Handbook of Research on Trends in Product Design and Development:
Technological and Organizational Perspectives

Arlindo Silva
Instituto Superior Tecnico, Portugal

Ricardo Simões
Polytechnic Institute of Cavado and Ave, and University of Minho, Portugal

BUSINESS SCIENCE Reference

BUSINESS SCIENCE REFERENCE
Hershey · New York

Director of Editorial Content:	Kristin Klinger
Director of Book Publications:	Julia Mosemann
Acquisitions Editor:	Lindsay Johnston
Development Editor:	Christine Bufton
Publishing Assistant:	Travis Gundrum
Typesetter:	Keith Glazewski
Production Editor:	Jamie Snavely
Cover Design:	Lisa Tosheff

Published in the United States of America by
Business Science Reference (an imprint of IGI Global)
701 E. Chocolate Avenue
Hershey PA 17033
Tel: 717-533-8845
Fax: 717-533-8661
E-mail: cust@igi-global.com
Web site: http://www.igi-global.com/reference

Library of Congress Cataloging-in-Publication Data

Handbook of research on trends in product design and development: technological and organizational perspectives / Arlindo Silva and Ricardo Simoes, editors.
p. cm.
Includes bibliographical references and index.
Summary: "This book provides a detailed view on the current issues, trends, challenges, and future perspectives on product design and development, an area of growing interest and increasingly recognized importance for industrial competitiveness and economic growth"--Provided by publisher. ISBN 978-1-61520-617-9 (hardcover) -- ISBN 978-1-61520-618-6 (ebook) 1. Industrial design--Research--Handbooks, manuals, etc. 2. New products-- Research--Handbooks, manuals, etc. 3. Production engineering--Handbooks, manuals, etc. I. Silva, Arlindo, 1968- II. Simoes, Ricardo, 1976- III. Title.

TS171.4.H36 2010
658.5'75--dc22

2010000039

British Cataloguing in Publication Data
A Cataloguing in Publication record for this book is available from the British Library.

List of Contributors

Table of Contents

Detailed Table of Contents

The goal of an inclusive society with equal opportunity for all is increasingly recognised and promoted worldwide. However, disabled people still face many difficulties in their daily lives, as well as some degree of exclusion from the wider society. Over the past few years, public design has considered ways to benefit disabled people, with an increasing trend to consult with and take advice from disabled people themselves. Taking aircraft lavatories as a case study, this chapter explores ways to help visually impaired people have a better quality of life, with minimal assistance from other people. This chapter discusses how the public design process can use consultative, participatory and inclusive approaches to generate understanding, so that the outcome of applied research fits the needs and preferences of disabled people. This chapter is not just a report on a particular case study. It also hopes to arouse designers' awareness of user-participation as an important trend in public design. Disabled people — users — should be invited to participate actively in the design process in order to bring real benefits to the disabled community, thereby contributing to an inclusive and harmonious society.

The multiple shortcomings of the current Design paradigm manifest the need of its modification. Our objective was to find out an appropriate mechanism. But such a mechanism could not be revealed

end user being an active partner in the designing process, a co-creator of knowledge. The concepts of *empathy*, *empathic horizon*, and the *material landscape* are illustrated through the discussion of a pilot design course within a university context. Using this approach, students with physical (visible) disabilities and product design students worked together designing everyday products. This empathic approach highlights research strategies that can support more effective design outcomes.

A new design process modeling approach focused on the information flow through design tools is discussed in this chapter. This approach is applied to three long term mechanical engineering design projects spanning 24 months, 12 months, and 4 months. These projects are used to explore the development of the new modeling approach. This is a first step in a broader effort in (1) modeling of design processes, (2) establishing case study research as a formal approach to design research, and (3) developing new design process tools. The ability of engineers to understand the dynamic nature of information throughout the design processes is critical to their ability to complete these tasks. Such understanding promotes learning and further exploration of the design process allowing the improvement of process models, the establishment of new research approaches, and the development of new tools. Thus, enhancing this understanding is the goal of this research effort.

Full traceability of products is extremely difficult, although it has been sought after for as long as production, distribution and sales chains exist. Electronic traceability methods, such as RFID technology, have been proposed as a possible solution to this problem. In the specific case of RFID, the number of applications that promote innovative solutions in retail and other areas has been continuous growing. However, RFID tags are mostly placed externally on a surface of products or their packages. This is appropriate for logistics, but not for other applications, such as those involving user interaction. In those, not only is the placement of the RFID tag more complex, but it is also necessary that the tag is not visible or not directly accessible, to prevent accidental damage and intentional abuse. This certainly imposes challenges to manufacturing, but mainly creates new challenges to the development of new products and re-design of existing ones. This chapter presents some insights and what we consider to be the two main approaches to incorporating RFID technology into consumer products.

Section 2
Supporting Technologies

This chapter presents the rapid prototyping and manufacturing concepts applied as means to reducing time between jewellery designs and manufacturing process. Different processes on jewellery modeling production are presented. Nowadays, the use of technologies as CAD/CAM - Computer Aided Design and Manufacturing in high production companies are very disseminated. However, the implementation of these resources at the design and manufacturing processes of jewels and fashion accessories, in small and medium size businesses, is still insipient. As reference, is presented the situation observed in small and medium companies located in Minas Gerais, Brazil.

The chapter in the broader sense will look in the area of creativity, creative process, and creative product. Specifically we shall look at the aspects of a creative product, discuss on thought process of the designers with focus on creative and visualization tools. Creative tools will be looked in context of changing mindset or assumptions, redefining problems, developing ideational fluency and bringing flexibility in thinking. Visualization tools such as sketches, storyboards, rough models, developing scenarios for the usage of proposed concepts etc will be discussed in parallel to creative tools serving as aids to externalize thought processes. Creative and visualization tools complement each other in enhancing the designer's creativity as well as help them come out of the stuckness feeling that they encounter while addressing design problems. Practicing the tools rather than being aware of them is important and the chapter will demonstrate the application of these tools with examples.

3D digital models are an essential tool that supports every step of the industrial product design process. For example, during the design review step photorealistic models are used as virtual prototypes for aesthetic and ergonomic evaluation of new products, in order to anticipate design evaluation and choices and to shorten the overall process. For this reason, industrial design enterprises usually have special rooms equipped with a projection based large display, for the realistic visualization of digital

This chapter outlines an innovative and collaborative design research project that connects New Zealand SME manufacturers with advanced design thinking about affective design. This project was developed and implemented by the centre for affective design research (Affect) at Massey University. Design researchers and five NZ companies have collaborated to foster 'design-for-desirability' thinking and develop capabilities by means of knowledge sharing, enterprise training and individualised projects. This created visionary product concepts utilizing the perceptual product experience framework (Warell, 2008). The chapter provides a novel model for collaboration between industry and academia that focuses on implementing 'design-for-desirability' thinking in SME companies with the aim of improving their international competitiveness. Well-designed, functional products are expected in today's competitive global markets. Gaining success in global markets requires a step beyond this level of usability in order to develop products that are desirable and appeal to the users on emotional, social and intuitive levels.

The world population is ageing. At both the individual and collective levels, ageing causes several changes in people's lives that influence their needs and the way in which they interact with products. This chapter aims first, to provide a literature review on different aspects of ageing in order to better understand how people will respond to products as they grow older. Second, it describes how the physical and cognitive decline of older users increases rather than decreases the need to focus on all aspects of their experience, including their physical, sensory, cognitive and affective responses. Having considered these issues, suggestions are made in order that designers might best focus their attention as they design for a population that is rapidly ageing, and where members of that population are encouraged to maintain full participation in society.

Section 7
Case Studies

"Madame Butterfly" is an industrial robot produced by Euroimpianti Spa. It is the result of the unusual cooperation between a cutting-edge robot factory and Fontanatelier, an unconventional architecture studio. Although it had all started as a common machinery project, and it was almost at the end of its development phase, the need for an iconic design led the designer to involve the architect Luisa Fontana in the project. The contrasting opinions between the engineers and the architect forced everybody to change points of view, in such a complicated and challenging field as robot production. The final design conveys a strong image of the product, very uncommon for the automation industry, maintaining the functionality of a standard robot, with improvements for some specific usage.

Singapore, a city state of 4.8 million people, located at the tip of the Malaysian Peninsula, was founded in 1819, when Sir Stamford Raffles of the East India Company established a trading settlement in Singapore. The meeting point for Chinese, Malays, Indians, Arabs, Europeans and others on their journey through the southern seas, Singapore achieved its initial economic success through international trade as a free port and free market. Given the status of an independent country in 1965, Singapore suddenly found itself in a struggle to survive. It's small population and scarce resources meant that regional and world markets were larger than the domestic market, presenting the government and its policymakers with distinctive economic challenges and opportunities. This chapter tries to recount the policies and subsequent actions put in place in Singapore from the 1960s till the present, promoting the creative industry, including product design, in order to transform a market dependent economy into a service centered economy. This chapter also discusses if and how such a 'planned intervention' played an important role in building up the resources and infrastructure within Singapore and in attracting multinational companies to locate their R&D and design facilities in Singapore, pointing to where it has succeeded and where it has not.

Contemporary product designers are increasingly attempting to utilize the latent potential of the product narrative and the impact it can have on the end-user and their relationship with products. Storytelling in

Foreword

In recent years, the practice of product design and development has faced numerous challenges, undergone significant changes, and seen many process improvements. In the near future, as always, firms will continue to face new competitive threats and the need to further improve their own capability to develop innovative products. Some of the most striking current trends include the following:

- **Development speed:** We are learning that many design processes are in fact procedural and amenable to lean principles and continuous improvement, yielding much faster product development cycle times.
- **Digital processes:** We are adopting new digital design, analysis, and collaboration tools to create better products with less time and effort.
- **Platform flexibility:** We are using modular product architectures to provide more product variety to customers.
- **Complexity management:** We are better able to engineer complex systems through analysis of interaction networks.
- **Outsourcing and offshoring:** We can take advantage of supplier skills and capacity, international operations, and new markets.
- **Customer involvement:** We are using the Internet to bring customers' ideas into the product development process much more quickly and easily than ever before.
- **Sustainability:** We are learning how to reduce the environmental impact of the industrial world in order to protect and sustain the natural world.

Recognizing these trends, researchers in many technical and social science fields study the associated challenges of product development and help to create improved methods and practices.

This handbook is a compilation of articles on a wide variety of the current trends, emerging technologies, important research, notable achievements, and key challenges in the field of product design and development. The range of topics covered here is as broad as the field itself: from product development methods and technologies, to process and project management, to some of the latest ideas about creativity, design for environment, and complex system development. Researchers, instructors, and practitioners alike will benefit from the insights and examples presented in this important handbook.

Steven Eppinger
Massachusetts Institute of Technology

Steven D. Eppinger *is Professor of Management Science at the Sloan School of Management of the Massachusetts Institute of Technology. He also holds the General Motors Leaders for Global Operations Chair and has a joint appointment in MIT's Engineering Systems Division. Dr. Eppinger served as Deputy Dean of the MIT Sloan School from 2004 to 2009 and as Interim Dean during 2007. From 2001 to 2003 he served as faculty co-director of the Leaders for Manufacturing and the System Design and Management programs, and from 1999 to 2001 as co-director of the Center for Innovation in Product Development. He received S.B., S.M., and Sc.D. degrees from MIT's Department of Mechanical Engineering before joining the MIT faculty in 1988. Prof. Eppinger has co-authored a leading textbook entitled Product Design and Development published by McGraw-Hill. His research is applied to improving complex design processes in order to accelerate industrial practices. He is a pioneer in development of the widely used Design Structure Matrix method for managing complex system design. He has authored over seventy articles in refereed academic journals and conferences. He has received MIT's Graduate Student Council Teaching Award, the Sloan School's Award for Innovation and Excellence in Management Education, and received twice the ASME Best Paper Award in Design Theory and Methodology.*

Preface

INTRODUCTION

Product design and development (PDD) is an area of increasing importance to industrial competitiveness. In a global market, the competitive advantage of companies will lie not only on mastering existing processes and methodologies, but most of all on their ability to pursue different directions, with increased value to their customers. This can only be achieved with an up-to-date knowledge of the field, but also with the inherent risk associated with innovative PDD processes and methodologies.

Each of the main topics covered by this handbook is perceived as contributing to an improved knowledge of PDD. Although each chapter will present possible approaches and solutions, there are no recipes for success. Each reader will find his/her balance in applying the different topics to his/her own specific situation. Case studies presented throughout will help in deciding what fits best to each situation, but most of all any ultimate success will come out of the interplay between the available solutions and the specific problem or opportunity the reader is faced with.

The main mission of the proposed book is to provide a snapshot of the current issues, trends, challenges, and future perspectives on product design and development, an area of growing interest and increasingly recognized importance for industrial competitiveness and economic growth. Product design and development is affecting not only industry, but society in general, as new and innovative products shape our way of life. The book intends to disseminate the latest approaches, practices, solutions, and pitfalls recognized by experts in this field.

A very important aspect of the book is to approach product design and development not only from the technology side, but to consider the business and social aspects of this area. For the past few years, the need to have a systems view on product design and development has caused many changes in the previously established methods and practices. It is now necessary to provide a fresh perspective on current trends in this field.

The overall objectives of the present Handbook are:

- To discuss the importance of product design and development and the most recent trends and approaches.
- To introduce state-of-the-art tools and technology to support PDD.
- To introduce the non-technological aspects, namely the most recent organizational and business aspects of PDD, but also the social perspective.
- To discuss the challenges associated to systems integration along the entire product development chain, and present success approaches.

- To tackle the important educational issues associated with the teaching of new product design and development.
- To address environmental issues associated with new product design and development.
- To present practical case studies that demonstrate pivotal solutions and pitfalls of relevance to the topic of this book.

To achieve the objectives stated above, contributions were accepted from 58 authors in 15 countries from all the continents: Australia, Belorussia, Brazil, Denmark, Finland, Hong Kong, India, Italy, The Netherlands, New Zealand, Portugal, Singapore, South Africa, UK, and USA.

ORGANIZATION OF THE BOOK

The book is divided in seven sections, each section encompassing several chapters:

- Section 1: Design Methodologies
- Section 2: Supporting Technologies
- Section 3: Organization and Process Management
- Section 4: Enhancing Creativity and Innovation
- Section 5: Social Sciences and Environment
- Section 6: Systems Integration
- Section 7: Case Studies

Section 1 is devoted to design methodologies. Within this section, six chapters explain different methods used in PDD. Chapter 1, "*Trends in Public Design for the Disabled: A Case Study on Public Design for Visually Impaired People*", looks at the ways in which product developers can design for an inclusive environment in which people with disabilities can live feel less dependent from others. To that aim, the authors present a methodology for PDD and a case study of aircraft lavatories design. Chapter 2, "*New Design Paradigm: Shaping and Employment*", presents a new proposal for a universal design theory, based on modularity principles and a scientific grounding, encompassing the design problem, design process and design representation. Chapter 3, "*Interdisciplinary Interaction for the Early Stages of Product and Service Development*", presents a different experience for structuring the multidisciplinary front end innovation with the INNOstudio® concept that facilitates innovation sessions. Chapter 4, "*Empathic Design Research Strategies: Designing for, with and by People with Disabilities*", presents its authors' view of a world designed by people with disabilities for users with and without disabilities. The authors call this design approach as empathic design, and state that it is a strategy that relies on the end user being an active partner in the design process, a co-creator of knowledge. Using this approach, students with (visible) physical disabilities and product design students worked together designing everyday products. Chapter 5, "*Tool and Information Centric Design Process Modeling: Three Case Studies*", proposes a new modeling approach for the process of product development, since the ability of engineers to understand the dynamic nature of information throughout the design processes is critical to their ability to complete these tasks. Chapter 6, "*Embedded RFID Solutions: Challenges for Product Design and Development*", presents an exploratory study on the implications of embedding electronic traceability devices in the product development process and the manufacturing process, especially if these devices are hidden within the product's body.

Section 2 is dedicated to the supporting technologies for PDD, containing three contributions. Chapter 7, "*Implementation of Rapid Manufacturing Systems in the Jewellery Industry in Brazil: Some Experiences in Small and Medium-Sized Companies*", describes how rapid prototyping can be used in small and medium-sized companies to assist in developing new products, focusing on the jewelry industry. Chapter 8, "*Creative and Visualization Tools in Context of Design*", looks at several aspects of a creative product, discusses the thought process of designers with a special focus on creative and visualization tools: sketches, storyboards, rough models, developing scenarios for the usage of proposed concepts, etc. These tools are discussed in parallel with creative tools serving as aids to externalize thought processes. Chapter 9, "*Virtual Reality Systems for Industrial Design Application*", investigates different CAD-3D software interaction techniques and outline their benefits and drawbacks compared to the interaction tasks required in the specific field of product design.

Organization and process management are the subjects for Section 3.This section contains two contributions from an educational perspective. Chapter 10, "*Integrating 'Designerly' Ways with Engineering Science: A Catalyst for Change within Product Design and Development*", suggests a new approach to engineering – product design engineering. The authors combine knowledge from design education and engineering education to build a degree program on product design engineering to educate engineering professionals with increased skills on creative design, social and environmental awareness, and a human-centered approach. Chapter 11, "*Rediscovering Design Education as a Social Constructivist Foundation for Innovative Design Thinking*", discusses design education, concluding that design education must change to adapt to the increasing pace at which different social groups are evolving new ways of communicating and living.

Section 4 has four chapters with a common background in the general topic of enhancing creativity and innovation. Chapter 12, "*Concept Naming: Exploratory Methods in the Development of Product Design and Brand DNA*", offers a unique alternative for integrating product personality in brand development. A new method, a specific derivation of product personality assignment referred to as 'Concept Naming,' is offered in combination with other more standard research methods. Chapter 13, "*PDD Trends: Research Driven by Laws of Product Evolution*", presents a framework that uses an evolutionary process to study the emergent behaviour, architectural innovation and organization redesign that can result from the development of radically new products and services. In a context of increasing awareness by companies to user needs and preferences, chapter 14 "*Customer Involved Open Innovation: Innovation of New Products with End Users and Customers*" provides guidelines for the involvement of end customers in the product development process, namely when, in which phases, which customers to involve, and which tools to use to support this involvement. Chapter 15, "*Stimulating Creativity and Innovation in and around Organizations: Co-Creation Experiments from Ongoing Research in a Bank*", outlines different approaches and developments related to co-creation, emphasizing its capacity also as a design approach. The chapter further defines the challenges of co-creation and reports experiences from an ongoing research in a service environment.

Four chapters compose Section 5 on social sciences and environment. The section is mainly devoted to the implications of considering non-technical issues when developing new products. Chapter 16, "*Research Project 'Future of the Present': The Process and the Importance Signs Observation into Fashion*", brings up a study of observation, analysis, and interpretation of signs and how new trends are possible to comprehend when developing new products. Understood as a language, the exploration of the signals issued by the society is a cross-behavioral study guide which enables tangible recommendations for all levels of products development. Chapter 17, "*Design for Desirability: A Collaborative Innovation-Initiative between New Zealand Design Academia and Industry*", outlines an innovative and collaborative design research project that connects New Zealand SME manufacturers with advanced design thinking about affective design. University and academia have collaborated to foster 'design-for-desirability' thinking and develop capabilities by means of knowledge sharing, enterprise training and individualized projects. The chapter provides a novel model for collaboration between industry and academia that focuses on implementing 'design-for-desirability' thinking in SME companies with the aim of improving their international competitiveness. Chapter 18, "*The Influence of Ageing on User Experience*", brings to analysis the topic of ageing. At both the individual and collective levels, ageing causes several changes in people's lives that influence their needs and the way in which they interact with products. The chapter describes how the possible physical and cognitive decline of older users increases the need to focus on all aspects of their experience when developing products directed at this target market. Chapter 19, "*The Contribution of Ergonomic Analysis in the Product Design for Recycling*", takes on the topics of Design for Environment (DFE), specifically Design for Recycling (DFR) and Design For Disassembly (DFD), in a particular social and economic situation, and the ways in which product designers need to take this into account when designing products that have to be disassembled or recycled.

Section 6 on systems integration has two chapters that aim to understand the ways in which a systems approach can be used to develop products. Chapter 20, "*Understand Complex Design Problems Using Systems Thinking*", provides a brief overview of systems theory and suggests that product designers could use systems theory and systems dynamics models to improve the understanding of complex Product Design research problems, anticipating how and where changes in these dynamically evolving systems might occur and interact with the current system in which the product is bound to be used. Chapter 21, "*Integrated Approach to Product and Process Design Based on Life Cycle Engineering*", introduces an integrated approach to product and process design and development based on Life Cycle Engineering principles. The aim is to propose a structured framework to drive the design team in their discussions and analysis towards the creation of design concepts and process alternatives and in their global evaluations towards informed decisions on a life cycle perspective. A simple case study is presented using the proposed integrated methodology.

Section 7 will present case studies of product design and development that are perceived as good examples in their fields. Six chapters compose this section. Chapter 22, "*The "Madame Butterfly" Robot: A Case Study in Product Design and Development*", is a case study of an industrial robot produced by Euroimpianti Spa. It is the result of the unusual cooperation between a cutting-edge robot factory and Fontanatelier, an unconventional architecture studio. The contrasting opinions between engineers and architects on this project forced a radical change in points of view, in such a complicated and challenging field as robot production. Chapter 23, "*Transformal Role of Product Design in Singapore's Transition to a Service Economy*", brings the reader into an environment where Chinese, Malays, Indians, Arabs, Europeans and others on crosses their paths in Singapore on their journey through the southern seas. The chapter recounts the policies and subsequent actions put in place in Singapore from the 1960s until the present day to promote the creative industry, including product design, in order to transform a market dependent economy into a service centred economy.

Chapter 24, "*The Value of Storytelling in Product Design*", outlines the end-user's relationship with products and the designer's role in this emotional and conceptual exchange of storytelling in product design through a framework defined by the authors as Narrative, Manufacture, History, and Interaction.

A case study will support this approach. Chapter 25, "*Deploying and Adapting an Indoor Positioning System in the Clinical Setting*", presents an innovative project at Massachusetts General Hospital where industrial design, operations research and outcomes research are weaved together with emerging technologies to provide a means for objectively and reliably measuring service time in the primary care setting. The RFID in Clinical Workflow Project aims to provide a tool with which to understand resource allocation and to shape appropriate and effective policy where the multidisciplinary team used a hybrid of design techniques sourced from the different disciplines represented. Chapter 26, "*Designing Toys, Gifts and Games: Learning through Knowledge Transfer Partnerships*", presents the Knowledge Transfer Partnerships (KTPs), a programme in the United Kingdom, administrated on behalf of the Technology Strategy Board (a Government Agency), where academics give their expertise to provide a solution for a particular problem, helping organizations improve their competitiveness and productivity. This case study investigation exemplifies some of the strengths and pitfalls of this type of knowledge transfer in the context of the global toy, games and gift markets. Chapter 27, "*Product Form Evolution*", brings about the notion that function and form are complimentary factors in improving the user's experience of a product and competing in saturated consumer goods markets. Through different methodologies with a case study on mobile phones, the authors have developed theories about the continuities that occur in product forms over time, and the forces that can disrupt this behaviour, and suggest how this view of form can benefit the design of future products.

Acknowledgment

Editing a book represents considerable work. However, more than anything, it embodies team work. The authors, editorial team, editorial advisory board, and external reviewers are all part of a multidisciplinary team. The work of editing a handbook on product design and development is in itself a product development process. It involves keeping track of contacts of authors and their contributions, exchanging information and ideas, managing the review process, feeding back review to the authors, managing conflicting perspectives, and integrating contents into a reasonable structure, with the ultimate goal of developing a product that adds value to the readers' body of knowledge.

As team leaders we, the editors, have to thank our team members for the effort involved in this initiative. This handbook is primarily supported by the team of professionals from IGI Global, most notably Ms. Jan Travers and Ms. Christine Bufton for the opportunity and constant support in editing the book, timely suggestions, prompt feedback and friendly reminders about deadlines.

To the Members of the Editorial Board, our gratitude for sharing with us their knowledge and experience in the support of the decision-making processes inherent to the project, for assisting in the review process, and for their help in shaping the handbook: António Augusto Fernandes (FEUP, Portugal), António M. Cunha (University of Minho, Portugal), Carlos Aguiar (FEUP, Portugal), Cees de Bont (TUDeflt, Netherlands), Chris Magee (MIT, USA), Dan Whitney (MIT, USA), David Wallace (MIT, USA), José Rui Marcelino (AlmaDesign, Portugal), Manuel de Freitas (IST, Portugal), Michael Tovey (Coventry University, UK), Paulo Ferrão (IST, Portugal), and Raul Cunca (FBAUL, Portugal).

As the contributions were extremely broad in scope, aside from reviews performed by the Members of the Editorial Board, we had to call on external experts to serve as reviewers for some of the chapters. To them, our sincere thanks for this entirely voluntary work: António Gonçalves Coelho (FCT-UNL, Portugal), António Pontes (University of Minho, Portugal), Carlos Tavares Ribeiro (Academia Militar, Portugal), Deana McDonagh (Univ. Illinois at Urbana-Champaign, USA), Jeremy Gregory (MIT, USA), Joaquim Mendes (FEUP, Portugal), Joshua Summers (Clemson Univ., USA), Kin Wai Michael Siu (Hong Kong Polytechnic University, Hong Kong), Lia Patrício (FEUP, Portugal), Mirja Kälviäinen (North Karelia University of Applied Sciences, Finland), Paulo Peças (IST, Portugal), Renato Bispo (ESAD-IPLei, Portugal).

We acknowledge all the authors, without whom there would be no Handbook in the first place! Some of the authors played a double role also as blind-reviewers for some of the chapters written by others. The process of editing a handbook of this kind brings the community together from academia to industry, making it an enriching experience for all.

We also have to acknowledge all the authors who submitted chapter proposals that were not included in the final version of the handbook. Many contributions were not considered, despite their merit, either because they were out of the scope for this handbook, time limitations, or other constraints.

A special word to our home institutions, the Instituto Superior Técnico of the Technical University of Lisbon and the Polytechnic Institute of Cávado and Ave, for providing the infrastructure, material resources, and logistics required for our work.

Thank you!

The Editors,
Arlindo Silva
Ricardo Simões
Portugal, December 2009

Section 1
Design Methodologies

Chapter 1
Trends in Public Design for the Disabled:
A Case Study on Public Design for Visually Impaired People

Kin Wai Michael Siu
The Hong Kong Polytechnic University, Hong Kong

ABSTRACT

The goal of an inclusive society with equal opportunity for all is increasingly recognised and promoted worldwide. However, disabled people still face many difficulties in their daily lives, as well as some degree of exclusion from the wider society. Over the past few years, public design has considered ways to benefit disabled people, with an increasing trend to consult with and take advice from disabled people themselves. Taking aircraft lavatories as a case study, this chapter explores ways to help visually impaired people have a better quality of life, with minimal assistance from other people. This chapter discusses how the public design process can use consultative, participatory and inclusive approaches to generate understanding, so that the outcome of applied research fits the needs and preferences of disabled people. This chapter is not just a report on a particular case study. It also hopes to arouse designers" awareness of user-participation as an important trend in public design. Disabled people, users, should be invited to participate actively in the design process in order to bring real benefits to the disabled community, thereby contributing to an inclusive and harmonious society.

INTRODUCTION

Visual impairment is a serious problem. According to the World Health Organization's statistics (2009), there are about 45 million blind people in the world, including 1.4 million children under the age of 15. In addition, there are 135 million people (that is, about one thirtieth of world's population) who have serious visual impairment (see also ORBIS, 2009; United Nations, 2009; World Health Organization, 2003). These numbers indicate that we should be preventing visual impairment; they also point to the many visually impaired people (VIP) whose daily lives are a constant struggle. Action on their behalf begins

DOI: 10.4018/978-1-61520-617-9.ch001

with a serious consideration of their needs, as *they* perceive them.

People with different types and levels of visual impairment face a variety of difficulties in their daily lives when they need to access public places and facilities with which they may not be familiar (United Nations, 2009). In recent years, policymakers, researchers, sociologists, planners, designers and engineers have tried to improve this situation. Among the many projects (with a variety of objectives, perspectives and targeted beneficiaries) the approach has generally been to apply technologies that provide convenience (or overcome existing "barriers") for VIP. However, as many of the studies in European and American countries have shown, the situation is still unsatisfactory. This is not because there is something wrong with the technologies, but because often they neither fit the actual wants and needs of VIP, nor function as they were planned and intended to do.

There are three consequences of our failure to help VIP to access public places and facilities. First, VIP tend to minimize the number of times that they go out to places with which they are not familiar. Second, if it is absolutely necessary for them to go to unfamiliar places, VIP need to spend much time (sometimes to an unreasonable degree) preparing to go out. Third, VIP need the assistance of people who are not visually impaired (or who are visually impaired but already familiar with the places). All three consequences are disadvantageous. They seriously diminish the normal daily lives of VIP, and other people as well.

When VIP leave their homes or environments with which they are familiar, public toilets are one of the most difficult places for them. Consequently, toilets are places where VIP are least willing to go. For VIP, using a strange public toilet is an unpleasant, stressful and even dangerous event. Even if they are familiar with a particular facility, using a public toilet is repulsive and distressing for VIP, because toilets are generally recognized as dirty places where it is possible to contract infectious diseases. VIP frequently complain about the inaccessibility and unpleasantness of public toilets, even in developed countries with well-established policies and better resources, such as the United States and European countries (Siu, 2008).

From 2004 to 2008, the Public Design Lab of The Hong Kong Polytechnic University conducted an in-depth exploration of public design for disabled people, as an applied research project aimed at improving the daily lives of VIP. The specific objective was to explore the needs of VIP in order to generate design solutions that would help them access aircraft lavatories. The broader objective was to explore how to help VIP have a better quality of life independently, that is, with minimal assistance from other people. This paper uses the project as a case study to show how active, participatory and inclusive approaches to disabled people can be incorporated into an effective design process. It is not just a report on a particular case study. It also hopes to arouse designers' awareness of user-participation as an important trend in public design. Disabled people, users, must be invited to participate actively in the design process in order to bring real benefits to the disabled community, thereby contributing to an inclusive and harmonious society.

CASE STUDY: AIRCRAFT LAVATORIES FOR VIP

To promote and develop public toilets that are more accessible, an applied research project on existing designs of public toilets and how VIP access them was conducted from 2004 to 2008. The project was funded by The Hong Kong Polytechnic University and the Asian Scholarship Foundation, and assisted by the Hong Kong Toilet Association, the Hong Kong Blind Union and several other non-governmental organizations (NGOs) that provide services to VIP.

After about six months of background study on VIP access to public toilets, aircraft lavatories were selected as a case study involving further exploration, serious investigation and analysis. No in-depth study on the topic had ever been done before. The focus of the study allowed for a manageable research scope leading to practical applied research. In addition to the standard quantitative approaches and methods (such as a questionnaire and survey), the project also took a qualitative approach to investigating the needs and preferences of VIP with regard to public toilets. The major reason for this approach was that the findings and results from a review of research reports about the special needs of disabled persons using public toilets (and other public facilities such as transportation) were too general to generate significant improvement in these environments and facilities. This was confirmed in our background study when we interviewed VIP and NGO social workers (from the Hong Kong Blind Union). They all agreed that there had been no improvement in public toilet facilities, because the findings and resultant "improved" designs did not meet the actual wants and needs of VIP (see also Siu, 2005a, 2008). Therefore, in order to more comprehensively investigate what VIP want and need, our project took a different direction. The project established a solid foundation based on VIP needs and wants by first understanding user experience, and then by generating directions and insights from them, which lead to more comprehensive studies in the future.

The project adopted three simultaneous approaches: (a) action, (b) user participation, and (c) inclusive design (Figure 1). The action approach implied that investigation, implementation and evaluation were run in cycles in order to better understand the issues and to improve the concepts (Calhoun, 2009; McIntyre, 2008). Sometimes, new problems would occur or old questions reappear in a new cycle of action. Nonetheless, these cycles proved worthwhile, since they are the only route to a final outcome that is satisfactory to VIP. The participatory approach, which involved VIP users, guaranteed that they would have opportunities to

Figure 1. Research framework and model of the case study

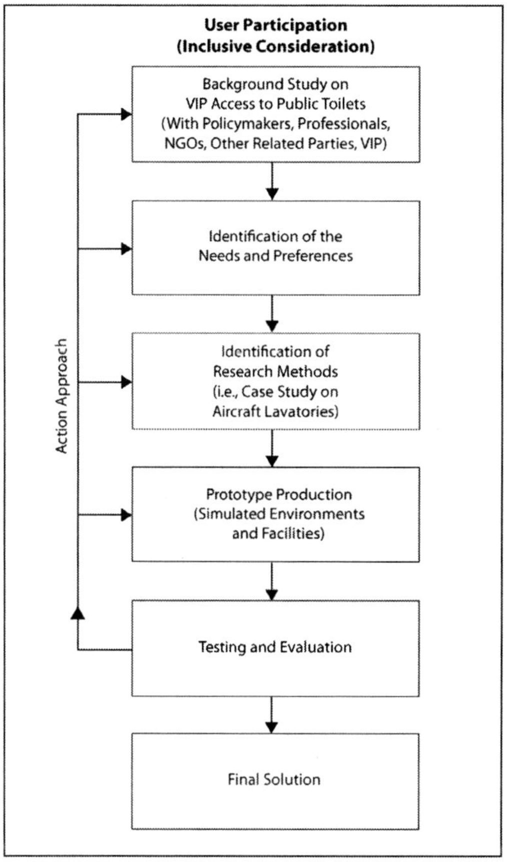

express what they actually needed, preferred and expected (Kwok & Ku, 2001; Kwok & Siu, 2002; Mehrhoff, 1999; Sanoff, 2000). The inclusive design approach pointed to solutions that allow people without visual impairment to benefit, or at least to avoid inconvenience or negative effects. Therefore, during the project, the project team spent a large portion of their time carrying out participatory research with VIP to ascertain daily practices through intensive observations and direct interviews

The detailed research methods included:

- A review of the existing policies and provisions of public facilities for VIP, including issues related to equal opportunity.

- Interviews with government officers responsible for the overall planning, design and management of public toilets.
- In-depth interviews with the NGO representatives (including social workers and visually impaired volunteers) who provide services to the VIP of Hong Kong.
- In-depth interviews with VIP having different types, levels and degrees of vision impairment.
- Observations of how VIP access and use public toilets, in particular aircraft lavatories.
- Arranging visits by VIP to aircraft, where they were invited to use aircraft lavatories. Observations were conducted on how they use lavatories. Their comments on the design of lavatories were collected.
- Product analysis of the facilities available in aircraft lavatories.
- Construction of a full-scale mock-up of an aircraft lavatory as a physical simulation for collecting the comments of VIP.
- Conducting workshops based on the above study, where proposed designs were generated for the comments of VIP. NGO social workers were invited to give assistance during the workshops.

From the beginning of the project, through comprehensive background study and intensive interactions with VIP themselves, the project team tried to understand the daily practices of VIP, and how they communicate. Research focused on the particular needs of VIP, and their preferences, expectations and practices with respect to accessing public spaces. The project team spent about one year conducting observations and interviews, in particular to ascertain how they access public toilets, and the different types of private toilets available for public access.

At the final stage, the project team spent about 18 months focusing on a research-and-design project on aircraft lavatories for VIP. In this phase of research, the most time-consuming part was the action process of exploration, prototype production and evaluation involving VIP in different environments and situations. Initially, unstructured and semi-structured interviews were conducted to understand backgrounds, experiences, needs and preferences of VIP. Brainstorming and discussion sections at different stages of the design development provided ideas from their special perspectives. These brainstorming activities with VIP were particularly important, because their creative and sometime unexpected ideas were based on their daily experiences, which sighted people do not encounter and therefore do not think about.

From time to time throughout the project, verbal communication between the (sighted) team and VIP did not work satisfactorily, particularly with respect to three-dimensional concepts (Kelley, Sanspree & Davidson, 2000; Prevent Blindness America, 2002; Seidman, 2003). Therefore, the project team used models and prototypes instead of abstract concepts to communicate more effectively with VIP (Lovelock, 1995; Hong Kong Blind Union, 2007). After exhaustive user tests involving different models and prototypes, the team constructed a new facility that allows VIP to independently access an aircraft lavatory. These results were so satisfactory that an airline recognized the particular needs of VIP and will propose the design concept to an aircraft production company.

VIP Needs in Using Public Toilets

The findings of the research project, as well as its applied research output, highlight several important aspects of design quality in public toilet environment and facilities for VIP. First of all, there have been very few investigations of, or design developments focused on the special needs of VIP users of public toilets. Worldwide, most designs only focus narrowly on the needs of physically disabled people. Effectively, "disabled toilets" simply imply toilets for physically

Figure 2. Accessible lavatory provided in aircraft

Figure 3. Accessible lavatory provided in aircraft

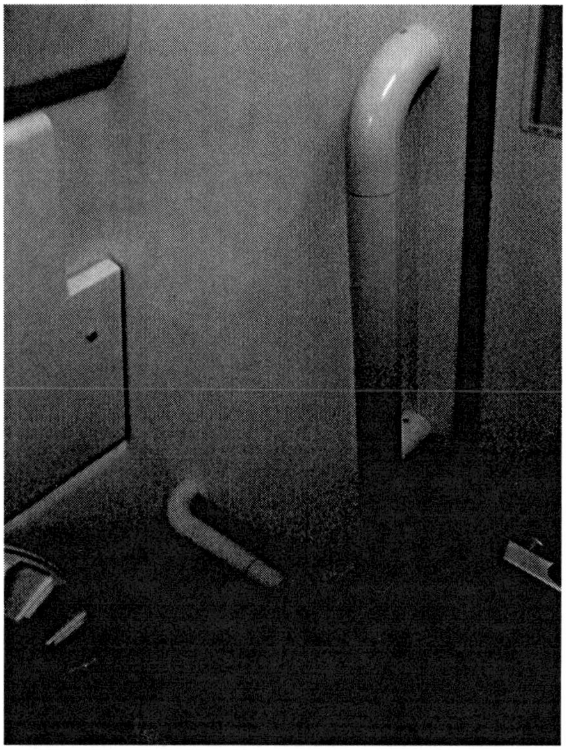

disabled persons, and even more narrowly for wheelchair users.

Some airlines have started to provide accessible lavatories for disabled people (Figures 2-3), by increasing interior space and providing some special facilities. However, all of these lavatories are specifically for physical disabled people, while the needs of VIP (as well as people with other sensory and cognitive disabilities) are not considered.

With the exception of the restrictions caused by limited or no vision, VIP have the same needs as sighted people when they access public spaces in their daily routines. The major difference is that, because facilities do not cater to non-sighted people, VIP generally prefer to go only to the places with which they are familiar. Or rather, they prefer not to go to the places with which they are *not* familiar. VIP interviewed for this study made clear that they do not lack the curiosity, interest or need to go to new places. However, it is not easy for them to overcome the difficulties, and sometimes embarrassments, of trying to access *unfamiliar* places. This is how we know that current designs do not meet the needs of VIP. This situation applies to aircraft lavatories. It is not easy for VIP to find aircraft lavatories, because they do not have a standard location in all aircraft. The lavatories themselves are unfamiliar places for VIP. The facilities inside the lavatories are difficult to for VIP to use. The flushing mechanisms are a typical example. Nearly all of the activation buttons have a flat surface and lack tactile information; in addition, they are located in different non-standard locations in different aircraft (Figures 4-5). Consequently, VIP are embarrassed to call aircraft attendants or to ask their friends for help in flushing the toilets. As a result, VIP report that they prefer not to drink water in order to minimise the need to go to lavatories.

Figure 4. Flushing mechanisms lack tactile or other VIP-friendly information. In addition, the button has no standard location.

Figure 5. Call-for-assistance buttons lack tactile or other VIP-friendly information. In addition, the button has no standard location.

If they need to go to a new place, VIP have to spend time preparing; however, it is very difficult for them to collect the information they need. Information suitable for VIP is limited. Even countries such as the United States, Germany and Japan are not able to provide public information by using advanced technology. Moreover, it is difficult if not impossible to provide detailed and updated information about public access facilities such as the interior every public toilet. The fact remains that even changes that are insignificant to a sighted person can significantly affect a VIP.

Specifically, there is no up-to-date information to help VIP access aircraft lavatories, other than individual assistance from aircraft attendants. Some VIP in the participatory design workshops suggested that a simple tactile map on paper, made available in the aircraft, would help most VIP locate lavatories. They could keep the map for their whole journey and use it in case of necessity, and they would not need to call for further assistance. VIP also commented on the fact that some airline companies request information about VIP disability when they reserve their tickets. However, all the companies do is remind aircraft attendants to give special attention to the VIP. There have been no VIP-oriented improvements in aircraft interior design.

When VIP travel by air, calling for assistance is neither their expectation nor their preference. Lavatories in aircraft are so small that there is not enough space for someone to enter and help–excepting only the 2% designed for physically disabled people. It is complicated, impractical and also embarrassing for an attendant outside the lavatory to shout instructions to someone inside, particularly as other passengers are sitting close by, and may be curious about what is happening.

VIP sometimes need assistance from sighted people, or from another VIP who is already familiar with the place in question. However, the VIP and NGO social workers who we interviewed stressed that, in common with people with physical limitations, VIP do not like to ask for help. They prefer to be independent and in going about their daily routines. They also expect to have more freedom. They do not want to be labelled as disabled. Asking for care and assistance generates a perception that VIP are functionless, or worse, a burden to society. People with physical limitations–not only VIP–often refuse assistance because people show pity when they give help. It is also the reason VIP prefer to stay at home or in familiar places. Or, as already mentioned, VIP shorten their time in public places and drink less water to avoid using public toilets.

FISH: Design Concept Applied in Aircraft Lavatories for VIP

To improve the design quality of public toilets and facilities, the project applied the FISH concept to the design of aircraft lavatories for VIP. This design concept is inclusive in nature, in that people without visual impairment can also benefit from the design, or at least avoid inconvenience and negative effects. The FISH concept includes four major directions (or design considerations): *friendly*, *informative*, *safe* and *hygienic* (for the details of FISH, see Siu, 2008).

Friendly

The participatory workshops made clear that a user-friendly aircraft lavatory should respect psychological as well as physiological factors (for different areas of user-satisfaction and pleasures, see Jordan & Green 1999; Siu, 2005a). In other words, the user must be the focus of high quality design, not the designer or the design itself. Fortunately, user experience is increasingly being recognized as crucial to creating a design with a high degree of user-friendliness (Kuniavsky, 2003; Rehabilitation Alliance Hong Kong, 2005; Siu, 2005a, 2008; Wiklund, 2005).

- Many designers lack both experience of and knowledge about vision impairment. They consider and propose designs from their own experiences, perspectives and imaginations, sometimes with very limited understanding and no experiential awareness of VIP needs and preferences. It is crucial that designers understand how VIP search for information, and to consider collateral problems such as injuries from sharp edges, or gaps between moveable parts such as door jambs and hinges, because when they are inside an aircraft, VIP primarily use their sense of touch to search for the lavatories and to use the facilities within.
- People (not only VIP) are unwilling to touch the facilities in lavatories. However, VIP need to use physical contact to search for and use the sinks and toilets, which can be dirtier than other public environments. User-friendly design minimizes contact by installing automatic flushing mechanisms, water taps and facility position systems (toilet seats). The Beijing and Belfast World Toilet Summits 2004, 2005 and 2008 indicated that with sufficient resources, public toilets, including aircraft lavatories, should provide automatic facilities that offer convenience to all users. Implementing this idea and finding an appropriate level of automation are issues that require further investigation, discussion and tests. In the case of aircraft lavatories, it will be difficult to implement automatic devices, since the spaces are so small.
- Friendly design implies standard and convenient locations for information. Since VIP must search for information by touch, standardised locations and information formats can minimise unnecessary and distasteful contact.
- User-friendliness includes psychological comfort and satisfaction as well as physiological, functional necessities. Nearly all the VIP we interviewed indicated that they were unwilling to use what are known as disabled toilets. One of the major issues they raised is that they do not have a physical disability, as do wheelchair users. Most of the time, VIP can take care of themselves without special physical facilities. VIP are willing to use *common* public toilets with *common* facilities. This expecta-

tion reflects VIP's dislike of being labelled as disabled.

- Providing a special lavatory for VIP in each aircraft is impracticable. User-friendliness implies a user-fit environment that works not only for VIP, but also for other users (Imrie & Hall, 2001; Siu, 2005b). VIP do not want to cause any inconvenience to the public due to their disability. During the interviews, VIP told us that they do not want to convey the image to the public that VIP are always asking for help and for more benefits and care from society. They also do not want their convenience to bring them into conflict with other users (Anogianakis, Bühler & Soede, 1997).
- User-friendliness also implies the best use of the strengths and specific abilities of different groups of users (Willems, Vlaskamp & Knops, 1997). The VIP we interviewed maintained that they have superior abilities than sighted people with respect to orientation and memory of physical locations and the forms of physical objects. They told us that their major difficulty in accessing aircraft lavatories is that nearly all of the aircraft interior settings and lavatories are new to them. "New" implies differences, changes, and divergences. Standardized aircraft lavatories would minimize the difficulties VIP encounter, and make good use of what they do better than sighted people. Standardization of the interior of aircraft lavatories is recommended. (Stephanidis, Akoumianakis & Ziegler, 1997).
- Another difficulty in designing user-friendly aircraft lavatories is that convenience for one group of users may inconvenience another. For example, many VIP do not want to access wheelchair-accessible lavatories, because VIP are easily trapped and hurt by the handrails provided for physically disabled persons in these special lavatories. These practical concerns are in addition to the psychological reasons mentioned above.
- User-friendliness is a design approach that promotes quality by designing to fit a larger population (Clarkson, 2003). The approach considers not only the functional factors but also the psychological factors. Inclusive and universal designs are therefore highly recommended in the design of aircraft lavatories.

Informative

- "Informative" implies information provided at the right time, in the right place, and in a form that is easily understood, without confusing different users. Good design provides VIP with tactile information to access aircraft lavatories and the facilities inside. (Figures 6-7).
- In the participatory workshops VIP indicated that their major concern is to avoid embarrassment while accessing information, particularly when sighted people do not have to do the same things as they do. For example, it is sometimes embarrassing for them to use guiding sticks or hands to search for lavatories or to ask for information in an aircraft compartment that is full of passengers and busy attendants.
- "Informative" also implies a high degree of independence in getting and understanding information. Because aircraft lavatories have limited space, it is impossible for a person inside to ask for support and information. Asking other people for assistance, for example, to flush the toilet, is both inconvenient and embarrassing.
- Information must have a high degree of accessibility and reliability. Aircraft are used by people who speak a variety of languages. Thus, a Braille version of words in a single language may not be appropriate. On the other hand, too many different ver-

Figure 6. The information provided in an aircraft lavatory is meaningless to VIP if they cannot "read" it by touch

Figure 7. The information provided in an aircraft lavatory is meaningless to VIP if they cannot "read" it by touch

sions of Braille are also impracticable and may easily cause confusion.

- The effectiveness of the information varies with individuals and groups. For example, fewer and fewer younger VIP can read Braille. Some older VIP are unable to read Braille due to the deterioration of their sensory capabilities. Additional and modified systems, such as easy-to-read and modified symbols, are good alternatives provided in aircraft lavatories for these VIP. Such alternatives can also prevent the language problem discussed above.
- The participatory workshops discovered that because the interior space of aircraft lavatory is so limited, information for VIP may sometimes inconvenience other users. Balancing different users' wants and needs, and avoiding assistance overload are both important issues (Siu, 2008).

Safe

- Safety is the most important consideration. Ensuring a safe as well as effective interaction between VIP and the lavatory environment and facilities is essential for quality design (Figures 8-9), since VIP have no (or limited) visual capability, which is the sense that enables the fastest response.
- Because many passengers use aircraft lavatories, the facilities can and do change, which may cause inconvenience and can sometimes be dangerous to VIP who are unable to observe the changes. For example, water may be spilled on the floor, or other users may have moved facilities and consumable items. Unless they have sufficient training, aircraft attendants may be unaware of these risks to VIP.
- Minimizing the unpredictable is one approach to making aircraft lavatories safe for VIP. Because they are *unfamiliar* places for VIP, standard designs and provisions are highly recommended (Stephanidis, Akoumianakis & Ziegler, 1997).
- Another approach to overcome the unpredictable is to notify VIP; that is, to make them aware of difficulties. This could include providing printed Braille information or a verbal reminder to VIP when they enter aircraft, even though this idea may not be related to the lavatory design itself.

Figure 8. Nearly all of the safety precaution warnings in aircraft lavatories are not accessible by VIP

Figure 9. Nearly all of the safety precaution warnings in aircraft lavatories are not accessible by VIP

Hygienic

- During the participatory workshops, VIP repeatedly emphasised that hygiene is their major concern when visiting any kind of toilet, in particular public toilets which are used by many people they do not know.
- Besides using their ears and guiding sticks, VIP need to use their bodies, hands, fingers and feet, to collect information and interact with the environment and facilities. This situation requires VIP to have direct contact with the facilities in aircraft lavatories. As has been mentioned with respect to all public toilets, aircraft lavatories are also commonly considered to be dirty. They can cause VIP to be more likely to get infections, particularly as most of these lavatories have a high rate of usage. The VIP we interviewed stated that not using or minimizing use of lavatories is their safe choice, even if it is not preferable. Consequently, VIP seldom get into trouble in either aircraft lavatories or other kinds of public toilets because VIP seldom go into lavatories with which they are not familiar. Hence, the design direction that respects VIP needs is to minimize contact with lavatory facilities which have been touched by other people, or which have a high chance of being infected (Figure 10).
- It is very difficult and unrealistic in terms of cost and management to aim at a "zero-contact" approach. Therefore, a good direction is to provide appropriate minimal contacts with facilities that have a high chance of public contact and infectious disease transmission (Siu, 2005a). Good design should provide sufficient information with minimal contact.

Figure 10. VIP need to use their bodies to collect information and interact with the environment and facilities

Figure 11. A proposed design for VIP to access aircraft lavatories (1). The universal twin-horizontal tactile guide bars provide internationally recognised information for VIP to locate different facilities in a lavatory. The user-friendly bars are installed separately from the facilities so that they remain safe and clean.

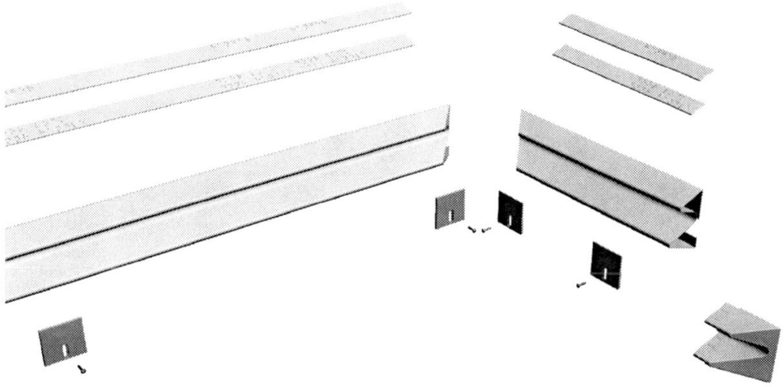

- A possible direction is to separate the facilities that must be touched from those that have a high potential to be infectious. For example, instead of providing Braille information on the surface of facilities that are touched by other people, independent information devices exclusively for VIP are a good design direction. A tactile information device for VIP could be installed in a position and location convenient for VIP, but also apart from those facilities that can easily be infected, such as water taps and basins (Figures 11 – 16)

Guiding Principles for Designers Addressing Problems of the Disabled

When we began our applied research on aircraft lavatories, "respect" and "understand" were two key words that were always on our minds. The project team maintained that only initial respect could direct the research in an acceptable way

Figure 12. A proposed design for VIP to access aircraft lavatories (2)

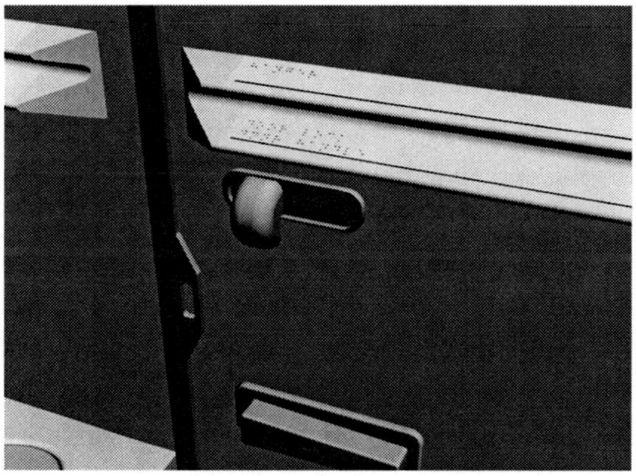

Figure 13. A proposed design for VIP to access aircraft lavatories (3)

Figure 14. A proposed design for VIP to access aircraft lavatories (4)

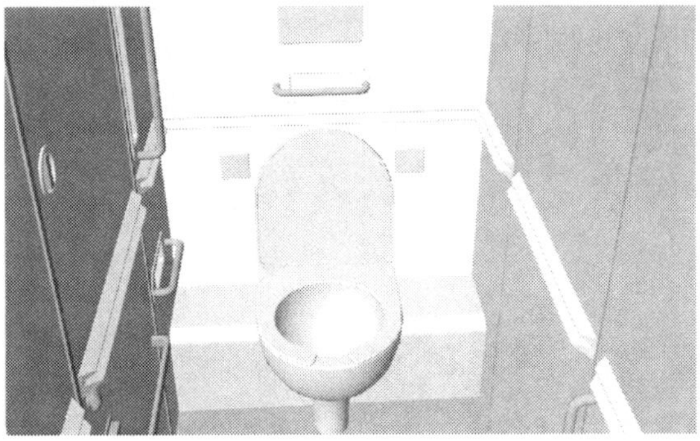

Figure 15. A proposed design for VIP to access aircraft lavatories (5)

that would lead to a better understanding of VIP's needs, preferences and expectations. The team also believed that only in-depth user research could bring understanding of how and why to respect VIP's needs and practices.

Obviously, improving the design of aircraft lavatories does not make any major change in the difficulties that VIP face every day. Most of the time, VIP, as well as other disabled people, still face challenges ranging from the unintended results of public policy to the intrusion of small objects in their daily lives. As many sociologists state, misunderstanding and neglect still exist today (Ferguson, 2007; Lovelock, 1995; Oliver, 2001; Seidman, 2003). Intentional or unintentional discriminations and exclusions are also still significant (Rehabilitation Alliance Hong Kong, 2004; United Nations, 1997, 2008, 2009; Siu, 2008).

Figure 16. A proposed design for VIP to access aircraft lavatories (6)

The United Nations has stated (2009) that one of the vital missions of human beings in the 21st century is to eliminate inequality in society. This mission applies not only to deprived and developing countries, but also to developed countries that claim to reach a high living standard. Disabled people all over the world face many difficulties, however, many people are unaware of how they go about their daily lives (Cooper, 2000; Rehabilitation Alliance Hong Kong, 2004; Siu, 2006a, 2006b, 2007; Treqaskis, 2004; United Nations, 2009). It is not well known that many VIP do not know how to read Braille; similarly, most people are unaware that there are several versions of Braille used in Chinese communities. Even in the English language world, Braille is not standardized, and there are many short-forms of Braille (Siu, 2008).

To improve the situation, designers can and should contribute more. First of all, it is important to respect the needs and preferences of disabled people, and also their limitations and difficulties. As Seidman (2003) states, disabled people should not be considered as a group with problems. Instead, they should be seen as groups of people with some special needs, who are commonly and normally living as members of our society, not strangers or aliens (Hong Kong Blind Union,

2007; Oliver, 2001; Seidman, 2003; Siu, 2008, 2009; United Nations, 2009).

Respect without action is insufficient. To obtain good quality design for disabled people as well as other people inclusively, attention to the whole design process must be shifted from the designer and the design to the user. Designers must recognise that they should not make decisions for users. Designers should not impose their value judgments and preferences on users, particularly not on disabled people or minority groups, especially when the designers know little about them (Luck, 2000; Scrivener, Ball & Woodcock, 2000; Siu, 2008). Instead, designers should embed user participation into the design process. As more and more researchers and practicing designers agree, participatory design is one of the best of the various design approaches, in that it allows users to engage in the design decision-making process (Kumar, 2002; Kwok & Ku, 2001; Kwok & Siu, 2002; Mehrhoff, 1999; Sanoff, 2000; Schuler & Namioka, 1993; Siu & Kwok, 2004). The opportunity for disabled people to participate in the design process not only results in better user-fit solutions, but also empowers the users by letting them influence the design decision-making process. This leads to an increased awareness of the consequences of these decisions (Kwok & Ku, 2001; Sanoff, 2000; Siu & Kwok, 2004).

A high degree of user participation does not imply that designers do not need to do anything or that they should be ignored (Siu & Kwok 2004). On the contrary, in participatory design, designers should begin with an understanding of disabled people, users, as well as extensive background research on the design topic (Siu, 2008). That is, designers need to conduct more empirical research and have more direct communication with disabled people in order to explore their diverse backgrounds, needs and preferences. When they have completed this preparation, designers can give direction to the participatory design work and communicate effectively with the disabled people who will participate in the design process.

In addition, designers should work as coordinators who gather together interested groups and professionals, and then as facilitators who assist users with different capabilities, disabilities, in participating, modifying, experiencing, creating, producing, and actualising the design. The aircraft lavatory project team spent a large amount of time communicating with NGOs, gathering VIP and coordinating the participatory workshops. At the outset, many disabled people are passive when it comes to voicing their opinions; and they cannot be relied upon to initiate and coordinate any movement to improve the designs that affect their lives (Rehabilitation Alliance Hong Kong, 2004; Siu, 2008; Siu & Kwok, 2004). It is necessary for designers to work closely with the various interested and potentially affected groups, and facilitate a supportive environment for working together. During the participation process, designers can give disabled people the opportunity to identify their needs and preferences, set goals, voice ideas and opinions, make decisions, and continuously evaluate the outcomes.

CONCLUSION

The goal of an inclusive society with equal opportunity for all is increasingly recognised and promoted worldwide. During the last few decades, there has been increasing recognition of the importance of research and design for disabled people. However, disabled people still face many difficulties and some degree of exclusion (Disability Rights Commission, 2009; Rehabilitation Alliance Hong Kong, 2004). Disabled people complain that their needs and preferences have still not been seriously addressed and satisfactorily met (Hong Kong Blind Union, 2007; Siu, 2008).

Taking aircraft lavatories as a case study, the project discussed in this paper explored how to help visually impaired people have a better quality of life with minimal assistance from other people. The outcome of the project was very satisfactory, as measured by not only the international invention and design awards received, but also by positive comments from visually impaired people.

When trends in public design for the disabled people are considered, the designers' job is not to produce finished and unchangeable products, but rather to develop solutions arising from continuous two-way communication among the different users who will use the final design in different ways. The energy and imagination of designers should be directed towards raising disabled people's level of awareness about design choices. Using the aircraft lavatory as an example, this means that the final design of the lavatory should arise from opinions and ideas exchanged between designers and VIP with different backgrounds, needs and expectations. The designers provided opinions and advice, and discussed the consequences of various alternatives; the VIP gave their opinions, and contributed their practical experience.

As stated by the airline company representatives interviewed, an aircraft lavatory is necessarily a small room in an aircraft. It is also a public place used by a large number of people with different backgrounds, capabilities, needs and expectations. It is not appropriate to cater only to the needs of a particular group of people. Thus, the lavatory should neither be a place exclusively for those who *can* see, nor a place for those who *cannot* see. Instead, it should be a genuinely *public* place, with *public* facilities. It should not be the intention or goal of good design to exclude any group of people, no matter whether the subject is a worldwide communication system, a city, an aircraft lavatory, a table, or even a pencil or a warning message. It is perhaps too idealistic to require that all public designs serve everyone's needs, preferences and expectations. However, when action, user participation, and inclusive design are persistently attempted, society can be bought closer to the ideal.

ACKNOWLEDGMENT

I would like to acknowledge the resources extended by The Hong Kong Polytechnic University. The Hong Kong Research Grants Council's Public Policy Research Grant support part of the study (ref. no.: RGC 5007-PPR-09). I would like to thank the Hong Kong Blind Union as well as other Hong Kong NGOs and volunteers (visually impaired persons) for their help in providing information. I would also like to thank Jiangyan Lu, Wai-lun Chan and Wan Pak-hong Wan for their collecting the data and design implementation work. Part of the findings of the project discussed in this paper was presented in 7th International Conference on New Directions in the Humanities in Beijing, PRC, and 12th ISSAT International Conference in Chicago, USA.

REFERENCES

Anogianakis, G., Bühler, C., & Soede, M. (Eds.). (1997). *Advancement of assistive technology*. Amsterdam: IOS Press.

Calhoun, E. F. (2009). Action research: Three approaches. In Schmuck, R. A. (Ed.), *Practical action research* (pp. 61–67). Thousand Oaks, CA: Corwin Press.

Clarkson, J. (Ed.). (2003). *Inclusive design: Design for the whole population*. New York: Springer.

Cooper, J. (Ed.). (2000). *Law, rights and disability*. London: Jessica Kingsley Publishers.

Disability Rights Commission. (2009). *Disabled people*. Retrieved June 22, 2010, from http://www.direct.gov.uk/en/DisabledPeople/RightsAndObligations/DisabilityRights/DG_4001070

Ferguson, R. (2007). *The blind need not apply: A history of overcoming prejudice in the orientation and mobility profession*. Charlotte, NC: Information Age Publishing.

Hong Kong Blind Union. (2007). *Ten tips to get along with visually impaired persons*. Retrieved June 22, 2010, from http://www.hkbu.org.hk/44.0.html/

Imrie, R., & Hall, P. (2001). *Inclusive design: Designing and developing accessible environments*. London: Spon Press.

Jordan, P. W., & Green, W. S. (1999). *Human factors in product design: Current practice and future trends*. London: Taylor and Francis.

Kelley, P., Sanspree, M., & Davidson, R. (2000). Vision impairment in children and youth. In Silverstone, B., Lang, M. A., Rosenthal, B., & Faye, E. E. (Eds.), *The lighthouse handbook on vision impairment and vision rehabilitation* (*Vol. 2*, pp. 1137–1151). New York: Oxford University Press.

Kumar, S. (2002). *Methods for community participation: A complete guide for practitioners*. London: ITDG Publisher.

Kuniavsky, M. (2003). *Observing the user experience: A practitioner's guide to user research*. San Francisco, CA: Morgan Kaufmann.

Kwok, Y. C. J., & Ku, H. B. B. (2001). *We are alive: Photonarrative of the everyday life of the elderly in Hong Kong* (Chinese ed.). Hong Kong: The Hong Kong Polytechnic University.

Kwok, Y. C. J., & Siu, K. W. M. (2002). *Participatory research for the designing of children and youth integrated services centres*. Hong Kong: The Hong Kong Polytechnic University.

Lovelock, R. (1995). *Visual impairment, social support: Recent research in context*. Aldershort: Avebury.

McIntyre, A. (2008). *Participatory action research*. Los Angeles, CA: Sage.

Mehrhoff, W. A. (1999). *Community design: A team approach to dynamic community systems*. Thousand Oaks, CA: Sage.

Oliver, M. (2001). Disability issues in the postmodern world. In Barton, L. (Ed.), *Disability, politics and the struggle for change* (pp. 149–159). London: David Fulton Publishers.

ORBIS. (2009). *ORBIS Hong Kong: Saving sight, rebuilding lives*. Retrieved June 22, 2010, from http://www.orbis.org/Default.aspx?cid=4870&lang=1

Prevent Blindness America. (2002). *Vision problems in the U.S.* Schaumburg, IL: Prevent Blindness America.

Rehabilitation Alliance Hong Kong. (2004). *Research on the accessibility of public disabled toilets*. Hong Kong: Rehabilitation Alliance Hong Kong.

Sanoff, H. (2000). *Community participation methods in design and planning*. New York: Wiley.

Schuler, D., & Namioka, A. (Eds.). (1993). *Participatory design: Principles and practices*. Hillsdale, NJ: L. Erlbaum Associates.

Scrivener, S. A. R., Ball, L. J., & Woodcock, A. (Eds.). (2000). *Collaborative design*. New York: Springer.

Seidman, K. R. (2003). Understanding visual impairments. In Axel, E. S., & Levent, N. S. (Eds.), *Art beyond sight: A resource guide to art, creativity, and visual impairment* (pp. 50–57). New York: American Foundation for the Blind Press.

Siu, K. W. M. (2005a). Userfitness in design: A case study of public toilet facilities. In H. Pham & S. Yamada (Eds.), *2005 proceeding: Eleventh ISSAT international conference: Reliability and quality in design* (pp. 41-47). New Brunswick, NJ: International Society of Science and Applied Technologies.

Siu, K. W. M. (2005b). Pleasurable products: Public space furniture with userfitness. *Journal of Engineering Design*, *16*(6), 545–555. doi:10.1080/09544820500273383

Siu, K. W. M. (2006a). Inclusive design: Public toilet designs for blind persons. *Annual Journal of SOPHEN*, *2*(1), 75–77.

Siu, K. W. M. (2006b). FISH: Design quality of public toilets for visually impaired persons. In H. Pham & S. Yamada (Eds.), *2006 proceeding: Twelfth ISSAT international conference: Reliability and quality in design* (pp. 122-128). New Brunswick, NJ: International Society of Science and Applied Technologies.

Siu, K. W. M. (2007). *Accessible design: Devices for blind persons to access buses*. Hong Kong: The Hong Kong Polytechnic University.

Siu, K. W. M. (2008). Better design quality of public toilets for visually impaired persons: An all-round concept in design for the promotion of health. *The Journal of the Royal Society for the Promotion of Health*, *128*(6), 313–319. doi:10.1177/1466424008092801

Siu, K. W. M. (2008). *Inclusive design*. Hong Kong: The Hong Kong Polytechnic University.

Siu, K. W. M., & Kwok, J. Y. C. (2004). Collective and democratic creativity: Participatory research and design. *Korean Journal of Thinking & Problem Solving*, *14*(1), 11–27.

Stephanidis, C., Akoumianakis, D., & Ziegler, J. (1997). Universal accessibility and standardisation: New opportunities and prospects. In Anogianakis, G., Bühler, C., & Soede, M. (Eds.), *Advancement of assistive technology* (pp. 39–43). Amsterdam: IOS Press.

Treqaskis, C. (2004). *Constructions of disability: Researching the interface between disabled and non-disabled people*. London: Routledge. doi:10.4324/9780203299517

United Nations. (1997). *A summary of United Nations agreements on human rights*. Retrieved June 22, 2010, from http://www.hrweb.org/legal/undocs.html/

United Nations. (2008). *Rights and dignity of persons with disabilities*. Retrieved June 22, 2010, from http://www.un.org/disabilities

United Nations. (2009). *Rights and dignity of persons with disabilities*. Retrieved June 22, 2010, from http://www.un.org/disabilities/

Wiklund, M. E. (2005). Making medical device interfaces more user-friendly. In Wiklund, M. E., & Wilcox, S. B. (Eds.), *Designing usability into medical products* (pp. 151–160). Boca Raton, FL: Taylor & Francis/CRC Press. doi:10.1201/9781420038088.ch16

Willems, C. G., Vlaskamp, F. J. M., & Knops, H. T. P. (1997). Use of technology in the support of independent living of people with special needs. In Anogianakis, G., Bühler, C., & Soede, M. (Eds.), *Advancement of assistive technology* (pp. 125–129). Amsterdam: IOS Press.

World Health Organization. (2003). *Up to 45 million blind people globally – and growing*. Retrieved June 22, 2010, from http://www.who.int/mediacentre/news/releases/2003/pr73/en/

World Health Organization. (2009). *Visual impairment and blindness*. Retrieved June 22, 2010, from http://www.who.int/mediacentre/factsheets/fs282/en/

KEY TERMS AND DEFINITIONS

Direct Discrimination: Direct discrimination occurs when, on the ground of disability, a person with a disability is treated less favourably than another person without a disability in similar circumstances.

Disability: A disability is a physical or mental impairment that substantially limits one or more major life activities. A person may also qualify as disabled if he or she has had an impairment in the past or is seen as disabled based on a personal or group standard or norm. Such impairments may include physical, sensory, and cognitive impairments.

Discrimination: Discrimination is the unjust or prejudicial treatment of different categories of people or things, especially on the grounds of race, age, sex or perceived handicap. It is also sociological term that refers to treatment taken toward or against a person of a certain group that is based on class or category. The United Nations points out that discriminatory behaviours take many forms, but they all involve some form of exclusion or rejection.

Inclusive Design: Inclusive design is an approach to design that aims to be flexible enough to serve the broadest possible range of users. Inclusive design calls for adaptable interfaces to be built into the product (or other kinds of designs) early in the design phase, producing features easily used by everyone. It allows for customization based on user needs and preferences. It provides equivalent access to content based on user needs and preferences. It also provides access to users with disabilities and provides better usability for everyone.

Indirect Discrimination: Indirect discrimination occurs when a condition or requirement is applied to everyone, but in practice affects people with a disability more adversely, is to their detriment, and such condition or requirement cannot be justified.

Public Design: Public design is a discipline, an approach and area of design that aims to serve a wide scope of users, as distinct from a particular set of user(s). Public design aims to serve the different needs and preferences of the public. Most of the time, public design relates to public systems (e.g., telecommunication systems), environments (e.g., streets), facilities (e.g., rubbish bins), and graphical design images (public poster images).

Public Toilet: Toilets may also be called lavatories, washrooms or bathrooms in different functional, physical, social and cultural situations. A public toilet is a toilet built and/or managed by a government or public agency, or a privately managed toilet, which is available for public access.

Visual Impairment: A visually impaired person is blind or has impaired vision. Blindness is the condition of lacking visual perception due to physiological or neurological factors. Blindness is frequently used to describe severe visual impairment with residual vision. Those described as having only "light perception" can see no more than the ability to tell light from dark. Low vision is defined as visual acuity of less than 6/18, but equal to or better than 3/60, or corresponding visual field loss to less than 20 degrees, in the better eye with best possible correction.

Chapter 2
New Design Paradigm:
Shaping and Employment

Vladimir M. Sedenkov
Belarusian State University, Belarus

ABSTRACT

The multiple shortcomings of the current Design paradigm manifest the need of its modification. Our objective was to find out an appropriate mechanism. But such a mechanism could not be revealed without assistance of a Design theory. The emergent dilemma – to use one of the available theories or develop a new one – was resolved by choosing the third way: rearrangement of the material at hand on modularity principles with initiation of fundamental (systemic) Design theory module via identification of its paradigm. While doing this, we had to overcome a number of delusions ingrained in engineering design, concerned firstly with design problem, process and design representation. To push these efforts forward, a scientific base named Continuous Process Theory had been developed. Systemic module initiation enabled to define a paradigm of the second Design theory module – the sought-for Design practice paradigm. Discussion on the outcomes of this definition rounds off this chapter.

INTRODUCTION

The Way to Reveal an Adequate Design Paradigm

The title of the chapter declares generation of a new design paradigm – the set of practices that define and drive (implements and manage) design process (*DPR*). This assumes that a certain version of the paradigm is available. Indeed, it cannot be out of place because of many decades of everyday designing. Design paradigm ("applied theory") is crystallized by design practice and is needed for practitioners in each design field – whether it be mechanical engineering, information science, architecture, chemistry, nano-technology or something else. Design paradigm is the base of design methodology within a given domain and the benchmark for developers of computer-aided design facilities.

DOI: 10.4018/978-1-61520-617-9.ch002

In Design theory absence, Design paradigm formation was spontaneous: the values of its descriptors mirror a specialized empiricism, intuition, experience, borrowings, parts of theories possessed by other disciplines, etc. This has stipulated a number of weak points of the paradigm. With the reference to design in mechanical engineering, recall some of those:

- it employs semi-intuitive design language;
- supports mainly adaptation design;
- generates ill-observable, non-holistic *DPR*, which is equally insufficient for learning and teaching and for the most part implicit;
- has no ideas how to keep the *DPR* complexity to a manageable level;
- structure synthesis problem remains unsolved;
- the role of computer in designing is obscure and insufficient;

Thus, modification or replacement (radical modification) of design paradigm is anticipated and in demand. However, there is no a regular mechanism of paradigm improvement, which should concurrently be a mechanism of paradigm identification forgoing to its change. Let us try to find out such a mechanism.

To be analyzed, evaluated and modified, the paradigm should have a sort of representation. We associate with any Design paradigm representation a set of descriptors or *paradigmants* – the certain characteristics of a paradigm, which take on one or another value. Thus, paradigmants characterize via their values this or that paradigm during a certain period of time. For instance, Design paradigm is concerned with such paradigmants as *notion base of design language* (formalized ∨ intuitive ∨ semi-intuitive), *the mode of structure synthesis problem realization* (explicit ∨ implicit), *design system architecture* (an hierarchy of subsystems ∨ another) and others.

If paradigm representation is available, we define a paradigm modification as the change of values for one or a subset of its paradigmants. To regularize the way of paradigm change, we distinguish within its representation a minimal subset of paradigmants sufficient for a unique paradigm identification – call it *design paradigm signature* (*Sg*). Then change of values for paradigmants, which are beyond the signature, would signify the paradigm modification, while the change of value for at least one signature's paradigmant replaces the paradigm. A signature considered without its paradigmants values is called a signature platform or *meta-signature* (*mSg*). Choosing different variants of *mSg* attribution, the produced signature alternatives (paradigm identifiers) could be compared and paradigm assessed as a whole.

It should be noted that the outlined mechanism of paradigm identification and modification has a heuristic base – the choice of both paradigm representation and modification rests mainly on experience, intuition and experiment. This does not make the mechanism reliable. Besides, it generates only paradigm clones according to given representation – this deprives the mechanism of practical value. Generation of paradigm versions becomes possible after changing the course of identification for an opposite one. This means that paradigm representation initially is unavailable and its deriving begins with identification of Design paradigm meta-signature and signature. Then the paradigm representation will be obtained by deployment of its signature. But such a systematic way of Design paradigm handling needs for feeding it by a resource of paradigm *mSg* generation. Such a resource could be provided by a Design theory only.

TOWARDS DESIGN THEORY PARADIGM IDENTIFICATION

Requirements to Design Theory

Does the stated need for Design theory signify development its new version or sampling from the list of currently available Design theories is also possible? To make decision, let us lay down three clue, in our view, requirements to Design theory.

1. First of all, the theory has to have a base of scientific principles. "The teaching of engineering subjects such as thermodynamics, mechanics, and materials are based on established principles like the first and second laws of thermodynamics, Newton's laws, and atomic and molecular theories of matter… Engineering design lacks sufficient scientific foundation, and without an adequate base of scientific principles, engineering design education and practice are too much guided by specialized empiricism, intuition, and experience" (Dixon, 1988, p. 145).
2. To be accepted by practice, a theory should answer its questions, solve its problems, and uppermost provide a regular mechanism of Design paradigm identification (attribution of *mSg* paradigmants) and modification. In case of design practice, its *mSg* is invariant and fixed a priori: i) constructive design definition; ii) a priori described process that provides the sought-for design, i.e. *meta design process* (Sedenkov, 2008, p. 93); iii) a system that supports meta process run or *design system*. Combining the requirements 1 and 2, we assert that the base of scientific principles should be aimed firstly at beneficial identification of Design paradigm (a proper *Sg* of Design practice paradigm).
3. The requirement of a viable theory organization is the last one in the triple. The origin of this requirement needs for comments.

We have in focus a Design theory, which should have its own paradigm – a set of ideas that define a scientific discipline during a particular period of time. This theory should identify an adequate design paradigm that is the part of itself, as the identified paradigmants of Design paradigm signature will become paradigmants of Design theory paradigm. This separation of Design theory and Design paradigm in their unity gives rise to the idea of modular structure of Design theory where paradigms of two main modules – Design theory and Design practice – define accordingly fundamental and applied aspects of a single theory.

Indeed, distinct components of Design theory may have substantially different dynamics of development. For instance, there are highly dynamic divisions aiming at knowledge handling. Next, market dynamics of computer aided design facilities and value dynamics of Design paradigm paradigmants, which are beyond the signature, interrelated closely and stimulate each other. Besides, the status of interdiscipline accepted by designing (Cross, 2002) obligates it to interact with a dynamic list of contractor disciplines. In this situation, a single Design theory has been either substantially loose or inflexible. The first variant makes a theory be of little use for design practice while the second one entails its frequent revisions and disintegration into a number of short-lived mini-theories supporting particular design stages (DfX). Between one-piece ("General") and mini-theories, there is a class of "heuristic" theories for particular design fields.

In case of modular structure of Design theory, the paradigms of modules get the autonomy in the sense of possibility to vary within the values of their current signatures. The paradigm change of Design theory itself (replacement of its signature value) occurs either on a signal "underneath" (when the paradigms of modules are not sufficient any more), or "from above" (with gathering new knowledge about signature's paradigmants possessed by Design theory paradigm and/or with alteration of scientific principles). It seems of importance to reserve the space for paradigmatics variations in each of modules as it should ensure simultaneously both the theory stability and its progress without strong conflicts between them.

Choice Testing

Let us see, whether some of the available theories could be suited for our objectives. To avoid single analysis (it is beyond the scope of this chapter), we shall manipulate the two above mentioned classes testing those for compliance with the stated requirements.

The class name "heuristic" indicates not only the origin of theories "from practice" but the domination of heuristics in their statements as well. The prevalent method of such theories creation is induction, which begins with generalization of design experience. With the terminology introduced, one may state that this class of studies (Lossack&Grabowski, 2000; Eder&Hosnedl, 2008) is characterized by the attempt to elevate applied aspect of design theory to the rank of fundamental. In other words, Design paradigm signature should become a core of Design theory paradigm signature. But as initial component is not reliable, the result proves to be artificial as well. In so doing, the weak points of design paradigm (for instance, the transfer of concepts related to representation and analysis of available products to synthesis of nonexistent products) turn to be weak points of Design theory.

The alternative class of theories ("General Design Theories") stresses the fundamental aspect putting off the handling of applied one to the end of theory formation. But when the theory comes to the focus of practice, it turns out that the applied aspect of the theory does not suit for complete and consistent use (Yoshikawa, 1981; Hatchuel, 2003). Thus, though the maxim "a theory will help advance the practice of design" (Dixon, 1988, p. 148) became a generality, it does not follows from the theory itself *how* and *by which* the advancement should come.

Having compared the properties of Design theory classes with the triple of requirements (sound scientific foundations, aimed at Design paradigm identification, and modularity), we have to conclude that the choice in the space of available Design theories could hardly be resultant to our end.

What Then, a New Theory?

However, the conclusion about necessity of new Design theory creation would be premature as well. In the light of modular structure and relative autonomy of both – modules evolution and its rate, theory shaping and its validation is the issue of time and efforts of the entire design research community. On the other hand, odd parts of a theory are in existence (Design paradigm, design methods, design knowledge, etc.) but those are either inadequate (like Design paradigm) or technologically imperfect in practical use (as knowledge or design methods).

Under this situation, it seems rational to rearrange the available material on the principles of modular theory organization. Then the first module should be established. It has doubling the parts of an attractor for odd elements of the theory and a platform for creation and synergetic operation the rest of the theory modules. This gives grounds to consider not a theory creation but its *initiation*, that is shaping a module of fundamental theory aspect or, hereinafter, a *theory platform*.

The Platform Based Design Theory Organization

The Design theory platform should provide a mode of Design paradigm signature identification and so is initially aimed at establishment and support its applied aspect. To develop the platform means firstly to propose a meta-signature of its paradigms. Then we have to initiate an adequate signature via attribution *mSg*'s paradigmants. The signature of Design theory platform will ensure the Design paradigm signature identification. The latter also has *mSg*, which is known a priori and was mentioned above.

We complete to plan a modular structure of Design theory by introduction of its third module. It will be a meta knowledge base that involves design methods, design knowledge, computer software, contractor disciplines, multidiscipline models and will be provided with control facilities to support knowledge access, knowledge storage and knowledge maintenance. We call this module "Design knowledge management system". Its meta-signature consists of three paradigmants: knowledge structuring, knowledge storage and knowledge access.

Thus, the only way to obtain Design paradigm signature (attribution of *mSg* paradigmants) is the initiation of constructive version of Design theory via identification of paradigm signature for the main module of the theory. In turn, Design paradigm signature has been a trigger for shaping the third module of the theory via identification its signature. Such a "modular-trigger" concept of theory development removes from agenda the necessity of General Design Theory. The latter can be changed for the new entry – a unified Design theory platform. On the other hand, it has been doubtful to have internal Design theory for each particular design field – with the use of master module they can adjust two other modules from the triplet and get a required version of Design paradigm for the domain.

The reminder of this chapter is organized as follows. In the next part we briefly review the proposed science base for design – Continuous Process Theory and the mode of problems tackling when those have the same insolubility (of the second kind) as a design problem. The third part addresses to establishment of Design Theory platform that paves the way to adequate design paradigm identification. The proper design practice paradigm revealing is the content of the forth part, while the main outcomes are discussed in the fifth part.

TOWARDS DESIGN THEORY PLATFORM: THE BASE OF SCIENTIFIC PRINCIPLES

Continuous Process Theory

Dixon (1988) states that "design is a process, and processes are not the usual subject of theoretical formulations" (p. 146). A process is a continuous entity and therefore it should have a continuous representation. Discrete representation of a process – by an hierarchy of sub-processes, for instance – mostly inefficient. *Continuous Process Theory* (*CPT*) is just intended to serve for the part of adequate scientific principle for design theory development. Here are two of the most important applications: (1) with the help of *CPT*, we shall show a design problem membership of the class of problems with insolubility of the second kind and outline the way to get an answer to such problems; (2) structure synthesis problem realization needs for continuous representation of a design, and we shall get it using (continuous) representation of *operation process* of the required product.

The starting point of *CPT* is representation of any process *PR* by its scheme:

$$PR=(D, P), \tag{1}$$

where *P* stands for a *processor* that performs transformation of energy, raw materials, information or products entering its input (I^P), and *D* stands for a *procedure* that describes the function of *P* over its I^P. *D* and *P* are referred to as a process *object* and *subject* respectively. Thus, the *subject matter* of *CPT* is a representation of complex process in a continuous mode. The *CPT method* – the scheme technique of process – is characterized by the following:

- A set of process schemes is added with a number of binary relations;
- Process schemes linked by distinguished relations make up a *structure*;

Figure 1. First order process structures on d- and p-relations

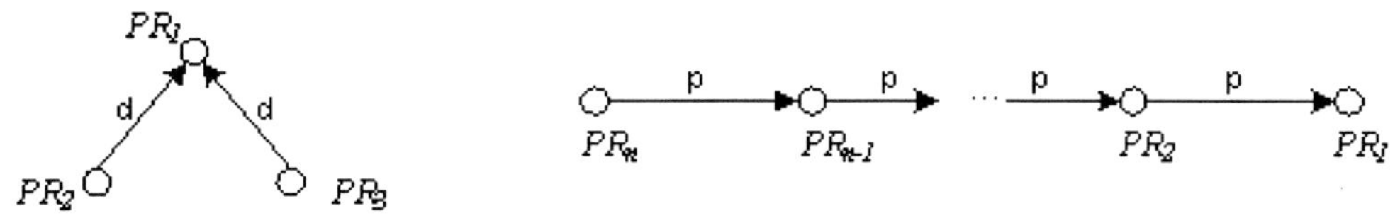

- The rules for structure shaping and conditions for its operability are stated.

The *CPT goal* is a constructive proof for a process (induced, for instance, by a declared problem and represented by the scheme) its runability through building up upon its scheme an operable continuous structure of processes.

There are two relations appropriate for making up the structures: providing relation or **p**-*relation* and relation of determination or **d**-*relation*. PR_1 and PR_2 are linked with **p**-relation $(PR_2 \xrightarrow{p} PR_1)$ if the output of PR_2 serves for the input of PR_1. If the output of PR_2 becomes a scheme component of PR_1 (*D* or *P*), these two processes are linked by **d**-relation $(PR_2 \xrightarrow{d} PR_1)$.

A set of processes (or their schemes) continuously linked by **d**- or **p**-relation forms an *elementary structure* of processes (or process schemes). This structure is represented by a plain graph, the nodes of which serves for the processes and each arc is a cross-linking relation. Elementary structures have an *order n*, equal to *1* (Figure1).

Each processes structure has a *core* (PR_1), a *tail* (PR_n) or *multitail* (PR_2, PR_3) and a *body* (the part between the core and tail). Elementary structures generated by one of the relations can form a new structure by the alternative (orthogonal) relation. The new structure gets *n=2* and is represented by a *super-tree* (*S*-tree) – an arc bichromatic (2D) tree where each node is an ordinary graph (Figure 2*a*). Relations between members within such non-elementary structure of processes are relations between the members' cores.

The structures with *n=2* may serve for the members in a structure of the next order. A usual motive for the third order structure generation is the synthesizing a required process from scratch. To this end, let us associate with each process

Figure 2. The second order S-tree by p-relation (a) and the third order S-tree by d-relation (b)

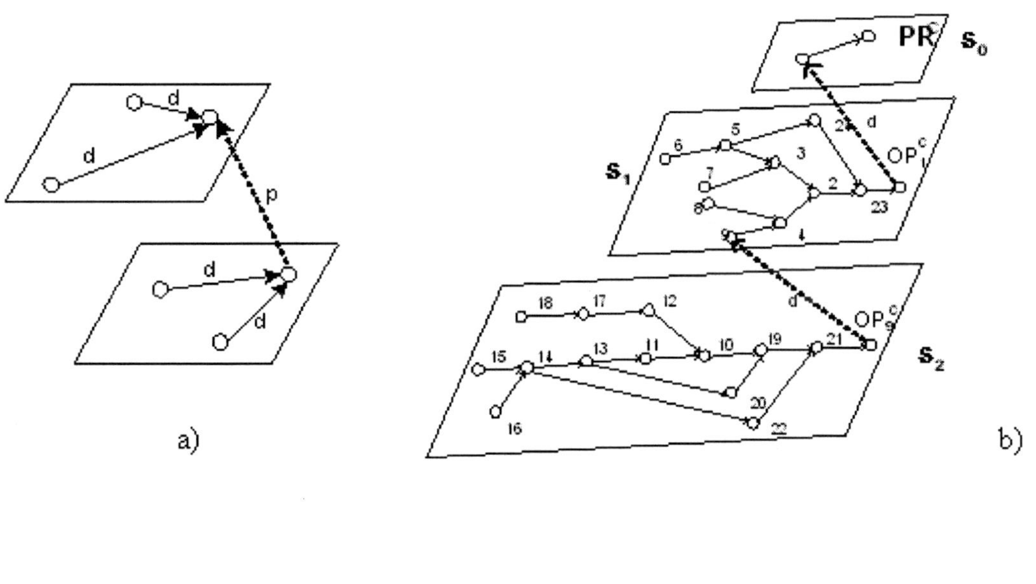

Table 1. Constructive proof of logical runability for a conditional process

Down stroke	Up stroke
$SD^V \xrightarrow{d} PR_1^C \xleftarrow{d} SP^V$ **1**	PR_1^L =(D, P) **8**
PR_1^V $p\uparrow$ PR_2^C **2**	PR_1^L $p\uparrow$ PR_2^L $\Uparrow$ I^P **7**
PR_1^V $p\uparrow$ $SD^V \xrightarrow{d} PR_2^C \xleftarrow{d} SP^V$ **3**	$SP^L \xrightarrow{d} PR_1^V \xleftarrow{d} SD^L$ $p\uparrow$ PR_2^L $\Uparrow$ I^P **6**
PR_1^V $p\uparrow$ PR_2^V $\Uparrow$ I^P **4**	PR_1^V $p\uparrow$ $SP^L \xrightarrow{d} PR_2^V \xleftarrow{d} SD^L$ $\Uparrow$ I^P **5**

scheme a level of its *uncertainty* (*UL*) as *UL* of the scheme's components.

- A process, which has $UL=0$, is called *physical*: its *D* and *P* are real.
- $UL=1$ corresponds to a *logical* process: its *D* and *P* have descriptions sufficient for their physical implementation.
- A *virtual* process has $UL=2$: its *D* and *P* exist only as mental images.
- $UL=3$ is assigned to a *conditional* process (PR^C): its result has been declared but its *D* and *P* are presented by their symbols only.

Constructive proof of logical runability for PR^C consists in stepwise reduction of its *UL*. A step of reduction is referred to as *determination* of conditional, virtual or logical process. Each step of $PR^C=(D, P)$ determination is modeled by the search on appropriate sets for its *D* and *P*. Thus, each search assumes the run of two processes – *search for D* (*SD*) and *search for P* (*SP*):

$$SD \xrightarrow{d} PR^C \xleftarrow{d} SP \qquad (2)$$

While the two-stroke determination of PR^C (Table 1), the objective of the *virtu al* (downward) determination is the reduction $UL=3 \rightarrow UL=2$; during the second or the stroke of *logical* (upward) determination, the reduction $UL=2 \rightarrow UL=1$ will take place.

This two-stroke cycle of PR^C determination results in *S*-tree (Figure 2*b*) where each node is the second order structure (2). Thus, while representing a complex process in *CPT*, the underlying principle is not a *D* decomposition with shaping a hierarchy of sub-processes, but making up a continuous structure out f process schemes.

Realization of Problems with Insolubility of the Second Kind

Let the structure (2) be elementary, i.e. when PR^C has a defined input (I^P) and does not need a generation of providing process:

$$SD \xrightarrow{d} \underset{I^{P\downarrow}}{PR^V} \xleftarrow{d} SP \tag{3}$$

Reduce (3) to the form (4) assuming **d**-relation between the contents in brackets:

$$P\ rB = < SD, SP > \xrightarrow{d} < \underset{I^{P\downarrow}}{PR^V} > \tag{4}$$

Following (Polya, 1965) while generic problem tackling, we distinguish a "*problem solution*" (the determined *D* and *P* from PR^L – the results of <*SD, SP*> performance) and an "*answer to the problem*" (the output of PR^L). The activity consisting in problem solving (<*SD, SP*> execution) and answer computing (PR^L run) is referred to as *problem realization*.

But expression (4) does not meet the structure precondition – the two processes in <*SD, SP*> are linked by empty relation. To restore for (4) the structure status, divide it into two schemes of processes where the relation between them is **d**-relation between their cores (the first scheme has no tail):

$$P\ rB = < SD > \xrightarrow{d} < SP \xrightarrow{d} \underset{I^{P\downarrow}}{PR^V} > \tag{5}$$

Next, making **d**-relations implicit, we get two expressions named *problem schemes*:

$$P\ rB = < SD > < SP, \underset{I^{P\downarrow}}{PR^V} > \quad \text{or} \quad P\ rB = < SP > < SD, \underset{I^{P\downarrow}}{PR^V} > \tag{6}$$

Thus, problem schemes like (6) get $n=2$ as each component scheme is the process structure with $n=1$. Incidentally, a structure made out of problem schemes (upon **p**-relation) should have $n=3$. Problem realization analysis in terms of problems schemes and their interrelations is named *problems schematics*. The use of this or that scheme from (6) depends on problem's specificity.

The problems that have both solution and answer are *soluble*. But, there are many more problems (posed usually directly upon a required result), which have an answer but have no solution. Those are referred to as insoluble of the *second kind*. (The problems that have neither solution nor answer have insolubility of the *first kind* – "perpetuum mobile" creation, for instance.) For all that, the answer availability to the problem with insolubility of the second kind indicates that instead of original (insoluble) some another problem is actually realized, the answer to which coincides with the answer to the initial problem. This another problem is said to be *conjugate problem* (*CP*) with respect to original one. Knowledge engineering gives the classic example of such a pair of problems: insoluble original – diagnosis problem, and soluble conjugate – knowledge inference problem. Answer deriving to a problem through realization of its *CP* is called *quasi-* or *q-realization* of the initial problem.

Turn now to design problem (*DP*), which has the reputation of ill-defined (Simon, 1970) or even wicked (Rittel & Webber, 1979), and write down its scheme:

$$P\ rB =< SD >< SP, PR^{V} > \quad {}^{\llcorner} I^{P} = \text{needs and requirements} \tag{7}$$

Any problem can be presented by "what is given" and "what is required". In case of *DP*, the "given" is a description of needs and a set of requirements, while the "required" is a sought-for product design. As the needs and requirements cannot be directly transformed into a goal design (the former and the latter belong to different conceptual worlds (Meijers, 2000), *DP* has no solution. Formally, there is no $D \in DPR$, so *DP* is a phantom problem. At the same time, an answer to *DP* (a design) is possible. Answer deriving to *DP* through realization of a conjugate problem (CP^{DP}) is referred to as *q*-realization of *DP*. Thus, the replacement, in respect to *DP*, of ambiguous "ill-defined" for constructive "insoluble of the second kind" means searching for CP^{DP}.

DESIGN THEORY PLATFORM: THE PARADIGM SIGNATURE IDENTIFICATION

A Conjugate Problem for a Design One

AXIOM 1

There is no a priori DPR: DPR shaping comes to an end concurrently with design obtaining

Let us begin the search for CP^{DP} with the search for a process conjugate to $DPR \in DP$ and preliminarily named "not *DPR*" ("$\neg DPR$"). Restore upon $\neg DPR$ the scheme of CP^{DP}:

$$CP^{DP} =< SD >< SP, \neg DPR > \tag{8}$$

Virtual (or top-down, see Table 1) determination of $\neg DPR$ by *D* requires generation of this or that virtual inputs for the process. For instance, any available design theory, which unavoidably employs some CP^{DP} by default, uses for the $\neg DPR$ input a future product representation. The latter may have different levels of abstraction – a product during redesign or its abstract idea – and was called "context description" (Braha&Reich, 2003); such input refines $\neg DPR$ to quasi-*DPR* (*qDPR*) or a process of product representation transformation. In this case the CP^{DP} problem scheme changes its view for (9) being ready for logical (bottom-up) determination of *qDPR*:

$$CP^{DP} =< SD >< SP, qDPR > \quad {}^{\llcorner} I^{P} = \text{"contect description"} \tag{9}$$

Explicit *qDPR* determination upon its object *D* (the result of *SD*>) is generally either impossible or restricted (*D* proposed in (VDI, 1987) or intended for Suh's model (1990, 2001)). Meanwhile, most of available design theories present this $D \in qDPR$ as a "*DPR* model" (*qDPR*=(*D*, *P*)=(*"DPR model"*, *P*)), though inexistent entity, particularly *DPR*, cannot have a model by definition. Next, the performance of <*SP*> results in a "human master processor" (*H*) because "*DPR* model" realization is only within human's depth. Thus, completeness and quality of CP^{DP} solution depend on what is entering the $\neg DPR$ input. How tear oneself away from the circle of mythic *DPR* modeling since a product representation could hardly be recognized as adequate input of $\neg DPR$?

Axiom 2

Just as a physical product is the outcome of product design processing, so a product design has been an outcome of design process design implementation (processing).

The stated axiom defines both generative process conjugate to *DPR* – call it *meta-DPR* or *mDPR* – and this process input, i.e. *DPR* design. *mDPR* should both generate *DPR* and produce a required product design through *DPR* performance. The new conjugate problem restored on *mDPR* is the problem of *DPR* design realization:

$$CP^{DP} =< SP >< SD, mDPR > \qquad \llcorner I^{P} =_{\text{DPR design}} \tag{10}$$

It remains to construct *DPR* design and proceed with the search of solution to CP^{DP} when the latter is represented by the scheme (10). In the course of this problem solving, we have to identify a signature for the platform (systemic module) of Design theory. The solution per se is required to be triply independent: of a master processor, design task and design domain. We call such independence "3D independence (3Di). 3Di solution to CP^{DP} will be called also *systemic* solution. To obtain the latter, the main input of *mDPR* – *DPR* design – is required to be represented in 3Di form as well. *mDPR* platform or $mDPR_3$ will undertake the realization of 3Di *DPR* design. The input data of initial design problem serves for the second input of *mDPR*. We have not recalled it after (7) but it will serve for control data for the sought-for $D \in mDPR$. Thereupon we refine the scheme of CP^{DP} under solution as in (11).

$$CP^{DP} =< SD >< SP, (D_{\in} mDPR, P_{\in} mDPR) > \qquad _{\text{needs and requirements}}= I^{P\lrcorner} \qquad \llcorner I^{P} =_{\text{DPR design}} \tag{11}$$

DPR Design Construction

Design Progress Concept

A product design does not emerge in a single-step action – it evolves over time. It is not desirable to have this evolving chaotic. Hence, any design process should realize some *design progress concept (DPC)*. For instance, the famous model of engineering design process (Pahl & Beitz, 1984) mirrors *DPC*, which may be called stepwise refinement of design abstract description (conceptual, embodiment and detail design). Another example of *DPC* is the *evolution of population* (selection, mutation, recombination), presented in numerous papers on evolutionary design, e.g. (Vajna et al., 2005). For our project, we have taken the most natural, in our view, concept named *evolution of individual* – adaptation of the current product design state to a new state of *operation environment design* related to the sought-for product. This *DPC* is fixed in (10) by I^P-input of *mDPR*. In compliance with this *DPC*, the (operation) environment, which will accept a product after its physical implementation, is under design concurrently with a product. The initial state of environment is mirrored in (11) by I^D-input of *mDPR*.

Dynamic Design Representation

Thus, we should have to deal with representations of three designs – a product design, environment design and design process design. The idea of some unification of these representations (via a platform, for instance) knocks at the door, as they say. But yet more earlier by the same door, another idea stands in the wings – dynamic design representation, which is integral to design synthesis. Lastly, the all design representations within systemic module of Design theory, inclusive of the unified, should be 3D independent.

In point of a (physical) design itself, the listed concepts are unrealizable. Though the following axiom brings about a possible solution to the problem.

AXIOM 3

Just as a product is obtained via implementation of a product design, so a product design should be obtained via implementation of product's design of design:

design of design → design → artefact. (12)

An extension of string (12) to the left (design of design of design, etc.) is replaced with the sequence of design of design maturity levels (ML_i, $i=1,2,...$). Thus, the possibility of a unified, dynamic and 3Di representation is associated now not with a design but with a design of design or, as we shall call it in what follows, *logical design* (*L*-design) of a product, environment or design process.

We shall say that the sequence of *MLs* names reflects a *diachronic* (*dh*) or "historical" structure of *L*-design. Componentization of a single ML_i will be referred to as *synchronous* or *sh-structure* of a respective item from the *dh*-structure. At the same time, diachronic structure reflects only design *quasi-dynamics*, if synchronous representations of *dh*-items (description of componentization) are static, e.g. drawings. If we manage to impart dynamics to *sh*-representations, we shall obtain the corresponding representation for the *L*-design in whole.

Dynamics signifies a process. Hence, the case in point is a representation of each *dh*-structure item in terms of processes. This is possible if the content of any such item, from the very first one, should be the determination of the process scheme – namely, the scheme of *operation process* of the sought-for product. In that case, each *sh*-component will be specified by the structure of processes. This opportunity is provided by continuous process theory. Thus, with the assumption that designer perceives the future product in terms of its operation processes, we shall try to display his reasoning by:

1. shaping the sequence of *L*-design states (design *dh*-structuring),
2. dynamic representation of these states (by the structure of processes),
3. choosing the number of states that provides designer with a comfort design increment ("the quantum of evolution").

Successful virtual designing of a product comes to an end with deriving of a final *L*-design state – dynamic representation of the last item's (within $ML_1, \ldots, ML_n$) *sh*-structure. This sequence is given the name of *approximate model* (*AM*) of *L*-design. Incident ML_i and ML_{i+1}, $i=\overline{1, n-1}$, are linked with *embedding relation* ($\prec$).

Let us divide the *AM* into its *structure St* (*dh*-structure of *L*-design) and *semantics Sm* (the *sh*-content of each *dh*-structure item). If *Sm* of each ML_i is determined, we have *approximation model* of *L*-design: $AM=(St, Sm)$. Otherwise, it will be a *quasi-approximation* model – $qAM=(St, Sm^*)$, where Sm^* is an abstract semantics. Then logical explicit designing is defined as *AM* construction for *L*-design or transformation of *qAM* into *AM* via refinement of $Sm^* \in qAM$. This *qAM* is called *L-design platform*.

Thus, the way to *DPR* design assumes construction of a unified *L*-design platform, refinement it till *L*-design and transformation the latter into a sought-for design. The role of objects to be unified will be performed by *dh*-structures of *L*-designs for a product and its operation environment, which design process should have to deal with. Those have to be obtained.

Primary dh-Structure for a Product L-Design

In compliance with the declared *DPC* (evolution of individual), we link by $\prec$-relation the following four product design states (*design goals*): *Prototype* (PRT), *Market version* (ITM), *Manufacturing version* (COM) and *Artefact* (ART).

$$\text{PRT} \prec \text{ITM} \prec \text{COM} \prec \text{ART} \qquad (13)$$

To smooth the discrecity of design representation in terms of its states, split the development of each state into four stages. Each stage results in *design subgoal*: *quasi-system* or qSYS (a minimal set of product units that validates the concept of

Figure 3. Primary dh-structure for a product L-design

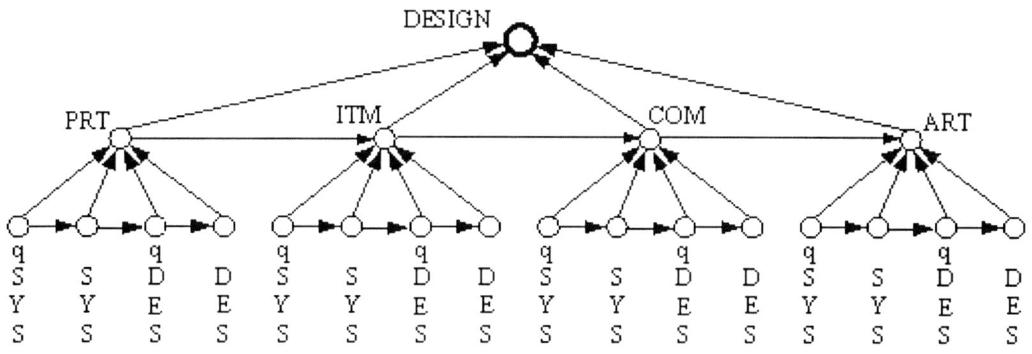

goal attainment), *system* or SYS (the extension of *q*SYS with control functions), *quasi-design* or *q*DES (space layout of the SYS constituents) and *design* or DES (the *q*DES components assigned with shape, materials, grades of finish and all necessary joints):

$$qSYS \prec SYS \prec qDES \prec DES \quad (14)$$

Combine all design states into its diachronic representation – the data structure named *quasi-hierarchy* (*q*-hierarchy) – by closing the nesting hierarchy, i.e. making the latter actual across horizontal as well. The resulted *dh*-structure is shown in Figure 3.

Primary dh-Structure for Operation Environment L-Design

OE is a family of processes $\{\mathbf{PR_q}\}$, which a product deals with throughout its lifecycle. There are only three kinds of relations between a product and a process from the family: a product can play the role of the *input*, *disturbance* or *processor* with respect to any process from $\{\mathbf{PR_q}\}$ (Figure 4*a*). Hence, we can break $\{\mathbf{PR_q}\}$ up to three constituent sets: $\{PR_{q1}\}$, $\{PR_{q2}\}$ and $\{PR_{q3}\}$ (Figure 4*b*).

Processes from $\{PR_{q1}\}$ place on their input a number of *requirements* (*Rq*); this set is shared by the product life cycle stages (six, as in our case, or more). Processes from $\{PR_{q2}\}$ impose *constraints* (*Cs*) on their disturbance. Processes from $\{PR_{q3}\}$ specify for their processor the *operation conditions* (*Cn*). In turn, each constituent set is divided into motivated (we drop the details) number of subsets – the hierarchy in Figure 4*b*. As product-oriented *OE* should also be under design, it needs *P*-independent diachronic representation (hierarchies are *H*-oriented structures). Having taken the members of each set of processes (Figure 4*b*) for the vector of 3D space, we shape *dh*-structure of *OE* design. The latter is referred to as *&-cube* (Figure 4*c*).

Embedding relation links incident cells of &-cube along the *track* of its scanning. This track leads from the starting cell of &-cube – $(X_0Y_0Z_0)$ – to its end cell – $(X_{max}Y_{max}Z_{max})$. Here are examples of tracks (brackets signify elements' iterations along the track up to their depletion):

- *(X(Z)(Y(Z)))* – layer-by-layer &-cube scanning along axis *Y*;
- *(X(Y)(Z))* – all *Z*-vectors, except the last one, are omitted in vertical layers (reduced layers);
- *(X)(Y)(Z)* – reduced &-cube (all *Y*-vectors are omitted, except the last one);
- *(X)(Z)(Y(Z))* – only the last vertical layer is taken wholly;
- *(X)(Z)(Y)(Z))* – the layer from the previous track is reduced.

Figure 4. OE design dh-structure construction

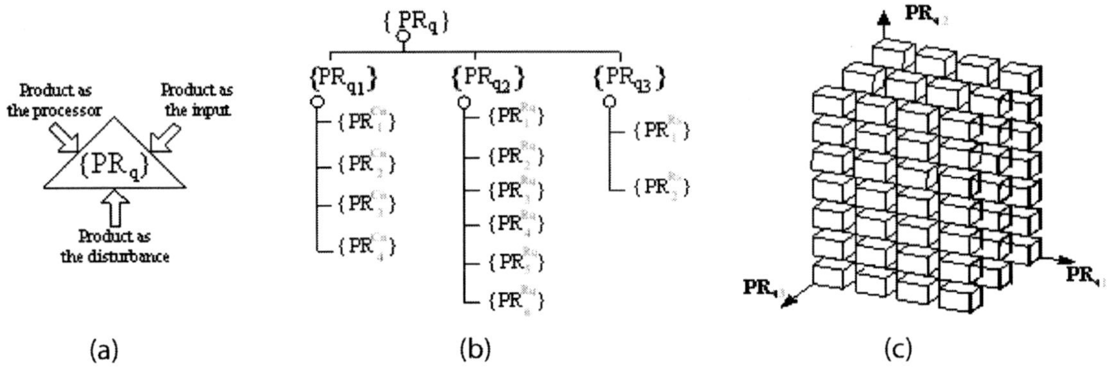

Thus, the choice of &-cube scanning track defines both relevant power of {$\mathbf{PR}_q$} and variants of its structuring. These variants are also valid for the sets {Rq} and {Cs}, generated by {$\mathbf{PR}_q$} members (*OE* processes), and consistent with the concurrently derived design properties and characteristics. Incidentally, the obtained structuring for properties and characteristics is more insightful than in (Eder & Hosnedl, 2008).

From Design Platform to DPR Design

The unified *dh*-structure is obtained by substitution of &-cube (Figure 4*c*) for terminals in the above *q*-hierarchy (Figure 3). After that, design *dh*-structure is valid for both product and *OE* designs, taking the status of $St \in qAM$. Whereupon, it takes only to refine $Sm^* \in qAM$ to come from qAM to approximate model of a product design (AM^P) and approximate model of *OE* design (AM^{OE}). The notion of design platform $qAM=(St, Sm^*)$ is completely pertinent for *DPR* design as well; the only difference is that "*L*-design" is used in this case without prefix because a physical representation for *DPR* design (a structure out of processes) coincides with the logical one.

Approximate model for *DPR* design is labelled as AM^{DPR}, and we come to it via refinement of $Sm^* \in qAM$. This refinement results in iterations of two processes (Figure 5) – synthesizing a new product design state (SPR^P, determination of the product operation process) and synthesizing a new operation environment design state (SPR^{OE}, changing for an increment the current set of conditions, requirements and restrictions, related to the sought-for product, and further paretization of this set).

Eventually, the obtained 3Di *DPR* design is represented by *AM*, in which every element of its *dh*-structure includes one iteration of the two processes mentioned above. Now when 3Di *DPR* design – the main CP^{DP} input (Figure 6) – has been obtained, we can proceed to CP^{DP} solving or determination of $mDPR \in CP^{DP}$ with respect to its subject (*P*) and object (*D*).

Systemic or 3Di Solution to CP^{DP}

P-Solution to CP^{DP} or mDPR Determination on P

To get *P*-solution to CP^{DP} means to determine *P* from $mDPR=(D, P)$. Thus, the search for CP^{DP}

Figure 5. AM^{DPR} semantics

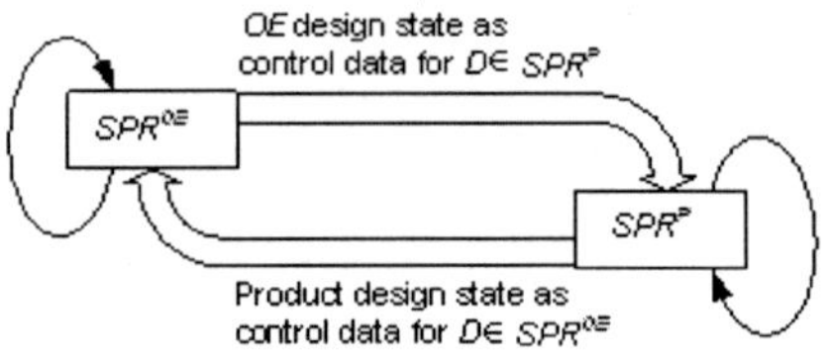

Figure 6. Input specification for the meta design process of 3Di

$$mDPR_3 = (D, P)$$

$$I^P = \text{3Di } DPR \text{ design}$$

solution on subject will proceed upon the expression (11) where <*SP*> means the search for $P \in mDPR$. To compile *P*, we have in general case: a) two types of processors – *H* and *C*; b) a list of members within a type; c) the kinds of relations between the types and members within a type (hierarchical subjection, cooperation and others).

We come to a virtual definition of the sought-for *P* in the following way. While $P \in PR$ is a *physical* processor of some *PR*, $D \in PR$ could be viewed as the *logical* processor with its own input – the processed control data (I^D). Then realization of any $D \in PR$ is performed in general case by a pair of processors – a *master* one (P_w), which deals with the input of $P \in PR$ (I^P), and *reference processor* (P_l), providing I^D for $D \in PR$. Let us consider the two as a single dual-core *P*. We call the latter a *diprocessor* (d*P*) – $P_w^{P^l}$. Diprocessor plays the role of action system (*AS*) platform, i.e. the base for shaping a real *P* for a particular *mDPR*. There are four variants for d*P* to be concretized: H^H, H^C, C^H or C^C. A process, which has the subject identical to one of the listed variants, will be referred to as *manual*, *computerized*, *automated* or *automatic* respectively.

Thus, CP^{DP} (virtual) solution on subject has been a d*P* with relation of interchangeability between *H* and *C*. This relation signifies that *P* of any type (*H* or *C*) is able to realize both work and reference function. Aiming at automated *mDPR*, we choose (logical solution) for the role of its *P* just $dP = C^H$. Then design automation gets under way from *mDPR* subject, and we have proposed instead today's intuitive design automation the more precise treatment of the notion. There are two ways of expanding d*P* to a real subject. First, each type of processors (*H* or *C*) has a number of instances that may be used. Secondly, $P^l \in dP$ serves for the gateway to descendants whose action systems can take for the base any of the four d*P* types. Lastly, we assign to each member of d*P* the equal ability to perform both functions (master and reference). Thereby determination of *mDPR* upon subject gets the status of 1Di *P*-solution of CP^{DP}.

D-Solution to CP^{DP}

3Di $D \in mDPR$ has been a procedure, which traverses the cells of &-cubes relevant to qAM^P and triggers in each of those the pair of processes (Figure 5). The corresponding meta-design process is designated as $mDPR_3$.

The Paradigm Signature for the Design Theory Platform

We call it Sg_l. The list of signature's paradigmants with their values is as follows:

1. Design Progress Concept = *evolution of individual.*
2. Conjugate Problem for a design one = *design process design implementation.*
3. Design (states) Representation = *dynamic.*
4. Fundamental definition of designing = *logical feedback synthesis.*

DESIGN PRACTICE PARADIGM IDENTIFICATION

Design process could not be other than domain-specific and task-oriented. At the same time, we retain it independent of a master processor type in d*P*, that is the aim of *mDPR* run is obtaining of 1Di design process.

In the light of the stated, the above obtained systemic solution to CP^{DP} (causing 3Di *DPR*) is out of practical sense as 3Di *mDPR* cannot transform entering its input 3Di *DPR* design into 1Di *DPR*. To get 1Di *DPR* as a result of CP^{DP} realization, we need the lDi status either for *mDPR* object and subject or for its I^P-input – *DPR* design. The choice of this or that variant will be defined by a strategy of design system construction. We shall dwell on the variant of successive concretization of 3Di solution to CP^{DP} that is deriving 1Di solution from a preliminary obtained 2Di solution. The corresponding *mDPR*s will be denoted as $mDPR_1$ and $mDPR_2$ respectively.

2Di Realization of the Conjugate Problem

2Di D-Solution to CP^{DP}

This solution consists in determination of $mDPR_2$ upon its subject (*P*) and object (*D*). Given the problem scheme – CP^{DP}=<*SP*><*SD*, $mDPR_2$> – let us feed the $mDPR_2$ input (Figure 7) with *DPR* design approximate model (AM^{DPR}) and design task (a sought-for product description and design goal).

CP^{DP} solution on subject, i.e. $P \in mDPR_2$, has been the above obtained *AS*. CP^{DP} solution on object (*D*) is a procedure that will allow to adjust the qAM^P structure to design goal and design subgoals. This adjustment consists in: i) selection of a required design goal (PRT, ITM, COM or ART) and terminal subgoal (qSYS, SYS, qDES or DES) for each inclusive goal; ii) elimination of irrelevant life cycle phases in subgoal's &-cube; iii) assignment of &-cube scanning path. Just the appropriate meta design process, designated as $mDPR_2$ with 2Di object (*D*) and 1Di *AS* in a subject (*P*) capacity, is taken as relevant for systemic module of design theory.

2Di CP^{DP} Realization

We assume here a logical realization of CP^{DP} – making arrangements for $mDPR_2$ rendering. The facilities for $mDPR_2$ implementation (a special purpose OS) have gotten the name of domain-independent *design machine* (Sedenkov & Guziuk, 2004).

Figure 7. Input specification for the meta design process of 2Di

$$mDPR_2 = (D,\ P)$$

$$\text{design task} = I^D \uparrow \quad \uparrow I^P = AM^{DPR}$$

1Di Realization of the Conjugate Problem: Design system

To obtain 1Di solution to CP^{DP}, there is a need for information about design domain. In the general case, such information is represented by a collection of domain-specific procedural and declarative knowledge: domain-specific SW libraries, theories, gained experience, design methods, etc. Acquisition, representation and assignment of such knowledge are the functions of the third module of Design theory, which gets the name of Domain Knowledge Management System (*DKMS*). A complete study of this module is beyond the scope of this chapter, while the outline of its signature looks like: 1) knowledge structuring; 2) knowledge storing; 3) knowledge access.

Starting 1Di solving of CP^{DP} (determination subject and object of $mDPR_1$), we claim the following input to $mDPR_1$ (Figure 8).

Then we take *P*=*DM* to be solution to CP^{DP} on subject, while the problem solution on object is taken of a procedure that couples in each &-cube's cell from qAM^P and qAM^{OE} structures the pair of processes (Figure 8) with *DKMS*.

The Paradigm Signature for Design Practice

Paradigm signature of the second design theory module (Sg_2) consists of the following three paradigmants:

Figure 8. Input specification for the meta design process of 1Di

$$mDPR_1 = (D, AS)$$
$$DKMS = I^D \circlearrowleft \quad \hookrightarrow I^P = qAM^P \;\&\; qAM^{OE}$$

1. Applied definition of designing = *transformation of design platform (qAM) into design.*
2. $mDPR_I$ = (Subject: *DM*, Object: *The procedure coupling DM with domain knowledge*).
3. Design system architecture = *Platform (DM) based Design System.*

SUMMARY: THE OUTCOMES OF NEW DESIGN PARADIGM REVEALING

Design Theory Rearrangement

In search of a new Design practice paradigm, it was imposs ible to avoid to refine some basic aspects of Design theory – design problem and process, design representation, constructive definition of Design. Available Design theories do not consider these issues in their close relationship. At the same time each theory gives its own definition of Design, that is claims a conjugate problem to solve. The adequacy of this problem is called by Design practice in question, as a rule. Besides, the proposed theories virtually do not take into account the participation in designing the formal processor (computer) though its expansion into the process will only grow. The next challenge is the steady increase of design process complexity. In sum, these circumstances cause considerable confusion, discordant opinions and chaotic tendencies in research (Hundal, 1990; Tomiyama, 1990). To take this situation for the reference point in search of adequate Design paradigm was as recklessly as to propose a regular new Design theory.

We decided to rearrange the current design theory through partition it into three conditionally independent units and their content revision. The unit of Design theory basic principles was named *systemic module* or a platform of the theory. Solutions to the problems related to this module are required to be independent of domain, design task and a type of processor – *C* or *H*. Systemic module is followed by a module of applied design aspects where the solutions produced at systemic level are coupled with design domain and design task. In turn, the applied module initiates the third module that provides designing with procedural and declarative knowledge. Each module was attached with the established signature of its paradigm.

Incidentally, the auxiliary (modular) concept of design theory reorganization has proved to be tempting as the concept of its initial arrangement as well. It seems a promissory alternative to both attempts to develop a General Design theory and tendency to have specific theory in every design field. Modular approach works in the best way also for theory consolidation, the calls for which sound ever more insistently (Birkhofer, 2006). This approach define three main lines of the theory development – fundamental, applied and design knowledgemation aspects – and distinct directions (upon signatures' paradigmants) within the bounds of each module. The relative autonomy of modules enables them to progress independently and at own rate.

In the course of systemic module establishment (design theory platform), we have considered the main, in our view, delusions ingrained in engineering design – design problem, design process, and product representation. The first among those is the design problem. It does not exist as a soluble problem. Instead of *DP* there is realized a conjugate problem (quasi-realization of *DP*). The choice of the latter will define the paradigm of the applied module, as well.

Design problem quasi-realization crosses the positivism of H. Simon (1970, designing as problem solving) and constructivism of D. Schön (1983; 1992, the study of designer's activity). In addition to the Simon's concept, we have changed the coined characteristic of design problem ("ill-defined") for the more promising one – "unsolvable of the second kind". This resulted in realization of the conjugate problem relative to the phantom *DP* – the problem of design process design implementation. In addition to the Schön's concept, we have proposed the proactive model of interaction between a product design and product operation environment ("evolution of individual") embodied in design process design.

The next modification was concerned with a design. Instead of misleading but widely discussed *product* representation (Ariyo et all., 2008), we have proposed continuous (dynamic) representation of a *product design*. This made possible explicit structure synthesis of both designs process and product.

It was now a design process turn. The latter a priori does not exist though it has been the current focus of design research community. Within the limits of Continuous Process Theory and its pivotal formalism – process scheme *PR=(D, P)* – the notion "(design) process model" becomes pointless. There are levels of uncertainty introduced for a process, pursuant to which it can be conditional, virtual, logical or physical. Hence, *DPR* model (Pahl & Beitz, 1988) has been an abstract procedure in the scheme of virtual process – *DPR=(DPR model, P)*, the refinement of which comes to end only when designing is over. In reality, there is in operation just a meta design process, which composes the design process proper part-by-part, supports these parts performance and, eventually, provides through this performance a required design. 3D independent meta-*DPR* (*mDPR* platform) is also defined within systemic module of design theory. Then a unified design platform had been constructed that enabled to represent design process design entering meta-*DPR*.

Thus, renunciation of the listed delusions shifts the focus from design problem to its conjugate problem, from design process to meta-design process, and from representation of a product to representation of a design – particularly continuous representation. But the all denoted innovations within systemic module would have been out of place without the proposed formal tools – Continuous Process Theory. While other design theories address to external facilities (which entail external problem as well) – for instance, issues in set theory or logic (Tomiyama & Yoshikawa, 1987; Hatchuel & Weil, 2003), we have used the ad hoc tooling. *CPT* meets a lack of scientific principles in engineering design and shifts the latter to a discipline in its own right.

The Impact of the New Design Paradigm on Design Practice

Design paradigm signature consists of three paradigmants. The names and values of the new one in comparison with the current design paradigm (based on German systematic model (Pahl & Beitz, 1984) are given in Table 2.

Within the new Design paradigm, *mDPR* deals with not a model but design of *DPR*, realizing it initially at systemic and then at applied level in line with constructive definition of Design for each theory's level – "Design as *DPR* design implementation" and "Design as transformation of design platform into design" correspondingly. Thus, with the account of Axiom 1, the quotation "Most researches and practitioners recognize that design process cannot be fully formalized mathematically" (Braha & Reich, 2003) gets a new sense: there is no such a necessity. Henceforth, exactly $mDPR_i$ $(i=1,2,3)$ is intended to attract main activities directed to resulted *DPR* perfecting.

In whole, new Design paradigm paves the way to new technologies in design computerization, design automation and design knowledgemation. For instance, the use of *design system platform* (design machine, *DM*) leads to the new efficient and effective technology of design computerization (automation): the compilation a wind range of domain-, product-, user-, and media-oriented design systems, via replication of domain-independent *DM*, and further extension of each specimen

Table 2. The comparison of two design paradigm signatures

Paradigmants / Design Practice Paradigm	Conjugate Problem under solution	Conjugate Problem Solution (*mDPR* input, its subject and object)	The system that supports ***mDPR*** **run** (Design System)
Current	Stepwise refinement of a prototype abstract description.	I^P = *(prototype, conceptual, embodiment) design description.* I^D = *design task.* *mDPR* subject = H^C. *mDPR* object = = *substantially implicit.*	Whole-cut and high-value item with long-term development but short durability. Supports Computer Aided Design (CAD).
New	Transformation of design platform into a product design.	I^P = $qAM^P\&qAM^{OE}$. I^D= *DKMS.* *mDPR* subject= *DM.* *mDPR* object= *D that supports interface between DM and DKMS.*	Platform based, holistic and product (domain, user, media)-oriented. Supports Computer Urged Design (CUD).

with domain-specific knowledge (design knowledgemation) via design knowledge management system. Design knowledge structuring within *DKMS* is guided by the structure of the unified design platform (*qAM*).

Due to *P*-independent theory of the systemic module, it is clearly stated the partner but not merely instrumental role of computer in Design: action system of $mDPR_I$ employs C^H platform, whereby we may identify the new paradigm as *Computer Urged Design* (CUD) that replaces CAD.

And one more benefit of the new design paradigm: it eliminates the problem of design process complexity: the realized $mDPR_I$ does not critically depends on complexity of a product under design.

CONCLUDING REMARKS

Any design theory should begin with propositions concerning the way of unavoidable design problem (*DP*) realization. To this end, it should have not involved but an ad hoc scientific base. In our case, the role of the latter is performed by Continuous Process Theory and Problem Theory.

(As for available design theories, most of those avoid to focus on design problem as such and begin just with design process modeling. The appropriate tooling is borrowed from other disciplines. This results in a doubtful resolving power of the tools and some extra problems for designing borrowed from the donor disciplines together with tools.)

Insolubility of design problem validated by *CPT* entails the search for a conjugate problem (CP^{DP}). For the role of CP^{DP}, Axiom 2 has identified the problem of design process design implementation. The process of such implementation was called *meta-DPR* (*mDPR*). *mDPR* dynamically generates design process, supports it performance and concurrently becomes the part of the process.

(Available design theories also realizes a conjugate problem but does this by default. Mostly it is one and the same problem of transformation a future product representation, which may have different level of abstraction. Thus, design theories with such CP^{DP} differ only in their input representation, involved tooling for the input transformation, and appropriate procedure. Schematic or concrete description of the latter is inconsistently called "*DPR* model", though inexistent entity, in particular *DPR*, cannot have a model – the only primary description of inexistent entity is a design. Due to the common CP^{DP}, these theories share such characteristics as catering for adaptation design, indispensable availability of a prototype, the luck of a structure synthesis problem solution, an obscure way of design automation, inability to manage the growth of design process complexity, and exclusive *H*-orientation as well. The latter prevents efficient usage in design a computer, reserving for it the role of instrument on demand. And though theory's human orientation is natural, we propose the trade-off decision – to have in the three-module theory structure a processor-independent systemic module.

Next, available design theories used to deal rather with *product models* than design representations. As inexistent entity cannot have a model, we considered the way of *product design* representation, particularly *dynamic* design representation required for explicit design synthesis. When designing, we have to deal with three designs – a product design, its operation environment design, and *DPR* design. So, the partial unification of their representations seemed both necessary and beneficial. The result of such unification has manifested itself in *design platform*. The latter gave rise to constructive definition of designing in the new paradigm: successive semantic attribution of each design platform cell.

Generally, the platform concept is also in the focus of the new design paradigm. Except design platform, it has defined generative *DPR* platform (*mDPR*), action system platform (diprocessor, d*P*), design system platform (design machine, *DM*) and design theory platform (systemic module). Besides, the bulk of facilities used for the platform technology description may shape one more platform – *design language platform*, which should allow to change from the current substantially intuitive notion base of designing to the more rigor one and thereby to serve for design coordination over distance and across professions. In whole, the platform-based reasoning paves the way for domain-independent approach to theory and research in design.

And one more point is noteworthy. There are an impressive number of questions, induced over time within design discipline, that still remain without answers (Sedenkov, 2008). The modular structure of Design Theory with identified signatures of modules' paradigms, the base of scientific principles in the form of *CPT* and platform-based technology enable to field these questions.

The main benefit of the new design paradigm for education is that it makes the primary design related processes ($mDPR_i$) observable, teachable and learnable. Incidentally, a pilot version of design system platform (design machine) is already in use for students training.

In whole, development of design practice paradigm contributes to shaping Design as "a coherent discipline of study in its own right, based on the view that design has its own things to know and its own ways of knowing them" (Cross, 2007, p. 3).

REFERENCES

Ariyo, O. O., Eckert, C. M., & Clarkson, P. J. (2008). Hierarchical decompositions for complex product representation. In *Proceedings of International Design Conference – DESIGN 2008* (pp. 737-744). Dubrovnik.

Birkhofer, H. (2006). The consolidation of design science – a critical review of the status and some proposals to improve it. In Vanek, V., & Hosnedl, S. (Eds.), *AEDS 2006 Workshop* (pp. 13–22). Pilzen, Czech Republic: University of West Bohemia.

Braha, D., & Reich, Y. (2003). Topological structures for modeling engineering design processes. *Research in Engineering Design*, *14*, 185–199. doi:10.1007/s00163-003-0035-3

Cross, N. (2002, February). *Design as a discipline*. Paper presented to assist discussion at the conference "The Inter-disciplinary Design Quandary", De Montfort University, Leicester, UK.

Cross, N. (2007). Forty years of design research. *Design Studies*, *28*(1), 1–4. doi:10.1016/j.destud.2006.11.004

Dixon, J. R. (1988). On research methodology towards a scientific theory of engineering design. *Artificial Intelligence for Engineering Design, Analysis and Manufacturing*, *1*(3), 145–157. doi:10.1017/S0890060400000251

Eder, W., & Hosnedl, S. (2008). *Design Engineering, A Manual for Enhanced Creativity*. Florida, USA: CRC Press.

Hatchuel, A., & Weil, B. (2003). A new approach of innovative design: an introduction to C-K theory. In *Proceedings of the International Conference on Engineering Design – ICED '03* (pp. 109-110, executive summary). On CD-ROM. Stockholm.

Hundal, M. (1990). Research in design theory and methodology in west Germany. In *Proceedings of the ASME Design Theory and Methodology Conference – DTM'90* (pp. 235-238), Chicago. New York: ASME.

Lossack, R., & Grabowski, H. (2000). The axiomatic approach in the universal design theory. First International Conference on Axiomatic Design. Cambridge, MA.

Meijers, A. W. M. (2000). The relational ontology of technical artifacts. In Kroes, P. K., & Meijers, A. W. M. (Eds.), *The Empirical Turn in the philosophy of technology*. Oxford: Elsevier Science.

Pahl, G., & Beitz, W. (1984). *Engineering Design*. London: Springer-Verlag.

Polya, G. (1965). *Mathematical Discovery. (VII)*. New York, London: John Wiley&Sons.

Rittel, H., & Webber, M. (1979). Dilemmas in a General Theory of Planning. *Policy Sciences*, *4*, 155–169. doi:10.1007/BF01405730

Schön, D. A. (1983). *The Reflective Practitioner: How Professionals Think in Action*. New York: Basic Books.

Schön, D. A. (1992). Designing as reflective conversation with the materials of a design situation. *Knowledge-Based Systems*, *5*(1), 3–14. doi:10.1016/0950-7051(92)90020-G

Sedenkov, V. (2008). An attempt to answer perennial design questions. In Rohatynski, R. (Ed.), *Design Methods for Industrial Practice* (pp. 21–30). Poland, Gdynia-Zielona Gora.

Sedenkov, V. (2008). Design problem *q*-realization and its outcomes. In Vanek, V., Hosnedl, S., & Bartak, J. (Eds.), *AEDS 2008 Workshop* (pp. 91–98). Pilzen, Czech Republic: University of West Bohemia.

Sedenkov, V., & Guziuk, Z. (2004). Design machine: theory and implementation. In I. Horvath & P. Xirouchakis (Eds), *Tools and Methods of Competitive Engineering* (Vol. 2), (pp. 1119-1120, executive summary). (On CD-ROM). Rotterdam: Millpress.

Simon, N. A. (1970). *The Science of the Artificial*. Cambridge, UK: MIT Press.

Suh, N. P. (1990). *The Principles of Design*. New York: Oxford University Press.

Suh, N. P. (2001). *Axiomatic Design*. New York: Oxford University Press.

Tomiyama, T. (1990). Engineering design research in Japan. In *Proceedings of the ASME Design Theory and Methodology Conference – DTM'90* (pp. 219-224), Chicago: ASME.

Tomiyama, T., & Yoshikawa, H. (1987). *Extended General Design Theory*. In Design theory for CAD. North-Holland: Elsevier Science Publishers B. V.

Vajna, S., Clement, S., Jordan, A., & Bercsey, T. (2005). The Autogenetic Design Theory: An Evolutionary view of the Design Process. *Journal of Engineering Design*, *16*(4), 423–440. doi:10.1080/09544820500267781

VDI-Guideline 2221/1987: Systematic Approach to the Design of Technical Systems and Products. VDI, Düsseldorf.

Yoshikawa, H. (1981). General Design Theory. In T. Sata & E. Warman (Eds), The IFIP WG 5.2-5.3 Working Conference (pp. 35-57). North Holland, Amsterdam.

KEY TERMS AND DEFINITIONS

Process scheme: A simple formalization for a process *PR* intended to represent the latter by the pair – a procedure *D* and processor *P*: *PR= (D, P)*. *D* describes the function of *P* over its input I^P. Process scheme is a basic unit for shaping process scheme structures – continuous representations of complex processes. To this end, a set of process schemes is provided with two initiating structure relations (**p**- and **d**-relation).

Problem Scheme: The structure of the second order made up of process schemes names by relation of determination: *PrB= <SD><SP, PR>*. Each delimiter (angle brackets or comma) indicates **d**-relation (from left to right) by default. *SD* – search for *D*, *SP* – search for *P*, *PR* – performance *D* by *P*.

Paradigmant: Distinguishing feature of the paradigm descriptive of its certain property.

Paradigm Signature (*Sg*): A number of paradigmants sufficient for a unique identification of a paradigm.

Problem Quasi-Realization: A situation when instead of initial (insoluble) problem another problem (conjugate) is under realization. The answer to the conjugate problem coincides with the answer to initial one.

Quasi Design Process (*qDPR*): The process intended to transform some initial representation of a future product design into a final state of the design.

Meta Design Process (*mDPR*): The process intended to implement design process design. Through this implementation, *mDPR* dynamically generates design process, supports its performance and supplies the final product design state. Splitting a virtual design process into the real *mDPR* and a pair of synthesis processes (for the product and environment designs) has proved to be a promising deed.

Chapter 3
Interdisciplinary Interaction for the Early Stages of Product and Service Development

Mirja Kälviäinen
North Karelia University of Applied Sciences, Finland

ABSTRACT

Saturated markets require user value through services and mass customised differentiation instead of mere products. This increases the significance of integrated innovation in the early stages of complex value offerings. Front end development combines the multidisciplinary professional perspectives and user insight in a cost effective way. Truly interdisciplinary interaction is reached through intrinsic motivation, shared goals and understanding. Experience for structuring the multidisciplinary front end innovation comes from the INNOstudio® concept created by the D'ART Design Resource Centre in the North Karelia University of Applied Sciences. This concept is about service and methods facilitating innovation sessions. Process support for communication, exploration, problem space definition and further development is provided by moving from abstract thinking into external observables – scenarios, sketches, or models. Innovative value concepts require both divergent, generative thinking and convergent, analytical thinking. Diverse methods support generative ideation, exploring future opportunities and user relevance or analysing the problem space.

INTRODUCTION

This article explains the significance of integrated, interdisciplinary innovation, especially in the early stages of complex value offerings. The experience for these development practices comes from the INNOstudio® activities, a service concept created by the D'ART Design Resource Centre in the North Karelia University of Applied Sciences. With the history of more than nine years of innovation camps in North Karelia, these were selected year 2005 as the best practice of the past period of EU projects in Eastern Finland. This led to the creation of the INNOstudio® service concept at the end of 2005. The INNOstudio® concept is about facilitating multidisciplinary innovative processes with

DOI: 10.4018/978-1-61520-617-9.ch003

structured sessions applying various innovation methods. The service has been created in practice of helping companies and organisations in their needs for new concepts. The approach of using people from diverse professional backgrounds, as well as user information and participation in innovation camps has proven successful.

The purpose of this article is not to explain in detail the executed development cases e.g. in the fields of wooden building industries, wellbeing services and products, new materials applications or in the real user participation. The experiences applying innovation sessions as a development tool have produced processes and methods that are mirrored here against the international professional practice and research. The INNOstudio® experience shows how these are applicable to regional development collaboration, especially with the local SMEs. The essential issue is not only to run an efficient development process but to turn the multidisciplinary approach into real interdisciplinary work by exploiting and synthesising the knowledge and skills of the diverse people. This development tool is currently being shared regionally through learning by doing approach via Multidisciplinary Innovative Environment for Product and Service Development 2008- 2010 European Social Fund project.

Figure 1. Services attached to mere products change the former producers of products as providers of complex product and service concepts. Summer kitchen design with an element based solution that allows mass customised end result and possibility both for delivery and built in place service. (© 2005, Harri Tuononen. Used with permission.). The customer value creation for this offering involves the delivery and building chain collaborators as vital stakeholders.

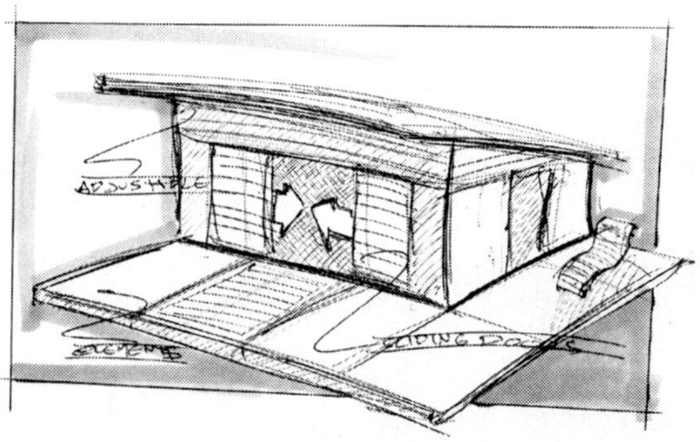

INTERDISCIPLINARY VALUE CREATION PROCESS

The current understanding of innovation is holistic reaching from new business ideas to the market success and can be described as the commercially successful exploitation of ideas (Stamm 2004, 11). u can be associated with an ideation process that has a value-creating outcome for the markets. Future orientation is required in a rapidly changing world predicting how the market will change through complex competition springing from any corner of the globe. As the changes offer new technologies, new processes and new approaches to living and working innovation has become a key to remain competitive. (Walton 2005, 6). New products and services draw their value from the networks they bring together including not only the product, but the whole venture. Developers will have to become proficient at aligning and managing the needs of multiple stakeholders: investors, suppliers, content-providers, distributors and others. (Hargadon 2005, 34). Complex product and service concepts need to be defined for the collaboration of these networks.

Concepts describe holistic business solutions, where extra value and desirability are created for the customer or end user. Concepts also involve the part of the different stakeholders in producing the solution. It is management information that illustrates what users expect from an improvement of situations in which a product or service might be acquired and used. Concepts describe the unique selling points valuable to the customer and differentiation from the existing offerings. They explain not only physical attributes (dimensions, colour, composition, features), but also quality (functionality, aesthetic, social/psychological attractions, value for money, safety, desirability, uniqueness, and reliability and mass customisation capabilities) (Jarvis 2008). As concepts define the strategic goals for the offering the brand, product and service concepts should be created at the same time for joint value often proposed through a complete service process from the user point of view.

The companies producing offerings can only build proposed value in their own processes. The perceived value for the end customer depends on the whole experience process including information about the offerings, choice and sales situation, delivery, initial use and usefulness, guidance, handling of the complaints, repair and disposal. Interaction in this whole user process builds positive experience constructed through networking with different stakeholders. All the process pieces must function to the end user's and the different stakeholders' satisfaction so that the last mentioned are willing to deliver the end user experience in a positive way. User needs for offerings are also highly heterogeneous even within a segment. People have specific, personal constraints and aspirations. (Hippel 2005, 33-43). In the global, saturated competition there are more offerings than demand, so mass customisation has become important trend in the markets. In addition to products and services the modular possibilities and executions through platform thinking exist also in marketing, delivery, installation and repair; in the whole experience process. The perceived value of the offering through the experience process is created in the situational interaction between the customer and the producers through the whole process.

Innovation is identifying and addressing the unmet customer needs beyond the existing demands. (Walton 2004, 7). Successful companies have understood that thinking consciously about the products and services in a real-use processes and situations is essential. (Stamm 2004, 11).

Figure 2. User process for perceived user value and experience

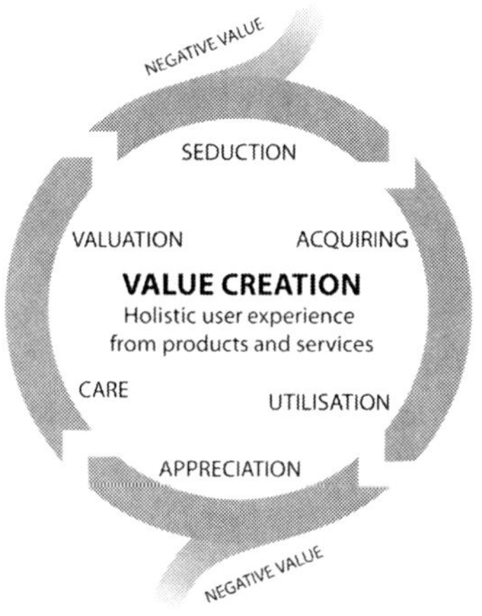

Building the experience means networking because the whole process means different stakeholders. All the pieces must function to user's satisfaction

Value can be created through supporting processes, production networks for relevant and desirable customer experience, providing platforms for mass customisation, diverse delivery channels and relevant communication or interaction with the customer. Development work for product and service designs is not only offering solutions, but also about challenging the strategy, business model and process for their production and marketing, innovating around holistic business concepts.

The demands of novel, appropriate business ideas, especially in entrepreneurial creativity, require greater competences and experience than few individual ones. (Leiviskä 2001, 62) Forming brand platform, service platforms, design platforms and technological platforms that work together requires different expert and stakeholder creation in the innovation process (Nieminen – Järvinen 2001). Different points of views from cross-disciplinary groups bring new information and ways of thinking into the development process. New points can lead to shortcuts such as solving a problem in one field by drawing an analogy to a problem solved in a different field. (Stefik - Stefik 2005, 11). Work groups with a greater mix of cultural backgrounds, mental models, experience and functional levels generate more innovative, complex and relevant offering solutions delivering appropriateness of products, processes and systems for different cultures.

But mere cross-disciplinary approach bringing a multidisciplinary group together is not enough. Multidisciplinary refers to diverse disciplinary specialists working side by side. The word interdisciplinary refers to high levels of integration, dialogue and interaction in the team process. (Leiviskä 2001, 12-13). Integrated collaboration is a constructive way to use the participant strengths: the knowledge, insights, comments, questions, ideas and new perspectives to add richness to the joint efforts. (Vidal 2006, Enhancing Your Creativity, 11). Out of the Box Thinking is one of the core concepts of innovation. Certain typical mental models, assumptions and patterns arise from the education and specialised work in different professions. We become efficient at thinking "inside the box". (Stefik - Stefik 2005, 11). The different professional boxes together form at least Big Box thinking as a wise alternative to out of the box radicalism. (Clarke – George 2005). Diverse disciplinary perspectives push to see out of the box and to solutions built as new combinations of the various boxes. They provide more and different sorts of thinking pathways and experience to the process to be combined in different ways.

Creative innovation process is inherently social instead of individual genius work. In collective creativity concepts emerge through participant and knowledge of the outside world interaction. The creative process flows easily through interaction and building on each other's thinking collaboratively. (CPH127 Design and Innovation). With more perspectives, there is a consistent finding that people working in teams generate more ideas. (Leiviskä 2001, 1; Sutton 2004, 45-46). Many ideation techniques help even homogenous groups spark innovation by broadening the range of differences. Cross-disciplinary work avoids drawing conclusions about situations and people based on very "thin slices" of experience. (CPH127 Design and Innovation). The right knowledge and timing in social meaningfulness, technical development and economic situation are required in breakthrough products. (Cagan – Vogel 2002). Specialised professionals are deep 'thin slice' experts, feature creeps in their own interest instead of experts in seeing the whole market situation and separating the unnecessary features from the critical ones. Product refinement requires editing out what is not essential for the simple user friendly experience. (Kelley – Littman 2001, 255-258). The multidisciplinary group can perform the task of the editor as the people from different disciplines are good critiques of the feature creeps from other specialised disciplines.

In a saturated, complex and proactive market situation customers and users are also active stakeholders and collaborative designers. Globalisation and the growth of Internet use have increased the consumer awareness. Consumers know a lot and they want to put their stamp on what they use. User driven approaches are also due to the more life style customised and interactive solutions in many offerings where the consumer's role is to make the individual choices. As personalisation is changing the assumption that the product or service is finalised before it goes to the consumer the challenge is to design interaction and this requires real user information and testing, methods that allow some play and interaction between the developers and users. The company has to make sure that they do not see the development question from their value points but from the way that customers and users perceive value in it.

The continuous, iterative feedback loop characterises many companies' focus on their customers and end users and the marketplace itself. (Walton 2004, 7). The Internet offers new possibilities to involve and interact with the customers and users on a global basis. Open source innovation describes participation on the Internet, on an open basis. (Hippel 2005). The development and use of social software for innovation community building is increasing, providing several aspects that are crucial for a good development processes: multidisciplinary input, open processes, ability to prototype, democratised dialogue, rapid development and improved timing in product launch (CPH127 Design and Innovation; Hippel 2005, 99-103). The new type of innovation networks call into question the former policy of keeping the company knowledge secret. If many organisations and users try to innovate together and find synergy and new middle-ground solutions, this happens only by sharing knowledge. Open innovation is a cultural change in business. (Vidal 2006, Fundamental concepts, 14). Stakeholders and ownership in networks means building a trust situation of the know-how that each participant possesses.

FUTURE VITALISATION OF THE NPD PROCESS WITH THE FRONT END INNOVATION ACTIVITIES

The traditional methods of New Product Development (NPD) are tools providing series of logical stage gate steps from the idea generation to the launch of the product onto a market. Each stage contains a defined series of professional activities with particular methods. The structured execution of the development has ensured efficiency and reduced time to market. The stage gate processes can, however, reduce the possibilities of ideation and exploration of complex solutions by separating different aspects of the process from each other and ruling out the users. Some NPD process stages also dismiss the important marketing process. Demands for changing the process arise as the whole business and the process of the consumption experience can be the source for innovation – not only reaching a neat step further in the offering.

Linear NPD processes with different stages are taking place under decreasing levels of uncertainty, which simultaneously means reduced flexibility (Innosupport 2005, 109). The NPD has started to develop in the direction of interdisciplinary value concept demands and user relevance, with multidisciplinary participants at the beginning and the customer's voice included. Front end development has emerged as an option tackling the complex problems with the participation of the user and later different stages already at the first concept phase. The first development steps should include a rich enough opportunity space, ideas to choose from that stem from a broad and deep understanding. It is also best to make the mistakes and do the testing as early as possible. In the front end of the process, the costs are comparatively low without the expensive production investments or big marketing campaigns. Through all this the front end situation is characterised by high flexibility.

The production process demands and getting the concept to the market cannot be separate processes from the offering planning. Complex problems and solutions need an interactive solutions network instead of separate linear functions that work in differentiated ways. The separation of planning and implementing comes from the conventional mathematical problem analysis -solution model. Yet, even in the mathematical problems you can carry out tentative calculations to find out the well working calculations. They may give you critical insights; therefore, no phase is pure analysis, as this activity is intermingled with synthesis. (Gedenryd 1998, 61-62). The process should have space for the wide interactive solutions network to communicate its thinking, make ideation, and explore and implement different solutions for successful concepts. The more separated and efficient stage gate process can be applied after the front end interdisciplinary work.

The often unspoken goal for innovation is to solve a problem. In a complex and fast moving market situation it is not, however, a question of a simple problem. Constant change and social complexity do not offer a definitive statement of the problem. This means that there can also be no

Figure 3. Front end innovation with early participation from the various areas of NPD

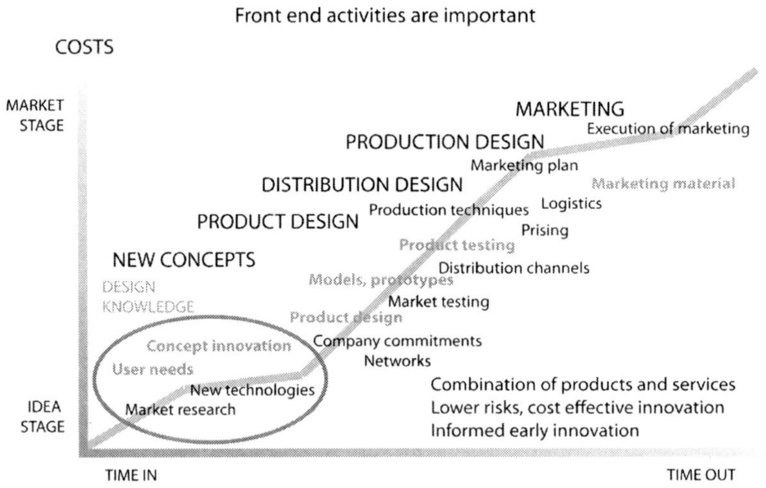

definitive solution and, therefore, no "stopping rule" for an optimum solution but competing solutions with subjective stakeholder perceptions and even discord among stakeholders. The development practice needs to devise solutions even before the problems and see how they further knowledge of the problem. Solutions are merely good enough or not good enough and they are born when there is wide enough vision with multidisciplinary professionals and closeness to customers to spot overall opportunities and threats at the edges. The explorative solutions shape the work forward at the same time as they provide the questions. (CPH127 Design and Innovation).

Setting the rich enough problem space can be described as the "Fuzzy Front End" and starting the process is about handling this ambiguous fuzzy front. (CPH127 Design and Innovation). Instead of the well defined and preset problem solving approach an innovation orientated approach is required. By having diverse people and perspectives to set the problems, to vary the problems and to test them is to set the rich enough problem space. The initial space search can be tentative, even arbitrary, since inquiry is to ensure that the framing will eventually be useful. (Gedenryd 1998, 91). Brainstorming or other ideation methods produce arbitrary results, but they still get the inquiry process going. Failing and having weak ideas is normal and useful but requires enduring risk and ambiguity.

The stage gate model guides to concentrate innovation-support resources on just a few preselected disciplines. The resource allocation to "the right people" can be hugely inefficient as the capability and the information needed to innovate can be widely distributed. (Hippel 2005, 123). Separation of different aspects of the process can easily lead to separating different disciplines from ideating and building exploration of complex solutions and finding the innovations between the core areas of different professional fields or for the user experience process building. The business stakeholders can be very diverse, different organisations and people that might have some production or distribution possibilities or other profits and that are forming the concept solution network. The extra value can be found by spotting the structural holes in between the different, even conflicting perspectives. (Uotila – Harmaakorpi – Melkas 2006). Practical examples of the new development teams are the ones around healthcare products. In addition to the typical engineering, design and business people the team should have participants form the healthcare professionals who are not at all professionals in product development processes but in wellbeing services and who are very much concerned about customer ethics and values behind the offerings. Also the user perspective can differ from the healthcare professional views as they might seek for pleasurable life, not only a healthy life.

PROCESS STRUCTURE AND FACILITATION FOR THE MULTIDISCIPLINARY FRONT END SESSIONS

The INNOstudio® service and facilitation concept from the D'ART Design Resource Centre North-Karelia University of Applied Sciences supports development projects by organising innovation sessions for company and organisational clients. INNOstudio® process structures and methods have particularly been developed for sharing ideas in the interdisciplinary front end development situations. The service helps to efficiently gather together multidisciplinary force of people and structure the work of this innovation network. With the INNOstudio® sessions development is conducted via intensive group work which produces lots of ideas in a short period of time. Innovation workshops have lasted from short (a few hours) sessions to up to 3 day camps and development support can also be a series of sessions. The participants do not necessarily come from the client organisation, so one preparation task is to gather a suitable group of participants from 10 to even 40 according to the task at hand. Efficient support can be provided and problems avoided by planning a suitable process structure and suitable methods in advance. So the service includes preparation such as planning the structure and methods, gathering the participants, smaller working groups and time use, knowledge and material gathering, and even accommodation and refreshment arrangements. The brief often comes from the companies, but planning, facilitation, documentation and reporting is done by the D'ART staff.

The innovation process facilitation and the organised sessions need to structure the work so that participant can orientate to the purposes of the project sharing understanding about visions, values, and goals. The facilitation of the interdisciplinary work should start with getting to know each other to the point where you can start communicating. In healthcare product development this has been, for example, done by letting the cross-disciplinary group choose everyday activity images to describe intuitively their views about health values in life. The mixing of participants form the different organisations and presenting them to each other in interesting ways is a vital starting point for the discussions. Intrinsic motivation for the joint task proceeds by a discussion of goals where everyone can express what they feel is important in the innovation case. This can also be a part of the participant presentation.

A communication problem in producing real interdisciplinary work is that each profession has its own vocabulary, concepts, theories and thinking styles. Successful innovation session requires an open-minded experience with participants willing to participate. To work in an interdisciplinary way means facing a challenge of interacting through even blocks and discrepancies with each other. This is important as controversy in groups seems to increase the range of ideas expressed and the achievements. Misunderstandings and disagreements of people with different backgrounds create negotiation situations, explanations and verbalisations that are important for learning and making new connections. (Leiviskä 2001, 115-116). The process of explaining a problem or an idea to someone, talking-it-out process, is most efficient when the second person is not an expert on the problem. Explaining a problem to a different-minded person reveals unconscious assumptions. (Stefik - Stefik 2005, 14-15). This helps thinking out of the box.

Trust is necessary for the co-creation process because the lack of it will prevent people from communicating when they find things unclear or doubt their interpretation. People in the innovative environment and process should let go of the need to be right and abandon the pretence of being superior (Walton 2005, 7). The expert opinion is not superior as the expert can define the problem and solutions according to the narrow discipline limits and tools. Citizen input cannot be sought on the expert's terms and it cannot be of secondary importance to the expert's opinion. (Vidal 2006, The Art of Facilitation, 23). Interdisciplinary achievements are reached only if everyone participating – experts and beginners, different professionals, customers and users – finds the atmosphere respectful. One joint work beginning example for a company developing automatic doors was an innovation session with elderly user participants and the company engineers and marketing people. The joint goals, scaffolds and democratic participation atmosphere were created through each of the participants presenting their practical problems with home doors which also the professionals had as well as the seniors.

The problems of mutual understanding have varied forms: cognitive, communicative, organisational, social, structural, cultural, functional and also geographical distance. (Uotila- Harmaakorpi – Melkas 2006). The distance spans from knowledge structures and qualities in habits, values and rules of acting or even developing. Boundary-spanning refers to the understanding that multidisciplinary participants operate from different values, knowledge backgrounds and cognitive structures with different languages and symbols. Even the same words can be understood in different ways as people interpret them in different contexts. An important task is to find a shared language to reach the accepted idea of the common goals and aims. The same distance issues apply also to international, multicultural development.

Figure 4. Different professionals do not share similar interests, although they would seem to discuss the same development issue

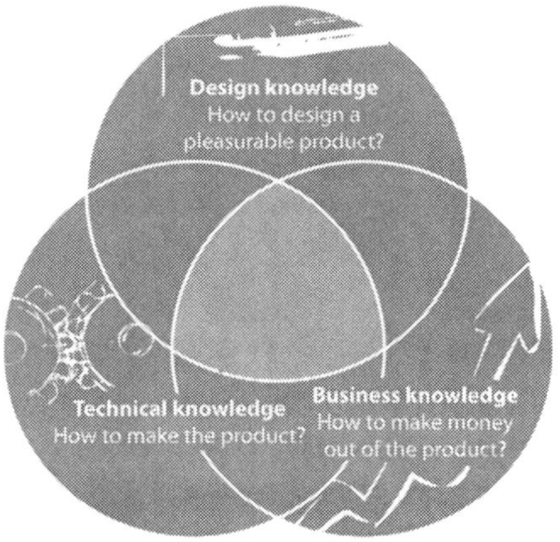

The various experts might approach the whole innovation question from very different points of view. The engineering approach is often about structural innovation organisations, stage gate management and production processes. The business approach is about business value, marketing strategies and quantitative market research, and the design approach is about ideation and exploration with the visual tools and user insights. Various research specialists are interested in their own specific research issues. The different professionals ought to be facilitated to produce the joint and meaningful goals for the collaborative work. Value creation in the business process through user understanding provides a common goal suitable for everyone. The change of roles, for example from expert to normal user, has been helpful in finding the joint user goals.

A distinction exists between mere creativity and innovation even when creative ideas are defined as being appropriate just as innovation includes the successful translation of the ideas into new offerings that serve market value. Creativity is one critical factor required producing the rich problem space, spotting opportunities in it and generating ideas for solutions. Getting loose from the normal, mind- shift type of activities support the diverse participants reaching the idea generating thinking mood, with feelings and intuition involved. This need is emphasised when you have to prepare the minds of diverse participants sharing ideas in a respectful way. The mind shift can be produced by various relaxation means. (Innosupport 2005, 42). In INNOstudio® applications external stimuli have been used for getting in the mood: change of place or environment atmosphere, art experience or exiting exercise.

Designers are comfortable in dealing with unfamiliar concepts, fuzzy problems, high levels of ambiguity and experimentation. They look to the future, imagine what might be and stress possibilities and opportunities. (Walton 2004, 7). Instead of mere thinking cognition should be regarded as an activity of inquiry with interaction between mind and environment. Designers explore the problem through a series of attempts to create solutions, articulating the problem space in a tangible way. (Gedenryd 1998, 11-13). This kind of design-based way of working can be used to externalise internal mental processes in a widely understandable way in the problem setting inquiry. The rate of exploration is important as most new ideas are bad. A feasible strategy is to search lot of alternatives quickly, without fear of failure, and then to throw away the bad ones. Exploration means letting go of perfectionism and developing confidence in a process of trial and error. (Stefik - Stefik 2005, 15). The teams must have permission to fail especially in the early stages when it is cheap, it helps idea generation and learning, and makes choosing from a wide range of solutions possible.

Articulation tools, for example sketching, can be used to test the problem setting. There is a simultaneous learning process about the nature of the problem and the range of possible solutions. (Gedenryd 1998, 93). Articulation helps understanding the development opportunities, what information is needed, and what particular circumstances and problems must be tackled. It is also a tool for reasoning, evaluating, and using the visual feedback as a basis for the next move. (Gedenryd 1998, 102-104, 107). Transformation of ideas into tangible results allow to see, feel and evaluate them, and helps the team in motivation and goal questions by creating a productive atmosphere as the results become real. So for the search and process support it is important to make rough presentations, visualisations and crude prototypes. The visual trying out process shows the feel of the variations and the features and allows evaluation for their fit with the customer value. (Kelley – Littman 2001, 108-110, 117, 232-235). Tangibles promote the iterative process structure with idea generation, feedback, analysing the situation and knowledge in constant variation. The consumer feedback and participation is also made possible by making the ideas and concepts tangible. Even services need to be described and communicated by creating scenarios, social pictures and touch points.

In multidisciplinary work, tangible outcomes show in synthetic reality the diverse mix of things that different sorts of people have brought to the discussion. Visual means can help in making even the individual embodied, emotional and unconscious mental images and models presentable for the collective interaction. Externalising the discussion reveals what the discussion was about and makes building on the ideas of others easier. (Kälviäinen 2002; Kälviäinen-Miller 2005). This externalising process meets the requirements of searching the structural holes in the knowledge building process by holistic visualisations that require the filling of the knowledge left untouched in the abstract discussions as certain areas or details can be missing. The synthetic visualisations present the holistic view of the issues in discussion with the relations of the smaller details and their weight in the whole solution. The visualisations also produce tangible details in concrete, one-off versions of the general signs or categories of things stated in words. This process points out the holes, helps to discuss how they should be filled in or if they seem to be filled in a right way, if the synthesis is the kind that different stakeholders thought of and if the details are the suitable ones and what is their relation and position in the solution.

Tangible outcome tools include sketching, situating strategies such as scenarios, action charts, simulation models, and storyboards, models made out of complex concepts or more advanced ideas of products and services, and prototypes as lifelike models of the design in progress. As a polished look can prevent the participants from changing the prototypes, for inquiry purposes it is better to apply work-in-progress type of paper or other cheap and easy prototypes. A future development question is to have new methods for prototyping the many different outcomes and relationships in multiple experience solutions, which will require possibilities for abstract network models.

In the INNOstudio® sessions, the results of the smaller groups' achievements have been gathered together from time to time with presentations joining all the session participants and their achievement for evaluation for further development. The results of the completed sessions have been organisational models and statements, design management plans, brand and product concepts, user models or offering ideas. All the materials produced in the sessions have been documented and memos or idea books have been produced from the session process and results. In many cases the innovation session can only produce vague ideas or idea scaffolds that have to go through modifications and more detailed work after the sessions. The tangible session outcomes serve the purpose of stakeholder dialogue, choices for further development and user testing.

METHODOLOGICAL TOOLS FOR INNOVATION SESSIONS

Methods suitable for supporting the communication, exploration, problem space definition, ideation and evaluation in the innovation sessions are numerous. In producing the pre-concept rich problem space and possibilities for complex concept solutions cycling through methods of generative divergent and analytical convergent thinking can be applied. Exploring a problem from different angles, generating ideas and concepts considering alternatives is divergent and selecting the problem and the concept is convergent. Evaluation means divergent monitoring of the use and convergent action. (Rieple 2004, 40-42). Creativity can be seen as an ability to see the situation in many and new ways or combining previously unrelated ideas and areas of life (divergent thinking) and to continue to question until an appropriate solution is reached (convergent thinking) and continuing these different thinking styles in an iterative process as a search for estimate solutions, testing, choosing and improving them through feedback from different sources. Iterative methods are especially useful for problems involving a large number of variables instead of direct methods of solving a simple problem.

Divergent thinking is rooted in the use of subjective judgements and intuition and requires soft or qualitative methods. Soft thinking processes include dreaming, humour, ambiguity, play, approximation, pleasure, fantasy, paradox, diffusion, hunches, generalisations and analogy. Unexpected combinations, recognising links among remote associates and transforming information into unexpected forms produces answers that vary among participants but are of alternative, equal value. Divergent thinking idealises rather than satisfies or optimises. Convergent thinking is oriented towards deriving the correct answer to a clearly defined question. It is hard: logic, reason, precision, consistency, algorithms, efficiency, results, reality, directedness, analysis, specificity and abstraction. Rational approaches produce the best possible or optimal outcome with the help of scientific or quantitative methods and accumulate information, reapply decision making techniques, and preserve the already known. (CPH127 Design and Innovation; Vidal 2006,

Fundamental concepts, 6-7). Both convergent and divergent thinking lead to the production of ideas, but with qualitative difference: divergent thinking generates variability, whereas convergent thinking generates orthodoxy without much space for ambiguity. As the idea of innovation is about new combinations and complex solutions defining the problem or the specific focus is not the suitable thinking style for the fuzzy front end. The methods required need to structure complexity within stakeholder networks and needs in the future lives of clients and users. To remember the complex field of opportunities when choosing the methods for the early stage is to emphasise divergent thinking for the wide enough problem space and alternative solutions. The next move is to switch to problem solving convergent thinking for the usefulness and relevance of the solutions.

People with different cognitive styles are tuned to these different approaches. "Innovators" prefer to do things differently, producing ideas with new elements in the problem context and by changing the relationship between the problem elements. "Adaptor" type of people prefer to do things better through order and accountability. They formulate ideas without changing very much of the initial data of the problem. (Innosupport, 2005, 39). The divergent thinking suits innovators but for the adaptors, convergent thinking is more suitable. Both sorts of thinkers are useful in a team, although the adaptive people could be a hindrance in the generative work and the innovative people in the analysis and evaluation. The use of different methods serves as a brokerage factor for facilitating different cognitive styles. In joint idea generation the cognitive differences produce positive or negative feelings among the participants. The very usual professional logic-solution participants easily see the generative thinking as unnecessary play, not as step to achieving feasible results. Within INNOstudio® experience this is one of the main attitudes that need to be tackled by the help of suitable methods for the divergent work.

Figure 5. Divergent and convergent phases of the process

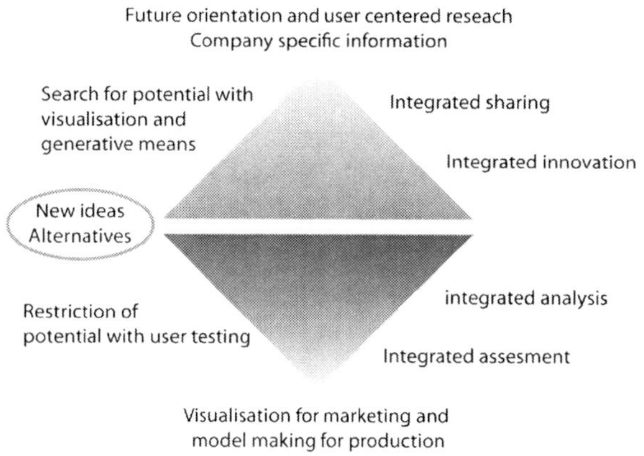

The session methods should promote the different expertise and skills of the participants. Participants can also be divided inside the group into smaller working units so that different strengths and capabilities can be included. Businesspeople are not used to drawing and technical people are not used to free brainstorming or expressing vague ideas visually. With many methods, each participant can have understanding scaffolds for the focus and the process of the ideation work. In the innovation sessions also the time available limits the scope of methods. The methods should strive to encourage participant action in articulating thoughts as tangible outcomes and thus sharing them in the joint interaction. In promising to participate, people feel intimidated about the tasks they have to do, such as skilful drawing. Drawing together is especially difficult. The facilitation can include specialised people helping with special skills. With visualisations, the use of readymade picture sources also saves time and frustrations connected to lack of skills.

Visual and other sensual work can be applied to all the methods as a means to support sharing values, feelings, experiences, ideas, mental images and maps, which are tools for the creative unconscious act, social discussion and interaction. For this, a wide range of stimulus materials, scaffolds or probes can be used. (Kälviäinen, 2002;

Kälviäinen-Miller, 2005). Using visual and tangible discussion tools is supported by psychological and sociological research methods for product perception, understanding and pleasure as they provide means to test the mutual understanding and solutions with the participants with differing mental images. (Kälviäinen-Miller, 2005). Mental images and maps are internal ideas, individual interpretations of something preserved in the mind in intuitive and multi sensory ways other than in verbal ways. Interpretation and interaction is influenced by this mental imagery world that presents individual values, attitudes, feelings and past experiences. (Kälviäinen, 2002). As visual stimulus provokes intuitive emotional experiences and mental understanding it stresses the freeness of association and prevents the influence of too much reason and thus promotes creative thinking. Visual information is not specifically limited by certain professional language and is useful in exploring values, creating meanings and prompting respondents to add ideas and details. Visual sorting tasks promote participants to show the richness of their thinking in how they structure and associate the world and make comparison with different mental maps possible. If the facilitator has chosen or pre-organised stimuli, it is vital to check what is missing or how the participants would organise the material. (Kälviäinen-Miller 2005).

The following methods provide tools for iterating between divergent and convergent thinking:

Ideation Methods

Typically connected to innovation work are the idea generation, ideation methods, that help to create imaginative ideas and to spot opportunities that might otherwise be missed. These methods help to find different potential solutions, reduce convergent thinking and break free from preconceptions. Idea generation takes into consideration mental imagery processes, associations, association chains, metaphors and metonyms and the processes of using different senses. The purpose of these methods is to enhance creative abilities such as fluency, flexibility, originality and elaboration.

Fluency is the production of multiple problems, ideas, alternatives and solutions (brainstorming approach). Flexibility is the ability to process ideas or objects in many different ways given the same stimulus (morphological forced joining together analysis methods, questioning methods, subject modifying metaphors or analogical processes). Elaboration is about structuring complex situations in expanding ways (mind mapping). In flexibility sense the creative tools can be divided into paradigm preserving tools where neither new elements nor relationships between the elements of the problem are introduced, into paradigm stretching tools, and paradigm breaking tools with new elements and relationships. Many free association (brainstorming) or situation analysing tools do not encourage diverting from the prevailing paradigm. Unrelated stimuli and lateral thinking methods stretch the paradigm but they do not necessarily break it. Paradigm breaking tools force us to think about the problem from different perspectives or totally alien starting points. Paradigm stretching techniques may be more useful in ill-structured problems whereas paradigm preserving tools in more structured problems. (Vidal 2006, Creative tools 12-15). The set information can be a disadvantage as it leaves no space for a free flow of thoughts and easily narrows down the perspectives. It provides categorisations and the ideation then occurs only inside this categorisation. The multidisciplinary participants can also be useful in the sense that their categorisations of issues are diverse. Forced ideation personas can be a way of building a team with diverse mental models and thinking styles to promote the discussion with different perspectives.

Narrative methods suggest using storytelling to build common understanding. The story can also be seen as the process description for the complex value offering. Practical examples are the whole business and customer processes that can serve as an ideation field for stage alternatives and various combinations. Strategy / Funding / Design / Production / Distribution / Communication / Acquiring / Customising / Installation / Repair / Disposal – different kinds of process stages have many different alternative solutions. Distribution has many alternative channels, routes and places, and the acquiring stage can alternatively be about buying, leasing, renting or acquiring joint ownership. These alternative stage solutions can be used building materials for the new combinations that form the processes stories.

Future Orientated Methods

Future orientated methods have their theoretical basis in futures studies where one is presenting multidisciplinary assessments of the future with the aim of building well-founded development paths and 'deviating', even critical alternatives. Future orientation in development work grows foresight for handling the changing competition or environment, spotting opportunities and business solutions that will add value in the future. For mapping alternative futures probable and desirable alternative changes in different aspects of society and nature are targeted. In future orientated innovation you can apply future casting to criticise the current situation, make future scenarios and then think how to reach the preferable situations. In back casting you dream about the future situation and then look backwards from it to estimate the steps to come from the desirable future to the current situation. (Vidal 2006, The future workshop, 2; Evans – Sommerville 2006).

Scenario work and opportunity ideation in innovation sessions is often based on trend lists for various purposes and at different level of influence (mega trends/ weak signals). Also future analysis models help to structure ideas about the future in the different sectors of society. The future tools can be used even for as short period as for the next year or anticipating the future changes in a suitable, longer time scale. As the future methods predict how society will change they present what opportunities these changes will provide for the business. Future methods often serve as an ideation sector lists that are used to build different layers of influence and opportunities. Typical sectors are business life, technological advancements, transportation, communication, social interactions, working life, health care, demographics, home life and hobbies. The purpose is to see how they all change, what kind of influence these changes impose, and how they affect the opportunities in a certain business field. Of course, the problem might be that some special or unexpected influential sector is missing from a model. In this the multidisciplinary group itself provides for informed, intuitive expert perceptions resembling the Delphi method, where experts' statements of the future arrive at a description of a plausible future. (Harmaakorpi – Melkas 2005, 652-653). So the session work has the effect of producing future insights with knowledge and expertise about trend influences from the various professional fields.

User Driven and Centred Methods

User centred methods aim at involving real user information and inspiration in the development. Understanding the customer is of vital importance for success in the saturated markets where engagement, deep emotional connections between the user and their experience with the offerings are required. The quality of experiences has been explained through the consumer lifestyles that help to make sense of what people do, and what doing it means to them and others (Chaney 1996, 4-12). On a deeper level, users' life worlds and life goals explain their emotional experiences, practical knowledge, and their intuitive understanding of a cultural way of life (Thompson 1998, 130-136). A positive experience means fitting the solutions into the identity and lifestyle of the user (Crozier 1994, 4-5). Positive experiences in products and services are about achieving your practical goal but also about intuitively and socially interpreted qualities such as emotional bonds, familiarity, aspirations, dreams, sentimentality, embodied aesthetics, general 'feel' and personality.

First hand learning about nuances of human behaviour and use context helps to open up opportunities from latent needs and underlying psychology of motivation and emotion with offerings. (Kelley – Littman 2001, 25-51). The purpose of user centred methods is about searching for empathy, latent needs and wants transferable to opportunities. Empathy requirements for the whole development group are essential because they discover the new opportunities and interpret user data into future products. Understanding the use situation and use action can offer opportunities of cross-pollination, taking processes and functions from remote areas of life to make the use experience better.

Many methods have been created to elicit user information and understand the users in great depth for business development purposes. Methods based on the future orientation such as the user trends predict what behavioural and social user lifestyles will be emphasised in the future. These trends provide possibilities of creating user scenarios as alternative future user and use situation descriptions. User segment information consists of demographic features and value segmentations that describe user groups with different inclinations of consumption based on life aspirations. Visual user scenarios or information about current segment groups is usable in making quick ideas for different segment groups and forcing out different ideas for individually customised choices. So instead of strict segment based development consumer segmentation can also be used as tool for mass customisation. Segmentation can also serve as a basis for searching features and concepts that would appeal to several segment groups.

For complex product and service solutions user process analysis is an effective tool. This is about depicting the story of the user involvement and experience with different goals, stages, touch points and interaction the user task has. Capturing the human experience as a scenario and with a stage, scene, plot, actors, products, touch points for services and user interaction provides a way of ideating the different solutions of the diverse user experiences. It also provides a way of analysing the user experience situation in all its complexity. If the scenario is constructed like a stage it provides the possibility to let the different user types have diverse strolling activities and searches important for the current customised and interactive experience.

For the innovation session purposes user information acquisition needs often be done beforehand. Triangulation, using different sources of information seems to be beneficial in providing a rich user insight. Observational methods are important for values and actions that are difficult for the users to describe. User personas create an emotional relationship with the user and with the real complexity of the user's life: their characteristics, life goals, experience goals and end goals. Persona design replaces the demand lists with a holistic perspective of the user and it should be built in an innovation session by the development before the offering. This helps the development team to meet the user goals at the same level of experience as the real living user. It then also serves as a practical tool for user systems analysis. Simple task analysis can be produced from observation material to show how the task is really done by the users and how it could be done more easily.

User participation is the tool for doing the user information integration in quick and dirty way but to have also deep insight, feedback and testing possibilities ad hoc. Involving users as co-designers in the innovation sessions is a possibility for gathering rich user vocabulary and spotting their possible tangible counterparts important for certain consumer experiences. Users highlight the important touch points and the desired feelings associated with them, which serve as a foundation for creating solutions with emotional connections, relevance, personalisation and delightfulness.

When the new solution is prototyped together with the user time is spared as the user testing for acceptance, desirability and interaction is conducted at the same time as the development process continues.

Probes and make methods help user participants express themselves in the sessions. Probe material is especially helpful in eliciting latent needs. This can be ambiguous visual stimuli that supports different people creatively express memories, thoughts, feelings, dreams and new ideas. The visual nature liberates people to make meanings beyond words and help to fill in what is unsaid or unseen. Visual and other building materials are tools for making the users envision the use situation, and the future, by identifying the important moments from their perspectives. Drawing on pictures or freehand drawing comes close to all sorts of prototyping. Building different levels of tangible outcomes as drawings, collages or prototypes is a way to reach synthesis from the discussion about the complex and abstract user conceptions that can in practice be applied to different solutions. (Kälviäinen 2007). Tangible participatory solutions are also a starting point for developing refined solutions that correspond to different notions about the enjoyable whole. In the user participation sessions, the development people gain the first hand understanding of the complex user experience and how it was constructed form the diverse probes.

Analysis and Evaluation Methods

The purpose of analysis and evaluation methods is to support the convergent phase of the process. Analysis is squeezing the richness of the possibilities into models or criteria. Evaluation is diminishing the alternatives to, hopefully, successful ones through some agreed protocol. Analysis methods can also be a starting point for the innovation and for sharing the current situation. The important thing to bear in mind is the convergent nature of these methods and how in some cases they can block the idea generation. There exists many information based ideation methods (future orientated and user centred) with embedded relevance and choosing criteria. They contain models and criteria of the future developments and users, and incorporate analytic thinking in the idea generation. The innovation process can begin with taking into account some previously acquired quantitative information such as marketing research results. The starting point can also be analytic qualitative information for successful offerings, such as future trend or user information, or feedback information from the markets. As the development criteria often stems from the strategic plans of the company in innovation networks this can mean joining criteria from several companies.

Figure 6. Senior user ideas transformed into tangible concept ideas for future door environments

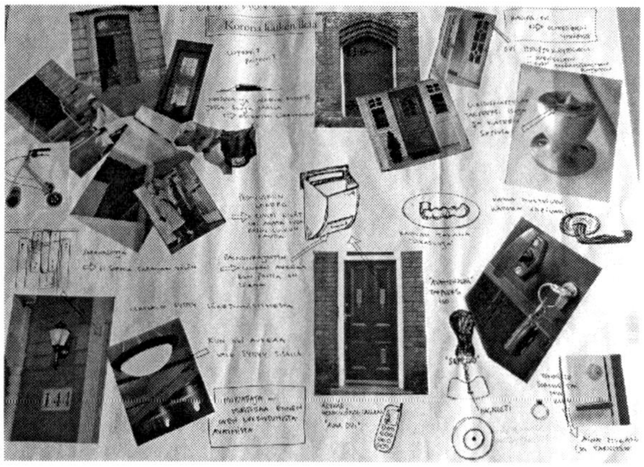

It is important to start sharing and understanding the information from the different stakeholders. This work can occur in the form of joint mental maps and continue by forming an analytical model of the rich problem space. Analysis of the information and evaluation criteria beforehand by some of the stakeholders is both difficult and dangerous. If the solutions are complex and look for new alternatives, the preset criteria might not even be suitable for complex concepts. For these reasons, it is wise to stick to the divergent innovation emphasis first using the analytic and evaluation tools with caution. Analytic methods in the beginning of the development can be useful if applied on a general enough level as a source for strengths and opportunities. Typical and useful alternatives for this are the traditional SWOT and TOWS analysis tools. Analytic future model tools look at the macro level changes and can provide for cross-disciplinary possibilities and lateral ideas from different specialisation areas. In the innovation sessions the evaluation of the ideation work has often been conducted by the multidisciplinary participant perspectives by assessing first best alternatives and thus providing the company clients just some choices continue.

As creativity includes the appropriateness of ideas, the methods can be formed so that they direct the relevance of the innovation results. Both the multidisciplinary knowledge from the participants and the pre-collected information (technical, market, user, future) can be part of an innovation session. The divergent ideation phase can, instead of supporting whichever wild ideas, support informed ideas. Informed innovation is using future tools and user tools that hold inside them information and perspectives. This further helps the convergent phase of choosing the options to continue with. The different types of methods often overlap (future user trends as a tool for ideation). This helps in using the methods in chains to produce the desired outcome. For example, the brainstorming kind of ideation techniques should be followed by using the alternatives in matrixes for combinations. The different methods can form the divergent convergent succession that builds up the route to certain goals, for example using user trends for idea generation and user empathy to build user value.

Design and product development emphasise the iterative process structure, where idea generation, analysing the situation and knowledge and feedback are in constant variation. This reflective variation in iterative thinking has been built into the structure of the innovation sessions with the help of the above described methods. Inside every session, the structure has been arranged in an order where the mind shifts, focusing, idea generation, analysis, assessment, and solution choice parts of the session change. Innovation sessions are about building communication and understanding, setting the rich problem space with wide enough questions, answering the questions in different ways, making the solutions tangible for everyone to understand, evaluating to find the suitable and optimal solutions, and implementing the solutions in a way that their further development and acceptance is assured.

CONCLUSION

The practical INNOstudio® experience has proved that innovation sessions provide a suitable tool for structuring the work of a multidisciplinary development network in the front end situation of the innovation process. This practice of truly interdisciplinary development has become common and necessary to serve the SMEs' networking, design, technological and marketing needs in an integrated way. Especially when supporting the SME projects help is required form several professional expertise areas and user information so that multidisciplinary approach and ideation can truly form the new business concepts.

The practical experience has also shown that to work efficiently all the innovation sessions have to be especially tailored to meet the requirements of the clients' case. The sessions cannot be a stable service so the INNOstudio® concept provides a customized platform to conduct innovation process support. INNOstudio® concept forms a living environment collecting the diverse stakeholders, structuring the development process, and facilitating the participants with various methods. For the structure and methods, INNOstudio® platform is collected and further developed in a virtual e-learning environment offering models of the innovation processes and timetables as well as methods. Virtual spaces can also be created for individual development cases and sharing the brief, process and documentation of the work done. The continuous search for good practices aims at developing the functionality and inspiration possibilities of sessions for a fast, intensive and integrated innovation service. Interdisciplinary sharing of thoughts in an understandable way, especially, requires new approaches. From the already executed sessions there is supporting evidence that the means of design are useful in producing shared understanding, especially by providing understandable visual probes and tangible synthesising results.

The factors in the INNOstudio® early stage interdisciplinary innovation applications noted as important to take into consideration are:

- Creating the wide network concept already in the beginning of the process (product and service offering with the user experience process)
- Multidisciplinary group of participants and strong user involvement
- Planning the joint sessions beforehand, but being flexible and responsive of guiding the sessions along the real team processes and their support
- Democratic atmosphere
- Emphasizing the team members endurance for ambiguity
- Building up joint goals
- Using diverse communication, ideation, user and future understanding tools and analysis and evaluation tools
- Documentation of the process and results for the evaluation and further development

The practical experience shows that cross-disciplinary sharing and ideation is by no means without problems. The main practical hindrance is the joint timetables aided through timetable planning and Internet environment. The building of trust and understanding including the development steps are critical and difficult to reach. The multidisciplinary approach brings the various thinking styles and working habits into the process. The goals, values and ideas how the development process proceeds vary for the different participants; this stems both from the professional and organisational backgrounds and user differences. Strong discrepancies arise when the basic research specialists are interested in their 'how things are now' research issues fulfilled and the application of research for real life complexities and development strives for what might be. The practical future offering is the main interest of the company representatives. The biggest hindrance for the joint development, however, has proven to be the different cognitive styles. Some participants are so well trained to only convergent thinking that divergent, generative and ambiguous work seems even painful to them.

The cognitive issues embedded in the front end development sessions concern wide research interests. Such issues are creativity, group processes and communication, not to mention innovation and innovation processes themselves. There is lots of cognitive research with diverse strands applied at least in business management, engineering, design, psychology and educational science research. The research available is often not business innovation related and when it is it often concerns large and influential companies and as such it is not applicable at the SME level. The practical experience has been of major value in creating the INNOstudio® service concept content despite the vast research field opening up behind it.

For the specialised SME managers even the basic question that in the current competition it is not adequate to innovate from one off disciplinary perspective can be hard to understand. The new demands of our interactive, fragmented knowledge society are changing the old structures. There are new approaches to what value is and new complexity to calculate this value. The agreement of the need for joint work can be hard to except but it builds up the respect and trust required in interdisciplinary development and is the cornerstone for getting the various participants to be real stakeholders with inner motivation. For the diverse stakeholders, value is different so for building value offerings the building of the joint value perceptions is a central one.

Evidently, the front end innovation sessions do not replace the whole development effort. Moving the front end solutions into the markets is essential with issues such as production planning, marketing and product and service launch. The innovation process can still go wrong at these stages. As the innovation sessions are used as a part of development processes leading to the markets also further offering specification, prototyping, testing, production, and marketing activities have been organised if required. As INNOstudio® work concentrates on the front end product and service development issues, so has this article explained the front end preconditions and practical experience. These experiences are supported by the international best practices and research that point out how the current market situation demands the diverse stakeholders' joint value creation and interaction in the early phases of the offering development.

REFERENCES

Cagan, J., & Vogel, C. M. (2002). *Creating breakthrough products: Innovation from product planning to programme approval*. Upper Saddle River, NJ: Prentice Hall, Inc.

Chaney, D. (1996). *Lifestyles*. London, New York: Routledge.

Clarke, P., & George, J. (2005). Big Box Thinking: Overcoming Barriers to Creativity in Manufacturing. *Design Management Review, 16*(2), 42–48. doi:10.1111/j.1948-7169.2005.tb00192.x

CPH127. (n.d.). *Design and Innovation*. Retrieved April 1, 2007, from http://www.cph127.com/cph127/html

Crozier, R. (1994). *Manufactured pleasures: psychological responses to design*. Manchester, UK: University Press Manchester.

Dant, T. (1999). *Material Culture in the Social World. Values, Activities, Lifestyles*. Buckingham, Philadelphia: Open University Press.

Evans, M., & Sommerville, S. (2006). Educating the future: embedding futures thinking in the design curriculum. In Rothbucher, Kolar, Ion, Clarke (Ed.), Educating designers for a Global Context? In *Proceedings of the 4th Engineering & Product Design Education Conference*, Salzburg, Austria, September 7-8 2006 (pp. 355-360). Basildon: Hadleys Ltd.

Gedenryd, H. (1998). *How Designers Work. Making Sense of Authentic Cognitive Activities. Lund University Cognitive Studies 75*. Lund, Sweden: Lund University.

Hargadon, A. (2005). Leading with Vision: The Design of New Ventures. *Design Management Review, 16*(1), 33–39. doi:10.1111/j.1948-7169.2005.tb00005.x

Harmaakorpi, V., & Melkas, H. (2005). Knowledge Management in Regional Innovation Networks: The Case of Lahti, Finland. *European Planning Studies, 13*(5), 641–659. doi:10.1080/09654310500139277

Hippel, E. (2005). *Democratizing Innovation*. Cambridge, MA: The MIT Press.

Innosupport: Supporting Innovation in SMEs. (2005). *Education and Culture. Leonardo da Vinci. Pilot Projects*. Retrieved September 16, 2006, from www.innosupport.net

Jarvis, C. (2008). *The Business Idea – Product and Service Concept*. Retrieved April 10, 2008, from http://www.bola.biz/businesses/busplan/product.html

Kälviäinen, M. (2002). Product design for consumer taste. In Green, W., & Jordan, P. (Eds.), *Pleasure with Products. Beyond Usability* (pp. 77–96). London: Taylor & Francis.

Kälviäinen, M. (2007). User experience capturing through Make Do innovation sessions. In *Proceedings of QMOD 2007 & First European Conference on Affective Design and Kansei Engineering*. Helsingborg, Sweden, 18-20 June. Lund University electronic publications.

Kälviäinen, M., & Miller, H. (2005, September 22-24). Visual research: means of producing shared meanings. In *Proceedings of Joining Forces International Conference on Design Research*. Helsinki: UIAH.

Kelley, T., & Littman, J. (2001). *The Art of Innovation. Lessons in Creativity from IDEO*. San Francisco: IDEO.

Leiviskä, E. (2001). *Creative Interdisciplinarity. IDBM Programme. Department of Teacher Education. (Research Report 227)*. Helsinki: University of Helsinki.

Nieminen, E., & Järvinen. J. (2001). *Teollisen muotoilun teknologiaohjelman esiselvitys. Teknologiakatsaus 119/2001*. Helsinki, Finland: TEKES.

Rieple, A. (2004). Understanding Why Your New Design Ideas Get Blocked. *Design Management Review, 15*(1), 36–42. doi:10.1111/j.1948-7169.2004.tb00148.x

Stamm, B. (2004). Innovation – What's Design Got to Do with It. *Design Management Review, 15*(1), 10–19. doi:10.1111/j.1948-7169.2004.tb00145.x

Stefik, M., & Stefik, B. (2005). The Prepared Mind Versus the Beginner's Mind. *Design Management Review, 16*(1), 10–16. doi:10.1111/j.1948-7169.2005.tb00002.x

Thompson, C. J. (1998). Living the texts of everyday life. In Stern (Ed.), *Representing consumers. Voices, views and visions* (pp.130-136). London, New York: Routledge.

Uotila, T., Harmaakorpi, V., & Melkas, H. (2006). A method for assessing absorptive capacity of regional innovation system. Fennia. *International Journal of Geography, 184*(1), 49–58.

Vidal, R. V. V. (2006). *Creative and Participative Problem Solving – the art and the science*. Infromatics and Mathematical Modelling, Technical University of Denmark, on line publication.

Walton, T. (2005). Innovation as a necessity. *Design Management Review, 16*(1), 6. doi:10.1111/j.1948-7169.2005.tb00001.x

KEY TERMS AND DEFINITIONS

Innovation: The commercially successful exploitation of new ideas creating business value.

Interdisciplinary: True integration of the understanding and achievements of a multidisciplinary group.

Front End Development: Flexible and cheap early phase of the offering development where the multidisciplinary stakeholders of the complex offering should meet to create the basic concept.

Complex Value Concept: Strategic definition and management information for a complex value offering created by diverse stakeholders stating the role of these participants in the value creation.

Product and Service Concept: Holistic business solutions for a product and service combination offering extra value and desirability for the customer or end user.

User Experience Process: Perceived value experience for the user through the different stages of using the offerings.

Chapter 4
Empathic Design Research Strategies:
Designing for, with and by People with Disabilities

Deana McDonagh
University of Illinois at Urbana-Champaign, USA

Joyce Thomas
University of Illinois at Urbana-Champaign, USA

Lydia Khuri
University of Illinois at Urbana-Champaign, USA

Susann Heft Sears
University of Illinois at Urbana-Champaign, USA

Feniosky Peña-Mora
University of Illinois at Urbana-Champaign, USA

ABSTRACT

Demographics are shifting. People are living longer and are expecting a higher quality of life than previous generations. Over a typical lifespan we will develop a range of disabilities, which are no longer perceived as a barrier to having a good quality of life. User expectation of products is growing which suggests a balanced approach to functionality is more important than ever. Rather than designing for the users, we need to be designing intimately with them to ensure that more intuitive design outcomes are generated. Ideally we hope to see people with disabilities designing for the wider population, which will push this model more towards design by. Empathic design research is a strategy that relies on the end user being an active partner in the designing process, a co-creator of knowledge. The concepts of empathy, empathic horizon, and the material landscape are illustrated through the discussion of a pilot design course within a university context. Using this approach, students with physical (visible) disabilities and product design students worked together designing everyday products. This empathic approach highlights research strategies that can support more effective design outcomes.

DOI: 10.4018/978-1-61520-617-9.ch004

INTRODUCTION

Creativity is an instinctive urge… that gives creators an unusual euphoria and generates an unmatched sense of satisfaction. Creativity is the core of new ideas. It's the source for new products, new designs, and vision to see the world in a renewed way (Canaan, 2003, p. 236).

Designers are traditionally and ostensibly able-bodied men and women. Design, of course, depends on the designers' creativity, which in turn is shaped by the material and historical dimensions of their lives. These conditions are often taken for granted and tend to be left unexamined. Along with individual differences in creativity, they provide the platform for designers to imagine what has not been imagined before. Able-bodied designers bring into prominence novel ways of thinking that nonetheless emerge against a backdrop of interacting with the material landscape which presupposes certain bodily qualities, a privilege not possible for individuals with disabilities or impairments. Although these delimiting conditions shape it, the very definitions of creativity denote pushing knowledge beyond things as they are accepted, defined, or known to exist – designers push beyond the boundaries of their empathic horizons.

The material and historical dimensions of their lives also shape the creativity of people with disabilities. Most often they are *not* in a position to take for granted what others accept without question – e.g., able-bodied people have an expectation that they will be able to reach door handles and elevator buttons, where a person in a wheelchair may not have the same assumption. A person's level of impairment can become magnified by the physical environment. How those who are able-bodied view the level of impairment of their peers with disabilities versus how people with disabilities view themselves can be entirely different, thus generating two realities. The conditions that provide the starting point for people with disabilities re-imagining the material landscape may or may not be radically different from their able-bodied counterparts. However, what they must contend with are the limiting factors of the environment, the nature of their disability and interpretation of it by others, and the seemingly universal phenomenon of devaluation (Vash & Crewe, 2004). In particular, people with disabilities face barriers in accessing the materials, skills, and facilities where product designers are educated and work.

Design is less about generating products and more about creating positive experiences for the user (Formosa 2006). This places design as a critical agent between the material world and the user. The material landscape impacts all people's daily experiences and affects personal perception of their independence in such basic areas as food packaging, mobility, personal hygiene, and communications. Effective product designers recognize they cannot rely solely on their own experiences to inform their designing process. More often than not they will be designing for people with experiences different than their own. Product design practitioners must be innovative problem solvers, studying and simulating life-expert-users in their environments to gain understanding of ways to create better user experiences with products.

Empathy "…is simply about achieving greater awareness, an extended imagination, and sensitivity to another person's world in a powerfully memorable way" (Fulton-Suri, 2003). It is the critical component that deepens the designer's understanding of users who may be very different from the designer. It allows the designer insight into intangibles such as feelings, emotions, dreams, aspirations, and fears that can provide the designer with critical cues, triggers, and inspiration that provide the foundation for more balanced functional and supra-functional products. Employing an empathic design research strategy enables the designer to expand his/her empathic horizon (McDonagh, 2008; Denton & McDonagh, 2003;

Laurel, 2003). It provides the designer with more relevant data and creative product outcomes as they apply what they learn and create together with the life-expert-users in the design process.

An empathic design research strategy was employed in an ongoing, design course-based project at the University of Illinois Urbana-Champaign (USA). The case study highlighted in this chapter describes the participants, course context, and implications for design education. It also examines project outcomes, including product opportunities, and explores their potential to be responsive to a shift in demographics in which people are living longer and are expecting a higher quality of life than previous generations. We will illustrate the engagement of life-expert-users and product design students in preparation for their professional practice. Advantages and challenges are discussed from the perspective of faculty and students involved, and how this will impact design education in the near future.

The roles of the life-expert-user within the design (research) process, the need for more empathic design, and the designer's boundaries to knowledge (empathic horizon) are critical elements in emerging design trends. By collaborating more closely with users, designers can begin to respond to more authentic needs, identifying "design moments" (product opportunities) within the users' naturalistic environments.

BACKGROUND

Barriers for People with Disabilities as Designers

The designing process typically involves two-dimensional sketches and three-dimensional models as a means of communicating and developing emerging concepts. Activities such as drawing and model-making can be significantly challenging for students who have disabilities and/or impairments. Many people with disabilities are unable to participate in design in an "equal manner" because the studio environment is not designed to meet their needs (Ginnerup, 2009). The physical limitations imposed by their disabilities may also be a barrier to experiencing empathy with some users (e.g., designing a running shoe, experiencing empathy with a runner – can you design a running shoe if you yourself cannot run or walk?). Users with "functional impairments" often are faced with serious difficulties with conventional product interfaces regardless of their disabilities (Keates *et al,* 2000).

By removing attitudinal and physical barriers, we can improve the lives of people with disabilities, giving them an equitable basis with the same opportunities as the able-bodied population (Models of Disability: keys to perspectives, 2001). To the extent that barriers are excluding people with disabilities from the creative process, we are faced with an under utilised resource and therefore missing important opportunities for novel and more creative solutions – not just in the realm of everyday objects for people with disabilities but in the realm of everyday objects for the wider population.

Demographic Overview of People with Disabilities

The United Nations estimates that more than 500 million people in the world are disabled because of mental, physical, or sensory impairment. In most countries at least 25 percent of any population is adversely affected by the presence of disability (United Nations Enable, n.d.).

A study found that 52 million people in the USA identified themselves as living with at least one disability (Kailes, 2002), which represents a significant portion of the US population (more than 15%). Moreover, due to the increase in the number of Baby-Boomers as a percentage of the total population, by 2030, there will be 70 million persons over the age of 65, with the fastest growing segment of that population in the US being those over the age of 85 (DOTCR, 2006).

In a study conducted in 1999 the National Center for Education Statistics in the US reported that 428,280 postsecondary undergraduate students identified themselves as having disabilities, representing 6% of the student body (Lewis & Farris, 1999). The types of disabilities reported by these students were:

- Learning disabilities (45.7%)
- Mobility orthopedic impairments (13.9%)
- Health impairments (11.6%)
- Mental illness or emotional disturbance (7.8%)
- Hearing impairments (5.6%)
- Blindness and visual impairments (4.4%)
- Speech or language impairments (0.9%)

Overview of the Campus Population of Students with Disabilities

In the 2008-2009 school year, the University of Illinois at the Urbana-Champaign (UIUC) campus had nearly 40,000 enrolled students. Out of that total number, 1081 students registered with the Department of Rehabilitation Education Services (DRES) as having visible or non-visible disabilities (e.g., depression, diabetes, and dyslexia). The University of Illinois Urbana-Champaign has one of the highest students populations in America that are wheelchair users (n=64).

Material Landscape

From the moment we wake to the moment we fall asleep we engage with products that fill our environments and that help us to communicate and construct who we are (Walker, 2006). We engage with these products that create our material landscape on both rational and emotional levels (Chapman, 2005). Such interaction should empower the individual, but unfortunately the experience often can be negative and erode one's sense of independence. Our material landscape sometimes can strip us of our dignity (e.g., difficulty opening basic food packaging, difficulty in operating television remote controls). Over time, individuals become aware of 'difficult' packaging and begin to avoid purchasing products that lead them to struggle. For people with disabilities (which also could include elders) this reaction can lead to a diminishing of food and other product range choice, thus eroding some of the key active ingredients of people's nutritional and sensorial experiences. Through more appropriate and sensitive design, people should be empowered.

Empathic Horizon

Babies are highly dependent on others to respond empathically to their limited, pre-verbal range of communication. They learn how to communicate hunger, discomfort, and enjoyment through body language, facial gestures, and vocal sounds. These modes of communication, the caregivers' consistent but not necessarily perfect attunement, and the incremental internalisation of this responsiveness, help babies to survive. The world exists from the perspective of how far they can reach, see, and hear. At this initial level, there is relatively little, or rather undifferentiated empathic awareness of others.

As a person grows a little older, he/she learns to engage with others, is introduced to the wider community (e.g., family, kindergarten), and slowly but surely becomes more socially and culturally integrated (stage 1 being the least differentiated, the final stage representing an expanded horizon). This empathic horizon is dynamic and ever changing (Figure 1).

As design is context sensitive, design understanding needs to be flexible because the user situation and cultural cues evolve. Expanding their personal empathic horizon to include the users they are designing *for* and *with* can take

Figure 1. The expanding empathic horizon

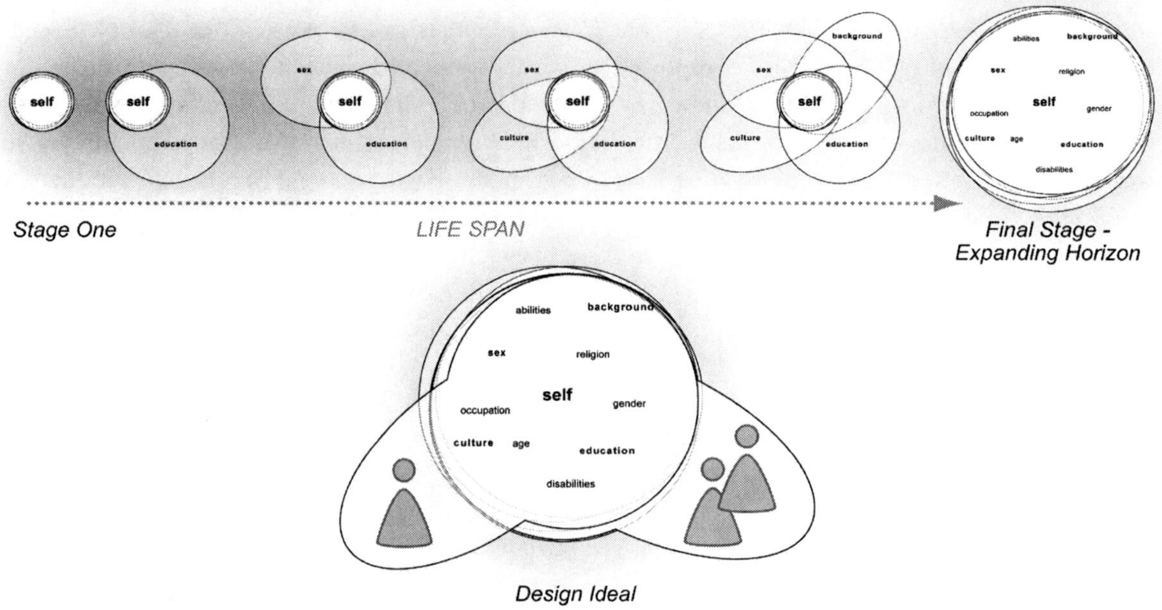

the designer outside their own comfort zone, but can lead to changes in thinking and in final design outcomes. This is the design ideal.

Supra-Functionality

Bearing in mind that user needs, expectations, and aspirations are vibrant and change frequently, the designer has many challenges to ensure that design outcomes are appropriate and relevant. Functionality alone will not necessarily satisfy a user. The experience needs to be enjoyable. Supra-functionality refers to the more ephemeral needs of the user that go beyond the utilitarian functionality of the product itself – the beauty, design, style; the colors, textures; the physical sensation of the product that makes the purchaser choose one product over another. When price points and functional needs are similar, supra-functionality can be the final, deciding factor that makes the customer pick it off the shelf in the store and unable to put it down. The elements that contribute to an enjoyable experience are often rooted in our emotional, social, and cultural desires. This has a significant impact on purchase decision-making, user-product bonding, and brand loyalty. In order to meet these needs designers must actively develop research methodologies that are specifically aimed at collecting design-relevant data which includes these often-difficult-to-grasp elements of the supra-functional (Weightman & McDonagh, 2003).

"Apple takes the view that its route to survival in a world dominated by Bill Gates's software, and Chinese hardware, is to use design as a lure to turn its products into aspirational alternatives to what its competitors are selling. It expects to sell fewer machines, but it charges more for them" (Sudjic, 2009, p. 15). "Design has become the language with which to shape those objects and to tailor the messages that they carry. The role of the most sophisticated designers today is as much to be storytellers, to make design that speaks in such a way as to convey these messages, as it is to resolve formal and functional problems" (Sudjic, 2009, p. 20).

Figure 2. Design Process and User Input (Highlighted area indicates the processes specifically targeted in this case study.)

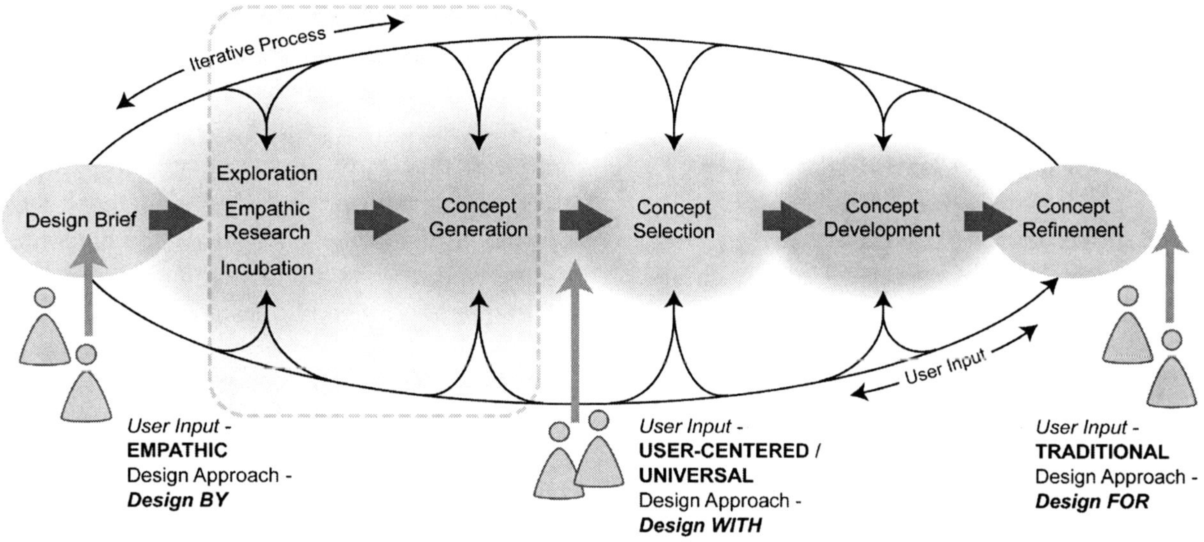

Product Design Process

Traditionally the product design process was a very linear activity with design followed by engineering, then tooling, manufacturing and production, each nearly completing its own phase before the process progressed to the next activity. User input was not sought until fairly late in the product development process, instead relying on the product designer to be the voice of the consumer in the early phases.

Designers have guided the consumer trends, the initiatives by which products were developed through their own intuitions and desires. As products become more commodity-based and users become more informed in their knowledge and desires, this paradigm of designers, engineers, and manufacturers guiding the trends is no longer a fiscally-or ecologically-responsible solution.

As a profession, product design has moved away from designer-centric approaches (e.g., the designer is removed from the user and professes good taste), to more user-centered ones (e.g., the designer acknowledges the need to study the user). The entire product development process has taken on a more collaborative approach, utilizing teams constructed of members from design, engineering, manufacturing, product management, and business from the very early stages of the process, engaging the user for input during the designing process (Figure 2).

"Universal Design is a strategy which aims to make the design and composition of different environments, products, communication, information technology and services accessible and understandable to, as well as usable by, everyone, to the greatest extent in the most independent and natural manner possible, preferably without the need for adaptation or specialized solutions" (Ginnerup, 2009, pp. 7-8).

By focusing on people different than themselves, designers are able to create more appropriate products and environments that benefit and accommodate the mainstream user/consumer. As the population in developed countries shifts toward more people with disabilities and longer life expectancies, it is beneficial for product designers to look beyond their own personal capital (e.g., background, physical abilities, and education) and embrace life-expert-users as co-creators to inform the designing process. Empathic design research relies on the user being an *active* and *participatory* partner within the information creation and designing process (Laurel, 2003; Preiser & Ostroff, 2001; Clarkson *et al*, 2003; and McDonagh, 2008).

Empathic Design Research Strategy

What differentiates participatory design from product-centered and user-centered design approaches is that the design work and process are conducted *with* the users rather than *for* the users (Iivari 2004). The participant has opportunities to interact with the design work in process and to express their feedback that often has a significant influence on the outcomes of the mockups and simulations (Clemensen *et al* 2007).

An important point in the participatory process is individual learning through increased awareness of a problem... It should encourage dialogue, debate, and collaboration (Sanoff, 2000, p. 10).

Empathic design research builds on this synergy of individuals developing relationships (Khuri 2004). "Developing empathy enables the designer to become closer to the user through respectful curiosity, genuine understanding, and suspension of judgment. Through this intimate, trusting relationship they create knowledge together" (McDonagh *et al*, 2009, p. 310). However, simply including the life-expert-user as a member of the team does not ensure communication, there must be a basis for understanding between these people with very different capitals to enable connections and interchanges. This dialogue allows the life-expert-user who engages with the world from an alternative perspective (*i.e.,* a person with disabilities) to become an integral part of the design process (Strickfaden *et al*, 2009).

Industrial designer Nancy Perkins writes "it is critical for women to work on designing products whose primary users are women. The experiences that women designers have in gendered roles can give them unique insights" (Perkins, 1999, p 121). Likewise people with disabilities as designers will bring unique perspectives and insights to products for themselves, for other people with disabilities, and for the general population.

Empathic design employs a range of research approaches that build on data and intuitive knowledge by expanding the horizons of those who practice it, thus enhancing the design outcomes - more empowering products and environments.

Employing empathic design strategies have resulted in the success of many commercial products. An example is OXO's Good Grips product range. "Designers interviewed consumers as well as professional chefs, examined competitive products, explored the physical effects of ageing and filmed arthritis sufferers" (Design Council, 2006). These products were developed for people with disabilities (specifically arthritis) but they do not attract any stigma. Reducing stigma reduces the risk of product abandonment, which is critical when the product is an assistive technology (e.g., walking cane). With their high style and excellent ergonomic design OXO's Good Grips have been successfully marketed to the mainstream population, winning design awards from the Industrial Designers Society of America, Arthritis Foundation, and Design Zentrum (Mueller, 1999).

CASE STUDY: DISABILITY + RELEVANT DESIGN COURSE

Course Structure

Disability + Relevant Design began as a pilot project[1] with the main goal of raising awareness of how product design can positively impact daily experiences through people's interactions with their material landscape. Graduate (n=6) and sophomore level (n=20) product design students[2] were partnered with students with disabilities (n=23), who were studying diverse subjects outside design[3]. Utilizing ethnographic

Table 1. Lecturers from the Disability + Relevant Design course

Dr. Deana McDonagh Associate Professor of Industrial Design		***Dr. Kathryn Anthony*** "Universal Design in Architecture" Professor in Architecture	
Eli Claire Author – Exile and Pride: Disability, Queerness, and Liberation		***Dr. Pat Malik*** "Ten things we should learn about disability" Director of Beckwith Residence Hall (for students with disabilities)	
Susann Heft Sears Disability Specialist DRES		***Professor Joyce Thomas*** "Disability + Relevant Design Project 2008" Professional Designer, Assistant Professor in Industrial Design	
Dr. Lydia Khuri Program Coordinator for Global Crossroads and Intersections Living Learning Communities		***Professor Carl Lewis*** "Accessibility Issues on Campus, Design for Disabled and Elderly" Visiting Assistant Professor of Architecture	
Dr. Brad Hedrick "Assistive Technology Abandonment" Director of DRES		***Professor Kevin Reeder*** "Anthropometrics and their Influence on Design" Associate Professor in Industrial Design	

and empathic research methods, the designers primarily developed products *for* their student with disabilities partner who was a volunteer participant and not enrolled in the class. This was not a traditional research relationship but one in which a meaningful relationship was developed (rather than researcher/subject, they developed designer/life-expert roles). The life-expert-users became co-creators of knowledge. The study included students with a variety of disabilities: Amputation, Cerebral Palsy, Dysreflexia, Muscular Atrophy, Muscular Dystrophy, Retinitis Pigmentosa, Sclerosis, Scoliosis, and Transverse Myelitis.

The projects undertaken emphasized the development of small, simple, insightful products which could have an incremental impact on the user, focusing mainly on improving the quality of life for the student with disability. The goal was to create products that did not carry a stigma and would visually integrate into the individual's lifestyle and personal environment. Products that were conceptualized included a pointing device to enable a law student with paraplegia to use his computer, accessible shelving for a wheelchair user, and a headset which uses puffs of breath to dial a cell phone for a student with Cerebral Palsy.

This project resulted in a public exhibition that illustrated the designing process and concepts, and celebrated the student collaboration. The exhibition enabled the school to share this project with the wider university community of students and faculty, as well as with the public. It provided the opportunity for critical conversations with colleagues outside design that further supported the efforts to develop the project into a more substantial and sustainable course that integrates the contribution of faculty and students across the university. The Disability + Relevant Design course welcomes design and non-design students with disabilities into the same learning space, including disciplines as diverse as product design, architecture, engineering, sport and tourism.

In the ensuing semesters of the Disability + Relevant Design course, the program has become more formalised. Students with disabilities enrolled in the course are not treated as a research subject, but rather as the designers' co-partners - creating *with* the designers, using the designers' expertise, hands, tools, and skills as they imagine and develop new concepts together. The students are immersed in video media, literature and guest lectures by stakeholders in the field of disability both from within and outside of the university (Figure 3), some of whom are people with disabilities themselves.

Key Insights of the Case Study: Employing the Strategy in the Design Atudio

For the designer and life-expert-user there are particular challenges and opportunities that this strategy offers:

Figure 3. Ethnographic observation in a naturalistic setting, a student simulating limited mobility while preparing meal, and students simulating Retinitis Pigmentosa

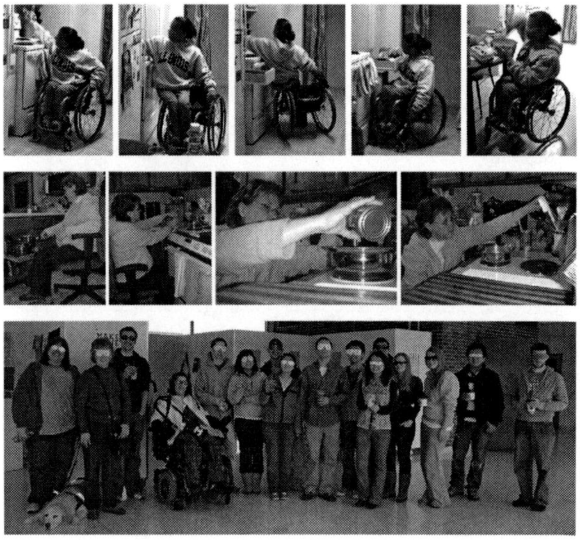

- Developing and nurturing a shared language between all members of the team is an essential element to enhance communication and collaboration whether it is simply between partners or across the larger classroom. A constant awareness and education of new team members or guests is required to maintain the shared language.
- Life-expert-users are partners in the designing process. They do not replace the designer but rather support more authentic data being retrieved and more relevant information being generated.
- Active collaboration between the designer and life-expert-user is crucial in the product development process. In this course, there was no room for the design student simply professing a knowledge and understanding of their partners. A partnership was required between the design students and their person with disabilities partner. The students needed to be focused and re-

sourceful, and to employ significant social skills.

- Taking designers outside their comfort zone to not only consider 'others' (e.g. elders, children, and people with disabilities) can require significant investment of time and emotion. For the typical student, it is very natural to design for someone just like themselves (e.g. educated, often able-bodied).
- Ethnographic shadowing by designers not formally trained in ethnography can provide unique insights and design moments for the product designer, but can be considered superficial research by experts in the field.
- Getting to know each other in non-structured activities, simple social conversations, and an "ongoing and significant dialogue" (Strickfaden *et al*, 2009) creates empathy between the designer and the life-expert-user.
- Empathic modeling can enhance the participant's ability to be open to understanding of other users needs.
- Realistic expectations need to be communicated to volunteer participants not familiar with the designing process. Design students were on fire with passion to reinvent the world. It was difficult to keep them focused on small, simple improvements for products that could be prototyped and used by their volunteer partners. Many of the students with disabilities partners were disappointed in only receiving a two- or three-dimensional representation of an idea rather than an actual product.
- Life experiences are particularly unique to each person; therefore, one person cannot generalize or 'speak' for others however small the target population. For example, one person's experience of living with a disability may be very different from another person's. Designers need to be aware that following an empathic route to eliciting data and creating information allows one of many perspectives. Equally, they need to be open minded to other people's 'unique' life experiences.

Shared Language

Collaboration tends to rely on natural respect, patience, tolerance and a shared goal. Within the process of collaboration, a shared working language develops that helps to define and sometimes redefine terms, language and processes. This takes time and effort to develop and nurture, and is particularly critical when working with individuals with disabilities. Synergy between collaborators is not automatically achieved. It is a product of time, investment, and mutual respect.

In the Disability + Relevant Design course, this fusion of able-bodied students and students with disabilities, this multicultural group – American, Asian and English, and ages ranging from 18 to 54 – had a diversity which challenged communication skills. Empathic methods needed to apply to communication as well as to research. This course emphasized usage of a "person first" language - e.g., "person with disability" rather than "disabled person." All students needed to learn a shared working language that incorporated culturally diverse meanings - e.g., empathy vs. sympathy, as well as discipline languages - e.g., non-design students may not be familiar with terms like concept generation, sketch model, or the designing process. As this shared language became common ground in the classroom, it was important for faculty and students alike to remember that new class members and guests would need to be introduced to this language.

Design Research: Ethnography and Empathic Modeling

Using ethnographic methods such as in-depth multiple interviews and observation of users, the teams gathered a range of textual, verbal, and visual data. The incubation period that followed provided them with opportunity to reflect, imagining a design solution/outcome that refined, developed, and/or matured what already exists. Incubation time allows the designer to make previously-unrelated connections (cross-pollination), identify emerging patterns, and ultimately create innovative design outcomes (Kelley & Littman, 2006). It required the student teams to have new ways of seeing, thinking, and experiencing as they developed more visionary ideas and concepts.

Some of the techniques that were used for eliciting authentic needs, both people with disabilities and able-bodied (Weightman & McDonagh, 2003) included personal diaries - audio, video, and text based (Lifchez & Winslow, 1979), one-to-one interviews (Bruseberg & McDonagh-Philp, 2001), observation (Kuniavsky, 2003), cultural probes (Aldersley-Williams *et al,* 2000), ethnographic shadowing (Hammersley & Atkinson 2003), and empathic modeling – able-bodied students restrict their own mobility, hand dexterity, etc. and attempt to complete everyday tasks, people with disabilities experiencing life without disabilities, or with a different disability than their own (McDonagh, 2008).

As this project concentrated on the students' experiences on the University of Illinois Urbana-Champaign campus, the ethnographic fieldwork was conducted in their respective naturalistic environments (e.g., students' kitchens, dining halls, and classrooms). Once the fieldwork was completed, the design teams focused on the concept-generation process, concept selection, development, and refinement, while maintaining regular contact with their life-expert-user partners for further dialogue, concept evaluation, and feedback.

One limitation encountered was the lack of formal training in ethnography (design education has only recently begun to introduce these type of research methods into the curriculum). Forsythe (1999) critiques the application of these methods as "do-it-yourself ethnography." She professes that this could be a form of superficial social research that confers the illusion of increased understanding when in fact no such understanding has been achieved. The concern noted by Forsythe is that the use of ethnographic methods is in name only, with little understanding of the actual methods used by ethnographers. These passive activities, however, often provided the design teams with valuable insights into the user experience.

Designers can experience a brief introduction into another person's life through empathic modeling. It is impossible for a designer using a disability-simulating device for a brief moment in time to authentically understand the life experience of a person *with* physical disabilities. It does, however, sensitize people to the boundaries of their knowledge and own personal experiences, helping them to expand their empathic horizon to embrace the life-expert-user.

Design students who used wheelchairs for a few hours expressed concerns of social responsibility to maintain the outward appearance of a person with disabilities as other able bodied people observed their movement unaware that the student was simulating a disability. Several students voiced feelings of being a fake, recognizing they could simply stand up and lift the wheelchair over the curb when crossing the street, or stand up and push the chair through a doorway.

Briefly simulating the experience of Retinitis Pigmentosa, the entire Disability + Relevant Design class wore "disability goggles" which impaired their peripheral vision by restricting their focal area in each eye through a narrow cylinder to a cone of about 2 degrees (Figure 3). They carefully walked down two flights of stairs and across a gallery to a coffee shop where they purchased beverages. Frustrated with difficulty seeing the sugar, cream, and coffee lids, a number of students abandoned the efforts to sweeten their coffee.

Uncertain of their footing and balance with this instantaneous loss of peripheral vision, students walked slowly and gingerly. These students noted a heightened sense of hearing with this loss of vision, recognizing the approach and passing of footsteps, but unable to catch sight of the people as they passed by through their newly narrow field of vision. Returning to the classroom, the students continued to wear their "disability goggles" as they wrote about their experiences.

A wheelchair bound student with Cerebral Palsy (CP) experienced an unexpected result of this empathic modeling. In her daily life, she frequently experiences body spasms due to her CP. As she approaches doorways, fear of crushing her hand between her chair and the door frame actually causes her hands to uncontrollably extend outwards, exacerbating the problem. The narrowed focal area of the "disability goggles" limited her field of vision to the door opening itself, and she was able to enter the doorway without fear, and without a body spasm that would put her hands at risk.

CASE STUDIES: DESIGNING BY, FOR AND WITH PEOPLE WITH DISABILITIES: DISABILITY + RELEVANT DESIGN COURSE PROJECTS

These representative students with profound disabilities exhibited tremendous enthusiasm when engaged in the creative design process. As active members of the design team, they brought new insights to the table that will ultimately benefit the wider society by adding perspectives otherwise not represented in design. For the product design students, seeing through the lens of students with disabilities provided valuable insights into their personal daily experiences.

The following three student experiences help to highlight student-to-student engagement with increasing levels of involvement and participation.

Design For: Bita

Bita's attitude is "I can do anything I want." She does not perceive herself as a person with a disability. An energetic and physically powerful person, Bita was a graduate student studying Human Resource Education. She worked as a Graduate Assistant and was a member of the elite US Paralympics track team. Her busy schedule included social events with friends, work, study, training, and travel.

The graduate student team of two women designers entered into their design research journey with Bita with an intentional degree of naiveté, purposefully open and suspending judgment in order to ensure they experienced new ways of seeing and thinking. This allowed them to see Bita with fresh eyes and to ask questions that they had no notion of how she might respond.

Using the creative exercises of observation, inspiration, research, and gathering in their studies the designers strove to be free of pre-conceived notions of how Bita's physical disabilities could affect her life. Initially focusing towards the authentic needs of their partner, they incorporated elements of Bita's personal environment, as well as a view of her social, educational, and work environments into their design thinking. This process required them to step outside of their personal comfort zones as they attempted to experience the ways in which Bita interacts with her daily environment. In methods similar to ethnographic research, the designers watched, listened, and absorbed – without interfering in her actions – as she worked, trained in the gym, and cooked. Using photographic ethnography, shadowing, and informal conversations in these natural settings, they began by asking questions about her life, dreams, goals, and aspirations – hoping to understand her personal needs, both physical and emotional.

As part of their empathic research the designers attempted to view the world through Bita's

Figure 4. Bita moves from her wheelchair to her car

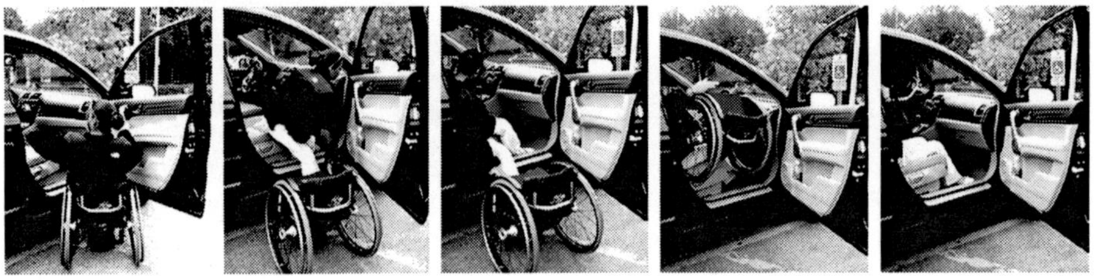

eyes, from her physical viewpoint, to become aware of frustrations and challenges related to being in a wheelchair. This part of the journey was particularly revealing to the able-bodied designers as they recalled the ease with which she dealt with activities of daily living compared to the difficulty and exhaustion they felt after only a few minutes as they momentarily experienced life in a wheelchair.

Moment of Inspiration: Bita told the design team that while she was studying for her first master's degree in Speech and Hearing Science, she was at class in a clinical / hospital setting. Although the hospital was wheelchair-friendly for patients, it was not so for the caregivers. Beds and examination tables placed the patients higher than what a caregiver in a wheelchair could easily access or even see; Bita could not see what the instructor was doing with the patient. Although wheelchairs with standing lifts are made, they are heavy and expensive. This frustrating situation was one element that caused Bita to choose to change her major from Speech and Hearing Science to Human Resource Education.

In her daily life, the kitchen table and the desk for her computer are generally higher than the appropriate ergonomic heights, causing significant additional stress to her body. With her positive attitude, Bita makes even these challenging situations work for her "because it's what I'm used to and I make them comfortable." Discussing these issues with the designers, she demonstrated how she increases her reach by sitting on the wheelchair's arm.

The design team's initial concept was to develop a mechanical adjustment section embedded in her wheelchair to change the seat height in reference to the floor to meet her needs. However, the designers quickly discovered the additional weight of this product would actually diminish its appeal. Bita's manual wheelchair must be light enough for daily travel and easy for her to lift into the car multiple times in a day (Figure 4).

From the perspective of functionality and practicality, with Bita's input the designers generated several versions of height adjustments on the wheelchair, concentrating on developing a cushion height adjuster which could raise her seated position in a localized setting without compromising her center of gravity when she moves about.

Design With: Jin

'Design for people with disabilities' was not a meaningful concept for us; only 'Design with Jin'. (Graduate Student Design Team, 2008)

Jin has a degenerative disease that has left her lower body unable to support her weight and requires her to use a wheelchair. Jin fully adopted this project as her own, which she illustrated by calling it "our work." Her collaboration extended beyond providing information and access to her environment - she was an active team member who brainstormed with the product design graduate students and made significant suggestions during the concept generation and refinement stages. Jin complained about discrimination in aesthetics, pointing out typical images of the products and facilities for people with disabilities: "Cold, boring… looks like a hospital." Access ramps for people with disabilities tend to be at the back of the buildings and often next to the rubbish containers. Jin is offended by the implications of this. Jin is sensitive to products designed specifically for people with disabilities because they tend to carry stigma.

The designing process with Jin had many differences from the designers' previous experiences in design research. Interviews and shadowing with Jin were not just to elicit her existing unresolved needs or problems with her material landscape, but also to develop consensus and empathy, and to discover the high impact of relatively incremental improvements. Using indirect questioning which did not specifically focus on Jin's material landscape, the design team tried to understand her life and her thinking rather than focus on her disability.

Understanding each other and obtaining familiarity were this team's first steps in their empathic designing process. Both members of the design team and Jin herself are Korean, which helped them find common ground as they talked about difficulties of living in the US. The designers regarded Jin as someone who has different kinds of difficulties. This team observed that this form of equalized relationship is fundamental for empathy, calling it "a perfect type of spiritual equality."

This step in developing a common ground allowed the design team to move naturally to substantive interviews and shadowing. The intent was to comprehend influential factors in her life including what is emotional. Sharing emotion in diverse stories, they developed an understanding of Jin's perspective toward each situation. This empathy in emotion helped them to establish which factors should be considered with Jin's perspective, including many challenging situations that needed functional and supra-functional improvements. Jin's emotional needs were very characteristic and clear and were primarily exhibited as aesthetic needs. Her shoe collection, including high heels,

Figure 5. Ethnographic observation of Jin's shoe collection, the movement from her wheelchair to her shower chair, and low level storage in her room

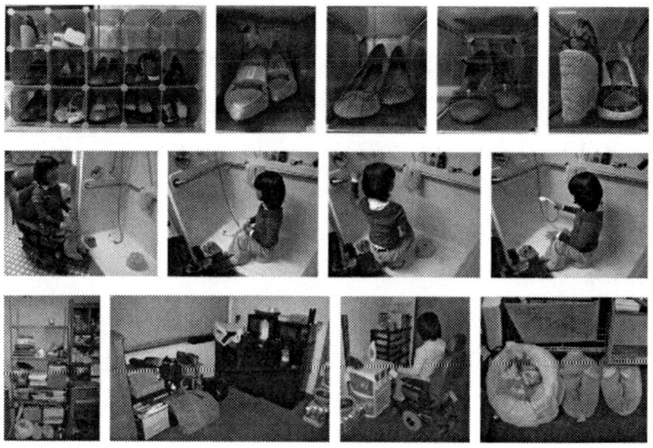

provides a compelling example of this characteristic preference, expressing her personal identity, gender, and sexuality (Figure 5).

Designing *with* Jin, the design team concentrated on her residence hall room to identify design moments which offered the potential for positive change in her life, focusing not only on improving functionality, but also on emotional satisfaction without a "special product look."

The first design opportunity was in her shower room during her simulation of her shower process. While Jin takes a shower, an armrest is needed for her safety. However when she moves from her wheelchair to the shower chair, the armrest impedes her movement. The team concentrated on developing a shower chair that was both functional and aesthetically pleasing without stigma.

Their second design opportunity was recognized in the methods Jin uses to store her books and shoes. Jin naturally uses the lower spaces in her room for storing her belongings because from her wheelchair she cannot reach the upper shelves. However, in the space which is closest to floor, Jin's line of sight and ability to reach these belongings is limited by her position in her wheelchair. To provide her with greater convenience, the designers developed an angled storage shelf which gave her better accessibility from her wheelchair and improved her range of sight of belongings stored at this low level. They concentrated on creating contemporary design concepts while considering accessibility for Jin, eliciting her feedback on sketches and mock-ups which they integrated into additional concepts. Jin also provided quantitative data in her evaluation of functionality and beauty.

Design By: Leigh

Leigh is sophomore studying Recreation, Sport, and Tourism. She is an honours student and a member of various leadership and honours organizations on campus. She has Athetoid Cerebral Palsy. Cerebral Palsy is an injury caused by a lack of oxygen to the brain while the baby is in utero or during the birthing process. It impairs a person's ability to control their muscles. It is not a progressive condition (Learn About Cerebral Palsy, n.d.). Athetoid Cerebral Palsy is characterised by involuntary, purposeless movements, especially in the arms and trunk (About Cerebral Palsy, n.d.). These movements tend to increase during moments of emotional stress, which can even include talking. Leigh cannot control her muscles to either cause or prevent movement of her fingers, hands, arms, and legs. She relies on her wheelchair as a mobility and support device, and on her personal assistant to help her eat, dress, complete her homework, and all other functions of life. Cerebral Palsy makes her daily life much more challenging.

Leigh's Story: "Almost one year ago I received my first power wheelchair. After many consultations with physical therapists it was determined that I could drive most accurately by using a head controlled switch. Prior to getting my power chair I had to have an aide push me manually wherever I wanted to go. I was so excited to get my new chair because it would give me a whole new sense of independence. Unfortunately I still was not able to fully gain my independence. One factor we did

Figure 6. Leigh Muscle spasms cause her hands to extend outward as she goes through doorways

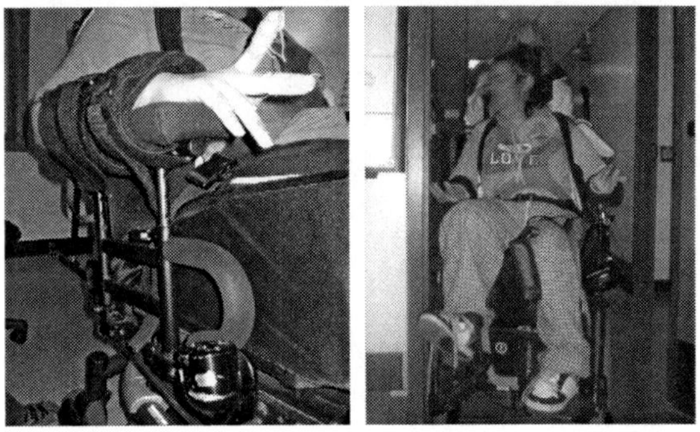

not consider was the uncontrollable movements of my hands due to Cerebral Palsy. My hands have a tendency to go into an extension and thrust out sideways (Figure 6). When this happens while going through doorways my hands and fingers get jammed backwards. To make matters worse, the fact that I am nervous about going through doorways causes my hands to extend out at the mere thought of it. This has resulted in several injuries and trips to the doctor. I now need to have an aide travel with me just so they can hold my hands in when I'm going through a doorway. After keeping track, for one week, of how many times I encounter doorways I was amazed to see it was over 240 times. During the past year I have had several physical and occupational therapists, physicians, and wheelchair technicians try to find a solution to protect my hands, all to no avail. Together with my design partner I hope to design a protective covering for my hands."

Collaborating closely with her design partner, Leigh learned to articulate her needs, create mood boards, define the product criteria, and brainstorm methods to solve the problem using techniques of design thinking. A basic overview of existing products designed specifically for wheelchairs was conducted, providing product awareness prior to concept generation. Her design partner's hands drew and created prototypes to communicate the ideas as Leigh directed the solution. They worked together to visually communicate (presentation form) her concept, review concepts and solutions, and reflect on the outcomes.

Awesome! From the moment I attached the prototype models and headed for a doorway I felt a sense of relief. It was amazing how my body reacted just because my hands and fingers were protected. We may still have some design issues to work out but we are well on our way to creating exactly what I need to feel safe and gain some independency. – Leigh

Informal conversations created an intimate relationship between Leigh and her designer partner. The trust between them broke the boundaries generated by their physical differences and allowed them to communicate more deeply, to seek solutions for independent living and to lead to the desired user experience. Leigh's participation in the designing process was a significant influence on the product development.

DESIGN OUTCOMES: DISCUSSION AND REFLECTION

Design Students Hopes and Fears

As they entered into this course, design students were asked to express their hopes and fears. Students hoped "to further my understanding of empathy...," "to better understand how (product design) can help people with disability…," "to help create a product to positively touch someone…," "to learn universal design in depth research…," and "to broaden my ability to empathise with other people…" They feared "political correctness…," "the sensitive situation…," "making anyone uncomfortable…," "accidently say(ing) something or do(ing) something that is not sensitive to a person with disability," and "not pulling through and (being) unable to create something useful."

This project encouraged the design students to go outside their personal and professional comfort zones. The design students generally were not familiar with people with severe physical disabilities, and the students with disabilities in turn were not familiar with being shadowed and observed in this way. The non-design students were unfamiliar with being a member of a product development team. Though research projects are often rated and valued by the outcomes, this particular project created *design moments* of student interaction that demonstrated a shift in thinking, practice, and designing. Not only did the collaboration and interaction between students with disabilities and the product design students have impact for the group, but also personally for each student.

Design Students Perspective

This experience was demanding but extremely rewarding for the design students, giving them insights into a world that they may not have had access to otherwise. The project provided unique opportunities to apply design research in a tangible way that added depth to their design projects. The design students have become more sensitive and aware to their use of language and now prefer the person-first terms.

Students with Disabilities Perspective

One of the more critical outcomes for the students with disabilities was the realization that their material landscape could (and should) be adapted to their needs (both functional and supra-functional). This shift in thinking was significant, as it gave the design projects an energy, direction, and real purpose.

In the pilot project, the volunteer students with disabilities were active participants, investing time and energy to ensure that each project developed and progressed. The faculty acknowledged early on that all of the students needed to receive credit for their input. The course now encourages participation and enrollment of both design and non-design students (with disabilities).

Design Educator's Perspective

The original project and the ensuing course required significant planning (e.g., organizing student meetings, gaining University approval for research activities). Integrating design research into the product design curricula and the personal student experience is very important. The fact that the undergraduate students became less grade-obsessed and more project-involved was an unexpected outcome. The lessons learned were less traditionally studio-oriented (drawing, model making) and more about developing and maturing the designers' as well as the life-expert-users' empathic horizons.

The undergraduates had a different journey than the graduate students and found the process of participatory design to be *time consuming*. Empathy takes time to understand and develop. It often takes the individual outside their comfort zone. We must acknowledge that it is more natural for some people to develop empathy than others. People with disabilities are equal collaborators - they need to go outside their own comfort zones as well. Educators must recognize that students with disabilities may need a personal assistant, which requires accommodation of non-students in the classroom that can impact on the dynamics of the setting.

It also could be considered *intrusive* in that the research needs to be conducted intimately in terms of space and time, rather than from a distance (e.g., relying on already-published material). Design students tend to want to dive into what they enjoy, what comes naturally to them, namely drawing, sketching, rendering, and model making. This is their language, their form of communication. For them to invest time in conducting design research can initially be considered as *postponing their designing process*. This perception changed once the students become aware of the relevance of the research and the impact it could have on their final design outcome. Research *takes the designer away from the studio*; the designers are outside their comfort zones and material landscapes. Initially this can be daunting, but relatively quickly the design students adapt and begin to acknowledge the design triggers and visual cues that are available to the designer. *Design research data tends not to be generalizable* (research that can be repeated locally or across cultures with similar results), as in the more traditional scientific (positivist) approaches. Design research has a huge cultural context that impacts the results. This form of research is NOT generalizable. For example, design research related to Chinese culture collected in China or Chicago would be similar, but not the same. Moreover, generalizable data tends not to be that helpful when we are dealing with visceral material and personal experience (as designers focus on particular users, tasks, and environments which are generally context specific). One concern is that such projects are just *paying lip service* and are not sustained over time and may just be touching the surface of much deeper social and cultural issues. It took time and patience for all involved to gain awareness of the appropriate use of language and terms when communicating with each other. There were times when students felt *verbally awkward and clumsy*. But with the support and guidance of the instructors and students with disabilities, the design students gained a more appropriate awareness of language and terminology.

This project provided the opportunity for *rapid immersion* and intimate contact with individuals who are outside the design students' experiences. It also introduced *design research as a dynamic and relevant skill* for designers, with additional value in gaining empathy with others to *expand one's empathic horizon*. Design students were given access to the *naturalistic environments* that they would not necessarily have been able to experience outside this project. This project also makes a *valuable contribution to the student's design portfolio.*

Though the data are not generalizable, they certainly are *transferable.* The design students will be able to *develop their own protocols and approaches* to design research based on this project. The students shifted their focus from their own design solutions to problems "out there" to the *authentic experience* supporting relevant design outcomes.

Future Trends: Social Inclusion

Expanding our horizon to a larger worldview, this project was an exercise in curriculum development with tangible outcomes. Education needs to be more socially inclusive. As design educators, we have the responsibility of preparing and equipping our students for jobs and technologies that may not yet exist. This is both a challenge and an opportunity for design educators.

This course has created a different educational experience for the students involved, one where they have become critical, active participants in developing the curriculum along with the faculty. As faculty and students both gain insight into issues of disabilities, more appropriate research approaches and design pedagogy develop.

Looking forward in the Disability + Relevant Design course, we are enabling product designers to develop more empathic approaches for designing FOR and WITH and also to push away the barriers and create methods and opportunities for design BY people with disabilities. We plan to expand on this growing body of knowledge and expect to develop products that will be *manufactured by* people with disabilities in sheltered workshops such as the Lighthouse for the Blind. In this exploration using empathic research strategies, we expect to engage students in new ways of thinking and developing more universally designed products and environments which are usable by all people, to the greatest extent possible, without the need for adaptation or specialised design (Mace, 1997).

In the near future, we will experience increased life expectancy, increased populations of people with disabilities and increased expectations of "quality of life" for all. There is no place for the "average" person being perceived as healthy and able-bodied. This will not reflect the consumers/life-expert-users that our future designers will be designing for. Through more empathic design research approaches, we can integrate the marginalised voice more effectively in preparation for our changing global demographics.

CONCLUSION

Just as energy is the basis of life itself and ideas the source of innovation, so is innovation the vital spark of all human change, improvement, and progress. (Levitt, n.d.)

In order to facilitate the full participation of people with disabilities in the designing process, we need to focus initially on the authentic needs of the individual, gathering information through a variety of methods including observation and recordings. Evaluation of the changing cultures and processes unique to this group of people with disabilities provides insight for greater exchange of ideas and teamwork in co-creating solutions that can provide benefits to the entire population (i.e. curb cuts that were initially designed to provide easy access for street crossing for people with disabilities have resulted in easier crossing for bicyclists). These steps provide the foundation for increasing economic development and competitiveness by promoting innovation in creativity education and practice in both people with disabilities and able-bodied population by removing the barriers that exclude people with disabilities from the creative process. With the current shifts in demographics, the increased expectations of quality of life for all, this approach to new product development is both beneficial and timely.

I have been impressed with the urgency of doing. Knowing is not enough; we must apply. Being willing is not enough; we must do. (Leonardo daVinci)

REFERENCES

About Cerebral Palsy. A resource for Parents and Families (n.d.). *About Cerebral Palsy*. Retrieved February 21, 2009, from http://www.about-cerebral-palsy.org/definition/athetoid-cerebral-palsy.html

Aldersey-Williams, H., Bound, J., & Coleman, R. (Eds.). (1999). The Methods Lab: user research for design. In Hofmeester K & de Charon de Saint Germain E. (Eds.), *PRESENCE: new media for older people*. Amsterdam: Netherlands Design Institute.

Bruseberg, A., & McDonagh-Philp, D. (2001). New product development by eliciting user experience and aspirations. *International Journal of Human Computer Studies, 55*(4) 435-452, October. ISSN 1071-5819

Canaan, D. (2003). Research to fuel the creative process. In Laurel, B. (Ed.), *Design Research: Methods and Perspectives* (p. 236). Cambridge, MA: MIT Press.

Chapman, J. (2005). *Emotionally durable design: objects, experiences and empathy*. London: Earthscan.

4Child, M. Help and Hope for Life, Learn About Cerebral Palsy (n.d.). *4 My Child*. Retrieved February 21, 2009, from http://www.cerebralpalsy.org/what-is-cerebral-palsy/

Clarkson, J., Coleman, R., Coates, S., & Lebbon, C. (Eds.). (2003). *Inclusive design: design for the whole population*. London: Springer-Verlag.

Clemensen, J., Larsen, S., Kyng, M., & Kirkevold, M. (2007). Participatory Design in health sciences: Using cooperative experimental methods in developing health services and computer technology. *Qualitative Health Research, 17*, 122–130. doi:10.1177/1049732306293664

daVinci, L. (n.d.). Leonardo daVinci quotes. *ThinkExist.com Quotations Online*. Retrieved March 15, 2009, from http://thinkexist.com/quotation/i_have_been_impressed_with_the_urgency_of_doing/214030.html

Denton, H., & McDonagh, D. (2003). Using focus group methods to improve students' design project research in schools: drawing parallels from action research at undergraduate level. *International Journal of Technology and Design Education, 13*(2), 129–144. doi:10.1023/A:1024149703800

Design Council. User Research – Case Study: OXO Good Grips (2006). *Design Council*. Retrieved September 14, 2008, from http://www.designcouncil.org.uk/en/Case-Studies/All-Case-Studies/OXO-Good-Grips/User-research/

Formosa, D. (2006). *Social responsibility through design: Smart Design calls for a wide-angle view of Universal Design*. Retrieved September 14, 2009, from http://www.smartdesignworldwide.com/news/article.php?id=98http://www.smartdesignworldwide.com/news/article.php?id=98 (accessed 14th September 2009)

Forsythe, D. E. (1999). It's just a matter of common sense, ethnography as invisible work. *Computer Supported Cooperative Work, 8*, 127–145. doi:10.1023/A:1008692231284

Fulton Suri, J. (2003). Empathic design: informed and inspired by other people's experience. In Koskinen, I., Battarbee, K., & Mattelmäki, T. (Eds.), *Empathic Design: User Experience in Product Design*. Finland: IT Press.

Ginnerup, S. (2009). *Achieving full participation through Universal Design* (pp. 7–8). Strasbourg: Council of European Publishing.

Hammersley, M., & Atkinson, P. (2003). *Ethnography: principles in practice*. London: Routledge.

Kailes, J. (2002). *Emergency evacuation preparedness taking responsibility for your safety: a guide for people with disabilities and other activity limitations*. Center for Disability Issues and the Health Professions.

Keates, S., Clarkson, P. J., Harrison, L. J., & Robinson, P. (2000). *Towards a practical inclusive design approach. In 1st ACM Conference on Universal Usability* (CUU 2000), Arlington, VA, USA, 45-52. Retrieved April 13, 2009, from http://www-edc.eng.cam.ac.uk/cgi-bin/publications.cgi?740

Kelley, T., & Littman, J. (2006). *The ten faces of innovation: strategies for heightening creativity*. London: Profile Books.

Khuri, M. L. (2004). Working with emotion in educational intergroup dialogue. *International Journal of Intercultural Relations*, *28*, 595–612. doi:10.1016/j.ijintrel.2005.01.012

Kuniavsky, M. (2003). *Observing the user experience: a practitioner's guide to user research*. San Francisco, CA: Morgan Kaufmann.

Laurel, B. (2003). *Design research: methods and perspectives*. Cambridge, MA: MIT Press.

Levitt, T. (n.d.). Ted Levitt quotes. *ThinkExist.com Quotations Online*. Retrieved September 20, 2009, from http://thinkexist.com/quotation/just_as_energy_is_the_basis_of_life_itself-and/161705.html

Lewis, L., & Farris, E. (1999). *An institutional perspective on students with disabilities in post-secondary education. Greene, B. project officer*. Washington, DC: U.S. Department of Education, National Center for Education Statistics.

Lifchez, R., & Winslow, B. (1979). *Design for independent living: the environment and physically disabled people*. Berkeley, Los Angeles: University of California Press.

Mace, R. (1997). *The Center for Universal Design, College of Design, NC State University*. Retrieved 12 April 2009 from http://www.design.ncsu.edu/cud/about_ud/about_ud.htm

McDonagh, D. (2008). Do it until it hurts!: empathic design research. Design Principles and Practices. *International Journal (Toronto, Ont.)*, *2*(3), 103–110.

Models of Disability: keys to perspective (2001). *Models of Disability*. Retrieved September 13, 2007, from http://akmhcweb.org/ncarticles/models_of_disability.htm

Mueller, J. (1999). *J. L. Mueller, Inc: Universal Design in products: case study - Oxo Good Grips*. Retrieved September 14, 2008, from http://home.earthlink.net/~jlminc/cases_oxo.html

Perkins, N. (1999). Women designers: making differences. In Rothschild, J., & Cheng, A. (Eds.), *Design and feminism: re-visioning spaces, places, and everyday things* (p. 121). Piscataway, NJ: Rutgers University Press.

Preiser, W. R. E., & Ostroff, E. (Eds.). (2001). *Universal Design handbook*. New York: McGraw-Hill.

Sanoff, H. (2000). *Community participation methods in design and planning* (p. 10). New York: Wiley.

Strategies in emergency preparedness for transportation-dependent populations (2006). *Strategies in emergency preparedness*. Retrieved September 20, 2007, from http://www.dotcr.ost.dot.gov/Documents/Emergency/Emergency%20Preparedness%20Strategy%

Strickfaden, M., Devlieger, P., & Heylighen, A. (2009). Building empathy through dialogue. In Malins, J. (Ed.), *Design Connexity: Proceedings of the Eighth International Conference of the European Academy of Design* (pp. 448-452). Aberdeen, Scotland: Gray's School of Art

Sudjic, D. (2009). *The language of things: understanding the world of desirable objects* (pp 15, 20). London: W W Norton.

United Nations Enable. (n.d.). *World Programme of Action*. Retrieved April 12, 2009, from United Nations Enable - Rights and Dignity of Persons with Disabilities: http://www.un.org/esa/socdev/enable/diswpa04.htm

Vash, C. L., & Crewe, N. M. (2004). *Psychology of disability* (2nd ed.). New York: Springer Publishing.

Walker, S. (2006). *Sustainable by design: explorations in theory and practice*. London: Earthscan.

Weightman, D., & McDonagh, D. (2003). People are doing it for themselves. In Forlizzi, J., Hanington, B., & Ayoob, E, (Eds.), *Proceedings of the International Conference on Designing Pleasurable Products and Interfaces*, (pp34-39). Pittsburgh, PA: ACM Press.

What is ID? (2004). *Industrial Designers Society of America*. Retrieved August 24, 2008, from http://www.idsa.org/webmodules/articles/articlefiles/what_is_id_brochure.pdf

KEY TERMS AND DEFINITIONS

Design Education: The conventional route through Higher Education (University) that students take in preparation for professional design practice.

Disabilities: Physical or mental impairments that prevent or hinder regular achievement.

Empathic Design Research: A strategy that relies on the end user being an active partner in the designing process, a co-creator of knowledge.

Empathic Horizon: The ability to acknowledge limitations to knowledge, experience, and understanding.

Empathy: The ability to appreciate another person's experience.

Industrial Design: The profession that determines the form of a manufactured product, shaping it to fit the people who use it and the industrial processes that produce it (Industrial Designers Society of America, 2004).

Material Landscape: The products that fill our personal and private environments.

Person-First Language: Referring to "a person with disabilities" rather than a "disabled person," and thus focusing on the person before the disability.

Product Design: The area of the visual arts that responds to user experience and generates products that resonate and relate to their needs.

Product Development: The process that supports the development of new products.

Shared Language: The working language (visual and verbal) that develops as partners work together.

Supra-Functional: Attributes that satisfy the user beyond the utilitarian functional needs.

ENDNOTES

1 This research within the Disability + Relevant Design course strictly adhered to the protocol and guidelines of the Institutional Review Board administered by the University of Illinois Urbana-Champaign.

2 Design students involved in this study were all enrolled as undergraduate or graduate Industrial Design students in the School of Art + Design, within the College of Fine and Applied Arts at the University of Illinois Urbana-Champaign. For the purpose of this discussion, the authors have purposively chosen to use the term "product design" in place of "industrial design."

3 Students with disabilities involved in the Disability + Relevant Design pilot project and the ongoing course are registered with the Division of Disability Resources and Educational Services (DRES). All have a physical and/or sensory disability, are over 18 years of age and are studying and living on the University of Illinois Urbana-Champaign campus.

Chapter 5
Tool and Information Centric Design Process Modeling:
Three Case Studies

William Stuart Miller
Clemson University, USA

Joshua D. Summers
Clemson University, USA

ABSTRACT

A new design process modeling approach focused on the information flow through design tools is discussed in this chapter. This approach is applied to three long term mechanical engineering design projects spanning 24 months, 12 months, and 4 months. These projects are used to explore the development of the new modeling approach. This is a first step in a broader effort in 1) modeling of design processes, 2) establishing case study research as a formal approach to design research, and 3) developing new design process tools. The ability of engineers to understand the dynamic nature of information throughout the design processes is critical to their ability to complete these tasks. Such understanding promotes learning and further exploration of the design process allowing the improvement of process models, the establishment of new research approaches, and the development of new tools. Thus, enhancing this understanding is the goal of this research effort.

INTRODUCTION

This chapter begins with a general discussion on the design process and approaches to capturing and modeling the design activities. It is not intended to be a comprehensive review of the literature, but to provide a frame of reference with respect to the critical issues associated with design processes. This is followed by a discussion of case study research in engineering design. Again, this section is used to provide the reader with a broad understanding of how current case study research has been undertaken in engineering design and how it can contribute to our understanding of the design process. Next, the reader is introduced to the relevant issues associated with representing the information flow through the tools used throughout the design process. Finally, three industry focused case studies are used to illustrate the application of this tool, demonstrating how it can be used to

DOI: 10.4018/978-1-61520-617-9.ch005

highlight issues such as iteration, information dead-ends, and tool function duality.

THE DESIGN PROCESS

The design process is a flexible, high level, logical network of activities to be performed and/or design tools to be used for the entire act of designing an artifact, formed by choosing desirable candidate(s) from a set of viable activities/design tools based on certain objectives (Hazelrigg, 1998). The design process is the collaboration of scientific "know how" with mental and physical steps being taken toward the goal of arriving at a satisfying solution (Simon, Kotovsky, & Cagan, 2001). It is a social activity which allows the generation of physical and intellectual property from mental organization and physical tasks (Leifer & Tang, 1988). Engineers perform design processes often with varying degrees of success. The goal of the research presented here is to enhance the ability of designers to understand and therefore complete design processes.

Design Process Modeling

Several approaches have been proposed in the literature for modeling design processes and activities. This section is not intended to review exhaustively the different approaches to modeling engineering design processes, but to highlight a fundamental limitation that appears to be common in most approaches. This limitation is their inability to easily capture the activities executed by the engineering designers in larger design projects outside of artificial academic environments. The goal of this research is to overcome this limitation by focusing on the actual engineering design tools that are used and to model the flow of information through these tools. This modeling approach will be discussed in the following sections.

One example of a design process modeling scheme is to view engineering design as a series of transformative steps that can be modeled as state transitions (Reymen, et al., 2006). In this approach, each state of the design, captured as the complete collection of the description of the design product, is modeled with a transition activity linking the initial and final state. A limitation of this approach is that the granularity of defining the transition activities is extremely coarse and there is no clear demarcation between states, as these are an artifact of the process modeler, rather than the design process itself.

Ullman presents a different approach to modeling the design process based on his empirical studies in the early 1980's (Ullman, Dietterich, & Stauffer, 1988). This approach seeks to define the task episode accumulation (TEA) model as a sequence of tasks that are essential information processing activities. The different activities are classified to include such actions as gather, synthesize, and decide. In this manner, the TEA model looks at sequencing the actual actions taken by individual designers as opposed to capturing the transition states between the entire collection of design product information. This model has been used to capture designer thought processes in controlled environments, but not in larger design projects as it would quickly become too unwieldy to process each task.

Another transformation based modeling approach to engineering design is that of the Function-Behavior-Structure model (Gero, 1990). In this model, there are eight basic steps that transform a function objective into a structure to realize that objective. While this model suggests that there is a distinction between different types of information and defines different design activities as formulation, synthesis, analysis, evaluation, documentation, and three types of reformulation, this model does not adequately capture all of the details of the design activities and it does not include other design information types, such as requirements.

Requirements tracing and design rationale capture is the basis for the design process model that has been proposed by (Burge & Brown, 2002). Essentially, this model creates a decision tree where design solution options are explored. At each branch, a criterion for success is selected and a branch is chosen. The rationale for selecting the option branch is defined. These three steps form a design task that is sequenced together to form a design plan. These tasks are used to define the value of the specified engineering design variables, thus linking the design requirements with the final design product. This model of engineering design provides for explanations as to why and how decisions are made. However, in large design projects, this level of documentation is not typically available thereby causing gaps in the modeling of the complete design process.

A different approach to modeling engineering design processes is to view the entire "space" of the possible design activities that can be used to achieve the artifact (Grabowski, Lossack, & Weis, 1995). In this model, the order of the design decisions leads to a path through this process space to the concretion of a design artifact. The activities, or directions, that can be chosen to explore this space include: concretion, abstraction, detailing, combining, varying, and limiting. These are similar to the activities defined by Ullman. This model provides, perhaps, an interesting theoretical view of the design process, it is not clear how this model could be constructed to capture a large scale design project.

Finally, Ostergaard and Summers present a model of engineering design from the perspective of the flow of engineering information through a series of design resistances that are the collaborative design situations (Ostergaard & Summers, 2007). In this model, the design information is represented as the current and each stage where information is passed between human and computer reasoning agents is treated as a resistance to this flow. This resistance is overcome through the application of external knowledge, termed design voltage. While this approach begins to capture the concept that information is accumulated throughout the design process through transformations, it has proven difficult to quantify and model actual engineering design projects.

These examples of different approaches to modeling the design process have been developed for many different purposes, including understanding how designers think, defining how computational reasoning support tools can be developed, how states of engineering design products evolve, and the flow information through collaborative situations. Unfortunately, none of these approaches have been easily applied to capturing the activities within realistic industrial engineering design projects. A model engineering design activities that can be used for studying large scale design projects.

Metrics of Design Processes

The design community has several measures of design success (Yang, 2007). One of these, product functionality, relates to how well the design solution accomplishes the tasks it has been given (Pahl & Beitz, 1996). Another common metric is customer opinion, a subjective measure. This metric can vary with aesthetics, comfort, appeal, or market trends (Kirschman, Fadel, & Jara-Almonte, 1996). Another more quantified success metric is the designer efficiency (Atkinson, 1999). This measure is used to find the time and cost to design. This could be measured as the time the designers take in the design process to develop the final product. It is recognized that many different methods of success evaluation in design exist (Ulrich & Eppinger, 2008). However, discrepancies exist in the academic community about which success measures are valid (Sobek II, 2007). Ultimately, only a handful of success metrics for design processes exist and there is no consensus which should be used.

All success metrics can be used by a variety of users. Designers, managers, and researchers all use success metrics of processes. Naturally, each user has their preferred success metric which caters

Figure 1. Design process hierarchy

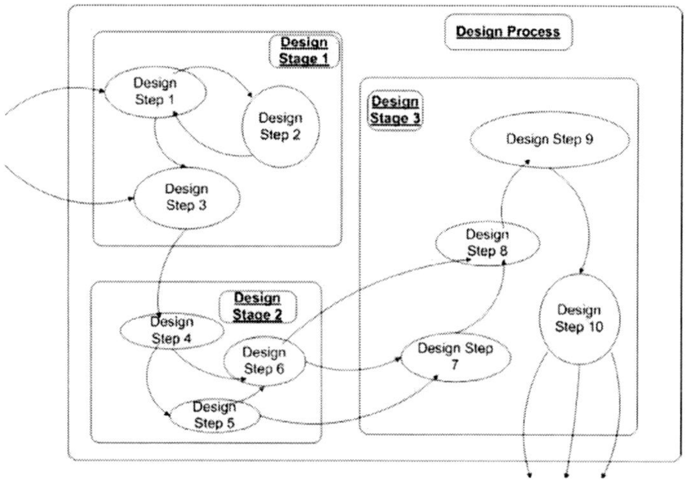

best to their particular area of interest. However, each category of the metrics can be evaluated independently and used to highlight the most desirable traits within the options given. An absolute metric would be beneficial over a relative metric because any ambiguity about the results would be eliminated. A direct measure of the design process is desirable because directly measuring the process leaves less room for interpretation errors that are present when the products are evaluated.

For design processes to be comparatively evaluated through different metrics, the processes and their goals must be clearly defined and then related to each other. To fully explain the process, it should be broken down into sections or stages which can give more detailed understanding of the change of information throughout the process. Doing so not only gives designers understanding of what the process accomplishes, but also how it is accomplished. Each stage can be further analyzed to determine the individual steps taken within that stage to achieve the deliverables and how those interact with subsequent stages. Each of these steps can be evaluated to determine the information that enters from the previous step and the information that exits to the next step. The tools used within the design step can be identified as well as their fundamental function.

Decomposition of the Design Process

Design steps can occur in loops, repeating and converging on information vital to success. Each design tool or method receives information from other sources, and transmits its exiting information outward to another step. Steps can possess singular and multiple information units which can enter and exit specific process steps. The information produced by any design step should have value greater than the sum of its components. The same can be said regarding the design process and its subordinate stages. Figure 1 illustrates a design process that has design stages decomposed into design steps. This notion of decomposition of the general design processes into constituent elements has been the focus of research efforts that yield ontologies of design activities at the atomic level (Kumar & Mocko, 2007; Sim & Duffy, 2003; Whitfield, Duffy, Coates, & Hills, 2002). From a pragmatic perspective, engineers and practitioners would likely prefer the granularity of resolution that focuses on design tools rather than atomic

Figure 2. Low detail resolution

cognitive and information manipulative activities. This view through the design tools is the focus of this research.

In order to quantify how well a design process has been executed, the process itself must be analyzed (Lockledge & Salustri, 2001). Analyzing processes can be done with various levels of detail resolution. A low detail resolution of the process allows the researcher to retain focus on the overall process goals. It does not lend detailed insight to information interactions within the process. Figure 2 shows a low resolution approach to process analysis. This type of analysis builds on the establishment of what is entering and exiting a specific boundary to the process. The result is a coarse representation of the process.

Conversely, in Figure 3, a higher detail resolution analysis of processes consuming more time and effort in detailing each step is shown. The work becomes tedious and can go into increasingly deeper detail, which may or may not be needed. This intermediate resolution of detail shows how

Figure 3. Intermediate detail resolution

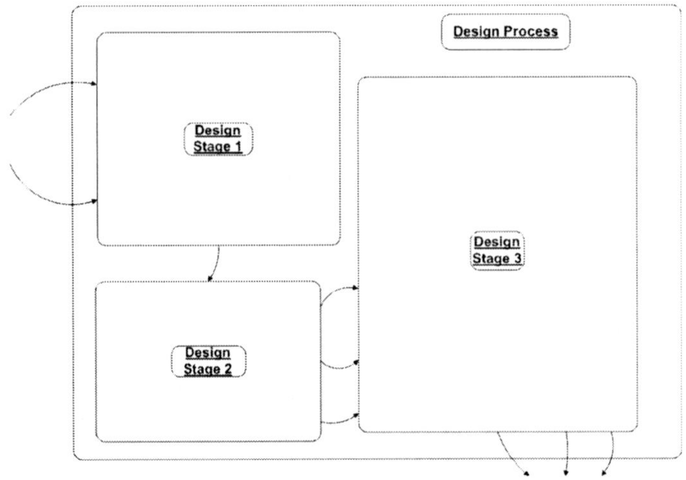

Figure 4. High detail resolution

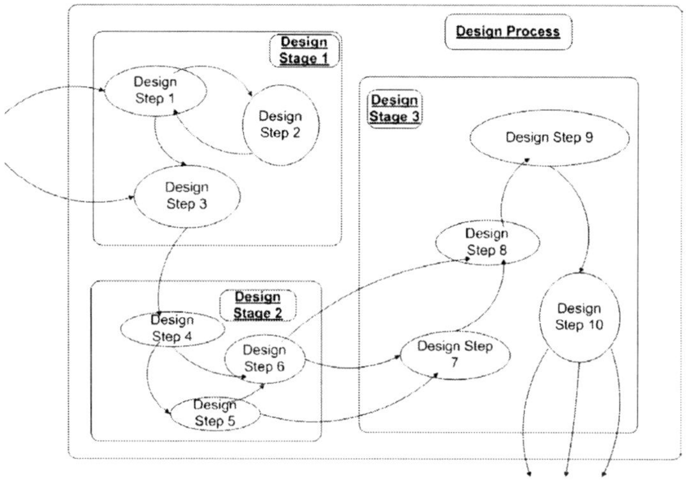

increasing detail requires increasing effort as well, because the entities being considered increases, thus increasing the work required to analyze.

This type of analysis permits the researcher to identify information interactions within the process, thereby identifying additional information within the process that was initially overlooked. A high detailed resolution analysis of the design process is shown in Figure 4 which is an extension of the hierarchy defined above. This breakdown of the process is considerably more "incremental" than a low detail resolution analysis can give. It is the designer, or as previously discussed, manager or researcher who must be capable of specifying appropriate detail to consider in the evaluation that will not consume enormous processing power yet still yield a sufficient analysis.

Design Process Tool Support

Design tools can exist in two different forms; product support and process support. Product support tools are used within the design process to complete and achieve the design solution. Process evaluation tools are used to evaluate and observe the design process itself, thus giving some measure of goodness. These tools may or may not be used to arrive at a solution, but they are frequently used to manage and monitor the design process either en route or after a solution has been achieved.

Analysis of design can be accomplished with several existing tools, albeit with limitations. Program Evaluation and Review Technique (PERT) is a design tool which is used in scheduling and planning processes (Battista, Pietrosanti, Tamassia, & Tollis, 1989). PERT is best used to evaluate singular project management entities, such as time. Gantt Charts are another project or process management tool. They are used to present time related information of processes in relevant timeline domains (Maylor, 2001).

The ability of the user to modify the evaluation tool for their specific needs does exist, but is often tedious due to evaluation tool restrictions of formatting and application. A process evaluation and analysis tool is needed to allow the users to track the critical information that is important to their individual research. The ability to decompose processes and evaluate each sub-system should prove beneficial for designers, by increasing their understanding of the design process.

Such a tool should facilitate information tracking and evaluation for each step within the process and should display the information produced by performing the specified tasks within that process. If units of information, such as design documents, are generated by a design task, they must either be useful or considered waste. Wasteful information means that the information is not useful in producing a solution and should not be created. If information is created and used later on, but not part of the original intention of that step, the requirements should be revised to include the newly discovered information. This analysis should enable researchers to better understand design processes. Researchers can analyze design projects to use the information available to learn about the process while using the process to learn about the applied case as well.

CASE STUDY RESEARCH IN ENGINEERING DESIGN

One of the most significant tools for design process analysis is case studies (Yin, 1994). Remarkably, case studies are often misunderstood when used as a design process analysis tool (Ahmed, 2007). Case studies are the empirical extraction of data from real world events that are used to view relationships and examine results about design (Teegavarapu & Summers, 2008). They provide relative, fact-based results to qualitative questions (Eisenhardt, 1989). Case studies focus on real world practices to develop theories and methodologies (Teegavarapu & Summers, 2008). They generate straightforward data from results which apply to a group of design variables, but cannot lend themselves to distinguish the effects of each variable individually.

Collecting data from case studies can laborious and can take place in a variety of manners, but tradeoffs of collection cost, collection effort, collection time, and influence on the results exist (Yin, 1994). The researcher must select what information is needed as well as how to extract it and do so without skewing the data. To extrapolate the conclusions to a theoretical application, properly constrained studies must be established (Yin, 1994). Developing such a bank of studies is difficult and resource consuming because the similarity of multiple cases is subjective and not easily compared.

Using case studies is said to be an all encompassing method which covers the problem definition, hypothesis formation, and collection and analysis of data stages. Case studies enable the designer to answer how and why questions about the specific occurrence. They do not allow the user to control variables, however, which would classify the study as an experiment. Case studies possess tremendous power in analyzing design processes, but wielding such power requires care. Despite the informational gain potential, case studies are easy to execute incorrectly. This reduces validity of the results as well as the beneficial experience gained by the researcher who executed the case study.

In order to enhance the use of case studies within design, one should be able to visualize the design process, without affecting the products of the process. This would allow the researcher to know how to construct the case study without corrupting the data. The visualization, as well as the results generated from it must be easily observed and understood. The information gained from visualizing the design process should not require exhaustive effort to generate or comprehend thus reducing the probability of erroneous case study design and extrapolation. The identification of multiple similar cases should not require exhaustive efforts but should rather be an observation of the two cases. By improving the use of case studies, engineers can focus more on what can be learned rather than doing tasks correctly.

The Need for a Case Study Research Tool

A design process model is needed. If a design process model existed which could communicate the information that the user needed, universally, understanding and manipulating design would require much less work. A way to communicate requirements of design processes as well as the way information transforms and flows throughout the design process is needed. Designers have product support design tools at their disposal. Using this as a starting point, a design tool based approach to modeling the design process could prove beneficial. Since design tools cannot compose a design process alone, the information which flows throughout the tools must also be considered. Using this approach, the information that enters and exits design tools could be connected to plot the design process evolution as it progresses.

Representing design processes with both product support and process evaluation tools could lend an advantage over other process models when representing design processes. With a model of design processes showing the importance and function of design tools, researchers could increase their understanding of design processes. Furthermore, the design process can be analyzed in individual portions as well as a whole to reveal fundamental relationships that are hidden within the design process.

A visual design tool based representation of design processes would enable researchers to modify the design process. When the design process is completely disassembled, the components and relationships within the process can be studied, thus allowing thorough understanding of the effect of each part of the process. This deep understanding could expose quantitative data about the effect of each design parameter and allow designers the ability to critique the design process. Having this concrete data would allow design process construction to be done on rules and facts which are easily quantified.

Observing and measuring precise rules and facts about design processes has been subjective for decades, but with the ability to "see" design, one could predict performance of designers engaged in design processes (Tufte, 1986). Engineers and designers are typically visually oriented people (Henderson, 1999). It is logical that if design processes could be illustrated with visual images, then the designers and engineers who use them so frequently would be able to better understand and manipulate the process to suit their needs.

Such a visualization of design should be capable of representing many if not any design process model. Researchers should be able to construct such a representation in context with their own domain of information. The users should be able to locate and follow any specified entity through the entire design process with ease. This model would enable performing visual and content based comparisons of multiple design processes. With such enhancements in design analysis, future researchers can build experiments to control design process outcomes. Such experiments would prove useful for design education as well as research projects. All of this could be applied to case study research, giving engineers a more powerful tool to advance the understanding of design. By observing the information flowing through each activity within the design process, bottlenecks can be spotted and addressed to streamline efficiency.

Fully understanding the design process and what happens within the context of design would allow engineers to improve the design process to suit the resources that are available to the designer. The design process itself could be manipulated intentionally to control the design process and learn about the effects that each component has on the final solution, thus blurring the line between experiments and case studies. With this improved understanding, the process can be used to improve the use of case studies in analyzing design processes. By implementing the aforementioned visualization method, case studies can become commonly used by many researchers and something that can work effectively to guide the design processes for engineers.

Thus, the development of a new process visualization tool is warranted. Here, we propose a visual representation scheme to illustrate information within design processes. This visualization method would enable the designer to see information that is of interest throughout the design process. As the process representation is constructed, the information of interest will develop from the first step of design to the final delivery of the product. Once complete, the process maps will allow researchers to follow critical information through the process as well as plan ahead to ensure that the current design process will yield something of sufficient value to the customer.

Development of a Design Process Visualization Tool for Case Study Research

A survey of process models is used to compare the differences between current design process model approaches and the limitations of each with respect to value gained by the executor. Inspired by IDEF0, an altered process model is presented in an attempt to increase the information captured by the designer when using and constructing these process models. The graphic representation proposed is applied to three case studies to illustrate the benefit of case study use in design tool development. Through applying this suggested representation to the case studies, it is shown that the Design Enabler Information Map (DEIM) allows researchers to create a visualization of design processes that can illustrate the needed information and be easily readable. By doing this, it is the goal of the representation to enable designers to understand, execute, analyze, and modify design processes in more effective ways than were previously available. By enabling design, the means of understanding, representing, modifying, and completing the process of design is permitted, or in this case, enhanced. Thus, this tool will be called Design Enabler Information Map (DEIM). Since information is the critical element of design, and the representation shows how information flows into and out of specific entities, these representations are called information maps. A design tool information map for the welding cell for metal frame construction project is shown below in Figure 5. Design activity sequencing is captured in the labeled arcs, along with the type of information that is flowing between design tools and activities. The design activities are grouped together into general design phases, such as *clarification stage*. The design activities are represented as formal tools (bold), general activity (thin), and unstructured (dashed).

Visualizing the design process traditionally falls into one of a handful of formats. Each of these formats has benefits and limitations to its use. Understanding what each can and cannot do is critical to proper application within the design process model. Fundamental methods to presenting the information contained within design process models exist and are commonly used widely; IDEF0, PERT, and Network schemes are typically the most commonly used (Dorador & Young, 2000; Shenoy, 2000).

Representation schemes for design processes are methods which present the information contained in the model to the reader. However, not all design process models are represented similarly. Some are represented via text and others are graphic with some being capable of using both. Words can be formed and sentences constructed to that communicate the information needed (Nanard & Nanard, 1995). While the text representation can give accurate and descriptive detail, the text is often lengthy and laborious to edit. Ideally, the detail will be maximized when writing the description and then filtered after the reader has deemed a specific portion unnecessary (Lockledge & Salustri, 2001).

A graphic representation may or may not give the same explicit detail as a text document, but can allow the user to observe the information that is pertinent to their present work. This can greatly reduce both the size of the representation as well as the volume of information that must be processed in order to obtain the required description of the image. A variety of graphic representations exist, and each performs a specific function appropriate

Figure 5. Generic design enabler information map (WMP)

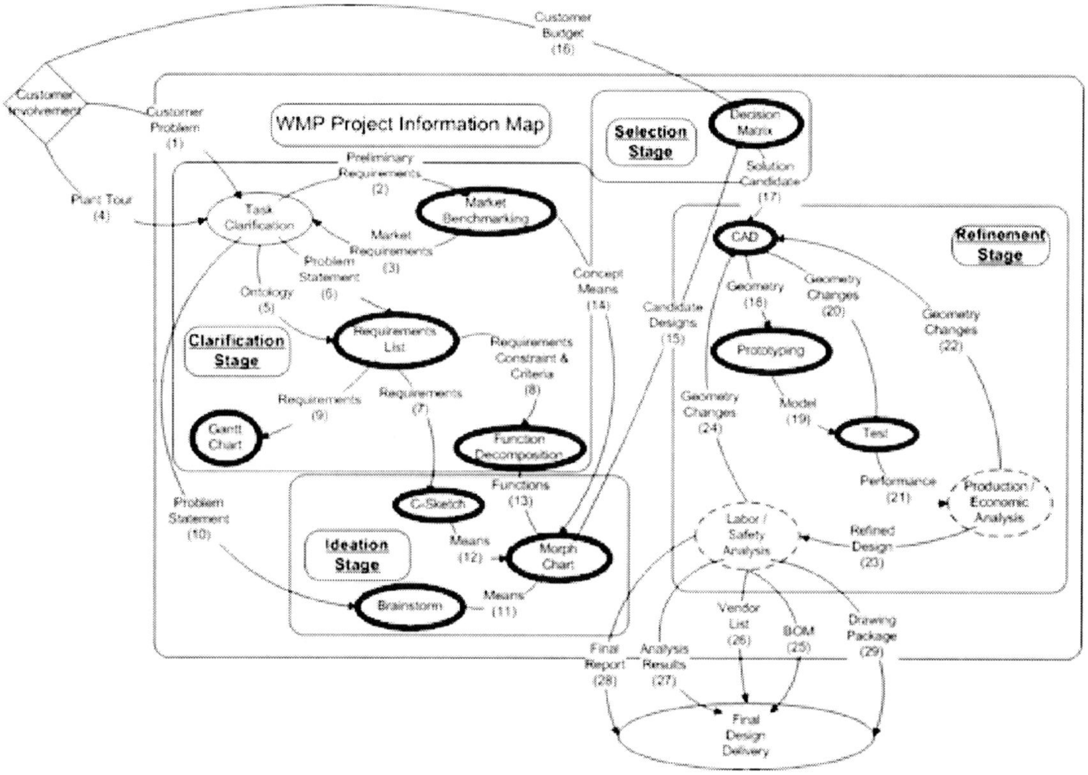

for its own use. Graphics such as diagrams, pictures, and icons are the most abstract. These have the least structure and can appear in almost any different form. Other forms of graphic representations are charts, graphs, and plots. These often, but not always, relate directly to mathematical entities allowing great detail to be communicated in a relatively small form. Charts are typically used for comparison of few characteristics. Matrix or table representations also utilize visual stimulus to communicate. However, these combine the visual principal with traditional text to relay information. Understanding a matrix requires the reader to associate the row and column of the specified entity to the characteristics of that entity.

A comparison of Design Structure Matrices (DSM) to an image diagram in representing design processes has been evaluated against design research criteria (Keller, Eckert, & Clarkson, 2006). In this evaluation, the DSM is compared to a Node Link diagram. Both of these methods are used to predict changes in process modeling and are a form of connectivity modeling which in turn introduces coupling complexities (Summers & Ameri, 2008). In order to generate a method of representing design processes, a form of graphic representation must be selected; being either a matrix or node-link graph.

In evaluating matrices against node-link diagrams, measures needed to be established to give a level of goodness. The ease of locating a specific information unit or node is of interest. Likewise, identifying and locating a specific relationship, or link between two information units is also needed. The ability to count the number of links entering and exiting a node should be possible. The identification of any adjacent nodes should be available to the user with minimal effort.

It was discovered that matrices lend favor to sorting information, thus aiding in selection

(Keller, Eckert, & Clarkson, 2006). Matrices are organized tables which can present large quantities of data in a clear fashion. However, matrices do make selection and counting of connecting links difficult. Identification of a relationship between two items is tedious to recognize meaning that identifying nodes adjacent to the current one is troublesome. Identifying the shortest process path via matrices is laborious if being analyzed manually.

Node-link diagrams present different benefits and challenges. They allow for identification of specific links easily. Determining the direction of the process links is easier with node-link diagrams, as is locating adjacent nodes. Following process paths is simple with node-link representations thus making the identification of the shortest possible path easy (Keller, Eckert, & Clarkson, 2006). The act of following paths, and shortest path identification were proven to be better represented with node-link diagrams for all datasets. This is believed to be due to the physical display of a path, which makes following easier. Since path following is critical to understanding information generation, flow, and change, node-link diagrams will be used to represent design processes.

IDEF0 is used to model functions of a system (Dorador & Young, 2000) and is similar to node-link diagrams. It represents the relationships and data which support the connection of those functions. The models are composed of hierarchal diagrams that show increasing levels of detail as the hierarchal level is reduced. IDEF0 is traditionally used to relate function to information of systems. The functions are traditionally shown as the nodes or entities. The information units or data which combine functions are shown as arrows leading from one function to another.

While IDEF0 diagrams allow designers to improve processes by describing critical information relationships, the traditional form of the tool is not optimal for representing process flows due to difficulty in tracking link flows. IDEF0 diagrams are suitable for showing activities and some connecting information. However, showing multiple domains of information flowing through nodes is something IDEF0 cannot easily communicate to the diagram reader.

An ideal scheme would combine the beneficial qualities of each of these into a single representation that researcher can use to illustrate design processes. The researcher should be able to specify the fundamental information of interest which is related to the specific information that flows through the process. The designer should also be capable of specifying the modifying bodies, or nodes, that exist so that the representation will be applicable to their specific area of work. The goal is to be able to determine the appropriate steps and expectations of design processes given the available information. The use of text should be available but not required and should not be limited to specific formats. This representation would be flexible to the researcher to use how each particular process needs it to in order to communicate the information that the designer desires to see.

The Case Studies

Three independent design projects were modeled after completion using the DEIM approach. The first project, designing a robotic welding cell, was a four month industry sponsored undergraduate design project for a small manufacturing firm. The second project was sponsored by another small company that is developing a combined trash and recycling collection and processing vehicle. The final project was to develop a tire testing system and was sponsored by a larger company. A brief comparison of the projects is found in Table 1.

These case studies are used to demonstrate the feasibility of using the DEIM to capture the type of information flow in the design projects. This quick and qualitative design research tool can allow engineers to visualize how different information is related throughout the design process. Coupling this visualization tool with formal ontologies for

Table 1. Case Study Comparisons

	EAI	WMP	Michelin
Duration	24 Months	4 Months	14 Months
# of Participants	6	6	8
Review Frequency	4 Weeks	1 Week	1 Week
Professors	1	0	2
Graduates	4	0	5
Undergraduates	1	6	1
Deliverable	Revised Vehicle CAD	New Work Cell	New Equipment and Testing Procedure
Time from project to mapping	5 Months	20 Months	1 Month

describing design information and activities is the next step in creating a formalism for conducting engineering design case study research.

For each of these projects, a visual design process map will be constructed representing the actual information and actions taken within the real life design project to develop a process representation and analysis tool. It is believed that throughout the development of these maps, the ability to represent design processes can be enhanced. Through these projects, the maps will be developed and tested, to show that visualization of design processes is possible and beneficial to researchers.

The graphic representation proposed is applied to three case studies to illustrate the benefit of case study use in design tool development. Through applying this suggested representation to the case studies, it is shown that DEIM allows researchers to create a visualization of design processes that can illustrate the needed information and be easily readable. It is the goal of the representation to enable designers to understand, execute, analyze, and modify design processes in more effective ways than were previously available. By enabling design, the means of understanding, representing, modifying, and completing the process of design is permitted, or in this case, enhanced. Thus, this tool will be called Design Enabler Information Map or DEIM. Since information is the critical element of design, and this representation shows how information flows into and out of specific entities, these representations are called information maps.

By intent, DEIM is to promote illustrative flexibility in what is represented as well as how it is represented for the case studies modeled. The main consideration when building a DEIM is to structure each connection in the context of what is being tracked, the information of interest. However, some formalism is required to enable logical perception by readers (Kim & Jang, 1999). The ability to freely position nodes and links within the map remains, but the net flow of the process should extend from the top and left to the bottom and right of the map. Additionally, the method of representing information types cannot be shared by multiple types. These are requirements that permit the process to be modeled flexibly yet with organization.

DEIM's consist of a few components which are connected together to create the representation of the process. These are the information links, the accompanying nodes, and any specific borders which are an extension of the nodes. Information is represented in the form of arrow links, which may have specified thickness, color, line style, length, labels, and direction (Kim & Jang, 1999). The nodes can consist of various shapes, sizes, colors, and labels. Any graphical property of the maps can be defined as a characteristic key of the map, thus representing some specific information unit. Borders shown within DEIM perform the same function of nodes, establishing detail limits to the investigation of the process. Borders are shown on the highest level as the design process boundary allowing initial information to enter and the solution information to exit.

Each component in DEIM represents some sort of information that is important to the overall representation. Information links represent the most elemental component of the process representation. The nodes represent the transformation of the information from one state to another. The borders that exist in the DEIM are expansions of higher level nodes. They serve to group information links and action nodes of specific relation together.

The goal of constructing a DEIM is to sufficiently track the initial information given to the designer through each step it is transformed through until the desired exiting information is reached. This will allow the designer to understand how information changes, how information flows, and how information is related to specific operations within a process.

DEIM can be used to represent actual complex design processes, consisting of a multitude of information domains, and including an assortment of active entities which work together to illustrate a design process in a clear and concise manner. The information contained within a DEIM can be low if the mapper desires, resulting in a DEIM showing a single type of information throughout the process. Additionally, DEIM can show many different types of information while still remaining readable by the user. Ultimately, they show what happens in a process as well as how it happens by illustrating the connectivity and transformation of links through nodes.

To test the use of and verify the benefit from DEIM, three design projects were represented using the DEIM method in a manner similar what others have used in validating their case study work (Hernandez, et al., 2001). For the DEIM case studies, the author was involved with each of these projects thus lending firsthand experience to the events as they occurred, giving the author the ability to map the events of the design processes into DEIM. These DEIM models were constructed after the project was complete, although the time span between project completion and map development varies for each example shown.

For each of the examples shown, the information links are generic design process information that may range from Requirements Lists to Concept Performance. A formal ontology for the type of information exchanged between the design tools has not yet been adopted. The nodes selected are the combination of design methods and tools (instantiations of tools). The author recognizes the difference in these two, but the distinctions are not of interest, therefore they will be used collectively. These serve to transform the information from its entering state to some other form via work.

In these maps, several generic design stages exist such as, Clarification, Ideation, Selection, and Refinement. These were chosen based on a traditional systematic engineering design process (Pahl & Beitz, 1996). Each respective map may or may not show the explicit stage boundaries, depending on the type of work being done. An example of this is the EAI project which consisted of three separate and concurrent detail refinement stages. The baler, trash compactor, and structure systems all were designed independently, at the same time. Rather than show a single refinement stage, the EAI map shows each separate system refinement stage of the design process.

THE EAI PROJECT

The Environmental America Inc. (EAI) project was a privately sponsored endeavor. The company, EAI, had an idea to patent and prototype an integrated recycling center on board a common trash collection truck. EAI approached designers with a fourth generation prototype and asked the

Figure 6. Final EAI DEIM

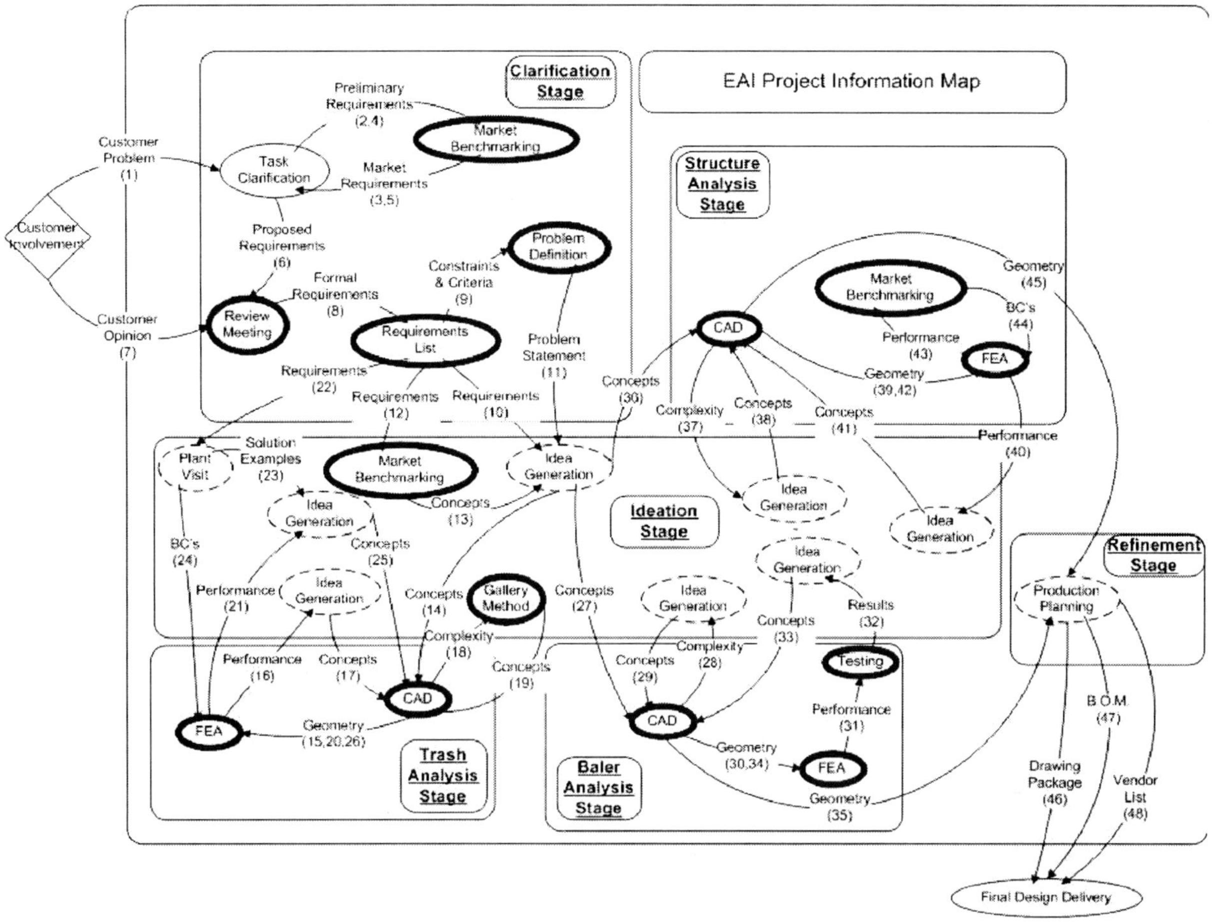

researchers to "streamline" the truck by reducing cost, mass, and complexity. This project relied heavily on the task clarification stage of design. Significant work was done to justify implementation of such a product into the infrastructure of local cities and residual plans had to be made that would affect the performance of the end product. Once appropriate justification of such a device was found, work began on ideation of concepts and product breakdown. The fundamental requirements of the truck were developed and used to eliminate excess systems on the current prototype. Once certain systems were eliminated, work began to improve efficiency and performance of the needed systems to conduct the truck's tasks. This work was segmented into three areas; Trash collection, Baler Design, and Structure Design (Smith, Johnston, & Summers, 2007).

The first of these is the design of the Trash handling equipment. This involved conceptualizing equipment to compact the trash and eject the load once at the land fill. Significant difficulty was encountered in modeling the behavior of trash, therefore, the project funding ceased before sufficient information about this could be collected. The baler design was more productive. The testing phase of this design proved instrumental in the generation of both concept ideas and Loading Conditions for the simulation. The superstructure design was also completed. Specification information was gathered from benchmarking existing vehicles and reproducing the functionality

Figure 7. Back propagation step 1

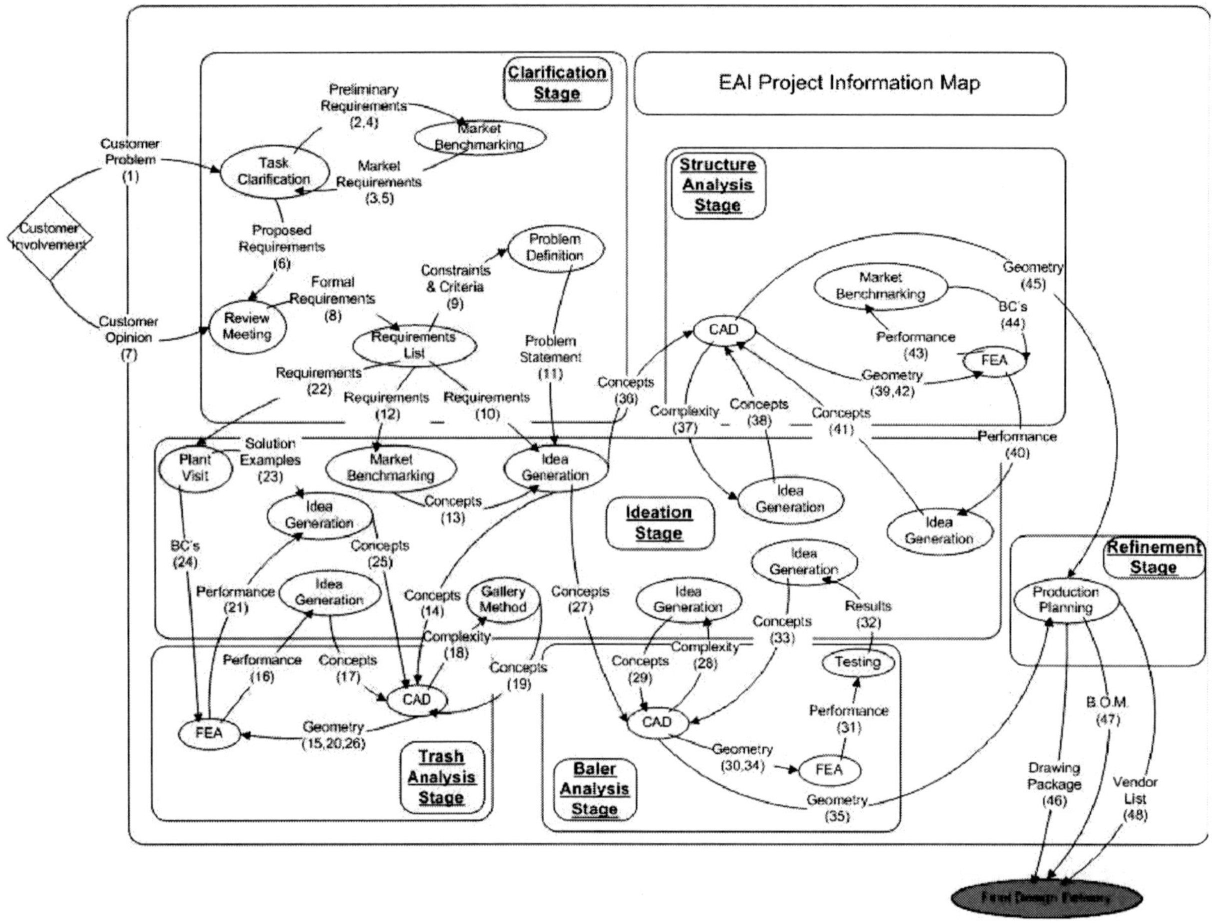

of those designs. The safety of the concepts was checked via FEA and the loadings gathered from market research.

Representation of the EAI Project

The DEIM of the EAI project was completed five months after the design project was finished. The DEIM representation of the EAI project is shown below in Figure 6. The information links vary in content but are primarily generic process information. Notice that each of the nodes is a design tool or design method which was used to transform the information to some other form. The line style of the node signifies what that node is. A bold node outline represents formal design tools. A thin node outline represents an informal action which can possibly be a design tool or method. Finally, the dotted outlines represent a fuzzy definition of that particular node. For these, the intent may be common, but the way in which the action was taken may constitute some modified version of a standard task definition. Thus these types of nodes are classified as "Fuzzy" and shown with a dotted line.

For this example, the information links can be related back to physical design documents or files. The Requirements List is a document that was printed and presented to the client. The geometry generated from the solid modeling (SolidWorks) step is the *.prt file for each respective part that was then analyzed in finite element analysis (Cosmos).

Figure 8. Back propagation step 2

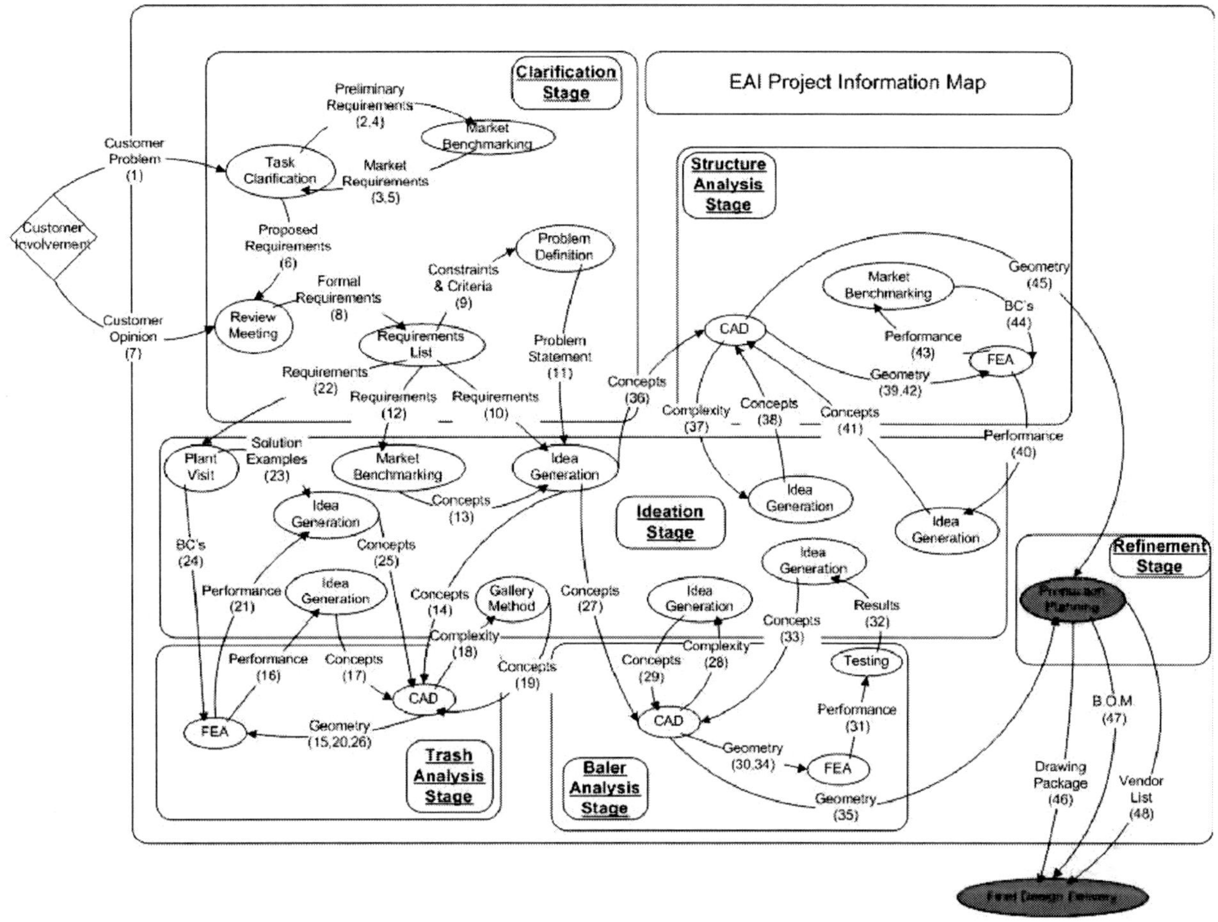

Observations of EAI DEIM

By constructing and examining the EAI DEIM, it can be concluded that:

- DEIM construction iteration improves readability
- Existing labeling of DEIM does not support parallel node sequencing
- DEIM readability can benefit from using cognitive tendencies
- DEIM construction iteration improves process content accuracy
- DEIM enables researchers to identify non-value adding nodes within a process
- DEIM enables researchers to identify information critical to a specific process

DEIM can allow the researcher to identify the critically needed steps within a process. By starting at the final design delivery, each subsequent step can highlighted. This can propagate all the way up to the initial customer input. By doing this, connection of each part of the DEIM to the final design delivery can be determined. In doing so, non-value adding steps are identified, revealing to researchers the wasteful steps within the process.

By studying the following maps, Figure 7, Figure 8, Figure 9, and Figure 10, it can be seen that the Trash Analysis Stage along with Idea

Figure 9. Back propagation step 3

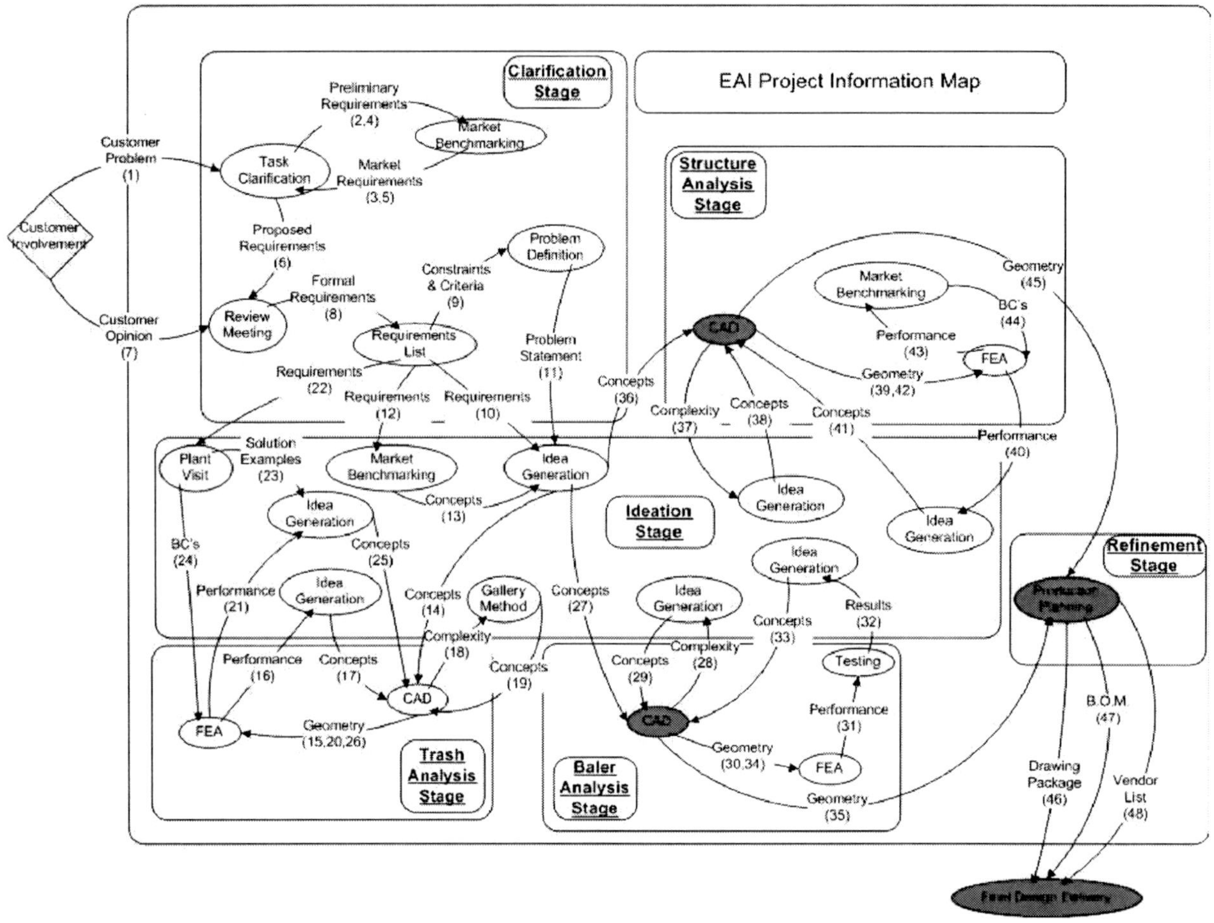

Generation Steps connected to it produced no work that contributed to the final design. This occurred because the effort required to model compacted trash behavior is extensive and the client decided to purchase an existing unit rather than consume resources to develop a new one. The identification of non-value adding nodes begins in Figure 7, with the final product. The required input information to that node is identified and then the source nodes of that information are then highlighted. This process propagates back through the process until the initial step is reached. Once a complete path of value adding nodes reaches from the final solution to the initial link, the process evaluation is complete. At this point, any remaining nodes are non-value adding.

THE WMP PROJECT

The Wright Metal Products (WMP) project was a four month long, sponsored design project. The client had purchased a welding robot, and wanted the designers to develop an efficient work area to use the robot in their facility. The client manufactures steel shipping crates out of pre-cut parts which are then welded manually in the facility. The goal of this project was to improve efficiency and production by using the robot along with manual workers. The client gave freedom to the designers to specify all needed layout, equipment, and fixtures that would be needed.

Figure 10. Back propagation complete

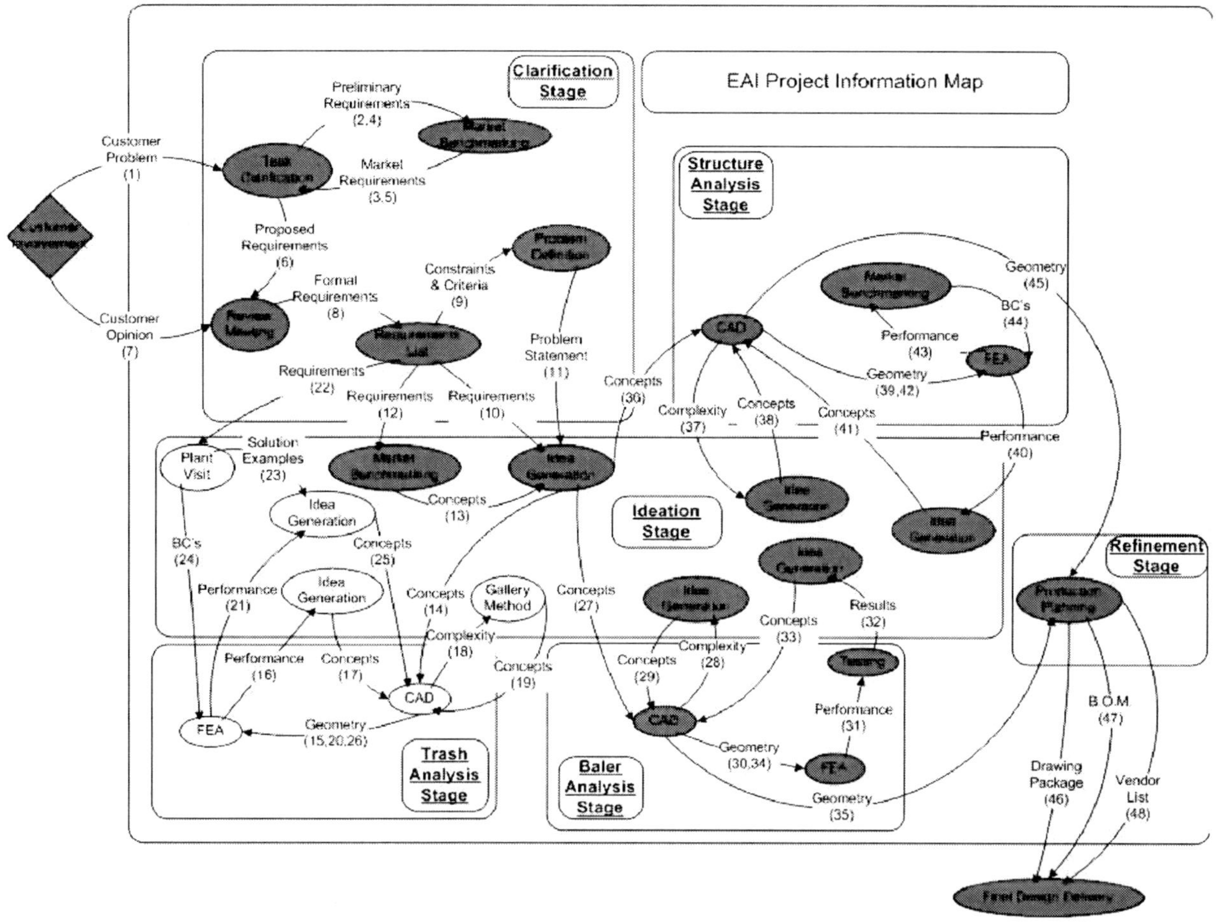

Representation of the WMP DEIM

The DEIM for the WMP project is shown in Figure 11.

It can be seen that this map does not contain the bold, thin, and dotted node outlines. That representation scheme was used in the EAI map to differentiate the different classifications of nodes that were used. For the WMP map, most of the nodes were formal design tools, therefore adding the different node formats would only serve to add information to the map which is not desired.

This project consisted of four main stages, starting with the Clarification, proceeding to the Ideation, then to the Selection, and completing the process with the Analysis. The Clarification stage served to properly define the problem and develop forma Requirements that must be met by the solution. The Ideation stage encompassed three idea generation techniques and the formation of a Morphological Chart to aid in concept organization. The Selection stage was the shortest in duration and consisted of the use of a Decision Matrix to rank solution candidates allowing the designers to choose the better concepts from all the ones generated. The final stage was the Analysis stage, where the solution was tested, analyzed, refined and validated into a suitable solution.

Figure 11. WMP DEIM

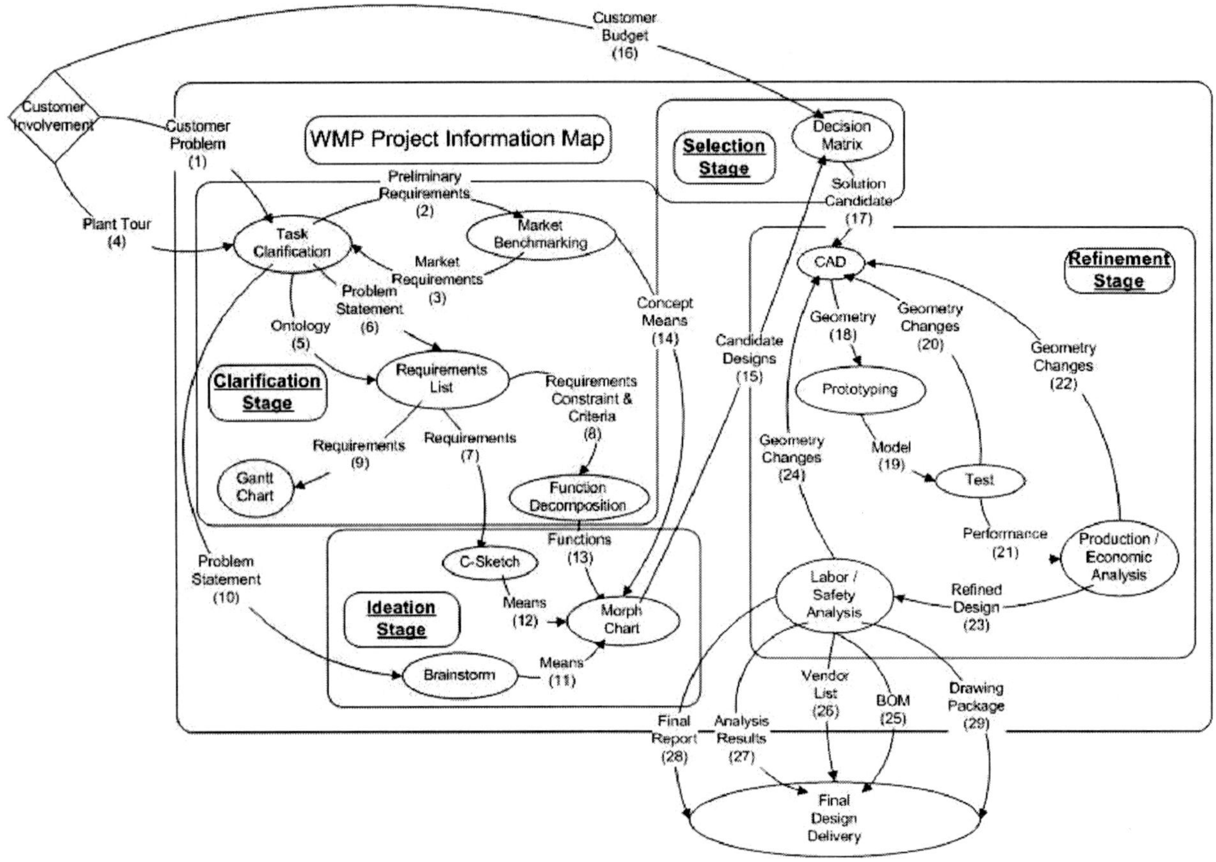

Observations

By constructing and examining the WMP DEIM, it can be concluded that:

- DEIM is flexible with respect to node classification and formality
- DEIM is flexible with respect to node grouping
- DEIM can enable researchers to identify non-value adding nodes
- Nodes can alter information and channel information

Some of the nodes in the WMP map accomplish dual functions. The first is that the node functions as the modifier of the information. Initially, this feeds back into the previous steps for iteration. However, when the information again reaches the same node, a different "check" may occur, thus allowing the information to "pass" through the node on to the next. An example of this occurrence is shown in Figure 12.

THE MICHELIN PROJECT

The Michelin project was a twelve month long project that began when representatives from Michelin Tire Co. approached the engineers with the topic of test procedure development. They were interested in understanding a soil and tire interaction and wanted to investigate the possibility of

Figure 12. Dual function nodes

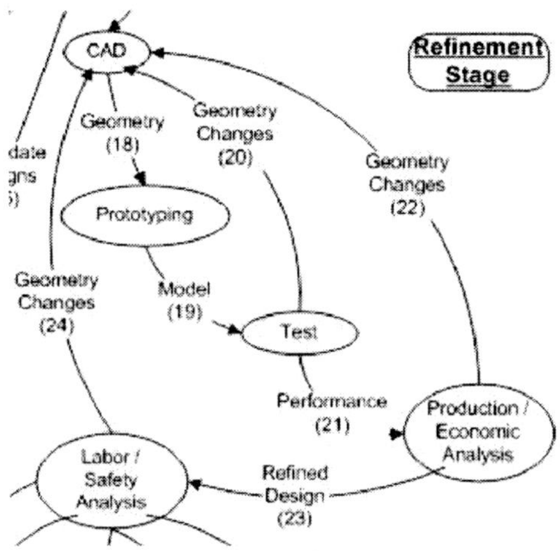

the designers developing an economic solution to their need of testing.

Representation of the Michelin DEIM

The DEIM of the Michelin Project is shown in Figure 13.

Tires are a complex machine, and the construction and simulation of their behavior is not a simple task. To test specific behaviors, tires are generally constructed, and used to examine the performance of certain parameters. The engineers will then specify changes, and will need

Figure 13. Michelin DEIM

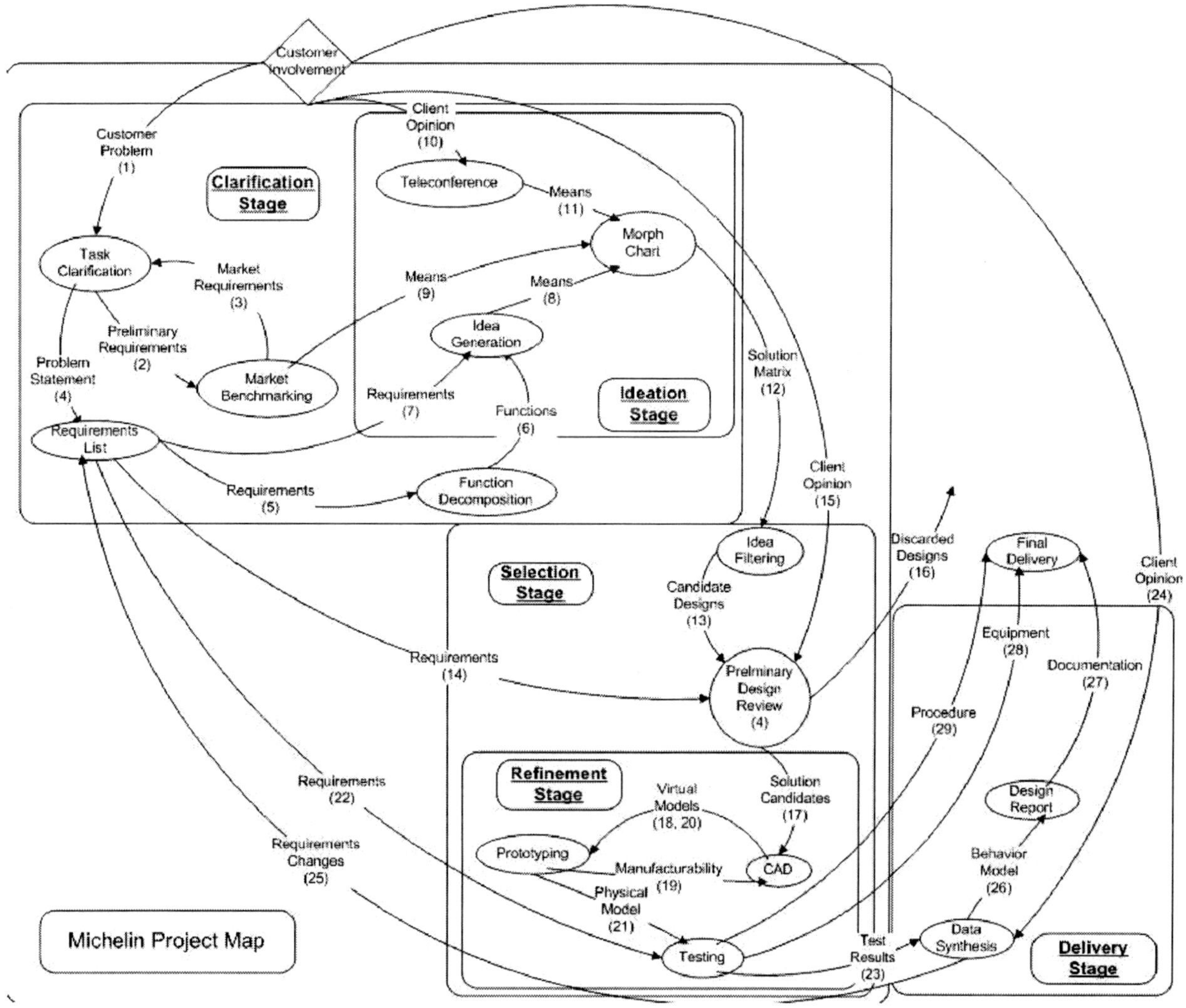

Figure 14. Michelin DEIM link length

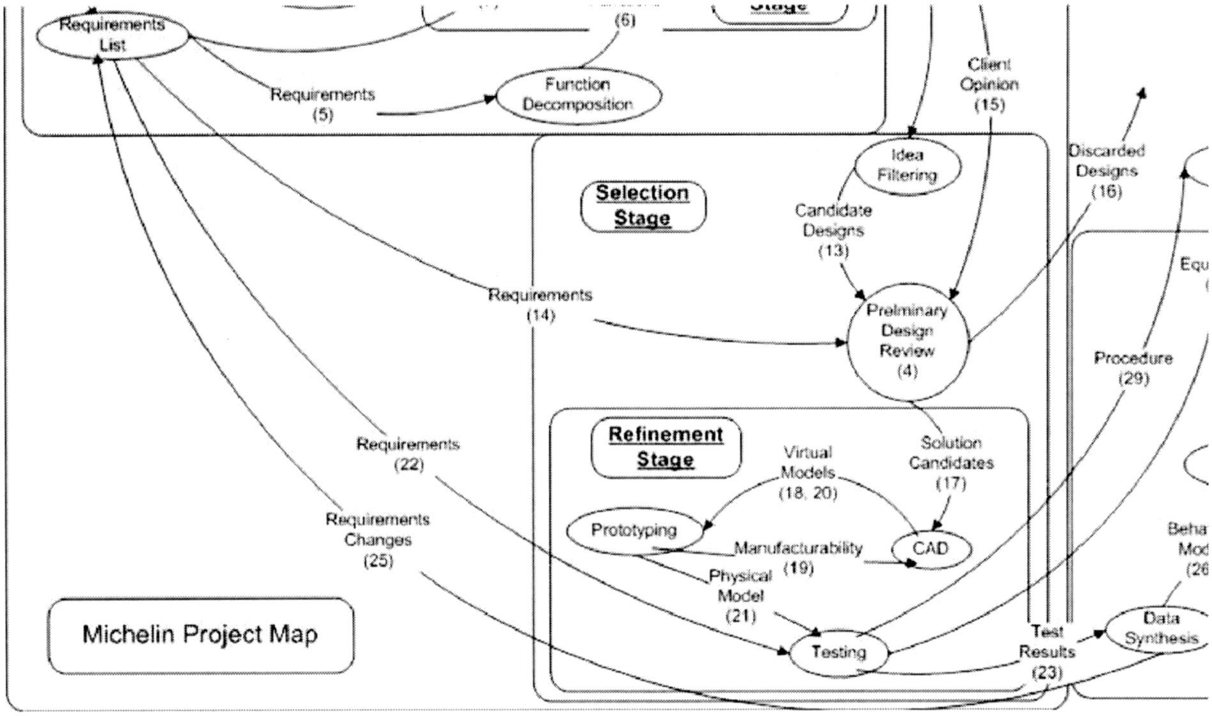

to manufacture a new set of tires. This is costly due to tire mold manufacturing cost. The testing cost limits the ability of the tire engineers to learn about the behaviors, so the need of a simulation protocol with more economic operational cost is warranted. This was the goal of the designers, as specified by the customer.

Observations

By constructing and examining the Michelin DEIM, it can be concluded that:

- DEIM can show non-critical information if the mapper wishes
- DEIM is flexible with respect to node grouping
- Link length may be proportional to specific information properties
- Node connectivity may be proportional to specific information properties
- DEIM can show the result of insufficient information
- DEIM can show the effect of making long iterations

It can be seen with this map (Figure 14) that the length of the links is proportional to the goodness of the design process. As mentioned above, early and often iterations make for better design processes, whereas the links representing those iterations would be spaced close together and in stages rather than in long sweeping curves. The link length in this DEIM varies greatly from node to node. This communicates some information about the process such as time required to iterate and percentage of time wasted by the iteration.

The Michelin DEIM shows relationships of node connectivity to the difficulty of that part of the process. If a node is highly connected such as the Requirements List or the Customer Involvement, then accurate and precise information from that source is needed for the design process to proceed properly. Accordingly, incomplete or inaccurate information from highly connected sources works against the productivity of the design process. As mentioned above, the lack of exhaustive requirements by the customer hindered the progress of the design project. If the significance of the Requirements List been known initially, the designers may have urged further clarification and development of the Requirements List before proceeding on to the next step of the design process.

The Michelin DEIM also shows the effect of insufficient information being used in a process. Fault of this deficiency cannot be given, but the effect of completing most of the process without it can be shown with the map. As the design data is being synthesized, discoveries were made that amended the requirements of the process. DEIM shows this by connecting a late node to an early node, thus initiating an almost complete process repetition.

CONCLUSION

By representing information within design processes, researchers can model complex processes graphically. Doing so gives researchers a better mental connection between what is perceived and what is real. With better understanding comes improved ability to analyze design processes with case studies. These improvements were shown through the application of a newly developed design process model in three case studies. Each case consisted of different deliverables, different experience levels of the designers, and project durations. The use of case studies to develop design tools and conduct design research has been proved beneficial by the successful application of and evaluation with DEIM to the case studies.

Through conducting these case studies, observations about the DEIM representation of design processes can be made and are:

- DEIM construction iteration improves readability
- Existing labeling of DEIM does not support parallel node sequencing
- DEIM readability can benefit from using cognitive tendencies
- DEIM is flexible with respect to node classification and formality
- DEIM is flexible with respect to node grouping
- DEIM can show non-critical information if mapper wishes
- DEIM is flexible with respect to node grouping

These observations illustrate ways that researchers can benefit from a graphic representation of design processes and that design processes can accurately be represented graphically while containing sufficient detail and promoting readability.

Observations about the cases studied through the application of DEIM were also made here are:

- DEIM construction iteration improves process content accuracy
- DEIM enables researchers to identify information critical to a specific process
- DEIM can enable researchers to identify non-value adding nodes
- Nodes can alter information and channel information
- Link length may be proportional to specific information properties
- Node connectivity may be proportional to specific information properties
- DEIM can show the result of insufficient information
- DEIM can show the effect of making long iterations

Information paths can be easily traced with DEIM. From the initial problem of the client to the documentation or artifacts that are physically delivered to the client, the type, quantity, and change of the information present at any given stage of the process can be displayed. This gives designers an ability to determine "Next Steps" for each process as well as determining information that may be needed to complete a specific task.

The critical path of information flow can be determined from reading DEIM. This can be done via back tracking the preceding nodes or any other means desired. The representation of links and nodes allows a holistic understanding of the process to advance into a detailed understanding of information that enters and exits any specific node.

The connectivity of the nodes within a DEIM can also be seen. It was shown that DEIM nodes that are highly connected tended to be significant contributors to the design process. The concept of this metric can be used to build DEIM in respect to other characteristics such as production dependencies, or information management. In the case studies shown, the connectivity relates to both the affect that the particular node has on other nodes as well as the difficulty that can be encountered when constructing DEIM containing that specific node.

More importantly, the development, application, and evaluation of new design tools and methods through the use of case studies has been proven. The three cases used show the benefit of the developed representation through the information revealed about the cases. The cases served to test the new representation, and allow the author to discover information about the representation that was previously unknown. These observations have been mentioned in the previous section. Furthermore, the application of the new tool to case studies has also shown that case studies can be used to evaluate design processes, in attempt to understand the process meaning that future endeavors of modifying and improving the process are possible.

Certain limitations were encountered when using DEIM to represent tools used in processes. These difficulties came from constructing the DEIM as well as reading and understanding the process being represented. The first difficulty is the difficulty to search for specified objects (links or nodes). With other representations such as DSM, readers can search for a specified entity in a one dimensional, linear manner by scanning either a column or row heading. DEIM expands the representation into a full two dimensional space, thus making searching more difficult. This was addressed by Keller et al as well in their Matrix vs. Node Link evaluation (Keller, Eckert, & Clarkson, 2006).

The layout of the nodes is also a cumbersome and tedious process. It is difficult to position the nodes of highly connected processes in a manner such that the links do not intersect. In fact, sacrifices must sometimes be made when constructing the maps to show some information and alter others. DIEM is a user defined representation of the process. It could be beneficial to allow link intersection sometimes, thus creating another entity which can represent additional information about the process.

The work presented in this chapter leads to some areas of future investigation such as the exploration of domain representation methods. They ways in which information domains can be shown within a map cannot currently be counted. The ability of DEIM to illustrate the sequence of operations within a design process is also of value to researchers. Future work could be focused on how to represent time and process sequence within the DEIM. DEIM contains complexity that could greatly benefit from computer integration. Implementing the theory presented in this chatper into a software package could prove to generate a useful and powerful tool for designers. Design disciplines such as project management could use such a program in researching and developing their own principles. More case studies should also be completed using DEIM to model the processes. These case studies give specific context to issues revealed in each map and allow researchers to relate the fundamental elements of DEIM to the details of design processes. The theory of information mapping reveals issues that could prove beneficial to process representation schemes. The application of the Conservation of Energy principle into information mapping could prove beneficial. By theorizing that "information cannot be created nor destroyed, but can only change form", researchers may be able to enhance the way that design processes are understood and completed. Similar benefit may also be had by investigating the integration of Boolean Logic into information maps. By identifying non-value adding steps, DEIM opens opportunity of integration with lean manufacturing process models. This can also be an extension of a software package. By exploring the theory of both lean manufacturing and DEIM, researcher may be able to discover similarities, thus enabling the design community to gain a broader and applicable model to understanding processes.

Finally, the expansion of design process models away from verbal descriptions into visually grounded renderings should be continued. This research has formed two dimensional representations of complex and real world design processes. A lofty future goal of this research is to enable designers to use complex and detailed scientific visualization tools similar to CAD software packages to represent design processes in a multi-dimensional virtual environment. This would give greater capability and freedom in information mapping and process modification to the researcher by representing design processes in virtual three dimensional spaces.

REFERENCES

Ahmed, S. (2007). An Industrial Case Study: Identification of Competencies of Design Engineers. *Journal of Mechanical Design*, 709–716. doi:10.1115/1.2723807

Atkinson, R. (1999). Project Management: Cost, Time, and Quality; Two Best Guesses and a Phenomenon, Its Time to Accept Other Success Criteria. *International Journal of Project Management*, 337–342. doi:10.1016/S0263-7863(98)00069-6

Battista, G. D., Pietrosanti, E., Tamassia, R., & Tollis, I. G. (1989). *Automatic Layout of PERT diagrams with X-PERT* (pp. 171–176). Washington, DC: IEEE.

Burge, J., & Brown, D. (2002). Integrating Design Rationale with a Process Model. *Workshop on Process Modelling, AID'02: International AI in Design Conference*. Cambridge, UK.

Dorador, J. M., & Young, R. I. (2000). Application of IDEF0, IDEF3, and UML methodologies in the creation of information models. *International Journal of Computer Integrated Manufacturing*, 430–445.

Eisenhardt, K. M. (1989). Building Theories from Case Study Research. *Academy of Management Review*, *14*(4), 532–550. doi:10.2307/258557

Gero, J. S. (1990). Design Prototypes: A Knowledge Representation Schema for Design. *AI Magazine*, *11*(4), 26–36.

Grabowski, H., Lossack, R. S., & Weis, C. (1995). A Design Process Model based on Design Working Spaces. In *T. T., M. M., & S. Finger, Knowledge Intensive CAD-1* (pp. 26–29). New York: Springer.

Hazelrigg, G. A. (1998). A framework for decision based engineering design. *Journal of Mechanical Design*, 653–658. doi:10.1115/1.2829328

Henderson, K. (1999). *On Line and on Paper: Visual Representations, Visual Culture, and Computer Graphics in Design Engineering*. Boston: MIT Press.

Hernandez, G., Allen, J. K., Woodruff, G. W., Simpson, T. W., Bascaran, E., & Avila, L. F. (2001). Robust design of families of products with production modeling and evaluation. *Journal of Mechanical Design*, 183–190. doi:10.1115/1.1359786

Keller, R., Eckert, C. M., & Clarkson, P. J. (2006). Matrices of node-link diagrams: which visual representation is better for visualizing connectivity models? *Information Visualization*, 62-76.

Kim, S.-H., & Jang, K.-J. (1999). Designing Performance Analysis and IDEF0 for Enterprise Modelling in BPR. *International Journal of Production Economics*, 121–133.

Kirschman, C. F., Fadel, G. M., & Jara-Almonte, C. C. (1996). Classifying Functions for Mechanical Design. *Design Engineering Technical Conference.* Las Vegas, NV: ASME.

Kumar, P., & Mocko, G. M. (2007). Modeling and Analysis Of Engineering Design Activities Ontology Using the Design Structure Matrix. *Design Engineering Technical Conference.* Las Vegas, NV: ASME.

Leifer, L. J., & Tang, J. C. (1988). *A Framework for Understanding the Workspace Activity of Design Teams*. New York: ACM.

Lockledge, J. C., & Salustri, F. A. (2001). Communication Using a Variation of the Design Structure Matrix. *International Conference on Engineering Design.* Glasgow: ICED.

Maylor, H. (2001). Beyond the Gantt Chart: project management moving on. *European Management Journal*, 92–100. doi:10.1016/S0263-2373(00)00074-8

Nanard, J., & Nanard, M. (1995). Hypertext design environments and the hypertext design process. *Communications of the ACM*, 49–56. doi:10.1145/208344.208347

Ostergaard, K., & Summers, J. (2007). Resistance Based Modeling of Collaborative Design. *Concurrent Engineering Research and Applications*, *15*(1), 21–32.

Pahl, G., & Beitz, W. (1996). *Engineering Design: A Systematic Approach*. New York: Springer-Verlag.

Reymen, I. M., Hammer, D. K., Kroes, P. A., Aken, J. E., Dorst, C. H., & Bax, M. F. (2006). A domain-independent descriptive design model and its application to structured reflection on design processes. *Research in Engineering Design*, *16*(4), 147–173. doi:10.1007/s00163-006-0011-9

Shenoy, P. P. (2000). Valuation network representation and solution of asymmetric decision problems. *Eurobean Journal of Operational Research*, 579-608.

Sim, S. K., & Duffy, A. H. (2003). Towards an ontology of generic engineering design activities. *Research in Engineering Design*, 200–223. doi:10.1007/s00163-003-0037-1

Simon, H. A., Kotovsky, K., & Cagan, J. (2001). *Formal Engineering Design Synthesis*. New York: Cambridge University Press.

Smith, E., Johnston, P., & Summers, J. D. (2007, September). Applying Lean Manufacturing Principles to Revolutionize Curbside Equipment and Collection Process. In E. Smith, P. Johnston, & J. Summers (Eds.), ASME-DETC2007, DFLM-35615. Las Vegas, NV: ASME.

Sobek, D. K., II. (2007). From Journals to Validated Tool: Results of Empirical Research on Student Designers. *International Conference on Engineering Design.* Paris: ICED.

Summers, J. D., & Ameri, F. (2008). *An algorithm for assessing design complexity through a connectivity view. Tools and Methods for Competitive Engineering*. Izmir, Turkey: TMCE.

Teegavarapu, S., & Summers, J. D. (2008). *Case Study method for design research. Design Engineering Technical Conferences*. New York: ASME.

Tufte, E. R. (1986). *The visual display of quantitative information*. Cheshire, CT: Graphics Press.

Ullman, D. G., Dietterich, T. G., & Stauffer, L. A. (1988). A Model of the Mechanical Design Process Based on Empirical Data. *Artificial Intelligence for Engineering Design, Analysis and Manufacturing*, *2*(1), 33–52. doi:10.1017/S0890060400000536

Ulrich, K. T., & Eppinger, S. D. (2008). *Product Design and Development* (4th ed.). New York: McGraw-Hill.

Whitfield, R., Duffy, A., Coates, G., & Hills, W. (2002). Distributed Design Coordination. *Research in Engineering Design*, *13*(4), 243–252.

Yang, M. (2007). *Design Methods, Tools, and Outcome Measures: A Survey of Practioners. ASME International Design Engineering Technical Conferences (pp. DETC2007-35123)*. Las Vegas, NV: ASME.

Yin, R. K. (1994). *Case Study Research: Design and Methods* (2nd ed.). Newbury Park, CA: Sage Publications.

Chapter 6
Embedded RFID Solutions:
Challenges for Product Design and Development

Álvaro M. Sampaio
Polytechnic Institute of Cávado and Ave, Portugal & University of Minho, Portugal

António J. Pontes
University of Minho, Portugal

Ricardo Simões
Polytechnic Institute of Cávado and Ave, Portugal & University of Minho, Portugal

ABSTRACT

Full traceability of products is extremely difficult, although it has been sought after for as long as production, distribution and sales chains exist. Electronic traceability methods, such as RFID technology, have been proposed as a possible solution to this problem. In the specific case of RFID, the number of applications that promote innovative solutions in retail and other areas has been continuous growing. However, RFID tags are mostly placed externally on a surface of products or their packages. This is appropriate for logistics, but not for other applications, such as those involving user interaction. In those, not only is the placement of the RFID tag more complex, but it is also necessary that the tag is not visible or not directly accessible, to prevent accidental damage and intentional abuse. This certainly imposes challenges to manufacturing, but mainly creates new challenges to the development of new products and re-design of existing ones. This chapter presents some insights and what we consider to be the two main approaches to incorporating RFID technology into consumer products.

INTRODUCTION

RFID technology has attracted an increasing interest from companies and R&D institutions around the world in the past few years. Some of the application areas of this technology can be seen in the automotive, retail, logistic and health industries. The benefits can go from increased productivity and cost reduction, to more indirect factors– difficult to quantify – such as improved post-sales consumer service (Hodges & McFarlane, 2005).

DOI: 10.4018/978-1-61520-617-9.ch006

This technology is seen as a potential new phase in the development of the information society, in which the overall day-to-day objects will be inter-connected to other objects, data, etc. This will lead to a considerable impact in some value chains, resulting in a plethora of new applications (Gellersen, Schmidt & Beigl, 2000).

However, this trend represents a challenge that has limited the feasibility of applying RFID technology. The fragility of these electronic devices, associated to the great variety of products and the processes to which these are submitted through their life cycle, restrict the adoption of these technologies (Deavours, Ramakrishnan & Syed, 2005), Moreover, the development of solutions to incorporate an RFID tag in a product and/or packaging in molded components with the minimum impact on the manufacturing process still remains a difficult task (Hodges & McFarlane, 2005).

Despite recent evolutions in design and development processes, that cover the necessary modular product architecture (e.g., Grady, 1999; Ulrich & Eppinger, 2007), these present themselves as processes for new or re-designed products. None of these new processes have been optimized for the development of products that need to perform their function coupled with another product that already exists, being therefore conditioned by that existing product's form, size, and other physical and functional characteristics.

Thus, the challenges posed by introduction of RFID technology into consumer products can be divided into two types: those related to the manufacturing process (the ability to embed the RFID tag into the product), but also those related to the required changes to the product form and possibly its architecture (where in the product should the RFID tag be placed and how much does the product need to change to accommodate it). For the former, we can expect industrial processes to improve to meet the demands of the market. As RFID technology becomes more widely adopted, advanced processing technologies (such as non-conventional injection molding) will enable new solutions for embedding the tags into products during the manufacturing process. However, for the later, which is the focus of this work, it will clearly be necessary that appropriate design methodologies be identified and, if necessary, developed, for seamlessly adding RFID tags to products.

In this chapter, we discuss that to successfully incorporate RFID technology, the design and development of products will demand an effective analysis of existing products, following a modular product architecture philosophy, which will create a significantly different development process. We also discuss what those changes may imply in the overall process and which type of product architectures offers competitive advantages for such solutions.

RFID TECHNOLOGY OVERVIEW

RFID is a wireless tracking technology that allows a reader to activate a transponder on a radio frequency tag attached to, or embedded in, an item, allowing the reader to remotely read and/or write data to the RFID tag (Das, 2009). Essentially, an RFID system comprises an RFID tag, also called transponder, a reader, also called transceiver, and the supporting IT infrastructure. An RFID tag embodies a built-in antenna connected to an electronic microchip. These tags carry on a unique identifier that relates the tag with the precise tagged object, allowing unique product identification. A tag receives and retransmits signals on a set of predetermined frequencies, in other words, in response, a tag transmits a predetermined message to a predefined received signal.

RFID tags can be read-only or read-write. A read-only tag includes a programmed identification code, recorded at the time of manufacture or when the tag is allocated to an object. Once programmed, the data cannot be modified but may be read multiple times. Read-write tags can have

Table 1. A general overview of RFID tags

	Passive	Semi-Passive	Active
Power Supply	External	Internal battery powering only processing units and sensors	Internal battery also powering radio chip
Typical read Range Requirement	1-3 meters	Up to 30 meters	Up to 120-150 meters
Typical Type of memory	Read-only	Read-write	Read-write

their memory changed, or written many times; in this way, they offer greater functionality. RFID tags can be categorized in three general varieties: passive, semi-passive and active tags (Table 1).

A passive RFID does not contain an internal power supply. This tag transforms the energy of radio-frequency it acquires from an RFID reader to answer by sending back information. Drawing power from the reader makes them cheaper and smaller than active tags, which have a battery used to broadcast the signal to the reader (Teixeira & Piçarra, 2007). Active tags are also more reliable than passive tags, because they have an internal read and write capability, and can transmit their signals over a longer distance. Depending upon the size of the tag and the frequency used, the current range of reception, or read range, of the reader is limited. Semi-passive tags use an internal battery to ensure data integrity, however the signal sent from the reader generates the power to transmit the signal from the tag.

The fast, automatic, pervasive and ubiquitous identification of objects is one of the challenges of today's business corporations. RFID's ability to read objects in motion and without a direct line of sight gives it the edge over traditional bar-coding methods. It can be said that the difference between RFID technology and bar-coding comes with the sensing capabilities that RFID offers in an increasingly dynamic, mobile physical world (Teixeira & Piçarra, 2007). Even if the main differences between these technologies may be seen as minor, other differences are opening the path for new applications. If we imagine that any product could have an RFID tag and that we could track this product with a computer-reader, this can be translated in an incredible number of applications in various domains, such as access control, logistics, or animal tracking. Some of the most interesting applications of this technology are: (a) the counterfeiting battle, which is huge threat to global business and concerns all kind of products and companies (e.g. pharmaceutical, luxury goods), as RFID technology allows the products to authenticate themselves to the user; (b) the reducing of crime, under the Design Against Crime program (Learmount, Press & Cooper, 2000; Design Council, 2002), which has been producing some interesting examples since 1999, such as the *Bike-off* project (Bikeoff, 2009) and *The Chipping of goods Initiative*, to show how property crime can be reduced through RFID technology (Adams & Hartley, 2000); (c) e-Government, as there is a growing demand for higher security standards and governments requiring secure identification of individuals, such as electronic passports; (d) e-Health, with healthcare being one of the most enthusiastic sectors adopting this technology as it offers multiple possibilities of tracking medical devices and; (e) Waste management. It is precisely this last example that emphasizes another important factor in the development and implementation of RFID technologies. Nowadays, as sustainable development is increasing in importance and every company must worry about their products throughout the end of the product lifecycle. It is very difficult to create a disassembly system for any kind of a product. It would be better if every company could trace its products and find out where they are at the end of their lifecycle. As stated, while being an innovative solution itself, it can be said that RFID triggered further innovation in products, services as well as in development processes.

RFID IN CONSUMER PRODUCTS

Incorporating RFID tags in products might seem like a trivial task. To support that hypothesis, one can even showcase the large number of products which already include RFID tags. However, a careful analysis shows that this is not the case. First, the large majority of current RFID-tagged products are containers (e.g. boxes, palettes) with very simple geometries. Second, in almost every case of currently tagged products, it is actually the product package which is tagged, and although packaging is a major application of RFID, and many products have and will continue having a package, it is important to be able to place tags in products that are commercialized without a package. Due to these two factors, until now there has not been a serious need to develop and implement efficient methods to place RFID in a wide variety of products. With the massive increase in RFID deployment and expansion to new applications/products, this situation must change very soon.

The two main reasons why it is important to start studying in much greater detail the placement of RFID tags in products is due to: a) technological issues related to the RFID technology itself, and b) the variety in products to be tagged. Other very important aspects of RFID incorporation into products are cost and sustainability, but these are outside the scope of this work and will not be explored.

The main RFID technological issues/limitations are signal attenuation and electromagnetic interference. Signal attenuation in RFID is related to losses in the energy reflected by the tag back to the reader which sends the signal. If less energy is able to reach the tag, then the tag must be closer to the reader to be read. The reflected signal decreases as the inverse fourth power of the distance between tag and reader. Since most tags are passive, the energy levels are already small to start with, and if losses occur, read distances can become a problem. Attenuation can occur from the environment or product characteristics. Water (thus, also products with high water content, including fruits, soft drinks, or moist goods) and carbon will absorb ultra-high frequency RFID signals (the most common passive RFID tags). In the case of metals (canned goods and metal objects), the RFID signal will bounce off the metal surface, typically causing the signal to be blocked. EMI (electromagnetic interference) is essentially noise that will affect the RFID signal travelling back to the reader. This can result from interference from motors, conveyors, old wireless networks, and other sources. In most cases, the source of the interference needs to be shielded in order to allow RFID system to work properly.

Signal attenuation requires particular attention due to the large variety of materials currently used, both in the product itself as well as its package. The development of consumer goods has evolved to take advantage of materials engineering, and thus contemporary products use materials such as wood, plastics, metal, paper/cardboard, composites, laminates, glass/ceramics, and often include several parts made of different materials. Without understanding the influence of the materials on RFID performance, it is not possible to ensure adequate implementation of this technology.

The second reason listed above for studying RFID placement in greater detail was the large variety of existing products. The diversity of materials has already been discussed, but product shape and other characteristics are also an issue. For example, a material might be relatively impervious to the RFID signal but only up to a certain wall thickness. Regarding shape, qualitative and quantitative criteria must be defined for where to place the tag on a product, and where along the production line the tag should be incorporated (possibly, but not highly likely, at the end). Trends towards more complex or more organic product shapes will affect where a tag can be placed, and specific inner features might be required to accommodate a tag. This becomes even more important due to the fact that specific design solutions can solve some of the signal attenuation issues listed above. For example, although metals can create serious problems for direct tag placement, introducing an air gap between the tag and the metal surface can increase its performance, in some cases, bringing it close to the level of performance expected in materials to which the RFID signal is impervious.

Other issues can also affect the RFID performance, for example the proximity of neighboring tags and the angle between the reader antenna and the tag. If RFID is being used for automated inventory upon delivery of goods to a store, adequate planning of the position of the tag in the product and the way multiple products stack inside a box can optimize the performance of the system.

It is also interesting to notice that different sectors of activity have been employing RFID technology in very different degrees. In retail, one can assume this to be highly related to considerable differences in the type of product of each sector. For example, textiles, consumer electronics, shoes, food packaging, all require considerably different implementation solutions, and a common platform is difficult to achieve. Due to this, the massive deployment of RFID in consumer products that has been advocated in the past decade is taking much longer than expected to come to pass. This situation results mostly from the inexistence of adequate methods to plan RFID tag placement in products, prevent knowledge-based development.

How to incorporate RFID tags in products in the most efficient way is a problem that needs to be solved. Two main approaches can be used for this. Externally attach the RFID tag to the product in some way, or make the RFID tag part of the product itself.

THE PRODUCT DEVELOPMENT PROCESS: TOWARDS THE INCLUSION OF RFID TECHNOLOGY

In the general trend towards wider application of RFID technology in products, we can envision two different roles that product development will assume. If we can think that tagging will move from being something attached to an object to become part of the object itself; for this to happen we will inevitable see changes take place in the design and product development process. Nevertheless, tagging any product is not a simple task and, thus, the shift from tagged to RFID-embedded products will be a necessary evolution. As stated before, product design and development processes present them as process for new or re-designed products. However, none of the traditional processes meet the requirements of developing products that support the incorporation of an RFID tag. During the product development process that absolutely requires the inclusion of RFID technology, we need to develop secondary products that have an embedded RFID and can be coupled to existing products. For this we need two product development processes: one that can add RFID technology to products that already exist (product development process – A), and a different process that has the incorporation of RFID technology in the product as one of its primary concerns (Product development process - B). These are schematically represented in Figure 1.

Figure 1. The two possible RFID product development process. NP stands for "New Product"

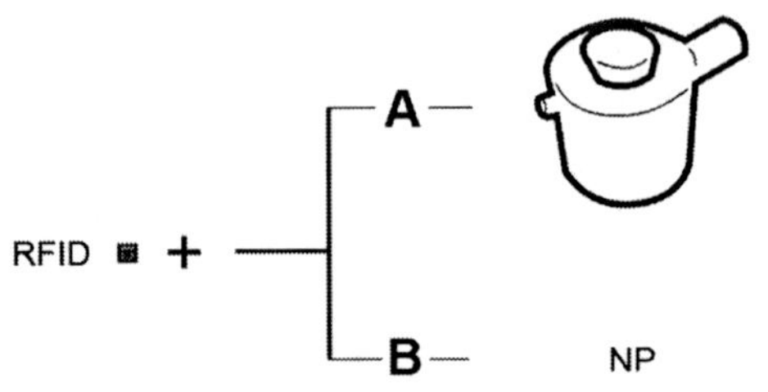

Defining the development process as the specific series of events and methods by which a procedure or set of procedures are followed, in order to achieve an intended purpose, goal or outcome (Best, 2006), the development process consist in a series of activities and methods that are grouped in phases that tackle a complex problem by decomposing the overall complexity in smaller, more manageable sub-problems (Simon, 1969). Dividing a problem into simpler sub problems is called decomposition and there arc many schemes by which a problem can be decomposed. One of them is functional decomposition (Ulrich & Eppinger, 2007). In product development, a part of the problem-solving activities consists of discovering unknown types of interdependencies between different design solutions. A decomposition of the product into nearly independent functional components is a way to clarify the overall problem and test the feasibility of different design solutions. A product can be thought in both functional and physical terms. The functional elements of a product are the individual operations and transformations that contribute to the overall performance of the product and the physical elements of a product are the parts, components, and subassemblies that implement the product functions (Ulrich & Eppinger, 2007).

Allocating a function of a product to a physical component is called product architecture. Roughly formulated, product architecture concerns the design and layout of the physical building blocks of a product. Product architecture is an abstract way of looking at a product and is usually determined before a product is actually designed into detail (Ulrich & Eppinger, 2007). Likewise, abstract terminology is used to refer to properties of components in products and the way these components interact. Ulrich and Eppinger (Ulrich & Eppinger, 2007) specified product architecture more precisely as: (a) the arrangement of functional elements; (b) the mapping from functional elements to physical components, and (c) the specification of the interfaces among interacting physical components.

The importance of problem decomposition and the subsequent product architecture to the inclusion of RFID technology is very high. Once again, if we need to place an RFID in existing product and further start to embed this technology in new products, we need to understand the existing products, their shapes, forms and functions to design a way to incorporate the RFID tag. Rethinking the product architecture is the best strategy to attach a new physical component, with a specific function, to an existing product.

In other words, when we need to attach an RFID tag to an existing product, and we begin with functional product decomposition, the existing product becomes one functional element itself. What is new in that architecture is the RFID tag. To attach thc RFID tag to the product we need to develop a new complementary product (NP) which contains the RFID tag (process A in Figure 2). This new complementary product will attach itself to the existing product, which ensuring adequate physical support to the RFID tag.

Figure 2. Functional product decomposition

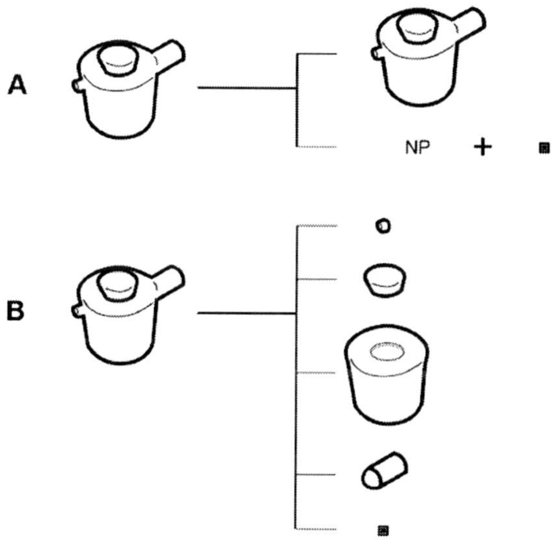

On the other hand, process B in Figure 2, illustrates the functional product decomposition of a product which is being developed and in which we need to incorporate the RFID technology. The fact that we have a considerable degree of control over each of the physical elements, with the exception of the RFID tag, will allow us to embed the RFID tag directly in the product. Independently of the product development process employed, one of the physical components will have the embedded RFID tag, ensuring traceability for the entire product. This, of course, will lead us to modular product architecture.

THE MODULAR APPROACH

According to a widely accepted definition (Ulrich & Eppinger, 2007), a product architecture is modular when components are functionally independent. A modular architecture includes a one-to-one mapping from functional elements in the function structure to the physical components of the product, and specifies de-coupled interfaces between components (Ulrich & Eppinger, 2007). Functional independence has a deep impact on the supply chain, since components may efficiently be developed and manufactured separately by different organizations, as well as on marketing, since product customization may be achieved with greater ease by simply swapping components (Cantamessa, 2005) One of other characteristics of product modularity is the ability to mix and match independent and interchangeable product building blocks with standardized interfaces and products in order to create product variants.

Modularity is an effective means to control complexity. Once the modular architecture is fully specified, the uncertainty in product development is confined to problems of identifying feasible solutions that will improve the functionality of the product, allowing at the same time the firm to minimize the physical changes required to achieve a functional change (Ulrich & Eppinger, 2007). Furthermore, it enables the production of variety while facilitating the achievement of economies of scale. Modularity is vital to responsive manufacturing and creates the opportunity of producing a huge variety of products that satisfy various customer requirements. A firm that offers several variants of a product can, at the same time, postpone the differentiation of a product until the very end of the supply chain. This is because products manufactured with a modular architecture can be easily modified without causing significant changes to the manufacturing system, because of the possibility of sharing components amongst various products.

The provision of product variety is often based on the concept of product platform. Since the effort, in terms of time and money, to implement a product platform is larger than a traditional development process, it is not possible for any new project to necessarily be a platform project. Product platforms are closely related to the technology development efforts of a determined firm, and to decisions about which technologies to employ in new products (Cantamessa, 2005). The technologies and the variable products needed, or possible, are expected to be planned in the beginning of the development process. This is the case of the inclusion of RFID technology in products. With the product platform being the set of assets shared across a family of products (Robertson & Ulrich, 1998), RFID technology is one of the assets shared, and needs to be planned in a way that achieves the overall performance of the technology. In other words, RFID technology could be the base for a new product platform, independently of the strategic development process chosen.

Notice that in the framework of the two distinct product development processes discussed above, the practical implementation of the integration of RFID technology in products can range anywhere from process A to process B. The process which can be implemented more rapidly and more straightforwardly is Process A, since it does not imply changes to the product itself. However, on

Figure 3. Possibilities of product development in the scope of Process A

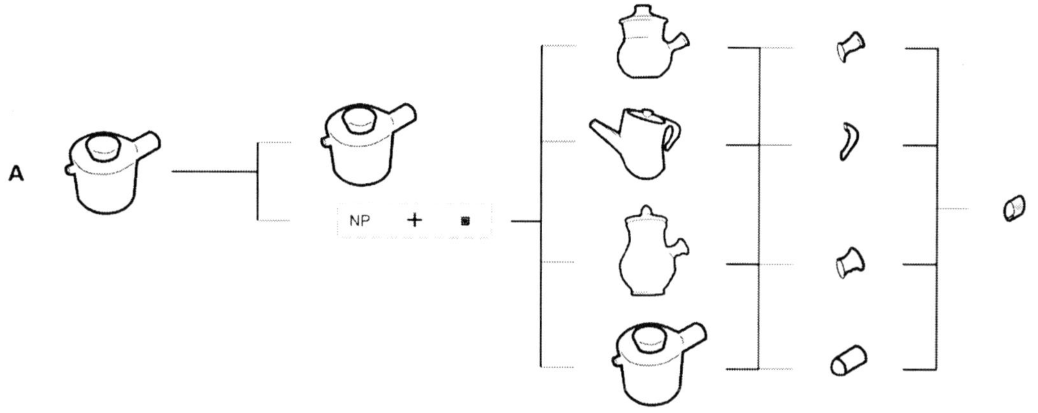

a longer time scale, Process B will most often be employed, due to its inherit advantages.

If we assume that RFID technology will be the support for one or several product platforms, the planning of these platforms will necessarily involve managing a basic trade-off between distinctiveness and commonality. These trade-offs can be formalized in the differentiation plan and the commonality plan (Ulrich & Eppinger, 2007). The differentiation plan explicitly represents the ways in which multiple versions of a product will be different from the perspective of the market or the customer. These differences are normally diverse specifications of the final product with the objective of satisfying the needs of the customers in a specific market target. Conversely, the commonality plan explicitly represents the ways in which the different versions of the product are the same physically and, sometimes, functionally. It is clear that in both product development processes discussed in this chapter, establishing the architecture of the product and its inherent modularity is a primary concern in the overall process. Nevertheless, turning to a more specific modular philosophy will necessary inevitably lead us to planning a product platform around RFID technology. In the case of the product development process A, it is necessary to take into account that developing a complementary NP that only works for one existing product is (aside from exceptions) unreasonable in terms of an economy of scale. So the inclusion of an RFID tag through this product development process will require a profound analysis of the shape of a variety of products (probably in the firm internal product catalogue) with the purpose of finding similarities that make it possible to develop a unique NP with an embedded RFID that can attach itself to as many products as possible (Figure 3).

From Figure 3, one can see that one of the possible similarities in the four products illustrated is the handle. Thus, to insure the possibility of incorporating RFID technology, it is necessary to create the NP with the embedded RFID in such a shape as to allow it to attach to all the handles of the considered products. This could be done with, as an example, a flexible polymeric bracelet.

On the other hand, product development process B (Figure 4), considering the development of new products from start, it is not restrained to any shape, form or style, but only to the functional elements needed. As in process A, several possibilities can be identified to achieve the goal of developing a new product that already features an embedded RFID. After decomposition of the functional elements, one needs to identify which functional element is going to have the embedded

Figure 4. Possibilities of product development in the scope of Process B

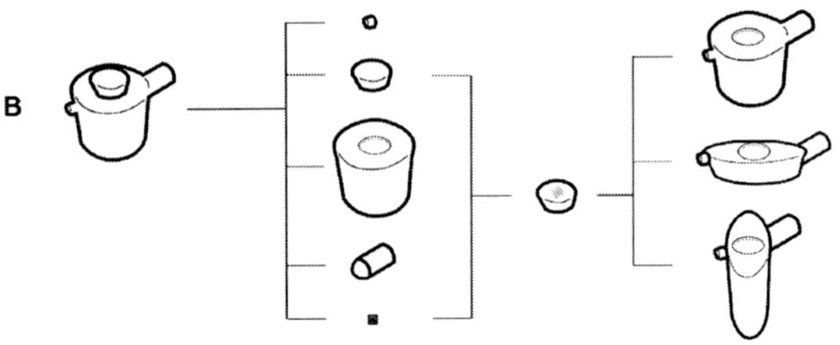

RFID; in the case shown in figure 4, that element was the cap.

Thus, in this case it is possible to talk about a functional element (asset) common to all products, and which already has an embedded RFID tag. It can also be planned to perform its function in other product families.

CASE-STUDIES

To better illustrate the two concepts being the product development processes discussed in this chapter, we shall use two case-studies of commercial products. The two products are mass-produced by the Daily RFID Co, Ltd. (2009).

The first example (shown in Figure 5) is a bracelet-like product that was designed to attach RFID tags to existing products, following the approach of product development process A.

The shape of the existing product to which this device will be attached must include some feature around which the bracelet will go (e.g. a tubular or quasi-cylindrical feature). It is a one-off seal, and once attached, cannot be removed without damaging the seal and leaving evidence of tampering.

As is well known, the most pervasive current application of RFID technology, and the one that still represents the large majority of tags in use, is logistics. Essentially, RFID tags are placed in boxes, cases, pallets, carts, and more recently, envelopes, books, magazines, and many other objects which have very simple shapes, and, almost without exception, offer a flat surface for placing the tag. For all those applications, the typical tag offers an adhesive layer so it can be stuck to the surface. The type of device in Figure 5 (which is only representative of a large number of devices with encapsulated RFID tags), appears from the need of placing tags in more complex products. If a retail store wishes to track its products that are not enclosed in a box (e.g. bicycles, toys, kitchen accessories, etc), they must find a way to ensure

Figure 5. RFID embedding solution for external coupling to existing products

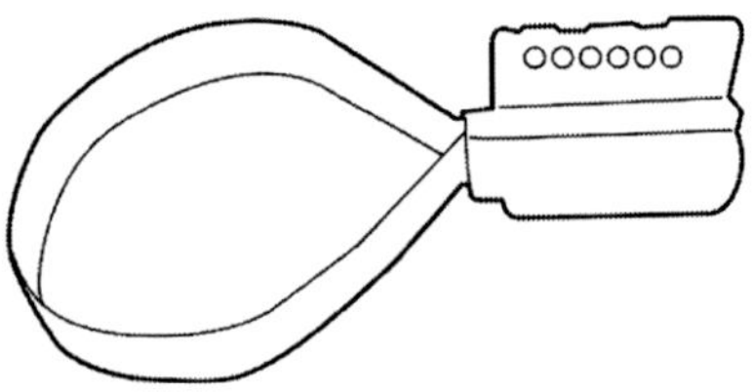

Figure 6. Key-chain with embedded RFID

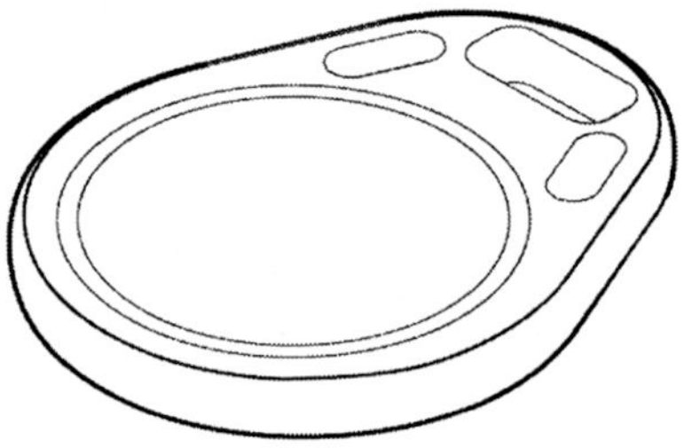

the tag can be safely placed on the product and that it will not come off when colliding with other objects. When there are no flat surfaces or when it is not allowed to leave residues of the adhesive on the object, the type of device in Figure 5 can be an excellent solution. It also prevents malicious users to simply peel off a tag placed externally on a surface of the product.

The second example is a key-chain (Figure 6). Designed from the start to include an RFID tag (following the approach of product development process B), using adequate materials and some planar surfaces, it becomes possible to use the key-chains for access control, security, and many other applications. The key-chain can even allow information (e.g. a logo) to be printed on its surface.

In this case, the fact that the RFID tag is not detectable by users is a vital advantage. At the early design stages, aside from all the product design specifications that one would typically consider for such product, the shape of the key-chain must be set taking into consideration physical requirements of the RFID tag as well as its placement. This implies finding the optimum way to enclose the tag inside the key-chain.

There are many technological options (fabrication solutions) to enclose the tag inside the key-chain, such as making the keychain in two parts that are locked together only after placing the tag between them in an appropriate hollow space, or finding a way to place the tag actually inside the polymeric material during processing. Notice that, in practical terms, the mechanism through which the tag is embedded in the product is not critical, since the end result is essentially indistinguishable by users.

CONCLUDING REMARKS

RFID technology is at a crucial point, in terms of the adoption exceeding current applications. As this technology becomes ever more deeply embedded in everyday life, it can no longer be seen as a simple appendix to increase traceability in the supply-chain, but as a technology that can be embedded in every existing product. The technical issues of tag reliability and the effect of liquids and metal surfaces are challenges and not true problems. One of the most important issues in the implementation of this technology is that the RFID will necessary transform the way in which we develop products and the way we identify potential advanced features of products. One can say that very little RFID novelty will be achieved from changes in the technology itself. Instead, it will arise from innovative use. When it comes to finding new ways to use RFID to create and build product value, user experimentation will inevitably bring original interactions.

In a trend towards a more effective use of this technology in product design and development, RFID tag incorporation needs to be an important specification of the development process. This, of course, will be an evolutionary process, and will range from the previously described development process A, where no changes are made to existing products, through development process B, where products are developed from start considering embedding the RFID tag. The endgame in any of those processes will be the increasing use of RFID technology in products, increasing the interaction of products with their environment and their users. A time will come when RFID will vanish from the mind of the consumers, becoming impossible to distinguish it from the product itself. As espoused by Weiser (Weiser, 1991, p. 94), "the most profound technologies are those that disappear. They weave themselves into the fabric of everyday life until they are indistinguishable from it".

Although the discussion about the examples provided in previous sections illustrates the need for manufacturing processes optimized for embedding RFID technology, processing technologies exist that can be adapted for that purpose, and the real challenges still clearly reside on appropriate product design and development methodologies.

It is also important to mention that RFID is only one example of the growing number of microelectronic devices incorporated into products, following the current trend of technology-driven product development (Simoes & Sampaio, 2008). However, many of the challenges are the same, and conceptual solutions can be reused for different scenarios.

ACKNOWLEDGMENT

We acknowledge the Foundation for Science and Technology, Lisbon, through the 3° Quadro Comunitario de Apoio, and the POCTI and FEDER programs.

REFERENCES

Adams, C., & Hartley, R. (2000). *The Chipping of goods initiative. Property crime reduction through the use of electronic tagging systems. (Police Scientific Development Branch, Publication N° 23/00)*. London: Home Office.

Best, K. (2006). *Design Management: Managing Design Strategy, Process and Implementation*. Lausanne: AVA Publishing.

Bikeoff. (n.d.). *Bikeoff.* Retrieved April 9, 2009, from http://www.bikeoff.org

Cantanessa, M. (2005). Product portfolio management. In Clarkson, J., & Eckert, C. (Eds.), *Design Process Improvement – A review of current practice* (pp. 404–435). London: Springer.

Daily RFID Co, Ltd. (n.d.). *Daily RFID*. Retrieved April 9, 2009, from http://www.rfid-in-china.com

Das, R. (2002). *An Introduction to RFID and Tagging Technologies*, Cambridge, UK: White Paper, IDTechEX

Deavours, D., Ramakrishnan, K. M., & Syed, A. (2005). *RFID Performance Tag Analysis (Tech. Rep.)*. University of Kansas, Information and Telecommunication Technology Center.

Design Council. (2002). *Evidence Pack. DAC case studies*. London: Design Council.

Gellersen, H. W., Schmidt, A., & Beigl, M. (2000). *Adding Some Smartness to Devices and Everyday Things*. Monterrey, USA: IEEE Workshop on Mobile Computing Systems and Applications.

Grady, P. (1999). *The age of Modularity*. Iowa: Adams and Steele Publishers.

Hodges, S., & McFarlane, D. (2005). *Radio frequency identification: technology, applications and impact*, AutoID Labs white paper.

Learmount, S., Press, M., & Cooper, R. (2000). *Design Against Crime. Report to the Design Council, Home Office and Department of Trade and Industry. ODPM (Office of the Deputy Prime Minister) (2004). Safer places. The planning system and crime prevention*. London, UK: Thomas Telford.

Robertson, D., & Ulrich, K. (1998). Planning for product platforms. *Sloan Management Review, 39*(4), 19–31.

Simoes, R., & Sampaio, A. M. (2008). Effect of technology-driven products in the future of product design and development. In *Proceedings of RPD 2008* (paper 8048). Oliveira de Azemeis, Portugal.

Simon, A. (1969). *Science of the Artificial*. Cambridge, MA: The MIT Press.

Teixeira, F., & Piçarra, S. (2007). *Estado da arte em RFID*. Sybase. Retrieved February 20, 2008, from http://www.portalrfid.net

Ulrich, K., & Eppinger, S. (2007) *Product Design and Development*. New York: McGraw-Hill international edition.

Weiser, M. (1991). The computer for the 21st Century. *Scientific American*, *265*(3), 94–104. doi:10.1038/scientificamerican0991-94

KEY TERMS AND DEFINITIONS

RFID: Radio-frequency identification is the use of an RFID tag applied to or embedded into a product, animal, or person for the purpose of identification and tracking using radio waves.

Product Architecture: The conceptualization, design, and description of a product, its components, the interfaces between components, and the relationships with internal and external entities, as they evolve over time.

Modular Architecture: Architecture in which the number of functions is approximately similar to the number of components, where usually the interactions (as well as the interfaces) between components are relatively simple, and where each module is intended to be the main (often the only) responsible for providing a given function or sub-function of the product.

Integral Architecture: Architecture in which the number of functions is considerably larger than the number of components, which implies that some components are involved in delivering multiple functions.

The EPCglobal Network: A computer network used to share product data between trading partners.

EPC: Electronic Product Code, is the basis for the information in the network. Each RFID tag has a unique EPC.

Section 2
Supporting Technologies

Chapter 7
Implementation of Rapid Manufacturing Systems in the Jewellery Industry in Brazil:
Some Experiences in Small and Medium-Sized Companies

Juan Carlos Campos Rúbio
Universidade Federal de Minas Gerais, Brasil

Eduardo Romeiro Filho
Universidade Federal de Minas Gerais, Brasil

ABSTRACT

This chapter presents the rapid prototyping and manufacturing concepts applied as means to reducing time between jewellery designs and manufacturing process. Different processes on jewellery modelling production are presented. Nowadays, the use of technologies as CAD/CAM - Computer Aided Design and Manufacturing in high production companies are very disseminated. However, the implementation of these resources at the design and manufacturing processes of jewels and fashion accessories, in small and medium size businesses, is still insipient. As reference, is presented the situation observed in small and medium companies located in Minas Gerais, Brazil.

INTRODUCTION

This chapter presents an analysis of the use of CAD/CAM systems (Computer Aided Design and Computer Aided Manufacturing) aiming to improve the connection between design and production in small and medium size jewellery companies located in the state of Minas Gerais, Brazil. In addition to that, the changes caused by this technology in the professional relationships are also discussed. The principal point is related to the influence of the use of CAD/CAM system on the designers' skill (competence) and design methods in an interactive design process and a Rapid Manufacturing (RM) system implementation.

In a global and highly competitive market, it is critical for the companies to develop distinctive products, joining aesthetic features with the effective control of the production processes,

DOI: 10.4018/978-1-61520-617-9.ch007

in order to ensure improved results. Within this scenario, the designer's activities are becoming more important for the jewellery industry; nevertheless, the design process is frequently regarded as an artistic activity dissociated to the industrial process. Improvements in jewel manufacturing process are extremely important, because the jewels and fashion accessories presents a high aggregate value. It is most important when raw material is abounding too. Brazil is a precious metal and stones products country, namely gold, silver, diamonds, emeralds, etc. Thus, better products imply in leadership of the market and consolidate designer's names in global market.

The intensive use of rapid manufacturing and CAD/CAM systems brings new opportunities to the integration between design and manufacturing process. Through a better assessment of the products during the preliminary phases of the project it is possible to predict mistakes that could result in additional labor, as well as increased production time and cost. In spite of technological advances and the use of machines at several stages of the industrial production of jewels (conception and execution of the first model), the process still maintains essential handcraft features. The intrinsic limitation of the manufacturing processes used in this phase for designer and goldsmith, results in mistakes which are transferred to the next phases, impairing process control and manufacturing phase. This is an important aspect to be considered in the product development, especially in small and medium size companies, which are typical of the jewellery industry in Brazil.

Changes in the communication methods and technical tools, such as CAD/CAM systems, must be carried out simultaneously with changes in the designers work method, aiming to adjust their projects to the needs of other production sections (including the goldsmith), with the purpose of obtained a finished product with the required quality. The RM concept can be used when the development of moulds or tools for moderate volume parts or products is required. Among the various RM processes, computer numerically controlled (CNC) machining is recognized as the technology most widely used in industry for the application of CAD/CAM systems. The production of jewels is based on the investment-casting process, in which the moulds and casting pattern can be produced in plastic or wax using an RM technology, thus reducing time and cost as well as ensuring the production of goods with high quality.

Different researches (as Rocha, 2007, Fernandes et al., 2005 and Han Boon et al., 1992) indicates that the use of Computer Integrated Manufacturing (CIM) systems can offer a substantial assistance to the solution of issues related to the interface between design and manufacture of jewels. However, some drawbacks associated mainly with the lack of qualified professionals (especially designers) were identified (Siu & Dilnot, 2001). The use of these technologies is only possible by employing three-dimensional CAD systems and trained professionals able to handle this task are currently not available in the jewellery industry. Moreover, the application of these tools alone is not enough to ensure the satisfactory integration of all phases of the project.

In jewels international market the Brazilian designer and manufacturer are traditionally competitive at producing small stones fashion jewellery and their skills in gem setting. These products attend the needs of its customers in terms of originality, creativity and fashion. However, the Brazilian Jewelers needing to search for ways to increase the efficiency and improves the productivity. Thus, this Chapter is concentrated on study of the rapid manufacturing innovative technique and their use in jewellery production for Brazilian small-size companies.

DESIGN AND MODELING PROCESS IN THE JEWELLERY COMPANIES

The first step of jewellery industry production process is the concept and design of a new jewel,

Figure 1. A final jewel design (including details)

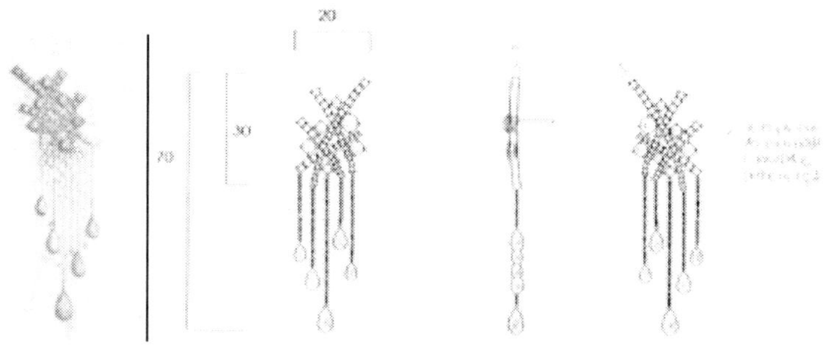

normally by a specialized industrial designer. After design process conclusion, another professional, the goldsmith, creates a prototype that will be the base for all subsequent industrial process. The interaction difficulties observed between designers and goldsmiths in the design and production stages are various and bring many problems to production results.

The deficiencies observed in the information transference at initial stages of jewel design and production results in important consequences for next phases, including serious imperfections in the final product, re-work in the production process, increasing of operational costs and delays in delivery product. In the same way, the improvement in the initial phases of jewel production represents a global development of all process quality results. Special attention is necessary on the design product process, for the reason that in this phase are elaborated most important specifications to an adequate and balanced production process. The design decisions are fundamental to a better quality in the final product and the success in a highly competitive market.

The Design Process

The designer activity begins with the creation of a piece or collection jewel. Ordinarily, the project includes specifications about costs, market share, piece weight, materials etc. From this information, the first ideas and freehand sketching drawings are doing on paper. After that, an adequate solution selection between different proposals is selected and refined in another development phase, a detailed and formal design plan. This result is sent to production phase (Figure 1). Conventionally the jewel design includes a perspective view (in colour, with an "artistic presentation", when is possible to anticipate the formal result, close to the real appearance of jewel to manufacturing) and a technical drawing (with the essential view to execution, in most times, superior and lateral views) that includes measure and materials used, finishing, enlarged details etc.

However, this formal representation, including perspective and artistic representation, technical drawings and others technical information, sometimes is not completed. In majority, the designs are presented to goldsmith using partial or alternative draws (Figure 2), unstudied techniques in different forms:

- Colour draw with one side of the piece, accomplished with measure-simplified representation;
- Technical drawing, using ISO norms;
- Technical detail with material and dimension description;
- Sketches and drafts;
- Prototypes using alternative materials (clay, paper etc.)

Figure 2. Jewel Design Examples. (Left: Design including only a colour draw, with material indication. Right: Design with only a 3 views draw, including fittings solutions. Source: Rocha, 2007.)

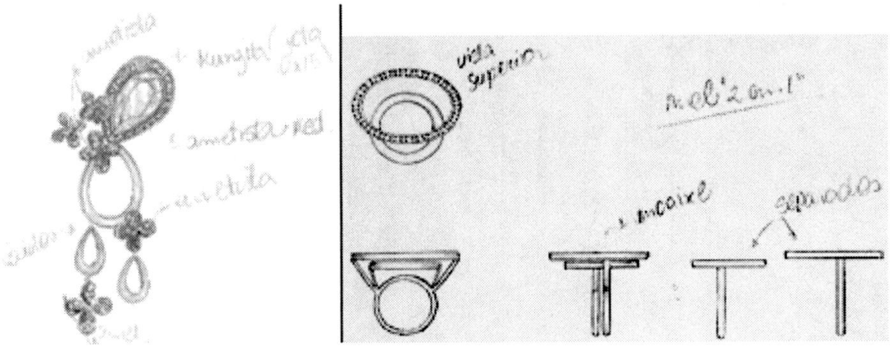

About representation ways, Noguchi (2003) presents a designer interview: "Some pieces are illustrated using technical draw, surfaces, cuts etc. A clay model is used when the goldsmith is not a high skilled professional. Sometimes, when I am tired of drawing, I do only a mock-up, the "direct form". Other times I do only a "sketch" on paper. When the goldsmith is highly skilled, he almost knows the future result, if it will be satisfactory". The Figure 2 presents some examples of "unconventional" designs, released by jewel designers.

The most important communication element between the conception and production steps is the jewel design. The quality and rules used in design process are important, however sometimes the information included in design documentation are not sufficient to goldsmith comprehension and don't trust the consistence in the process, requiring others representations (and communications) resources. Eventually, some design characteristics are "deduced" by goldsmith, because cannot be presented on paper as like as in a three-dimensional model. Movements, fitting systems, volume and weight of pieces are items not always adequately represented in the conventional draw. Moreover the characteristic limits on the representation process, there is the problem of the ignorance by designers about the jewel production process, its possibilities and restrictions. This fact can cause a series of incompatibilities between design and production, bringing delays and changing necessities in the pieces.

For this reason, it is usual that modelling goldsmith modify some design characteristics, adapting it to product process limits and avoiding problems in the final model. These modifications can respect and maintain the original design aspects, when are about specific adaptation as plate thickness, measurement or fitting adjust etc. However, in some cases the goldsmith change important aspects of design pieces, owing to impracticability in the production process or about difficulties in the design interpretation.

The Wax Model

After conclusion of design work, the draws are sending to goldsmith, which start the construction of a "model" (or prototype) of jewel piece, in plastic or special wax. This model can be made by metal (gold, silver etc.), plastic or a special wax (used in jewel industry). The choice depends fundamentally of the techniques used in the piece confection. In this process are used specific tools like files, saws, drills etc. in a handcraft process, until the final form of the piece.

In the traditional process, the model construction is a fundamental step at industrial process. In cases of exclusive pieces, made to order, the goldsmith finalizes his work producing the first piece directly in metal. But in the industrial process, thousands pieces can be produced from the wax model. This step (model construction) is a handcraft work and depends on highly of the goldsmith skill. For this reason, problems can be caused by limitations in the work conditions and tools. Some design pieces propose details and symmetries impossible to handcraft techniques. In the same way, complex geometric parameters turn impossible an adequate control by goldsmith about some aspects like exact weight and thickness in all points of piece, in order that the final product will be different that described in the design.

REVIEW ON JEWELL RAPID PROTOTYPING

In recent years, the use of computers throughout all engineering disciplines has obtained an increase importance. This is particularly true in activities that span the life cycle of discrete products development, namely in products design and in manufacturing. Computer-aided manufacturing design (CAD) and computer-aided manufacturing (CAM) software programs, computer numerical control (CNC) machining and Rapid Prototyping (RP) process has certainly gained momentum in the marketplace.

Several works have been published aiming to evaluate the RP process and computer technologies on the manufacturing mechanical and medical complex pieces. However, a few studies have been making of the RP techniques and CAD/CAM/CNC tools in jewellery industry. Leong et al (1998) showed the effect of abrasive jet deburring processes on the surface finishing of jewellery models built by stereolithography apparatus (SLA). The investigation aims to determine the significant parameters of the deburring process, to determine a practicable range of settings of these parameters for effective deburring, and to establish the optimum settings to the jewellery models built by stereolithography.

Traditionally, a human process planner performs the manufacturing process planning, with acquired expertise in machining practices that determines from a parts engineering drawing what the machining requirements are. Wong et al (1999) developed a decision support system to assist in the operations of typical jewellery manufacturer. The paper describes the development of a management information system with a specially designed decision support system to meet the specific requirements of Hong Kong's jewellery industry. It aims at assisting a jewellery manufacturer to make decisions in various areas of operation, including price quotation, sales analysis, materials requirement planning, vendor selection and job assignment.

Using the conventional design and production system in the Hong Kong jewellery industry as a case, Siu and Dilnot (2001) investigated the reason for the high degree of failure of design representations in the jewellery context, through illustrating the difficulties and problems (in maintaining quality of craftsmanship) when introducing CAD/CAM/RP technologies in production of craft-based object as jewellery in modern factory production. In this study, the authors affirm that the tacit-format attributes, which were separately contributed by the jewellery designers and goldsmith, can be extracted, recaptured, recorded, integrated and finally coded into CAD database.

Many commercial CAD systems have been developed for the purpose of designing jewellery (e.g., JewelCAD, JewelSpace, Rhinoceros, Matrix, ArtCAM Jewelsmith, etc). They provide graphical interfaces with excellent rendering capabilities. The majority of these systems provide built-in libraries of settings and cut gems and stones and advanced feature-based design tools. Also, the majority of these systems have the capability of exporting models to RP Machines. However, none of these systems is appropriate for designing and creating pierced Byzantine jewellery or other jewellery of this kind of detailed engraving. Stamati and Fudos (2005) shows a parametric

CAD system for the design of pierced medieval jewellery, which is jewellery created by piercing, a traditional Byzantine technique. For other hand, Soo et al. (2006) applied fractal geometry for artistic product design in a CAD platform. Based on the proposed data structure, the geometrical information of the fractal can be extracted in a more effective way and the fabrication of a fractal shape for an artistic product design can be realized. The CAD platform has been employed to fabricate a physical prototype in RP technology.

Recently, in the paper titled "I want and now", Langer (2008) affirm that consumers demand for individualized products. This individualized series production, or call it mass customizations, is driving a paradigm shift for manufacturing and electronic manufacturing (e-manufacturing) with its flexible and fast response times is a major factor in the drive for change. The e-manufacturing allow delivers end products, functional parts and tools directly from computer aided design (CAD) data. A laser heats and melts powdered metals or plastics layer by layer, until the build is complete. Among the novel products with highly complex and sometimes filigreed structures and forms that are unthinkable geometries for conventional series production is jewellery, clothes, lamps, chairs or others functional parts for industry.

NEW TRENDS ON CAD/CAM IMPLEMENTATION IN JEWERLY INDUSTRY

As far as concerned on computer technologies, the resources are today very popular and used in different industrial sectors since 1960 decade. At first in the aerospace and automotive industries, the CAD/CAM (Computer Aided Design and Manufacturing) has been expanded the application on jewellery. CAD Software, CNC (Computer Numeric Control) machines and Rapid Prototyping equipments are improving the jewellery sector results, bringing new competitive advantages. On the other hand, some questions obstruct an accelerated diffusion of new technologies between the users, especially in the medium and small companies. These questions are highly related to the difficult in the learning process by designers and the conflict between using the new CAD/CAM technologies and the traditional work process, including the relationship between designers and goldsmiths.

The improvements on results of new technologies application are only possible, in practice, with an intensive use of Three-dimensional CAD Systems (CAD3D). However, the professionals with a high knowledge about CAD are almost rare, and can be observed a resistance by traditional designers in front of new tools, in general more complex and (initially) slower than the traditional sketch process. The sub-utilization and unfamiliarity with the real potential of CAD are, as a rule, the main reason of the problems observed. In fact, many designers put questions about use of CAD technologies instead of traditional sketches and manual draws. The CAD application needs a high detailed solution, which can need more time in the initial steps of design process. However, the reduction of mistakes in the next steps and the improvement of a global result, owing to a better evaluation of design alternatives, can result in a time and resource economy in a long time. These aspects justify an effort in learning and application of these systems in the product development and production process. Notwithstanding the apparent complexity, after the software learned, the total time necessary to design process can be significantly reduced with facilities to generation of new alternatives, for example.

In the midst of new exigencies of consumers, quality improvement, reduction of products life cycle (time in market) and technical innovations, the jewellery industry has slowly introducing resources as CAD/CAM. By means of its characteristics, these systems present potential benefits in a crucial problem in jewel design: the communication interface between design and production

Figure 3. An example about a jewel development process: (a) handmade sketches; (b) first studies using a CAD system; (c) wire frame perspective; (d) different views in CAD; (e) jewel perspective; (f) jewel materials simulation in CAD3D; (g) and (h) model production;(i) final model in special resin.

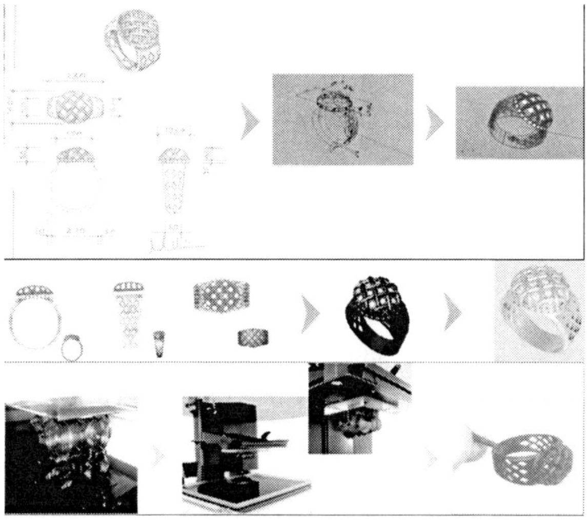

sectors. The use of CAD system associated to CAM tools and rapid prototyping technologies turn possible a total integration in the initial steps in jewel production. A CAD3D system using solid modelling is able to simulate a virtual (and three-dimensional) model, showing forms and dimensions exactly as a real piece, including elements as colour, stones and construction details. These characteristics can be anticipated and visualized in the conception stage, avoiding problems like inaccuracy interpretation, improving quality in final design and reducing fails in the production process (Figure 3).

Using a CAD3D virtual model, the designer can visualize the final appearance of jewel, with a "rendering" resource. It is possible to simulate a change of the material (for example, testing the results in same piece applying gold or silver as material basis). The modern CAD technologies present simulations very close to the real product. Different evaluation versions can be generated quickly, without wax models or different draws. The CAD systems bring great benefits in highly detailed pieces, when the exact dimensions and mechanisms descriptions are essential. Fitting and fixation mechanisms in these models are, in practice, impossible to a goldsmith. Also in the repetition of components in different pieces (as a collection), is possible by achievement in computer memory. At long last, CAD system can automatically generate all technical information essential to production and directly send to CAM machines, evicting mistakes interpretation.

After design process conclusion by CAD, nevertheless is possible a traditional production process, with the goldsmith work. In this case, the model can be printed, in different views, including technical and "artistic" information. However, the CAD resources are better used in connection with CAM technology, especially Rapid Prototyping machines to models construction. The main advantage of this technology is the possibility of a controlled result; with sure of the model resultant have the same characteristics of designer conception. Another fundamental advantage is a possibility of adoption of complex and symmetric forms, impossible to a traditional process production, involving a goldsmith work. Finally, it is observed a sensible reduction in the time-to-market, since the design and production process are integrated and the time between conception and production is progressively reduced.

METHODS USED IN JEWELLERY MANUFACTURING

Several techniques have been employed to make jewellery from gold and silver and its alloys are Investment Casting, Die casting, cuttle fish casting and hand fabrication. Many jewelers use CAD/CAM to make jewellery. CAD (Computer Aided Designs) and CAM (Computer Aided Manufacturing) to reproduce a model piece of jewellery that can be mass-produced. Making jewellery requires knowledge and expertise in gold smiting, stonecutting, engraving, mold making, fabrication, wax carving, lost wax casting, electroplating, forging, and polishing. These are the various steps needed to make jewellery. The first step in making a detailed piece is making of a mold (Ray, 2009), it can be made in desk or computer aided.

Mold

A mold is the exact and perfect replica of the piece to be made, copied either from a design or a piece/object. A mold is shaped around the shape/figure with the help of casting process. Molds are made from many materials, including plaster compounds. Some different methods of casting include the lost wax process, centrifugal or investment casting, and sand casting. In jewel manufacturing the process begins with a mold. The casting process involves a number of steps. There are two methods of casting using in jewel and fashion accessory manufacturing; investment casting or die casting each with its own advantages.

The Investment Casting

The "Investment Casting" or "Lost-wax Casting", also called by the French name of *cire perdue* or the Latin *cera perduta*, is the process by which a bronze, brass or other special alloys are cast from an artist's sculpture and jewels. The casting process is one of the oldest known metal-forming techniques, was first used during the period 4000-3000 B.C. Lost-wax casting was widespread in Europe until 18th century, when a piece-mold process came to predominate. Believed to have been developed by the Mesopotamians, it remains as the most popular process of making gold (metal) jewellery and forms the basis of modern investment casting process. Nowadays, with the appearance of the high-technology waxes, refractory materials and specialist alloys, the castings allow the production of components with accuracy, repeatability, versatility and integrity in a variety of metals and high-performance alloys, thus, in industrial uses, this modern process is called investment casting.

Kalpakjian (1995) said that this process involves dipping a mold into a ceramic mix. Sometimes new materials like plastic or polystyrene foam is used instead of wax. This process has a number of steps involved. The figure 4 shows a brief description of the Investment Casting process begins by the injection of wax in a metallic mould that has previously built up or plastic hand-made pattern.

a. A primary model can be made in hard alloy like nickel silver or just silver. Surrounding this primary model, using sheet rubber in a mold frame, makes a rubber model. Placing it in a heated press then vulcanizes it. On cooling, it is cut with a scalpel into halves or more and removing the primary model. This rubber mold is used to make many copies of the primary model on wax. Molten wax is then introduced into the mold cavity by using a wax injector. On cooling the wax is removed to get an exact copy of the primary model in wax. Other way is wax or plastic models hand-made by craftsmen.
b. Then the wax or plastic models are welded to a structure also in wax, to set up an assembly (tree form).
c. The ceramic mould or negative of the wax pattern is formed up by immersing the assembly (tree) in a slurry coating (babbitt), keeping the finest details as the wax pattern.
d. The assembly is then surrounded by refractory materials (e.g., fine silica) till that a ceramic shell is obtained, also called stucco coating or investment plaster. Thus, form stiff and sturdy molds are produced.
e. The wax is removed in an autoclave by a pressure-temperature process to pattern melt

Figure 4. Steps involved in jewellery manufacturing by investment xasting process

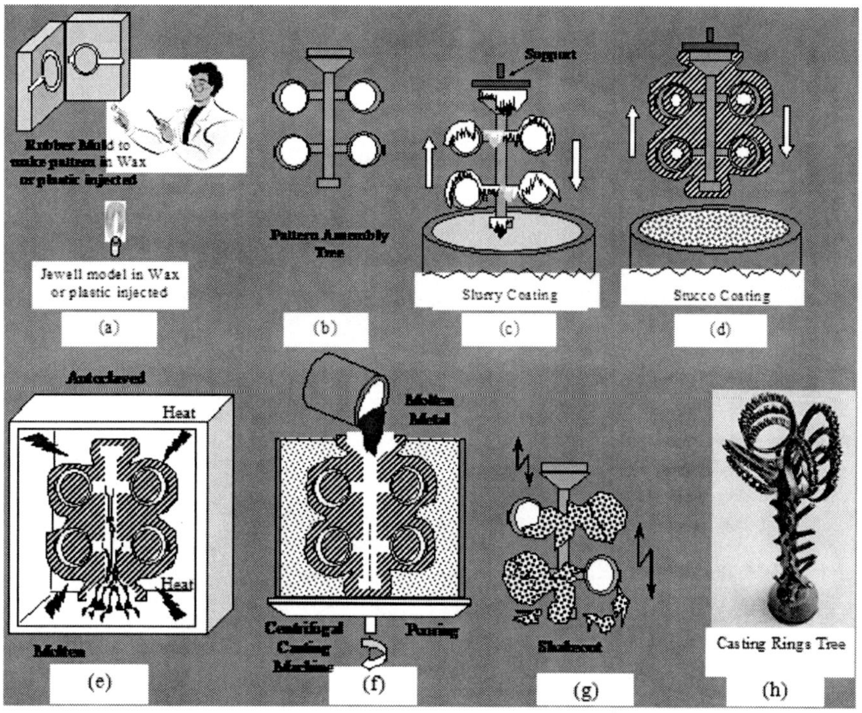

out. The flask is inverted and placed in a kiln/furnace. The wax is melted by steam or air to remove all the wax. The furnace is set in stages and the maximum temperature reached is 750 degrees centigrade. The melting process takes about 12 to 16 hours. This melting down of the wax is called the 'lost wax process'.

f. The ceramic mould is pre-heated at 1000°C and the metallic material is poured into at temperature higher than fusion temperature (e. g., 1550 °C steel or 1100 °C copper, etc) Two types of casting machines can be used; the centrifugal casting machine which is the older technique or the modern technique of static vacuum assist machines.

g. When the assembly is cold the ceramic shell is separated from the steel bye vibration process.

h. The steel parts are then separated for the assembly by mechanical process.

i. The process finishes with all kinds of controls, heat treatments and machining operations to achieve quality and finishing required.

As advantages of casting investment in jewel manufacturing can be said that it is an old age proven method. It allows the jewellery flexibility to create complex designs. The details can be copied perfectly. The control of colour is better. The finished product can be highly polished. It results in very fine surface finish. The metallurgical properties are also excellent. However, some disadvantages of investment casting also can be cited, as that this process can result in porosity. Also the dimensions may not be as accurate as the die struck method. Nevertheless, this process can and is used for almost all gold jewellery and remains a favourite with jewellers even after 6,000 years later.

Figure 5. Applications of computer technology in manufacturing

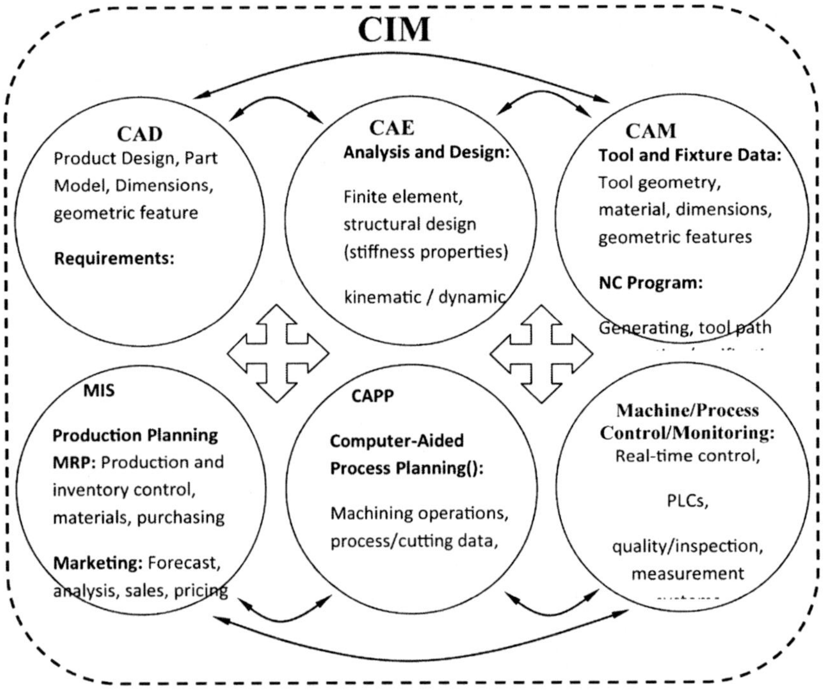

Die Struck Method

Die struck method is a casting method where the metal to be cast is forced under pressure into a mold which is usually made out of metal. This is an excellent method of producing complex shapes. The earliest recorded history of die casting by pressure occurred in 1800's. Using a plunger or compressed air, molten metal is forced into a metallic die and the pressure is maintained until the metal settles and solidifies. The pressure reaches 300 MPa. The intense pressure causes the atoms in the metal to move closer together and solidify to dies form or molds. Using compressed sheet metal and dies steel mountings are formed with metal parts mechanically stamped out. Each part is matched and fitted into the correct portion of die halved and stamped and shaped. A hydraulic press is used (Ray, 2009).

MODERN MANUFACTURING SYSTEMS

A number of steps are involved in manufacturing a part from its conceptualization to production. They include product design, process planning, production system design, and process control. Computers are used extensively in all these stages to make the entire process easier and faster (Mehrabi et al, 2002). Potential benefits of using computers in jewel manufacturing can include reduced costs and lead times in all engineering design stages, improved quality and accuracy, minimization of errors and their duplication, more efficient analysis tool, and accurate control and monitoring of the machines/processes, etc.

As showed in Figure 5, the applications of computers in manufacturing systems allow development products by computer-aided design (CAD), using computers in the design and analysis of the

Figure 6. Products life cycle phases and computer technology applications

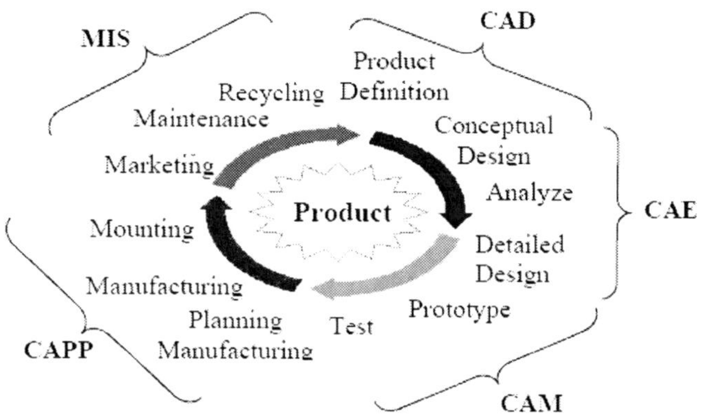

products and processes. The computer technologies play a critical role in reducing lead-time and cost at the design stages of the products/process. Also, computers may be utilized to plan, manage, and control the operations of a manufacturing system: computer-aided manufacturing (CAM). In CAM, computers are used directly to control and monitor the machines/processes (in real-time) or used off-line to support manufacturing operations such as computer-aided process planning (CAPP) or planning of required materials. Materials Requirement Planning (MRP) determines a schedule for the operations and raw material purchases shows the production requirements for the whole production system.

At higher levels, computers are utilized in support of management. They play a critical role in all stages of decision-making and control of financial operations by processing and analyzing data and reporting the results (Management Information Systems, MIS). Also, computers facilitate integration of CAD, CAM, and MIS (Computer-Integrated Manufacturing, CIM). They provide an effective communication interface among engineers, design, management, production workers, and project groups to improve efficiency and productivity of the entire system. (Mehrabi et al, 2002).

Product Development Life Cycle

Many enterprises have new product development methodologies that provide a standard framework for planning and managing development efforts. It can be discriminated four phases to this life cycle: concept (preliminary ideas), production (including product development and purchasing), use and disposal. The Figure 6 shows the life cycle phases and computer tools that can be using in improvements of the process.

Improvements in jewel manufacturing process are extremely important. Because the jewels and fashion accessories presents a high aggregate value. It is more important when raw material is abounding too. Brazil is a precious metal and stones products country, namely gold, silver, diamonds, emeralds, etc. Thus, better products imply in laddering of the market and consolidate designer's names in global market.

An Integrated Method for Jewel Manufacturing

For the last few years, the process of manufacturing jewellery has seen lots of innovation, and particularly in term of Design & Manufacture. CNC machine are adapting to the jewellery industry,

Figure 7. Time saving opportunities in the product development life cycle

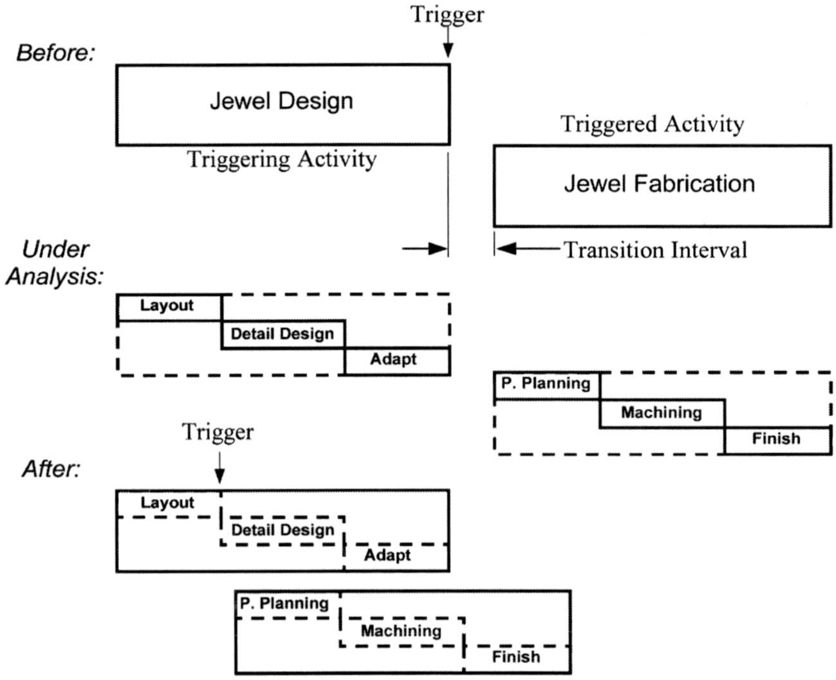

as for the laser welder and market, and Design & Manufacture software. For the last one, it has come much slower, and the available solutions offered on the market are not affordable to the majorities of the Brazilian jewelers.

Rapid Prototyping, RP, techniques has a vast application on several fields, such as architecture, art, engineering and industrial design. Using these technologies, and from a CAD3D geometrical file, it is possible in a few hours to build functional prototypes of components or even production tools, known as Rapid Tools or Rapid Manufacturing (RM). Hence, there is no need to build moulds to produce some components, e.g. in the development stage, that are constantly changed by the design teams. The number of RP systems in use, as well as theirs typology, is increasing every day, using different building materials such as waxes, polymers, metals, paper, sand and mix materials. There are already some desktop solutions (Mascarenhas & Esperto, 2006). Using two (2) of these RP techniques it was created a few jewel prototypes, namely Fused Deposition Modeling (FDM) and CNC micro-milling machining.

Rapid Manufacturing is the embodiment of the concept that allows defining elements or modules with well defined interfaces and "hooks" in a way that the process of manufacturing can be: developed in an independent manner, customized based on installation specific data, based on formal models of controller behavior, amenable to automatic generation of control software modules, and reconfigurable to adapt to changes in the environment through changes to data and/or regeneration of control software. Smith (1998) shows that application of RM allows identify time saving opportunities in the product development lifecycle, namely in design and manufacturing phases. As showed in Figure 7, the jewel manufactures may be improved by reducing both the production time and costs by work's superimposing. It is possible by using the computer technology, namely Rapid Manufacturing (RM).

Figure 8. Diagram illustrating the various methodologies' phases used in jewel rapid manufacturing

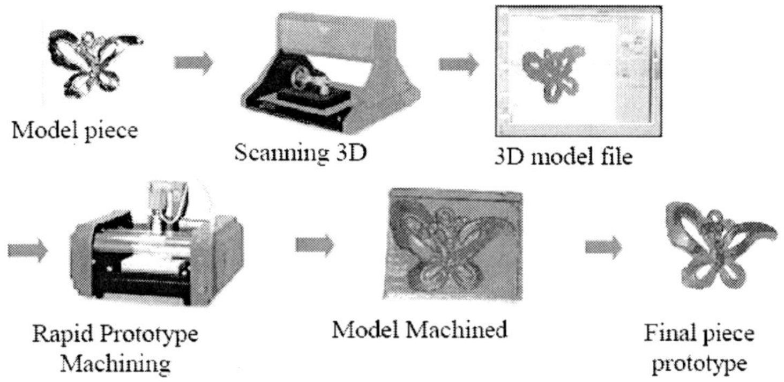

In last year's, artistic design & manufacture software has been developed to the jewellery industry, aimed to reduce production costs and hand process. This technology allows designers to the freedom to development quickly design, modify and view the models. Thus, designers can create new models with the benefit of a parametric construction tree which allows modifying the design at any stage, without going back countless steps or starting from scratch, reducing both the production time and costs. It can use the jewellery specific tools, such as the ring wizard to quickly build a band then change for example, the finger size while keeping the thickness constant.

Once the jewel model is completed, the designer can resize it, reconfigure its dimensions or use it as a component in another design by adding it to a pre-existing library of parts. Another problem is the cost of the jewel designed; the software can alert the designer if it goes under or over the desired weight. Allow a more accurate vision with the production costs. Most of materials are in the software, and designers can also easily create your own material for new jewel or fashion accessories.

Methodology Using in Jewel Rapid Manufacturing

The Rapid Manufacturing (RM) concept can be used when the development of moulds or tools for moderate volume parts or products is required. Among the various RM processes, computer numerically controlled (CNC) machining is recognized as the technology most widely used in industry for the application of CAD/CAM systems. The basic methodology for RM techniques can be summarized as follows:

a. CAD jewel model is constructed, and then converted to STL format. The resolution can be set to minimize stair stepping.
b. The RP machine processes the STL file by creating sliced layers of the model.
c. The first layer of the physical model is created. The model is then lowered by the thickness of the next layer, and the process is repeated until completion of the model.
d. The model and any supports are removed. The surface of the model is then finished and cleaned.

Different CNC machines can be used in rapid prototyping aided jewel manufacturing. In Figure 8, a small Roland desktop 3 axis machine was used as RM process, It can machining soft metal, wax or resin that can be vulcanized in rubber moulds. There are two work ways; from the scratch (start drawing) and 3D scanning and modelling (reengineering).

CAD/CAM IN JEWELL INDUSTRY

The use of latest trends in design tools and manufacturing technologies allows jewels design and manufacturing through CAD/CAM technology, from the basic concept to final productions. According to Osgrove (2005), can be cited three (3) advantages of this technology in jewel industry, namely accuracy, time and cost advantages. Among of the accuracy advantages that can be cited are the capability in provides a realistic image of the jewel even before it is made, thus aiding inspection to improve quality. Same, achieve exact symmetry of design, especially for left and right earrings. It can create identical looking designs of different diamond weights fasters. Likely, can be created ensembles will have same family looks, described precisely the metal dimensions for gemstone holding areas, no variation in consistency and accuracy, such as provides casting friendly even cross sections, etc.

Regarding to the time advantages in the application of the CAD/CAM technology in jewel manufacturing, can be said that CAD and CAM based technologies help to obtain a silver master models, faster than traditional wax or solid silver based carving methods. One can convert a complex design sketch into a finished silver model in few days time. Modifications in master model are faster and creating identical designs for different diamond weights, earrings and ensembles is possible.

The costs are extremely important in jewel manufacturing, and CAD/CAM technology allows obtaining the uniformity of scoop for maximum removal of excess metal, giving consistent and minimum metal weight. Same, the model making productivity increases many times without increase skilled manpower. As form a contract of high capital involvement, optimum utility of the capacity makes models affordable is allowed when CAD/CAM technology is applied in jewel industry. By using the CAD software in jewel development the designer might creates and produces a bigger range of jewellery models at the same time (family pieces) allowing creating and producing many finished models simultaneously.

In practice, CAD software is written in numerous different programming languages and can be used in many environments and platform (e.g., Windows, Linux, Apple Mac, etc). Some CAD using in jewellery are written in the Java computer, they can also run in any machine, and exchange data with other users. Jewellery's CAD software present friendly and intuitive interface, as well as help box dynamic. There are libraries of precious metals and gemstones with gemmological materials, which can also add too. Thus, a designer with few days of practicing would be able to design nice items, showing that working on the model can be very clear and the final rendering to be easy to use and extremely effective.

Kai et al (2000) presents a methodology to converting scanned data to.STL format. They said that the data modeller is the most important process for the success of jewel making. Its purpose is to convert the scanned image file to editable CAD format. The common converted file formats include IGES, DXF and ACIS. The conversion to CAD readable format is to facilitate correction and rework of the scanned image before converting to.STL format. Figure 9 illustrates the flow chart for converting scanned data to.STL format, followed by Figures 10a and 10b which present the steps to convert the input (scanned file) to the desired output (.STL file) of the ear mould, and finally can be used as a wax jewel model or vacuum cast mould.

All RP systems generally adopt the additive fabrication technique, which is a process that builds a part layer by layer. Each layer is represented and determined by its corresponding slice file. The slice file stores the geometric detail data of the model at that particular slice height. The slice file is created from the.STL file. After specifying the slice height, usually smaller 0.30mm, the RP attached software will slice the.STL file to a slice

Figure 9. Steps indicating the conversion of scanned data to.STL format

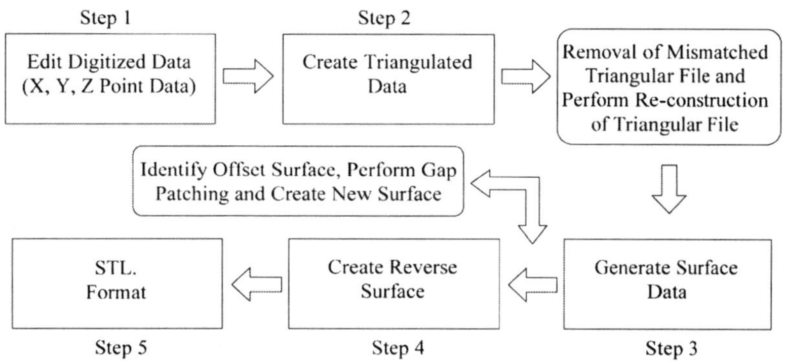

(.SLI) format. The.SLI file is then used to build the RP model (Osgrove, 2005).

JEWEL MODEL RAPID PROTOTYPING

Rapid Prototyping (RP) can be defined as a group of techniques used to quickly fabricate a scale model of a part or assembly using three-dimensional computer aided design (CAD) data. Rapid Prototyping has also been referred to as solid free-form manufacturing; computer automated manufacturing (CAM) and layered manufacturing. RP has obvious use as a vehicle for visualization. In addition, RP models can be used for testing, such as when an airfoil shape is put into a wind tunnel. RP models can be used to create male models for tooling, such as silicone rubber molds and investment casts. In some cases, the RP part can be the final part, but typically the RP material is not strong or accurate enough. When the RP material is suitable, highly convoluted shapes (including parts nested within parts) can be produced because of the nature of RP.

Why Rapid Prototyping?

The reasons of applied Rapid Prototyping (RP) in jewel manufacturing are:

- Increase effective communication.
- Decrease development time.
- Decrease costly mistakes.
- Minimize sustaining engineering changes.
- Extend product lifetime by adding neces-

Figure 10. Convert; (a) the scanned image file to (b) editable CAD format

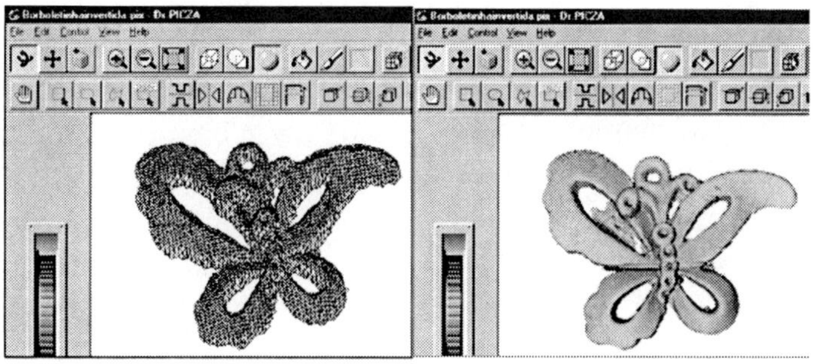

Figure 11. Jewel development process inside RM environment

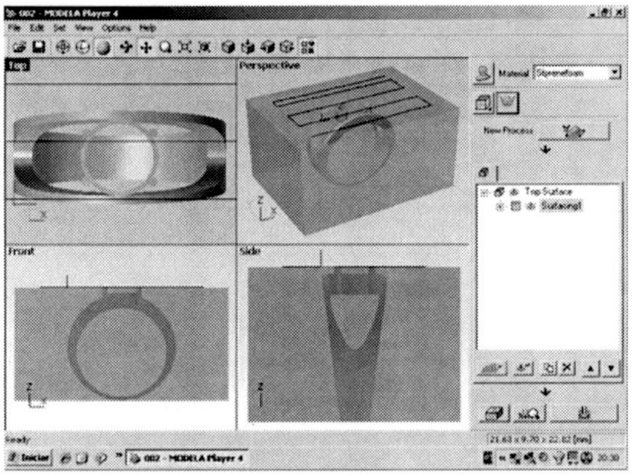

Figure 12. FDM extrusion head

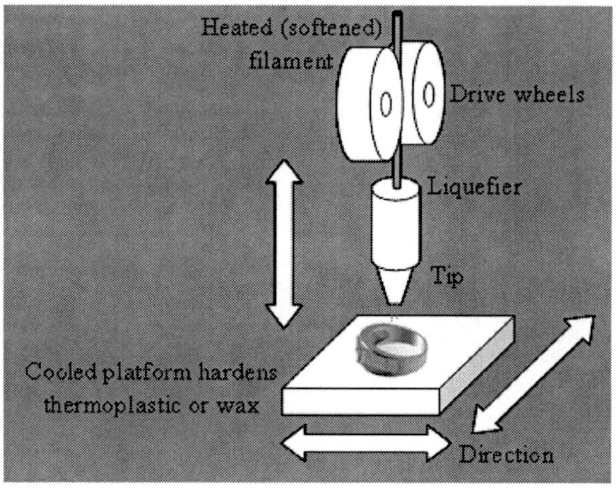

sary features and eliminating redundant features early in the design.

Rapid Prototyping decreases development time by allowing corrections to a product to be made early in the process. By giving engineering, manufacturing, marketing, and purchasing a look at the product early in the design process, mistakes can be corrected and changes can be made while they are still inexpensive. The trends in manufacturing industries continue to emphasize the following:

- Increasing number of variants of products.
- Increasing product complexity.
- Decreasing product lifetime before obsolescence.
- Decreasing delivery time.

Rapid Prototyping improves product development by enabling better communication in a concurrent engineering environment.

Different CNC machines can be used in rapid prototyping aided jewel manufacturing. In this work, two RP systems were used to fabricate the jewel and fashion accessories pattern for the silicon mould making or wax model. They are fused deposition modelling (FDM) and CNC Micro-Milling Machining. The first works by material deposition and the second works by material retreat. The two RP systems will be briefly discussed in the following. In addition, SLA and CNC will be used to make the RP jewel mould or model. This is to facilitate the evaluation on the patterns fabricated by each RP system. The evaluation will show which RP systems produce better-finished pattern suitable for investment casting. Figure 11 shows development process of the ring using software oriented to RP machine inside RM environment.

Fused Deposition Modeling (FDM)

The Fused Deposition Modeling (FDM) process was developed by Scott Crump in 1988. The fundamental process involves heating a filament of thermoplastic polymer and squeezing it out like toothpaste from a tube to form the RP layers. The machines range from fast concept modelers to slower, high-precision machines. The materials include polyester, ABS, elastomers, and investment casting wax (Osgrove, 2005). The overall arrangement is illustrated in Figure 12.

Figure 13. CNC micro-milling machine used in jewel rapid prototyping

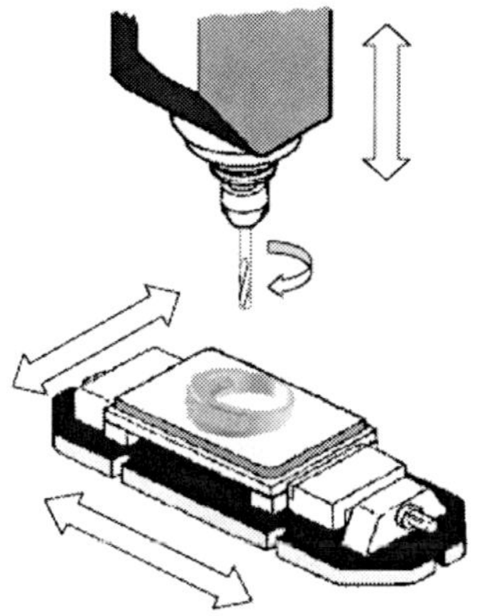

Highlights of Fused Deposition Modeling:

- Standard engineering thermoplastics, such as ABS or investment casting wax, can be used to produce structurally functional models.
- Two build materials can be used, and latticework interiors are an option.
- Filament of heated thermoplastic polymer is squeezed out like toothpaste from a tube.
- Thermoplastic is cooled rapidly since the platform is maintained at a lower temperature.
- Milling step not included and layer deposition is sometimes non-uniform so "plane" can become skewed.
- Make rapid progress in past few years and be used widely

CNC Micro-Milling Process

A wax model of a jewel can be produced using CNC machining techniques. The virtual model (CAD) of the jewel created from digital data obtained from the jewel model. The jewel model prototype is then fabricated from the wax model or other plastic materials. On the other hand, the CNC micro-milling process can be used in mold manufacturing. Thus, the virtual model (CAD) is used by coping the replica of the negative piece to be made, The Figure 13 shows CNC micro-milling machine used in jewel model rapid prototyping.

Different CNC machines can be used in rapid prototyping aided jewel manufacturing. In this work, was chosen a small Roland desktop 3 axis machine, It can machine soft metal, wax or resin that can be vulcanized in rubber moulds. There are two works ways; from the scratch (start drawing) and 3D scanning and modelling (reengineering). Figure 14 illustrates the stages from scanned jewel model to wax jewel model and finally jewel prototype.

This alternative offers quickly turnaround metal mold prototypes and rapid prototyping for metal parts. Depending on material, timing requirements, critical dimensions and post-machining operations, there are several different methods that can be used to satisfy metal prototyping requirements. CNC machining is the most common method to manufacture rapid metal prototypes. Figure 15 shows a 3D model file and CNC machine cutting tool path using CAD/CAM software.

Figure 14. Wax jewel model manufactured through CNC micro-milling machine

Figure 15. 3D model file and CNC machine cutting tool path using CAD/CAM software

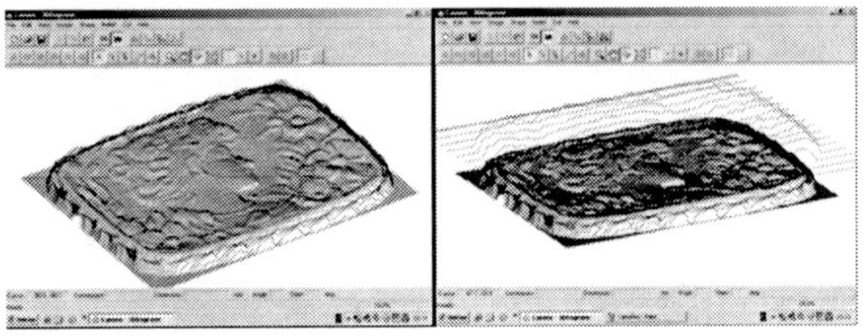

CONCLUSION

Everything considered, the benefits of CAD/CAM systems in front of the traditional design and production process are clear, and the use of these systems probably will increase in the next years. Notwithstanding the jewel industry modernization is only at the beginning. More and more references about this subject are presented, bringing incentive to entrepreneurs to implement new technologies in the jewel industry. However, the CAD/CAM is not a simple new tool, but it includes some difficulties to learn an adequate application. Problems like communication between designer and production sector, including goldsmith, can be solved with the use of CAD3D. In the same way, the technical characteristics of jewel are necessarily defined in the first stage of design process. This integration between design and production processes is the main benefit of the new technology. On the other hand, it is a great challenge to jewellery industry, especially small and medium enterprises.

REFERENCES

Fernandes, J. M., Romeiro Filho, E., Araki, L. A., Reis, L. P., Figueiredo, A. C., & Vaz, C. R. (2005). Cross-functional Teams and Concurrent Engineering: Contributions to the Development of Product Design Through Multidisciplinary Integration Using CAD Systems. *Product: Management and Development*, *3*(1), 5–9.

Han Boon, L., Ko, M. S. H., Gay, R. K. L., Kah, F., Leong, K., & Chua, C. (1992). Using computer-based tools and technology to improve jewellery design and manufacturing. *International Journal of Computer Applications in Technology*, *5*(1), 72–80.

Hilton, P. D., & Jacobs, P. F. (2000). *Rapid Tooling: Technologies and Industrial Applications*. New York: Marcel Dekker, Inc. doi:10.1201/9780203908020

Kalpakjian, S. (1995). *Manufacturing Engineering Technology*. Reading, MA: Addison-Wesley Publishing Company.

Kei, C. C., Meng, C. S., Ching, L. S., Tei, L. S., & Aung, S. C. (2000). Facial prosthetic model fabrication using rapid prototyping tools. *Integrated Manufacturing Systems*, *11*(1), 42–53. doi:10.1108/09576060010303668

Langer, H. J. (2008). I want it now - pushes manufacturers into rapid responses. *Metal Powder Report*, *63*(4), 8–9. doi:10.1016/S0026-0657(09)70035-9

Leong, K. F., Chua, C. K., Chua, G. S., & Tan, C. H. (1998). Abrasive jet deburring of jewellery models built by stereolithography apparatus (SLA). *Journal of Materials Processing Technology*, *83*, 36–47. doi:10.1016/S0924-0136(98)00041-7

Mascarenhas, J., & Esperto, L. (2006). Direct Metal Laser Sintering (in Portuguese). Ciência & Tecnologia dos Materiais. *Journal of Portuguese Materials Society*, *18*(3/4).

Mehrabi, M. M Galip, G., Ulsoy, A., & Koren, Y. (2002). Manufacturing Systems and Their Design Principles, In Osita D.I. Nwokah and Yildirim Hurmuzlu (eds.), *The Mechanical systems design handbook: modeling, measurement, and control* (Vol. 1 – Manufacturing) Boca Raton, FL: CRC Press LLC.

Noguchi, L. D. (2003). *The Goldsmith and the Skill: The Jewel Quality on the Interface Between Design and Production Process in the Handmade Jewellery*. (Thesis). Federal University of Minas Gerais, Belo Horizonte, Brazil. [in Portuguese].

Nwokah, O. D. I., & Hurmuzlu, Y. (2002). *The Mechanical systems design handbook: modeling, measurement, and control*. Boca Raton: CRC Press LLC.

Osgrove, B. (2005). *Comparison of Low Cost Rapid Prototyping Machines*. Retrieved from rpworld.net

Ray, N. (2009). *Jewellery Manufacturing*. Retrieved February 22, 2009, from Jewel Info 4 U, www.jewelinfo4u.com

Richards, T. R., Lorrain, H., & Hilton, P. D. (2000). The Role of Rapid Tooling in Investment-Casting Applications. In Hilton PD. Jacobs PF (eds.), *Rapid Tooling: Technologies and Industrial Applications*. New York: Marcel Dekker.

Rocha, V. C. (2007). *The Use of Information Systems on Jewellery Design and Production Process*. (Thesis). Federal University of Minas Gerais, Belo Horizonte, Brazil. [in Portuguese].

Siu, N. W. C., & Dilnot, C. (2001). The challenge of the codification of tacit knowledge in designing and making: a case study of CAD systems in the Hong Kong jewellery industry. *Automation in Construction*, *10*, 701–714. doi:10.1016/S0926-5805(00)00088-1

Smith, P. G. (1998). Leverage the Cycle-Time Capability of Your Rapid Prototypes. *Time-Compression Technologies. North American Edition, 3*(7), 36-43.

Soo, S. C., Yu, K. M., & Chiu, W. K. (2006). Modeling and fabrication of artistic products based on IFS fractal representation. *Computer Aided Design*, *38*, 755–769. doi:10.1016/j.cad.2006.04.003

Stamati, V., & Fudos, I. (2005). A parametric feature-based CAD system for reproducing traditional pierced jewellery. *Computer Aided Design*, *37*, 431–449. doi:10.1016/j.cad.2004.07.004

Stamati, V., & Fudos, I. (2007). A feature based approach to re-engineering objects of freeform design by exploiting point cloud morphology In *Proceedings of SPM 2007: ACM Symposium on Solid and Physical Modeling*. Beijing, China.

Wong, T. N., Cheung, C. H., & Lau, H. (1999). Decision support system for a jewellery manufacturer. *International Journal of Production Economics*, *60-61*(20), 211–219. doi:10.1016/S0925-5273(98)00142-X

KEY TERMS AND DEFINITIONS

Rapid Prototyping: A group of techniques, which has a vast application on several fields, such as architecture, art, engineering and industrial design. Using these technologies, and from a CAD3D geometrical file, it is possible in a few hours to build functional prototypes of components or even production tools, known as Rapid Tools or Rapid Manufacturing (RM).

Rapid Manufacturing: A computer technology, the embodiment of the concept that allows defining elements or modules with well defined interfaces and "hooks" in a way that the process of manufacturing can be: developed in an independent manner, customized based on installation specific data, based on formal models of controller behaviour, amenable to automatic generation of control software modules, and reconfigurable to adapt to changes in the environment through changes to data and/or regeneration of control software.

Computer Aided Design and Computer Aided Manufacturing (CAD/CAM): Systems are diverse software and hardware technologies, used in many applications in industries like automotive, aerospace, mechanical and others. One of the most important aspects of CAD/CAM implementation is the integration between different areas of enterprise, like product development, production and assembly.

Computer-Integrated Manufacturing (CIM): An engineering approach for manufacturing, where computer technology controls different functions in the enterprise. The use of CIM technology provides an effective communication among design, analysis, planning, purchasing, cost accounting, inventory control, and distribution. However, the CIM can be understood like a management strategy, using the computer resources to integrate functions and people in the corporation.

Fused Deposition Modeling (FDM): Process involves heating a filament of thermoplastic polymer and squeezing it out like toothpaste from a tube to form the Rapid Prototyping layers. The machines range from fast concept modellers to slower, high-precision machines. The materials include polyester, ABS, elastomers, and investment casting wax.

Mold: The exact and perfect replica of the piece to be made, copied either from a design or a piece/object. A mold is shaped around the shape/figure with the help of casting process. Molds are made from many materials, including plaster compounds. Some different methods of casting include the lost wax process, centrifugal or investment casting, and sand casting.

"Investment Casting" or "Lost-Wax Casting": also called by the French name of *cire perdue* or the Latin *cera perduta*, is the process by which a bronze, brass or other special alloys are cast from an artist's sculpture and jewels. The casting process is one of the oldest known metal-forming techniques, was first used during the period 4000-3000 B.C. Nowadays, with the appearance of the high-technology waxes, refractory materials and specialist alloys, the castings allow the production of components with accuracy, repeatability, versatility and integrity in a variety of metals and high-performance alloys, thus, in industrial uses, this modern process is called investment casting.

Chapter 8
Creative and Visualization Tools in Context of Design

Mamata N. Rao
National Institute of Design, India

ABSTRACT

The chapter in the broader sense will look in the area of creativity, creative process, and creative product. Specifically we shall look at the aspects of a creative product, discuss on thought process of the designers with focus on creative and visualization tools. Creative tools will be looked in context of changing mindset or assumptions, redefining problems, developing ideational fluency and bringing flexibility in thinking. Visualization tools such as sketches, storyboards, rough models, developing scenarios for the usage of proposed concepts etc will be discussed in parallel to creative tools + serving as aids to externalize thought processes. Creative and visualization tools complement each other in enhancing the designer's creativity as well as help them come out of the stuckness feeling that they encounter while addressing design problems. Practicing the tools rather than being aware of them is important and the chapter will demonstrate the application of these tools with examples.

BROAD OVERVIEW OF CREATIVITY

The dictionary meaning of creativity is the "ability to create". In the words of de Bono (1992) who popularized lateral thinking "Being *creative* means to *bring into being* something that was *not there before*". Lubart (1997) views "Creativity as the production of *original* (novel), *socially valued* (appropriate) works resonated by Jones & Murray (2003) as "A creative idea is generally considered to possess two main qualities: *appropriateness* and *originality*". Csikszentmihalyi (1996) defined creativity as the production of an idea, action or object that is new and valued, although what is considered creative at any point depends on the cultural context.

There are two aspects interpreted from the above statements:

DOI: 10.4018/978-1-61520-617-9.ch008

- of *novelty/newness/originality and socially valued/appropriateness* in *something* hinting about the quality of an idea or product to qualify as *creative idea/product* and it also hints that a creative idea /product being new means bringing change.
- words *ability* and *being creative* refer to the ability of an individual and we will refer to this as *creative thinking*.

Howard-Jones & Murray (2003) in view of the main qualities of a creative idea such as *appropriateness* and *originality* state that creative problem solving will involve *less focused generational process*, allowing access to remotely associated concepts (originality) and *more focused analytical processes* (appropriateness). Based on Freud's (1949) proposal of two modes of thinking primary process and secondary process, they further state that in *primary process* thinking is more concerned with defocused, more freely associative thinking whereas *secondary process* is concerned with conscious, focused and logical analysis. The generational thinking mode is referred to as *divergent* and the analytical as *convergent*.

Although both divergent and convergent thinking abilities are needed for creative thinking, divergent thinking mode is regarded as a key aspect of a person's creative thinking ability. Guilford (1950) regarded divergent thinking as an ability to generate multiple alternative solutions to a given problem. Divergent thinking is a key capacity underlying creative thought. To add to Guilford's views several others such as Christensen & Guilford (1958), Wallach (1970), Torrance (1974), Pellegrini (1984) have regarded ideational productivity, as a key sign of a person's overall creative ability. Mckim (1980) regarded fluency and flexibility as two important attributes of ideational productivity. He further states that *fluent ideation* is demonstrated by a thinker who generates many ideas; fluency being measured in terms of quantity and not quality. *Flexible ideation* is exhibited by a thinker who expresses diverse ideas, the measure in this case being variety. An example to demonstrate this concept is, looking at the task of generating alternative uses for a newspaper. The various uses stated are 'shade from the sun', 'clean a dirty surface', 'cover food items', 'shelter from rain', 'serve eatables', 'holding nuts', 'as a fan', 'hit a fly', 'hit someone', 'pretend reading to avoid someone', 'pack items', 'use as light fixture' etc. Uses such as cover food items or a lighting fixture, holding nuts, provide shelter from rain/sun shade belong to the same functional set (covering) and are regarded as less diverse when compared with one another. Such a set of ideas are rated low in terms of flexible ideation or diversity.

Boden (1998) viewing creative idea from the perspective of an individual and history states "A creative idea or product may be novel for an individual, referred as P-creativity (P for Psychological). This creative idea when found to be novel historically (unique and produced for the first time by anybody) is called H-creativity (H for Historical)". In the above example of alternative uses of a newspaper, an individual may feel excited that he/she has been able to list so many uses of newspaper getting the sense of P-creativity, but none of these uses are novel historically.

Sternberg(1999) has proposed several methods for measuring and developing creativity:

- The *pragmatic approach*, develops creativity with methods and tools like Edward de Bono's Lateral Thinking, Osborne's Brainstorming, Gordon's Synectics etc
- The *psychometric approach* measures creativity by means of Guildford and Torrance Tests etc.
- The *Cognitive approach* concerns the mental representations and cognitive processes that are involved in a generative and exploratory phase using methods like retrieval, association, synthesis, transformation, analogical transfer etc.

- The Social personality approach explores personality development in the context of setting up an environment that encourages creativity.

Corbolan-Berna (1992) investigated on cognitive processes and came to the conclusion that question generation was the most important factor in achieving a creative output.

The psychologist Amabile (1983) proposes a componential model for creativity. The three major components of creativity are domain-relevant skills, creativity-relevant skills and task motivation. The domain relevance refers to knowledge, skill and expertise in a given area. For a designer domain relevant skills could be visualization skills such as sketching, making models; good know how of material knowledge, manufacturing processes. Creativity-relevant skills are related to the way an individual views a given problem, related issues and uses his/her skills to come up with the best solutions. Task motivation includes both intrinsic and extrinsic motivation (external factors such as deadlines, monetary benefits etc) with intrinsic regarded as necessary to reach higher creative output.

Simonton (1984) studies the correlation between intelligence and creativity and found that creativity has to do with how a person's knowledge is accessed and applied for a given problem and not to do with how knowledgeable an individual is. Intelligence influences divergent thinking by providing a basis for generating not only original, but also high quality ideas (Vincent, Decker & Mumford, 2002). Although quality in terms of originality and usefulness, is essential in judging the final outcome of a creative problem, the ability to produce a number of ideas would mean some of them would be qualitative. (Howard-Jones & Murray, 2003).

In view of creative thinking or the ability of an individual to be regarded as creative, it is essential that one should focus on bringing in a large number of ideas. Bringing design ideas that are diverse would mean attempting to focus on quality of ideas. Domain relevant (knowledge) and intrinsic motivation of an individual play an important role in creative thinking.

THE CREATIVE PROCESS

Archimedes eureka moment, Kekule's discovery of benzene molecule structure, Poincare's description of how he arrived at Fuschian functions are instances of insights occurring during illumination phase of problem solving process. To explain these instances Wallas (1926) proposed a four-stage model namely *preparation*, *incubation*, *illumination*, and *verification*. *Preparation* involves defining and analyzing problems, a conscious effort that is related to a person's education, knowledge. Next is *incubation* when there is no focused conscious work on the problem, involves taking a break, relaxing etc during which mind continues to work on the problem forming a network of associations, rejecting a number of associations and if at this point one finds a promising association this leads to *illumination* phase which is identification of one or more creative ideas popularly referred as *insights*. This is followed by a phase of conscious work in terms of refining, developing and evaluating ideas which is called *verification*. This model is regarded as a classic model for depicting stages of creative processes. Wallas (1926) stated that one could return to earlier phases if the idea seems to be less effective during verification.

Martindale (1995) proposed that during preparation, attention becomes highly focused with a few nodes dominating thinking which exert strong lateral inhibition on other nodes, making them inactive. As attention is gradually defocused during the incubation period, inhibition caused by previously highly active nodes that encoded the problem decreases and other nodes that had remote association with the problem become more active, providing the experience of illumination. This brings forth the importance of alternating between less focused generational process and more focused analytical process during problem solving.

Hertzberger (1991) emphasizes the need to gain knowledge and experience for design students where more seen and experienced will be stored in memory and recalled whenever the mind seeks a connection with relevant problem. He underlines the importance of knowledge and experience for promoting associational thinking.

Amabile (1996) based on her componential model of creativity (Domain-relevant, Creativity-relevant & Task motivation) described the creative process as consisting of the following phases: (1) *problem or task identification* (2) *preparation* involving relevant information gathering (3) *response generation* to do with producing alternative solutions (4) *response validation and communication.*

CREATIVITY ACROSS PROFESSIONS

Lawson (2006a) views creativity as follows – "Most people would describe design as one of the most creative human pursuits. The so-called creative arts include musical composition, painting, sculpture and the various forms of two and three dimensional design. However creativity and creative thought can be applied just as much in science, medicine, philosophy, the law etc. In the creative arts including design, the whole point of the business is to *create something* which other people experience and which is in *some way or other is original and new*.

Creativity is not limited to any profession. The forms of creative behavior might vary for different professions, like in arts it is based on perceptual and subjective insights, while in science insights should be based on rational and objective behavior (Ludwig 1995). Lubart (2000-2001) based on Dudek & Cote (1994) states problem finding in art is internally oriented process to come to terms with a topic, express one's emotions, point to a new social reality or externalize an inner state and referring to work done by Ochse (1990), views problem finding in science as characterized by discovering gaps or discrepancies in existing knowledge, sensing difficulties when one's expectations are not met, when observations do not match with an existing mental model of a phenomenon.

It is common for a lot of practicing designers, architects and design students to view creativity as an artistic ability and attach a lot of importance to develop aesthetic sensibilities such as sensitivity to compositional aspects of design (look and feel). Although aesthetic sensibilities are to be built in a designer, it should be noted that they are not the only aspects of design as shown in the framework by Donald Norman and Patrick Jordan.

Donald Norman (2004) captures very beautifully the various design aspects (parameters) at three levels as *visceral* level, *behavioral* level and *reflective* level. At the *visceral* level, physical features – look, feel, sound dominate and shape, form, physical feel and texture matter and highly rated products may be turned down if they do not appeal to aesthetic taste. Visceral is all about immediate emotional impact or in other words catching attention. *Behavioral level* is all about performance with function, understandability, usability and physical feel playing an important role. This is the aspect of design that usability practitioners focus on where the inspiration behind design begins with user research. *Reflective level* is all about message, culture and about the meaning of the product where overall impact of the product comes through reflection i.e. in retrospective memory and assessment.

Another framework titled 'The Four Pleasures' by Jordan (1999) was derived from the theoretical work of Lionel Tiger (1992), addresses the experiential aspects of products. *Physio-pleasure* is to do with pleasures derived from the senses such as touch, taste, smell and sensual aspects. *Socio-pleasure* is the enjoyment derived from the company of others where products can facilitate social interaction in a number of ways. *Psycho-pleasure* relates to the extent to which a product can help in accomplishing the task (usability approaches) and make such tasks satisfying. *Ideo-pleasure* refers to the pleasures derived out of theoretical aspects such as the aesthetics of a product and the values that a product embodies (recycle, reuse, history) etc.

An important aspect in design which falls within the experiential zone is the feeling of 'surprise' invoked by a product which can be beneficial to both the designer and the user. As stated by Derbaix & Vanhamme (2003) "The designer benefits from a surprise reaction because it can draw attention to the product, leading to increased product recall, recognition and increased word-of-mouth." The product user benefits from the surprise because it makes the product more interesting to interact with meaning that a user can learn something new about the product (Ludden, Hendrik & Schifferstein, 2008).

Li Zuo (1998) has written about aesthetic sense in creativity as being beyond visual beauty alone and viewing aesthetic ability as being able to recognize promising problems (patterns) in their field and the ability to persist until a right solution is found. The uneasiness about a lack of fit, can be equated with aesthetic sensitivity that lead Einstein and Darwin to discovery of important problems in their domain. He further states that when good solutions are found they yield more aesthetic pleasure with the aesthetic criteria being simplicity and elegance. Darwin explained the entire theory of evolution in three simple words – generation, selection and preservation.

Norman's framework focuses on physical and functional aspects of a product and Jordan's on experiential aspects bringing to the forefront the various design parameters and context that can be considered while ideating creative products. While Li Zuo reflects on aesthetic sensitivity as the ability to intuitively seek for patterns in terms of harmony, order and simplicity as indicators of good solutions for any given problem. In the profession of design it is a common notion to state that a good design solution is one that brings simplicity from the complexity that existed in a given problem. Invoking surprise in a product can be benefitted by both the designer and the user. All these various frameworks and perspectives present a rich context a designer can address in a design problem by bringing in a number of qualitative ideas to make the product or a design solution to be regarded as novel or original.

ORIGINALITY OR NOVELTY AS SEEN IN A CREATIVE PRODUCT

People who are aware of '*Sydney Opera House*' in Sydney, '*Lotus/Bahai Temple*' in Delhi or other architectural wonders find them interesting as they are different and unique. The uniqueness could be the period and style of architecture (form, color, materials), massive scale, and the story-like imagery that they set within us. For example Sydney Opera House sitting on the harbor looks like a sailing ship to some, to others it may seem like a huge pearl oyster; in general it evokes imageries connected with sea very appropriate for its situational context and sets one on a journey of discovery.

Phillipe Starck's 'Lemon squeezer' for Alessi is loved by many as a collectible item and criticized by those who regard functionality as an important attribute in a product. It is famous for its value not as a squeezer but as a decorative object and is considered an icon of industrial design, leading to it being displayed in New York's Museum of Modern Art.

In contrast to the above two examples of iconic nature, if we look at the design of public facilities such as hospitals, shopping centers or products used in such places the originality will depend on understanding the users behavior (patients, doctors, staff, buyers) and their activities. This is evident in IDEO's shopping cart design part of a national level show that was telecast on TV (ABC, USA). This demonstrated to viewers the actual process of innovating a new product, based on their methodology which included insights from user's within the shopping environment.

Similarly Nokia's mobile phones are examples of research based design outcomes and are very popular when it comes to masses, whether in established or emerging markets. They are known for the choice of design options in terms of look and feel, as well as pricing. The most important aspect of Nokia product is easy learnability for various features, which are evolved incrementally and are intuitive due to their connection with precedent (past) design.

IKEA's furniture and lifestyle accessory products are appreciated worldwide because of their affordability and functionality. Their products are based on modular construction and require assembly by user's themselves, which gives the user an experience of assembling a product together, making this experience very unique and emotional.

Shanghai Tang's (lifestyle accessory and fashion house) collection are desired as their products carry a distinct touch of Chinese cultural aspects designed to suit a contemporary lifestyle. Animation films made by Hayao Miyazaki although are fantasy stories have strong depiction of Japanese flora, fauna and local buildings and are renowned world over for their excellent details and use of cultural aspects, which differs from other animation and science fiction stories.

Apple's ipod/ iphone are regarded as breakthrough products in the area of digital devices, because of their simple sleek form, a very different looking touch interface that departs from the prevalent design styles.

In all the above examples the products were liked for being original/unique or in other words novel due to differing aspects such as:

- *Surprising* or *discovery led* journey or aesthetically pleasing/decorative look and feel as in iconic buildings (Sydney Opera) or products (Alessi Lemon squeezer). In the context of iconic products/ buildings the *formal aspects* (form, color, materials) of product become very important.
- Ease of use and provision of features appropriate to *user needs* as in IDEO's shopping cart or Nokia's mobile phones become important in products for mass usage.
- Closeness to *cultural aspects* as seen in Shanghai Tang's products or Hayao Miyazaki's animation films.
- Apple's approach to design products that look different and set *a new trend.*
- IKEA's approach to cut down costs and pass the benefits to users by bringing out affordable and modular designs that users could progressively buy and build which is also related to formal aspects of a product but with constraint of *cost factor* and accommodating user's diverse needs through *modularity*.

Based on instances above we can say that we look at *originality* and *uniqueness* from various design aspects (parameters) such as form/style factors, being user friendly, associating with our history and culture, setting a new trend and other attributes for a product to be regarded as creative. To sum up a *creative product* is one which is unique, original and new, with there being several design aspects that play a role in bringing about this *uniqueness* and *originality* depending on the context (situation, users, economy etc) and the kind of experience it provides.

QUALITIES OF APPROPRIATENESS IN A PRODUCT

An important criterion for a product to be called creative is society's acceptance of that product and the value the society attaches with that product. A product ahead of its times, may not be much valued although being very original and new, Sydney Opera House an icon of Australia today is a good example. A product which is perceived as creative at a particular point of time may not hold the same value at another point, this holds true for many of the historical products such as typewriter, bicycle etc.

The context, environment, weather, usage, societal status etc. play an important role in acceptance and appropriateness of the product for example the formal aspects of Sydney Opera House based on sea imageries were appropriate to the site context as the building is located on the harbor.

In the examples above there is a sense of originality/uniqueness created by designers which are very appropriate to the contexts and users they address. Formal surprises and discovery-led design will not work in railway stations, hospitals, other public places whereas are appropriate for iconic buildings and collectible artifacts. Philippe Stark's "Lemon squeezer' was appropriate for the brand Alessi who value giving a status symbol to owners of their product.

Apple's products which look very sophisticated are designed for people who want to look different and set a new trend, even though it means learning a newer style of interaction. While Nokia which caters to masses is able to address their users by understanding their needs and bringing in incremental developments. Their products are robust and look very typical that link to the past products making them appropriate for their user segment.

The appropriateness of a product comes from the context and the users targeted. It also depends on the time factor in terms of societal values and aspirations.

DESIGN PROBLEMS, DESIGNERS AND PROBLEM SOLVING BEHAVIOR

Nature of Design Problems

Mayer (1999) defines "problem as a situation in which a problem solver has a goal but does not know how to achieve it." He further elaborates between well and ill-defined problems. Well defined problems has a clearly understood given state and well stated goal; while in an ill-defined problem the given state and the goal state both are not so well defined. Design problems are regarded as 'ill-defined problems'. Most design projects come with problem definition in terms of probable requirements, budget, and proposed site etc slackly defined (Cross 2006). Schon (1979) viewed that problems being human constructs are not fixed and are created as responses to complex and troubling situations.

Eastman (1970) viewing the behavior of subjects (architectural design)from his design protocol study stated " Instead of generating abstract relationships and attributes, then deriving the appropriate object to be considered the subjects always generate a design element and then determine its qualities." Lawson (1979) looked at the strategy adopted by two groups of students (one final year students of architecture and the other postgraduate science students) to solve design-like problems with rules and found that scientists focused their attention on understanding the underlying rules, a *problem-focused strategy*, while the architects were obsessed with achieving the desired goals or in other words *solution-focused strategy*. These studies hint that designers address any given problem by generating solution ideas and evaluating their effectiveness for the problem through trial and error method, rather than defining the problem.

Cross (2006) has stated that the task of problem formulation requires refined skills in gathering, structuring information and judging to move to solution generation phase. He quotes studies done by Christiaans & Dorst (1992) of junior and senior industrial design students where it was found that some students got stuck with information gathering and did not move to solution generation, while in another study by Atman et al (1999) of engineering students found junior student designers group spent more time defining problem but did not have quality designs; while the senior student designers group defined problem better which resulted in better designs. This means problem formulation is a requisite to achieve better designs.

Cross (2006) states "Solution-focused strategy adopted by most designers has drawbacks such as 'fixation effect'." He quotes the study by Jansson and Smith (1991) who proposed that fixation could stop the designer from considering the aspects of knowledge and experience for the problem at hand. Designers based on their previous experience and designs suggest ideas without looking into the given problem and generating new designs. Studies by Purcell and Gero (1991, 1993, 1996) conclude that industrial designers seem to be "fixated on being different" and that fixation in design may exist in a number of forms. The other type of fixation seen among designers is attachment to a single concept generated early in the design process and not generating alternatives (Cross 2006). There is also a need to generate alternatives during design exploration and the study by Fricke (1993, 1996) concluded that successful designers are those who have 'balanced search' for solution alternatives (not too many nor too less both regarded as weak design strategies). He also found that with a precise design assignment more alternatives were generated and vice versa.

All the research studies in the area of problem solving in design underline the importance of problem formulation, need to get away from being fixated with past designs or the urge to be different(with no basis), and not getting carried away with information gathering only while problem formulating. This makes creativity-relevant skills as described by Amabile important for solving design problems.

Creative Insights in Design Problems

Insights might occur while solving problems in form of associations as evidenced in examples below of Sony's Walkman connecting walking and listening, the discovery of Velcro by George de Mestral on noticing that his dog's coat and his pants were covered with cockleburrs, Utzon's architectural form for Sydney Opera house based on ocean-related forms, Starck's form for Lemon squeezer. Let's look at some design developments based on insights described by designers.

Joern Utzon architect mentions the *yachts*, *gothic cathedrals*, *Mayan temples* with strong horizontal lines and huge platform as his source of *inspiration* for his architectural creation, the Sydney Opera House one of the most renowned buildings in the world. Narrating on the problem, he states that being located on the harbor meant that it could be viewed from all sides including top and hence had to have a unique roof form. Evaluating the appropriateness of a design form for the harbor site, he formulated the problem rightly and visualized a solution with cues from chosen inspirations. In view of how these cues were taken to further develop this initial idea Utzon says "I shaped like a sculptor; I tried to shape the things and made many models. It came out of *model work*, more than *paper work*." Utzon was at the time criticized for not having detailed diagrams for implementation of his structure which was to build giant shells, all of differing sizes from wire-mesh frames and surface them with tiles. This idea was rejected by structural engineers as not feasible. Utzon took to solving this problem by looking at shells as part of a gigantic sphere and the answer to the problem came as a sudden breakthrough when he noticed a fruit bowl with the orange and began to slice the peel into segments.

When Philippe Starck created the 'Lemon Squeezer' one of Alessi's iconic products, his response was an *idea for an octopus like* lemon squeezer *sketched* on a napkin. It was inspired by squeezing a lemon over a squid in a sea food restaurant.

In the case of IDEO, as outlined by Kelly & Littman (2000) the focus was to understand the market, client in terms of current perceptions and challenging them during the later phases; gaining inspiration by observing people in real life. They externalized their ideas by *visualizing concepts through sketches, physical models, computer based rendering or simulation, scenario etc.*

Lindholm, Turkka, & Harri (2003) have explored a number of Nokia design projects and affirm that usability and user-centered design are considered as core competencies at Nokia with focus on incremental innovation specific to local cultures based on user research and *insights*. Nokia focuses on an evolutionary approach to UI development as small improvements are easier, faster, and cheaper to implement than large ones with possibility of turning back if the new design is not appreciated.

As Lucretius said "Nothing can be created out of nothing" something is required to approach a design problem. In all the above examples there are two aspects that need to be closely understood, one being the *source of inspiration* (cues from environment, users, earlier products etc.) for design ideas and the other being *methods for visualizing* some such ideas for further evolution, evaluation and development.

CREATIVE AND VISUALIZATION TOOLS

Broadly *creative tools* will look at methods to connect to *sources of inspiration* to build ideational productivity both in terms of quantity and quality; while *visualization tools* will look at methods to *externalize* ideas for further evolution and communication.

Creative Tools

Rao, Lo and Fung (2007) proposed a framework of objectives to nurture creativity as follows:

- *Getting rid of assumptions*: Starting with a fresh perspective.
- *Restating problems*: Redefining/reframing project briefs.
- *Ideational fluency*: Ability to generate many ideas
- *Bringing deliberate shifts*: Ability to change one's perception and frame of reference.
- *Recording and Reflection*: Promote recording and reflection of thought.
- *Experimental Evaluations*: Evaluating concepts through models/prototypes.

Fung, Lo & Rao (2005) suggested a set of creative tools within the framework of these objectives. There are many instances when design students as well as designers experience the feeling of stuckness or fixation. In such situations it is common practice to talk of abilities such as being imaginative, playful, and bringing shifts, to get oneself out of the feeling of stuckness and promote creative thinking but very difficult to have tangible and deliberate ways of experiencing such concepts (imagination, playful, bringing shifts). Practice of Creative Tools raises ones abilities to achieve and experience them and also facilitate divergent thinking by bringing attention one's to varying perspectives of a given design problem.

Visualization Tools

Many of us have experienced the fact that ideas seem to evaporate quickly unless captured. Noting ideas through textual and visual notations by maintaining a diary is a common practice among a number of designers. This visual diary could include sketches, writings, inspirational visuals from various sources such as magazines, photos, observations etc. Sketching (doodles, associational maps, thumbnails, scenario generation, storyboarding, 2D-3D product sketches), making models, simulation are some ways to externalize thoughts for evolution and communication.

Doodles are sketches that are less intentional, more abstract and are usually formed using bubble and line structures. *Associational maps* show a network of association using lines and shape

Figure 1. Thumbnails of poster ideas for traffic, empathizing a policeman and the traffic signal post. (sketches by Sachin Rathi and Pallavi Naik)

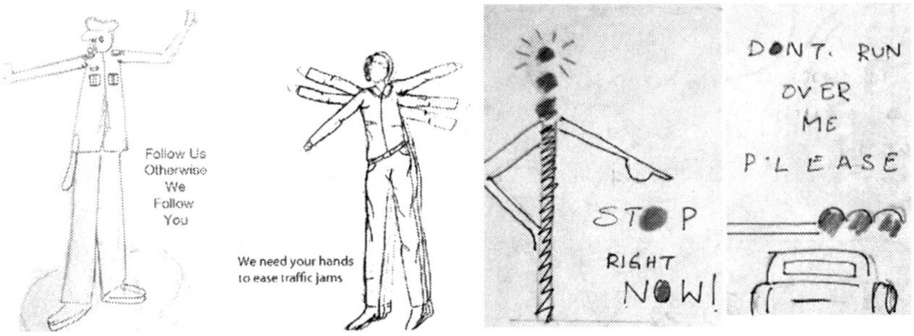

structures together with keywords. *Thumbnails* sketches are very small in size and can be quickly made to capture essence of an idea without much detailing. *Scenario generation* uses a series of sequential thumbnail sketches, visuals from other sources to narrate a situation. It uses the technique of *storyboarding* where a series of sketches are made to tell a story. *Two and three dimensional matrices* serve to hold a large amount of information within a grid structure. *2D-3D product sketches* focus on the look and feel as well as the functionality of the product.

By externalizing one's design ideas, one is also able to translate the inspirations to the problem at hand, by exploring the formal and functional aspects of the product, resolving complexities to attain simplicity, experience the notion of fit, aesthetically. Among the various methods to externalize thoughts sketching is regarded as a key tool for generation and exploration by designers. The creative and visualization tools to nurture creative thinking are discussed hereunder.

Getting Rid of Assumptions

With education and practice we get into set patterns (due to repetition in type of design projects) forcing us into routine ways of doing things. We become fixated with our approach to design and experience a feeling of stuckness. Tools such as *Empathizing*, *Experiencing products* can help us look at problems with imagination and a fresh perspective through quick primary research. Such research might look at target user, existing products, manufacturing processes etc with focus on using all our senses bringing about the tangible experience of imagination and playful attitude to the aspect of problem solving. This concept is closer to divergent thinking mode related to defocusing and activating remote associations to the related problem area and generational abilities. Observing through primary research also encourages questioning and interpretation in context of users, environment, materials and a number of other factors during the preparation phase of a design problem.

Empathizing is understanding people by being them for sometime through role play and attempting to experience some of their needs and aspirations. This is connected very closely to the Design research method Informance as stated by Johnson (2003) which begins with ethnographic research (observing people and acquiring data) and interpreting this data through empathy (seeing situations, uses and elements as the studied consumers).

For example if the task is to create awareness posters for traffic, one could look at visualizations as *thumbnail sketches, scenarios* based on role playing as a policeman, red-light etc (Figure 1). Watching, imagining oneself as a policeman, traffic signal post could bring interesting insights,

Figure 2. Experiencing products and scenario generation in context of proposing improvements in retail cart design. (Sketches by Kartik Gupta)

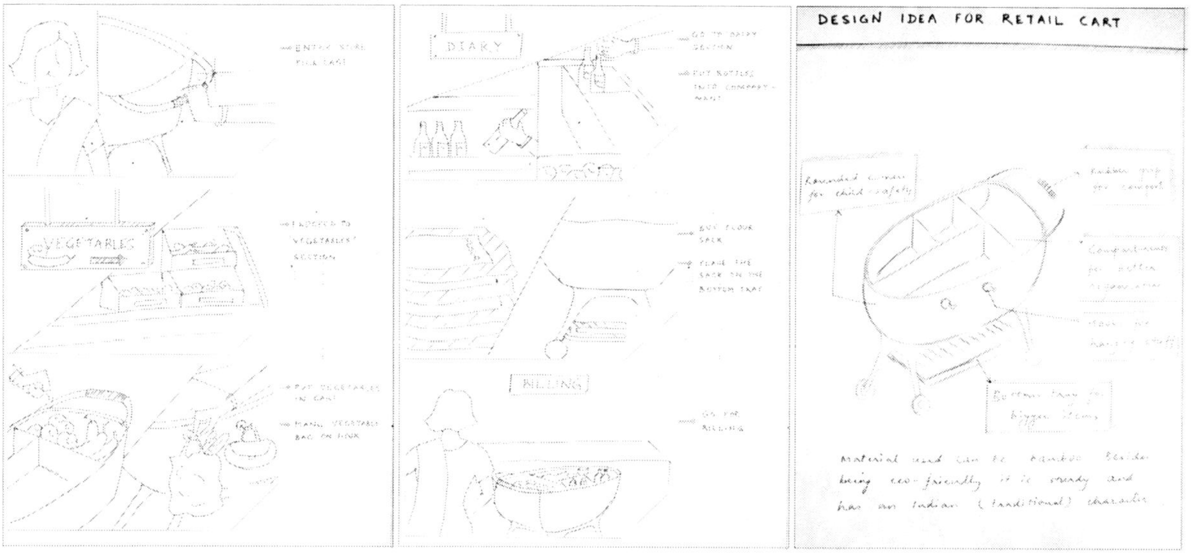

associations related to traffic, the various players and some kind of local, cultural flavor to design illustrations.

Lets now take an example related to product design that exists in market today – the problem is to design a thermometer for a kid. One can imagine being a child which might lead to revelation that younger kids are scared of a mouth-thermometer. Next is to look at thermometer concepts such as ear-based, strip-based etc.

Experiencing Products is about primary research of existing products before designing or re-designing such products. Hands-on experience with products helps one broaden focus areas of the problem and explore the performance of products by evaluating the product form & features, functionality, interaction modes etc. One can undertake certain tasks to evaluate the product or have users to perform tasks to look at how others interact with the product. One could also gain by observing the influence of local customs and habits on such product. The visualization method of *scenario generation* is very helpful in capturing some of the observations and evaluations of various tasks with a given product (Figure 2).

Restating problems is looking at problem statements or briefs from different perspectives. One can have varied statements for the same given problem. For example let's say the problem is to 'Design a memento'(Figure 3). This statement suggests and brings in visual images of existing mementos (past designs). It makes one think of a product as an object, making students to look at product forms, creates a fixation effect brought from past designs, or the fixation among designers to simply make a different looking product form. With the statement being very specific people forget to bring context and the mind is made to think about objects alone.

This problem could be restated as 'Design a gift'(Figure 3). In this statement as there are a number of gift options available connection to one or two strong visual images does not occur opening up a lot more possibilities. 'Design a memento' is a close-ended kind of problem and forces one to look at product details such as forms, structure, color etc. While the statement 'Design a gift' with

Figure 3. Sketches on product forms for 'Design a memento' and an associational map with keywords for 'Design a gift' showing the effect of problem statements on thinking. (Sketches by Girish Yelka and Ayush Srivastava)

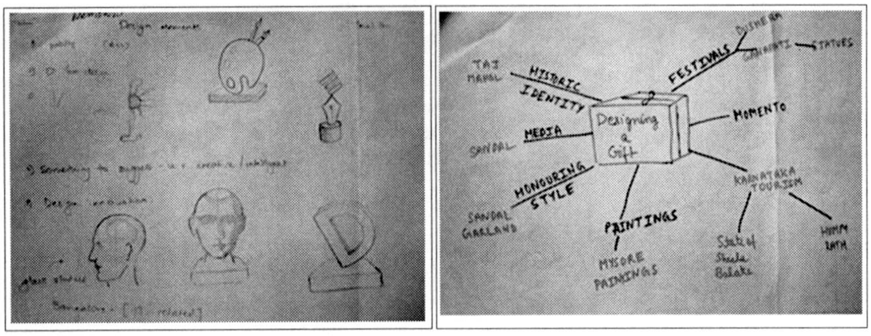

the absence of any particular visual clue makes one to ask questions like 'What are the types of gifts?', 'What is the significance of a gift?' etc to create a network of associated keywords related to context and helps one to broaden the focus. The statement being generic invokes broader level concepts or design strategies.

Any given problem statement can be restated in a number of ways under two broad categories:

1. Specific problem statement
2. Generic problem statement

Specific problem statement gives a visual cue to past practices invoking thoughts and concepts at object level such as product form, size, materials, and construction details. *2D and 3D sketches* related to product forms come naturally during visualization. Generic problem statement invokes thoughts related to the whole context and not much to do with a specific object. Such a statement brings visualization in the form of *associational maps* with keywords and textual references. It also brings in questioning, instigates information gathering due to generality of the problem statement and is very beneficial during the preparation phase of creative design process assisting problem finding and construction. In view of the problem of fixation – an inability experienced by an individual in getting away from a particular area to discover new associations (Jannson & Smith 1991), Finke et al. (1992) suggested that broadening the focus of attention might improve creativity and help one overcome the problem of fixation. Generic problem statement broadens the focus of attention for an individual. The specific design statement helps in the generational phase of the design process for specific product level ideas.

Ideational Fluency

The tools discussed earlier such as *empathizing*, *experiencing products*, and restating problems also bring in ideational fluency, but tools such as *brainstorming*, *mindmapping* and *morphological matrix* help one to deliberately create a number of ideas and network of associations triggering remote associations. The aim is to build on quantity, some of which would be qualitative.

Osborne's (1979) *Brainstorming* is a popular technique for quick generation of number of ideas without the constraint of any kind of judgment. The first phase is to aim for more quantity of ideas based on the fact that the first 20-30 ideas would be usual ones and asking for more quantity of ideas would mean triggering unusual ones. During the first phase one is to list ideas spontaneously without critical analysis of judging. The second phase is to shortlist qualitative ideas based on analysis. The visualization is mostly seen in the

Figure 4. Visual brainstorm on uses of newspaper (Sketches by Jai Shanker Singh Beban)

form of *textual listing* of ideas or as a series of *thumbnail sketches* when it is called visual brainstorming (Figure 4).

Buzan's (1995) *Mind Mapping* is a technique of communicating one's thinking by use of keywords and images connected with a web-like network of associated connections as spontaneously as possible. Mind map (Figure 5) categorizes associations and the use of word-images makes it very memorable for reflective analysis. It sets the mind free from linear thinking.

Morphological matrix (Figure 5) by Swiss astrophysicist Fritz Zwicky is a tool for structuring thinking within a *two dimensional matrix*. It helps one to bring the complexity of information into smaller manageable segments by placing them within a grid structure. It reveals patterns for combinatorial thinking by combining two or more ideas from each grid in newer ways. It serves as a map to chart all kinds of observations and thinking in a structured and crisp format.

Bringing Deliberate Shifts

There are a number of unwritten rules we observe in our regular activities which we tend to stick to. Such rules do help us in getting through our daily needs/activities with ease but fail to address changing trends in lifestyle. When a new problem arises these rules may not be enough to get to a

Figure 5. Morphological matrix charting the components associated with Digital Design and a mind map showing thoughts on retail cart design. Both reduce complexity through categorization (Sketches by Himanshu Suthar and Kartik Gupta)

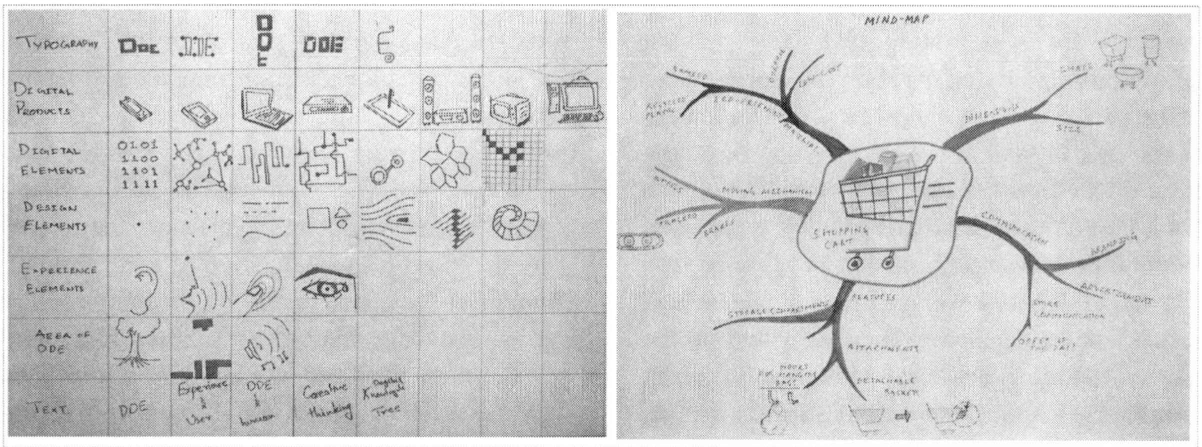

solution. *Analogical thinking* and *Break the rules* when used deliberately bring about greater shifts in thinking.

Analogical thinking is a thinking strategy wherein ideas are generated by making associations between objects that seem to be unconnected. Ball & Christensen (2009) believe "Analogical reasoning involves accessing and transferring previously acquired knowledge of objects, attributes and relations to support current problem solving and decision making activities".

Similies, metaphors and analogies are thinking strategies in which one seeks connections deliberately between a given problem and other objects or concepts. Analogical thinking involves drawing a connection from a known concept referred as *source* to the problem at hand where it could be applied referred as a *target*. All of them provide a comparative platform that can highlight similarities between the source and the target domain, thus bringing in unique insights for problem solving. With similies the term 'likeness' is used to draw connection between two objects or concepts, in metaphor the connection is such that one object is considered as if it were the second object being associated, making this connection seem very direct. Analogies seem to look at functional/structural resemblance to associated objects rather than looking at physical aspects.

Synectics proposed by Gordon (1961) is a thinking process where analogies are used to stimulate thinking for problem solving. This technique looks at gaining insights by making familiar strange and strange familiar. For example in the case of Sydney Opera House the architect Utzon had to look at the construction of a new form of shell roof, a case of strange problem or not a familiar problem. The solution which was inspired by orange peel came from a familiar object showing how a strange problem was solved by seeking connection to familiar object. In the case of the construction of Millennium Bridge, London the inspiration for the design of a bridge, a familiar object came from eyelid making it a strange bridge, an example for making the familiar strange.

While engaged in a design problem it is very difficult to imagine what might be a good source for drawing an analogical connection. It's a common practice among a lot of designers to browse through visual resources (magazines, books, visual diary, web resources, and surroundings) as preparation of ground work for design project, a natural urge to connect to probable stimuli. Gordon defined four types of analogies which help one to think about various sources. *Personal analogy* is when one can analogize by personifying the problem as though it was human, for example the design of a lighting fixture one can gain insights by imagining 'What is it like to be a lighting fixture?' *Direct analogy* looks at problems from one domain, for application in another domain like in the case of Georges de Menstral who invented Velcro noticing plant burrs on his dog's fur. Here nature, keen observations can act as sources. *Symbolic analogy* defines relationship where elements of the problem are seen as mental images. This looks at defining the problems by visual and pictorial thinking. Example imagine a blender that is super silent. *Fantasy analogy* looks to draw connections through visuals and scenarios that are not realistic or may be futuristic.

Analogical thinking tool is difficult to be practised unless one is deeply involved with the problem, only being aware of this tool will not achieve any connections. Initially deliberate efforts are needed on the part of an individual to use this tool and very soon one starts to sense insights from surrounding environment. One such deliberate exercise required picking words such as 'stuck', 'conflict', 'movement' and mapping a relationship based on experiences associated with the word for visualization of ideas through *thumbnail sketches* (Figure 6). In this case the given word is the target with visual action mapped by imagining the word as a being in action. This exercise leads to some kind of *force-fit* from the action to the word. *Force-fit (Forced Connection)*

Figure 6. Force-fit: mapping relationships visually for given words. (Sketches by Kartik Gupta and Jai Shanker Signh Beban)

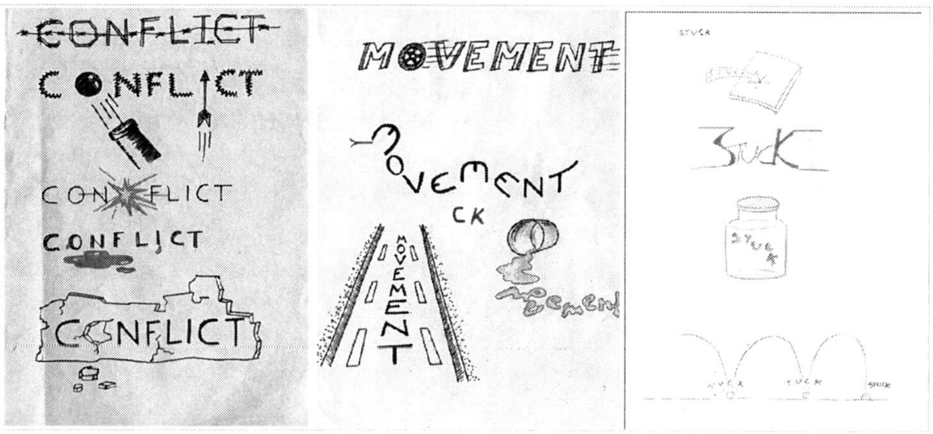

is combining two or more unrelated things to bring out a new message that could be surprising, and original as seen in the poster ideas for global warming and education and literacy (Figure 7).

Associations that are made using analogical thinking will happen only if one is totally immersed in the given problem area, when objects and concepts within our environment begin to suggest connections. For example let's say one is thinking about creating awareness on driving while drinking coffee in the cafeteria. One can draw a connection between a 'Human head' being as fragile as 'ceramic cup', and such a connection will only happen if a person has been thinking about the problem.

In terms of the distance of the source from the target domain, we have near or far analogies. In *Near analogy* the relationship is retrieved from a domain similar to target, while in *Far analogy* the retrieval is from a domain that is very different to target domain. For design ideas on lighting fixture an example for *Near analogy* could be a 'candle stand holder'; while the 'firefly' is an example of *Far analogy* as the distance from the domain of

Figure 7. Analogical thinking: Poster ideas for awareness on global warming, education & literacy (Ketches by Himanshu Suthar and Kartik Gupta)

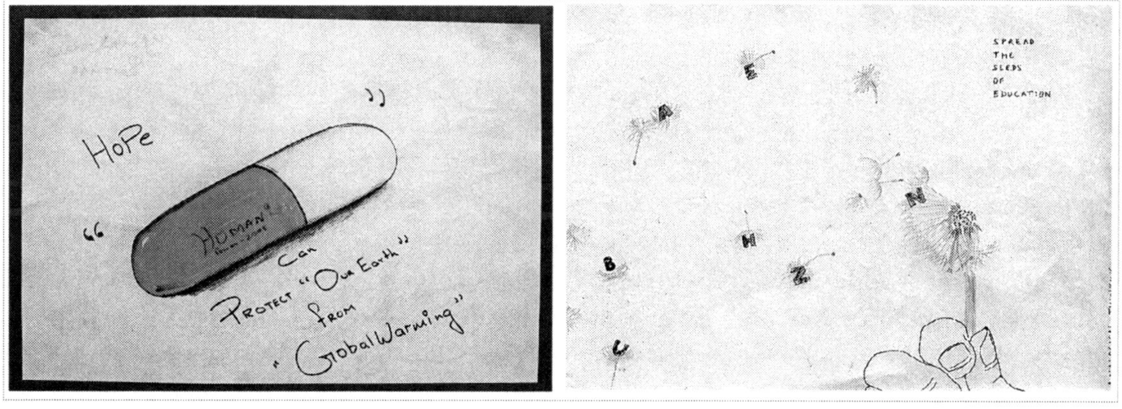

lighting is farther apart. Design ideas based on Far analogies are regarded to be more novel and surprising than those based on Near analogies. Tseng, Moss, Cagan & Kotovsky(2008) have found that Far analogy impacted idea generation more when given with an open goal to solve the problem.

Till now we have mostly looked at analogizing as a tool for solution generation and here it is found to be a difficult tool to apply. Christensen & Schunn (2007) have proposed that analogizing and mental stimulation can serve as strategies to resolve uncertainty in design problems. They have shown that analogies serve three purposes:

- Problem identification (noticing a possible problem in the emerging design)
- Solution generation (transferring possible solution concepts from source to target domain)
- Explanation (using a concept from the source domain to explain some aspect of target domain to members.

Use of metaphors in design needs a lot of skill in terms of translation from the source to the target. For example Sydney Opera house is based on ocean forms, but does not literally represent any of them. There are also examples of buildings which look like sheeps, ducks, etc and are not regarded as novel as they are lacking translation from the architect designer's side and seem too literal. It should be noted that metaphors are used commonly in our day-today language like "Computer is down", "She gave a cold stare" although our perception is that it is used by great writers and in literary work. Like analogies metaphors can assist designer as models that help us take our experience from familiar, concrete object onto unknown or abstract concepts or things giving the unknown concepts structure and meaning (Ericksom 1990). Saffer (2005) underlines the importance of metaphor for a designer where naming and framing often happen via metaphor and it can be used from the initial problem solving to delivery for the creation of products and services. He regards this as useful for a designer as designers are often thrust into domains where they have little or no background.

Coyne & Snodgrass (1995) emphasize the role of metaphors in design process as design models such as *Rationalism*, *Romanticism* and *Pluralism*. These models if kept in the mind are so powerful that they bring about a paradigm shift among designers and the design process. *Rationalism* is about clarity, formulating appropriate methods and models with a desire for simplicity, plain thinking and common sense and resistance to subjectivism. *Romanticism* is characterized by a dependence on imagination, intuition, emotion, feeling and in this tradition design is commonly promoted as an art. *Pluralism* is detachment from the above models with basis in philosophical ideologies.

Break the rules is another tool, the deliberate use of which can bring in radical approaches to addressing a problem and can free one from routine approaches, but is one of the toughest tools to be put into practice. To apply this tool one needs to"

- be informed about the current trends and practices,
- then challenge some such rules,
- decide on which one or more of these rules needs to be changed.

For example let us say the task is to come up with contact management application for a mobile phone. A review of existing products in this family will reveal that contact management is based on saving first, last name of people. One could challenge this rule based on the fact that people remember contacts not only by names but as well by their relationship (family, friend, friend's mother, service provider), location (place, organization). Such a vision will bring out a radically different contact management product in terms of interaction. *Break the Rules* does not mean doing something illegal and is a strategy to break away from routine modes of thinking.

To begin with one can break routine habits such as taking a new route, visiting places in the city one has not been to, watching a performance maybe of style and form not earlier watched, talking to people from different walks of life etc. This makes one notice new things and record a number of perspectives. It also motivates one to be more curious about any given task or activities. Apple's products which set a new trend can be stated as examples of such a tool being operated.

Analogical thinking, *Force-fit* and *Break the Rules* can help one generate ideas that are diverse, surprising i.e qualitative ideas. These tools are very difficult to be applied in the context of design problems and deliberate effort is needed in the beginning. But application of these tools to explain unfamiliar design problems using familiar concepts comes in very naturally and is helpful in terms of problem identification and construction.

Recording and Reflection

Ideas and associations come in very fast and disappear quicker as well. One needs to maintain a handy diary to quickly capture and record emerging associations that occur while observing people, browsing book, travelling etc. This could be the place for quick sketches, preparing image collage, explorations/experimentation, reflection on thoughts etc. This should help one to activate a number of nodes while thinking or in other words divergent thinking. Overtime one will be able to create unforeseen associations within our day-to-day environment.

Experimental Evaluations

Thumbnail sketches, scenario storyboards, drawings, making mock-up models are means of visualizing design ideas. It's a common practice among students to first sketch design ideas, then prepare drawings and models to communicate their ideas or present ideas without much evolution of ideas. Here sketches, drawings and models are used as methods to present an idea. As a number of practicing designers do, one needs to look at all these various methods for evolution and evaluation of design ideas as well.

CONCLUSION

Creativity is not limited to design profession alone. Creativity can be nurtured by being aware of characteristics of design problems, the various phases of creative process, use of creative and visualization tools. Creative tools act as aids for design ideas, while visualization tools help in externalizing these design thoughts for evolution and communication, with both (creative and visualization tools) being complementary to each other. It must be noted that inspirations from nature, observing users, culture serve as triggers for creative ideas. It is very important to capture creative ideas as many of them could be ethereal. Many of the creative and visualization tools will be internalized with usage over longer timeframe and across various design projects. It is essential to look at aspects of nurturing creativity by looking at thinking modes, creative tools, and creative processes as a whole in context of current design education where much focus is placed only on studio-based learning. Providing creativity-relevant skills and being aware of the close relationship between 'Design thinking' and 'Creative thinking' would be important for designers. There is further scope for research on 'creativity' within the context of design discipline, particularly in the larger interest of promoting informed design education and not looking at 'creativity' as either a talent born with or as a mysterious process.

ACKNOWLEDGMENT

I would like to thank the students of Design for Digital Experience, a PG programme offered at

National Institute of Design (NID), Research and Development Campus, Bangalore, for letting me use their sketches (individual names mentioned under figures) which came as a part of Creative Thinking and Design process course offered by the author. I would also like to thank the various faculty members of NID with whom I had fruitful discussions and gained insights while writing this chapter.

REFERENCES

Amabile, T. M. (1983). *The Social Psychology of Creativity*. New York: Springer-Verlag.

Atman, C. J., Chimka, J. R., Bursic, K. M., & Nachtmann, H. L. (1999). A comparison of Freshman and Senior Engineering Design Processes. *Design Studies*, *20*(2), 131–152. doi:10.1016/S0142-694X(98)00031-3

Ball, L. J., & Christensen, B. T. (2009). Analogical Reasoning and Mental Simulation in Design: Two Strategies linked to Uncertainty Resolution. *Design Studies*, *30*(2), 169–186. doi:10.1016/j.destud.2008.12.005

Boden, M. A. (1998). Creativity and artificial intelligence. *Artificial Intelligence*, *103*, 347–356. doi:10.1016/S0004-3702(98)00055-1

Buzan, T., & Buzan, B. (1995). *The Mind Map book*. London: BBC Books.

Christensen, B. T., & Schunn, C. D. (2007). The Relationship of Analogical Distance to Analogical Function and pre-inventive Structure: The Case of Enginerring Design. *Memory & Cognition*, *35*(1), 29–38.

Christensen, P. R., & Guildford, J. P. (1958). *Creativity/fluency scales*. Beverly Hills, CA: Sheridan Psychological Services.

Christiaans, H., & Dorst, C. (1992). Cognitive Models in Industrial Design Engineering: a protocol study. In Taylor, D. L., & Stauffer, D. A. (Eds.), *Design Theory and Methodology – DTM92*. New York: American Society of Mechanical Engineers.

Corbolan-Berna, F. J. (1992). Creativity as a cognitive style. *The Journal of Creative Behavior*, *26*(3), 163–164.

Coyne, R., & Snodgrass, A. (1995). Problem Setting within Prevalent Metaphors of Design. *Design Issues*, *11*(2), 31–61. doi:10.2307/1511758

Cross, N. (2006). *Designerly Ways of Knowing*. London: Springer-Verlag.

Csikszentmihalyi, M. (1996). *Creativity: Flow and the Psychology of Discovery and Invention*. New York: Harper Collins Publishers.

de Bono, E. (1992). *Serious creativity*. London: Harper Collins Publishers.

Derbaix, C., & Vanhamme, J. (2003). Inducing Word-of-Mouth by Eliciting Surprise – a Pilot Investigation. *Journal of Economic Psychology*, *24*(1), 99–116. doi:10.1016/S0167-4870(02)00157-5

Dudek, S., & Cote, R. (1994). Problem finding revisited. In Runco, M. A. (Ed.), *Problem finding, problem solving, and creativity* (pp. 130–150). Norwood, NJ: Ablex.

Eastman, C. M. (1970). On the Analysis of Intuitive Design Processes. In Moore, G. T. (Ed.), *Emerging Methods in Environmental Design and Planning*. Cambridge, MA: MIT Press.

Erickson, T. D. (1990). Working with Interface Metaphors. In Laurel, B. (Ed.), *The Art of Human-Computer Interface Design*. Reading, MA: Addison-Wesley.

Finke, R. A., Ward, T. B., & Smith, S. M. (1992). *Creative Cognition: Theory, research, and applications*. Cambridge, MA: MIT Press.

Freud, S. (1949). *An outline of psychoanalysis.* New York: Norton.

Fricke, G. (1993). Empirical Investigations of Successful Approaches When Dealing with Differently Precised Design Problems. *International Conference on Engineering Design ICED93*, Heurista, Zurich.

Fricke, G. (1996). Successful Individual Approaches in Engineering Design. *Research in Engineering Design, 8*, 151–165. doi:10.1007/BF01608350

Fung, A., Lo, A., & Rao, M. N. (2005). *Creative Tools*. Hong Kong: The School of Design, The Hong Kong Polytechnic University.

Gordon, W. J. J. (1961). *Synectics, the development of creative capacity*. New York: Harper and Row Publishers.

Guilford, J. P. (1950). Creativity. *The American Psychologist*, *5*, 444–454. doi:10.1037/h0063487

Hertzberger, H. (1991). *Lessons for students of architecture*. Rotterdam, UK: Uitgeverij 010.

Howard-Jones, P. A., & Murray, S. (2003). Ideational productivity, focus of attention, and context. *Creativity Research Journal*, *15*(2&3), 153–166. doi:10.1207/S15326934CRJ152&3_07

Jansson, D. G., & Smith, S. M. (1991). Design Fixation. *Design Studies*, *12*(1), 3–11. doi:10.1016/0142-694X(91)90003-F

Johnson, B. M. (2003). The Paradox of Design research – The role of Informance. In Laurel, B. (Ed.), *Design Research Methods and Perspectives*. London: The MIT Press.

Jordan, P. W. (1999). Pleasure with products: Human factors for body, mind and soul. In Green, W. S., & Jordan, P. W. (Eds.), *Human Factors in Design: Current Practice and Future Trends*. London: Taylor & Francis.

Kelley, T., & Littman, J. (2001). *The art of innovation: Lessons in creativity*. New York: Broadway Business.

Lawson, B. R. (1979). Cognitive strategies in architectural design. *Ergonomics*, *22*(1), 59–68. doi:10.1080/00140137908924589

Li, Z. (1998). Creativity and aesthetic sense. *Creativity Research Journal*, *11*(4), 309–313. doi:10.1207/s15326934crj1104_4

Lindholm, C., Keinonen, T., & Kiljander, H. (2003). *User interface styles. Mobile Usability: How Nokia Changed the Face of Mobile Phone*. New York: McGraw-Hill.

Lubart, T. I. (1997). Emotion, metaphor, and the creative process. *Creativity Research Journal*, *10*(4), 285–301. doi:10.1207/s15326934crj1004_1

Lubart, T. I. (2000-2001). Models of the creative process: past, present and future. *Creativity Research Journal*, *13*(3&4), 295–308.

Ludden, G. D. S., Schifferstein, H. N. J., & Hekkert, P. (2008). Surprise as a Design Strategy. *Design Issues*, *24*(2), 28–38. doi:10.1162/desi.2008.24.2.28

Ludwig, A. M. (1995). *The Price of greatness*. New York: Guilford.

Martindale, C. (1995). Creativity and connectionism. In Smith, S. M., Ward, T. B., & Finke, R. A. (Eds.), *The Creative Cognition approach* (pp. 249–268). Cambridge, MA: MIT Press.

Mayer, R. E. (1999). Problem Solving. In Runco, M. A., & Pritzker, S. R. (Eds.), *Encyclopedia of Creativity* (*Vol. 2*, pp. 437–447). San Diego: Academic Press.

Mckim, R. W. (1980). *Experiences in Visual Thinking* (2nd ed.). Monterey, CA: Brookes/Coles Publishing Co.

Norman, D. A. (2004). *Three levels of Design: Visceral, behavioral, and reflective. Emotional Design – Why we love or hate everyday things.* New York: Basic books.

Ochse, R. (1990). *Before the gates of excellence: The determinants of creative genius.* New York: Cambridge University Press.

Osborne, A. F. (1979). *Creative collaboration by groups. Applied Imagination: Principles and Procedures of Creative Problem Solving.* New York: Charles Scribner's Sons.

Pellegrini, A. D. (1984). The effects of exploration and play on young children's associative fluency. *Imagination, Cognition and Personality, 4*, 29–40. doi:10.2190/0W5N-X8E8-WKWF-9A1R

Purcell, A. T., & Gero, J. (1991). The Effects of Examples on the Results of Design Activity. In Gero, J. (Ed.), *Artificial Intelligence in Design AID91.* Oxford: Butterworth Heinemann.

Purcell, A. T., & Gero, J. (1996). Design and Other Types of Fixation. *Design Studies, 17*(4), 363–383. doi:10.1016/S0142-694X(96)00023-3

Purcell, A. T., & Williams, P. (1993). Fixation Effects: Do they exist in design problem solving? *Environment and Planning. B, Planning & Design, 20*(3), 333–345. doi:10.1068/b200333

Rao, M. N., Lo, A., & Fung, A. (2007). Nurturing students to think creatively in design education. In Katiyar, V. S., & Mehta, S. (Eds.), *Design Education, Tradition and Modernity.* Ahmedabad: National Institute of Design.

Saffer, D. (2005). *The Role of Metaphor in Interaction Design.* (Thesis), Master of Design in Interaction Design, The School Of Design. Pittsburgh, PA: Carnegie Mellon University.

Schon, D. A. (1979). *Generative Metaphor: A perspective on problem-setting as a social policy. Metaphor and Thought.* Cambridge, UK: Cambridge University Press.

Simonton, D. K. (1984). *Genius, Creativity and Leadership: Historimetric Enquiries.* Cambridge, MA: Harvard University Press.

Sternberg, R. J. (1999). The concept of creativity: prospects and paradigms. In Sternberg, R. J. (Ed.), *Handbook of Creativity.* New York: Reinhold.

Tiger, L. (1992). *The pursuit of pleasure* (pp. 52–60). Boston: Little, Brown and Company.

Torrance, E. P. (1974). *The Torrance tests of creative thinking.* Bensenville, IL: Scholastic Testing Service.

Tseng, I., Moss, J., Cagan, J., & Kotovsky, K. (2008). The Role of Timing and Analogical Similarity in the Stimulation of Idea generation in Design. *Design Studies, 29*(3), 203–221. doi:10.1016/j.destud.2008.01.003

Vincent, A. S., Decker, B. P., & Mumford, M. D. (2002). Divergent Thinking, Intelligence and Expertise: A Test of Alternative Models. *Creativity Research Journal, 14*(2), 163–178. doi:10.1207/S15326934CRJ1402_4

Wallach, M. A. (1970). Creativity. In Mussen, P. H. (Ed.), *Manual of child Psychology* (pp. 1211–1272). New York: Wiley.

Wallas, G. (1926). *The art of thought.* New York: Harcourt Brace.

Chapter 9
Virtual Reality Systems for Industrial Design Application

Alice Pignatel
Politecnico di Milano, Italy

Fausto Brevi
Politecnico di Milano, Italy

ABSTRACT

3D digital models are an essential tool that supports every step of the industrial product design process. For example, during the design review step photorealistic models are used as virtual prototypes for aesthetic and ergonomic evaluation of new products, in order to anticipate design evaluation and choices and to shorten the overall process. For this reason, industrial design enterprises usually have special rooms equipped with a projection based large display, for the realistic visualization of digital models. Usually, the interaction with 3D models is still based on mouse and keyboard, even if this kind of input is not fully adequate to the interaction needs concerning such special application. The aim of this chapter is to investigate different interaction techniques and outline their benefits and drawbacks compared to the interaction tasks required in this special field.

3D DIGITAL MODELS IN THE INDUSTRIAL DESIGN PROCESS

Today, Industrial Design is a key factor to get competitive advance over market competitors. In fact, the role of the ID is to understand the needs of the potential user in order to create more attractive and useful objects. In the global market, the fundamental issue for a successful product is the shortening of time-to-market.

In this context, 3D digital models are a powerful tool that can support all the steps involved in the Industrial Design process for new products development. In fact, digital models are quicker to prepare than physical *maquettes* and they allow faster changes to the original shape. Moreover, they are less expensive and easier to store. Thus, digital models can sometimes replace the physical models in the Industrial Design process. For example, 3D digital models can be used as *virtual prototypes* to check both technical and aesthetic qualities of

DOI: 10.4018/978-1-61520-617-9.ch009

the industrial product, in order to anticipate the evaluation step of the process.

A virtual prototype is a 3D model that reproduces the features of the real object that must be developed. When a specific issue is under evaluation, it is necessary to set up a system that allows to interact with the virtual object exactly in the same way as if it was real. For instance, during the so called *Design Reviews,* when the aesthetic of a new product is under evaluation, it is important to display the digital model as it would really appear, with realistic looking materials and with the possibility of moving all around the object to consider every detail. This activity is crucial to achieve high quality products, because all the issues related to the final user's perception of the product are discussed by the design project team. In fact, these discussions are generally more focused on generic subjects rather than on specific technical problems. Usually the chief designer, or project leader, shows to other people attending the design review the product details and features, in order to check the new product before proceeding to further development steps. People attending Design Reviews usually belong to different production departments, not only to the design workgroup. In fact, while discussing the shape and the object features, it is important to consider all the issues involved during the life cycle of the new product, in order to avoid or at least to correct design errors, saving time and money.

During the Design Review it is necessary to display highly realistic looking 3D models, and also to perform some basic interaction task in order to present the best visualization of the different product parts. For this reason the Industrial Design Companies attention towards real-time rendering software s always increasing. This kind of software allows not only to visualize digital objects in a realistic way, but also to show product and material variation in order to evaluate different product configurations. Moreover, the virtual camera can be interactively displaced, according to user's visualization needs.

For this reason, an always increasing number of private Industrial Design Industries built "Virtual Theatres" fully dedicated to design review activities.

TYPICAL VIRTUAL ROOM SET-UP AND RELATED ISSUES

A Virtual Room is expensive and space consuming, because it needs special equipments that must be installed in a large room fully dedicated to 3D models visualization purposes. Thus, virtual rooms are a convenient investment when dealing with highly complex products with a long life cycle.

Usually they are equipped with large screen that allows the visualization of the virtual object at the real scale. People attending to the review need to move freely in front of the screen; for this reason retro-projection techniques are used and projectors are connected to one or more workstations, according to the needed rendering performances, equipped with the high end real time rendering software. Moreover, additional devices allow stereoscopic visualization of the digital models in order to improve the realistic look of the virtual prototype.

Usually ten or fifteen people attend to the design review and their seats are placed right in front of the large screen, while the workstation is placed on their back, in a room corner, so that it is scarcely perceived during the work session.

Figure 1. A typical virtual room set up

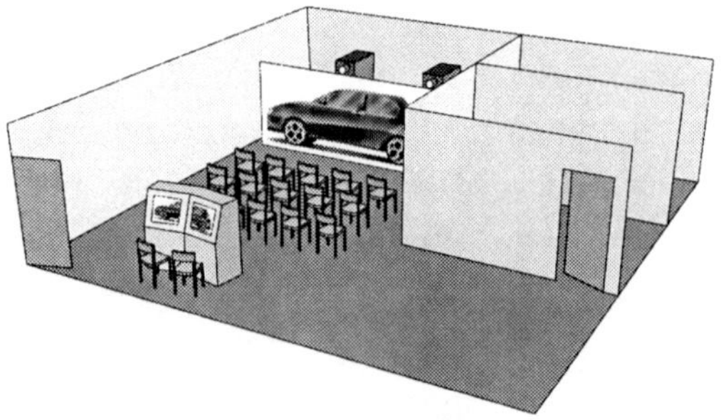

The input devices are usually traditional tools, such as mouse and keyboard, directly attached to the workstation far away from the screen. Advanced interaction systems, such as 3D mouse or haptic systems, used in Virtual Reality environments are not largely implemented. The main reason is that this special application of virtual reality requires a lower immersion feeling compared to the traditional virtual reality environments visualization. Indeed, a typical design review requires a highly photorealistic look of shapes and materials, but on the other hand, the interaction with objects displayed on the digital scene is very poor and restricted to a few product features. On the contrary, the main interaction task performed during the review is the manipulation of the virtual camera, to change the point of view in order to examine every detail of the product.

During the design review, the project leader stays near the screen, pointing out the product features and discussing with other contributors. Therefore, the design review in a virtual room requires an additional qualified technician to execute all the interaction tasks with digital scene as the review activities leader may ask for from his location near the display device.

In this scenario a direct and effective interaction with virtual prototypes would be a basic need, but the interaction issue remains an unsolved problem. In fact, this typical virtual room set up causes some "relation" problems that reduce the effectiveness of virtual design reviews: the project leader needs to interact at the same time with his audience, to discuss about design issues, and with the operator at the workstation, to give instructions for the right camera placement while asking for changes on the displayed scene. This situation is really annoying, because the project leader and the operator at the workstation are in opposite corners of the room, while other people attending to the review are right in the middle.

To provide a solution to this problem, it is possible to build an interaction tool that could be used directly by the project leader during the design review. Thus, it would be possible to decrease the flow of information not directly relevant to the project and improve the quality of communication. This could be achieved, for example, with an additional input device connected to the workstation, at least for the execution of basic interaction tasks.

CASE STUDIES

This chapter will discuss three different solutions to manage basic interaction tasks with 3D digital models on large display based systems. In general, the interaction device must be small and lightweight, so that it could be easily handled during the design review. Moreover, a wireless connection systems is required, in order to provide the necessary freedom of movement to the chief designer. Finally, the device must be very easy to use and to understand, in order to help the review activity and not to disturb it.

Interaction by Means of a Graphical Interface

The first case study describes an interaction system based on a small device with a graphical interface.

The whole system is based on two components: a Pocket PC that transmits data to the workstation and an accelerometer fixed to the handled device, that determines the user's hand movements.

Several issues influenced the choice of a Pocket PC as a key tool for 3D model interaction. First of all the small size and weight, that allow to easily handle the device and also to place it in a pocket when two-handed tasks are required. Moreover, it is a fairly popular device, well known and quite inexpensive. Another important characteristic is the large touch screen display that allows direct selection of the commands displayed by means of graphic interface. Finally, this device is built for easily exchange data with a workstation, thanks

Figure 2. The pocket PC and the accelerometer

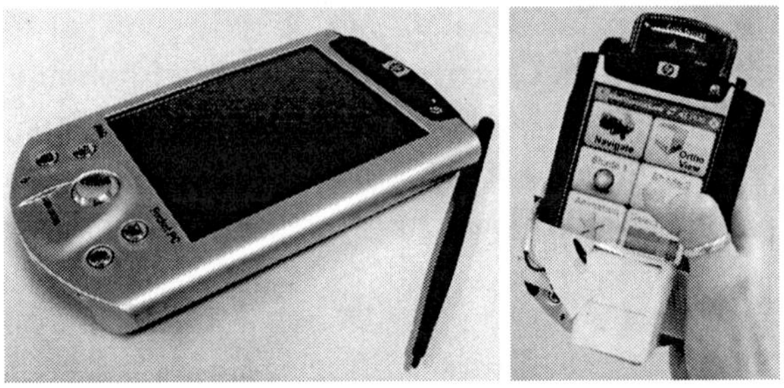

to data exchange protocols an the high operating system compatibility.

We used an accelerometer to extend the input of the handled device. The sensor is very small and lightweight and can be easily fixed on the top of the Pocket PC; in this position the sensor does not interfere with the user's activity and at the same time, it allows to obtain the best leverage length. We used the accelerometer to detect the inclination of the device compared to the vertical direction: this information is used as input to control the camera movement on the virtual scene. We ignored horizontal movements because the sensor reported high measure errors; thus, it was impossible to have an effective control over the device.

The whole system is based on the data exchange between the workstation, the Pocket PC and the sensor node. In order to avoid delays in the information exchange, all the incoming data are processed directly by the workstation, that is the device with the larger computing power. The sensor node sends a continuous stream of data via Bluetooth transmission protocol, with a speed of about 30 packages per second. For the Pocket PC a different communication protocol was used, to prevent interferences with the channel already used by the sensor node. In this case the device sends discrete packages of information via WiFi network, using TCP/IP protocol. The workstation receives both signals coming from the external devices and translates them into commands sent to the real-time rendering software. A *middleware* called Senie (Sensor Network Integration Environment) provides data interpretation thanks to specific drivers for decoding data stream: the input from the external devices are translated into shortcut keys for the visualization software that must be previously configured.

Table 1. List of commands implemented on the interaction device with graphical interface

Camera controls	
Rotate	●
Pan	●
Zoom	●
Focus on parts	●
Default views	●
Ortho views	●
1:1 scale	●
Model handling	
Geometry variation	●
Material/colour variation	●
Movements	○
Parts selection	●
Hide/isolate model parts	●
Others	
Link to external files	●
Light variation	●
Environment variation	○

This system reproduces the input coming from mouse and keyboard, but with a slightly different interaction paradigm. Thus, the system is not linked to a specific software and can be easily adapted to different visualization solutions.

Before setting up the whole system it was necessary to determine which features had to be implemented on the device interface. We decided to implement the commands to manage the virtual camera and some basic command for the control over the digital model. Complex features were not implemented at this time.

Once the main features were chosen, they were divided in groups of about 6-8 homogeneous elements in order to obtain a hierarchical organization for the whole system. The number of elements per each group was fixed considering two issues: on one hand the small dimension of the Pocket PC display and on the other hand the maximum information that could be recognized at one time by the user. In fact, the buttons on the display must be very easy to reach by user's finger and, at the same time, the user must be able to immediately recognize the right feature between the available commands. Thus, the system was organized with a main window with six buttons pointing to different sub-menus. Every sub-menu is organized with eight buttons: seven commands plus a "back" button on the lower right corner to return to the main window.

The graphic interface was build in order to maximize the ease of use and to minimize the time necessary to perform the requested tasks. First of all, all the buttons were placed in a regular grid, for a faster memorization of the commands position. Furthermore, we used a combination of text and icons to clearly identify features on the device display.

The interaction device was tested in the Virtual Room of the Virtual Prototyping Lab at Politecnico di Milano. A group of about ten users tested the device during simulations of typical design reviews: every user was asked to guide the review activity; afterwards, every user filled in a questionnaire for the device evaluation. The questionnaire is composed of two sections. The first part is focused on the general appearance of the device and its interface, while the second section is focused on the navigation tasks.

The test results show that the main problem of the device is the scarce control over the virtual camera, while the graphic interface is very helpful during the review activity. In fact, users showed to appreciate the graphic interface and had no problems in recognizing the commands. Moreover, the number of features implemented allows to have a complete and satisfactory control over the digital scene and the objects displayed on the screen. The only problem noticed by users is

Figure 3. Some results from the usability test: (a) overall evaluation of the interface; (b) learning time; (c) interaction mode evaluation

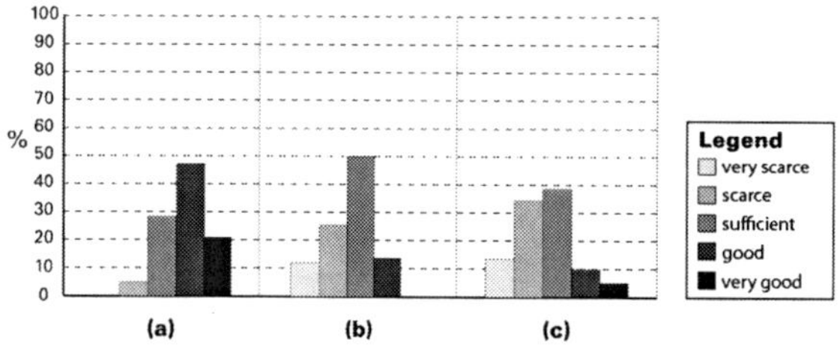

the lack of feedback on the command selection, that sometimes avoid to understand if the finger pressure over the touch screen was right or not.

The navigation tasks, on the opposite, gained very different rating. The ROTATE command got a high rating: most of the users were able to manage the camera rotation after few seconds, even if the rotation speed needs to be reduced. The high motion speed is the main problem also for the PAN command: users loose completely the control over the camera and they are forced to select a default view of the object. Anyway, the most difficult task to perform is to zoom the camera: users had great difficult not only in managing the ZOOM command, but also in understanding the right hand movement.

Thus, the overall evaluation of the navigation tasks is not satisfactory and the learning time is too long. The main problem lies in the continuous movement of the pointer on the screen, that confuses the user, preventing him to understand quickly how to master the movement of the virtual camera.

Interaction by Means of a Simple Interface

The second case study discusses a different interaction approach and it is based on a device with very simple user interface. This approach aims to reduce the learning time and to ease the interaction with digital models displayed on a large screen.

Figure 4. The wiimote: the accelerometer detects translation and rotation

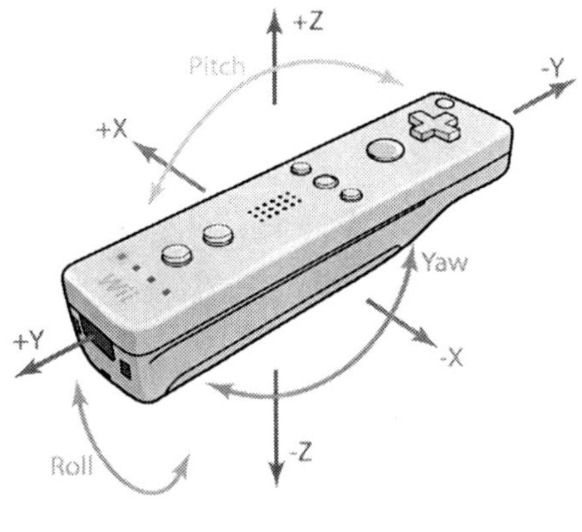

The system is based on a wireless controller of a commercial videogames console, called *wiimote*.

We decided to use this device for several reasons: first of all, the controller uses a wireless communication protocol based on bluetooth technology, so that it can easily exchange data with a common workstation. Second, the device is very small and lightweight and can be easily handled with a single hand; moreover, it can be placed in a pocket when needed. Furthermore, the device looks like a common remote controller; thus, everyone can easily and quickly understand the right way to handle it. Finally, it is equipped with an accelerometer that detects the controller orientation and transmits data to the gaming console. Moreover, the controller can be purchased separately at a very low price and no special equipment is required for the power supply.

The system is based on the data exchange between the *wiimote* and the workstation, that receives and translates the incoming information. To translate data we used a software that collects the input coming from the remote and changes into data coming from other input devices, such as mouse and keyboard. Thus, we reproduced the usual workstation input system with a different set of peripheral devices. The input from the external devices are translated into shortcut keys for the visualization software that must be previously configured, so that the system is not linked to a specific software and can be easily adapted to different visualization solutions.

Table 2. List of commands implemented on the interaction device with simplified interface

Camera controls	
Rotate	●
Pan	●
Zoom	●
Focus on parts	●
Default views	●
Ortho views	○
1:1 scale	○
Model handling	
Geometry variation	●
Material/colour variation	●
Movements	●
Parts selection	○
Hide/isolate model parts	○
Others	
Link to external files	●
Light variation	○
Environment variation	○

For this project we used GlovePIE5 (Glove Programmable Input Emulator), a free software distributed over the Internet. This application is based on a configuration script that associates the input parameters of a device with the input parameters of another device.

Before setting up the whole system it was necessary to determine which features had to be implemented on the device interface. To preserve the simplicity of the device interface, it was necessary to reduce the number of implemented features. In fact, the small number of buttons on the controller represents its main characteristic as well as the main reason why it is so easy to use. Thus, we implemented only the main commands that allow the management of virtual camera and some command for a basic level of interaction with the digital model. Moreover, a button on the controller was configured with a command that immediately stops the data stream, restoring the usual input system configuration based on mouse and keyboard.

The control over the virtual camera can be obtained with the combined use of buttons and values coming from the accelerometer: the button activates the desired task, while the hand motion determines the camera movements. Also in this case study, as in the previous one, only wrist rotation in the vertical planes were considered, while the rotation on the horizontal plane was ignored. In fact, the sensor reports a high error level for the motion values on this plane. Moreover, this direction of wrist rotation produces an unnatural movement that is not easy to perform.

At this early stage of development of the prototype, we introduced no feedback from the device. Nevertheless, the device can produce auditory, visual and "force" feedback, offering a wide range of opportunities.

This device, as the previous one, was tested in the Virtual Room of the Virtual Prototyping Lab at Politecnico di Milano. A group of about ten users tested the device during simulations of typical design reviews: every user was asked to guide the review activity; afterwards, every user filled in a questionnaire for the device evaluation. The questionnaire is composed of two sections. The first part is focused on the general appearance of the device, while the second section is focused on the navigation tasks.

The test results were absolutely satisfactory, with high average scores in all the questionnaire areas. In fact, users appreciated the ease of use of the device and they were able to perform all the required tasks during the design review, despite the small number of implemented features. Instead, the buttons layout was scarcely appreciated: in fact, it is designed to have a good gaming experience, which requires completely different tasks from a design review activity. Nevertheless, all the users were able to manage the digital camera in some few minutes: all the navigation tasks were easily

Figure 5. Some results from the usability test: (a) overall evaluation of the interface; (b) learning time; (c) interaction mode evaluation

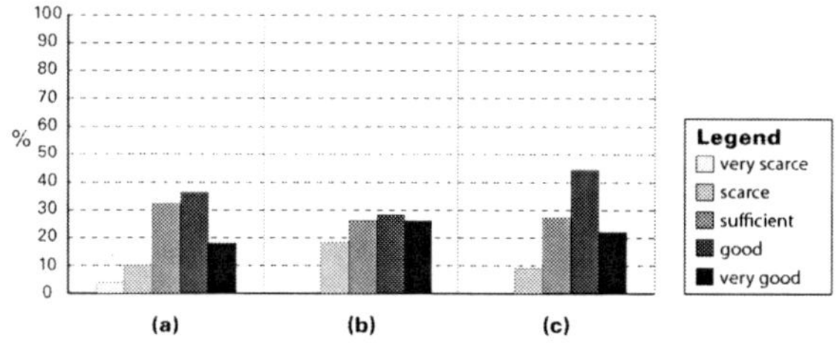

performed with high accuracy and with a very short learning time.

Interaction by Means of Hand Gesture

The third case study discusses an interaction approach based on direct interaction by means of hand gesture. The tool used for the implementation of the described system is a glove-like device, originally developed for game purposes. In this case, the glove was used to provide additional input data to the real-time rendering software that displays the digital scene on the large screen, in order to obtain a direct interaction with the virtual prototype.

Currently, there is a large number of different glove-like devices that use a variety of sensors for gesture recognition and with very different prices. For this project, we decided to reduce the overall complexity and we chose a device with a simplified architecture. We used the Reality P5 Glove, a sort of "exoskeleton" that is fastened to the back of the user's hand with a couple of rubber bands. This device can be easily adapted to fit different hand sizes, it is lightweight and can be worn for long time without causing any trouble.

The Reality P5 Glove has three independent input sets. First, the glove has a total of eight infrared position markers, used to determine the position and orientation of the user's hand: the

Figure 6. The glove-like device: The sensor detects both translation and rotation along the three axis

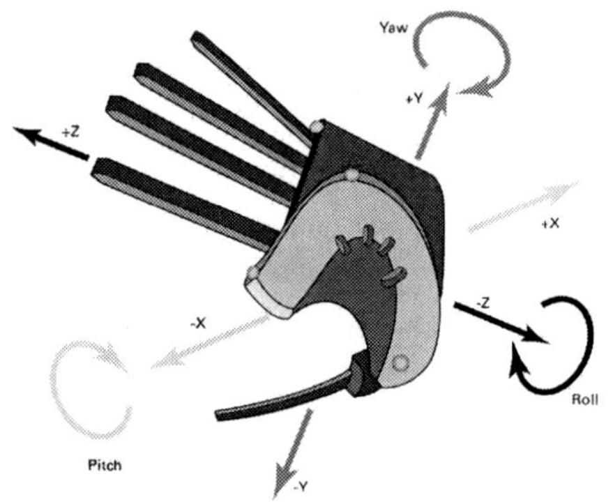

system calculates the devices position analysing the number of visible markers on the glove.

Second, the glove is equipped with bending sensors: every finger on the user's hand is covered by a rubber band that follows the finger movements. A sensor inside this element recognizes the displacement of a finger segment; thus, the system is able to detect the extension and the bending of every single finger.

These two sensors provide continuous input data flow: the information is sent to the system that is constantly updated with the current status of the device.

Finally, the glove is equipped with a set of buttons, placed on its back. These buttons are fully programmable, and they can be used to send a single information package to the system, associated with a specific action.

The system is based on the translation of input data coming from the glove into commands for the real-time rendering software. The overall structure is quite similar to the system architecture of the previous case study, so we used the same emulation software to translate incoming data into shortcut keys for the visualization software. We used the emulator GlovePIE5 (Glove Programmable Input Emulator) and compiled an "ad hoc" configuration script to obtain the proper input translation.

Table 3. List of commands implemented on the interaction device with hand gesture based interface.

Camera controls	
Rotate	●
Pan	●
Zoom	●
Focus on parts	●
Default views	●
Ortho views	○
1:1 scale	○
Model handling	
Geometry variation	●
Material/colour variation	●
Movements	●
Parts selection	○
Hide/isolate model parts	○
Others	
Link to external files	●
Light variation	○
Environment variation	○

The choice of features to implement on the interaction device was a crucial point of the process, because it was absolutely necessary to minimize the number of commands and of different gestures. In fact, a large number of gestures to remember is very difficult to manage, specially by the inexperienced user; thus, we preferred to use simple gestures that would be easier to remember, even if the device offered a wider range of opportunities. Moreover, some gestures are recognized by the system but they are really hard to perform: the ability to move every finger independently is different from person to person, therefore the capability of performing some complex gestures must be checked in advance.

For this reason, we implemented only the main commands that allow to manage the virtual camera and some command for a basic level of interaction with the digital model. Moreover, a button on the glove was configured with a command that immediately stops the data stream, restoring the usual input system configuration based on mouse and keyboard.

The buttons on the back of the hand were used to implement commands that does not require strong interaction with the virtual scene. For example, one of the buttons was used to restore the initial position of the camera and another one allows to shift from a detailed view to the general view of the whole object.

On the contrary, the commands for the interactive control of the camera were associated to a combination of finger gestures and hand motion: the system executes the desired action recognizing the finger gesture, while the wrist rotation controls the camera motion. In particular, camera rotation is activated by extending thumb and forefinger and bending all other fingers, and the wrist rotation controls the direction and amount of camera rotation. The PAN command, is activated by bending the thumb toward the palm while keeping the other fingers extended and the camera shifting is controlled moving the hand from side to side. Finally, the ZOOM command is activated by pointing the scene with the forefinger and the middle finger, while the user's hand shifting controls the amount of the magnification factor.

Once again, the device was tested in the Virtual Room of the Virtual Prototyping Lab at Politecnico di Milano. The test was organized and conducted in the same way than the previous case studies, in order to compare the three interaction devices. Also the evaluation questionnaire has the same structure: the first part is focused on the general appearance of the device, while the second section is focused on the navigation tasks.

The first test results showed interesting and unexpected issues.

Unless the small number of features implemented on the interaction device, most users were satisfied with the interaction experience level: in fact, all the main commands to control the digital

Figure 7. Some results from the usability test: (a) overall evaluation of the interface; (b) learning time; (c) interaction mode evaluation

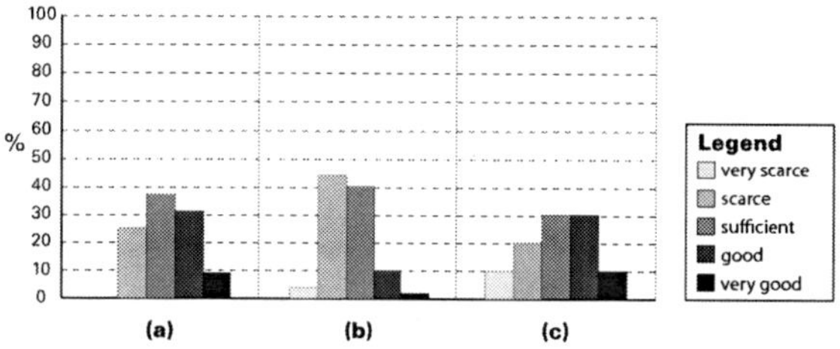

camera and to perform basic interaction tasks were available. Moreover, users reported no difficulties in understanding and remembering the gestures associated with the navigation tasks: most of them stated that the gesture associated to camera controls were absolutely natural and adequate to the connected task. However, most of the users experienced many problems in the gesture execution reporting great problems to control their hand movements. Part of the problem stays on the user's hand dimension: if the glove does not fit exactly the user's hand, the probability of inaccurate response from the bending sensor increases a lot. Moreover, even if a particular movement is easy to perform, it is still necessary to spend some time in training, in order to achieve a complete control over the virtual camera.

In conclusion, this interaction system proved to have an enormous potential; however, it showed many problems in the practical implementation. In fact, the necessary training time is definitely too long, specially for inexperienced users and it is absolutely inadequate for the basic level of tasks to be performed. Moreover, a preliminary calibration of the system would be necessary to improve the system accuracy, but sometimes it is impossible to perform calibration tests in advance. Finally, in order to prevent random camera movements, the wearable device must be continuously stopped and restarted, while using hands for different purposes. Thus, the user is forced to push repeatedly the "stop" button.

Table 4. Some results from the usability test: comparing the interaction devices

	Pocket PC	Wiimote	Glove
ROTATE: understanding	***	*** ½	****
ROTATE: perform	** ½	****	****
ZOOM: understanding	** ½	** ½	* ½
ZOOM: perform	* ½	****	**
PAN: understanding	**	*** ½	***
PAN: perform	** ½	****	*** ½
Navigation mode	***	****	** ½
Learning time	** ½	****	**
Overall evaluation	** ½	****	***

CONCLUSION

As mentioned, the evaluation tests were organized in the same way and the questionnaires intended to check about similar issues, in order to compare the different devices and to understand which one could give the better solution in the reference scenario.

The device with graphical interface got the highest rating in the questionnaire area that concerns the completeness and the clearness of the interaction device interface. In fact, the visual support in the selection task proved to be very effective and allowed users to speed up the learning process, specially for inexperienced users. Moreover, the possibility of grouping features and organizing the interface in different ways, allows to expand the opportunity to use the interaction device, even when more complex tasks are required.

As we can imagine, the device that proved to be easier to use is the interaction device with simplified interface. In fact, the buttons are always at hand and their position is easy to learn and remember. Moreover, all the available features are presented unambiguously on the device, and the user must not look for hidden features or unexpected combinations of actions.

The second section of the evaluation questionnaire checks about the ease of use for camera controls and the naturalness of the gestures required fro each interaction device.

A first interesting point emerging from the survey, is that users report similar problems in using the same feature on diverse devices, even if the satisfaction level is quite different. In general, the ROTATE command is easier to manage than the others and the associated gesture is always easy to understand. On the contrary, the PAN and the ZOOM commands are much more difficult to control, specially because the camera acceleration drives the digital model out of the user's field of view, forcing him to restore the previous view and to restart the learning process. In general, the device with the highest score in this evaluation area is the interaction tool with simplified interface. In fact, users noticed that the device based on the *wiimote* controller, thanks to its shape, is very easy to handle and the gesture of pointing and turning the wrist is very natural, specially when facing a large screen. Thus, also the learning time is very short and the user's satisfaction level became very high.

In conclusion, the interaction device that proved to adapt in the best way to the task required in the proposed scenario is the system based on the game controller. In fact, the simplified interface is very easy to learn, and the device is quite manageable, so that even inexperienced users are able to master it within a very short time.

NEW ISSUES AND FUTURE APPLICATIONS

The evaluation tests were useful not only to understand the single characteristics of the interaction devices, but also to outline their hidden potential for further application in a different scenario.

In fact, we considered a very special scenario, in which only the basic interaction tasks were needed. Nevertheless, during all the evaluation steps of the industrial design process, an effective interaction system would be a key point for shortening the time-to-market while improving the product quality.

The interaction system based on graphical interface showed some deficiencies in practical implementation. However, once problems identified by users are fixed, this system is very flexible and it offers many opportunities. First, the graphic support to command selection allows many user's type to easily access all the device information. Furthermore, the system is based on a device with autonomous computing resources; as a consequence, the interface can be easily changed and adapted to different situation, simply by introducing new features to the main structure. Thus, the same tool could be used by both trained and inexperienced users, to accomplish basic or complex interaction tasks according to the user's need.

The *wiimote*-based system proved to be perfect for a basic interaction whit digital models. However, the advantages that this device offers are also its main bonds. In fact, the small number of buttons limits the features that can be implemented on the interface. To solve this problem two strategies can be adopted: on one hand, the buttons on the device can be increased; on the other hand, combination of buttons can be used. However, both solutions increase also the complexity of the device interface, that will loose its main advantage. As a consequence, this interaction system would not adapt to a situation were complex interaction tasks are needed.

Finally, the glove-based interaction system showed a great potential, specially for the implementation of very complex interaction tasks. In fact, the most interesting issue is the complete absence of a tangible interface that gives to the user the feeling of a direct manipulation of the digital scene. Thus, also complex tasks would easily be implemented, simply by mapping them to appropriate hand gestures. However, this kind of interaction device requires a very long training time; moreover, a preliminary calibration is needed to have the best response to the user's action. For these reasons, a glove-based interaction device would be an interesting solution when the design review requires very complex tasks and it is conducted by an expert user.

We may be sure that some of the tested interaction methods perfectly adapt to the needs and tasks of the specific scenario we outlined here. Other methods, instead, proved to be more successful when applied to different fields.

REFERENCES

Assmann, E., & Ramezani, H. (2000). *The use of virtuality in car development*. San Diego: Houghton Mifflin Harcourt.

Beaudouin-Lafon, M. (2004). Designing interaction, not interfaces. In *AVI '04: Proceedings of the working conference on Advanced visual interfaces*. New York: ACM Press.

Berta, J. (1999). Integrating VR and CAD [Los Alamitos, CA: IEEE Computer Society.]. *IEEE Computer Graphics and Applications*, 19.

Bowman, D. A., Kruijff, E., LaViola, J. J. Jr, & Poupyrev, I. (2005). *3D user interfaces*. Reading, MA: Addison-Wesley.

Bowman, D. A., & Wingrave, C. A. (2001). *Design and evaluation of menu systems for immersive virtual environments*. Institute of Electrical and Electronics Engineers Computer Society.

Bretzner, L., & Lindeberg, T. (1998). Use your hand as a 3D mouse, or, relative orientation from extended sequences of sparse point and line correspondences using the affine trifocal tensor. In Proceedings of the *5th European Conf. on Computer Vision*. Freiburg, Germany.

Burdea, G. C., & Coiffet, P. (2003). *Virtual reality technology*. Hoboken, NJ: Wiley-IEEE Press.

Cabral, M. C., Morimoto, C. H., & Zuffo, M. K. (2005). On the usability of gesture interfaces in virtual reality environments. In *CLIHC '05: Proceedings of the 2005 Latin American conference on Human-computer interaction*. Cuernavaca, Mexico: ACM Press.

Ciger, J., Gutierrez, M., Vexo, F., & Thalmann, D. (2003). *The magic wand*. New York: ACM Press.

Dumas, J. S., & Redish, J. C. (1999). *A Practical Guide to Usability Testing*. Portland: Intellect Books.

Fairen (2004). MiniVR: a portable virtual reality system. *Computers graphics, 28*. New York: Pergamon Press.

Green, D., Stanton, N., Walker, G., & Salmon, P. (2005). *Using wireless technology to develop a virtual reality command and control centre, Virtual Reality*. New York: Springer.

Kim, H., & Fellner, D. W. (2004). *Interaction with hand gesture for a back-projection wall. Computer Graphics International*. Washington, DC: IEEE.

Kölsch, M., & Turk, M., (2002). Keyboards without Keyboards: A Survey of Virtual Keyboards, *UCSB Technical Report, 21*.

Kukimoto, N. (2005). Scientific visualization in collaborative virtual environment with PDA-based control and 3D annotation. [Tokyo: Japan Society of Mechanical Engineers.]. *JSME International Journal. Series B, Fluids and Thermal Engineering*, 48.

Leubner, C., Brockmann, C., & Muller, H. (2001). Computer-vision-based humancomputer interaction with a back projection wall using arm gestures. In *Proceedings of the Euromicro Conference*, Varsavia, Polonia.

Mazuryk, T., & Gervautz, M. (1996). *Virtual Reality: History, Applications, Technology and Future*. Vienna, Austria: Vienna University of Technology.

Milgram, P., & Kishino, F. (1994, December). A Taxonomy of Mixed Reality Visual Displays. *IEICE Transactions on Information and Systems, 77*(12).

Reitmayr, G., & Schmalstieg, D. (2005). OpenTracker: A flexible software design for three-dimensional interaction. *Virtual Reality (Waltham Cross)*, 9.

Sutcliffe, A. (2004). Heuristic evaluation of virtual reality applications. *Interacting with Computers*, 16.

Ulrich, K. T., & Eppinger, S. D. (2003). *Product design and development*. New York: McGraw- Hill.

Wang, Q. H., & Li, J. R. (2004). A desktop VR prototype for industrial training applications. *Virtual Reality (Waltham Cross)*, V7.

KEY TERMS AND DEFFINITIONS

Digital Models: Digital representation of objects obtained by means of 3D modelling software.

CAID Models: 3D models produced by CAID (Computer Aided Industrial Design) software;

Virtual Prototype: Digital model used to validate product features during the design process, before committing the physical artifact.

Virtual Room: Special rooms equipped with a projection based large display, for the realistic visualization of digital models.

Virtual Reality: Technology which allows a user to interact with a computer-simulated environment.

Interaction Device: tool that enables users to move and perform interaction in 3D environments.

Industrial Design: Creating and developing concepts and specifications that optimize the function, value and appearance of products and systems for the mutual benefit of both user and manufacturer.

Design Review: Mechanism for ensuring design standards, alignment, and diligence throughout the course of the product design process.

Section 3
Organization and Process Management

Chapter 10
Integrating 'Designerly' Ways with Engineering Science: A Catalyst for Change within Product Design and Development

Ian de Vere
Swinburne University of Technology, Australia

Gavin Melles
Swinburne University of Technology, Australia

ABSTRACT

The fields of design and engineering both contribute to product design and development. Increasingly design teams require an integrated approach in environments where mutual understanding and respect replace traditional professional rivalries. These new synergies both enhance communication and understanding between designers and engineers and lead engineering into new areas of professional activity. Engineers are integral to the product development process, but change in product development and manufacturing requires new responsibilities; design engineers must assume a greater role to achieve successful product realisation. However, to be effective engineers must develop new skills; creative design ability, understanding of societal and environmental impacts and a human-centred approach. These themes, not typically addressed by engineering curricula are evident in a new approach to engineering education - product design engineering. This chapter addresses issues confronting product design and development and examines the emergence of this new engineering professional in response.

INTRODUCTION

Product design and development (PDD) responsibilities have changed. There is greater focus on sustainable design, socially responsible design and design for need. Opportunities exist for design teams to make a positive commitment to the welfare of global communities whilst advancing technologies that support sustainable development. It is no longer appropriate for designers and engineers to serve solely the interests of business; instead PDD teams must understand the potential for design to make a greater contribution to lives and society.

DOI: 10.4018/978-1-61520-617-9.ch010

As the roles and responsibilities of product design and development teams are reformed, so too are their professional composition. The single discipline purity of the traditional industrial design consultancy has evolved into an inter-disciplinary team, where designers and design engineers collaborate harmoniously to provide an extended palette of services. Product design teams require an integrated and collaborative approach in environments of understanding and mutual appreciation. The product design and development process is enhanced by these new synergies between engineers and designers, as is the progression of the engineering designer into new areas of professional activity.

Whilst design engineers have always been an integral part of the product development process, their roles have traditionally been confined to working within constraints and defined parameters to achieve closure to the product realisation stage. However, emerging trends in manufacturing and revised professional responsibilities require design engineers to have a greater role in product design and development, particularly in the conceptual design and embodiment stages. Yet to be effective, they require an extensive palette of new skills; creative design ability, a thorough understanding of the societal and environmental impacts of their professional activities and a human-centred and responsible approach. These attributes are not characteristic outcomes of traditional engineering curricula, but are evident in product design engineering courses.

This new engineering discipline results from the integration of two traditionally disparate professions; mechanical engineering and industrial design. It responds to the need for interdisciplinary professionals and a greater participation in design teams by engineers conversant, indeed accomplished, in the product design and development process. These new engineering pedagogies support the changing role of the engineering designer and are catalysts for significant change in product design and development through greater team synergy, interdisciplinary understanding and communication. "Times of great flux call for those who can cross disciplines, who can see and understand the big picture." (Akay, 2003)

BACKGROUND

It has been observed that the boundaries between the design and engineering can inhibit both innovation and successful product realisation, particularly in the product design and development milieu. "These two mindsets often clash as one seeks to broaden the scope of the problem, while the other is working to achieve closure." (Fry, 2006)

In Engineering Design Methods: Strategies for Product Design, Nigel Cross notes that "the increasing competition for consumer markets and the growing awareness of the importance of design for the market has led to reinforcement of the view that successful design can only be accomplished by an integration of the skills of both engineering and industrial designers." (Cross, 2000)

This trend is evident both in the traditional industrial design consultancy and in the manufacturing sector where there is increased demand for engineers who can operate effectively in a variety of environments within global multidisciplinary teams. Engineers, particularly those in product design and development, are now expected to creative, flexible and adaptable, responsible and human-centred designers. "In this evolving world, a new kind of engineer is needed, one who can think broadly across disciplines and consider the human dimensions that are at the heart of every design challenge." (Grasso & Martinelli, 2007)

Conflict between disciplines has been common in product design and development, where differing professional approaches can destabilise the progression of design ideals. A long-standing cultural conflict exists between designers and engineers, one that can be traced to the foundation of their approach to problem solving and design. Fry (2006) defines industrial designers as "looking for new contexts and opportunities for innovation" whilst engineers "predominately work to define a set of parameters and target values up front that would define a specific, successful solution within a narrow range."

It is important to acknowledge that the approach of engineers to design activity results from the attitudinal emphasis of their education. Many engineering curricula offer limited exposure to the design process and often graduates are inexperienced in open-ended problem solving and fearful of the uncertainty of the design process. Engineers trained only in the 'science' and 'methods' of engineering are often hesitant to engage in intuitive or creative processes that may lead to an unexpected outcome.

This absence of cohesion between engineers and designers can lead to misunderstanding, a lack of appreciation, and subsequently, professional distrust and discord within the PDD team. Professional disparity between engineers and designers can result in inefficiencies in the PDD process and can be detrimental to successful product realisation. To facilitate developments in product design and development, the disparities between the professions must be addressed through engineering educational programs that seek to emphasise design, encourage creativity and innovation, resulting in improved cross-disciplinary communication, understanding and mutual appreciation.

We need "a new generation of adaptable, flexible, well rounded and innovative professionals" (Stouffer, Russell, & Oliva, 2004). However innovation and creativity does not readily emerge from typical engineering process. To be creative one needs to seek the unexpected through a process of divergent thinking. Unfortunately engineering design and industrial design follow differing methods; "convergence is at the core of the engineering process...divergence is at the core of the industrial design process" (Fry, 2006) It is therefore apparent that to develop creative engineers, their educational experience must not only include rational applications of science, but also experience in open-ended problem solving and familiarity with what Cross describes as 'designerly ways'. (Cross, 2001)

'Designerly ways' suggests a way of thinking that is common in designers, but which differs significantly from those with a techno-scientific approach, such as engineers. Cross observed that scientists problem-solve by analysis, whereas designers problem-solve by synthesis. This constructive and often intuitive approach uses non-verbal thought and communication, translates abstract requirements into concrete objects and contributes to the solution-focussed approach designers take to solving ill-defined problems. Engineering education can learn much from design pedagogy if engineers are to develop creative problem solving skills early through education, rather than later in the workplace.

The engineering profession must progress from the role of technical service provider, to a profession that leads change through understanding of the human, environmental, societal and cultural challenges and the consequences of professional activity.

ISSUES CONFRONTING PRODUCT DESIGN AND DEVELOPMENT

Design is a fundamental building block of innovation and a critical enabler of competitive advantage. Product design is essential in addressing the challenges that modern societies face in the 21st Century; whether it be enabling the delivery of technological, scientific and medical advances, mobilising public opinion, empowering communities, or simply improving the experience of a product. (Design Victoria, 2008)

However there are many challenges facing the product design and development industry as we move into the 21st century. These range from the decline of local manufacturing and the impact of a global economic downturn, to new opportunities afforded by emerging markets, the integration of new disciplines and the redefinition of existing professional responsibilities. Sustainable practice and socially responsible design will continue redefine the landscape and the PDD industry must take responsibility for its activities and show leadership as drivers of change and innovation.

The fortunes and growth potential of the product design industry are too closely allied to the manufacturing sector, an area of recent considerable upheaval and changing alliances. The transfer to offshore manufacturing has had unprecedented impact in the design sectors of many countries, as have the competitive forces of a global economy and the reduction in Government protection of local industries. Many product design environments have witnessed a rationalisation of product lines, a reduction of local content and a shift in outsourcing towards 'design and supply' contracts.

In response product designers must take responsibility for their destiny through business diversification, entrepreneurship and the creation of new opportunities through flexible and adaptable approaches. PDD must reduce its reliance on the manufacturing sector (as a sole source of revenue); in turn this will lead to greater diversification of services into new sectors, including interdisciplinary activities which are solution, rather than market focussed.

Socially responsible design initiatives and sustainable solutions are imperative and offer PDD teams a diversity of new opportunities. This will need to be complimented by design-led initiatives as PDD teams redefine and create new avenues of activity, and respond to societal needs on a global basis.

A new economy is evolving and it needs products and services that have been designed to eliminate environmental impacts"(Kel Dummet, Sustainability Victoria (Design Victoria, 2008)).

As the manufacturing industries adjust to a changing economic and social environment, new markets are emerging with diversified consumer behaviour and expectation. It is therefore essential that PDD teams develop interdisciplinary attributes that successfully integrate design, engineering, management and marketing in a flexible milieu of understanding and respect, with a high level of creativity and consideration for environmental, societal, cultural and economic demands.

Typically, poor product design decisions can adversely affect manufacturing costs, assembly schematics, quality control, product to market lead-times and overall product success. This can be addressed through consideration of manufacturing issues earlier in the PDD process, in particular design for assembly, with user-centred design addressing not only the end user, but also production, transportation, and where appropriate, installation personnel. This early integration of manufacturing consideration into the design process can be facilitated by the inclusion of design-skilled engineers into PDD, bringing production concern to the forefront of product embodiment decisions.

Product design is at a turning point, one which will redefine the industry and allow significant societal contribution to be made through new directions of activity. The key issues facing the PDD industry; sustainability, social responsibility, design for need, global distribution of design and manufacturing processes, new and emerging markets and shifting economic power will be the focus of the next generation of product designers, who must be sufficiently informed and skilled to lead the inevitable reform of industry practice.

In Australia, the Design Victoria report "Five Years On: Victoria's Design Sector 2003-2008", (through a survey of 340 design service providers and 1253 businesses) identifies design as making an essential contribution of economic growth, yet also identifies that graduates lack appropriate skills including business acumen and production and manufacturing knowledge. (Design Victoria, 2008) This trend is apparent in both the design and engineering sectors. Typically design curricula have become generic and holistic whilst engineering curricula continues to focus on the science rather than preparing graduates for engineering practice. Clearly the challenges faced by industry must be met firstly by the educational community through revised, industry relevant and progressive curricula: industry reform should be led by the next generation of designers and engineers.

So how do we define the roles and responsibilities of future PDD teams and what will be the challenges they face and the contributions they make? Firstly we must anticipate the impact of trends in product design and development, respond to the emergence of new technologies and constraints, and most importantly we must involve engineering education in the conversation. We cannot hope to foresee all of the challenges that our graduates will face throughout their careers; however we can prepare them for significant change through educational programs that deliver flexible, adaptable and responsible engineers who are lifelong learners and creative problem solvers.

Tertiary education should be the foundation of improved and redefined professional outcomes for product design. New paradigms such as product design engineering (discussed later in this chapter) are catalysts for significant change and directly respond to industry needs and societal demands. The integration of disparate disciplines (industrial design and mechanical engineering) facilitates greater professional synergies, and thus, efficiencies in the product design and development process. It is essential that all PDD team participants, particularly the engineers, are fully engaged and cognisant in all aspects of the product's 'function-aesthetic-experience' and demonstrate an eagerness for innovation. The challenges of future product design and development will be greatly facilitated through teams which include these 'integralists' (Eekels, 1987) who can articulate the overall product vision in an collaborative environment devoid of traditional professional rivalry.

The product design engineering discipline emerges from industry-informed curricula which develop well prepared, highly skilled and adaptable practitioners vital to the security and future competitiveness of the PDD industry.

DEFINING THE ROLE OF ENGINEERING IN PRODUCT DESIGN

What is the role of this new discipline, the product design engineer, in the PDD process? Before we address this, it is important to define the roles of engineers and industrial designers and examine disparate claims regarding who 'owns' the process.

Ulrich and Eppinger in their definitive text "Product Design and Development" (Ulrich & Eppinger, 2004), devote a chapter to Industrial Design (ID) where they define the role of ID in the PDD process. In addition, they endeavour to quantify the effectiveness and utilisation of industrial designers and the timing of their involvement in the development process. They state that "typically, ID is incorporated into the product development process during the later phases for a technology-driven product and throughout the entire product development process for a user driven product." They also declare that the ID process is a "subprocess of the product development process" and that "the technical nature of the problems that confront engineers in their design activities typically demands more development effort than do the issues considered by ID".

The reader may discern two things from the position taken by Ulrich and Eppinger. Firstly, that the product design process is not directed from a design perspective and secondly, that ID is a non-technical and potentially optional phase of PDD which is not integral to product development. With their statements, they are reinforcing a position not uncommon in production engineering circles, but one which devalues not only the contribution that industrial designers make to product design and development, but also the potential for designers to contribute significantly to product forecasting and planning.

With due respect to Ulrich and Eppinger, whose writings on PDD are exemplary and extremely valuable to students and practitioners seeking to develop a thorough and 'validate-able' PDD process, the industrial design profession may wish to present an alternate stance.

The early involvement of designers greatly informs the product design and development process and can actually drive technology development through earlier consideration of user needs and more appropriate and targeted product specification. Indeed it is essential that the early stages of all projects (regardless of whether it is for user or technology led products), involve those personnel with a human-centred focus and a divergent thinking approach.

It has been our professional experience that all too frequently Industrial Design input is integrated far too late in the project, often at a convergent stage of development, where the product's technical specification has been established. This limits the scope and potential of the product and forces the industrial designers to adapt and package often unsuitable or inappropriate technology and components that have been developed for poorly defined user needs. The impact of the designer's understanding of market, expertise in socially responsible design (SRD), sustainable design, ergonomics, user needs, product interaction, aesthetics and product differentiation combined with their intrinsic quest for creative solutions and innovation cannot be underestimated in product scoping and planning.

Accordingly, design practitioners must be central to the product planning process, not introduced retrospectively when parameters are pre-defined and existing constraints restrict innovation. This can be addressed in two ways, with the establishment of a collaborative team comprising both industrial designers and engineers early in the product planning stages, or with the inclusion of a new type of engineering professional; the product design engineer.

These new engineering professionals though the integration of 'designerly ways' into the practice of engineering are well placed to drive a user-centred and creative design philosophy in the planning stages of PDD, particularly in technology driven products where typically industrial designers may not be included in the discussion.

New Directions for Engineering

It is apparent that as product design and development evolves to meet new challenges that its engineers will require:

- improved design emphasis (including sketching ability),
- a creative approach,
- a socially responsible design philosophy,
- an emphasis on sustainable product design and manufacture and
- advanced open-ended problem solving ability

An Emphasis on Design: Integration of Designerly Ways

Most engineering faculties have been reticent to incorporate any significant design course components, at least until the final or capstone years. Faculties prefer to build a solid foundation of engineering science, but without supporting design activities that provide opportunities for this knowledge to be applied. "We do not teach the language of design, preferring instead the language of mathematics". (Dym, 1999) This view is supported by a perusal of typical engineering curricula that finds that negligible design projects and a notable absence of divergent thinking or creativity agenda.

Design has always been fundamental to the practice of engineering, and thus should be a key constituent in engineering education. Unless design and creative skills are a focus of learning activity, graduate engineers will continue to be technically competent, but will not prepared for the practice of engineering or the challenges they will face professionally. Both Fox and Cross have identified that there is 'an educational justification for design' as a means to develop cognitive skills and real-world problem solving abilities. (Fox, 1981) (Cross, 2001)

Yet it appears most engineering curricula continue to focus on the theory-based science model, denying students the learning opportunities afforded by design projects, particularly in the application of engineering science to real world practice. Design needs to move from the periphery to a central role in engineering education if we are to graduate innovative and adaptable engineers, and enable more valuable engineering input in the PDD process.

"One of the consequences of design-focused education is that students learn that they are applying knowledge in differing forms to serve different ends, which means that they can become fluent translators of engineering languages." (Dym, 1999)

Fostering Creativity

It appears that many in engineering faculties do not value creativity or do not readily acknowledge its importance in engineering practice. This despite many calls for a focus on fostering creativity. "Creativity is the essence of engineering. Yet creativity is neither explicitly taught nor promoted in the engineering curriculum." (Santamarina, 2002)

Creativity is frequently misunderstood, and is often viewed with an element of mistrust, especially by those outside the design community. This is due to creativity being viewed as resulting from 'ex nihlo' (something from nothing), rather than from less confronting processes involving structured problems and expected or predefined solutions. However teaching engineers that creativity involves "a non-linear, unstructured and flexible approach to solving problems and generating ideas" (Pappas, 2002) can facilitate acceptance and willingness. Engineering students must become comfortable with the creative process, if they are to embrace unexpected outcomes and pursue innovative solutions. "Making the strange familiar – accepting creativity as a desirable mindset and attribute of engineers – is a tangible and realisable goal." (Stouffer et al., 2004)

So what is creativity and why is it important in an engineering educational context?

Creativity emerges from a divergent thinking process that allows the designer unconstrained exploration, the use of intuition and reflection to respond to a problem with a solution-focussed approach. Creativity seeks to generate new, unique and unexpected solutions.

"Creativity is of paramount importance in engineering for it endows one with insight and discipline to seek out and address problems from the boundaries of different engineering disciplines" (Ghosh, 1993)

Creativity involves having unusual ideas, tolerating the unconventional and seeing unexpected implications. (Cropley & Cropley, 2000) This can be rather challenging for engineering students who are more comfortable working within defined parameters and tackling constrained (rather than open-ended) problems, and who typically, have a tendency to fixate on prior solutions. "Students must be aware that instruction in creative thinking will not provide the certainty offered them in most engineering tasks. Creative thinking is a more ambiguous endeavour than most engineers are used to or skilled in...there are no right answers" (Pappas, 2002).

Yet without a creative methodology, the potential for engineers to contribute to the PDD process is limited and innovative engineering solutions are less likely. Creativity results from a flexible and open approach to problem solving, an approach dependent on confidence and a willingness to take risks and trust intuitive ideas. Typically, these attributes have not been developed through the engineer's education, but have been gained through extensive industry experience. To better prepare graduates for the demand of industry and to enable more innovative engineering practice, we must foster creativity in students. To achieve this, curricula must facilitate significant experience of design problems, in an environment that appreciates unexpected solutions, tolerates failure and nurtures students through the framing and resolution of ill-defined problems.

Initial attempts at fostering creativity focussed on teaching specific techniques, for example Osborn's brainstorming. However this ignored the non cognitive aspects of creativity including motivation and self confidence (Cropley & Cropley, 2000), which are integral to student engagement in the creative process. Engineering education must focus on developing problem solving skills, in particular, to reduce anxiety about unexpected solutions, develop confidence in the student's creative potential, and instil a desire for seeking creative and innovative solutions. It is no longer sufficient for students to be engaged in simple problem solving activities; experience in the 'practice of engineering' must occur through experiential learning projects that allow science to be applied to real world problems.

The Importance of Sketching

There is significant research that links sketching to creativity, yet many engineering curricula fail to develop the sketching skills essential for design conceptualisation and communication and resultantly limit the ability of engineering graduates to explore design possibilities. Sketching enables abstract idea development and is not just a documentation and communication process; rather it facilitates the creative process through contextual citing of the design intent. Without sketching ability, many engineers struggle to find the unintended consequences that stimulate the design exploratory process in what Schon and Wiggins (1992) called the 'reflective conversation with the situation'.

Figure 1. Differing roles of conceptual sketching: Exploration of form (left) and abstraction of technical details (right)

It appears that the very design process is limited by the ability to use graphics as a cognitive extension. This implies the need for training... not only in the standard drafting skills, but additionally in the ability to represent concepts that are more abstract and best represented as sketches (Ullman, Wood, & Craig, 1990).

Sketching is fluent, flexible and inaccurate and because sketches are 'rough' there is less reluctance to discard them in favour of alternative designs. Sketches are fast, implicit, inexact and abstract, require minimal commitment and serve both analysis and simulation requirements. (Lipson & Shpitalni, 2000) Goldschmidt in his definition of sketching dialectics, refers to a dialogue between reflective criticism ('seeing that') and analogical reasoning and reinterpretation of the sketch ('seeing as'). (Goldschmidt, 1991)

Most importantly, sketches allow the designer and engineer the opportunity to explore as many concepts as possible before moving into the detailed design stage. That sketching is integral to the creativity process, is reinforced by studies of industrial design students which compared skilled and unskilled sketchers (Verstijnen & Hennessey, 1998) who found that those with sketching ability were advantaged by externalisation of mental imagery.

Yet many engineering faculties rely solely on CAD for design, neglecting the possibilities that sketching offer in developing creative solutions. As sketching facilitates "the creative shift to new alternatives" (Goel, 1995) it is imperative that engineering students are competent sketchers, capable of articulating unformed ideas and design intent.

SOCIALLY RESPONSIBILITY DESIGN (SRD)

It is imperative that engineering serve the community in a socially responsible, sustainable and culturally sensitive manner. "Because of the intrinsic connection of engineering design with values, the engineer as designer shall not only be answerable for his/her engineering capabilities, but also and always for his/her ethical conceptions and behaviour as a moral person." (Eekels, 1994) Not only do product design teams need to be aware of the consequences of their professional actions, they must also be cognisant of the opportunities afforded them to lead positive change.

Designing for our complex societies requires anticipation of future needs and cultural sensitivity, but there exists a greater need, that of those at the base of the pyramid (the other 90%). Product design teams must address the societal needs of developing nations through designs that are indeed life supporting, sustainable and empowering. Urgent action is required in many critical areas including clean drinking water, sanitation, renewable energy, and healthcare and disease prevention.

A new engineering conscience is needed, one that emboldens 'design for need' over the material needs of first world consumers and repositions product design and development not as a functionary of business, but as the leader of change for greater social good. Engineering education is well positioned to make a significant contribution to determining attitudinal change amongst the student cohort and thus developing social conscience and sustainable and ethical practice in its graduates. The next generation of engineers can contribute to appropriate product design and development through responsible, ethical and culturally and environmentally sensitive design practice.

The engineering profession must progress from the role of technical service provider, to a profession that leads change through understanding of the human, environmental, societal and cultural challenges and the consequences of professional activity. This can only be achieved through a systematic and thorough grounding in the principles of socially responsible design (SRD) combined with opportunities to apply sound engineering practice to real world problems throughout the educational journey.

Sustainability

The UN has estimated that 80 percent of all product-related environmental impacts are determined during the product design stage. If issues of resource depletion, energy usage, material reuse and recycling, environmental degradation and climate change are to be addressed in sufficient time, environmental and social considerations need to be integrated early into the education of design engineers. Our responsibility as educators cannot be just to our students; we must be answerable to societies in general for the consequences of our graduates' professional behaviour. Our engineering graduates need not just awareness of the issues, but an embedded ethical philosophy and the tools to effect reform through sustainable design and product engineering.

The decade 2005-2014 has been declared the United Nations 'Decade of Education in Sustainable Development'. This initiative aims to "encourage changes in behaviour that will create a more sustainable future in terms of environmental integrity, economic viability and a just society for present and future generations." We must respond to this call through a targeted program that teaches not only awareness, but also the tools to affect behavioural change and to lead reform in design practice and manufacturing. The next generation of product designers and design engineers must direct initiatives to address the challenges faced by the manufacturing sector. Energy and material consumption and carbon emissions must be reduced, environmental degradation must be halted and alternative technologies developed. But most importantly the needs and aspirations of those currently outside the first world consumer societies must be addressed; not only are they the unwilling participants in relentless and unsustainable development, they also lack the basic elements that constitute a safe, healthy and equitable existence.

"Sustainability cannot be conducted on the sidelines. It can only be achieved through a paradigm shift which results in sustainability becoming part of everyday life, directing the way in which communities and individuals make decisions that contribute to the development of broad social goals." (Hammer, 2007) Engineering and design educators are ideally placed to make a positive contribution to reducing the impact of product manufacture and life cycle through the development of pedagogy that facilitates this paradigm shift.

Many of these goals can be addressed through innovative curricula that inform, encourage and require design project outcomes that include sustainable manufacturing processes and material selection, life cycle analysis (cradle to grave), design for assembly and disassembly and a holistic approach to product design that questions not only the impact of the product but also the 'need' for the product to exist at all. Designers and engineers must be taught to question established practice and to be sufficiently confident to make strategic calls that may defy established practice in search of a more sustainable product solution. "Designers have responsibility over the choices they make in design processes." (Papanek, 1985) Sustainable design practice must merge as an entrenched design methodology that is at the core of all professional activity. This will require attitudinal change and will require product design teams to serve not only their immediate stakeholders (clients and customers), but also the needs of global societies, including those outside their own market.

Sustainable design addresses the 'triple bottom line', a reporting framework that considers ecological and social aspects, in addition to the traditional financial measures. These goals demand that corporate responsibility be to its stakeholders not just its shareholders; consequently anyone who is influenced or affected, either directly or indirectly, must be considered. Appropriate design, fair trade initiatives, reduced ecological footprints and true accountability must be paramount in the minds of product design teams if we are to reduce the negative impact of our activities in product design and manufacturing. Victor Papenek famously declared that "there are professions more harmful than industrial design, but few of them." (Papanek, 1985) In response, product design and development teams must take the initiative to secure a sustainable lifestyle for future generations through well considered and responsible design and engineering practice; the role of engineering education is to provide the knowledge, aptitude and attitude required to lead this paradigm shift.

Problem Solving: 'Wicked' and Open-Ended Problems

It is evident that the 21st century design engineer will be confronted with ill-defined problems; design problems that will not be solved solely through the appliance of engineering science. These 'wicked' problems (Rittel & Webber, 1973) will require a new engineering approach, one that incorporates divergent thinking, creative and intuitive processes, and a willingness to embrace unexpected solutions; a designers approach to problem solving.

As discussed earlier, design is a divergent activity, whereas engineering thinking is mostly convergent. It is no surprise that mechanical engineering students struggle with open-ended problem solving when compared with design students. Most engineers are more comfortable with constrained problems, where the parameters are set and the outcome is, if not immediately apparent, determined by constraints afforded by material and processes. By contrast designers 'live' for the creative moment: they thrive on the pressure of uncertainty, exult in the quest for uniqueness and triumph with innovative solutions.

Design is an intensely personal pursuit where one makes a commitment to the problem that involves significant personal vulnerability and a willingness to face uncertainty and failure head-on. It has been said that a designer is only as good as his/her last design, as such designers must constantly expose themself to intense scrutiny as their outcomes are measured against unforgiving criteria, and judged by a market often driven purely by economic rationalism. Successful design practice requires a degree of self examination and confidence in the face of critical scrutiny that is not required by other professions; in design mere competency is insufficient.

However, many of the problems facing design engineers in product design environments will also be poorly defined and their first challenge will be problem definition; this will need experience and confidence when faced with uncertainty. Students need a learning process of construction and confrontation rather than memorisation. Project-based learning allows students to define and analyse the problem, develop alternative strategies to problem solving and to build, enhance and practice their expertise. But not all project-based learning will afford intense exploration and challenge students to work without constraints or parameters, as will an open-ended problem.

Open ended design problems "force students to think creatively and ultimately foster in them an appreciation for developing creative solutions." (Ghosh, 1993) This requires students to tolerate the unusual, unconventional and unexpected, and to become comfortable with divergent thinking processes. Open-ended problems require problem finding, evaluation, definition and resolution skills together with ideational fluency. They take students away from a problem 'solving' role into a position of exploration and synthesis that enforces a more holistic reasoning approach with critical analysis and creative expression. Experience with ill-defined design problems (that are not amenable to the techniques of science and engineering), is invaluable in engineering education. Students must develop flexible and divergent dichotomies and eliminate any tendency to fixate on prior solutions; this will require a new approach to the teaching of engineering design.

To address ill-defined problems, engineers must be confident seeking solutions outside their traditional fields of expertise, and comfortable using intuition to solve problems, rather than relying on solely on mathematical and scientific solutions.

The Emergence of a New Engineering Discipline within PDD

Industry trends indicate that role and responsibilities of the engineering designer in product design and development have changed from technical resolution to a more central role in the planning and design process. Design engineers are often engaged in creative 'front end' design activities (typically the domain of industrial designers) in addition to systems engineering and manufacturing resolution roles. "Current trends in technology and our increasingly complex society and workplace require engineers to have a wider variety of skills and a broader understanding of engineering as a discipline if they are to be successful." (Pappas, 2002)

Unfortunately it appears that traditional engineering curricula has not responded to the repositioning of engineering in PDD environments, and continues to focus on the 'theory' rather than the 'practice' of engineering, neglecting the development of design and creativity in students. "Historically, engineering curricula have been based largely on an 'engineering science' model." (Dym, et al, 2005) However, curriculum developments, particularly during the last decade, have led to the emergence of new engineering pedagogy that support integrated design environments. These new educational models show evidence of an attitudinal shift that leans "toward a more explicit recognition of design as a distinguishing feature of engineering practice and a motivating factor in the learning of engineering." (Dym, 1999)

Product design engineering (PDE) has created a new style of engineering designer; creative, flexible and adaptable, sustainable, ethical and responsible. Product design engineering whilst a relatively new engineering discipline, is gaining support with global distribution of curricula. These pioneering curricula integrate the traditionally disparate disciplines of industrial design and mechanical engineering to develop a creative 'next generation' engineer more adapt at the changing roles of product design and development.

Swinburne University of Technology PDE graduates have already made a significant impact in local product design and development. From observation of (and discussions with) leading industrial design consultancies in Melbourne, Australia, it is apparent that the product design engineering model is leading a trend in product design and development team constitution, with these interdisciplinary professionals defining a new approach to product design and breaking down traditional professional rivalries.

All of the major consultancies (and most companies in the manufacturing sector) are employing PDE graduates as they provide extended services to their clients and engage in brand building, global manufacturing and sustainable design solutions. Upon investigation, it appears that in the three leading industrial design consultancies examined closely, Swinburne product design engineers represented approximately 20-30 percent of the design staff, the balance being industrial designers. This is a significant achievement for a single course that has only seven years of graduates and that is lacking in public awareness as a distinct design discipline.

Ensuing conversations revealed that when employing a graduate, consultancy directors are more inclined towards employing a product design engineer for a traditional ID role, on the basis of their greater technical ability. It is felt that product design engineers 'reflected the industrial design graduates of previous generations', were more industry ready and could make a valuable contribution immediately upon employment. By contrast most ID graduates lacked technical ability and an understanding of professional practice, as a result of an education process less rigorous and more generic in nature. In consultancies whose activities were defined as approximately 20 percent front end creative, 60 percent detail design and 20 percent project management, the PDE's were a better investment with several consultancy directors declaring that they prefer product design engineers when hiring junior staff.

It was said that ID graduates were still slightly better at purely conceptual front-end design and that at present the most preferred employee was an industrial designer with 5-10 years industry experience; although this may change as PDE graduates reach professional maturity.

CASE STUDY—THE SWINBURNE PRODUCT DESIGN ENGINEERING CURRICULA: A CATALYST FOR CHANGE

Overview

Product design engineering represents a different approach to the teaching of engineering. It is an exemplary example of interdisciplinary education that not only develops a creative and human centred engineer, but understands the need for a more integrated approach to product embodiment. Without neglecting the science of engineering, it seeks to develop graduates who utilise the intuitive creativity of the design practitioner, united with sound engineering design method, to create a new approach to product design and development.

This educational paradigm is evident across a range of educational institutions worldwide as engineering faculties seek to address shortcomings in traditional curricula; in particular the notable absence of embedded design activities and negligible fostering of creativity. There are many models emerging either through targeted product design activities in mechanical engineering curricula (e.g. the product design major at Stanford), double degree courses or fully integrated programs such as the PDE courses offered in the United Kingdom, Europe and Australia. But all of these courses address the identification by Dym, Pappas, Akay, Santamarina and others, that engineering education needs reform if we are to develop engineers who are 'creative thinkers'.

It is important to clarify that the product design engineering paradigm discussed here is the product of fully accredited engineering courses with an engineering qualification (e.g. BEng, MEng) outcome and should not be confused with variations of the industrial design paradigm; PDE graduates are a new category of engineering designer.

However this new engineering curricula is not the only educational response to develop greater synergies between engineering and industrial design. The industrial design engineering (IDE) programs prevalent in Europe and the UK respond to the same industry requirements but through an expansion of industrial design curricula to incorporate engineering content.

Product design engineers are capable of operating with distinction in both professions, they utilise the divergent, reflective approach of designers, but engineer with the disciplined thoroughness and scientific method of engineers. More importantly, their integrated approach to product design incorporates a level of engineering proficiency not evident in industrial design, and their engineering methods benefit from the creative, flexible and human centred approach of the designer. This allows a valuable contribution to the PDD environment as they design from the 'inside out'; firstly engineering the functionality and then designing the user experience.

Product design engineers with their interdisciplinary attributes utilise intuitive creativity with sound engineering design method, to develop innovative solutions to complex, ill-defined problems. This results in a more proactive role in product design and development as these new engineers engage in initial problem framing activities, share responsibilities with industrial designers, and produce user-centred design solutions.

Background to PDE

Product Design Engineering originated as a new engineering discipline in Glasgow, Scotland in the late 1980s as a changing manufacturing environment demanded a professional fluent in both engineering and design. The first product design engineering course resulted from collaboration between Glasgow School of Art, (Industrial Design at the School of Design) and the University of Glasgow, Department of Mechanical Engineering and was soon followed by similar pedagogy at the University of Strathclyde. It intended to alleviate the gulf between the design and engineering professions and address disparate technical and cultural issues. It also aimed to introduce many of the educational topics raised by Donald Schön in 'Educating the Reflective Practitioner' including reflection-in-action and joint experimentation on open-ended problems. This new engineering curriculum integrated the design and engineering disciplines to create one model of reflective practicum (Green & Kennedy, 2001) with "design a continuous thread running through the teaching", an attribute preferred by the Institution of Mechanical Engineers, in the Grant Report, The Formation of Mechanical Engineers: Present and Future Needs.(IME, 1985)0

The importance of this interdisciplinary educational approach is evidenced by the emergence of similar pedagogy within the last decade, particularly in the United Kingdom and Western

Europe (PDE is now offered by least twenty-five Universities worldwide), and the success of the Swinburne University of Technology PDE program in Melbourne, Australia, established in 1997. As industry enthusiastically embrace these 'new' engineering graduates, these innovative curricula appear to be feasible alternatives to traditional methods for teaching engineering, particularly in areas where design skills, creativity and problem solving ability are valued professionally.

The intent of the product design engineering curriculum is best described by a sample of course mission statements from institutions offering this new curriculum, as follows:

"There are increasing pressures, both from existing and emerging world marketplaces, for products which not only respond to the needs of function, user and society, but which can be brought to market ever more rapidly through state-of-the-art development and manufacturing processes. Industries which develop, manufacture and market today's products need high calibre graduates equipped to handle these processes with management skill and creative drive, and the PDE programme develops graduates with these skills." (Glasgow School of Art, 2009)

"The Product Design major concerns itself with the conception and design of products, services and experiences for the benefit of society. The program teaches a design process that encourages creativity, craftsmanship and personal expression and emphasises brainstorming and need finding to discover latent or un-served human need." (Stanford University, 2009)

"This course bridges the worlds of design and engineering, producing true hybrid professionals who will move immediately into companies as design engineers and product development specialists. This rare degree format has proved to be highly regarded by industries worldwide for producing valuable employees." (Brunel University, 2009)

From these course descriptors, it is apparent that, regardless of the country or institution, the product design engineering paradigm intends to develop a creative and human-centred design approach to the practice of engineering within the PDD environment.

This new engineering curricula is dependent on a number of distinct and crucial elements which are integrated throughout all areas of the learning process. These include:

- interdisciplinary education and skill integration
- an emphasis on design – integration of designerly ways
- the fostering of creativity
- sketching
- social responsible design
- sustainability
- experience of 'wicked' and open-ended problems
- experiential project based learning

Project Based Learning: Learning from Design

"Innovation requires creativity. The easiest vehicle for promoting creativity and for developing the student's decision making ability is the design project." (Eekels, 1987)

Developing product design aptitude requires experience; opportunities to learn, to explore, to experiment and to fail. Traditional engineering pedagogy, reliant on a theory-based scientific model does not offer many such learning opportunities. Students are not required to apply their newfound knowledge to the design of product solutions for real world problems; it is more likely that students will be tested on their science through examination and through solution-focused problem solving activities that lead to predetermined outcomes. This leads not to creative and innovative engineering practitioners, but rather, to pragmatic and inflexible professionals who readily fixate on the security of known solutions.

On the other hand, a design curriculum is 'learning in action' where students learn theory, and develop creativity and design experience through project-based or experiential learning processes. Students are encouraged, indeed required, to explore the limits of their imagination, to push the boundaries of technology, processes and materials: to investigate the possibilities and extend their own potential. Project based learning affords opportunities for the application of the known, the science of engineering to be applied to the unknown, often resulting in an unexpected solution. The joy of teaching design projects is that every brief will result in unique and innovative designs from each and every student; there are no correct or incorrect answers, only creative and unexpected solutions.

Project based learning is constructive, participatory and problem driven; an exemplary pedagogical model of 'learning by doing' and the most effective way to educate creative engineers for product design and development.

Creativity and Innovation

Felder (Felder, 1987) notes that creativity can be fostered through exercises that focus on:

- ideational fluency
- poorly defined and open ended problems requiring divergent thinking
- synthesis of material outside normal boundaries
- evaluation – technical decisions vs. social and ethical considerations
- problem finding and definition (not just problem solving)

This model is intrinsic to the PDE educational approach where creativity and innovation are emphasised. Engineering education that focuses on problem solving results in graduates who can resolve a technical issue, but who are not solution focussed. Product design innovation does not result from solving a problem; it emerges from a creative, explorative and reflective process that expects only the best, a unique and creative solution.

Figure 2. Innovative product solutions (Device to apply chemically-impregnated tape to grapevines to eradicate environmental contamination (left) Shared ownership urban 2020 vehicle concept utilising a sustainable propulsion system (right))

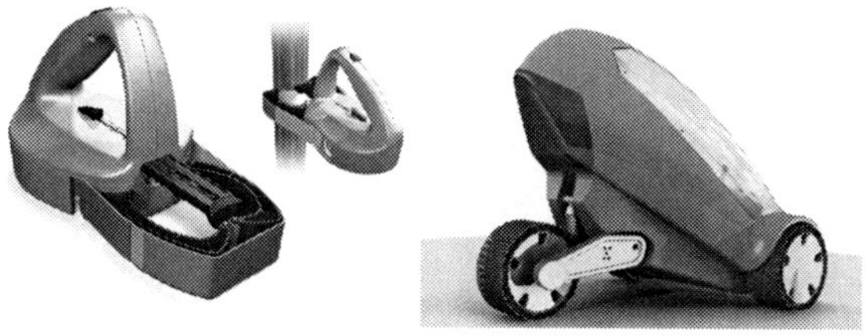

Sketching

The product design engineering pedagogy through its integration of engineering with 'designerly ways' ensures that these new engineers are fluent at creative exploration and critical reflection and can 'handle different levels of abstraction simultaneously." (Cross, 2000) This is achieved through the requirement for aptitude in perspective sketching, which is taught from the first semester, and developed continuously throughout the four year undergraduate program. Students are required to utilise sketching in all stages of a design project; beginning with initial conceptualisation, then concept development (Figure 2) and progressing to detail design and technical resolution (Figure 2).

In the PDE curriculum, perspective sketching and rendering are integrated into the learning process at all levels of the course experience and students are expected to become fluent communicators of form and function through non-digital drawing media.

Socially Responsibility

To address these goals, the Swinburne product design engineering program has established several

Figure 3. Socially responsible designs. (Mobility aid/shopping trolley for elderly users (left), Ceramic pressurized sterilization unit for Nepal designed for local production (right))

Figure 4. 2nd year project work – sustainable electronic products. (Solar-powered soil analysis device (left); Hand-powered communication/banking access device for remote communities (right))

strategic relationships with humanitarian aid agencies with the view to providing proficient design and engineering services in areas of desperate need, whilst providing students an opportunity for real world learning and experience in design for good. Product design projects have been conducted addressing scenarios provided by World Vision (a leading humanitarian aid agency) (Figure 3) and liaison with Engineers without Borders has been established to engage in long term infrastructure building activities.

Final year 'capstone' projects are required to address a social need through a product outcome. Students are encouraged to collaborate closely with partners from aid agencies, medical research organisations, disability support organisations and education and healthcare specialists to provide product solutions to real world needs (Figure 3). Such projects have developed an ethical and responsible approach to the practice of engineering design amongst the student cohort, and many altruistic endeavours post-graduation.

Sustainability

The product design engineering program focuses on sustainable design from the second year where the students are introduced to the principles of sustainability, the importance of life cycle analysis (cradle to grave), and ethical and eco-design methods. Students examine the impact of e-waste and the WEEE directive and then utilise environmental design tools to develop environmentally sustainable electronic devices (Figure 4). Eco-design is integrated from the early conceptual stages to ensure appropriate and well considered designs; band-aid solutions and products without a 'sustainable purpose' or genuine need are not accepted. Design outcomes must adhere to the "Ten Golden Rules of Eco-Design" (Luttropp & Lagerstedt, 2006), life cycle analysis (using Eco-Indicator and other software) and 'end of life' scenarios must be considered.

Open Ended Problems

It is important that students are set open–ended challenges where they must address a real world scenario without clear definition of potential outcomes. A typical project might be the design of a water transportation device to enable remote communities to convey water across either rugged or flooded terrain (Figure 5). These 'wicked' problems have numerous potential outcomes and are not constrained by known entities such as materials or manufacturing processes. Instead the students must consider the user and environment, define the problem and then design the product solution.

These projects are invaluable tools to develop creativity, and as students become familiar with

Figure 5. Open-ended problem solving exercise: water transportation for remote communities (Modular water transportation system for use in flooded landscapes (left); Water transportation system utilising readily available discarded materials (right))

uncertainty and confident in their design ability, true innovation emerges.

Future Trends in PDD: A Changing Environment

Industry trends indicate that product design and development will increasingly require an inter-disciplinary professional capable of operating with distinction in both design and engineering roles. It is already apparent that successful design results from teams that integrate the skills of both engineering and industrial designers through non-combative collaboration built on mutual respect.

The challenges of the next few decades are immense; design teams must lead behavioural change through well considered, appropriate and sustainable products that meet the future needs of all stakeholders, not just their clients and customers. Confronting climate change inflicted by environmental degradation, reducing the impact of products throughout their life cycles, and enhancing the health, life expectancy and aspirations of those at the base of the pyramid will require real innovation in product design and development. This innovation will not occur naturally, a paradigm shift is required; one that will challenge expected conventions in PDD and revolutionise product design and manufacturing activities.

Advances in solid freeform fabrication (SFF) or rapid manufacturing (RM) have the potential to dramatically reshape the existing manufacturing sector. New design protocols and working methodologies will be developed to maximise the potential of these emerging manufacturing processes. The advent of rapid manufacturing will profoundly impact existing global manufacturing systems and may precipitate a move to local, or even customer production, using SFF technologies. As components and products are freed from the constraints of traditional manufacturing processes, fresh and innovative product architecture will emerge: we are at the dawn of a new product vernacular. Without the restrictions of centralised manufacture, transport and storage considerations will no longer critical as localised 'just in time' or 'buy and make' production of 'end use parts' will drive new directions in product design.

The implications of RM for the developing world are immense; previously unheralded access to products and services will drive rapid societal change. The removal of transportation from the product supply chain would dramatically reduce costs and aid in greenhouse gas reduction and fossil fuel consumption. Manufacturing processes not reliant on immense tooling and infrastructure costs will allow product design teams' far greater freedom to explore product potential and even to customise products for individual markets and/or customers. With this emerging technology, the traditional shackles of design for manufacturing will be removed or at least will be much less constraining, and this will revolutionise the product design and development environment.

The existing system of product manufacture, which has its origins in the Industrial Revolution, will be rapidly transformed and these changes may have similar socio-economic consequences and cultural impact, to that of mechanisation in the late 18th century.

This new environment will see the emergence of a new creative PDD professional; adaptable flexible, multilingual in the languages of engineering and design, responsible and culturally sensitive. The product design engineering paradigm which integrates engineering and design to develop an interdisciplinary professional, aims to produce graduates for these challenging and uncertain times; design engineers who are well equipped to adapt quickly to shifting environments and new methodology, in the pursuit of socially responsible and sustainable product innovation.

CONCLUSION

This chapter has dealt with the challenges facing product design and development teams in the near and immediate future. PDD needs to respond to existing and emerging challenges with a new paradigm that incorporates fresh approaches to practice that respond to global societal and environmental needs, empower individuals and cultures and embrace emerging technologies that will impact significantly on established protocols and the definition of professional disciplines.

The role of engineers in the product design and development process is shifting and will continue to be redefined as technological, environmental and societal considerations impact on professional practice. It is felt that these challenges are best addressed with a complex and multi-skilled practitioner, who bridges traditionally rival professions and facilitates a new era of collaboration, understanding and respect. Both the engineering and design professions need redefinition, a paradigm shift to address the rapidly changing climate with which we are faced. Engineers have always been integral to the PDD process, but their roles are shifting and the importance of their contribution in the near future, should not be underestimated.

The product design and development process can be enhanced by innovative engineers, but often is hampered by inflexibility, poor understanding of user needs and cultural insensitivity. The PDD environment will be the beneficiary of a more enlightened (and less rigid) approach to engineering education that seeks to impart skills that result in responsible practice and creative outcomes. And both society and environment will benefit from an engineer who is creative, innovative, adaptable and responsible; who can contribute to appropriate product design through new professional synergies and guide attitudinal change.

The role of engineering education in appropriate product design and development cannot be underestimated. Educators have a responsibility to global communities to produce engineering graduates with embedded ethical and creative philosophies who are well prepared and eager to make a significant contribution to environmental and societal issues. Early indications are that the product design engineering paradigm (emerging in direct response to the needs of industry) contributes to future global needs through a new engineering pedagogy that imbues its students with a strong understanding of the importance of their future contributions to society and the impact of appropriate professional behaviour.

The world is expectant of product design and development to address the future needs of our societies and environments. This will require new skills, enhanced communications and new synergies between stakeholders. The next generation of engineers, properly prepared, can direct the product design and development process confidently in new strategic directions.

REFERENCES

Akay, A. (2003). The renaissance engineer: educating engineers in a post-9/11 world. *European Journal of Engineering Education*, *28*(2), 145–150. doi:10.1080/0304379031000078979

Brunel University. (2009). *Product Design Engineering - Course Summary*. Retrieved in 2009, from www.brunel.ac.uk/courses/ug/cdata/p/ProductDesignEngineeringBSc

Cropley, D. H., & Cropley, A. J. (2000). Fostering Creativity in Engineering Undergraduates. *High Ability Studies*, *11*(2), 207–218. doi:10.1080/13598130020001223

Cross, N. (2000). *Engineering Design Methods: Strategies for Product Design* (3rd ed.). New York: Wiley.

Cross, N. (2001). Designerly ways of knowing: design discipline versus design Science. *Design Issues*, *17*(3), 49–55. doi:10.1162/074793601750357196

Design Victoria. (2008). *Five Years On: Victoria's Design Sector 2003-2008*. Melbourne: Retrieved from http://www.designvic.com/~/media/PDF/FiveYearsOn.ashx

Dym, C. A. (1999). Learning Engineering: Design, Languages and Experiences. *Journal of Engineering Education*, *88*(2), 145–148.

Dym, C. A., Agogino, A. M., Eris, O., Frey, D. D., & Leifer, L. J. (2005). Engineering Design Thinking, Teaching, and Learning. *Journal of Engineering Education, January 2005*, 103-119.

Eekels, J. (1987). Guidelines for Engineering Teachers Concerning Educating the Engineer for Innovation and Entrepreneurial Activity. *European Journal of Engineering Education*, *12*(3), 259–270. doi:10.1080/03043798708939369

Eekels, J. (1994). The Engineer as Designer and as a Morally Responsible Individual. *Journal of Engineering Design*, *5*(1), 7–23. doi:10.1080/09544829408907868

Felder, R. M. (1987). On Creating Creative Engineers. *English Education*, *77*(4), 222–227.

Fox, R. (1981). Design based studies: an action-based 'form of knowledge' for thinking, reasoning and operating. *Design Studies*, *2*(1), 33–40. doi:10.1016/0142-694X(81)90027-2

Fry, R. (2006). *Defining the Obvious: Explaining Creativity and Design Thinking to Nondesigners*. Paper presented at the 2006 IDSA National Education Conference.

Ghosh, S. (1993). An Exercise in Inducing Creativity in Undergraduate Engineering Students Through Challenging Examinations and Open-Ended Design Problems. *IEEE Transactions on Education*, *36*(1), 113–119. doi:10.1109/13.204828

Glasgow School of Art. (2009). PDE Program - Relevance to Industry and Society. Retrieved from www.gsa.ac.uk/gsa.cfm?pid=1421

Goel, V. (1995). *Sketches of Thought*. Cambridge, MA: MIT Press.

Goldschmidt, G. (1991). The Dialectics of Sketching. *Creativity Research Journal*, *4*(2), 123–143. doi:10.1080/10400419109534381

Grasso, D., & Martinelli, D. (2007). Holistic Engineering. *The Chronical Review*, *53*(28), B8.

Green, G., & Kennedy, P. (2001). Redefining Engineering Education: The Reflective Practice of Product Design Engineering. *International Journal of Engineering Education*, *17*(1), 3–9.

Hammer, J. (2007). Sustainability - an Introduction from the National President. Retrieved November 12, 2008, from http://www.engineersaustralia.org.au/ieaust//quicklinks/sustainability.cfm

IME. (1985). *The Formation of Mechanical Engineers: Present and Future Needs* (The Grant Report).

Lipson, H., & Shpitalni, M. (2000). Conceptual design and analysis by sketching. *Artificial Intelligence for Engineering Design, Analysis and Manufacturing, 14*, 391–401. doi:10.1017/S0890060400145044

Luttropp, C., & Lagerstedt, J. (2006). EcoDesign and The Ten Golden Rules: generic advice for merging environmental aspects into product development. *Journal of Cleaner Production, 14*, 1396–1408. doi:10.1016/j.jclepro.2005.11.022

Østfold University College. (2009). Study Program for the Engineering Education: Product Design Engineering - course description. Retrieved from www.hiof.no/index.php?ID=1760&lang=eng

Papanek, V. (1985). *Design for the Real World: Human Ecology and Social Change* (2nd ed.). Chicago: Academy Publishers.

Pappas, E. (2002). *Creative Problem Solving in Engineering Design*. Paper presented at the ASEE Southeast Section Conference

Rittel, H. W. J., & Webber, M. M. (1973). Dilemmas in a general theory of planning. *Policy Sciences, 4*(2), 155–169. doi:10.1007/BF01405730

Santamarina, J. C. (2002, February 21-22, 2003). *Creativity and Engineering - Education Strategies*. Paper presented at the International Conference/Workshop on Engineering Education Honouring Professor James T. P. Yao, College Station, Texas.

Stanford University. (2009). *Product Design - course description*. Retreived from http://design.stanford.edu/PD/undergraduates.html

Stouffer, W. B., Russell, J. S., & Oliva, M. G. (2004). *Making the Strange Familiar: Creativity and the Future of Engineering Education*. Paper presented at the 2004 American Society for Engineering Education Annual Conference and Exposition.

Swinburne University of Technology. (2009). *Product Design Engineering - course description*. Melbourne, Australia: Swinburne University of Technology.

Ullman, D. G., Wood, S., & Craig, D. (1990). The Importance of Drawing in the Mechanical Design process. *Computing and Graphics, 14*(2), 263–274. doi:10.1016/0097-8493(90)90037-X

Ulrich, K. T., & Eppinger, S. D. (2004). *Product Design and Development* (3rd ed.). New York: McGraw-Hill/Irwin.

United Nations, W. C. o. E. a. D. (1987). Our Common Future. *Report of the World Commission on Environment and Development, General Assembly Document A/42/427*

University of Strathclyde. (2009). *Product Design Engineering-course description 2009*. Retrieved from www.strath.ac.uk/dmem/studying/undergraduatecourses/choosingtherightcourseforyou/

Verstijnen, I. M., & Hennessey, J. M. (1998). Sketching and Creative Discovery. *Design Studies, 19*(4), 519–546. doi:10.1016/S0142-694X(98)00017-9

KEY TERMS AND DEFINITIONS

Product Design Engineering (PDE): Product design engineering is an educational pedagogy and professional engineering paradigm, which seeks to integrate two disparate and traditionally rival professions; mechanical engineering and industrial design. Product design engineers are a new breed' of design engineer, interdisciplinary, creative and human centred, and highly skilled in the product design and development arena.

Base of the Pyramid (BoP) - the Other 90%: The base of the pyramid refers to the largest but poorest socio-economic groups; the four billion people who survive on less than $2 per day. These groups are not only isolated from the affluent materialistic lifestyles of first world Western populations, but are adversely affected by our development, manufacturing and quest for diminishing resources, and are the earliest victims of environmental degradation. Central to assisting those at the BoP are the principles of understanding, empowerment and respect, participatory development, mutual value and co-creation. Many of 'the other 90%' lack access to clean drinking water (1.1 billion), adequate sanitation (1.4 billion), healthcare and education. There exists many opportunities for design to contribute to improving the life expectancy and aspirations of those in need, through socially responsible design.

Socially Responsible Design (SRD): The use of design to address social, environmental, economic issues and focuses on a move to move beyond first world consumer demands towards a more holistic and responsible approach to product design that embraces ethical, cultural and humanitarian values; simply termed 'design for good' or 'design for need'. SRD responds to all stakeholders (not just clients and customers) and examines the consequences of design activity and the potential for design to contribute positively to societal aspirations and expectations, health and lifestyle.

Sustainable Design: Sustainable design addresses the 'triple bottom line', addressing the principles of economic, social and environmental sustainability. It aims to counter negative impacts of products throughout their lifecycle with 'cradle-to-grave' philosophy that employ low impact materials and processes, and considers energy usage, embedded energy, design for disassembly to facilitate reuse or recycling, resource renewability, bio-mimicry and design impact measures (such as life cycle analysis). Sustainable design aims to support development "that meets the needs of the present without compromising the ability of future generations to meet their own needs." (United Nations, 1987)

Rapid Manufacture (RM): Rapid Manufacture utilises processes originally known as rapid prototyping (RP) and employs solid free-form fabrication (SFF) to deliver three dimensional components. It is an additive fabrication method that utilises sequential delivery of material and/or energy to specified areas to construct a particular form, layer by layer. As this technology evolves, increasingly complex forms are possible, including shapes and assemblies not achievable through traditional fabrication processes. This 'next level' technology has the potential to revolutionise manufacturing with particular regard to localised or even customer manufacture through advanced '3D-printing' processes.

Pedagogy: Refers to strategies of teaching or principles and methods of instruction, sometimes referred to as 'the art of teaching'. It is effectively an array of teaching strategies that support the learning experience, through a well considered approach to curriculum development that empowers the teacher, student and learning experience. In this chapter, pedagogy is used as a broad descriptive term when describing the teaching strategies and curricula of the product design engineering paradigm.

Project Based Learning (PjBL): An experiential learning process that involves the use of classroom projects to stimulate deep learning through enquiry and experimentation. In the context of design education it involves the students undertaking a design process in response to a detailed brief usually in a way that replicates real world professional practice. Project bases learning requires students to engage in design activities including conceptualisation, reflection, problem solving and exploration of possible solutions and detail resolution. The learning process is defined partly by the final artefact (design) and the success of the methods applied. In the context of teaching engineering design, PjBL allows students to integrate scientific knowledge into a solution-focussed approach encouraging 'learning through doing' which entails increased knowledge retention, and facilitates development of skills in the 'practice' of engineering' rather than relying on memorisation of theory.

Creativity: A process that involves generation of new ideas and concepts; resulting in unique, innovative approaches and solutions. Creativity is an essential element of design thinking and depends on confidence to explore the unfamiliar, tolerate the unexpected and encourage the unknown. Creativity is an expressive and personal process that involves divergent thinking, exploration and reflection and at times relies on intuitive action.

Open-Ended Problems: Facilitate the development of creative problem solving skills. They are typically ill-defined problems that do not have an obvious single correct answer, and require a unique solution that emerges from a creative process. Open ended problems often require intuitive responses from students and the acceptance of the unexpected, which can cause discomfort for those with underdeveloped critical thinking or creative ability. It is critical that designers and engineers experience many of these 'wicked' problems throughout their education in order to develop problem framing skills, confidence in divergent thinking, and lose the tendency to fixate on prior solutions.

Chapter 11
Rediscovering Design Education as a Social Constructivist Foundation for Innovative Design Thinking

Johann van der Merwe
Cape Peninsula University of Technology, South Africa

ABSTRACT

Design has been described by Bruno Latour as the missing masses, and tellingly as "nowhere to be said and everywhere to be felt" (2005: 73). Traditionally, not only objects, but design's presence in general has gone largely unnoticed by the public, but that is changing, due, in considerable part, to the ubiquitous presence of computing technology. Design, as representative of unnoticed and neutral objects, is no longer feasible, but design, as a participative presence in the lives of its users, is fast gaining ground in our complex society. Designers are no longer fully in control of the design process, meaning design practice, and as a result design education must change to adapt to the increasing pace at which different social groups are evolving new ways of communicating and living.

INTRODUCTION

Until quite recently, it seemed as if the prevailing notion of design was based on a linear cause and effect process that relies on logic, rationality and scientific rigour, a very orderly practice that guarantees control and defined outcomes. Unfortunately, this can result in fixed structures protective of design 'truths' and hence restrictive of thought patterns, and by concentrating on what is being designed and not reflecting on why these objects are being designed, we seem to have created a design crisis in self-conception. On the other hand, much has been written about the new era of design that acknowledges the shift in focus from the object itself to the process of design, with a further shift to what is being called service design, which takes the concept of process even further (especially in Scadinavian and Finnish publications). We find that in many fields of design the importance of a user-centred methodology has been emphasised, with this general approach further evolving to include the significance of the users' experience. However, any approach

DOI: 10.4018/978-1-61520-617-9.ch011

to design research, theory and practice that deliberately includes the social element has to relinquish at least part of the 'guaranteed' control leading to defined outcomes that it depended on previously. Consider, then, this question: How do designers know when they are on the right track if there are no guarantees in design, if there are no set formulas to follow? How is it possible to design anything at all, let alone be innovative? Too many designers are trying to adapt to a new and complex world environment with a mindset that was formed during their educational phase, based largely on set formulas and ways-of-doing that just about guaranteed the outcomes.

Once a designer starts off on this road of social inclusion, however, the situation becomes worse. How does any designer progress to the point of being able to handle situations in our contemporary and complex socio-technical world? How is it possible for any one individual to be able to deal with a multitude of complex and seemingly opaque socio-technical situations that are encountered in the contexts that need design solutions? In order to know which direction (the process of) design needs to take at any given stage, where do we look to for answers? There seems to be far more questions than there are answers, and design is about finding answers, is it not? Isn't it?

Wrong. Design is not about finding set-answer solutions, as if the socio-technical relationship does not matter and will not interfere with the design process. If we only had to concentrate on the technical, the mechanical, that would be easy, and engineers used a methodology called Cybernetics (now recognised as First Order Cybernetics), a system of control that depended on negative feedbacks, hence the guarantees of defined outcomes. When this same system was transplanted to the study of the social, however, it was soon found that a negative feedback system, used to control the outcomes, simply was not possible, or indeed, desirable. People do not like to be controlled as if they have no choice, and hence no control over their own actions and decisions, and so the adaptation to cybernetic thought brought about Second Order Cybernetics, a system of positive feedbacks that studies thinking and acting human beings, a system in which the observer of that system (in our case the designer) becomes an integral part. A positive feedback system inevitably means that whatever is fed back into the system changes the system itself, the way it thinks and the way it operates (as AI researchers have found), an iterative process that amounts to a learning process. Seen against this background, if design were to be seen as a process of simply finding solutions to defined problems, it would have to use a system of negative feedbacks, and hence control over the whole process, leaving no or little space for the involvement of the social. Accept the positive feedback approach to design thinking, and the involvement of the social becomes not just indispensable, but crucial to the designer's ability to find direction, and knowing if the design process is on the right track or not; all of this depends on the total context and not on any single individual.

Design is not initially about finding answers or solutions, but about observing the system that needs design intervention, and *then* design becomes an investigative system to redefine the problem space, for, as John Chris Jones (1988: 224) stated, "the 'right' requirements are in principle unknowable by users, customers, or designers at the start". Observing the system also means redefining the so-called problem, for not only is it wrong to assume the stable nature of the problem requirements, but pre-empting the solution in this way means nothing much will be learned from the whole process, and hence very little innovation becomes possible. We all work with what most people would either call (professional) intuition, 'know-how' or tacit knowledge, the knowledge base we have built up over the course of a design career. Paradoxically, this working method can be the worst barrier to product innovation, because "At the start one's intuition is likely to be wrong, informed by what IS, but not by what is to be conjured into existence" (Jones, 1984: 136).

Designing more of what already exists, or derivatives of the same, is not what design trends should be about, not in our contemporary and very complex world with its quite specific and socially-based, and wide-ranging, *problem environments* that need addressing by people capable of systemic thinking. I will argue (below) that there are too many designers now having to work with complex world environments while using the learned thinking tools of the industrial age, set formulas that bring about guaranteed outcomes. For this reason they need rigid operational requirements, not realising that these same initial requirements can be very misleading, for the real requirements emerge from the collaborative design process (other designers, clients, users) that is, at the same time, a learning process (Jones, 1988: 223).

A good case in point is MIT's Engineering Systems Design (ESD, 2008); they have reconsidered their educational approach to design thinking, and with this process they obviously wish to influence design practice in the real world. That, I agree, can best be achieved by re-thinking design education to focus on the benefits of socio-technical relationships, thus my educational approach of focusing on the individual and his / her interactions with the world. Another department at MIT (Aeronautics and Astronautics), as part of the CDIO™ Initiative, also focuses on the individual via conceptual understanding (using constructivism as a learning tool), but they do so within the individual/group/real-world-needs relationship (Darmofal, et al., 2002). It is in this sense that the individual, as a critical thinker, can use design as a type of innovation intelligence, as Clark and Smith (2008) believe, and since design "innovation's terrain is expanding" it would be wise to use systemic solutions to solve complex real-world problems (Brown, 2008).

The purpose of this chapter is therefore to offer a version of a design educational methodology based on radical social constructivism, while at the same time exploring the emergent effects of an ontological phenomenological process that each individual (whether student or professional designer) has to experience before being able to systemically become part of a group network for design. Additionally, to more fully understand the relationship between the social and the technical, we need to look at the insights that Actor-Network Theory can bring to any discussion on user-centred design processes; many people tend to 'forget' that the word *user* presupposes the presence of a designed object, anything from a match box to the biggest industrial piece of equipment we have yet devised, and that this 'presence' is often anything but neutral and / or under our full control.

But, how can individuals think as if each constitutes a group? The individual, in an ontological sense, comes before the group, yet, seemingly quite in contradiction, the individual-self cannot come into being without the input from the group, from others external to this individual-self. Radical constructivism takes the stance that social reality is constructed, following the lead set by Dewey, Vygotsky, and Piaget, among others. Even though individual subjectivity is part and parcel of this process, it is because the formation of each individual-self depends on similar processes occurring in other subjective selves that the larger picture of social construction and re-construction can be studied, its actions investigated, and, in line with second-order cybernetics, these observing systems (each human, thinking being) can themselves become aware of their own observations. "Ceccato spoke of operational awareness, deliberately leaving implicit that he was focusing on the construction of knowledge" (von Glasersfeld, 2005:9), which, for educational purposes, means that students can learn how to learn, in the same way that Jones (1988) stated the designer has to learn what the requirements are, from the process of investigative design thinking. Students of design can thus also begin to learn that designed objects and systems absolutely depend, for their coming-into-existence, on many external-to-'design', and quite complex, factors that emerge during the socially constructed process of daily life.

Before tackling the systemic / constructivist and phenomenological aspects of design education, let's look at some of the trends that seem to be emerging in the world today.

BACKGROUND

Designers have been described as change agents, and change is definitely in our contemporary air. The question is, who determines what these changes are, what they are about, and what they mean for design as a profession? For designers to function as change agents – since 'design' needs to be about action and change – they have to examine the nature of contemporary design, and from within that viewpoint look at the trends in design.

Some of the larger companies seem to be reversing the product design trend that has had a rich economic base as its non-thinking (in sustainable design terms) *raison d'être* for many decades now, in fact, a trend that has existed ever since the notion of planned obsolescence in design was first discussed when mass production became possible. It would seem as if General Electric, Nokia, and Proctor & Gamble are "creating entry-level goods for emerging markets," but then also 're-importing' these 'new products' (at a fraction of the cost of the original design) for the benefit of first-world markets (Jana, 2009). This would seem a good idea, because in these first world economies design has traditionally been brought into being for a mass and economically stable market (with its associated buying power) without taking the real people who make up that market into account: a consumerist market presupposes a large and captive audience willing and able to listen to the product-speak of the large companies. But perhaps this reverse trend has less to do with a change in how companies and manufacturers regard their users, as it has to do with the economic recession. Such a quick change of heart regarding user expectation has less to do with altruism than with economic expediency, since "Not making a purchase was the most powerful impulse in the global economy ... Last year [2008] there was ample evidence that the muscle consumers were flexing most was that of constraint" (Hockenberry, 2009). As Jana (2009) emphasizes, the first-world customers are in need of bargains, and these companies can 'score again' by means of their 'trickle-up' innovative thinking, hence the re-importation of goods originally meant for emerging markets.

An example of this is the MAC 800, a portable ECG system that weighs a mere 3 kg instead of the nearly 30 kg for a standard machine. It comes with a laptop-type battery and a cellphone type keyboard, and is meant to improve the "Connectivity within a physician's work environment [that] is key to coordinate care and improved diagnoses" (medGadget, 2009). A pity, then, that the target market physician was a Chinese one, and that the device was not originally designed (i.e. planned) for the first-world market. As Jana's (2009) article bears out, past-their-prime products used to be passed on to emerging markets "as if they were unloading fleets of used cars", but this 'new reverse trend' is proving much more profitable.

So what is a product designer to do? Why design cheaper products for first-world markets that can well afford more expensive equipment? What does the term *affordance* mean, and who are these *users* you are designing for? Take an even bigger step back, and dare to ask "What is good design?", because this question "goes further than designers' personal need for recognition and reveals a much deeper cultural anxiety about consumerism" (Hall, 2009). To attempt an answer to the last question first, good design is not what the MoMA exhibition of 1951 portrayed it as being (*if it sells well it must be 'good'*). Hall assumes a moral dimension to this question, and we may well ask with him, "*Whose 'good' are we talking about?*" A further elaboration on this first answer would be to turn in the direction of the 'social good', and ask what can these users *afford*, in terms of buying power, but more importantly, in terms of Gibson's notion of *affordance*, by asking, can we afford to allow this mechanical / technical device to enter into our network of social relations, given the possible (some foreseen but others unforeseen) results of its presence; will convenience help us do our work more efficiently, or will convenience become the enemy (words borrowed from Victor Papanek) of an existing social and work fabric? Even though Norman differed from Gibson, in the former's initial translation and use of the term *affordance*, Norman (1999) now acknowledges the broader definition of affordance as reflecting "the possible relationships among actors and objects: they are properties of the world", and it is these relationships between user-actors and designed objects that should be the core business of design education.

So where is a product designer to find direction? Change is definitely in the air, in terms of an enforced economic re-think that makes all types of consumers, from individual users to large institutions (cf. the medical health example, above), think twice about buying-as-usual, but, more importantly for design as a discipline, change has been in the air for some considerable time now. Lancaster and Northumbria Universities jointly hosted a Design PhD Conference (2009) based on the interrelationships between sustainability, innovation and design; "The changing nature of our activities - reflecting changes in society, technology, the environment and business – demands the development of new creative directions ... and new kinds of designers and creative professionals." Findeli (2001) has warned designers about this transformative paradigm shift, and he called upon them to "open up the scope of inquiry ... and push back the boundaries of our system in order to include other important aspects of the world in which design is practiced." This call to action was addressed at the Cumulus Kyoto 2008 Conference, entitled *[Cu:] "emptiness" Resetting Design – A New Beginning*, and a Declaration was signed which stated that all the people of the world live in interdependent systems for living, a veritable groundless and perfectly cybernetic field for design investigation. This Declaration calls for the merging of the sciences and humanities, technology and the arts, and puts it clearly that design thinking places itself in the midst of this important paradigm shift, and must therefore redefine itself (Kyoto Design Declaration, 2008). These calls to action are based on real world challenges, and in *Brighton 05-06-07* (Boddington et al., 2008) a number of international designers urge the rest of the design community to accept the challenges of contemporary design, which serves as a reminder that this challenge has been taken up by design thinkers at, e.g., NextD, Doors of Perception and dott07 (the UK Design Council's *Designs of the time* 2007), Boddington et al. are asking designers to seriously look at ways to transform society through the powerful influence of design.

Calls for action and redefinition are not sent out by a few isolated individuals, but by *movements* based on design thinking and design sensibility, One of the keynote speakers at the Design PhD Conference (2009), Alistair Fuad-Luke, and Ezio Manzini, who organised the *Changing the Change* 2008 conference, are both advocates for Slow Design, a mode of interaction/living that promotes sustainability both material and metaphysical, These designers are representative of changes in design thinking, aligned with real-world paradigm shifts, finding, for instance, that in many different cultures around the world "an obsession with things is being replaced by a fascination with events" (Manzini, quoted in Thackara, 2006:6). Calls for action and redefinition also mean that, as Thackara (2006:7) puts it, the solo designer (who relies on stable problem requirements to guarantee the outcome, above) that acts as a 'celebrity' more than a facilitator of design solutions is no longer needed, being consequentially not part of the real-world context that harbours the complex systems that are shaped by the actors who initiate, construct and maintain these systems (cf. Actor-

Network Theory, below). Designers are changing their roles from individuals essentially outside the system under observation (the sole author of objects, *vide* Thackara) to being simply one of the actors among the larger group of actors that need the design facilitation; design thinking changes from an individual supplying the group with a solution, to a new individual/group relationship that allows (*affords*) design solutions to emerge from particular social contexts.

It is in this sense that individuals need to learn how to 'think like a group' (Introduction, above), and how to learn and experience, systemically, what becoming part of a group network for design can mean for (individual) design thinking. Not only is social reality constructed, but also the complex social contexts that design thinking has to function in, a living and working context that is in no small part due to the interrelationships between user-actors and designed objects, and the best way to assure a new and sustainable future for design as a human activity is to redesign design education itself.

DESIGN EDUCATION

As I have written elsewhere (van der Merwe, 2007), design education still concentrates too much on styling and form-giving, but a curriculum that does not challenge students, one that makes it easy to move from logical idea to the logic of the finished form, is not conducive to development and innovation. Despite the increase in the complexity of social, economic and political structures on a world-wide scale, and the consequent increase in the complexity of designed objects, we have inherited design as an effect of the machine age c. 1851, and as a guild-oriented arts and crafts activity, something the modern world simply does not recognise anymore. Design, both as an activity and as education, must be approached from fresh vantage points to rethink and to broaden its character, and to do so a new educational structure is required. Not only do we need an understanding of the objects we design, we also need this same instructional understanding of the users of those objects and the contexts within which both function (Beucker, 2004; Formosa & McDonagh, 2005; Gedenryd, 1998; Jonas, 1997; Kapustin, 1998; Kolko, 2005; Michl, 2002; Pombo & Tschimmel, 2005; Restrepo, et al., 2004). Not all design education curricula hold to this type of outmoded view, and one of the new design trends today is to focus on process instead of product, but the mere fact that these observations can still be made is cause for concern. "Perhaps design today is a reductionist parody of what should be a truly systemic activity" (Broadbent, 2005).

Much has been written about design methodology and the process of design, but too often, with hindsight, it can be said that to design in such a prescribed manner simply does not work, and that a much more sense-making and authentic practice is called for (Gedenryd, 1998). I would rather follow Jonas (2004) in declaring design, as a groundless field of knowledge, free from such prescription and allow design to re-create itself as a systemic activity. To do so design education will need to concentrate on the learning process itself, which means un-learning conditioned and uncreative habits, moving towards authentic, competent and cognitive design processes and practices, generic and holistic skills and understanding, and the promotion of life-long learning and development. New educational models must allow students to construct their own learning capabilities, thus structuring their own modes of knowledge acquisition, in order to liberate themselves from programmed knowledge and, instead, allow for emergence as a creative input (Albers et al., 2004; Basadur, in Van Patter, 2002; Beucker, 2004; Dowlen & Edwards, 2004; Gedenryd, 1998; Giaccardi & Fischer, 2005; Pombo & Tschimmel, 2005; Overbeeke et al., 2004).

A design methodology for future design development begins with the individual and ends with the group, but to do that we have to set aside, for now, the practical design of objects and, instead, concentrate on how and why designed objects come into being in the first place. If we do not forget that modern design is much more about the process than it is about the product, that means we can acknowledge the human element in the design thinking process, and so make sense of this statement: "It is ourselves … that are the real purpose of designing. The biggest mistake is to take the product alone as the aim. It is always secondary" (Jones, 1988: 224). Now we can start to ask questions such as, *how do designers design* (synonymous with, *how do designers think?*), and *how do they know when designing becomes innovation?* Anne Richards (2008) makes an important connection between the work of Elinor Gadon (2006) and Gerard de Zeeuw (2005), in the sense that we have to pay attention to the relationship between the individual and the group. Gadon (2006) found through self-observation what many researchers believe to be the case: quality research is often driven by personal curiosity, with the focus on "a certain quality of mind and personality", and, for good measure, Richards also found that this type of quality research "is not just about publishing for career advancement but for the social good". De Zeeuw (2005) speaks of acquiring a high quality experience within one such "social good", which he describes as 'stable collectives', making a point, however, of warning that we have to focus on the difference between knowledge and experience: "people may have knowledge, while stable collectives are high quality experience." As part of their social constructivist education my students learn to deal with the ramifications of systems theory, firstly, to prepare them for actor-network theory, the ideal vehicle for observing a stable (albeit temporary) and sustainable 'collective' in the making, an emergent situation only made possible because of a co-design collaboration between a social group with specific needs, and the group of (student) designers who wish to create designs for a civil society: design for the 'social good'. Both 'personal' knowledge and group 'knowledge' can be engendered at the same time, and the 'design situation' of the group dynamic provides the high quality experience, something that many professional designers cannot seem to observe while it is happening.

Secondly, in re-thinking design education to focus on the benefits of socio-technical relationships, it means my approach to this situation is to not regard designed objects as of primary concern, but to begin with, as well as to observe, the smallest element in the design / artefact / social / world relationship: the individual and his / her interactions with all things external to the self. It also means that students may benefit from the insights of a study of phenomenology, and specifically the ontological phenomenology of Heidegger, which not only deals with the socio-technical relationship, but sheds light on the seeming dichotomy of the individual / group problem. It is a contemporary given that designing for the social good, while dealing with a formative interaction between the individual and the group, is a matter of complexity-in-action.

The world that the designer has to deal with, be *concerned* with in Heidegger's terms, is a complex one, and we are not educating students for meaningful work once they enter this new world, according to Davis (2008). This author believes that design education treats the *fact* of complexity as a *problem* to be 'designed away' through positivistic reductionism, an old scientific treatment of anything that 'lacks' control. Davis deals head-on with the increasing distance between the real-world circumstances of design practice and what is being taught in our design schools, citing this (negative) relationship as 'disorienting', a term that is very apt in the light of the discussion to follow. Design education, as I believe large sections of design practice still demonstrate, avoids the complex problem of the making of meaning in the first place, and the still greater problem of 'designing' that meaning into an artifact / system, in the second place. Not only is design as educational subject, and design as practice, a complex relational system instead of a straightforward how-to model of simplistic form-giving, but students can, very early on, learn to deal with design basics while becoming familiar with the expressive qualities of the complex and the applied (Davis, 2008).

Looking forward to what the designer of 2015 should be capable of, a recent AIGA research project highlights a number of trends for the future of both design education and design practice (Davis, 2008), the gist of the report being corroborated by the *Design for Future Needs* (ICSID, 2001) project, run by European designers in collaboration with the business establishment, on behalf of the European Commission, while the ICSID/Danish Designers (2008) research project concentrated on design's vital role in society and the impact this should have on education:

- Design is being described as a 'weak' discipline *because* it has to deal with increasingly complex real-world challenges, which means design is seen as interpretative instead of prescriptive, forming a social discourse about reality in which hard facts are subordinate to perception and experience.
- Design is in need of new sources and a new basis of knowledge that will act as a new foundation for meeting the needs of users of the future, necessitating new techniques and practices that can responsibly and sustainably deal with emerging issues.
- Being a 'weak' discipline (cf. 'weak' nuclear forces), design operates on the strengths of a many-method platform while taking account of new and hybrid technologies; each new project for social interaction uses this new platform to blend such methods as are locally suitable, resulting in multiple future visions applicable globally.
- This approach leads to underscoring the importance of community, impossible without regarding the people we design for as co-designers in some form of collaborative and participatory process; design methods are now user-centred and dialogue-driven, fostering understanding between diverse groups.
- With the original meaning of the word 'design' changing from form-giving to design-as-planning we can look at the value-added potential of design as an innovation tool
- All this does mean that students have to learn how to work in large, interdisciplinary teams in order to deal with a complex and largely unknown future while acquiring the capability to effect transformation. Working within the socio-technical sphere and mining data-into-information from the users themselves will enable future designers to generate rich results from what appears to be poor and scattered inputs.

One thing, above all else, is non-negotiable: "We have to kill the myth about the solitary designer, who can see through the complexity of this world alone" (ICSID/Danish Designers, 2008), and nowhere is this changed vision for design made any clearer than in developing countries. Reminiscent of the statement that Danish design students "possess a large degree of humanistic understanding" (ICSID/Danish Designers, 2008), Tim Brown, the CEO and president of IDEO, states that innovative design thinking for developing countries can supply the necessary alternatives to our Western, developed modes of practice, and for this to happen we need to take a systems view, the many-method platform capable of accounting "for the vast differences in cultural and socioeconomic conditions" (Brown, 2008:90-91). The socio-technical example he cites is the case of India's Aravind Eye Care System, a company that "built a systemic solution to a complex social and medical problem" by not just simplistically lowering their prices and providing cheaper products, but by sourcing ideas for quality and equivalent products from the communities they serve. The preferred (yet to be designed and manufactured) products had to comply with two constraints: the poverty of the social groups being designed for, and the fact that Aravind did not have easy access to the (expensive) solutions available to developed countries. The innovative design-for-the-social-good solution, forced upon them by the circumstances of their clientele, was to design and manufacture their own intraocular lenses, using alternative technology (cf. new and hybrid technologies, above), to be able to offer their users the equivalent of $200 Western quality lenses for $4 a pair.

Taking a systems view of design's development also means that ideas come directly from the community, and are thus user-centred and dialogue-driven. During 2005 the Bank of America commissioned IDEO to create a new product, one that I recognise as corresponding to a pattern of behaviour from my own daily life, and a product designed to establish and maintain brand loyalty. Working as a collaborative design / industry team, they called this new savings account "Keep the Change", since that focuses on consumer behaviour that the consumers themselves would willingly transform into something more positive, since we all know that the change we receive from a cash sale somehow 'disappears' because nobody really keeps track of what happens to it. Now you can 'siphon off' the small change left over from a credit card transaction and transfer these amounts directly to your electronic piggy-bank account (Brown, 2008:92), and it really does amount to a substantial figure at the end of a year's worth of indirect saving. This same basic idea has been put into effect in a very practical and sociocultural way, and it is called a *stokvel*, a community-driven solution to poverty that was 'designed' by ordinary black people in South Africa. Stokvels are very 'low-technology economic instruments' in first-world-speak, and means that a number of people get together to create an informal, mutually supporting, savings society, every member paying a fixed amount into the stokvel on a monthly basis, with, usually, each of the members being handed the full amount, each month, in turn. "Despite the high level of unemployment in SA ... poor people do save [despite unconventional methods, because] they realise the importance of saving for funerals, education and emergencies" by forming stokvels (Shezi, 2005). On a par with the Bank of America idea of creating a product sourced directly from an activity inside a community, First National Bank (FNB) designed the Stokvel Account, targeting the poor who could not or would not make use of any form of banking facility in the past. No fees are payable for cash withdrawals or deposits, all transactions are physically recorded, two designated signatories from the group are able to access the funds, and all movement of monies are regulated by a constitution drawn up by the members of the group (FNB, 2008): in other words, a product was designed that incorporated the best features of a working system already in place, making it more transparent and safer.

In the West we hardly need more of the same products we have been offered, by design, for so long. What we do need is something else, hybrid 'products' that will make more sense of our lives on a personal, on a social group, and on a sustainable future level, and these hybrids "will be complex combinations of products, services, spaces, and information" (Brown, 2008:92), in fact, hybrid products that make full use of the possibilities inherent in a social constructivist approach to design thinking, based on innovative knowledge of the socio-technical interactions of our daily lives.

RADICAL SOCIAL CONSTRUCTIVISM

In the constructivist classroom teaching and learning must be approached with the assumption that the teacher's personal constructs of the world cannot be made readily available to the students, as if these constructs could be found in a textbook on design. Each student must of necessity construct his or her own version of the world, as well as the information in and of that world, as it appears to them; at best they can compare notes on what they have discovered, and so correlate their constructs (their 'designs' of what the world looks like to them). Here, however, von Glasersfeld (2005:5) makes an important point from which I extrapolate the following: in correlating your constructs with those of your fellow constructionists, you are, in fact, valuing your 'new' construct against another construct that not only already exists, but a construct that is based on previous interactions between other constructs and the world, as it appeared to *them*. Our knowledge of the world are interpretations of information already interpreted by other actors-in-the-world, and while it is accepted that knowledge is an individual construct, it is also known that this process is based on the communal interpretation-of-interpretations of the world and its interactions with all the actors in our information sphere.

The point made by von Glasersfeld is that the term *environment* has two meanings, depending on your observational viewpoint, and the capability we have to acquire is to observe how these two can merge into one:

On the one hand, when we speak of ourselves, environment refers to the totality of permanent objects and their relations that we have abstracted from the flow of our experience. On the other, whenever we focus our attention on a particular item, environment refers to the surroundings of the item we have isolated, and we tend to forget that both the item and its surroundings are part of our own experiential field, not an observer-independent objective world (von Glasersfeld, 2005:5).

To individual students interacting with the world, their total environment consists of all manner of formal and informal teaching environments, which include other students, teachers, family and friends, as well as information from texts, objects and conversations / discussions, images from popular culture and entertainment, and (to a student, *really* 'external' to much of their 'world') news from the rest of the world's events. It could seem as if information from these disparate sources is simply readily available to be picked up by the individual, take-aways to be transformed into knowledge by the act of assimilation, as if we can unthinkingly consume them instead of having to experience them. The *flow of experience* means we have to 'tell' ourselves what we are experiencing, and that 'telling' means interpretation, leading to the making of meaning: we abstract from these sources what we need, and this *need* will vary from context to context. We do not, and cannot, allow all of the possible information coming from these sources, which constitute our experiential field at any given moment, to enter into our meaning-making process, since that would simply create a sensory overload. We abstract and select, turning the flow of experience into a filtering system at the same time, thus designing a process of individuation based on the collective.

A non-constructivist approach sketches a different scenario. As human knowing organisms, ones that can 'draw' information towards ourselves and turn this into knowledge, we assume the same process is applicable for all other organisms and objects outside our own environment. As von Glasersfeld points out (above), when we focus our attention on a particular item outside our own environment - an object, a book, a person – we tend to see these as discrete items each with their own environments that supply information to be turned into knowledge, which is in turn embedded in that item. What can be easier than to extract this knowledge from the book, the object, even the person (i.e., the teacher)? There is one small problem, however. There are various means of extraction, but any process used turns individual knowledge back into mere information, because knowledge cannot be transferred *intact*, it has to be constructed, or, rather, re-constructed, by the individual receiving that information. My environment and your environment may look different from each person's observational viewpoint (which includes literal *and* figurative perception, as well as personal experience and memory), but when I instigate the interaction, *your* environment becomes another item for *my* environment, and must be treated as such, as another item / object / event in my experiential field that can add to the information mix I need for my flow of experience to be activated.

There is only the one, larger, experiential field of possibility that all human systems can make use of, to varying degrees, depending on the particular context, and the processes we use to make meaning, to construct knowledge *for understanding*, these processes need to be *holistic* views of the complete network of relations created by people, activities, and technologies, instead of treating any item (especially technological devices) as if it could be understood in isolation from the rest of the network (Winograd and Flores, 1988:6). This approach, of course, correlates with the views taken by Richards (2008), Gadon (2006) and de Zeeuw (2005), discussed above. There will always be a relationship between an individual and the rest of the group, who are, lest it be forgotten, all individuals themselves, and we must become capable of discerning the web of relations between all the individuals in any grouping around human action issues. Knowledge is supplied by each of these individuals to the group, but from the viewpoint of an other-than-self individual, the pool of 'knowledge' accessible is at best quality information based on narratives of individual experience. Only *participation* within the *concerns* of this group (to borrow a term from Heidegger) will give me access to what this quality information / 'experience database' can offer and afford the individual-self who wishes to tease out the emergent properties within this new group database, also known as a collective memory.

A mobile library, for example, with its smaller database in the form of a limited number of books, connects a group of individuals to a central and much larger database, thereby creating a relational network between people and the potential knowledge in a couple of hundred books. But the 100s of mobile library books are supported and enhanced by 1000s in the central library, while these in turn are backed up by as many other depositories of cultural memory as our contemporary electric culture can offer, making this network essentially endless, without boundaries. But the question has to be asked, does all of this simply represent *storage space*? What is being stored? Not knowledge, for only data and low level information can be stored and kept. All the texts in the Alexandrian library were not much more than cellulose and keratin crumbling to dust, if it were not for the capability of human memory, and shared memories within the group. "In this concept memory is not a 'container' for remembrances, not adherence to the past, but rather a prerequisite for communication" (Halter, 2005:81). As thinking human beings we are who we think we are only through negotiated, interactive, and therefore constructed, communication, the most important culturally designed 'product' human beings have ever devised. As an essential 'technology', communication in its many forms is, really, all we have, leading Luhmann (2002:156) to state, "only communication can communicate" in the networks we have to build, and the resultant actions we decide upon are the result of our understandings generated within these (actor) networks of communication. Memories, and especially collective memories, are socially constructed narratives that are re-told with each interaction within the group, and with each contact between different groups.

For the purposes of design, as part of the socially constructed process of the reassembling of parts to form new wholes, the notion of a collective memory is worth exploring, particularly to combat the comparison between the human brain and the computer, as storage space, "yet another anthropomorphic technology model of the Modern Age" (Halter, 2005:79). Citing the work of Esposito (2002), who in turn based her sociological theory on Luhmann's theories of social communication, Halter (2005:82) points to the contrariness of memory's strength: what is 'stored' has more to do with *forgetting* than with what is in fact 'there', with *remembering*. Not only does memory not store contents as such, but it makes much of what contributes to 'memory' quite redundant, a quality without which memory could not help us identity differences, and certainly without which variety, which leads to the recognition of innovative ideas, would not be possible.

As human thinking beings we are pattern recognition organisms, and are used to dealing with what might appear to be fragments (parts) instead of wholes. Condensing information as pattern in memory is simply a cost effective way to deal with the world, specifically in our complex modern world with its ever-threatening information overload (always at hand, like a brooding presence that has to be taken notice of). Our patterns of perception, in the first place, are mediated by thought, not based on the Cartesian split between mind and body, nor based on the Kantian disinterested thought as objective observation, but thought processes that are constructed because of our interactions with everything external to our self as individual, thus making thought socially constructed. The seeming paradox, that the information quanta obtained through our senses can be 'mysteriously' increased in the image-schema areas of the brain, can be explained through the contrariness of memory that 'forgets' and makes large sections of information 'redundant'. When we use an expression such as 'having something in mind', are we referring to just what went into our memory banks and was stashed away, unadulterated? Not only is social reality constructed instead of corresponding to an objective reality 'out there' in the *real* world, but our memories are similarly and socially constructed, and worse still for positivistic realists, what we 'remember' about so-called reality is nothing but abstractions. Social constructivism is *radical* in this sense only, that what we perceive, feel, think and remember, are constructed image-schemas devised by us, the *collective*, even though we would call the image schemas we live by uniquely, individually, ours. I stated (above) that, following von Glasersfeld, we take our *flow of experience* to mean we 'tell' ourselves to ourselves, and this is applicable in both directions – perception *and* memory. This interpretative process allows the making of meaning, and abstraction plays a vital role; we abstract from our sources what we need for any particular task at hand, and if the source is our own memory, we still cannot allow all of the possible information coming from our sources, which constitute our experiential field at any given moment, to enter into our meaning-making process. We abstract and select, turning the flow of experience into a filtering system at the same time, thus designing a process of individuation based on the collective, and designing a process of remembering based on the collective and the forgotten, except that the proper name for this 'forgetting' is 'convergence'.

Whenever you retrieve a 'memory' you are, in fact, 'retrieving' a convergent pattern that is also an augmented version of what *you* may call an individual memory, but which is anything but. When we perceive something, we do not *see reality*, but patterns created by this mode of interaction, and this act of quite considerable abstraction focuses on just the essential features of what is perceived. Call this convergence-from-divergence, or less from more, but we appear to lose information in this redundancy process, were it not for the fact that (abstracted) convergence loses nothing at all, not only requiring less storage space but, as mentally compressed 'zip files' the convergent process carries within itself the possible bounty of the divergent circumstances from which it was abstracted. Socially and constructively thought-into-being, as it were, every time you remember something, because what emerges is an augmented version of what went in, interpretations of previous interpretations, created by personal knowledge expressed via the individual memory fragment recalled, but now (as happened previously) combining with the new data obtained from information in the world, to create a new memory fragment to put before the group. What we need to bear in mind is that all perceptual inputs are stored in what appears to be a fragmented process, parts of the 'whole picture' stored here, parts way over there. The smell of the coffee here, the impression of the café setting over there. You may or may not recall the intact picture, because the first augmentation happens on the first pathway followed to the storage spaces, and the second augmentation happens on the way out, as a remembered 'fact' that is, really, 'contaminated' by other memories along the same pathway. All memory parts are combined or associated with information 'already there', which makes it possible to retrieve more information than that stored, because retrieval requires those same associanistic pathways to different loci in the brain, and this second augmented production (reinterpreted mindful artifice) has the possibility of enhanced complexity because of the specified pathways (Bohm and Peat, 1989:64; Dembski, 1997; Paivio, 1977:48-51; Pribram, 1977:172). The very possibility of clear, innovative thought and 'flashes of insight' is created by this *new* flow of experience, first because of your need to abstract from this, your own memory-as-source, and secondly, in the group dynamic, and according to Kelly (quoted in Bannister and Mair, 1968:2-4) this version (of 'the truth') has as much relevance to us as the objective and so-called *real* world out there.

"But how can a concept of memory that is simultaneously and exponentially characterized by forgetting be relevant for design [if] The point is … to mediate between man and machine to enable meaningful communication"? (Halter, 2005:83). One theory that investigates these mediated relationships, and especially the more difficult (for being 'missing' in the sense of unnoticed) socio-technical relationship between human actors and the non-human actors called designed objects, is Actor-Network Theory.

ACTOR-NETWORK THEORY

The paradox of *remembered redundancy* can be explained in another way. In our dealings with other people, other actors, we can be seen to operate as if on automatic pilot, for how do we *know* how to act, what to say, what *not* to say, how to respond adequately or even just within the bounds of propriety? What is appropriate behaviour, and do we just follow the rules that we were taught as if reading from a text book? How, then, would we deal with new and unforeseen circumstances, and would we be able to 'think on our feet' in an emergency? Apropos of this, how do we 'know' how to operate the many and varied mechanical and electronic designed devices that fill our homes and work places? Can all this knowledge be put down to training and / or following the rules, and would this obedience (really a form of obeisance) suffice when things go wrong?

Actor-Network Theory (ANT) is a way of observing the relationships that develop between human beings and all manner of designed objects and systems, and it treats people as human actors while treating machines, designed artefacts, systems and even events as non-human actors that play roles within this created relationship cycle. It also investigates how and when people acquire this knowledge to act, but can also focus on actions and decisions that are obviously out of line with the relationship cycle that was established by any grouping at the start of the relationship, whether that grouping consist of a single user and a designed object, or a number of people attracted to and by a collective interest. The latter situation can normally obscure the fact that many non-human actors are involved, and that the very issues around which the human actors are grouped can themselves serve as actants in the total relationship. ANT, as a way of seeing and understanding the world, is thus normally concerned with the human / machine interface, but it can also shed light on the relationship between two communicating human actors, and begin to answer the question, how do we *know*?

The paradox of *remembered redundancy* is twofold, and we dealt with one aspect, in that memory, as source material for action, can be augmented on two counts, in order to deal with everyday situations and the designed objects that play non-human actor roles. How do we *know* how to use / operate designed objects, especially in the case of interface design? We 'remember', and bring this knowledge, twice augmented, to the new situation involving the designed object. The larger the gap between the designer's field of experiential knowledge and that of the user, the more chance there is that the design will be boring (got that, done it before), too new (how *do* you turn this thing on?), or dangerous (let's see, if I push *this* button …). How do you as designer *afford* the user the accessible knowledge to operate your device (i.e., design into the device the *information* to be turned into *knowledge*, on which to base a *decision* that can lead to an *action*)? The users will socially construct knowledge of the situation, including how to deal with your designs, *as non-human actors*, by following their own paths of discovery, as always. Do I, as user, recognise the role that your design is playing, or not? Make this role part of my flow of experience, especially if your 'new design' has to rely on my *new* flow of experience (above), and I will be able to construct the knowledge I need to use your end product.

As a designer you can also rely on the second aspect of *remembered redundancy*, which is simply a part of our communicative makeup. After writing *Les Misérables* the author went on holiday, but wanted desperately to know how the publication was faring. So Victor Hugo sent this abstracted letter to his publisher, containing only one symbol, namely "?". In reply he got this answer, "!". How do we *know*, and where does this information come from? Much of it is already there, and only *seems* to be forgotten or made redundant. Norretranders (1999: 92; 109) calls this contextually relevant information that we can have access to, *exformation*, socially constructed information that both parties, above, still had access to. In another context and at another time Hugo's "?" would have been meaningless, but after writing his latest book the abstracted *in*formation made sense because it was backed up by the still present *ex*formation. Design a 'new' device and users can still access the information they will need because of this phenomenon; to paraphrase Norretranders, a design will have meaning to users if it contains a large quantity of exformation. "We do not consider the fact that there is more information in an experience than in an account of it. It is the *account* that we consider to be information. But the whole basis of such an account is the information that is discarded" (Norretranders, 1999:109). The individual can gain information leading to knowledge from the group high quality experience (de Zeeuw, 2005), and much of it is based on the construction of exformation. Do not rely on the *account* (the actual, surface design) to convey all the information that is needed, but communicate with the user via this largely untapped and 'discarded' basis of knowledge, in experience. Let the role that the design can play in a person's life be part of this construction of knowledge.

As a way of sharing this knowledge among a group of like-minded people, ANT is a methodology that works well for designers, but we must remember that, as a theory, ANT is not applicable to anything (Latour, 2005 141). Any good methodology, as a good theory, is not a method for doing anything, and it is not a recipe that guarantees an outcome. Latour's (173-74) description of how ANT can be useful includes "specific tricks to help resist the temptation to jump to the global", because, in the very act of communication, in forming networks (even if only between one local actor and a second), we are moving ('jumping') from the local (the individual) to the global, to the many, to the away-from-us. "A ploy has to be found to make the two social theories diverge, letting the sociology of the social go its own way while the sociology of associations should be able to keep drawing more and more accurate maps." ANT is, simply put, [1] a way of looking at the roles that human and non-human actors play in any one situation, with non-human actors identified as designed objects, rocks in the road, the snow in winter that will not let your car start, the window that sticks and will not open. Non-human actors are all the external-to-being-human objects (even events) we notice when they do not work as they should, and, more seriously, objects that intrude in our (un)consciousness and our decision-making processes even though we are not aware of the fact. ANT looks at the relationship we have with these objects, and considers the observation of social systems as of necessity having to include both. More to the point, but more difficult to 'see', is the ANT approach to the context of the social. There is no such thing as 'the social' in ANT terms, because "In the alternative [ANT] view, 'social' is not some glue that could fix everything including what the other glues cannot fix; it is what is glued together by many other types of connectors" (Latour, 2005 5). When ANT speaks of the social, [2] what is meant are the associations entered into by the members of that 'social' – which can be a single conversation between two people, or a whole group of like-minded designers coming together in order to decide on the way forward, for the sake of a sustainable, living future. The 'social' is a temporary concept, which might last for many years, or centuries even, but change it most certainly will, and in our contemporary age change is not only inevitable, but quick (van der Merwe, 2008).

So, what has all of this to do with design education, and, more to the point, with design practice? What can help people who wish to adopt a way of life that focuses on long-term sustainability? ANT proposes a number of phases any individual (in joining a like-minded group) has to go through for the theory (of looking at the world differently) to be given any chance of changing your mind, about anything. ANT can tell you what to look out for, but it will not, and cannot, help you make up your mind. Because ANT is also known as the sociology of translation, everything we touch (everything / everybody we reach out to, whether physically or with our thoughts, we 'touch' and explore phenomenologically) we simultaneously translate. That would normally mean we interpret the phenomena confronting us, in order to understand, not the phenomena themselves, but to understand our own selves, in translation. Everyday contact with the other, every interpretation of the external, is a translation of our old selves into our new selves. We are not really interpreting or translating the other, but moving our old understanding from its previous position to its new position, in relation to this new association. As far as I understand the ANT process, most descriptions seem to agree that there are four major phases to go through, and, it needs to be said, I am appropriating this theory to the extent that my understanding is a design-based sociological translation of the actant texts I work with, and we will only deal with the first two phases:

Phase 1: *Problematization*, which simply means you, the focal or main actor in the system, identifies something outside your system / knowledge environment that does not fit into your scheme of things (who, what, and how you understand, anything), and, this is so important, you must suspend disbelief, must imagine, for a moment, that you do not know all there is to know. That is the first moment of translation, the first obligatory passage point (OPP), when you 'recognise each other', when you recognise that there is something external-to-the-self that you need; where design knowing would start to veer off is in ANT's recognition of the focal actor's role, not so much in this first phase, but the implications for the second phase, and, particularly, with Callon's (1986) notion of this focal actor "rendering itself indispensible".

Phase 2: *Interessement*, the second moment of translation (and here understanding will start to founder if we do not acknowledge that translation must be admitted into the process *as* transportation, which in turn, creates the environment for transformation), is usually the persuasive phase of enrolling others in your network of belief / understanding. However, here is a second point on which the usual ANT process, as described (a little too pre- and in-scribed?) by certain texts, differs from design / systemic thought. Interessement, as a second phase of translation ... stop. What are you 'translating'?

In the first phase, *Problematization*, the focal actor (you, as the individual 'I') identifies that, as a system, something is lacking, something that is vitally necessary for the continuation of the (your) system. Your gaze upon the world-out-there, the other systems external to yours, is initiated because of a need within the system itself. It would be a *falling back into* (cf. Heidegger's *Verfallen,* below) the trap of concepts that ANT can replace (in terms of constructing a human as opposed to a social world viewpoint), in the sense that these concepts manufacture meganarratives, or stories of conduct that become inscribed, in no uncertain terms, in the social book of conduct, the very notion that Latour says is false! So, what are you translating, with a weather eye kept on transformation, if what you need for the survival of your system depends on the acceptance, by the other systems, of your narrative, or system of survival? Allied to radical social constructivism (RSC), the notion of autopoiesis (self-generating systems) and second order cybernetics (observation of observing systems) position our understanding on the knowledge plane where each system reacts 'selfishly', because "it" needs to survive. This, on the face of it, seems to support Callon's contention that the focal actor should render itself indispensible, but that would be a one-sided interpretation with no transportation at all. An autopoietic system has no interest in rendering itself indispensible to its environment, but it does have an interest in rendering its conduct – its interaction with the external environment – congruent with a sustainable future: a system cannot draw from its environment to the extent that this environment becomes unstable and therefore unsustainable (to its own survival). Cybernetics tells us that, as a designer, you want to keep the second phase of *Interessement* as neutral as is humanly possible, therefore you let the 'translation' circulate where it will (along with its shadow, meaning your specific version of 'it') to, not 'persuade' others into your circle of knowing, but to offer your new-self-understanding of/an object, a person, an event/ to whomever needs it and can use it. That is called metadesign thinking, which is nothing, and potentially everything (new).

For a design understanding of ANT we need not go further, and it is my contention that Callon did *not* state the necessity of the focal actor's indispensability, on the contrary, what he described were "attempts by these researchers to impose themselves and their definitions of the situation on the others", in clear violation of the first ANT principle of *agnosticism*, i.e., impartiality between actors (Callon, 1986). The designer, as the first focal actor, must deal with this issue of impartiality, and be willing to become just one of the actors in the design relationship with the users *and* the potential product designs-in-the-making as developing non-human actors. Callon (2005:181) later confirmed that one of the strengths of ANT is the fact that a theory of the actor is an unstable one, and *not* fixed; biological systems need *stable* environments (above) for symbiotic survival, but social systems need *unstable* environments for learning and development; "rather, it [ANT] assumes the *radical indeterminacy* of the actor" which approach breaks with orthodox social science viewpoints, but agrees with that of RSC, which also supports, as does ANT, the dissolution of "the sterile individualism / holism dichotomy" (Callon, 2005:182). The quality of a user *and* a designer who can construct the capability of moving between the individual and the group, of answering the question, *how can individuals think as if each constitutes a group?* is dealt with by social constructivism, by ANT, and, as we shall see, by an ontological phenomenological approach. In focusing on the after-ANT scenario, Law (2005:10) makes it clear that, in dealing with complexity issues, the diasporic nature of ANT means it transforms itself, perfectly adaptable to the design contexts it can help illuminate, and like RSC the framework for understanding is constructed *inside* contexts, using the elements *from* the grouping, in the relational context surrounding the design problem space. In concord with the issue of memory storage and more from less, Law (2005:12) describes the understanding of the after-ANT scenario as one that recognises neither singularity nor multiplicity or plurality; "this is something that is indeed *more than one and less than many*", refusing the easy single / plural opposition in favour of a relational emergence, much the same way that Giddens (1986: 3; 14; 24) refused the opposition between agency and structure. Understanding, as a framework-for-action, has to be constructed somewhere in-between the single and the multiple, the individual and the affordances for structuration to be found in the 'quality experience' of the group.

When we consider the importance of communication, it has to be borne in mind that what is of essence, for understanding human relations, is the associations being entered into in the very flow, in the act of true communication. If there can be a glue that holds everything together, it is this. New associations mean new ways of augmenting existing knowledge, in fact, it means that learning takes place, and that means the new individual is at the heart of this association, or reassembly, as Latour prefers to call it. ANT, it must be remembered, is also known as the sociology of translation, and it is well to remind ourselves that the act of communication (any relationship with the external-self) is at the same time a transportation of both what we know and of the information we are dealing with, a flow from one contact point to the next, and in this flow transformation takes place. Giddens (1984, xx), in focusing more on 'ontological concerns' rather than on epsitemology, meant by this statement that his structuration theory of the social "should be concerned first and foremost with reworking conceptions of human being and human doing, social reproduction and social transformation". To designers, these 'ontological concerns' should mean, very simply, an ongoing and constant becoming (Heidegger 1962, 359), the crucial point, to both Latour and to Giddens, being that the people making up the grouping, and the designed objects and systems we surround ourselves with, cannot be separated as if they do not structure each other. In this duality of structure, the human need to voice what cannot be seen (human-world experience) nevertheless demands shaped expression, which becomes literature, art (Steiner 2002, 279-280), or becomes design, an outward expression or manifestation of who we are, and what we want or need to become (van der Merwe, 2008). The necessity of rethinking design education and the role of design in most peoples' lives is thus of paramount importance.

ONTOLOGICAL PHENOMENOLOGY

That, I believe, can be achieved by making use of, amongst other 'like-minded' theories of being-human, Heidegger's notion of ontological phenomenology, or at least my ANT 'translation' of the term. Phenomenology is the study of how we look at and experience the world, and it is a study of subjective experience, following Husserl, although some have used this approach to deny any experience beyond the concrete objectivity of what is. Others have contrasted phenomenology to ontology, which studies the essence, the very nature of existence. Put the two together, and Heidegger's ontological phenomenology starts to make sense, in that individuals can study themselves as developing structures, coming-into-being through a movement away from the *what is* existence of a concrete world, towards a possible existence in renewal. Ontological phenomenology allows each individual to map their own progress, even (especially) from within the group, of like-minded people, of course (van der Merwe, 2008).

Heidegger should never be read as if one could interpret his texts too easily, and never literally, since, very much in the vein of Merleau-Ponty, he writes about the visible only in order to make us see the invisible, something which has an intelligence of its own, as Findeli (2001:12) suggested, in advising design education to understand human social systems from within, using a qualitative, phenomenological approach to do so. This makes the work of Heidegger wonderfully relevant to design thinking, because you simply cannot 'see' the process and the experience of design directly, but have to describe it in a circumspect and indirect way, via the actions, intentions, and (more difficult still) the emotions of the people involved. When a designer thus claims that a designed object does not exist in phenomenological terms, it means the design is not interactive in any sense and does not contribute anything to the flow of experience of the user. Heidegger (1949) maintains that only man exists, but not rocks, trees or horses. This does not mean that human actors are the only real entities, but it does mean that we are the only 'designs' that are emergent and ongoing, while rocks, trees, housing estates, cars and animals are all 'finished designs' as far as 'existence' is concerned. That marks a difference that is crucial – ontological phenomenology studies the ongoing and emergent nature of peoples' *existence*, not simply *what is*, and not the everyday lives we lead in Dasein's *Verfallen*, but a future and better existence that is being 'designed' as we speak. The world designs us and we design the world, e.g., the effect we have on our environment (including other people) through our actions and our interpretations results in changing that very environment, and this change, in turn, again affects us as we interact with our (new) environment in Dasein's everydayness. Heidegger used the term *Dasein* to indicate a person going about his or her ordinary business, in a state of *Verfallen* or falling back into the world of things and events, so much so, that nothing at all can change; we simply do not notice, being far too busy 'living'. For Heidegger, *existence* means Dasein can and should become aware of the *Being*ness of a better and other-self, something not to be found in appearances, but in action and change, and therefore something that can be 'found' when things 'break down'.

I am not so much concerned with physical breakdowns as with an interruption of concerns, e.g., when we do not 'notice' a piece of equipment, or do not 'notice' the way we drive our cars, then we are in a state of *Verfallen*, so busy 'using' the equipment that these designs become 'missing' in Latour's sense of the term, and we take them for granted by not noticing, not so much the things themselves, as not noticing our own interactions and the possible consequences of say, bad driving habits, because the car has ABS brakes and power steering, designed devices that will keep us safe, as if these were real human actors 'in charge'. The design (phenomenological) 'breakdowns' are things such as beeps that sound when you are about to exit the car without switching off the headlights. These designed interventions can become intrusive, or they can become part of the lifeworld system within which they can prompt a change in action, for the better. The design of industrial products can benefit greatly by our studying not only the phenomenal user worlds of experience, but also the gap-creating breakdowns in the use of the most ordinary tools. I have yet to see the new designs for a one-handed chainsaw, and my challenge to students to design a one-handed power drill has not yet been taken up. Tree fellers use chain saws in situations that were never designed for such use, and I have used a power drill two meters off the ground, clinging on to a ladder with my knees. Users will always interact with your designs in ways that you cannot foresee, and you cannot design for all possible circumstances, but you can use the notion of breakdown to innovatively redesign equipment so that the gap between knowing and not-knowing can be bridged.

Truly innovative designs will emerge from this qualitative, phenomenological approach, because, as human, biological systems we are not different from the other "complex adaptive systems in the biosphere [that] evolve to a natural state between order and chaos, a grand compromise between structure and surprise" (Kauffman, 1995:15). Breakdowns in the normal flow of experience, and specifically those that involve devices more complicated than a screwdriver, can be a source of innovative information to not only improve designs, but that lead to new adaptations, and from there be a springboard for totally new devices. All we need to remember is that one of the Blackberry's grandfathers used to be an old bakelite landlocked telephone, and that the *ontological* flow of experience (of dialogic communication) experimented, as it were, with these swings between structure and surprise, between what seems to us to be order (while it works) and what could turn into chaos, when it either physically breaks down, or we experience a 'breakdown' in the flow of experience because we want something else, something that *adds more value*. As Prigogine (1980:95) reminds us, to be alive as a biological (hence an *adaptive*) system, to *exist* in Heidegger's sense of a constant *becoming*, means that feedback mechanisms (feedback loops) are of prime importance, and it is more than tempting to see structure, in particular one that flirts with Kauffman's structure / surprise *indeterminacy*, as a breakdown of the *stability of systems*, not something that we want to avoid, since this surface appearance of instability simply points to "a situation of non-obviousness, in which recognition that something is missing leads to unconcealing … some aspect of the network of tools that we are engaged in using" (Winograd and Flores, 1988:165).

Willis (1999) calls ontological designing a radically different approach to the practice of design and to the understanding of the objects of design, which viewpoint would correspond to the insights of Heidegger. However, she also states that ontological designing involves equally different ways of looking at ourselves, who and what we think we are, and our 'place' in the world. That this makes ontological designing sound like a cross-over between social constructivist and actor-network theories is no mere coincidence, since to Willis this essentially describes the relational network between people and the spheres they inhabit, between human actors and everything in their lifeworlds, including Heidegger's views on the ontology of equipment. Winograd and Flores's reference to *unconcealing* is derived from Heidegger's use of the Greek term *aletheia*, thus equating 'truth' with *unforgetting*, and when one *unforgets*, something is brought forth from a *space of hiddenness*, thereby equating truth with discovery (Dreyfus, 1991:270). The term 'truth' is, of course, used in the sense of an *ontological development* that is emergent in human action, supported by reflection throughout. Heidegger (1962:62) made it clear how, in a process of ontological understanding, an alethic truth may be discovered that can serve both human and non-human actors ('some aspect of the network of tools'). In this very process of *understanding-towards*, we strive to uncover both the beingness and the structure (the ontology) of Dasein *and the ontology of equipment*, because, *in so doing* we create the very conditions that can further uncover both the beingness and the structure of entities that are not-Dasein, i.e. non-human actors, conditions that are described by Willis (1999) as "the breakdown [that] opened up a space for interpretation, an opportunity towards disclosure of the thing itself", that design space we can create and where "We are concerned with what happens when new devices are created, and with how possibilities for innovation arise" (Winograd and Flores, 1988:177).

Heidegger (1962:98) named this ontology of equipment, *readiness-to-hand*, and it is in this space of becoming that new and innovative product designs will be forged, in opposition to run-of-the-mill copies of existing (as in *merely occurring*) products. The beingness of tools that we chance upon through *readiness-to-hand* "can be exhibited phenomenologically if we take as our clue our everyday Being-in-the-world, which we also call our '*dealings*' *in* the world and *with* entities within-the-world" (Heidegger, 1962:95), a 'hindsight' endorsement of Actor-Network Theory. Heidegger also named this readiness-to-hand aspect of equipment by using the term *Gestell*, in his own inimitable way, of course. This new interpretation of *Gestell* works along the lines that the after-ANT approach would have us look at the world, by transporting the 'original' meaning to a different space of meaning-making, and thus transforming it without relinquishing the whole of the old meaning, only adding value to the new hybrid (very much like the design process, really). *Gestell* now denotes both a reunion or reassembly (*Ge-*) and place, as in location, or placing (*Stelle*), and Ciborra (1998:318) rightly asks, "what has such a reunion to do with technology?"

On the face of it, nothing, because it has nothing to do with the equipment itself, but everything to do with the human actor who uses that equipment, and how, in what *sense* the interaction takes place. Expressing the ontology of equipment through *Gestell*, Heidegger is focusing on our interactions with the world, a space where we do not *regard* (as in seeing, observing, and reflecting) designed objects as mere collections of things, but a specific space where we come to look upon designed objects "as 'in order to' devices endowed with functions" (Ciborra, 1998:318). This is the phenomenological space-of-reflection that design students and professional designers alike have to learn to create, and then to occupy for long enough, in order to emerge with the interpretative 'disclosures' of equipment, created by the breakdown effect, that Willis (above) mentioned. For Heidegger (Ciborra, 1998:318) the essence of technology was anything but technical, a sentiment echoed by Winograd and Flores (1988:179) in considering our own transformation as a non-technical evolution, within the relational network we form with other, and non-human, actors, in this ontological space where "we continue becoming the beings that we are".

CONCLUSION

Product design is increasingly moving towards the field of interactive design, and as a natural result of user expectation and user experience-driven innovation, a merger between product design and human-computer interaction (HCI) design is inevitable. The exact proportions do not matter that much, whether it is the latest model car, washing machine, or any industrial design product that is 'live', i.e., that uses some form of electronic circuitry, or whether it is a designed object that runs almost exclusively on digitally enhanced binary blood, e.g., cellphones, televisions, computers, and all forms of electronic, computerised gadgetry that we find so indispensable today; these products become part of the contemporary lifeworlds of users, and they do have a formative effect, as non-human actors, because we *allow* them to play these roles in our lives, and at times we have no choice in the matter, e.g., when a device is introduced to the workplace, and the worker / user has to accommodate this new 'actor' in the network of communications and associations that make up a 'normal' work environment. This state of affairs is a given, and as designers we have to make use of what is becoming an environment for innovation, rather than a liability. Product design practice has to change its very nature and become an adaptable hybrid, and to do so design education will have to change fundamentally, and become an interdisciplinary endeavour, to start with, and change very basically, meaning we have to teach students to teach themselves, by helping them to acquire the capability of reflective, constructive, and if necessary, radically critical thinking.

REFERENCES

Bannister, D., & Mair, J. M. M. (1968). *The Evaluation of Personal Constructs*. London: Academic Press.

Boddington, A., Brown, B., Chapman, J., Cooper, R., Doordan, D., & Garl, K. (2008). Brighton 05-06-07. *Design Issues*, *24*(1), 91–93. doi:10.1162/desi.2008.24.1.91

Bohm, D., & Peat, F. D. (1989). *Science, Order and Creativity*. London: Routledge.

Brown, T. (2008, June). Design Thinking. *Harvard Business Review*, ▪▪▪, 84–92.

Callon, M. (1986). Some elements of a sociology of translation: domestication of the scallops and the fishermen of St Brieuc Bay. In J. Law (ed.), *Power, action and belief: a new sociology of knowledge?* London: Routledge. Retrieved February 23, 2009, from http://www.csi.ensmp.fr/Perso/Callon/MCarticlesavant1987.html

Callon, M. (2005). Actor-network theory – the market test. In Law, J., & Hassard, J. (Eds.), *Actor Network Theory and After*. Oxford, UK: Blackwell.

Ciborra, C. U. (1998). From Tool to Gestell: Agendas for managing the information infrastructure. *Information Technology & People*, *11*(4), 305–327. doi:10.1108/09593849810246129

Clark, K., & Smith, R. (2008). Unleashing the Power of Design Thinking, *Design. Management Review*, *19*(3), 8–15.

Cumulus Kyoto. (2008). *Cumulus Kyoto*. Retrieved on April 11, 2008, from http://www.kyoto-seika.ac.jp/cumulus/e_prospectus/index.html

Darmofal, D. L., Soderholm, D. H., & Brodeur, D. R. (2002). *Using Concept Maps and Concept Questions to Enhance Conceptual Understanding*. Retrieved December 5, 2008, from http://www.cdio.org/papers/papers.html

Davis, M. (2008). *Toto, I've got a feeling we're not in Kansas anymore ... AIGA Design Educators Conference Massaging Media 2*, (April 4-6, 2008, Boston). Retrieved March 29, 2009, from http://wwwmassagingmedia.org/keynote-address-meredith-davis

de Zeeuw, G. (2005). The Acquisition of High Quality Experience. *Journal of Research Practice*, *1*(1).

Dembski, W. A. (1997). *Explaining Specified Complexity*. Retrieved from http://inia.cls.org/~welsberr/evobio/evc/ae/dembski_wa/19990913_explaining_csi.html

Design PhD Conference. (2009). *Design PhD Conference*. Retrieved March 20, 2009, from http://imagination.lancaster.ac.uk/conference2009/

Dreyfus, H. L. (1991). *Being-in-the-World. A Commentary on Heidegger's Being and Time*. Cambridge, MA: The MIT Press.

ESD. (2008). *ESD Releases New Strategic Plan*. Retrieved December, 2, 2008, from http://esd.mit.edu/HeadLine/esd101408.html

Findeli, A. (2001). Rethinking Design Education for the 21st Century: Theoretical, Methodological, and Ethical Discussion. *Design Issues*, *17*(1), 5–17. doi:10.1162/07479360152103796

FNB. (2008). *FNB's Stokvel Account, making saving more attractive*. Retrieved May 6, 2009, from http://www.fnb.co.za/news/archives/2008/20080413stokvel.html

Gadon, E. W. (2006). My life's journey as researcher. *Journal of Research Practice*, *2*(2), Article M1. Retrieved December 8, 2008, from http://jrp.icaap.orh/index.php/jrp/article/view/40/53

Giddens, A. (1984). *The Constitution of Society*. Berkeley, CA: University of California Press.

Giddens, A. (1991). *Modernity and Self-Identity: Self and Society in the Late Modern Age*. Stanford, CA: Stanford University Press.

Hall, P. (2009). A Good Argument. *Metropolis*. Retrieved March 3, 2009, from http://www.metropolismag.com/story/20090318/a-good-argument

Halter, R. (2005). Remembering the Future: Memory and Interaction. In Buurman, G. M. (Ed.), *Total Interaction: Theory and practice of a new paradigm for the design disciplines*. Basel: Birkhäuser.

Heidegger, M. (1962). *Being and Time. John Macquarrie & Edward Robinson (trnls.)*. New York: Harper Collins.

Hockenberry, J. (2009). Within the Product of No Product. *Metropolis*. Retrieved March 20m 2009 from http://www.metropolismag.com/story/20090318/within-the-product-of-no-product

ICSID. (2001). *Design for Future Needs*. Retrieved December 10, 2008, from http://www.icsid.org/resources/library.htm

ICSID/Danish Designers. (2008). *The increasingly vital role of design*. Retrieved December 10, 2008, from http://www.icsid.org/resources/library.htm

Jana, R. (2009). Innovation Trickles in a New Direction. *BusinessWeek*. Retrieved March 20, 2009, from http://www.businessweek.com/magazine/content/09_12/b4124038287365.htm

Jones, J. C. (1984). *Essays in Design*. Chichester, UK: Wiley.

Jones, J. C. (1988). Softecnica. In Thakara, J. (Ed.), *Design After Modernism: Beyond the Object*. London: Thames & Hudson.

Kauffman, S. (1995). *At Home in the Universe*. New York: Oxford University Press.

Kyoto Design Declaration. (2008). *Cumulus Kyoto 2008 Conference*. Retrieved April 11, 2008, from http://www.cumulusassociation.org/

Latour, B. (2005). *Reassembling the Social*. Oxford, UK: Oxford University Press.

Law, J. (2005). After ANT: complexity, naming and topology. In Law, J., & Hassard, J. (Eds.), *Actor Network Theory and After*. Oxford, UK: Blackwell.

Luhmann, N. (2002). *Theories of Distinction*. Stanford, CA: Stanford University Press.

medGadget. (2009). MAC 800 Portable ECG System. *Internet Journal of Emerging medical Technologies*. Retrieved March 13, 2009, from http://medgadget.com/archives/2009/03/portable_ecg_system_with_sms_keyboard.html

Merwe, J. van der (2007). The magic of Three. *Kybernetes*, (Cybernetics and Design), 36(9/10): 1436-1457.

Norman, D. (1999). *Affordances, Conventions and Design (Part 2)*. Retrieved March 27, 2009, from http://www.jnd.org/dn.mss/affordance_conv.html

Norretranders, T. (1999). *The user illusion: Cutting consciousness down to size*. New York: Penguin.

Paivio, A. (1977). Images, Propositions, and Knowledge. In Nicholas, J. M. (Ed.), *Images, Perception, and Knowledge*. Boston: Reidel.

Pribram, K. H. (1977). Holonomy and Structure in the Organization of Perception. In Nicholas, J. M. (Ed.), *Images, Perception, and Knowledge*. Boston: Reidel.

Prigogine, I. (1980). *From Being to Becoming: Time and complexity in the physical sciences. San Fransisco: W*. H. Freeman.

Richards, A. R. (2008). The Art of Rhetoric as Self-Discipline: Interdisciplinarity, Inner Necessity, and the Construction of a Research Agenda. *Journal of Research Practice*, *4*(1). Retrieved from http://jrp.icaap.orh/index.php/jrp/article/view/115/104.

Shezi, A. (2005). Under the mattress or into the stokvel, SA's poor put money away for rainy day. *BusinessDay Online*. Retrieved May 24, 2009, from http://www.cssr.uct.ac.za/media/stokvels1.pdf

Steiner, G. (2002). *Grammars of Creation*. London: Faber and Faber.

Thackara, J. (2006). *In the Bubble: Designing in a Complex World*. Cambridge, MA: MIT Press.

van der Merwe, J. (2008, July 10-12). Changing a phenomenal change: Reassembling the self through a new ethics of negotiation. In Proceedings of *Changing the Change*, an international conference on the role and potential of design research in the transition towards sustainability, Turin, Italy.

von Glasersfeld, E. (2005). Thirty Years Radical Constructivism. *Constructivist Foundations*, *1*(1), 9–12.

Willis, A.-M. (1999, May). Ontological Designing. In *Proceedings of the European Academy of Design conference, Design Cultures*, Sheffield.

Winograd, T., & Flores, F. (1988). *Understanding Computers and Cognition*. Reading, MA: Addison-Wesley.

KEY TERMS AND DEFINITIONS

Radical Social Constructivism: Believes that 'reality' is a social construct, and that we cannot have objective or direct access to 'a real world out there'. With the design's focus shifting from product to process and experience, social constructivism becomes vitally important.

Actor-Network Theory: As a way of seeing the world, and specifically the relationships and associations that people enter into, ANT regards all entities that play a role in these relationships as equal actors in an unfolding narrative, including designed object as non-human actors.

Socio-Technical: Allied to both RSC and ANT, this term implies that a working relationship has to exist between users and all manner of technical devices, since a reciprocal designing process is at work between the two.

Ontological Phenomenology: Phenomenology studies all aspects of the world as they subjectively appear to us, not 'as they are', objectively. An ontological approach, in Heidegger's terms, studies the coming-into-being type of 'existence' that is enhanced by our interactions with these phenomena.

Complexity: As a scientific theory, complexity has much to offer the study of design, in terms of communication, e.g. when two communicative wavefunctions collide (person A's communication structurally coupling with person B's communication). Not to be confused with complication, complexity (the structural interplay between order and chaos) has always been part of life.

Exformation: The world runs on exformation, those unsaid, sometimes taboo, and very large areas of knowledge that still exists, but are not present in fact, yet present in reference mode. The explicit information we are offered relies for its effectiveness on the implicit exformation knowledge held by a particular audience.

Section 4
Enhancing Creativity and Innovation

Chapter 12
Concept Naming:
Exploratory Methods in the Development of Product Design and Brand DNA

John H. Takamura Jr.
Arizona State University, USA

ABSTRACT

The primary research method described in the case studies for this chapter offers a unique alternative for integrating product personality in brand development. A new method, a specific derivation of product personality assignment referred to as 'Concept Naming,' is offered in combination with other more standard research methods. The Concept Naming method in particular asks respondents to assign personalities to particular areas or features on a product rather than the entire product itself in order to explore consumer/user perceptions, motivations and attitudes towards specific product/brand attributes and the individual meanings and values they place on them. Concept Naming in combination with other qualitative and quantitative data collection methods facilitates the development of a product character vocabulary based on form, color, material, texture, details, and brand. This product character vocabulary when analyzed leads to the development of product and environment design based on Brand DNA (a combination of the tangible and intangible aspects of the product brand).

INTRODUCTION

This chapter describes 3 case studies (a master's thesis research study, a funded research collaboration that involved an industry sponsor, and a funded research collaboration with a city government department in the U.S involving a transdisciplinary team made up of sponsor representatives, university faculty, graduate students, and undergraduate students. This chapter explains the Concept Naming research method in combination with other more standard methods and how its use can capture latent user preferences that can be used to generate design and development of products and environments.

This chapter begins with a discussion on the metaphor of DNA and its use in Brand DNA. A conceptual framework that puts into perspective

DOI: 10.4018/978-1-61520-617-9.ch012

the interrelationships between the tangible and intangible aspects of a product/environment is discussed and expanded upon in terms of the brand experience, product experience, and brand DNA within the context of the overall consumer experience. This chapter then goes into a discussion of the unique Concept Naming method; its background, development, and its use in combination with other standard qualitative and quantitative methods.

This chapter finally focuses on three case studies that employ the Concept Naming method in the development of brand DNA for a theory-based research study on an existing high profile product and brand, an industry sponsored human-centered research and design program, and a government sponsored brand identity and environmental human-centered action research and design program. The objective of this chapter and each case study within it describes in detail the usage of the unique Concept Naming method and the obstacles and challenges as well as the overall benefits and outcomes of its use as evidence of its efficacy.

BACKGROUND

Defining the Term 'Brand'

David Aaker (1996, p. 68), noted marketing strategist, defines brand identity as; "a unique set of brand associations that the brand strategist aspires to create or maintain. These associations represent what the brand stands for and imply a promise to customers from the organization members." Knapp (2000, p. 7) states that, "In order to be a brand, a product or service must be characterized by a distinctive attribute in the consumer's mind." Knapp defines a 'genuine brand' as comprised of an internalized sum of impressions, a distinctive position in the mind and the perceived functional and emotional benefits a product or service provides (Knapp, 2000).

A brand is made up of the associations consumers make with it on four levels: perceiving a brand as a product, as an organization, as a personality, and as a symbol (Aaker D., 1996). In the context of this chapter, brand is defined as the sum total of consumer experiences including both the intangible attributes of a brand (voice, mission, promise and proposition) and the tangible attributes (form, color, material and texture) that make up that experience. In essence, brand is seen as a system of beliefs the truth-value of which is obtained through consumer experiences. This definition of brand best addresses the complexities of the consumer mind in that it takes into account all the touch points of a brand both tangible and intangible.

Brand/Product Attributes and Conceptual Knowledge

Product designers hold a certain portion of the responsibility for managing the conceptual knowledge and meanings attributed to form, color, material and texture from the earliest beginnings of the product design process (Swann, 2002). Jane Darke (1979, p 181) states that there exists a "primary generator" that functions as a "visual image" that arises early in the design process and that it is "...a group of related concepts rather than a single idea." Post and Cramer (1989, p. 1) define conceptual knowledge as, "... knowledge that is rich in relationships. It can be thought of as a connected web of knowledge, a network in which the linking relationships are as important as the discrete pieces of information. By definition, a piece of information is part of conceptual knowledge only if the holder recognizes its relationship to other pieces of information (Hiebert & Lefevre, 1986)." Conceptual knowledge is also perceptually or visually represented in the mind as opposed to being verbally represented (Kan, Barsalou, Soloman, Minor, and Thompson-Shill, 2003, Darke, 1979). The connection between conceptual knowledge and design lies within the visual relationships which consumers create in their network of knowledge regarding products.

Figure 1. Conceptual framework

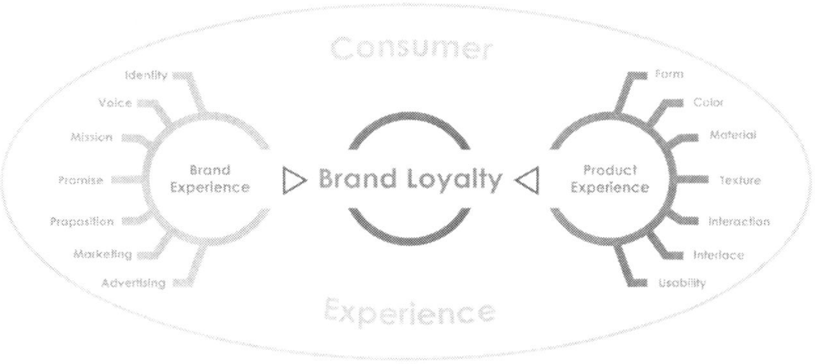

Consumers' conceptual knowledge of products and brands can be seen as the sum total of the links and interconnections they make regarding the various aspects of a product and its attributes. In the way that Post and Cramer describe conceptual knowledge, it is the linking that consumers' make between the pieces of conceptual knowledge (brand and physical product attributes) that are every bit as important as the attributes themselves.

The conceptual framework for this chapter is based on the general interrelationship between what is referred to as the product experience (comprised of the formal attributes of a product or environment) and the brand experience (comprised of the formal attributes of a brand). Product experience is defined as the consumer physical and cognitive response to the formal attributes of form, color, material, texture, interaction, interface and usability, terms that readily fall into the realm and responsibility of product design and product designers (Cagan & Vogel 2002, 2003, Jordan, 2000, Schmitt, 2003). The formal attributes of the brand experience: identity (Aaker 1996), voice (Walker 1999), mission (Knapp, 2000), promise (Knapp, 2000), value-proposition (Aaker, 1996), marketing (Foxall, 1996) and advertising (Berger, 2004, LaSalle and Britton, 2003) were based on the key features of the expression of a brand. Figure 1 illustrates the conceptual framework (Takamura, 2007) for the general interrelationship between the brand experience attributes and the product experience attributes and how together they represent conceptual knowledge in the mind of the consumer.

Jonathan Cagan and Craig M. Vogel in their book Creating Breakthrough Products suggest that within the consumer realm of experiences a brand is basically "articulated" through a product's attributes (p. 101). The core of this framework is based on the premise that brand loyalty is maximized in the consumer experience through a specific blending or integration of the brand and product formal attributes. In other words, brand loyalty is fully realized when both product experience and brand experience reinforce each other. This conceptual framework suggests the existence of fundamental building blocks in branding, otherwise known as brand DNA (Upshaw, 1997, Kapferer, 1998, Marsden, 2000, Gad, 2001, Greenberg, 2003, Cagan & Vogel, 2003) and postulates that brand DNA is the vehicle through which sustained brand loyalty is achieved.

DNA in Design

DNA Metaphor, Brand DNA, and Teleosemantics

The completion of the human genome in 2003 has changed the way modern society views the world on a fundamental level. The DNA metaphor, now quite prevalent in the hard and soft sciences, has recently made its way into the business-consulting world where terms such as 'Organizational DNA' and 'DNA Management' have come into use. Design as well has embraced the DNA metaphor in brand development known as 'Brand DNA.' Brand DNA can be defined as the words and perceptions of users contained in memory over time (Greenberg, 2003, Marsden, 2000). In other words, brand DNA can be interpreted as how consumers think and describe their experience of a product. As Figure 1 suggests, these experiences are facilitated by the brand and product that function as the vehicles for embedding information.

As product and brand design further evolve into a single transdisciplinary field within the information network economy, the DNA metaphor will begin to better address the complexities of embedded information in the designing of artifacts. In the study of how information is transferred through the gene Ruth Millikan, well-known philosopher of science and author of the book Variety of Meanings, uses the terms 'producer' and 'consumer' in explaining the complex relationships within the gene. In her theory, information is produced by the gene and consumed by the organism or parts of the organism. Millikan refers to this theory of information content and transfer as 'Teleosemantics.' Millikan (2004) states that "'Teleosemantics,' as it is sometimes called, is a theory only of how representations can be false or mistaken..." In other words teleosemantic theory recognizes that intentional information from a 'producer' can be misinterpreted by the 'consumer' during transfer. This brings to bear the idea that information regardless of the intentions can be misinterpreted. In terms of product design, misinterpretation of product designers' intentions can create new meanings in the minds of the consumer leading to new uses of the product. In essence, the challenge of product design today is to express designers' intentions in such a way that misinterpretation and reinterpretation is seen as a positive avenue for new product development.

The DNA metaphor is most appropriate in describing how designers transfer information or brand meaning through product designing due to the fact that it provides a teleosemantic as well as socio-economic lens through which the relationship between designers (producers) and users (consumers) can be reviewed, analyzed, and potentially changed. Perhaps 'Product Teleosemantics' will be the next stage in the natural evolution of Product Semantics (Takamura, 2007). Figure 2 shows the expanded conceptual framework (Takamura, 2005, 2007) for the transfer of brand intentions through the design of artifacts and the interpretation or misinterpretation of those intentions by consumers ultimately leading to brand beliefs through consumer ascribed meaning.

This transfer of brand intentions through the design of artifacts is the realization of the ideas of Darke, Post, and Cramer regarding the before mentioned conceptual knowledge of a brand (brand attributes) contained in the experience of the artifact (product attributes). Richard Dawkins, noted biologist and author of The Selfish Gene, coined the term 'meme' which is now defined in the American Heritage Dictionary of the English Language Fourth Edition Copyright 2000 as "A unit of cultural information, such as a cultural practice or idea, that is transmitted verbally or by repeated action from one mind to another." WordNet® 2.0, Copyright 2003 Princeton University defines a meme as, "a cultural unit (an idea or value or pattern of behavior) that is passed from one generation to another by non-genetic means (as by imitation); 'memes are the cultural counterpart of genes'." Dawkins (1976) states, "Examples of memes are tunes, ideas, catch-phrases, clothes fashions, ways of making pots or of building arches. Just as genes propagate themselves in the gene pool by leaping from body to body via sperms or eggs, so memes propagate themselves in the meme pool by leaping from brain to brain

Figure 2. Expanded conceptual framework

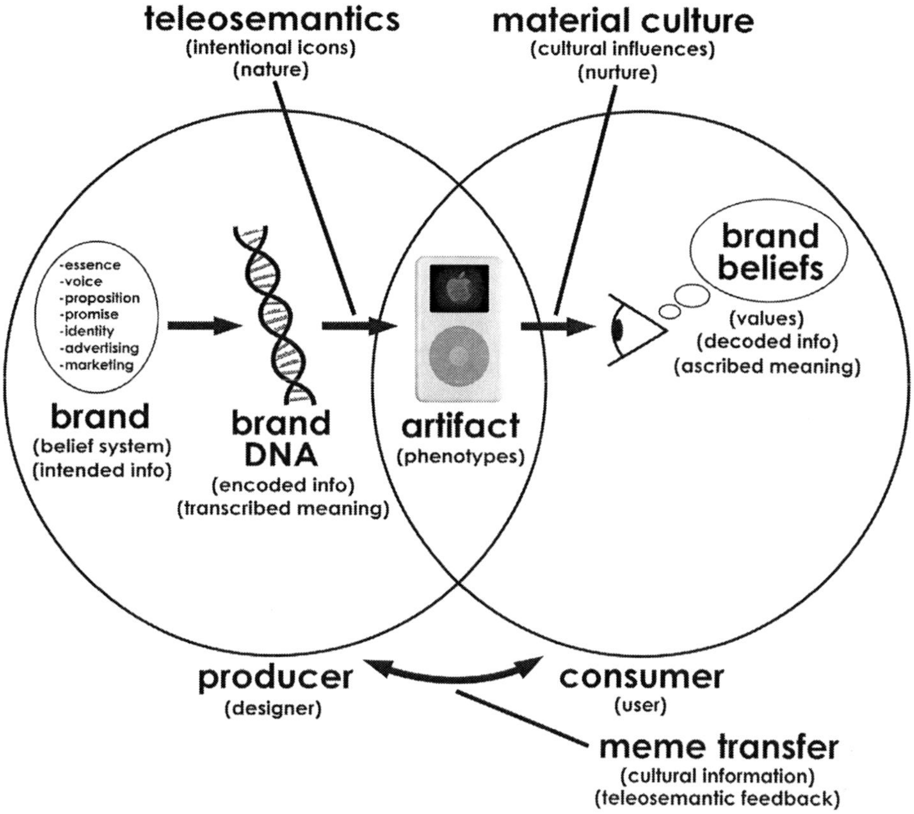

via a process which, in the broad sense, can be called imitation" (Dawkins, 1976, Chapter 11, paragraph 10). Dawkins draws out the parallels between the transfer and propagation of knowledge through memes to the propagation of traits of an organism through genes. Figure 2 illustrates the brand meme transfer from the producer (designer) to the consumer (user) through user interaction with the artifact (product). In essence, product design is depicted as the vehicle through which brand memes (brand values and beliefs) are transferred to consumers. It must be noted that there are a variety of other approaches regarding material cultural studies and consumer psychology that also deeply influence the producer/consumer relationship and the notion of brand DNA within the previously discussed expanded conceptual framework as noted Figure 2. Noted cultural theorist Paul du Gay (1997, p. 21) in his book Doing Cultural Studies: The Story of the Sony Walkman, states that "the role of the producer and the so-called consumer of culture are becoming much more interchangeable. Consumption is becoming more of a personal act of 'production' in its own right." Regarding the notion of Brand DNA as consumer perceptions of memory, Christopher Jones and Russell Fazio (2008) in their chapter in the Handbook of Consumer Psychology entitled Associative Strength and Consumer Choice Behavior, corroborate Greenberg (2003) and Marsden's (2000) ideas of how products and brands are represented in memory. The authors state that "many different kinds of associations in memory may prove relevant to consumer behavior. A brand, for instance, might be associated with usage situations, particular attributes, previous experiences and so on." (Jones & Fazio, 2008, p. 437)

Assigning Product Meaning

Brand Personality

Jennifer Aaker (1997) in her article entitled Dimensions of Brand Personality, states that brand personality "refers to the set of human characteristics associated with a brand." Aaker (1997) goes on to say "it is argued that the symbolic use of brands is possible because consumers often imbue brands with human personality traits." In her research, Aaker (1997) refers to the 'Big Five' human personality traits in psychology; Sincerity, Excitement, Competence, Sophistication and Ruggedness. Aaker's research in product personality dimensions led to the development of the subset of fifteen product personality traits based on the 'Big Five' human personality traits. These traits are listed as; Down-to-earth, Honest, Wholesome, Cheerful, Daring, Spirited, Imaginative, Up-to-date, Reliable, Intelligent, Successful, Upper class, Charming, Outdoorsy, and Tough. In a painstaking analysis of her data Aaker concludes by stating; "...brand personality information, used as a heuristic cue, might influence consumer attitudes and attenuate the processing of brand attribute information under low motivation." In other words, companies can potentially use brand personality traits to stimulate user interest in their products and brands. By incorporating the personality traits as defined by consumers or users within the design process companies and organizations can potentially strengthen the viability of their brands within the ever-increasing competitive marketplace.

Product Personality

Patrick Jordan (2000) in his book Designing Pleasurable Products defines a product's 'personality' as a product's experiential properties. Jordan corroborates Aaker's suggestion that consumers imbue products with human personality traits and states: "Research has shown that people tend to imagine products as having personalities and that they tend to express a preference for products that they perceive as reflecting their own personalities (Jordan, 1997)." In their book Pleasure with Products, Patrick Jordan and William Green (2002) describe the twenty personality dimensions of: Kind/Unkind, Honest/Dishonest, Serious-minded/Light-hearted, Bright/Dim, Stable/Unstable, Narcissistic/Humble, Energetic/Unenergetic, Violent/Gentle, Complex/Simple and Pessimistic/Optimistic used in a 1997 study on Product Personality Assignment conducted by Jordan that studied the link between product aesthetics and product personality.

Product Personality Assignment

Product Personality Assignment (PPA) was originally used by Philips design as an internal tool to assess product design efforts (Jordan, 2000). Additional studies in the Netherlands by Lenau and Boelskifte (2004, p. 6) tested "sensory and symbolic product attributes" using words in order to delineate product character and personality (p. 1). A 1997 study by Patrick Jordan gave product designers seventeen of the before mentioned twenty personality dimensions to assign personalities to products. Jordan concluded that there was evidence that products within a specific brand were perceived similarly (Green & Jordan, 2002). When looking at PPA from the larger brand-building context of companies and manufacturers, it would appear to have great benefit as a method towards ensuring the intentions of corporate designers in designing the experiences of their company's products. Product personality assignment offers a tool to be able to understand how the 'experiential properties' of products relate with the 'formal properties' of products (Jordan, 2000, p. 187). In other words, personality assignment as a method for designing offers a lens through which to see the interplay between the brand experience (formal attributes of the brand) and the product experience (the formal attributes of the product).

Concept Naming

Concept Naming (Takamura, 2005), a unique method based on a derivation of traditional PPA (Product Personality Assignment), was originally developed to conduct brand-profiling studies. Concept Naming proved to be a useful tool in understanding the nature of the relationship between a product's physical or formal attributes (form, color, material, texture, and detail) and the way consumers interpret or perceive the product's brand (Takamura, 2007). In traditional PPA, respondents place a single overarching personality on an entire product based on a set of human personality traits. The Concept Naming method asks respondents to place personalities on a variety of locations (displays, keypads, gauges, switches, handles, etc.) on a single product or environment. These multiple personalities are then analyzed and aggregated into word lists to create a product character vocabulary that represents how users/consumers transcribe product attributes into their individual perceptions of the brand: the personality traits of the product or environment. Concept Naming extracts much more granular detail regarding consumer perceptions of a brand by allowing respondents to focus on the individual design aspects, details, and components of a physical product or environment. In short, Concept Naming helps to develop a deeper understanding of the perceptions of consumers/users by providing actionable insights that can be used to generate new design concepts.

Multi-Method Combination

Concept Naming (CN) is best used in combination with other standard qualitative and quantitative research methods for data collection and data analysis. The use of multi-methods is usually a safer research strategy in terms of minimizing researcher bias. Triangulation of methods helps to ensure higher levels of objectivity when conducting any research study and Concept Naming (CN) studies are no exception. One of the most effective qualitative methods used in combination with Concept Naming (CN) would be the semi-structured interview (Robson, 2002). Semi-structured interviews provide the 'grounded' approach to eliciting subject response regarding consumer/user attitudes and perceptions and functions as a reflective device for the word lists (personality traits) generated in CN.

Contextual research methods such as user observational studies are also appropriate qualitative methods to use in conjunction with CN. Observational field studies open new avenues for further understanding the consumer/user context by providing rich 'thick description' (Geertz, 1973) of the consumer/user experience of both the brand and the product or environment. Observational studies help to create a well-rounded foundation for understanding the context of the specific personality traits developed through CN by providing the 'why' to the 'what.' Understanding why respondents say what they say is essential to understanding consumer/user motivation.

In addition to semi-structured interviews and observational studies, open-ended questionnaires are an excellent qualitative method to partner with CN. Open-ended questionnaires (Robson, 2002, O'Leary, 2004) allow subjects the opportunity to respond to topics more freely leading to potentially unique, unexpected, and sometimes more pertinent lines of thought. Open-ended questionnaires also add to the richness of the data collected adding deeper dimension to the word and trait lists provided by CN.

Quantitative methods should also be combined with Concept Naming. Likert scale or 5-point ordinal surveys based on the key issues and or personality traits that arise while conducting CN may help in setting metrics when analyzing each issue or trait. It is important to flesh out each personality trait but without a system to judge, rank, or assess each trait in relation to one another it may be difficult to draw conclusions on CN data alone.

Another quantitative method, 5-point ordinal Concept Sorting (Rugg & McGeorge, 2003), may also be used with CN. 5-point ordinal concept sorting is a very effective means of collecting data similar to that of the 5-point ordinal surveys but in a more visual manner. The actual sorting of concept cards aids in engaging respondents in the research by immersing them in a kinesthetic activity. Much like the Likert scale or 5-point ordinal surveys, concept sorting helps to define the hierarchy of consumer (user) preferences or needs.

PRODUCT BRAND PROFILE DEVELOPMENT CASE STUDY

Project Background

The first research application of Concept Naming to be discussed is a Master's degree thesis study based on the idea that the consumer experience of a product is the key to the consumer experience of its brand (Kapferer, 1998, Cagan & Vogel, 2002, Schmitt, 2003). The main premise of this study is that companies can gain brand equity by developing brand loyalty through the practice of product design and the use of product design methods. The following research case study focuses on the use of Concept Naming in the context of brand profiling.

A collective of instrumental case studies (Stake, 1994 p. 237) was determined as the most appropriate means by which the phenomenon of product design could be studied due to the fact that broad applicable knowledge was sought regarding the extrinsic research concerns of external influences on brand loyalty. Regarding instrumental case studies, Stake (1994) states that "The case is of secondary interest; it plays a supportive role, facilitating our understanding of something else. The case is often looked at in depth, its contexts scrutinized, its ordinary activities detailed, but only because this helps us pursue the external interest."

The study centered on products designed and manufactured by the Apple™ Computer Company and the brand loyalists that used them between 2004 and 2005. Apple was chosen because of the avid following of its products and its reputation as a powerful brand (Kahney, 2002). Much has been written and documented regarding the brand and products of Apple. Apple products are quite commonly used on university campuses and so the university setting was selected, as it was ideal for targeting respondents. Students as well as faculty commonly use Apple products and the university setting offered an opportunity to investigate a specific community of Apple product users that encompassed both computer and MP3 products. Secondary sources of respondents were solicited via one of the largest online Apple user groups on the West coast called Planetmug.org, a local university user group called MacHelp, and by word of mouth introductions from avid Apple consumers.

Data Collection

Brand loyalists were selected as subjects because the study sought to observe the correlation between product design and brand loyalty. It was believed that a study of brand loyalist perceptions, actions, and attitudes would begin to clarify any relationships between brand loyalty and product design. Brand loyalty was confirmed verbally for questionnaires, interviews, concept naming activities, and observations. Brand loyalty was confirmed via email for questionnaires and surveys given over the Internet. The subjects were segmented into the three age groups of y-generation ages 18 to 25, x-generation ages 26 to 39, and baby boomers ages 40 to 59 years old. An overall script was written for each test to ensure consistency of delivery amongst the three age groups. A total of 82 respondents were solicited for the entire study.

Five-Point Ordinal Survey Questionnaire

The researcher conducted correlational research using a survey questionnaire on a five-point ordinal scale. (Groat & Wang, 2002, Sinclair, 1995) Correlational research methodology (Groat & Wang, 2002) was employed in this study though two survey questionnaires. One survey had twenty-three and the other had thirty-two items that the subjects rated on an ordinal scale of one to five. On the five-point scale, one was indicated as not important and five was indicated as extremely important. The items on the survey questionnaire ranged from attitudes towards shape, color, material, texture, weight and size. Test subjects were asked to rate each of the thirty-two items in terms of what they expect from the Apple™ brand regarding product design. The five-point ordinal survey was selected for its usefulness in causal-comparative studies (Groat & Wang, 2002). The researcher was interested in observing the effect of the physical attributes of a product on a user's sense of brand loyalty. The researcher designated the physical attributes of a product as the "independent variables" and user's perceived sense of brand loyalty as the "dependent variable" (Groat & Wang, 2002). A total of 12 respondents participated in the five-point ordinal survey questionnaire.

Concept Card Sorting

In order to verify and cross-reference data taken by the survey questionnaire the researcher employed Knowledge Elicitation methodology (Shadbolt & Burton, 1995) through a concept sorting exercise. Test subjects sorted twenty-three cards with words and images printed on them representing form, color, material, texture, size and weight issues regarding the Apple brand into 5 categories. The study designated category one as least important and category five as most important. Each of the twenty-three concept sorting cards directly coordinated with the twenty-three items on the first survey questionnaire so that later cross-referencing could occur. Subjects sorted the cards in terms of their expectations of the brand and in order of importance (value) to them as Apple brand loyalists. Subjects generally took about ten to fifteen minutes to complete the concept sort and once their responses were documented they were then debriefed on the test and what the researcher was observing during the test. The researcher selected concept sorting as an additional method to test brand loyalist attitudes that could easily be coded and cross-referenced with the five-point ordinal survey questionnaire in the first test.

Open-Ended Questionnaire

A twenty-two-question open-ended questionnaire (Robson, 2002) was written and distributed via the Internet through two user groups. The Planetmug.org and the university Machelp user groups provided an avenue into the vast Apple community of users as well as an opportunity to gain knowledge regarding Apple user attitudes and values. A total of 13 respondents participated in the questionnaire.

Concept Naming

The method of Concept Naming was used in order to study how brand loyalists transcribe product attributes into their individual perceptions of the brand. This was done to facilitate the development of a product character vocabulary that could be used as a reference for the respondent's attitudes to the physical attributes of a product, namely the product's form, color, material, texture, details, and the brand. Respondents received single sheets each printed in color with four views of the same product (see Figure 3). The respondents then assigned adjectives that described the personalities of parts, locations and areas on the product. Each respondent assigned personalities to four out of a total of eight separate Apple products within a fifteen-minute time period. A total of 13 respondents participated in the Concept Naming exercises.

Figure 3. Concept naming sheet

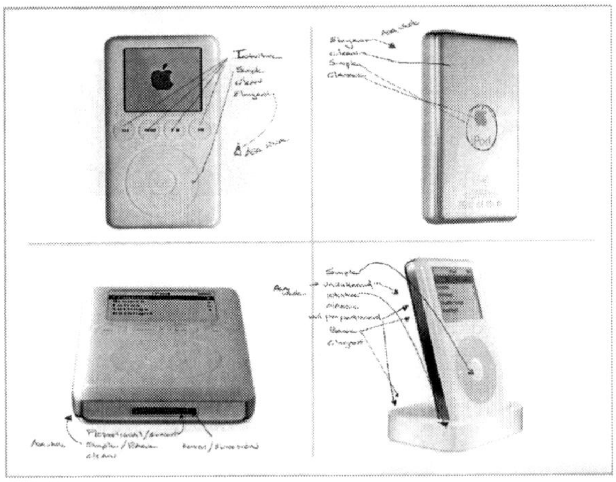

Participant Observation

The study employed observational methods, specifically participant observational methods (Webb, Campbell, Schwartz, and Sechrest, 1966) as a means to enter the 'symbolic world' of Apple loyalists (Robson, 2002). Participant observation (Robson, 2002, p. 317) through the joining of Apple user groups and the participation in the various on-line dialogues with other Apple enthusiasts allowed the study to delve into the inner world of the brand loyalist. Several respondents who expressed a willingness to be video taped in their specific environments for field study observations (Robson, 2002, p.316) were solicited. Observations and video recordings taken on site revealed how Apple loyalists use their products, how they coddle or regard their Apple products, as well as what other types of products they use. Additional observations were conducted to understand the Apple loyalist's general lifestyles and personal artifacts. Participant observation was selected for its 'directness' in obtaining data related to the subject of study (Robson, 2002) that could later be supported by the data from the other methods employed in the study. A total of 25 respondents participated in the observations.

Semi-Structured Interviews

The fourth research test utilized semi-structured interviews (Robson, 2002). The subjects were pre-selected from Apple user groups, by referral, and solicited from the university campus. The video and audio recordings taken at each interview allowed the researcher to concentrate on creating a dialogue with the interviewee rather than being absorbed by note taking. The interviews were treated as 'conversations with a purpose' (Robson, 2002), based on each individual interviewee's willingness and ability to address the questions. Because of the Grounded Theory approach (Strauss & Corbin, 1990) taken in this study, it was believed that semi-structured interviews would allow the opportunity for unexpected and somewhat research-bias-free data to emerge. A total of twenty-five interviews were conducted in conjunction with the previously described twenty-five observations.

Data Analysis

As data was collected it was transcribed in Microsoft Word prior to analysis. Analysis began by a thorough review of field notes, interview transcripts, concept sort results, ordinal survey results, and the completed Concept Naming sheets. Analysis began at the onset of data collection. Mixed methods (Robson, 2002, p. 43) used where both quantitative and qualitative data analysis approaches were adopted. Throughout the analysis, patterns referencing brand and product design were sought and general themes were developed based on those patterns. Through content analysis of the data, parent codes were assigned as well as any child codes that arose throughout the process (Huberman & Miles, 1994, pp. 55-69). The 'thick description' (Geertz, 1973) provided by the field notes and semi-structured interviews were broken down into separate units of information (memos) so that they could be organized in a variety of categories. Axial coding (Morrow, S. L. and Smith, M.L., 2000, p. 213) was used to understand any relationships that formed between the categories and the subsequent units of information (incidents). Although the actions and activities observed in the study were unique and temporally bound, the grounded theory approach of the study intended to develop a general theory or theories regarding product designing and brands that could have broad application in the industry practice. Paradigmatic-type grounded theory analysis (Morrow, S. L. and Smith, M.L. 2000, p. 204) provided a deductive phase from which to develop a general theoretical statement about product design and branding, bounded by data units, that had broader applications.

Unique visualization methods were used in order to better discern the emerging patterns in the data. Word lists and word matching posters (Figure 4) were created to visually depict the data resulting from the Concept Naming exercise and the semi-structured interviews. The graphic layout of these words into posters allowed for the matching up of words leading to the visualization of data patterns once several posters were completed and compared. The color-coding of words within the posters provided a means of differentiating the topics of form, color, material, and texture. Words that related to form, color, material and texture were color-coded in blue, green, orange and purple respectively.

Master personality trait posters were also created to begin to visualize patterns in the data collected. These posters proved to be valuable in comparing the personalities given to the various products in the study. All the words used to describe specific areas on the products were collected and placed on the poster illustrating the entire lot of responses to the product.

Product Brand DNA

During the analysis of the lists of words and phrases accumulated in the various studies (semi-structured interviews, surveys, concept sorts, and Concept Naming exercises) several re-occuring patterns were detected. Respondents generated over 330 different words and short phrases within the 25 semi-structured interviews, 13 open-ended questionnaires, and 13 concept naming exercises for the 2004 to 2005 set of Apple products. The words were analyzed for patterns then reduced down to 58 by their general meanings. The resulting 7 categories of simplicity, dependable, technology, professional, creative, group, and design were chosen to best represent the groups or categories of words in the analysis process through word equivalencing.

The largest group of words entitled 'Professional' contained 14 words pertaining to the professional, elite, and high-end quality of the brand. The second largest of the word groups entitled 'Simplicity' contained 12 words relating to the simple and easy to use aspects of the Apple brand. The third largest of the word groups 'Depenable' contained approximately 10 words relating to dependablity, quality, and trust. The 'Technology' group followed with 8 words that covered the wide spectrum from technology to innovation. Next came the 'Creative' group which had 5 words that related to creativity and individual expression. This was tied with the group entitled 'Design' which was comprised of 5 words pertaining to design, aesthetics, and form. The last group entitled 'Group' followed with 4 words having to do with the Apple community of users.

Each group of words were arranged by the researcher in terms of their individual root relationships with each other. The 'Dependable' group was linked to the 'Technology' group because data paterns suggested that technologies are often seen as dependable or trustworthy. The 'Technology' group was situated next to the 'Professional' group due to the associations respoondents made between professionals and technology. The 'Professional,' 'Creative,' and 'Group' groups were linked together due to the reoccuring theme that creative professionals were seen by respondents

Figure 4. Word matching posters

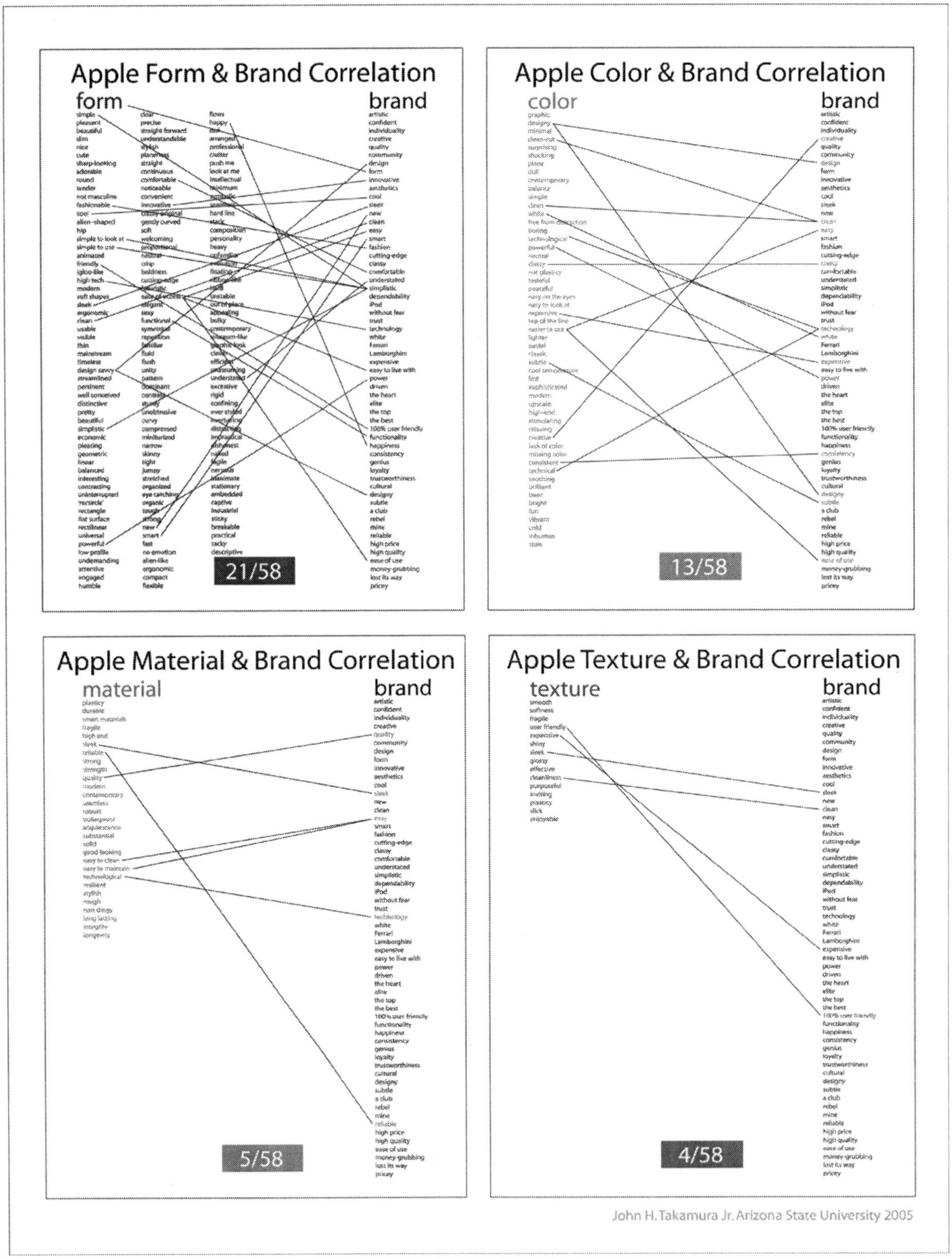

Figure 5. The apple brand DNA molecule

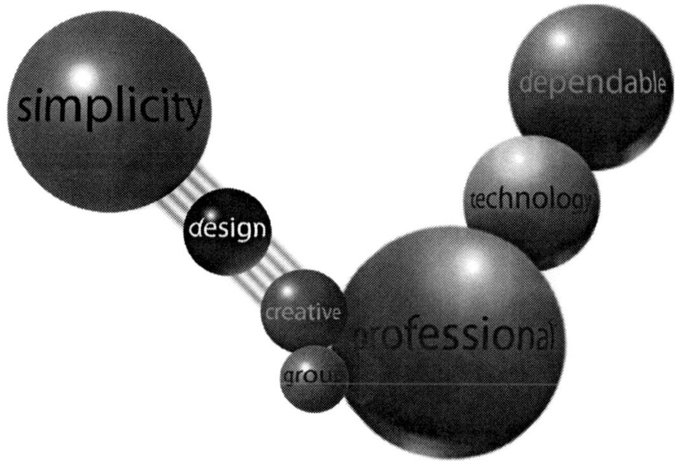

to be the primary users of Apple products and that many loyalists viewed creatives as a 'niche group' targeted by Apple™. Lastly, important bonds were observed through the respondents' associations between the 'Creative' group, the 'Design' group, and the 'Simplicity' group.

It is hypothesized that the 'Design' group of words represents the fusing element in bridging brand loyalist perceptions of 'Simplicity' with the rest of their perceptions of the Apple brand in what could be considered the rudimentary Apple brand DNA molecule (see Figure 5). This molecule, not unlike the human gene, is a complex configuration of elements that can be altered by the adding of new elements. The resulting brand profile based on the results of this study and the subsequent development of the brand DNA molecule led to a better understanding of the nature of the relationship between product design attributes and the sense of brand loyalty of consumers (users).

Benefits and Challenges of the Concept Naming Method

Concept Naming within the context of this brand profiling case study, in combination with a variety of other qualitative as well as quantitative research methods, proved to be of value in directly mapping the physical attributes of products with consumer/user beliefs and interpretations of the brand. CN filled in the knowledge gaps between surveys, observations, and interviews by eliciting responses directly referencing the physical attributes of the product experience namely the form, color, material, texture, details, and logo placement of the product. As a result, more direct associations between the product experience and the brand belief system of consumers could be observed and analyzed.

The main challenges with the use of Concept Naming involved the difficulties faced by respondents in generating a number of adjectives that could express their individual interpretations of the brand through the product. The placing of personalities on the separate components of a product was something respondents had never done before and had no previous experience to reference. This lack of a familiarity with the task slowed the first round of Concept Naming but by the second or third round most respondents appeared to be able to master it. Those respondents who had the greatest difficulty with the task often pressured the researcher to provide words and special effort by the researcher had to be made not to bias the test by providing any. Although quite convenient in terms of task set up, implementation, and speed of data collection, the use of printed sheets with only four views of the product proved to be slightly limiting due to the fact that the minute physical aspects of the products could not be viewed in detail.

PRODUCT SUB-BRAND DESIGN DEVELOPMENT CASE STUDY

Project Background

The next case study that utilized the Concept Naming method to be discussed in this chapter involves an industry sponsored research and design project with faculty, PhD, graduate, and fourth year undergraduate industrial design students at an American university. The sponsor, a high-end gasoline-powered generator company based in America, sought out the services provided by a university with the intention to completely re-design their existing line of products which were originally designed and manufactured in China. The sponsor wished to take advantage of the exploratory research capabilities available at a research one level American university with the belief that the university would provide a fresher and more innovative approach to understanding their user base.

The project was structured very much like any common industrial design consulting contract where kick-off meetings, team meetings, brainstorming sessions, and presentations were scheduled at specific intervals with the sponsor throughout the project with the expressed goal to design, patent, and ultimately manufacture a new product. The team of faculty and students met frequently through the five phases of the project (research, brainstorm, ideation, selection, and prototyping). The four fourth year undergraduate industrial design students were treated as separate teams paired with the PhD. and graduate student researchers on the project. The two faculty members on the project functioned as primary investigator and co-investigator and managed each of the teams.

Data Collection

The industry sponsor provided the initial specifics on market target for the gasoline-powered generator as well as user email lists from which some of the respondents were drawn for the study. A basic user profile was determined and other respondents were solicited from the parking lots of large home improvement outlet stores as well as from city and town communities across the United States from basic home gasoline-powered generator users, to fire fighters, local public event retailers, construction workers, mobile home owners, and university researchers. The regions of the United Stated covered in the study varied from dry desert climates to extreme weather climates throughout the Midwestern to the Southern and Southwestern U.S. states.

Concept Naming

Concept Naming (Takamura, 2005) was designated as the primary data collection method with the understanding that the research data would be supplemented by other methods. Respondents were offered cash incentives and were asked to complete a Concept Naming Exercise, Descriptive Product Personality Rating Survey, and a Prescriptive Product Personality Rating Survey in a single session that lasted approximately one hour and thirty minutes.

An actual generator of the existing design was placed in front of the respondent and respondents were given a black marker pen and a pad of sticky notes. Respondents were then asked to walk around the generator and were allowed to touch and interact with it (pushing buttons, pulling levers, holding handles, etc.). They were then instructed by the researcher to write down adjectives or short phrases they felt described the personality of the specific details on the product

Figure 6. Concept naming product personality assignment exercise

and to place them directly on the product with the sticky notes (see Figure 6). The researchers sometimes facilitated this activity to ensure that respondents did not skip areas on the product and that there was a thorough product user overview by the respondent. This also helped to keep a consistent account of the respondents' reactions to the various interaction points on the product. Twenty-five Concept Naming exercises were conducted in total for the project.

Descriptive Product Personality Rating Survey

A descriptive product personality rating survey was given sometimes prior to and sometimes after the CN exercise. The order of the CN exercise and the surveys were conducted randomly to minimize the bias or influence they may have caused during data collection. The descriptive product personality survey was based on eight of the fifteen personality traits as described by Aaker (1997). These were honest/down-to-earth, exciting, daring/imaginative, up to date, competent, reliable, sophisticated, and rugged/tough. The research team determined that these eight traits were the most appropriate and applicable to the gasoline-powered generator product, brand, and user interaction. Respondents were asked to rate the sponsor's product as well as two of the sponsor's closest competitor products using the eight personality traits on a Likert scale from one to seven (where one was indicated as 'not at all' and seven as 'very'). Respondents were given a color printed hand-out with the images of the two competitors and the sponsor's products and were then asked to rate each one regarding the before mentioned eight personality traits by placing an 'x' on the appropriate box on the survey that best expressed their opinion.

Prescriptive Product Personality Rating Survey

Respondents were given a prescriptive product personality rating survey where they were basically asked to rate each of the eight personality traits of honest/down-to-earth, exciting, daring/imaginative, up to date, competent, reliable, sophisticated, and rugged/tough in order of importance to them as gasoline-powered generator users again using a Likert scale from one to seven where one was indicated as 'not important' and seven as 'very important.' In addition to the twenty-five Concept Naming exercise sessions this survey was also sent via email as well as given in person across the United States. A total of forty-five surveys were collected for this study that spanned five U.S states.

Contextual Observations

Contextual observations were conducted in order to understand the specific user contexts of the various types of users in the research study. These observations revealed several insights on user motivations regarding their activities and interactions with artifacts within the specific atmosphere (both physical and psychological) in which those activities and artifacts existed. The lead researcher as well as the designers on the teams participated in these observations taking field notes, digital photographs and digital video. Respondents were asked to show and describe their particular use context with their generator and their surrounding environment. This would entail walking around with the respondent in their environment as well as demonstrations of how they used their generators.

Data Analysis

The data analysis was conducted in parallel with data collection in order to discern data patterns as quickly as possible that in turn aided in effectively managing the collection of data. Early data analysis (Huberman & Miles, 1994) was adopted as a means of ensuring quality data collection throughout the data collection phase where the researchers could reflect on the data at a real time pace allowing for small changes in how the data collection methods were administered.

Figure 7. Master concept naming posters

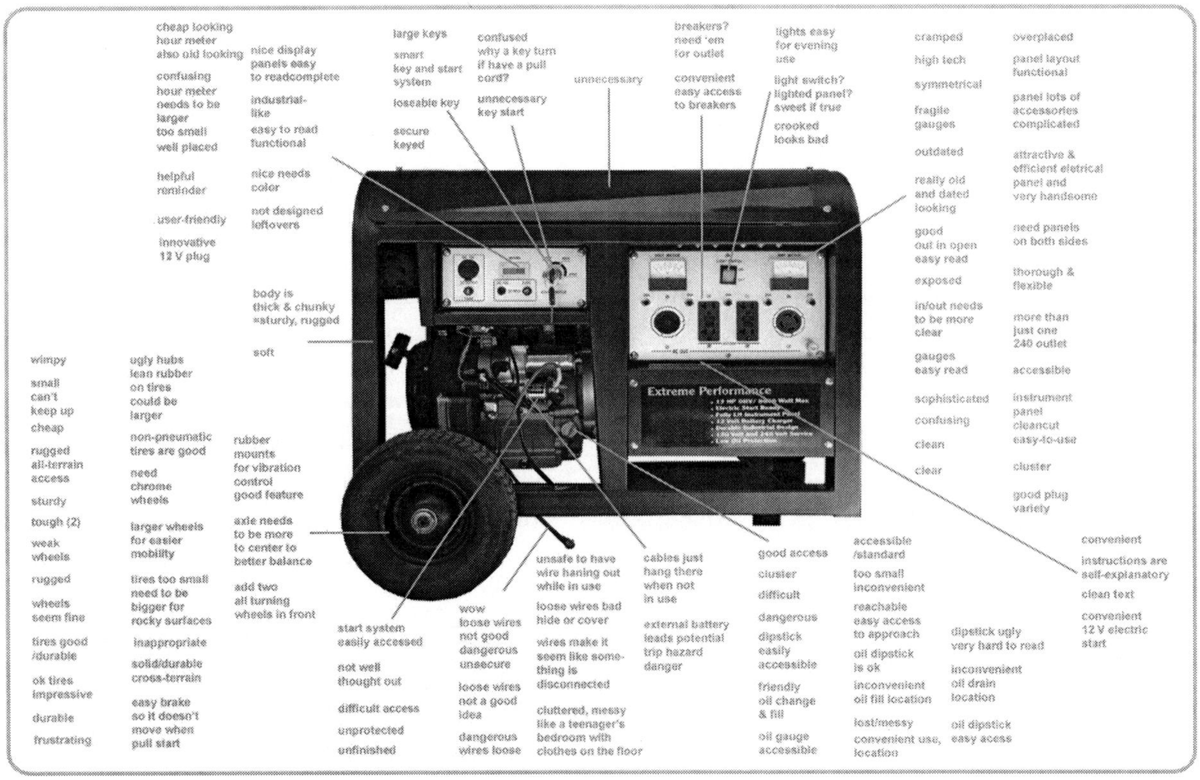

Concept Naming Posters

As with the previous product brand profiling case study on Apple, unique data visualization techniques were used to aid in the data analysis. Concept Naming posters were generated for each of the twenty-five respondents in the CN exercise. These posters helped to organize the various notes and photographs from each session into one set of visual data that could be compared between respondents.

Once all of the CN exercise data was collected and each individual poster analyzed for patterns, master CN posters were created for each of the orthographic views of the generator (see Figure 7). These posters functioned as a master snap shot of all the personalities as described by respondents for every interaction point on the product and aided in the brainstorming and ideations sessions that followed later on in the development process.

Both data sets from the descriptive and prescriptive product personality rating surveys were entered into spreadsheets in Microsoft Excel and output as bar graphs. The data from the descriptive product personality rating survey was broken down into the eight personality traits (Aaker J., 1997) for each of the two competitors' and the sponsor's products respectively.

The prescriptive data was output in Microsoft Excel and was over laid on the descriptive data to provide a respondent ranking of each of the eight personality traits per brand in order of overall importance and value to the user (see Figure 8). At face value one might conclude that each brand out performed in areas of relative low importance to users. Another way of assessing this data suggests that this 'out performance' could be attributed to the specific performance of each individual brand.

Generator Brand DNA

The resulting data from the CN exercises led to the construct of word clusters (bubble diagrams) formed by word equivalencing. The number of words within each cluster determined their relative size. Twenty five (25) words relating to 'rugged' comprised the largest of the clusters followed by nineteen (19) words relating to 'accessible,' nineteen (19) words relating to 'easy,' eight (8) words relating to 'convenient,' and five (5) words

Figure 8. Descriptive / prescriptive assessment

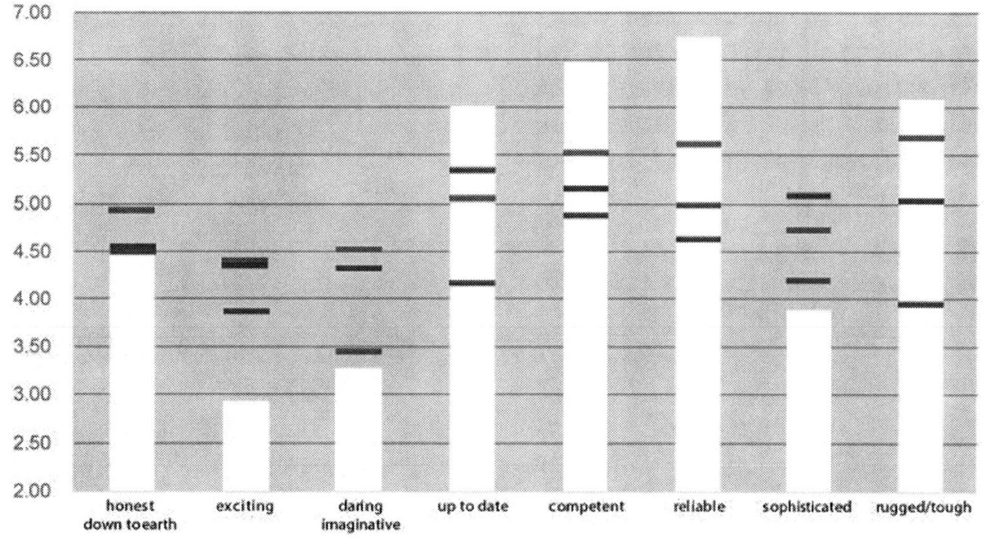

relating to 'friendly.' A basic analysis formula (see Figure 9) was used to calculate the relative cluster sizes for each set of word groups.

The final brand DNA molecule was derived from the word cluster relationships and visually 3-dimensionalized for use by the design team in the brainstorming and ideation sessions conducted in the product development process (see Figure 10). The DNA molecule was used as a guide and filter for any design ideas that came up through the process. Essentially all product performance and function discussion had to be assessed through the brand DNA molecule to ensure that the ideas and issues in the developmental meetings stayed true to the research findings. The DNA molecule functioned as a reflective tool to judge and assess design ideas and design sketches later on in the ideation phase of the project.

Benefits and Challenges of the Concept Naming Method

Concept Naming in terms of brand and product development represented a core activity for both the researchers and designers on the team. CN established itself as a vital tool in assessing user interaction on a micro level as well as in the development of the core set of product values as indicated by users: the Brand DNA molecule. The resulting Brand DNA molecule based on CN data functioned as a critical filter to ensure that design activities remained on task and true to the research results. Little question was left regarding both the brand and product performance issues of the gasoline-powered generator primarily due the high level of detail at which CN operates regarding user feedback. The CN data clearly defined both the positive as well as negative traits of the product as embodied by the brand and its physical expression in the functionality and experience of the product. The large size of the product actually facilitated the use of sticky notes and allowed the user direct access to the areas on the product they were assessing.

The main challenges faced by the research team in conducting the CN exercises once again focused on the difficulties of respondents in coming up with appropriate adjectives that best represented their experience with the product. Researchers had to make special effort not to bias the data by providing words to respondents. In order to maximize data collection efforts the CN exercises were done in combination with the descriptive and prescriptive surveys. This combination although quite convenient for time saving for the collection of data may have also made the overall data collection sessions longer than what most respondents had expected especially

Figure 9. Word cluster analysis formula

$$x = \frac{k\,y}{z}$$

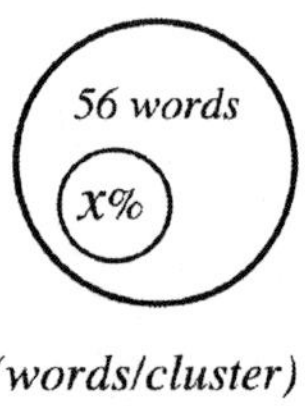

$$x\% = \frac{100\ (words/cluster)}{56\ words}$$

Figure 10. Generator brand DNA molecule

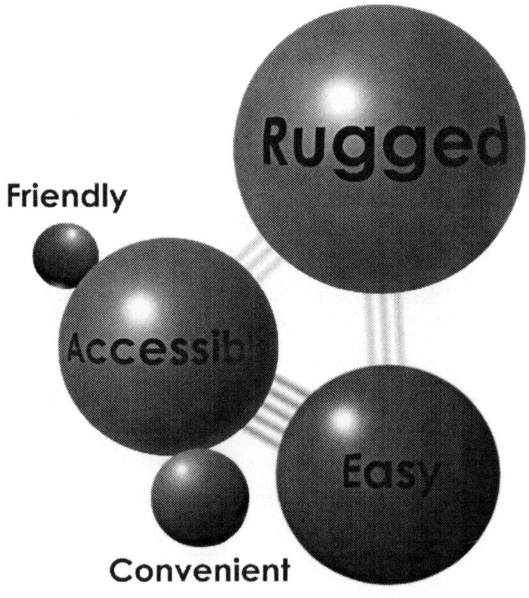

for those respondents who had difficulty and took more time to complete the CN exercise.

ENVIRONMENTAL/ ORGANIZATIONAL BRAND DESIGN DEVELOPMENT CASE STUDY

Project Background

The final case study to be discussed in this chapter regarding the use of Concept Naming involves a sponsored brand research and revitalization program with a local government department in the United States. An American research one state university was sponsored by the state's Department of Economic Security (DES) to assess its current brand in order to begin brand revitalization efforts that would eventually be expressed through the design of a new brand and its implementation in the environmental design of the DES offices state-wide. The DES is made up of several departments each with separate and related areas of service, expertise and knowledge ranging from social welfare to child support to adults with disabilities services. The history of the DES of the state in this case study had been plagued by a severe lack of integration and cross-pollination between its various departments and case-workers. The DES sought the research one level university for inspiration and benchmarking in fostering internal initiatives towards brand revitalization and optimum facilities design with the hopes of fostering departmental integration, employee retention/recruitment, and to better the overall customer experience of DES clients (the general public).

The objective of this sponsored action research and design program was to research DES customers and employees regarding the DES brand and its expression through facilities design. The ultimate goal of this project was to revitalize the DES brand by providing a well managed blend of research and design utilizing a transdisciplinary team made up of Visual Communication, Industrial design, and Interior design students and graduate student researchers led by a single faculty as primary investigator. The final deliverable for the project was a revitalized DES brand identity and a branded facilities lobby design concept.

Data Collection

The data collection process began with several meetings with the director and deputy director of the Department of Economic security and several meetings with the DES executive team manager charged with directing the sponsored program. Once a general understanding of the problems and issues faced by both employees and clients of the DES was established from the point of view of the DES a full research study was constructed.

Semi-Structured Interviews

As discussed in the previous case studies the semi-structured interview method was chosen with the intent of allowing respondents to voice the unexpected issues that were believed to exist. These issues dealt with the dynamic relationships between DES internal divisions, their case-workers and management, and the individual employee attitudes and experiences within the individual DES office environments. Fifteen semi-structured interviews were conducted with the DES managers and case-workers in three separate office sites within the metropolitan city as designated by the DES executive manager directing the program. The semi-structured interviews were conducted in the meeting rooms and office cubicles located at each of the three sites and were audio taped for later transcription. Field notes by the researchers and design team were also taken at each interview. A single interviewer gave the interview although there were up to 3 other team members present in the back taking notes.

Bilingual Surveys

As the research for this program was conducted in a Southwestern state in the U.S it was determined by the research team that a bi-lingual survey be crafted to fit the population under study. Although many languages are served at DES offices it was determined that Spanish was the second major language used by case-workers and clients next to English and so a Spanish version of the original English survey was developed. The DES provided Spanish-speaking interpreters for several of the thirty-three surveys collected in the study so that immediate translation could be captured on the audio tape for transcription.

Environment Open-Ended Questionnaires

Specifically constructed open-ended questionnaires (Robson, 2002) were written for key personnel at the DES offices that addressed the issues faced with the current design of the lobby environment. More general questions were given to the pool of DES clients however these questions were much more specific in terms of the operational functions and limitations of the current lobby in terms of daily case-work from the point of view of a DES employee. It was believed that these confidential and anonymous open-ended questions would liberate the respondent to reveal their true opinions about their work environment without risking their employment status with upper management.

Non-Participant Observation

The team made up of the faculty primary investigator, the lead graduate researcher, assistant researcher, and the four multi-disciplinary undergraduate design students conducted several non-participant observational studies (O'Leary, 2004). The team spent several hours observing how DES case-workers interact with clients in the lobby and several sets of field notes were taken as well as digital photographs and physical measurements of the three locations. This data would lead to a design narrative that documented the environment and a separate 2-dimensional layout of the existing space.

Concept Naming Exercises

Thirty Concept Naming exercises were conducted in the DES lobby space. Respondents made up of both DES employees (case-workers and management) and DES clients (various private individuals seeking DES services) were given sticky note pad and black markers. Each respondent was instructed by the researchers to walk through the lobby environment placing adjectives on the sticky notes that best described the personality of that particular area or artifact in the lobby (see Figure 11). Researchers walked along side of the respondent digitally photographing the sticky notes while also taking notes. The respondents were encouraged to speak aloud while participating in the Concept Naming exercise allowing the researcher to take notes and sometimes audio record the session.

Facilitated Executive Brainstorming Sessions

Twenty-eight respondents participated in the executive brainstorming sessions conducted for the study. These high-level brainstorming sessions conducted at the Department of Economic Security headquarters helped to form a clearer understanding of the DES management team's point of view regarding the DES brand and the issues of the branded lobby design. Specific word exercises were constructed to help the DES executive team define the department's brand essence, brand tagline and brand promise. Again using sticky notes respondents were asked to write down single words or short phrases that described the DES as a brand and government department and to place them on large newsprint sheets. These

Figure 11. Concept naming in an environment

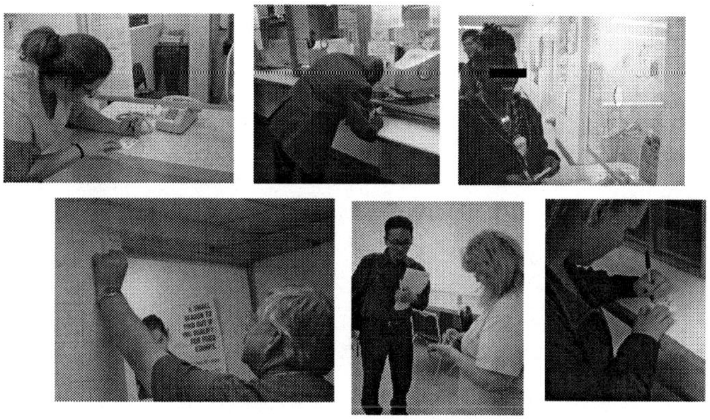

sheets were digitally photographed and kept for later analysis.

Data Analysis

Since the data collected in the study came from a variety of research methods management of the data became absolutely critical. Each research method data set had to be organized separately and analyzed and distilled into a set of separate word clusters. Word clusters from each set of data were later combined through equivalencing of the words to a final set of words that would later become the final DES DNA molecule.

Data from the facilitated executive brainstorming session activities and two of the brand related questions from the interviews were coded into groups and later combined to form larger word groupings. These word groups or clusters were then used to develop the final DES brand DNA. A word cluster analysis formula similar to that shown in Figure 9 was used for the eight hundred and eighteen words collected describing the

Department of Economic Security Brand

The survey and interview data sets were transcribed in Microsoft Word and analyzed for patterns through word searches. Since this part of the research was primarily focused on both the brand and design issues regarding the DES lobby environment, CN data was also analyzed and integrated into a hierarchal list of key issues or what was referred to as the DES lobby personality traits.

The data patterns began to lead to the counter area as a primary feature of the DES lobby and so the data was localized on the same scale of word frequencies as the overall lobby personality traits. Both the overall DES lobby personality traits and the localized lobby counter traits help fuel the transdisciplinary design team with actionable insights towards the development of the final branded environmental space.

Environment/Organization Brand DNA

CN data was once again used to generate word clusters (see Figure 12) representing the types of words used to describe the Department of Economic Security's overall personality. These word clusters were depicted in terms of their relative size based on the numbers of words within each cluster group using the word cluster analysis formula (Figure 9). The clusters were then arranged in position and proximity that reflected the overall data in terms of how the groups related to one another.

Figure 12. DES concept naming word clusters

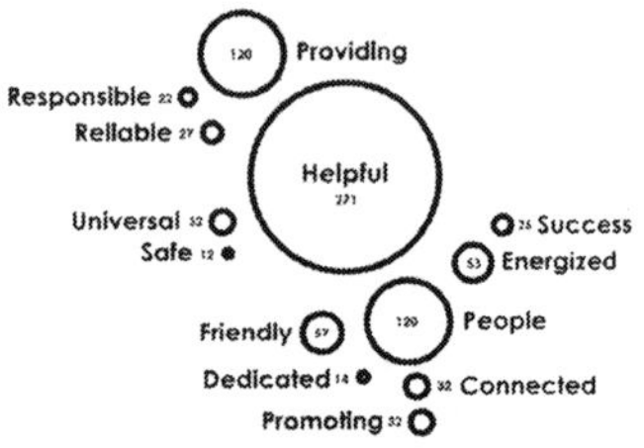

Figure 13. Department of economic security brand DNA molecule

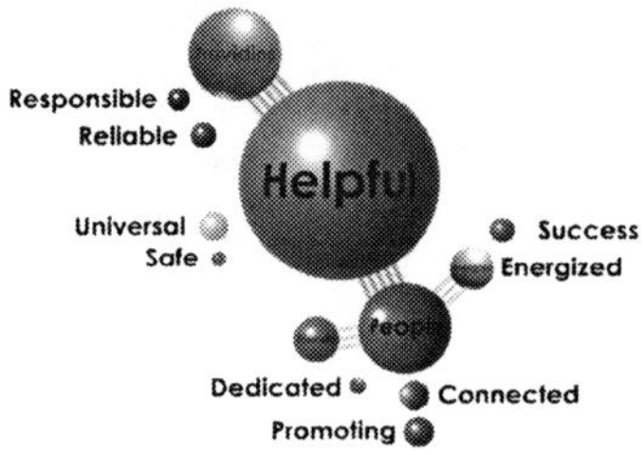

The word clusters were then translated into a final 3-dimensionalized illustration of the brand DNA molecule (see Figure 13) again to show the relative positions of each of the molecular elements in the brand as well as the weak and strong forces between them. Due to the complexity of this brand DNA molecule, an actual 3-dimensional virtual model was developed and a short animation (see Figure 14) was created to allow the design team to virtually experience the true molecular positioning of the brand personality elements. Unfortunately due to budget and time constraints this virtual modeling of the brand DNA molecule was not further explored in the program.

Benefits and Challenges of the Concept Naming Method

Concept Naming exercises, in terms of a government environmental and organizational brand design development program, provided a powerful tool for action research. Concept Naming helped to unveil the socio-economic, political, and anti-bureaucratic sentiments that characterized the politically charged landscape of the DES. Concept Naming aided in the study of the internal workings of a government department by providing respondents with a hands-on level action research activity that facilitated a deeper understanding of the 'personality' of the organization.

Some of the challenges faced with using CN in a large environment dealt with the management of sticky notes and their digital photo documentation during and after each CN session. Most of the CN exercises were fast paced and were conducted in sometimes extremely crowded working environments. This was done to get an accurate response from participants regarding the actual lobby environment experience. The effect of onlookers and the curious although minimal may have had some effect in terms of subject anxiety in that they may have felt like they were the center of attraction.

FUTURE TRENDS

Prior to the advent of Concept Naming (Takamura, 2005) the term 'brand DNA' was merely an esoteric metaphor verging on 'buzz word' status used to describe the fundamental nature of a brand. With unique research methods like CN brand DNA takes on much deeper meaning and literal dimension. Research has revealed the complexities of the brand and consumer relationship (Takamura, 2007). As the complexities of brands are better understood, 3-dimensional visualization, modeling, and analysis methods will clearly set the stage for the future of Brand DNA development much as it did for the mapping of the human genome completed in 2003. It is hypothesized that like the human genome, that involved an international consortium working together for several years, a product design genome could one day be researched and mapped with an international consortium made up of de-

Figure 14. Virtual modeling of 3-dimensional brand DNA

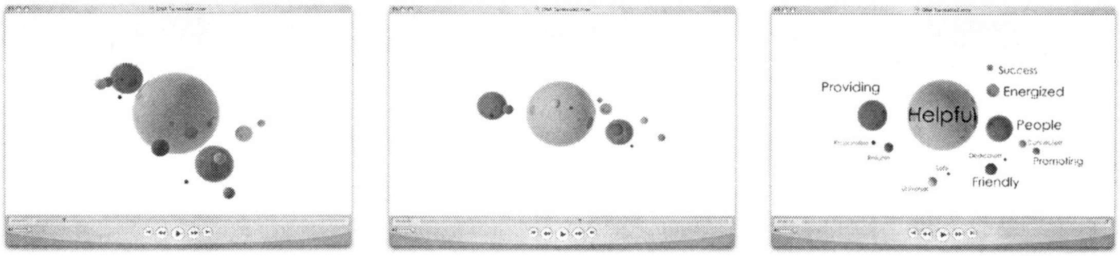

signers, design educators, and design researchers (Takamura, 2007). Deeper understanding of the brand/consumer relationship through the lens of brand DNA will have a profound influence on the future of products, environments, services, and the brands that encompass them.

CONCLUSION

The Concept Naming method is helpful in understanding a physical product or environment because it asks respondents to assign personalities to particular areas or features of a design rather than just the overall design itself. Concept Naming allows designers and human-centered design researchers the ability to explore consumer/user perceptions, motivations, and attitudes towards specific product/brand attributes and the individual meanings and values they place on them. Concept Naming coupled with a multi-method data collection and analysis approaches leads to actionable insights for product, environment, and brand design. Unlike other more common Product Personality Assignment methods, Concept Naming offers a more granular collection of data regarding latent user needs and has proved valuable in the design of brands, branded products, and branded environments because it facilitates the development of a product character vocabulary needed in understanding the fundamental building blocks of brands otherwise known as Brand DNA.

REFERENCES

Aaker, D. A. (1996). *Building strong brands*. New York: Free Press.

Aaker, J. (1997). Dimensions of Brand Personality. *Journal of Marketing Research, Los Angeles, August*, (pp. 347-356).

Berger, A. (2004). *Ads, fads and consumer culture: advertising's impact on American character and society. Lanham, MD*. Oxford: Rowman & Littlefield Pub.

Cagan, J., & Vogel, C. M. (2002). *Creating Breakthrough Products*. Upper Saddle River, NJ: Prentice Hall Inc.

Darke, J. (1979). The Primary Generator and the Design Process. In Cross, N. (Ed.), *Developments in Design Methodology*.

Dawkins, R. (1976). *The Selfish Gene. (1989 edition)*. Oxford, UK: Oxford University Press.

duGay, P., Hall, S., Janes, L., Mackay, H., & Negus, K. (1997). *Doing Cultural Studies: The Story of the Sony Walkman*. London: Sage.

Foxall, G. (1996). *Consumers in Context*. London: Routledge.

Gad, T. (2001). *4-D Branding: Cracking the Corporate Code of the Network Economy*. New York: Financial Times Prentice Hall.

Geertz, C. (1973). *Thick Description: Toward an Interpretative Theory of culture. The Interpretation of Cultures*. New York: Basic Books.

Green, W., & Jordan, P. W. (Eds.). (2002). *Pleasure with Products: Beyond Usability*. London: Taylor & Francis.

Greenberg, K. (October 20, 2003). Rock on, folks: Caddy's Back. *BrandWeek*. Retrieved November 6, 2003, from http://O-web.lexis-nexis.com.library.lib.asu.edu:80/universe/printdoc

Groat, L., & Wang, D. (2002). *Architectural Research Methods*. New York: J. Wiley.

Huberman, A. M., & Miles, M. B. (1994). Early Steps in Analysis. In Norman K. Denzin and Yvonna S. Lincoln (Eds.), *Handbook of Qualitative Research*. Thousand Oaks, CA: Sage. (pp.50-88, 428-444).

Jones, C. R. M., & Fazio, R. H. (2008). Associative Strength and Consumer Choice Behavior. Haugtvedt C. P., Herr P. M., & Kardes F. R. (Eds.), *Handbook of Consumer Psychology*. New York: Taylor & Francis.

Jordan, P. W. (2000). *Designing Pleasurable Products*. London: Taylor Francis.

Kahney, L. (2002). *Apple: It's All About the Brand*. Retrieved December 18, 2003, from http://www.wired.com/news/mac/0,2125,56677,00.html

Kan, I. P., Barsalou, L. W., Soloman, K. O., Minor, J. K., & Thompson-Shill, S. L. (2003). Role of mental imagery in a property verification task: fMRI evidence for perceptual representations of conceptual knowledge. *Cognitive Neuropsychology*, 2003, *20*(3/4/5/6), 525-540.

Kapferer, J. N. (1998). *Strategic Brand Management: Creating and Sustaining Brand Equity Long Term*. London: Kogen Page Limited.

Knapp, D. (2000). *The Brand Mindset*. New York: McGraw-Hill.

LaSalle, D., & Britton, T. (2003). *Priceless: turning ordinary products into extraordinary experiences*. Boston, MA: Harvard Business Press.

Laurel, B. (2003). *Design Research Methods and Perspectives*. Cambridge, MA: MIT Press.

Lenau, T., & Boelskifte, P. (October 1, 2004) Soft and Hard Product Attributes in Design. *Semantic & Aesthetic Functions in Design*. (Working Papers 28). Helsinki, Finland: University of Art and Design Helsinki.

Marsden, P. (2000, March). Brand Selection, Naturally Mapping Brand DNA, Engineering Mind Viruses and Adapting to the Evolved Consumer. In *Proceedings of the Annual Market Research Society conference*. Brighton, 24-27. Retrieved December 5, 2004, from http://www.brandgenetics.com/archive/bsnat.htm

Millikan, R. G. (2004). *Varieties of Meaning: The 2002 Jean Nicod Lectures*. Cambridge, MA: MIT Press/Bradford Books.

Morrow, S. L., & Smith, M. L. (2000). Qualitative Research for Counseling Psychology. In Brown, S. D., & Lent, R. W. (Eds.), *Handbook of Counseling Psychology* (3rd ed., pp. 199–230). Thousand Oaks, CA: Sage.

Neander, K. (2004). Teleological Theories of Mental Content. In Edward N. Zalta (ed.), *The Stanford Encyclopedia of Philosophy* (Summer 2004 Edition). Retrieved March 1, 2005, from http://plato.stanford.edu/archives/sum2004/entries/content-teleological/

O'Leary, Z. (2004). *The Essential Guide to Doing Research*. Thousand Oaks, CA: Sage Publications Ltd.

Post, T., & Cramer, K. (1989). Knowledge, Representation and Quantitative Thinking. In Reynolds, M. (Ed.), *Knowledge Base for the Beginning Teacher - Special publication of the AACTE* (pp. 221–231). Oxford, UK: Pergamon Press.

Robson, C. (2002). *Real World Research*. Malden, MA: Blackwell Publishers Inc.

Rugg, G., & McGeorge, P. (2003). Sorting and Categorization: Case Study. In Drake, M. A. (Ed.), *Encyclopedia of Library and Information Science: Pub-Zoo* (2nd ed., *Vol. 4*). Boca Raton, FL: CRC Press.

Shadbolt, N., & Burton, M. (1995). Knowledge elicitation: a systematic approach. In Wilson, J. R., & Corlett, E. N. (Eds.), *Evaluation of Human Work* (pp. 406–440). London: Taylor & Francis.

Sinclair, M. A. (1995). Subjective assessment. In Wilson, J. R., & Corlett, E. N. (Eds.), *Evaluation of Human Work* (pp. 406–440). London: Taylor & Francis.

Stake, R. (1994). Case Studies. In Denzin, N., & Lincoln, Y. (Eds.), *Handbook of Qualitative Research* (pp. 236–247). Thousand Oaks, CA: Sage.

Strauss, A., & Corbin, J. (1990). *Basics of qualitative research: Grounded theory procedures and techniques*. Newbury Park, CA: Sage.

Swann, C. (2002). Action Research and the Practice of Design. *Design Issues*, *18*(2), 43–61.

Takamura, J. H. (2005). *Brand X: Product Design's Influence on Brand Loyalty.* (Unpublished Master's Thesis), Arizona State University.

Takamura, J. H. (2007, December). *Molecular Branding: A Study on the Nature of the Relationship Between Product Design and Brand Loyalty.* In *Conference Proceedings of DeSForm*. Newcastle, UK: Northumbria University.

Takamura, J. H. (2007). The Product Design Genome: Mapping the Mysterious Interrelationships Between Products and Brands. In *Proceedings of IDSA Innovation,* (Summer 2007).

Upshaw, L. B. (1997). Transferable Truths of Brand Identity. *Design Management Journal, 8*(1).

Walker, J. (1999). Writing to Build Brand Voice. *Public Relations Strategist*, *5*(1), 34–35.

Webb, E. J., Campbell, D. T., Schwartz, R. D., & Sechrest, L. (1966). *Unobtrusive measures* (pp. 112–141). Thousand Oaks, CA: Sage.

KEY TERMS AND DEFINITIONS

Brand: Brand for this chapter is defined as the sum total of consumer experiences including both the tangible and the intangible attributes of a product, environment, service, company or organization. In essence, brand is defined as a system of beliefs the truth-value of which is obtained through consumer experiences. This definition of brand best addresses the complexities of the consumer mind in that it takes into account all the touch points of a brand.

Brand DNA: Brand DNA can be defined as the words and perceptions of users contained in memory over time (Greenberg, 2003, Marsden, 2000). The brand DNA metaphor describes the interaction of fundamental elements (the combinations of the essential building blocks of a brand) that lead to the evolution of the brand as a living organism.

Teleosemantics: Teleosemantics is a theory of information content and transfer. The term "teleosemantics" is used to refer to the class of theories of mental content that use a teleological notion of function. Teleosemantics is best understood as a general strategy for underwriting the normative nature of content, rather than any particular theory (Neander, 2004, paragraph 33). Teleosemantic theory recognizes that intentional information from a 'producer' can be misinterpreted by the 'consumer' during transfer.

Product Personality Assignment: Product Personality Assignment (PPA), originally developed at Philips Design is a research method that defines 'personality ' as the 'experiential properties of a product' (Jordan, 1997). In Product Personality Assignment, products are seen as personalities and this research method is based on the notion that there is a link between product personality and user needs and preferences. PPA is concerned with how users perceive products as personalities and the formal attributes of a product as associated by users to be a product's 'personality traits' (Jordan, 2000).

Product Attributes: In this chapter Product Attributes are defined as the physical or formal properties of a product or environment that being their form, color, material, texture, layout, configuration, and details.

Meme: A meme is 'a unit of cultural information, such as a cultural practice or idea, that is transmitted verbally or by repeated action from one mind to another' (American Heritage Dictionary of the English Language, Fourth Edition Copyright, 2000).

Chapter 13
PDD Trends:
Research Driven by Laws of Product Evolution

José Manuel Ferreira Gaspar
Instituto Superior Técnico, Portugal

Arlindo Silva
Instituto Superior Técnico, Portugal

ABSTRACT

One of the most important research issues in the product design and development field is concerned with the emergence of organizational undesirable behaviors caused by product architecture innovations. These undesired effects occur when the architectural innovation requires, for example, the training and acquisition of new skills. If those are not accomplished in the right time the product quality will deteriorate, affecting negatively the organization. For this reason research is being developed to predict and mitigate the undesirable behaviors, but this is not an easy task, because it requires the creation of an evolutionary process applicable on complex and large engineering systems, such as the product design and development organization. The objective of this chapter is to propose a framework for the development of this evolutionary process.

INTRODUCTION

The view of product design and development changed in the past few years. Researchers started to migrate to a system's view of the product design and development process that integrates the domains of product architecture, development process and development organization, and began to study the interactions within and across them to understand how they affect each other. However, this is a static view of the problem, because there are architectural changes that have a cumulative effect in the future that can be harmful to the product development organization. For example, organizational adaptation to technical change requires training and acquisition of new skills, which interfere with the product quality if not accomplished on time.

These cumulative effects or emergent behaviors are a consequence of the system evolution.

DOI: 10.4018/978-1-61520-617-9.ch013

Because of this researchers are looking for an evolutionary process to predict and to mitigate the undesirable behaviors. Even if it seems, this is not an easy task, because the product development systems are large and complex, particularly in the cases where dispersed teams are used, i.e. Global Product Development (GPD). The absence of a starting point to develop the evolutionary process also increases the problem difficulty. Some researchers may argue that it can be developed from a theory of heuristics or algorithms that must be found and described (Whitney, 2004b).

The work explained in this chapter was done around this research clue. The research questions were: where we can find a theory of heuristics to apply on the engineering systems? What kind of similarities can be found between the two conceptual structures (heuristics and engineering systems)? Is it possible to describe the architectural evolutionary process? Can it be tested? What are the expected impacts on the engineering systems research?

During the research, a theory of heuristics was found that can be adapted to develop and test an evolutionary process for large and complex engineering systems. The finding of an evolutionary process that takes into account the relations within and across the three domains of product design and development provides a direction to create a research framework.

If the test of this framework is successful, it will bring benefits to the academic and technical communities in terms of research, teaching and application of new methodologies of product development, reorganization management, innovation management, product and organization innovation, and enhanced creativity. All of these benefits are given in a system's view of PDD – where possible – to study the influence of product architecture on the development organization.

BACKGROUND

Architectural Evolution in Product Design and Development

The system's view on product design and development led the researchers to a study of the interactions within and across the product architecture, development organization and development process domains. The research procedure used consisted on (Sosa, 2003; Eppinger, 2002; Gulati, 1996): (1) capturing the product architecture by recording the design interfaces (see Table 1 for a detailed explanation of the concepts, (2) capturing the development organization by recording the team interactions and (3) comparing design interfaces with the team interactions. All the recordings were translated into matrices adapted from the design structure matrices (DSM's). This procedure was applied on a one-to-one mapping basis, among the product development domains. Usually this kind of structure is found in large and complex systems.

The data was then analyzed in order to answer the following research questions: "does the organization properly execute the development process?", "Is the development process effectively implementing the product architecture?", "Are the architecture interactions driving the organizational communications?" (Eppinger, 2002, p. 2). Some important conclusions were found, and presented as hypotheses for further research work:

- Domain evolution: The density of the interactions within each domain varies with the domain maturity and the pattern of those interactions change over time;
- Co-evolution: The interaction patterns in the three domains are coupled and coevolve together.

Table 1. Key terms and their definitions (adapted from ESD terms and definitions 2002, Goldenberg, 2002, Altshuller, 2005)

Term	Definition
Component	Part of a system relative to that system; A component can be a system if it contains components.
Module	A part of a system that is built to have minimal, standardized interactions with the rest of the system or a part that provides a given function / sub function of the system.
Integral architecture	An architecture in which the number of functions is significantly larger than the number of components, i.e. components with multiple functions.
Modular architecture	An architecture in which the number of functions is comparable to the number of components and each function is carried out by one module;
Interfaces	An interconnection between systems or their components; interfaces may be intended or unintended.
Attribute independence template	Find two variables (product or service attributes) that are independent (i.e., a change in one does not cause a change in the other), and make a new connection between them. The new functional dependency can be a degree, discontinuous or complex function. Note: A variable is an element that is subject to measurable change. The measure may be exact (quantitative) or categorical. An internal variable is under the full control of the manufacturer. An external variable is a variable of a component within the immediate environment of the product and in direct contact with it, but not under the manufacturer's control. External and internal variables can be independent variables. Only the internal variables can be dependent variables.
Replacement template	A resource (material, energy, and phenomenon) or component existing in the system (internal component) or in its immediate environment (external component) is replaced in order to fulfill a necessary role (or the same needed function). In some cases two existing components are connected. Both cases differ on the achieved level of resource savings (system compactness). Note: A component can be a self-standing part or sub-system. An internal component is one whose variables the manufacturer or the service deliverer has full control. External component is one that comes in contact with the product at a certain point in time, but is not controlled by the manufacturer or the deliver of the service.
Displacement template	Remove an intrinsic component from the system configuration including its functions, in order to cause a qualitative change in the configuration
Component control template	A new link between two previously unconnected components is introduced. The link can be with an existing (internal and external) component and does not necessarily imply uniting or bringing components closer. It can also be used to separate, block, neutralize or at least limit the impact of one component over the other. A negative link between an external component and the product configuration is solved by a new functional link between the external and internal components. If no internal component is able to make the necessary link a new component must be included in the configuration. A positive link delivers an unknown function that can be used to take advantage in the market. It does not require a new component into the configuration. A link may exist between two components only when a change in one component will be directly responsible for changing the parameters of the other component. A link may exist between two attributes only when a direct dependency exists between the attributes.
Division template	A component is split into two, and each new component becomes responsible for a different function.

One might infer from these findings that changes done in one domain can create desirable or undesirable changes on the other, because they are coupled, and this problem can increase in dimension for global new product development or GPD (Tripathy, 2007; Mohrman, 2001). This explains why there is a concern to "balance the social and technical goals, and long term aims with short terms benefits" (Baird, 2000, p.353) in this kind of collaborative projects. This behavior is common in complex and large systems and is characterized by a dichotomy between desirable and undesirable effects. They emerge from the system's evolution and must be managed.

A way to manage these behaviors is by studying the relation between architectural evolution and emergent behavior on large and complex engineering systems. Research clues are given by Whitney (2004b). He says that architectures emerge and evolve from an evolutionary process guided by market pressures on functions or pressures on the ilities. However, the evolutionary process is unknown and it should come from a theory of heuristics or algorithms. Then, where can we find them to be used in the engineering systems context? What kind of similarities can be found between both conceptual structures? Is it possible to describe the architectural evolutionary process? Can it be tested?

Laws of Product Evolution

Answers to the research questions can be found in the field of product innovation, where some methodologies present characteristics that fulfill the requirements of an evolutionary process – product evolution driven by market pressures. These methodologies represent a new approach to product innovation that views the information encoded in the product as a result of its evolution during the struggle to fit or adapt to the past market pressures. This information is used to formulate laws of product evolution that are systematically applied to create new products from the existing ones. As the following paragraphs show these methodologies are good candidates for an evolutionary process guided by laws of product evolution (this is the first major point of this chapter). They are the TRIZ and Creativity Templates theories.

The TRIZ theory (Altshuller, 2005) was the first to materialize this new view. It provided an analytical approach to solve technical contradictions found in products. A technical contradiction is defined by a deterioration of a product's parameter as a result of the improvement of another parameter. For example, the increase – positive improvement – of an engine power requires an increase – negative effect – in the engine size. These parameters and the causality between both are then used to find an algorithm to solve the technical contradiction. The algorithms and the procedures to apply them to the product are provided by the TRIZ methodology. Presently TRIZ is being integrated into the axiomatic design process (El-Haik, 2005), also a systematic and scientific methodology for product innovation.

The Creativity Templates theory (Goldenberg, 2002) is another innovation methodology that uses product information. It is consistent with preceding theory because it uncovers the logical patterns applied to the technical problem solving, but in a smaller number when compared with the huge number of principles and standards provided by the TRIZ. It is characterized by being: (1) identifiable, objectively verifiable and generally observable, (2) applicable to any product or service without market restrains, product maturity or even fabrication process, (3) market independent but including codes of product evolution that correspond to the market evolution, (4) used to improve, create and forecast new products and services and (5) applicable with only minor changes in the technology.

The creativity templates are also heuristics. With their characteristics, it is possible to provide quick and effective strategies that guide problem-solving and invention (Abbot, 2004; Amabile, 1996; Vincenti, 1990). Or in other words, are frugal, fast and have a high correspondence (Magee, 2006) with the product success.

The application of these heuristics is done through the following process (see Figure 1 and Table 1 for concept definitions). At an early stage the product's attributes and components, and the external components that have a direct impact during the product operation, are gathered. The components and the attributes are then categorized as being internal or external, depending whether they are controlled or not by the product manufacturer. The next stage consists on creating links within components and among the product attributes characteristics, i.e. the product's configuration. The configuration of the product is then presented graphically. The last stage is the manipulation of

Figure 1. The creativity templates process (adapted from Goldenberg, 2002)

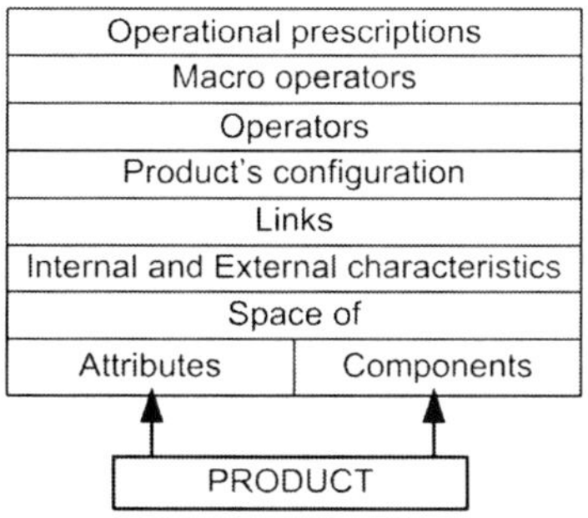

the product configuration by operators in a given sequence. The five macro operators or templates are explained in the following paragraphs with examples to better understand this methodology (for a detailed explanation of the methodology see Goldenberg, 2002).

The replacement template is defined as being "...based on the replacement of a resource or component existing in the system or in its immediate environment in order to fulfill a necessary role" (Goldenberg, 2002, p. 104). The component must be local or included in the "content world of the system", and have similar visual, physical or functional characteristic of the one it is meant to replace. The same happens for the resources that can be a material or energy. Usually the application of this heuristic reduces the resources and components used by the product, increasing at the same time its compactness. For example, for safety reasons there is a need to close the car doors automatically during the car motion. Instead of using a safety system that increases the car weight and cost, we can apply air pressure on the doors. That's why the doors do not open towards the exit of the passenger making it easier to access the vehicle, but quite in the opposite way.

The displacement template prescribes a suppression of an intrinsic component and its functions from the product's configuration. An intrinsic component by definition is the one that will eliminate a large part of the product's advantages, making it useless when removed. Nevertheless, it provides the opportunity for a qualitative change in the configuration of the product. For example, the Bicycle Lefty Fork® was created from a suppression of the bicycle's right side of the fork. The new solution provides added benefits over the competition, namely a smooth travel without stiction, light weight, precision tracking and braking control, strong and durable, intuitively responsive and balanced bicycle.

The control template prescribes the creation of a new link (see Table1) between an internal component and one in the external environment. The link controls the union, separation, approximation and impact limitation between the components. In some cases negative links are detected and limited, or neutralized, by a new link followed by the discovery of new benefits and needs for the product. For example, the Post-It Notes® resulted from the creation of a new link between the desk and the note. Prior to this solution the notes got an insufficient contact with the desks, and usually were blow away by light wind. This negative link was then solved by adding a new one, the weak glue (Goldenberg, 2002).

The division template divides a component (and its function) into two components, each one responsible for a different function. For example, in the 3 Wheel Scooter®, the front wheel was replaced by two wheels, leading to greater stability, traction in wet and rough road conditions, far less braking space and unprecedented riding experience.

The fifth and last heuristic is the attribute dependence template. It prescribes the creation of new dependences (described by a degree, discontinuous or complex function) among the product independent attributes. The used variables are the product's internal and external ones. For example, the F-14 combat aircraft has a wing geometry that varies accordingly to the operational needs. The wing is configured for combat (supersonic flight) and landing in an aircraft carrier (subsonic flight). In this case the wing parameter turned to be dependent on the aircraft speed.

The Role of Knowledge in Product Design and Development

The use of product innovation methodologies requires the creation of new knowledge. For example, Altshuller (2005) said that knowledge must be created from different areas within an industry relevant to the technical system, to solve a technical contradiction with the TRIZ methodology. However, the coupling between the technical and knowledge domains is also the source of organizational barriers to product innovation (Markides, 2006; Christensen, 1997; Utterback, 1996), because the knowledge intersection (Johansson, 2004) done within the design members brings together different practices, skills and viewpoints (Fong, 2003; Carlile, 2002; Vincenti, 1990). Therefore, only the creation of a base for communication and understanding will solve this problem.

The base for communication and understanding was already found by Carlile (2002), when he proposed a "boundary infrastructure" with boundary knowledge and boundary objects. The boundary knowledge is the "place" where different practices meet to create new knowledge and the boundary objects provide the individuals a shared and stable syntax to represent and transform their knowledge. Typical boundary objects or models are the sketches, assembly drawings, parts, prototype assemblies, mockups and computer simulations. Nevertheless, boundary objects per se are not enough for successful team collaboration. The most important factor is the mutual understanding between the interacting members. Each team member must be receptive to change his or her knowledge, but also to influence or transform the knowledge used by other members.

In this context, there are several studies that provide insight for the management of social interactions in design teams affected by culture diversity, ages, personalities (Gratton, 2007; Ogot, 2006; Kurtzberg, 2005; Carrillo, 2002), geographic dispersion (Kratzer, 2006; Cherbonneau, 2005; Wilczynski, 2005; Fruchter, 2003; Lenss, 2001; Tang, 1989), creativity assessment (Redelinghuys, 2006; Busseri, 2000; Valkenburg, 1998), communication between its members (Kratzer, 2004; Carlile, 2002; Stemple, 2002; Austin, 2001; Eckert, 2000) and finally, the team leadership (Kratzer, 2008; Hastings, 2004; Herb, 2001). However, the role of technical problems on team management is rarely referred in this literature and must be considered. Vincenti (1990), for example, said that: "Engineering knowledge is... the product of communities committed to doing and having a sense of collective identity fostered by complex interaction based in part on a shared problem" (p. 239).

From the literature research one might infer that technical contradictions are shared problems, coupled with the laws of product evolution and knowledge creation. Can this relation be used to build a research framework for an evolutionary process? Can this framework be developed and tested in large and complex engineering systems? The answer is given in the following section.

THE RESEARCH DEVELOPMENT FRAMEWORK

As presented in the previous sections a literature research was carried out to determine if the heuristics and algorithms for product evolution can be borrowed (Abbot, 2004) and used in the engineering systems (ES) framework (see Table1 for a detailed description of the ES concepts and terminology). Basically, what was done was to find similarities between the three conceptual structures (TRIZ, CT and ES). The results (see Table 2) showed that a theory of heuristics: (1) can be adapted for an evolutionary process of large and complex systems; (2) can be related to causes of architectural innovation and (3) can be

Table 2. Engineering systems, creativity templates and TRIZ: similarities between conceptual structures

Engineering System	Product Innovation: Creativity Templates	Technical Innovation: TRIZ
Product architecture	Product configuration	
Design dependencies: spatial type transfer type (M, E, C) symmetric, asymmetrical	Attribute dependencies: reciprocal, non reciprocal	
Design interface matrix: required design dependency (+2) desired design dependency (+1) neutral design dependency (0) undesired design dependency (-1) detrimental design dependency (-2)	Forecasting matrix: dependency (+1) no dependency (0)	
Conflicting design interfaces: Positive design dependence Negative design dependence	*Inference:* Positive link Negative link	Technical contradictions: Parameter improvement Parameter deterioration Physical contradiction: Opposed properties
Integral architecture	Replacement heuristic	Ideality law
Modular architecture	Division	
Interfaces	Component control	
	Displacement	
Positive effects: Intended interface, interaction, Interdependence or property. Desirable emergent property ('ilitie')	*Inference:* Positive link	*Inference:* Parameter improvement
Negative effects: Unintended interface, interaction, Interdependence or property. Undesirable emerg. property ("ilitie").	*Inference:* Negative link	*Inference:* Parameter deterioration
Evolution: the patterns of design interfaces and team interactions within each PD domain change over time Co-Evolution: the interaction patterns in the three domains change in coupled ways.	Product evolution laws with universal application to other fields.	Laws of technical evolution applicable to other fields by analogy.
Knowledge management: Alignment of the different practices in the knowledge boundary through negotiation and use of common boundary objects.		Knowledge required: From different areas within the industry relevant to the industry From different other industries

Note: empty rows correspond to similarities not identified between the three conceptual structures.

tested. A description of the results is given in the following paragraphs.

Conflicting Design Interfaces (ES) vs. Technical Contradictions (TRIZ)

As said before, technical innovations occur when *contradictions* are found amongst two characteristics or parameters of the technical system. For example: "Increasing the power of an engine [positive improvement] requires an increase in the size of the engine [negative effect]. So, an inventor considers increasing the power partially in order to reduce the *negative effect* [compromise solution]" (Altshuller, 2005, p.17). Another example can be found in the engineering systems field, but in this case for a large commercial aircraft engine: "An example of a *negative dependency* would be the *undesired* transfer of energy [vibration] from the LPT [low-pressure turbine] vanes to the LPT blades. However, the blades and vanes are *positively codependent* for proper inlet and exit gas conditions to achieve optimum aerodynamic efficiency" (Sosa, 2003, p. 242).

As showed above, the examples are similar but use different concepts and terms. They are classified – including the CT conceptual structure – in:

- Positive effects: improvement of a parameter or characteristic of the technical system (TRIZ), a positive link between two components or attributes (CT), or a desired, required or positive design dependency (ES);
- Negative effects: deterioration of a parameter or characteristic of the technical system (TRIZ), a negative link between two components or attributes (CT), or an undesired, detrimental or negative design dependency (ES);
- Opposed effects or pairs of: improvement – deterioration parameters (TRIZ), or positive –negative dependencies (ES), also known as conflicting design dependencies.

Design Interface Matrix (ES) vs. Forecasting Matrix (CT)

The design interface matrix (DSM) of a technical system is filled with components, and the design dependencies that exist between them. Each matrix cell is filled with five types of dependencies: "+2" for a dependency necessary for functionality, "+1" for dependency beneficial, but not absolutely necessary for functionality, "0" for a dependency that does not affect functionality, "-1" for a dependency that causes negative effects, but does not prevent functionality, and "-2" for a dependency that must be prevented to achieve functionality. Each box on the DSM diagonal represents a subsystem and includes all the design dependencies within the components. The design dependencies within components of different subsystems are given by marks (-2 to +2 dependencies) outside the diagonal boxes.

The forecasting matrix (CT) is built from a list of internal and external variables, and the type of dependencies that exist between them. Each cell is filled with a "0" for no dependency and a "1" for a dependency. Its function is to provide guidance for the application of the dependency template on the product attributes.

The search for similarities between the two matrices above mentioned gave the following results:

- **Internal variables (CT):** those that are related to the components within the subsystem (DSM diagonal box);
- **External variables (CT):** those that are related to the external components (outside the DSM diagonal box) and come into direct physical contact with the subsystem at a certain time and space;
- **Dependencies between the variables (CT):** are the marks found in the design interface matrix (DSM).

Integral Architecture (ES) vs. Replacement Heuristic (CT)

The following example illustrates integral architecture and its relation to modularization:

...here the panels [of the car body] are spot welded together and some of them, especially the panel that extends from the rear door area back over the rear fender, comprise a mix of interior ribs and exterior finish surfaces all within a single part. In the sense that structure and appearance are normally separate, their inclusion in a single part marks this design as being somewhat integral. In addition, the exterior portions of some of these panels provide some structural rigidity as well, a function that is provided in the aluminum body

exclusively by the frame...integral designs are favored when performance is the highest priority. Such designs are likely to be more efficient in their use of space, weight, and energy because they can be optimized to a known combination of chunks and can contain their own interfaces (Whitney, 2004a, p. 346).

There are similarities between this example and the ideality law (TRIZ). The ideality law states that " any technical system, throughout its lifetime, tends to become more reliable, simple, effective...it costs less, requires less space, wastes less energy, etc...reflects the maximum utilization of existing resources, both internal and external to the system..." (Altshuller, 2005, p. 16). This is also true for the replacement template. However, these operations go beyond the optimization of chunks and interfaces during the system integration, i.e. they reduce the number of existing components and interfaces.

Modular Architecture (ES) vs. Division Heuristic (CT)

The division template can be seen in the last example by reversing the integral operation. The continuous use of component division will lead to a limit, where each component can be designed and manufactured independently of all the others and the product can be produced simply by plugging these elements together at their interfaces (or links, by similarity). However, the flexibility achieved by using more interfaces that are not contained in the same part implies a deterioration of the system's efficiency (more space, energy and weight) and reliability, but potentially extreme cost savings.

Interfaces (ES) vs. Component Control Heuristic (CT)

The system's reliability is related to the number and quality of the interfaces used across its modules. The interfaces are the main sources of failures and when used in great number the result will be a less reliable system. The reason for this problem is related to the impossibility to contain, control and optimize the interfaces, as in the case of a purely integral design, because it is difficult to know which modules might find themselves in the same product unit. The degree of uncertainty, responsible for unexpected failure modes also increases when the design is outsourced, for example, in GPD projects.

In the ES framework the quality of each interface depends on critically the level of the dependency, i.e. how a given component depends on the others. The control template has a similar approach that eliminates or mitigates the negative links and explores the positive ones. Then, there exists a similarity between the two conceptual structures:

- **Negative links (CT):** are the (-1) and (-2) negative design dependencies (ES);
- **Positive links (CT):** are the (+1) and (+2) positive design dependencies (ES);

Creativity Templates as Boundary Objects (ES)

The engineering systems provide a platform and syntax that can be shared among different individual functions in an organizational structure. It also identifies and provides information about the individual's dependencies across a given boundary. This information is useful for the process of knowledge creation.

The creativity templates also provide a simpler syntax, language, and methodology that can be easily shared among all the functions. Each creative template changes the product configuration (or product architecture by similarity) and consequently it will be possible to forecast the team interactions required to create new knowledge at the boundary.

FUTURE TRENDS

As shown in the last section there are enough similarities between the TRIZ, CT and ES conceptual structures to advance the hypothesis of developing from them an evolutionary process for large and complex engineering systems.

There is a debate in the engineering systems field about the testability of an evolutionary process based on a theory of heuristics such as the proposed one. Some researchers advocate that physical evidence and rigorous mathematical proofs must be found to test the theory, others argue if it is possible to achieve such rigor or if it is necessary in practice (Whitney, 2004b). However, the CTs are robust enough to overcome these controversies, because they are systematic, identifiable, objectively verifiable and generally observable.

There are three aspects that will characterize the evolutionary process: evolution causes (design dependencies), laws of product evolution (heuristics) and the effects on the product architecture, and developing organization domains. The relation between these three aspects of the evolutionary process is shown in Table 3. This table was built by applying the results from Table 2 into a PDD system characterized by a one-to-one mapping between the product architecture and the developing organization domains. In the first column the causes of product architecture evolution, are presented. There are different causes, characterized by a null (0), positive (+1; +2) and negative (-1; -2) dependencies, and pairs of conflicting dependencies: (+1,-1; +1,-2; +2, -1; +2,-2). Each one is associated to one or more laws of product evolution presented in the second column. The evolution effects of these laws on product architecture are explained in the third column. In the fourth column the effects that those architectural changes have on the team interactions (each team has only one component assigned) are presented. As can be seen in the Table 3, the level of product architecture change increases in the following order: +2; +1; -1; -2; (+1, -1; $\Delta = 2$) to (+2, -2; $\Delta = 4$).

The null (0) dependency can be changed to other kind of dependencies, because according to the dependency heuristic, product evolution occurs when new dependencies are added. The introduction of new dependencies usually calls for the remaining heuristics, introducing team movements inside the developing organization. For example, one replaced team can be useful in another project, i.e. organizational dynamics. In conclusion, the Table 3 can be used for product architecture innovation, as well for the redesign of the developing organization.

The laws of product evolution can be applied directly on the product configuration (or by similarity, into the product architecture) without looking for the design dependencies. As a result, these laws can be seen as an analogy of natural selection where the product adapts itself to the environment (causes) or has a mutation that will give it an advantage against competing products. It is expected that changes on product architecture will propagate into the organization structure, because they are coupled. However, the reversal can be true.

If the test of this framework is successful, it will bring benefits to the academic and technical communities in terms of investigation, teaching and application of new methodologies of product development.

CONCLUSION

Research done in product design and development shows that product architecture, development process and development organization domains evolve, but also are coupled and co-evolve together. The study of these interactions is one of the most important research issues, because, for example, desirable changes done in product

Table 3. Relation between the design dependencies, heuristics and PDD domains (one-to-one mapping)

Type of depend.	Heuristics or laws of prod. evolution	Product Architecture	Team interactions (Developing organization)
(0)	Dependency	A new dependency can be introduced for technical innovation. Involves the application of the remaining heuristics.	
(+1) (+2)	Control Control	Unchanged. Changes only on the module interfaces. Unchanged. Changes only on the module interfaces.	Unchanged. The knowledge boundary continues to evolve without the need to introduce new practices (teams). Idem.
(-1) (-2)	Control Control	An additional component can be introduced. Changes only on the module interfaces. Idem	One team is created. The knowledge boundary is changed by the addition of a new practice (team) related to the additional component. Idem
(+2,-2) (+2,-1) (+1,-2) (+1,-1)	Replacement Division Displacement Control Dependency All All All	A component is replaced by an internal or external one for the same function. A component is divided in two components, each one with different functions. A critical component is removed An additional component can be added. Changes only on the module interfaces. Involves the application of the remaining heuristics.	One team is "replaced" by other. A practice in the knowledge boundary is replaced by a new one. Two new teams are created. The original one "disappears". A practice in the knowledge boundary is replaced by two new practices. One team is "removed". The knowledge boundary lost one practice. One team is created. The knowledge boundary gains a new practice.

Note: +2 = dependency necessary for functionality, +1 = dependency beneficial, but not absolutely necessary for functionality, 0 = dependency that does not affect functionality, -1 = dependency that causes negative effects, but does not prevent functionality, and -2 = dependency that must be prevented to achieve functionality.

architecture can create undesirable behaviors on the development organization.

These undesirable behaviors are generated from the system evolution and for this reason researchers are looking for an evolutionary process to predict and mitigate them. The goals of this chapter are aligned with these intentions. A literature research was performed to determine if there are methodologies that can be adapted, developed and tested to structure and study the evolutionary process. The results indicated that there are methodologies that present characteristics that fulfill the requirements, i.e. architectural evolution driven by market pressures. A research framework was then created and proposed in this chapter for development. It integrates new methodologies of product innovation with the engineering systems methods.

The research framework consists on: (1) evolution causes or design dependencies, (2) laws of product evolution and (3) effects on product architecture, and developing organization structure. If successfully developed, the evolutionary process will be useful to study emergent behavior in large and complex systems, such as product design and development system, and will help decision making regarding architectural innovation and redesign of the development organization.

REFERENCES

Abbott, A. (2004). *Methods of Discovery: Heuristics for the Social Sciences*. New York: W.W. Norton & Company.

Altshuller, G. (2005). *40 Principles Extended Edition: TRIZ keys to Technical Innovation*. Worcester, MA: Technical Innovation Center.

Amabile, T. M. (1996). *Creativity in Context: Update to the social psychology of creativity*. Boulder, CO: Westview Press.

Austin, S., Steele, J., Macmillan, S., Kirby, P., & Spence, R. (2001). Mapping the conceptual design activity of interdisciplinary teams. *Design Studies*, *22*(3), 211–232. doi:10.1016/S0142-694X(00)00026-0

Baird, F., Moore, C. J., & Jagodzinski, A. P. (2000). An ethnographic study of engineering design teams at Rolls-Royce Aerospace. *Design Studies*, *21*, 333–355. doi:10.1016/S0142-694X(00)00006-5

Busseri, M. A., & Palmer, J. M. (2000). Improving teamwork: the effect of self-assessment on construction design teams. *Design Studies*, *21*(3), 223–238. doi:10.1016/S0142-694X(99)00024-1

Carlile, P. R. (2002). A Pragmatic View of Knowledge and Boundaries: Boundary Objects in New Product Development. *Organization Science*, *13*(4), 442–455. doi:10.1287/orsc.13.4.442.2953

Carrillo, A. G. (2002). *Engineering design team performance: quantitative evidence that membership diversity effects are time dependent*. (Unpublished doctoral dissertation). Stanford, CA: Stanford University

Cherbonneau, G. (2005). A framework for improving the effectiveness of distributed project teams. (Unpublished doctoral dissertation). Cambridge, MA: Massachusetts Institute of Technology

Christensen, C. C. (1997). *The Innovator's Dilemma: When New Technologies Cause Great Firms to Fail*. Boston, MA: Harvard Business School Press.

Eckert, C. M., Cross, N., & Johnson, J. H. (2000). Intelligent support for communication in design teams: garment shape specifications in the knitwear industry. *Design Studies*, *21*(1), 99–112. doi:10.1016/S0142-694X(99)00006-X

El-Haik, B. S. (2005). *Axiomatic quality. Integrating axiomatic design with six-sigma, reliability and quality engineering*. Hoboken, NJ: John Wiley & Sons.

Eppinger, S. D. (2002). *Patterns of Product Development Interactions*. (Working paper ESD-WP-2003-01.05). Cambridge, MA: MIT Engineering Systems Division.

ESD Symposium Committee (2002). *ESD terms and definitions*, (version 14). Retrieved September 1, 2009, from http://ocw.mit.edu/OcwWEB/engineering-systems-division/ESD-342Spring-2006/readings/

Fixson, S. K. (2001). *Three Perspectives on Modularity – A Literature Review of a Product Concept for Assembled Hardware Products*. (Working Paper ESD-WP-2001-06). Cambridge, MA: MIT Engineering Systems Division.

Fong, P. S. W. (2003). Knowledge creation in multidisciplinary project teams: an empirical study of the processes and their dynamic interrelationships. *International Journal of Project Management*, *21*(7), 479–486. doi:10.1016/S0263-7863(03)00047-4

Fruchter, R., & Townsend, A. (2003). Multi-Cultural Dimensions and Multi-Modal Communication in distributed, Cross-Disciplinary Teamwork. *International Journal of Engineering Education*, *19*(1), 53–61.

Goldenberg, J., & Mazursky, D. (2002). *Creativity in Product Innovation*. Cambridge, UK: Cambridge University Press. doi:10.1017/CBO9780511674464

Gratton, L., Volgt, A., & Erickson, T. (2007). Bridging faultlines in diverse teams. *MIT Sloan Management Review*, *48*(4), 21–29.

Gulati, R. K., & Eppinger, S. D. (1996). *The Coupling of Product Architecture and Organizational Structure Decisions*. (Working Paper #3906-96). Cambridge, MA: MIT Sloan School of Management.

Hastings, D. (2004). *The future of engineering systems: development of engineering leaders. Engineering Systems Monograph*. Cambridge, MA: MIT Engineering Systems Division.

Herb, E., Leslie, K., & Price, C. (2001). Teamwork at the top. *The McKinsey Quarterly*, *2*, 32–43.

Johansson, F. (2004). *The Medici Effect: breakthrough insights at the intersection of ideas*. Boston, MA: Harvard Business School Press.

Kratzer, J., Leenders, R., & Engelen, J. (2006). Managing creative performance in virtual environments: an empirical study in 44 R&D teams. *Technovation*, *26*, 42–49. doi:10.1016/j.technovation.2004.07.016

Kratzer, J., Leenders, R., & Engelen, J. (2008). The social structure of leadership and creativity in engineering design teams: an empirical analysis. *Journal of Engineering and Technology Management*, *25*, 269–286. doi:10.1016/j.jengtecman.2008.10.004

Kratzer, J., Leenders, R. T., & Engelen, J. M. (2004). Stimulating the potential: creative performance and communication in innovation teams. *Creativity and Innovation Management*, *13*, 63–71. doi:10.1111/j.1467-8691.2004.00294.x

Kurtzberg, T. R. (2005). Feeling creative, being creative: an empirical study of diversity and creativity in teams. *Creativity Research Journal*, *17*(1), 51–65. doi:10.1207/s15326934crj1701_5

Lenss, V. G., & Pumpuni, G. A. (2001). A systems engineering approach to managing communication in globally dispersed teams. (Unpublished doctoral dissertation). Cambridge, MA: Massachusetts Institute of Technology.

Magee, C., & Frey, D. (2006). Experimentation and its role in engineering design: linking a student design exercise to new results from cognitive psychology. *International Journal of Engineering Education*, *22*(3).

Markides, C. C., & Geroski, P. A. (2005). *Fast Second: How Smart Companies Bypass Radical Innovation to Enter and Dominate New Markets*. New York: John Wiley & Sons.

Mohrman, S. A., Klein, J. A., & Finegold, D. (2001). *Managing the Global Knowledge-Creation Network: A Sense-Making perspective*. (Working Paper ESD-WP-2001-05). Cambridge, MA: MIT Engineering Systems Division.

Ogot, M., & Okudan, G. (2006). The five-factor model personality assessment for improved student design team performance. *European Journal of Engineering Education*, *31*(5), 517–529. doi:10.1080/03043790600797335

Redelinghuys, C., & Bahill, A. (2006). A framework for the assessment of the creativity of product design teams. *Journal of Engineering Design*, *17*(2), 121–141. doi:10.1080/09544820500273136

Sosa, M. E., Eppinger, S. D., & Rowles, C. M. (2003). Identifying Modular and Integrative Systems and Their Impact on Design Team Interactions. *Journal of Mechanical Design*, *125*, 240–252. doi:10.1115/1.1564074

Stemple, J., & Schaub, P. (2002). Thinking in design teams – an analysis of team communication. *Design Studies*, *23*(5), 473–496. doi:10.1016/S0142-694X(02)00004-2

Tang, J. C. (1989). Toward an understanding of the use of shared workspaces by design teams. (Unpublished doctoral dissertation). Stanford, CA: Stanford University.

Tripathy, A., & Eppinger, S. D. (2007). A System Architecture Approach to Global Product Development. (Working Paper #4645-07). Cambridge, MA: MIT Sloan School of Management.

Utterback, J. M. (1996). *Mastering dynamics of innovation*. Boston, MA: Harvard Business School Press.

Valkenburg, R., & Dorst, K. (1998). The reflective practice of design teams. *Design Studies*, *19*(3), 249–271. doi:10.1016/S0142-694X(98)00011-8

Vincenti, W. G. (1998). *What engineers know and how they know it*. Baltimore, MD: The Johns Hopkins University Press.

Whitney, D. E. (2004a). Product Architecture. In *Mechanical Assemblies, Their Design, Manufacture, and Role in Product Development*. New York, NY: Oxford University Press.

Whitney, D. E. (2004b). *The influence of architecture in engineering systems. ESD Symposium*. Cambridge, MA: MIT Press.

Wilczynski, V., & Jennings, J. (2005). Creating Virtual Teams for Engineering Design. *International Journal of Engineering Education*, *19*(2), 316–327.

Chapter 14
Customer Involved Open Innovation:
Innovation of New Products with End Users and Customers

Marcel Weber
Altuition BV, The Netherlands

Simone A.M. Geerts
ABNAMRO, The Netherlands

ABSTRACT

Because of both technological developments on the internet and user preferences, user or customer involvement in NPD is becoming more popular and receives broad attention from both academia and businesses. One of the major advantages of involving the customer in product development is that user needs and preferences can be implemented at an early stage of NPD, therefore enhancing the chance on acceptance and adaptation of the new product in the user market. This advantage leads to firms and companies wanting to involve their customers in the innovation process at every possible moment. But then questions arise on when, in which phases, which customers to involve, and which tools to use to support the involvement. This chapter addresses these questions by providing guidelines for customer involvement in NPD.

INTRODUCTION

Industrial society has changed a lot since the development of the service economy and information technology. We have entered an era in which customers take part in the value creation processes of firms, either on their own initiative or on invitation by these firms, and thus creating and producing their own goods and services. Initiatives from companies like Adidas, LEGO Factory, Starbucks and Dell are only a few of the many examples where consumers are invited by firms to supply ideas for new products and services, sometimes to even co-design and co-develop them, and, in a few cases, co-produce the goods and services. The "co-" in these terms indicates that it concerns activities, conducted by two or more actors together. This togetherness does not necessarily imply that these activities take place

DOI: 10.4018/978-1-61520-617-9.ch014

in the same time or at the same location; they can be asynchronous and remote. As actors, we mean, in this particular chapter, customers or end users and the firm.

However, there are still a lot of organizations that think that customer involvement is just a fad, a hype. A recent survey among Dutch marketers exhibited that about 70% of researched companies has never made the step to invite customers in co-creation, half of them state they fear to do so, because the company is not ready for the step, but also because of fear of an assertive and empowered customer (Helkema, Vugt, & Bosveld, 2008). Other surveys show that organizations think it is costly to involve customers (Chan & Lee, 2004). And many R&D driven organizations also are reluctant because of IP-breaches or disclosures (West & Gallagher, 2006). These fears and uncertainties can be taken away, when proper guidelines can be provided on how to harness these customers' potential.

Based on *research synthesis* (Denyer, Tranfield, & van Aken, 2008) we will explore this phenomenon which is often called *co-creation* (Prahalad & Ramaswamy, 2003), and will provide guidelines for companies that want to make use of the 'wisdom' of their customers in innovation projects. These guidelines are for selecting the customers to involve, identifying the innovation phases or activities that can benefit from customer participation, and applying techniques and tools best suited to support the process of customer involvement in innovations.

(THEORETICAL) BACKGROUND

Prosumerism

Until the industrial revolution it was very common for production and consumption to be fused into a single life-giving function; the producer consumed his own goods. The industrial revolution changed this situation dramatically and created a civilization in which almost no one was self-sufficient any longer. Everyone became almost totally dependent upon food, goods or services produced by someone else. The industrial revolution also led to specializations in the production function, accompanied by the rise of professionalism, thus fragmenting and concentrating product and production knowledge to the emerging professions. The difference between producer and consumer became manifest (Toffler, 1980).

In the last two decades of the 20th Century, individual, societal, technological and economic changes have led to a decrease of this difference between producers and consumers. We can distinguish a tendency towards a market as existed before the industrial revolution, where consumers are producing their own food, goods and services, the so called prosumer (Toffler, 1980). Mass customization, do-it-yourself and self-service are some of the ways this development has resulted in. The consumer becomes a co-producer by performing a part of the production process.

As consumers become more informed, connected and active, with the ability, means and motivation to take control of their interactions with companies, companies move toward co-creating and co-designing unique experiences at critical points of interaction with consumers (Prahalad & Ramaswamy, 2003). The traditional division between producer and consumer no longer holds. In order to survive this movement, firms can no longer rely on their own ideas and paths to market, but must make use of external ideas and paths to market as well, where ideas do not only come from other firms or knowledge facilities, but users as well (Chesbrough, 2003).

Is co-creation with or involvement of customers a realistic alternative? Many studies on the success of innovations have concluded that to develop successful new products or services, firms must – aside from knowing customers' needs and requirements – involve users in the development process (Brown & Eisenhardt, 1995; Cooper, 1999; Gardiner & Rothwell, 1985; von Hippel, 1986).

Companies may benefit from customer involvement by developing new products or services with better value or which better meet customers' demands (Biemans, 1991; Franke & Piller, 2004), a reduced cycle time (Biemans, 1991; Karagozoglu & Brown, 1993), educating the customer in using the new product/service (Alam, 2002), a more rapid diffusion and acceptance of the new product (Biemans, 1991), and an improvement of the long-term relationship and loyalty (Akamavi, 2005; Alam, 2002; Ancarani & Shankar, 2003; Comer & Zirger, 1997; Ennew & Binks, 1996; Friesen, 2001).

Customer involvement goes by many names in literature. In literature we encounter *co-development* (Anderson & Crocca, 1993; Mullern, Wildeman, & White, 1993; Neale & Corkindale, 1998), *customer participation* (Martin & Horne, 1995), *user involvement* (Kaulio, 1998), *partnership* (Campbell & Cooper, 1999), *co-creation* (Prahalad & Ramaswamy, 2003), *value co-production* (Lusch & Vargo, 2006; Normann, 2001; Normann & Ramirez, 1993; Ramírez, 1999; Vargo & Lusch, 2004) and *customer involvement* (Alam, 2006). Although all these names differ in a way from the others, they also seem to overlap substantially in a sense that customers or users are being involved in the value creation process of firms, instead of being regarded as mere destroyers of value created by producers (Ramírez, 1999; Vargo & Lusch, 2004) – e.g., Lusch and Vargo (2006) define co-creation and value co-production as two separate constructs, but they acknowledge that the two are linked as "nested concepts with the co-production being a subordinate concept to that of co-creation of value". Elaborating on the differences between all these names goes beyond the intention of this chapter. We will suffice by stating that we prefer the term *customer involvement in innovation,* as it captures all actions where product manufacturers and/or service providers engage with their end users or customers in (parts or phases of) innovation projects with the aim of increasing effectiveness and efficiency of the innovation process. Effectiveness refers to (1) the result of meeting users' and customers' needs and demands in a better way; and (2) increasing customer loyalty. Efficiency refers to (1) the reduction of research and development costs; and (2) the reduction of development time.

Some studies also warn for too much involvement of customers because customers can be protective or closed concerning their inventions or innovations, or simply can't be trusted concerning their commitment to participate in the innovation process, often leading to an early withdrawal from the process or a low productivity because of a lack of understanding what to do (Christensen & Bower, 1996; Leonard-Barton, 1992). And although there is no consensus on whether or not to involve users and consumers directly in innovations, a growing body of empirical work shows that customers are the first to develop many and perhaps most new products, both industrial and consumer (Foxall & Tierney, 1984; Hienerth, 2006; Lüthje, 2004; Olson & Bakke, 2001; Tietz, Morrison, Luthje, & Herstatt, 2005; von Hippel, 2005).

However, most research on the subject is limited to describing and explaining the phenomenon, and is diffused, leaving product developers and marketers that wish to make use of their customers' potential behind in confusion about what to do. Inviting customers to take part in the new product development is also not easy, or without risks and strategic consequences; firms have to consider such implications. Until this moment there is hardly any research that has completely addressed these aspects of involving users and customers in the innovation process, except for one attempt on involvement in *new service development*, which has been envisaged in a book on the topic (Edvardsson, Gustafsson, Kristensson, Magnusson, & Matthing, 2006).

Societal Benefits from Customer Involvement in Innovation

Is innovating by users a "good thing"? Henkel and von Hippel (2005) explored the social welfare implications of user innovation. User innovation is defined as the phenomenon of new products and innovations being developed by customers and end users, rather than manufacturers (von Hippel, 1988). They found that, relative to a world in which only manufacturers innovate, social welfare is very probably increased by the presence of innovations freely revealed by users. These findings are confirmed in other research (de Jong & von Hippel, 2009; Franke & Shah, 2003; Gault & von Hippel, 2009; Harhoff, Henkel, & Hippel, 2003; Henkel, 2006). There are three major reasons for this. First, user innovations tend to complement manufacturer innovations, filling small niches of high need left open by commercial sellers. Second, user innovation helps to reduce information asymmetries between manufacturers and users. Third, user innovations are more likely to be freely revealed and commercialized faster than manufacturer innovations.

We can also argue that user innovations have more value for the users because the product or service meets their needs (von Hippel, 2005). Aside from product related benefits, customers also gain psychological benefits - e.g. it is intrinsically attractive to produce something for yourself, or social benefits, like peer recognition, community belonging, status, and identity (Jeppesen & Molin, 2003; Lakhani & von Hippel, 2003; Schreier & Prügl, 2008). So, research seems to indicate that society, both on individual level as on community level, will prosper when customers participate in the innovation process of a firm.

NEW WAYS TO INVOLVE CUSTOMERS

Following the seminal studies from von Hippel, but also as a result from technological and societal changes in which firms became aware of users' potential to aid in the value creation, new product scholars and practitioners have proposed a range of successful techniques for obtaining customer input into product development processes, such as lead user analysis (Herstatt & von Hippel, 1992; Lilien, Morrison, Searls, Sonnack, & Hippel, 2002; Lüthje & Herstatt, 2004; Urban & von Hippel, 1988), mass customization (Berger, Möslein, Piller, & Reichwald, 2005), information acceleration (Urban, Hauser, Qualls, Weinberg, & et al., 1997), beta testing (Dolan & Matthews, 1993), consumer idealized design (Ciccantelli & Magidson, 1993), quality function deployment (Griffin, 1992), the ideal oriented co-design (Albinsson & Forsgren, 2004), participatory design (Greenbaum & Kyng, 1991) and user communities (Chan & Lee, 2004). We will discuss some of these ways that companies nowadays follow to get customer input in their NPD. The order in which we do this is of increasing activity from the participating customers, and an increasing external locus of customers' activities. Customers' contributions can be active (work, expertise, or information) or passive and even unknowing, e.g. behavioral data that is gathered automatically during a transaction or an activity. By activity locus we refer to where the actor, which is in control of the activities, resides. This can be within the innovating firm, on its boundaries, or even outside its boundaries, denoting the increasing autonomy that users and customers have in the activities.

Market Research

Although market research has almost always been an essential means of acquiring insights in customers' needs and requirements (Cooper, 1999; Holt, 1987; Holt, Geschka, & Peterlongo, 1984), we see that in the last decade this research has shifted from a passive involvement of customers and consumers – which is commonly referred to as 'traditional market research' in many studies – into a more active participation (Davenport, Leibold, & Voelpel, 2006; Prahalad & Ramaswamy, 2003; Senge, Carstedt, & Porter, 2001). This active participation usually takes place in a direct conversation between customer and firm representatives (Flores, 1993; Lundkvist & Yakhlef, 2004), preferably by or in the presence of those that are directly involved in product or service development (Tomes, Armstrong, & Clark, 1996). A very well known example of this active participation is the development of the Boeing 777 (Condit, 1994). This direct conversation can be conducted either physically, e.g. qualitative customer interviews (Buber, Gadner, & Richards, 2004; de Ruyter & Scholl, 2004), or online (Rey, 2008). Owing to the present internet capabilities of Web 2.0., such as social networks, weblogs and wikis, customer participation becomes even more active, not to say proactive; the rise of these technologies give way for people to express themselves in all possible ways, thereby exposing their needs, wants, feelings and thoughts, which are indispensable for NPD (Antikainen, Ahonen, & Mäkipää, 2006; Bonabeau, 2009; Li & Bernoff, 2008). Market research to elicit customers' needs and wants has thereby gone beyond the traditional form of companies asking questions and customers answering them, resulting in an actual dialogue between companies and customers.

Co-Design or Participative Design and Development

Initially conceived in information systems development, but finding application in other sectors as well, numerous techniques are being deployed to involve users specifically in the design and development phases of the product development process. The user participation usually entails active participation in requirements specification and design through user *experience design* (Norman, 1999), *empathic design* (Leonard & Rayport, 1997), *user centered design* (Beyer & Holtzblatt, 1998; Greenbaum & Kyng, 1991) and *co-design* (Albinsson & Forsgren, 2004). We refer to the literature for an elaboration on these ways of participatory design of IT. Each technique is supported by its own tools, which are usually applied in user group discussions to assess needs, and face-to-face interviews with users and managers to get a good picture of the use context and personal experiences with system failures or inadequacies. A particular tool or technique that is used at the end of the development phase is the *beta-testing* (Dolan & Matthews, 1993), in which users are invited to use and test the new software in their custom context. An important lesson that can be learned from these developments is that success is highly dependent on the direct dialogue between users and designers (Tomes, Armstrong, & Clark, 1996). As stated, these modes of customer involvement have also found application in other industries, e.g. the public sector (Brand, 2005; Sanoff, 2005) and library documenting systems (Anderson & Crocca, 1993).

Mass Customization

Mass customization refers to a customer co-design process of products or services that meet the needs of each individual customer with regard to certain product features. All operations are performed within a fixed solution space, characterized by stable but still flexible and responsive processes (Piller, Schubert, Koch, & Möslein, 2005). Mass customization differs from ad hoc customization in terms of costs. In order to obtain mass customization companies must modularize their goods and services, where a specific combination of modules is delivered for each particular customer.

In order not to overwhelm the customer with almost infinite possibilities of product combinations – *mass confusion* –, a configuration tool (Salvador, Forza, & Rungtusanatham, 2002) is applied, usually online (Franke & Piller, 2003; von Hippel & Katz, 2002). In developing online mass customization tools it is important to offer easy to operate configuration tools or facilitate customer collaboration in communities in order to prevent the aforementioned mass confusion (Piller, Schubert, Koch, & Möslein, 2005), enabling customers to easily create new combinations (Jeppesen, 2005). Applying these toolkits in a competitive setting for users gives firms the opportunity to access innovative ideas and solutions from users, shifting development and design from the locus of the firm to the user (Piller & Walcher, 2006; Thomke & von Hippel, 2002). We find mass customization appliances with design possibilities in all kinds of industries, both consumer and industrial (Berger, Möslein, Piller, & Reichwald, 2005). LEGO Factory is a well known example (Antorini & Schultz, 2007).

User Innovation

We refer to *user innovation* as innovations developed by customers and end users for their own benefit, rather than manufacturers. The driving force behind the theory on user innovation comes from von Hippel (1988), but other and earlier research confirms that this phenomenon has existed for quite some time (Enos, 1962; Freeman, 1968; Pavitt, 1984). VanderWerf (1992) studied samples of important industrial gas-using and plastics forming process equipment innovations, and found in both samples, that users were to be the most frequent developers of these innovations. Von Hippel demonstrated that organizations were not the only ones having the abilities to make valuable innovations and that the users also have innovative abilities. His findings show that the source of innovation varies across industries – users and manufacturers are the main contributors in innovation, even though they are motivated for totally different reasons (von Hippel, 1988). Manufacturers innovate in order to sell the innovation, to profit from it. Users innovate or develop an innovation because the existing products do not match their needs or requirements. Realizing this potential among users, companies can benefit from user innovations, requiring them to identify, locate, and subsequently, engage users to share their ideas or designs. Von Hippel identified a specific segment of users, which he called lead users (von Hippel, 1986). He defines lead users as "users who present strong needs that will become general in a marketplace months or years in the future. Since lead users are familiar with conditions which lie in the future for most others, they can serve as a need-forecasting laboratory for marketing research. Moreover, since lead users often attempt to fill the need they experience, they can provide new product concept and design data as well" (1986, p. 791). In a more recent publication lead users are defined as "members of a user population who (1) anticipate obtaining relatively high benefits from obtaining a solution to their needs and so may innovate and (2) are at the leading edge of important trends in a marketplace under study and so are currently experiencing needs that will later be experienced by many users in that marketplace" (Franke, von Hippel, & Schreier, 2006, p. 302). Lead users do not innovate to compete with the manufacturer, but rather because of *sticky information* - information that is hard to communicate and to transfer to the firm. This means that where information 'sticks' with users, that innovation will be most probably initiated by users, and the other way around. In this sense, 'stickiness of innovation-related information' could be an indicator in predicting the locus of innovation in several sectors. Unless an organization succeeds in motivating and enabling these lead users to submit their ideas, designs and concepts to the firm, lead users are likely to innovate themselves. So organizations have to develop systems based on the newest communication technology that motivate the users to share their sticky knowledge, and further facilitate that the sticky and tacit knowledge can be made explicit. Examples of (lead) user innovations in the sports sector are the mountain bike, kayak rodeo and kite surfing; in the industrial sector a well known example of an organization that involves lead users in its NPD is 3M (von Hippel, Thomke, & Sonnack, 1999). We conclude by stating that user innovation in itself isn't a mode of customer involvement in firms' innovations, unless companies succeed in harnessing their customers' creative potential into the development and commercialization of a new product or service.

Open Source Software

In general, co-creation implies a collaboration between a firm and its customers or users within the firm's boundaries. Open source stretches the design function and goes beyond the firm's boundaries. Open source software (OSS) is software of which the source code is available to the public and therefore can be used, changed, and redistributed according to specific licensing rules to make sure no one can appropriate the code. The code is developed by groups of thousands of voluntary developers who collaborate in online communities. Although many people did not believe in the concept at the start, - why would people do this work voluntarily, working with people they have never met before? -, there were many people who were willing to do the job, and they did it successfully. Several open source products have demonstrated that these projects can indeed lead to software systems with high functionality and quality. The success of OSS products such as Linux is widely acknowledged (Amant & Still, 2007).

An open source software community is a comfortable and exciting working environment, since the volunteers can choose their own tasks to work on, which usually are the ones they are best in, and can decide how much time to spend working on it. Participants are motivated by their own personal software needs, learning opportunities and social interaction (Bloem & van Doorn, 2007; Hertel, Niedner, & Herrmann, 2003). Research has shown that with the proper mechanisms in place, a group of intrinsically motivated, self-organizing volunteers can be as least as productive as a firm (Wendel de Joode, 2005).

Although we see open source mainly applied in software development, we see other areas where open source principles are applied to involve amateur participants in product development, including marketing (Pitt, Watson, Berthon, Wynn, & Zinkhan, 2006) and even space technology (Hancock, 2005).

User Generated Content

User-generated content (UGC), also known as Consumer Generated Media or User-created Content (UCC) (Wunsch-Vincent & Vickery, 2007), refers to various kinds of media content that are produced by end-users, as opposed to traditional media producers such as professional writers, publishers, journalists, licensed broadcasters and production companies. The term entered mainstream usage during 2005 after arising in web publishing and new media content production circles. It reflects the expansion of media production through new technologies that are accessible and affordable to the general public. These include forums, weblogs, podcasts, wikis and the posting of digital video and photographs.

Sometimes UGC can constitute only a portion of a website. For example on Amazon.com the majority of content is prepared by administrators, but numerous user reviews of the products being sold are submitted by regular visitors to the site. But in many cases the complete website thrives on user input, as is the case with Wikipedia and YouTube. Often UGC is partially or totally monitored by website administrators to avoid offensive content or language, copyright infringement issues, or simply to determine if the content posted is relevant to the site's general theme.

Most user-created content activity is undertaken without the expectation of remuneration or profit. Motivating factors include connecting with peers, achieving a certain level of fame, notoriety or prestige, and self-expression. Technical and content quality is guaranteed through the choice of the traditional media "gatekeepers". Relative to the potential supply, only a few works are eventually distributed, for example, via television or other media (Wunsch-Vincent & Vickery, 2007). In the UGC value chain, content is directly created and posted for or on UGC platforms using devices (e.g.

digital cameras), software (video editing tools), UGC platforms and an Internet access provider. There are many active creators and a large supply of content that can engage viewers, creating a potential competition for traditional media. Users are also inspired by, and build on, existing works as in the traditional media chain. Users select what does and does not work, for example, through recommending and rating, giving guide to visitors and leading to recognition of creators who would not be selected by traditional media publishers.

User-generated content has also come under fire from established media outlets such as the New York Times. Many claim that the quality of user-generated content is not up to par with the quality produced by formally trained writers and is contributing to the decline of standards in publishing, particularly with regard to news (Wunsch-Vincent & Vickery, 2007) or even 'truth' (Keen, 2007). Another concern often raised is related to privacy, with social networking sites encouraging users to share their personal information and messages in publicly viewable areas.

However, user-created content is already an important economic phenomenon despite its originally non-commercial context, e.g. iStockphoto.com which started as a creative outlet for amateur photographers and has become a real competition for professional photography. The spread of UGC and the amount of attention devoted to it by users appears to be a significant disruptive force for how content is created and consumed and for traditional content suppliers.

Crowdsourcing

Crowdsourcing is closely related to customer involvement in innovations since it also involves non-professionals performing some tasks for firms. These non-professionals are referred to as a crowd, since crowdsourcing stands for outsourcing to the crowd. Since there are several ways to use the crowd, crowdsourcing can be seen as a collection of business models. This means that although crowdsourcing is closely related to OSS and UGC, crowdsourcing is meant to result in a profit. The – for many unexpected – high quality products that resulted from OSS and UGC initiatives inspired other sectors to apply the principle of involving non-professionals in their activities and finding ways to make a profit from the crowd's efforts.

One of the ways to involve a crowd in a firm's innovation activities is the use of online user communities (Chan & Lee, 2004; Janzik & Herstatt, 2008; Nambisan, 2002). Generally these initiatives involve a forum where participants can posts their ideas and suggestions for new products or services, comment on each other's postings and vote for the best contributions. Participants generally do not get paid for their input, but are motivated by personal needs and the social benefits that such a community can provide. Many companies have found the ideas and the discussions a very useful input for their innovation process (Howe, 2008). Furthermore, firms benefit from the loyal customer base that results from the open way of interaction between customers and the firm (Li & Bernoff, 2008).

That co-creation does not always involve customers is shown by another type of crowdsourcing. In this business model, a specific challenge is broadcasted to the widest possible audience in the hope that someone will be able to solve it (Howe, 2008). The crowd at InnoCentive, where complex scientific challenges are posted and solved, consists of about 160,000 people who are intrinsically motivated by these challenges and compete for a prize. In exchange for the monetary reward, the firm receives the rights of the solution and can use it to make a profit. Despite the fact that these solvers can be considered hobbyists, they are able to solve some of the challenges that have puzzled the R&D departments of some prominent firms for months. The basis for the success of this business model lies in the fact that the more diverse the crowd, the higher the chance that someone will have a solution.

GUIDELINES FOR CUSTOMER INVOLVEMENT IN INNOVATION

Why Do We Need These Guidelines?

As has been demonstrated, management theory hasn't reached a sufficient level of completeness, comprehensiveness and applicability for firms to start engaging customers in innovations. It isn't that surprising that, even if firms and organizations acknowledge the importance of customer involvement, the inclination to act as such is still low and could use some boosting up. Most firms restrict their customer involvement at this moment to concept, prototype and product testing. Even fewer have deployed an online dialogue, from which they occasionally receive a bright idea from one or two customers, others without success. But we can decidedly state that many organizations do not recognize the potential that customers possess in supporting their innovation efforts. The good news is that we can observe a growing interest for the phenomenon, both in the academic field as among firms, making followers aware of the importance of engaging with the customer to innovate. Some of these firms may actually want to start one or more innovation projects with the involvement of their customers. They will be looking for a workable guide on how to do this, not for a participation in the academic discussion. To that avail we are developing a comprehensive, robust protocol for engaging customers and users in innovations. A protocol which addresses the following questions: How can firms engage and involve customers in the innovation process in an effective way? What kind of customers, process, procedures and methods should be engaged and followed, what are the tools to be used? What pitfalls or disadvantages exist in engaging customers in this innovation process, and how can they be overcome and avoided?

The development of this protocol will be based on research of both academic literature and practice; practice is included because of the incompleteness of academic literature. The protocol is currently in the phase of being applied and tested as an important part of the first author's PhD research, publication forthcoming in the beginning of 2010. Aside from this theoretical contribution to management science, we aim to support firm managers which are tasked with the management of NPD appropriate guidelines and suggestions to decide on customer involvement, to select the proper participants and to engage them with the proper tools. Because of this objective, and the fact that the protocol has not yet been tested in practice, we will not present the protocol as it will appear in the dissertation, but will address the aforementioned questions.

Choosing the Appropriate Mode for Customer Involvement

As we have seen in our review over the different modes of involvement, any sector can harness their customers' potential to co-create. In the Open Innovation approach it's now conventional wisdom that virtually no company should innovate on its own (Chesbrough, 2003). Firms nowadays have a myriad of potential partners and ways to collaborate with them, making the choice for the right partners and ways much more difficult. Should firms open up and share their intellectual property with the community? Should they nurture collaborative relationships with a few carefully selected partners? Should they harness the "wisdom of crowds"? There is no best approach to leveraging the power of outsiders. Different modes of collaboration involve different strategic tradeoffs. Firms who have already tried have become experienced in choosing the right partners, those who haven't should experiment, but should also take caution.

In principle, all innovation projects are suitable for customer involvement. As the Open Innovation Paradigm depicts, it is recommended to source the innovation partly or completely from outside the organization. However, the firm must make decisions on appropriateness, based on the following related considerations.

The Stage of the Innovation Process

The further the firm is in its innovation process, the less necessary or even appropriate it is to involve customers *for the first time* in the process, because customer involvement could lead to a declination of the reached results. Conversely, the earlier we are in the project, the more suitable customer involvement will be, because this will prevent the firm from developing the wrong specifications, requirements, prototypes and so on. Thus, to increase the success probability of the innovation, the company should aim to involve these customers in an early stage of the innovation process, preferably at the start of the project (Calantone, Di Benedetto, & Haggblom, 1995; Cooper & Kleinschmidt, 1986; Gruner & Homburg, 2000; Pitta, Franzak, & Katsanis, 1996; Urban & von Hippel, 1988; van Kleef, van Trijp, & Luning, 2005; von Hippel, 1986,, 1988; Yoon & Lilien, 1988).

The Impact on Customer Relations and Loyalty

Having customers contribute in innovations has a positive effect on customer relationship, as is concluded in several studies (Akamavi, 2005; Alam, 2002; Ancarani & Shankar, 2003; Comer & Zirger, 1997; Ennew & Binks, 1996; Friesen, 2001). Although we assume that firms will always contemplate loyal customers, their innovation activities do not have to be aimed at creating customer loyalty. More important objectives of an innovation can be, for instance, staying ahead of competition in terms of market share, being a first mover, initial penetration of new markets, rendering increase of customer loyalty a minor or less important objective. Thus, the company has to consider, whether customer relations and customer loyalty are important objectives of the innovation project. In such a case, customer involvement is recommended. Simply stated, at this stage, the company has to decide on being market oriented or innovation oriented (Berthon, Hulbert, & Pitt, 2004).

The Preferred 'Openness' to the Project (Pisano & Verganti, 2008)

In this case, openness is referred to as the amount of external participants. Firms may want to source as much ideas or customer input as possible to ensure that the great idea or input is obtained. Disadvantage of this approach, however, is that the cost of searching for, screening, selecting contributors and contributions grow as the network of participants becomes larger and can become prohibitive, e.g. IBM's Innovation Jam in 2006 (Bjelland & Wood, 2008). An open approach can also expose the project too much to competitors, increasing project risk and time pressure. Choosing an open approach with a large amount of participants implies that, to source the large amount of participants as efficiently possible, it is preferable that the tasking of the participants can be divided in small, well-defined discrete chunks of works that participants can work on autonomously, e.g. open source initiatives. A large amount of participants also implies involvement via an online approach, because of the mere fact that an offline approach is too costly or troublesome (travel, logistics, venues). Therefore, to have a large amount of (potential) customers to participate, it has to be necessary that the tasking of the participating customers is possible through the online channel. For instance, an online approach will be particular difficult to implement in case of the prototype testing of high tech or bulk products.

The Preferred 'Completeness' of the Contribution

Completeness refers to terms of 'raw ideas', 'market-ready ideas', or 'market-ready products'.

Nambisan and Sawhney (Nambisan & Sawhney, 2007) have organized the possibilities for the external sourcing of innovations by four variables: (1) the reach that firms have as they cast about for innovative ideas to assess; (2) the cost of acquiring and developing these ideas; (3) the risk involved in trying to turn the ideas in marketable products, and; (4) the speed with which the ideas can be brought to market. Nambisan and Sawhney argue that the more market-ready the idea or product is the firm is looking for, the higher the search costs, but the shorter the time to market will be. However, the chance (reach) of finding such market-ready offerings is low. In contrast, raw ideas, still have to be developed further, increasing project risk and time to market, but reducing costs and difficulty of finding an idea. Our view on this is that market-ready ideas or products are in a way user innovations, and can be expected mainly from professionals, inventors, experienced users or lead users, while ordinary, untrained or inexperienced users are expected to supply mostly raw ideas of which manufacturability has not been tested at all (Magnusson, 2009).

The Accessibility of Knowledge Concerning the Intended Innovation

As studies show (Un, Cuervo-Cazurra, & Asakawa, 2008; von Hippel, 1994) knowledge needed from the customer can be difficult to access, reducing the chance for success when involving customers. Other studies teach us that radical and disruptive innovations do not lend themselves to involve the customer from the beginning, because of the fact that customers can't be knowledgeable about things that do not yet exist (Callahan & Lasry, 2004; Christensen & Bower, 1996), unless companies can find a means to access customers' tacit and latent knowledge, or sticky information. It is not easy to depict when customers' knowledge is easy or difficult to access – even Un et al. (2008) remark that this variable cannot be measured in a direct way, and postulate that it is determined by several factors, like the acquisition costs, the difficulty of identifying the customers, IP-sensitivity, speed of access, and such. Von Hippel defines sticky information as information that is "costly to acquire, transfer, and use" (von Hippel, 1994, p. 429). In order to choose for customer participation, firms should therefore assess accessibility of customers' knowledge – as we will review further on, there are tools to accomplish this –, and calculate the expected costs and benefits for doing this.

The Specificity of the Problem or Question

The consideration is here whether the firm is looking for solutions to specific questions or trying to 'discover' on its own which solutions or applications exist in the market which it can fit in its innovation (Hansen & Birkinshaw, 2007). The first option requires a more active participation from customers – it is done by firm's request –, and can be directed towards pre-selected participants, while the discovery option requires the search of a wide and diverse range of possible contributors and sources. Looking for a discovery therefore usually results in higher search costs and more development time than requesting a solution to a specific problem. On the other hand, we expect that asking specific questions requires thorough assessments and screenings of potential innovation directions, in which the customer probably will not take part, unless they are about solving problems or acquiring ideas on existing products and services, or prototypes – the so-called re-innovation phase of NPD (Rothwell, 1986). In fact, customer complaints, suggestions, visits, user-to-user interactions in communities, user innovations or modifications may well contain specific solutions that firms are looking for.

The aforementioned considerations are very interrelated, implying that we cannot decide in a linear manner on when to and when not to choose for customer involvement. Decisions made at this stage depend highly on aspects like preferred speed (e.g. for firms in highly competitive industries speed can be a crucial), customer base (large or small, heterogeneous or homogeneous), reach (i.e. how many or which of these customers are we able to reach), available budget, and such. We also observe that the considerations aren't a matter of either one option or the other, but that mixed options are also possible, as LEGO proves by observing and involving communities, commercializing user innovations and engaging lead users in projects (Antorini & Schultz, 2007; Seybold, 2006). We also do not exclude the possibility for firms to create a mix of external sources in which other firms, like suppliers, or research institutes participate, confusing simple decision making on the matter. It is important also to denote that many decisions to be made in this step are also highly dependent on organizational characteristics and the experience that the company itself already has with involving customers or other external parties in innovations or any other organizational process (Hansen & Birkinshaw, 2007; Nambisan & Sawhney, 2007). Otherwise, the organization will expose some barriers that can be detrimental to customer involvement.

Firms should find their own approach in deciding, but we would like to propose to review all these considerations in a facilitated workshop for all concerned in this decision making. The combination of criteria – the innovation project phase, the impact on customer relations and loyalty, the intended openness, the preferred completeness of ideas, the accessibility of knowledge, and the specificity of the problem –, should lead to a discussion, followed by some kind of consensus on whether or not to involve customers, why to involve them, and expected contributions, by the participants.

Choosing Participants: Who to Involve

The previous expose, in a way, has already addressed the issue of which customers should be involved. We will elaborate more on this question in this section.

As one can derive from this chapter's title, our first proposition suggests that firms should engage the *end users* of their products and services in NPD or NSD, as has been suggested over the years by QFD-proponents (Griffin & Hauser, 1993; Mazur, 1994), but other research as well (Ancarani & Shankar, 2003; Seybold, 2006; Wind & Mahajan, 2002), since these end users are the keystone customers that determine the real value of the product or service.

As for the question whether to involve existing customers or potential customers, we can be very brief. Several studies show that existing customers are best involved when developing incremental innovations or line extensions, where potential customers should not be neglected when developing really novel products or services (Christensen & Bower, 1996; Christensen & Raynor, 2003; Danneels, 2004; Hamel & Prahalad, 1994; Slater & Narver, 1995).

Initial research indicates that not every (potential) customer or user is capable of contributing in the innovation process (Rothwell & Gardiner, 1983). Von Hippel (2005) suggests that companies should aim to involve its lead users, meaning that these users are usually professionals or experts in the field of work of the product or service being innovated. Gruner and Homburg (2000) found that for the best result in NPD, the characteristics of customers in order of the best to the least success should be: lead users, financially attractive customers (potential market), close customers (relations) and technical attractive customers, whereas the last group led to almost no success at all. To trace and locate such lead users, von Hippel introduced the Lead User Approach (von Hippel, 1986), which has been tested and applied in several subsequent studies (Herstatt & von Hippel, 1992; Lilien, Morrison, Searls, Sonnack, & Hippel, 2002; Olson & Bakke, 2001; Urban & von Hippel, 1988), showing that, although it is very well possible to locate lead users, the process in which to do this, is complex and time consuming. Additionally, application of the Lead User Approach seems to be easier and appropriate in industrial and technology sectors, than in consumer goods or service settings, or innovations where individuals are affected by the innovation (Intrachooto, 2004), since the approach requires the specification of lead user trends in the market by means of reviewing user modifications or innovations – such an inventory is easier made for industrial than for consumer goods. In the industrial goods sector, customers typically have a higher level of expertise regarding the product compared to those in consumer goods markets (Gruner & Homburg, 2000). And although von Hippel (1988) indicates that lead users' participation increases the chance of adoption, it has not been evidenced yet that this group can represent and speak for the majority who will actually use the goods and services that are being designed and developed. Lead users have been identified also to be innovators and early adopters (Schreier & Prügl, 2008), as Rogers (1995) designates people which usually are the first to adopt new innovations. Lead users may well be ahead of the market trend, but could have such a jumpstart from ordinary users, creating a 'chasm' between them and the following groups of adopters (Moore, 2002).

Later research indicates that ordinary, less experienced users, can generate more original or better ideas than professional or even experienced users (Kristensson, Gustafsson, & Archer, 2004). Training these ordinary users in advance in creativity techniques and process skills enhances users' contributions in better completing innovation tasks than professional designers do this (Magnusson, Kristensson, & Hipp, 2007). To give useful contributions, customers or users do not have to be technical experts in the field (Magnusson, 2009). Thus, customer involvement in NPD or NSD can be aimed at co-creating with any end customer. We believe that all people can be creative (Boden, 1990; Sanders & Stappers, 2008). But asking someone to be creative will probably not work, but teaching or training him how to do it, will surely help (Boden, 1990). There are certain prerequisites for individuals to become creative. People have to have curious and inquisitive minds, handle ambiguities and uncertainties well, and highly motivated to act (Boden, 1990; Chang, 2008). They also need to have some experience with the related problem – e.g., being a user – to be capable to participate in any kind of co-creating activity with the firm.

Our own practice (Brabant-Zeeuwse Werkgeversvereniging, 2004; Weber, 2008) and numerous cases, like LEGO, Fiat, Kraft, IBM, show that requirements regarding knowledge, expertise, technical skills, and such, indeed do not matter. As long as participants are a qualitative representation of a company's end user base, i.e. familiar with the company, its products, and use of its products or services, involvement may always pay off. And obviously, it can be expected that people who are expressive and socially communicative will provide better input for the process, but we believe that – because it is difficult to find such people – companies can better start with this representation of their customers, and build up experience in recognizing these qualities among them. People to whom it is clear what is expected from them, who feel they can contribute or have contributed previously, and are enthusiastic about that, will contribute, whether professionals, amateurs or pro-ams (Leadbeater & Miller, 2004; Li & Bernoff, 2008). And the more you can engage in participation, the greater the chance that someone or the crowd as a whole will deliver the solution or great idea you are looking for.

Contributions from Participants: Phases for Involvement

Some studies emphasize the importance of customer involvement in early phases of the development process (Calantone, Di Benedetto, & Haggblom, 1995; Cooper & Kleinschmidt, 1986; Urban & von Hippel, 1988; von Hippel, 1986,, 1988; Yoon & Lilien, 1988). Other studies highlight the importance of customer involvement throughout the length of the development process (Cooper, 1994; Cooper & Edgett, 1996; Donath, 1992) on the assumption that customer needs change throughout the design, development and purchasing activities. It seems that all phases or stages that are suited for involvement with customers as long as the interaction between firm and customers is intensive (Parkinson, 1982; Voss, 1985) (Holt, 1988; Nambisan, 2002; Sawhney & Prandelli, 2000). Gruner and Homburg (2000) found that a high intensity of firm-customer interaction during the stages of idea-generation, screening, prototype testing, and the launch significantly influenced the performance of new products. As for services, it was found that involvement during most of the development stages, but particularly in the idea generation, service design, and testing, were important for the success of the new service (Alam, 2002). Research is however very scarce, if not, unavailable concerning the commercialization of innovations. There seems to be ample experience on customer involvement in this stage. This involvement can take place either by involving customers in the marketing or advertisement development for the new product (Weber, 2008), selecting them as launching customer or providing customers with a sample, starting a word of mouth campaign (Eilander, 2008) or through user communities or social networks (Delre, 2007).

These findings supports our idea that all phases are suited for customer involvement, but that the role of the customer and the intensity of the interaction with the firm will differ along the stages. The more stages customers are participating in, the better this would be for the result. The different phases of the innovation process may require different skills and knowledge from the participants in the several phases, implying that one can not engage the same customers throughout the complete innovation process, because of these changing skill needs. Customers can also fulfill different roles in the innovation process (Nambisan, 2002), making it more difficult to find all these roles within one single person or small group. The aforementioned literature illustrate and emphasize the importance of customers in several, mostly all phases of the innovation process, but neglect to address the question whether these were the same customers for all stages. In our own practice (Weber, 2009) we have experienced that engaging with the same customers throughout all the stages, will present some objections, like a decreasing motivation as time goes by and the effect that people get carried away with their own ideas, thereby blocking new ideas. We therefore propose to try changing participants in each stage. This will imply extra effort to accustom and eventually train these new participants each time, but on the other hand it ensures new and different ideas and insights throughout the whole process, and also creates diversity among participants, increasing the chance for success.

In the Appendix we summarize the distinct stages in the NPD along with the activities in which customers can be involved.

Maximizing the Contribution: Tools, Techniques and Methods

As we've seen all kind of customers can be involved in all NPD stages - success is likely to depend on the (process) tools that are applied. Many studies about the use of tools in the NPD or NSD process only address what we've named 'traditional market research' methods where the input from customers is concerned (Holt, 1987;

Tidd & Bodley, 2002). The problem with these studies is that they do not elaborate on tools which are best suited for customer involvement. To tackle that, van Daelen (2005) identified in her master's thesis, 29 tools suited to support the customer involvement in NPD, and presented these to experts, through a Delphi study (Linstone & Turoff, 2002), to determine appropriateness of these tools for each stage of the innovation process. According to these experts, customer contribution seems to be relatively important in all innovation phases, especially for idea generation, concept development and testing. Customer interviews, group interviews, and some other qualitative and observation techniques were identified as best suited for the ideation, concept and test phase. Von Hippel (1988) shows that application of the Lead User method is preferred in such stages, since most users 'suffer' from functional fixedness (Adamson, 1952), disqualifying traditional methods like focus groups and customer surveys. However, there is ample evidence that modern qualitative techniques, like the Zaltman Metaphor Elicitation Technique or ZMET (van Kleef, van Trijp, & Luning, 2005; Zaltman, 2003), the Long Interview (McCracken, 1988; Mullins, 2007), and customer observation (Leonard & Rayport, 1997), are perfectly suited to access latent feelings, needs and wants from ordinary users as well. However, we will have to keep in mind that these methods require expert skills, making the use of a qualified assessor necessary. Additionally, the ZMET is a patented technique, requiring licensed researchers for assessment. Beta testing proved to be suited for the testing of prototypes and new products, while brainstorming can be used in ideation too. For the development and commercialization phase experts could not reach a consensus. Van Daelen attributed this to "the Dutch background of the experts (…) because enabling customers to contribute in the form of 'creation' during development and commercialization is not yet a very common practice in the Netherlands" (2005, p. 69). As has been observed in the previous paragraph, this is a universal problem.

Some research focuses on aspects like the communication and interaction mode between firm and users, concluding that intensive communication and interactions with customers about new products, designs, and prototypes is associated with NPD success (Atuahene-Gima, 1996; Biemans, 1991; Drew, 1995; Frambach, Barkema, Nooteboom, & Wedel, 1998; Gobeli & Brown, 1993; Gruner & Homburg, 2000). From other studies it becomes obvious that direct contact between NPD or NSD team members and participating customers is also beneficial to the success of the project (Alam, 2002; Anderson & Crocca, 1993; Comer & Zirger, 1997; Magnusson, Matthing, & Kristensson, 2003; Sanders & Stappers, 2008; Tomes, Armstrong, & Clark, 1996), while integration of these participants in NPD or NSD teams is recommended (Alam & Perry, 2002; Lagrosen, 2005). In the interactions between the actors it is crucial that team members really understand the latent and unarticulated needs and wants of the customers (Mullins, 2007), and communicate in a way that avoids misunderstandings between firm representatives and participating customers (Tomes, Armstrong, & Clark, 1996). To accomplish this, not only techniques like the ZMET, the Long Interview, and customer observations – which we mentioned earlier – are appropriate. It has been demonstrated that the use of metaphors and analogies, as a 'language', throughout the whole NPD, improves understanding and communication, increasing the success of the NPD (Albinsson & Forsgren, 2005; Herstatt & Kalogerakis, 2005; Sifonis, Chernoff, & Kolpasky, 2006; Teichert, von Wartburg, & Braterman, 2006; Zaltman, 2003).

It can be concluded that tools and techniques that facilitate a dialogue or an intensive interaction between participating customers and company's NPD-team, and the use of metaphors and analogies are suited to support the customer involvement because:

- customers might use metaphors and analogies to express their feelings (Zaltman, 2003)
- unspoken, latent needs can be elicited (Mullins, 2007)
- a common vocabulary can be created that can be used for communicating ideas and concepts (Sifonis, Chernoff, & Kolpasky, 2006), enabling outsiders or newcomers to understand what is intended – this way, changes in participants can be made without any loss of information, thereby hardly affecting the speed of the process.

We intentionally have not addressed tools and techniques that should be used to conduct and facilitate the innovation process itself, since these are usually dependent on sector, competitive pressure, firm size, cultural and organizational variables, as well as firm's budgets. We have focused our review on tools for interacting with participating customers, and propose without further ado that companies use their customary tools and techniques as much as possible, but take into account that interaction with the customer is conducted through the (basic) techniques incorporated in the Appendix. If customers are unfamiliar with these tools – which often will be the case – a simple training of these customers in advance will be of great help. At this stage we also observe that several innovation tools have been developed in software and online applications, enabling firm to customer interaction asynchronously and at distance. Two known examples are GroupSystems (van Zanten, 1998) and Virtual Customer (Dahan & Hauser, 2002).

Special Situations, Recommendations and Avoidances

In this last paragraph of the protocol we will discuss some do's and don'ts.

Inviting Customers to Participate

It goes beyond the intention of this chapter to fully discuss the steps that companies should take, in order to streamline the process of searching for, inviting, welcoming, accommodating, motivating, thanking, and dismissing participating customers in the innovation activities. We believe that most firms are capable and experienced to handle these aspects effectively. However, we would like to address some of these aspects.

One of the most important activities in the preparation phase for involvement consists of formulating the proper task, challenge, question or invitation to customers. This invitation should be uncomplicated and easy to understand, otherwise customers might not be motivated to participate or lose interest during participation. We propose to recruit participants in an open way (Pisano & Verganti, 2008) in order to acquire interested and motivated volunteers, unless the innovation objectives are to focus on a select group of customers. To place an 'open invitation', all media or communication channels can be used, like a website call, emails, newspapers, radio, television, depending on the amount of disclosure the company wishes to reach. Regardless of these media, the firm should make clear in advance to all prospects what is expected from them when applying for participation, e.g. objectives, required skills, selection and intake procedures, and needed training.

Another important aspect in the preparation for customer involvement concerns the allocation of proper resources, like a project manager, a customer interaction moderator, training of employees, training of participants. The firm should also consider the recording of meetings, session, correspondence, and such. We deem it unnecessary to elaborate on how to prepare for and conduct meetings or interactions, but suffice with pointing out that dates, duration, and venues should all be made well in advance.

Guidelines for Online Involvement (e.g. Crowdsourcing)

Online involvement is in essence not very different form offline involvement, described previously; we see evidence that even in online situations participants can become a part of NPD (Füller, Bartl, Ernst, & Mühlbacher, 2006). As in all user involvement, trust is an important issue. But since building trust can be more difficult in an online environment, this issue needs more emphasis. Communication aspects that are natural in an offline environment, such as welcoming participants and guiding the discussions, need some attention. Since online participation is more anonymous than offline participation, it also tends to be more informal. Therefore, firms should find a way to motivate users to participate, and ideally, keep participating and form a loyal customer base. Creating a critical mass is therefore crucial. To do this, the request towards the participants should be clear and simple enough to keep their attention. Even if all goes well, it takes about one year to create such a community (Li & Bernoff, 2008).

Rewarding Participants

Collaborative innovation bestows several advantages on companies. The payoff for customers is less clear. IP resulting from company-customer collaborations is typically owned by the company, and so are the profits generated by that IP. Today the most creative innovators may find more than one company competing for their time and ideas. So customers, understandably, are starting to ask questions: "What's in it for me?" (von Krogh, 2006).

Research shows that many users that participate in co-creation do this primarily for intrinsic reasons (self efficacy, recognition) rather than for monetary rewards (Füller & Bartl, 2007; Jeppesen & Frederiksen, 2004). One of the interesting perspectives is provided from the viewpoint of OSS (open source software) communities where people are working on a voluntary basis without receiving direct compensation. Although some of the participants are receiving their salaries from the companies, the basic idea of OSS has been traditionally based on free work and still often is. But, contrary to experimental findings on the negative impact of extrinsic rewards on intrinsic motivations (Deci, Koestner, & Ryan, 1999), it was found that being paid and feeling creative on OSS projects do not have a significant negative impact on project effort (Antikainen & Väätäjä, 2008). In the light of the rewarding models used in successful open innovation intermediaries, like InnoCentive, it seems reasonable to assume that multiple and varying types of motivations are present and members may also have multiple simultaneous goals behind their participation. If this is true, a combination of both monetary and non-monetary rewards would be optimal for members. The term "incentives" (Reeve, 2005) is therefore a more appropriate term. Examples of rewards or incentives are approval, paychecks, trophies, money, praise, attention, grades, scholarships, prizes, food, awards, honor-roll lists, public recognition and privileges.

Summary of Our Guidelines

We can now summarize all these findings in some general do's and don'ts:

1. It does not matter in which industry or sector you are employed; your customers possess knowledge and experience about the use of your product, which you can harness for your NPD-initiatives. This will enhance the success of your innovation, and your customers' loyalty. The earlier you involve them in a certain project, the better the results will be.
2. Carefully decide on this customer involvement by identifying in advance what benefits you want to get from this and whether your

customers are easy to locate, access and cooperative. Communicate your intentions and objectives upon invitation. Be transparent about these. Do not underestimate their power to neglect your invitation to participate.

3. You can co-create with your customers in every phase of the NPD process, you can use your own tools and techniques, as long as you do not make things too complicated for the participants. That means that you can train them in the use of these techniques – this will lead to better inputs from them. But keep in mind that it is not this toolbox that determines the success. It is the dialogue that you have with your customers that will do the trick. Therefore, use additional techniques that facilitate this dialogue. In that respect, choose participants that are willing and able to take part in that dialogue.
4. Do not rely solely on a few important, financially promising or lead customers. On the other hand, do not think that a large amount of participants will be helpful. Create a diverse participation by alternating customers in phase or activities of an amount you can handle. This will also prevent group thinking, the preference for one's own ideas or even boredom in a lasting process. In many online situations you might aim for a larger participation, but do this only when you do not have to handle their inputs simultaneously.
5. Finally, treat your customer respectfully. Reward him properly and treat him as one of the NPD-team, by including him in decision-making, be transparent about royalties and rights. But, be certain that your whole organization adopts this attitude. Any mistake can be catastrophic for you customers' willingness to participate.

CONCLUSION

Technological and social developments in society give way to the participation of customers in the creation of goods and services, especially in their design and development, thereby blurring the boundaries between firms and their customers. We call that co-creation. This co-creation can take several shapes, i.e. from passive to active participation from customers, as well as on initiative of the firm to full initiative of the customer, as has been demonstrated by examples like market research, mass customization, user innovation, open source developments, user generated content and crowdsourcing. Firms can harness their customers' potential to co-create by systematically applying the guidelines we have presented, with which they can identify the appropriate modes of customer involvement, the type of customers to involve, the phases for involvement and the appropriate methods, tools and techniques to support it. These guidelines have been developed through research synthesis of literature and practice, making it as robust possible for all kinds of industries and channels.

REFERENCES

Adamson, R. E. (1952). Functional Fixedness as Related to Problem Solving: A Repetition of Three Experiments. *Journal of Experimental Psychology*, *44*, 288–921. doi:10.1037/h0062487

Akamavi, R. K. (2005). A research agenda for investigation of product innovation in the financial services sector. *Journal of Services Marketing*, *19*(6), 359–378. doi:10.1108/08876040510620148

Alam, I. (2002). An Exploratory Investigation of User Involvement in New Service Development. *Journal of the Academy of Marketing Science*, *30*(3), 250–261.

Alam, I. (2006). Process of Customer Interaction in New Service Development. In Edvardsson, B., Gustafsson, A., Kristensson, P., Magnusson, P., & Matthing, J. (Eds.), *Involving Customers in New Service Development* (pp. 15–32). London: Imperial College Press. doi:10.1142/9781860948893_0002

Alam, I., & Perry, C. (2002). A customer-oriented new service development process. *Journal of Services Marketing*, *16*(6), 515–534. doi:10.1108/08876040210443391

Albinsson, L., & Forsgren, O. (2004, Feb. 23-24). *Ideal Oriented Co-Design of e-Services.* Paper presented at the First Scandinavian eGovernment Conference, Örebro, Sweden.

Albinsson, L., & Forsgren, O. (2005, June 19-20). *Co-Design Metaphors and Scenarios – Two Elements in a Design Language for Co-Design.* Paper presented at the Language Action Perspective (LAP) 2005, Kiruna, Sweden.

Amant, K. S., & Still, B. (2007). *Handbook of research on open source software: Technological, economic, and social perspectives*. London: Information Science Reference.

Ancarani, F., & Shankar, V. (2003). Symbian: Customer interaction through collaboration and competition in a convergent industry. *Journal of Interactive Marketing*, *17*(1), 56–76. doi:10.1002/dir.10048

Anderson, W. L., & Crocca, W. T. (1993). Engineering practice and codevelopment of product prototypes. *Communications of the ACM*, *36*(6), 49–56. doi:10.1145/153571.256015

Antikainen, M., Ahonen, M., & Mäkipää, M. (2006). *Customers as Designers, Sellers and Buyers: Utilizing Web 2.0 Business Models in Mass Customization.* (Unpublished Working Paper). University of Tampere, Finland.

Antikainen, M., & Väätäjä, H. (2008). *Rewarding in open innovation communities – How to motivate members?* (Unpublished Paper Accepted to be published in Proceedings of the 2008 ISPIM conference.). Tampere University of Technology.

Antorini, Y. M., & Schultz, M. (2007, May 31 - June 2). *Brand Revitalization through User-Innovation: The Adult Fan of LEGO Community.* Paper presented at the 11th International Conference on Reputation, Brand, Identity & Competitiveness Bergen, Norway.

Atuahene-Gima, K. (1996). Differential potency of factors affecting innovation performance in manufacturing and services firms in Australia. *Journal of Product Innovation Management*, *13*(1), 35–52. doi:10.1016/0737-6782(95)00090-9

Berger, C., Möslein, K., Piller, F. T., & Reichwald, R. (2005). Co-designing modes of cooperation at the customer interface: learning from exploratory research. *European Management Review*, *2*(1), 70–87. doi:10.1057/palgrave.emr.1500030

Berthon, P., Hulbert, J. M., & Pitt, L. (2004). Innovation or customer orientation? An empirical investigation. *European Journal of Marketing*, *38*(9/10), 1065–1090. doi:10.1108/03090560410548870

Beyer, H., & Holtzblatt, K. (1998). *Contextual Design: Defining Customer-Centered Systems.* San Francisco: Morgan Kaufmann.

Biemans, W. G. (1991). User and third-party involvement in developing medical equipment innovations. *Technovation*, *11*(3), 163–182. doi:10.1016/0166-4972(91)90032-Y

Bjelland, O. M., & Wood, R. C. (2008). An Inside View of IBM's 'Innovation Jam'. *MIT Sloan Management Review*, *50*(1), 32–46.

Bloem, J., & van Doorn, M. (2007). *Open voor business: Open-source-inspiratie voor innovatie.* Groningen: Uitgeverij Kleine Uil. (in Dutch)

Boden, M. A. (1990). *The Creative Mind: Myths and Mechanisms.* London: Weidenfeld and Nicolson.

Bonabeau, E. (2009). Decision 2.0: The Power of Collective Intelligence. *MIT Sloan Management Review, 50*(2), 45–52.

Brabant-Zeeuwse Werkgeversvereniging. (2004). Klantenleercentrum: Van klantgericht naar klantgedreven innoveren. *InterActie, jaargang 2 nr. 7,* p. 4 (in Dutch).

Brand, R. (2005). The citizen-innovator. *The Innovation Journal: The Public Sector Innovation Journal, 10*(1), 1–11.

Brown, S. L., & Eisenhardt, K. M. (1995). Product development: Past research, present findings, and future directions. *Academy of Management. Academy of Management Review, 20*(2), 343–378. doi:10.2307/258850

Buber, R., Gadner, J., & Richards, L. (Eds.). (2004). *Applying Qualitative Methods to Marketing Management Research.* New York: Palgrave MacMillan.

Calantone, R. J., Di Benedetto, C. A., & Haggblom, T. (1995). Principles of new product management: Exploring the beliefs of product practitioners. *Journal of Product Innovation Management, 12*(3), 235–247. doi:10.1111/1540-5885.1230235

Callahan, J., & Lasry, E. (2004). The importance of customer input in the development of very new products. *R & D Management, 34*(2), 107–120. doi:10.1111/j.1467-9310.2004.00327.x

Campbell, A. J., & Cooper, R. G. (1999). Do Customer Partnerships Improve New Product Success Rates? *Industrial Marketing Management, 28*(5), 507–519. doi:10.1016/S0019-8501(99)00058-9

Chan, T.-Y., & Lee, J.-F. (2004, April 3-7). *A Comparative Study of Online User Communities Involvement in Product Innovation and Development.* Paper presented at the 13th International Conference on Management of Technology (IAMOT), Washington D.C.

Chang, C. M. (2008). Engaging the creative minds - the ENGAGE models. *International Journal of Innovation and Technology Management, 5*(1), 149–165. doi:10.1142/S0219877008001254

Chesbrough, H. W. (2003). *Open Innovation. The New Imperative For Creating and Profiting from Technology.* Boston, MA: Harvard Business School Publishing.

Christensen, C. M., & Bower, J. L. (1996). Customer power, strategic investment, and the failure of leading firms. *Strategic Management Journal, 17*(3), 197–218. doi:10.1002/(SICI)1097-0266(199603)17:3<197::AID-SMJ804>3.0.CO;2-U

Christensen, C. M., & Raynor, M. E. (2003). *The Innovator's Solution. Creating and Sustaining Successful Growth.* Boston, MA: Harvard Business School Press.

Ciccantelli, S., & Magidson, J. (1993). From experience: Consumer idealized design: Involving consumers in the product development process. *Journal of Product Innovation Management, 10*(4), 341–347. doi:10.1016/0737-6782(93)90076-3

Comer, J. M., & Zirger, B. J. (1997). Building a Supplier-Customer Relationship Using Joint New Product Development. *Industrial Marketing Management, 26,* 203–211. doi:10.1016/S0019-8501(96)00114-9

Condit, P. M. (1994). Focusing on the customer: How Boeing does it. *Research Technology Management, 37*(1), 33–37.

Cooper, R. G. (1994). Perspective third-generation new product processes. *Journal of Product Innovation Management, 11*(1), 3–14. doi:10.1016/0737-6782(94)90115-5

Cooper, R. G. (1999). From Experience: The Invisible Success Factors in Product Innovation. *Journal of Product Innovation Management, 16*, 115–133. doi:10.1016/S0737-6782(98)00061-7

Cooper, R. G., & Edgett, S. J. (1996). Critical success factors for new financial services. *Marketing Management, 5*(3), 26–37.

Cooper, R. G., & Kleinschmidt, E. J. (1986). An investigation into the new product process: Steps, deficiencies, and impact. *Journal of Product Innovation Management, 3*(2), 71–85. doi:10.1016/0737-6782(86)90030-5

Dahan, E., & Hauser, J. R. (2002). The virtual customer. *Journal of Product Innovation Management, 19*(5), 332–353. doi:10.1016/S0737-6782(02)00151-0

Danneels, E. (2004). Disruptive Technology Reconsidered: A Critique and Research Agenda. *Journal of Product Innovation Management, 21*(4), 246–258. doi:10.1111/j.0737-6782.2004.00076.x

Davenport, T. H., Leibold, M., & Voelpel, S. (2006). *Strategic Management in the Innovation Economy. Strategy Approaches and Tools for Dynamic Innovation Capabilities*. Berlin Heidelberg, Germany: Springer Verlag.

de Jong, J. P. J., & von Hippel, E. (2009). *Measuring user innovation in Dutch high tech SMEs: Frequency, nature and transfer to producers*. (Unpublished Working paper). EIM Business and Policy Research, Zoetermeer, The Netherlands.

de Ruyter, K., & Scholl, N. (2004). *Qualitative Market Research. Theory and Practice*. Utrecht: Lemma.

Deci, E. L., Koestner, R., & Ryan, R. M. (1999). A meta-analytic review of experiments examining the effects of extrinsic rewards on intrinsic motivation. *Psychological Bulletin, 125*(6), 627–668. doi:10.1037/0033-2909.125.6.627

Delre, S. A. (2007). *Effects of social networks on innovation diffusion and marketing dynamics* (Unpublished Ph D Dissertation). University of Groningen, Groningen.

Denyer, D., Tranfield, D., & van Aken, J. E. (2008). Developing design propositions through research synthesis. *Organization Studies, 29*, 393–413. doi:10.1177/0170840607088020

Dolan, R. J., & Matthews, J. M. (1993). Maximizing the utility of customer product testing: Beta test design and management. *Journal of Product Innovation Management, 10*(4), 318–330. doi:10.1016/0737-6782(93)90074-Z

Donath, B. (1992). The Customer as Consultant. *Sales and Marketing Management, 144*(11), 84–87.

Drew, S. A. W. (1995). Strategic benchmarking: innovation practices in financial institutions. *International Journal of Bank Marketing, 13*(1), 4–16. doi:10.1108/02652329510075418

Edvardsson, B., Gustafsson, A., Kristensson, P., Magnusson, P., & Matthing, J. (Eds.). (2006). *Involving Customers in New Service Development*. London: Imperial College Press.

Eilander, E. (2008). Word of Mouth rond Omo Klein & Krachtig. [in Dutch]. *Tijdschrift voor Marketing, 42*(10), 52–55.

Ennew, C. T., & Binks, M. R. (1996). Good and bad customers: the benefits of participating in the banking relationship. *International Journal of Bank Marketing, 14*(2), 5–13. doi:10.1108/02652329610106872

Enos, J. L. (1962). *Petroleum Progress and Profits: A History of Process Innovation*. Cambridge, MA: MIT Press.

Flores, F. (1993). Innovation by listening carefully to customers. *Long Range Planning*, *26*(3), 95–102. doi:10.1016/0024-6301(93)90011-4

Foxall, G., & Tierney, J. (1984). From CAP 1 to CAP 2: User-Initiated Innovation from the User's Point of View. *Management Decision*, *22*(5), 3–15. doi:10.1108/eb001358

Frambach, R. T., Barkema, H. G., Nooteboom, B., & Wedel, M. (1998). Adoption of a Service Innovation in the Business Market An Empirical Test of Supply-Side Variables. *Journal of Business Research*, *41*, 161–174. doi:10.1016/S0148-2963(97)00005-2

Franke, N., & Piller, F. (2004). Value Creation by Toolkits for User Innovation and Design: The Case of the Watch Market. *Journal of Product Innovation Management*, *21*(6), 401–415. doi:10.1111/j.0737-6782.2004.00094.x

Franke, N., & Piller, F. T. (2003). Key research issues in user interaction with user toolkits in a mass customisation system. *International Journal of Technology Management*, *26*(5-6), 578–599. doi:10.1504/IJTM.2003.003424

Franke, N., & Shah, S. (2003). How communities support innovative activities: an exploration of assistance and sharing among end-users. *Research Policy*, *32*(1), 157–178. doi:10.1016/S0048-7333(02)00006-9

Franke, N., von Hippel, E., & Schreier, M. (2006). Finding Commercially Attractive User Innovations: A Test of Lead-User Theory. *Journal of Product Innovation Management*, *23*(4), 301–315. doi:10.1111/j.1540-5885.2006.00203.x

Freeman, C. (1968). Chemical Process Plant: Innovation and the World Market. *National Institute Economic Review*, *45*(1), 29–51. doi:10.1177/002795016804500104

Friesen, G. B. (2001). Co-creation: When 1 and 1 make 11. *Consulting to Management*, *12*(1), 28–31.

Füller, J., & Bartl, M. (2007, October 7-10). *What Consumers Expect from Virtual Co-Creation.* Paper presented at the Mass Customization & Personalization Conference 2007, Cambridge, MA.

Füller, J., Bartl, M., Ernst, H., & Mühlbacher, H. (2006). Community based innovation: How to integrate members of virtual communities into new product development. *Electronic Commerce Research*, *6*(1), 57–73. doi:10.1007/s10660-006-5988-7

Gardiner, P., & Rothwell, R. (1985). Tough Customers: Good Designs. *Design Studies*, *6*(1), 7–17. doi:10.1016/0142-694X(85)90036-5

Gault, F., & von Hippel, E. (2009). *The prevalence of user innovation and free innovation transfers: Implications for statistical indicators and innovation policy* (Unpublished Working Paper). Cambridge, MA: MIT Sloan School of Management.

Gobeli, D. H., & Brown, D. J. (1993). Improving the process of product innovation. *Research Technology Management*, *32*(2), 38–45.

Greenbaum, J., & Kyng, M. (Eds.). (1991). *Design at Work: Cooperative Design of Computer Systems*. Hillsdale, NJ: Lawrence Erlbaum Ass.

Griffin, A. (1992). Evaluating QFD's use in US firms as a process for developing products. *Journal of Product Innovation Management*, *9*(3), 171–187. doi:10.1016/0737-6782(92)90028-B

Griffin, A., & Hauser, J. R. (1993). The Voice of the Customer. *Marketing Science*, *12*(1), 1–27. doi:10.1287/mksc.12.1.1

Gruner, K. E., & Homburg, C. (2000). Does Customer Interaction Enhance New Product Success? *Journal of Business Research*, *49*(1), 1–14. doi:10.1016/S0148-2963(99)00013-2

Hamel, G., & Prahalad, C. K. (1994). *Competing for the Future*. Boston: Harvard Business School Press.

Hancock, T. (2005). Towards a free matter economy (Part 3). Designing the Narya Bazaar [Electronic Version]. *Free Software Magazine*. Retrieved September 10, 2006 from http://www.freesoftwaremagazine.com/free_issues/issue_09/free_matter_economy_3.

Hansen, M. T., & Birkinshaw, J. (2007). The Innovation Value Chain. *Harvard Business Review*, *85*(6), 121–130.

Harhoff, D., Henkel, J., & Hippel, E. v. (2003). Profiting from voluntary information spillovers: how users benefit by freely revealing their innovations. *Research Policy*, *32*(10), 1753–1769. doi:10.1016/S0048-7333(03)00061-1

Helkema, R., Vugt, T. v., & Bosveld, K. (2008). Grote Marketing Enquete 2008: Marketeer bang voor mondige consument. *Tijdschrift voor Marketing, 2008*(Juli-Augustus), 16-24 (in Dutch).

Henkel, J. (2006). Selective revealing in open innovation processes: The case of embedded Linux. *Research Policy*, *35*(7), 953–969. doi:10.1016/j.respol.2006.04.010

Henkel, J., & von Hippel, E. (2005). Welfare Implications of User Innovation. *The Journal of Technology Transfer*, *30*(1/2), 73–87.

Herstatt, C., & Kalogerakis, K. (2005). How to use analogies for breakthrough innovations. *International Journal of Innovation and Technology Management*, *2*(3), 331–347. doi:10.1142/S0219877005000538

Herstatt, C., & von Hippel, E. (1992). From experience: Developing new product concepts via the lead user method: A case study in a "low-tech" field. *Journal of Product Innovation Management*, *9*(3), 213–221. doi:10.1016/0737-6782(92)90031-7

Hertel, G., Niedner, S., & Herrmann, S. (2003). Motivation of software developers in Open Source projects: an Internet-based survey of contributors to the Linux kernel. *Research Policy*, *32*(7), 1159–1177. doi:10.1016/S0048-7333(03)00047-7

Hienerth, C. (2006). The commercialization of user innovations: the development of the rodeo kayak industry. *R & D Management*, *36*(3), 273–294. doi:10.1111/j.1467-9310.2006.00430.x

Holt, K. (1987). User-oriented product innovation in theory and practice. *Technovation*, *6*(1), 37–45. doi:10.1016/0166-4972(87)90037-X

Holt, K. (1988). The role of the user in product innovation. *Technovation*, *8*, 249–258. doi:10.1016/0166-4972(88)90028-4

Holt, K., Geschka, H., & Peterlongo, G. (1984). *Need Assessment—A Key to User-Oriented Product Innovation*. Chichester, UK: Wiley and Sons.

Howe, J. (2006, June 2006). The Rise of Crowdsourcing. *Wired, 14.06*.

Howe, J. (2008). *Crowdsourcing: Why the power of the crowd is driving the future of business*. New York: Crown Business.

Intrachooto, S. (2004). Lead users concept in building design: its applicability to member selection in technologically innovative projects. *The TQM Magazine*, *16*(5), 359–368. doi:10.1108/09544780410551304

Janzik, L., & Herstatt, C. (2008, September 21-24). *Innovation Communities: Motivation and Incentives for Community Members to Contribute.* Paper presented at the International Conference on Management of Innovation and Technology (ICMIT) 2008, Bangkok, Thailand.

Jeppesen, L. B. (2005). User Toolkits for Innovation: Consumers Support Each Other. *Journal of Product Innovation Management, 22*(4), 347–362. doi:10.1111/j.0737-6782.2005.00131.x

Jeppesen, L. B., & Frederiksen, L. (2004, January 5). *Innovative Users in a Firm-Established User Community: The characteristics of users co-developing "computercontrolled music instruments"*. Paper presented at the DRUID, PhD Winter Conference, Aalborg, DK.

Jeppesen, L. B., & Molin, M. (2003). Consumers as Co-developers: Learning and Innovation Outside the Firm. *Technology Analysis and Strategic Management, 15*(3), 363–381. doi:10.1080/0953 7320310001601531

Karagozoglu, N., & Brown, W. B. (1993). Time-based management of the new product development process. *Journal of Product Innovation Management, 10*(3), 204–215. doi:10.1016/0737-6782(93)90026-M

Kaulio, M. A. (1998). Customer, consumer and user involvement in product development: A framework and a review of selected methods. *Total Quality Management, 9*(1), 141–149. doi:10.1080/0954412989333

Keen, A. (2007). *The Cult of the Amateur. How Today's Internet Is Killing Our Culture*. New York: Random House.

Kristensson, P., Gustafsson, A., & Archer, T. (2004). Harnessing the Creative Potential among Users. *Journal of Product Innovation Management, 21*(1), 4–14. doi:10.1111/j.0737-6782.2004.00050.x

Lagrosen, S. (2005). Customer involvement in new product development: A relationship marketing perspective. *European Journal of Innovation Management, 8*(4), 424–436. doi:10.1108/14601060510627803

Lakhani, K. R., & von Hippel, E. (2003). How open source software works: "free" user-to-user assistance. *Research Policy, 32*(6), 923–943. doi:10.1016/S0048-7333(02)00095-1

Leadbeater, C., & Miller, P. (2004). *The Pro-AmRevolution. How enthusiasts are changing our economy and society*. London: Demos.

Leonard, D., & Rayport, J. F. (1997). Spark Innovation Through Empathic Design. *Harvard Business Review, 75*(6), 104–113.

Leonard-Barton, D. (1992). Core Capabilities and Core Rigidities: A Paradox in Managing New Product Development. *Strategic Management Journal, 13*, 111–125. doi:10.1002/smj.4250131009

Li, C., & Bernoff, J. (2008). *Groundswell: Winning in a World Transformed by Social Technologies*. Boston: Harvard Business Press.

Lilien, G. L., Morrison, P. D., Searls, K., Sonnack, M., & Hippel, E. v. (2002). Performance assessment of the lead user idea-generation process for new product development. *Management Science, 48*(8), 1042–1059. doi:10.1287/mnsc.48.8.1042.171

Linstone, H. A., & Turoff, M. (Eds.). (2002). *The Delphi Method: Techniques and Applications*. Newark, NJ: New Jersey Institute of Technology.

Lundkvist, A., & Yakhlef, A. (2004). Customer involvement in new service development: a conversational approach. *Managing Service Quality, 14*(2/3), 249–257. doi:10.1108/09604520410528662

Lusch, R. F., & Vargo, S. L. (2006). Service-dominant logic: reactions, reflections and refinements. *Marketing Theory*, *6*(3), 281–288. doi:10.1177/1470593106066781

Lüthje, C. (2004). Characteristics of innovating users in a consumer goods field - An empirical study of sport-related product consumers. *Technovation*, *24*(9), 683–695. doi:10.1016/S0166-4972(02)00150-5

Lüthje, C., & Herstatt, C. (2004). The Lead User method: an outline of empirical findings and issues for future research. *R & D Management*, *34*(5), 553–568. doi:10.1111/j.1467-9310.2004.00362.x

Magnusson, P. R. (2009). Exploring the Contributions of Involving Ordinary Users in Ideation of Technology-Based Services. *Journal of Product Innovation Management*, *26*(5), 578–593. doi:10.1111/j.1540-5885.2009.00684.x

Magnusson, P. R., Kristensson, P., & Hipp, C. (2007, October, 8-10). *Exploring Ideation Patterns among Ordinary Users: The case of mobile telecommunication services.* Paper presented at the Mass Customization and Personalization Conference 2007, MIT, Cambridge, MA.

Magnusson, P. R., Matthing, J., & Kristensson, P. (2003). Managing User Involvement in Service Innovation. Experiments With Innovating End Users. *Journal of Service Research*, *6*(2), 111–124. doi:10.1177/1094670503257028

Martin, C. R., & Horne, D. A. (1995). Level of success inputs for service innovations in the same firm. *International Journal of Service Industry Management*, *6*(4), 40–56. doi:10.1108/09564239510096894

Mazur, G. H. (1994, June 12-14). *QFD for Small Business. A shortcut through the 'Maze of Matrices'.* Paper presented at the 6th Symposium on Quality Function Deployment, Novi, Michigan.

McCracken, G. (1988). *The Long Interview.* Newbury Park, CA: Sage.

Moore, G. A. (2002). *Crossing the Chasm. Marketing and Selling Disruptive Products to Mainstream Customers* (Revised ed.). New York: Harper Collins.

Mullern, M. J., Wildeman, D. M., & White, E. A. (1993). Taxonomy of PD practices: a brief practitioner's guide. *Communications of the ACM*, *36*(4), 26–27.

Mullins, J. W. (2007). Discovering "Unk-Unks". *MIT Sloan Management Review*, *48*(4), 17–21.

Nambisan, S. (2002). Designing virtual customer environments for new product development: Toward a theory. *Academy of Management. Academy of Management Review*, *27*(3), 392–413. doi:10.2307/4134386

Nambisan, S., & Sawhney, M. (2007). Buyer's Guide to the Innovation Bazaar. *Harvard Business Review*, *85*(6), 109–118.

Neale, M. R., & Corkindale, D. R. (1998). Co-developing products: Involving customers earlier and more deeply. *Long Range Planning*, *31*(3), 418–425. doi:10.1016/S0024-6301(98)80008-3

Norman, D. (1999). *Invisible Computer: Why Good Products Can Fail, the Personal Computer Is So Complex and Information Appliances Are the Solution.* Cambridge, MA: MIT Press.

Normann, R. (2001). *Reframing Business: When the Map Changes the Landscape.* Chichester, UK: Wiley & Sons.

Normann, R., & Ramirez, R. (1993). From value chain to value constellation: Designing interactive strategy. *Harvard Business Review*, *71*(4), 65–77.

Olson, E. L., & Bakke, G. (2001). Implementing the lead user method in a high technology firm: a longitudinal study of intentions versus actions. *Journal of Product Innovation Management*, *18*(6), 388–395. doi:10.1016/S0737-6782(01)00111-4

Parkinson, S. (1982). The Role of the User in Successful New Product Development. *R & D Management*, *12*(3), 123–131. doi:10.1111/j.1467-9310.1982.tb00497.x

Pavitt, K. (1984). Sectoral patterns of technical change: Towards a taxonomy and a theory. *Research Policy*, *13*(6), 343–373. doi:10.1016/0048-7333(84)90018-0

Piller, F. T., Schubert, P., Koch, M., & Möslein, K. (2005). Overcoming mass confusion: Collaborative customer co-design in online communities. *Journal of Computer-Mediated Communication*, *10*(4), 1–8.

Piller, F. T., & Walcher, D. (2006). Toolkits for idea competitions: a novel method to integrate users in new product development. *R & D Management*, *36*(3), 307–318. doi:10.1111/j.1467-9310.2006.00432.x

Pisano, G. P., & Verganti, R. (2008). Which Kind of Collaboration Is Right for You? *Harvard Business Review*, *86*(12), 78–86.

Pitt, L. F., Watson, R. T., Berthon, P., Wynn, D., & Zinkhan, G. (2006). The Penguin's Window: Corporate Brands From an Open-Source Perspective. *Journal of the Academy of Marketing Science*, *34*(2), 115–127. doi:10.1177/0092070305284972

Pitta, D. A., Franzak, F., & Katsanis, L. P. (1996). Redefining new product development teams: learning to actualize consumer contributions. *Journal of Product and Brand Management*, *5*(6), 48–60. doi:10.1108/10610429610152831

Prahalad, C. K., & Ramaswamy, V. (2003). *The Future of Competition: Co-Creating Unique Value with Customers*. Boston, MA: Harvard Business School Press.

Ramírez, R. (1999). Value co-production: intellectual origins and implications for practice and research. *Strategic Management Journal*, *20*(1), 49–65. doi:10.1002/(SICI)1097-0266(199901)20:1<49::AID-SMJ20>3.0.CO;2-2

Reeve, J. (2005). *Understanding motivation and emotion*. New York: John Wiley & Sons.

Rey, J. D. (2008). Tapping the Community Pool Online Communities Let You See What's On Your Customers' Mind. *Inc*, *30*(7), 37–39.

Rogers, E. M. (1995). *Diffusion of Innovation* (4th ed.). New York: Free Press.

Rothwell, R. (1986). Innovation and Re-Innovation: A Role for the User. *Journal of Marketing Management*, *2*(2), 109–123. doi:10.1080/0267257X.1986.9964004

Rothwell, R., & Gardiner, P. (1983). The role of design in product and process change. *Design Studies*, *4*(3), 161–169. doi:10.1016/0142-694X(83)90025-X

Salvador, F., Forza, C., & Rungtusanatham, M. (2002). How to mass customize: Product architectures, sourcing configurations. *Business Horizons*, *45*(4), 61–69. doi:10.1016/S0007-6813(02)00228-8

Sanders, E. B. N., & Stappers, P. J. (2008). Co-creation and the new landscapes of design. *CoDesign*, *4*(1), 5–18. doi:10.1080/15710880701875068

Sanoff, H. (2005). Community participation in riverfront development. *CoDesign*, *1*(1), 61–78. doi:10.1080/15710880512331326022

Sawhney, M., & Prandelli, E. (2000). Communities of creation: Managing distributed innovation in turbulent markets. *California Management Review*, *42*(4), 24–54.

Schreier, M., & Prügl, R. (2008). Extending Lead-User Theory: Antecedents and Consequences of Consumers' Lead Userness. *Journal of Product Innovation Management*, *25*(4), 331–346. doi:10.1111/j.1540-5885.2008.00305.x

Senge, P. M., Carstedt, G., & Porter, P. L. (2001). Innovating Our Way to the Next Industrial Revolution. *MIT Sloan Management Review*, *42*(2), 24–38.

Seybold, P. B. (2006). *Outside Innovation: How Your Customers Will Co-Design Your Company's Future*. New York: Collins.

Sifonis, C., Chernoff, A., & Kolpasky, K. (2006). The Innovation Pipeline: Analogy as a tool for communicating about innovation. *International Journal of Innovation and Technology Management*, *3*, 1–19. doi:10.1142/S0219877006000697

Slater, S. F., & Narver, J. C. (1995). Market orientation and the learning organization. *Journal of Marketing*, *59*(3), 63–74. doi:10.2307/1252120

Teichert, T., von Wartburg, I., & Braterman, R. (2006). Tacit meaning in disguise: Hidden metaphors in new product development and market making. *Business Horizons*, *49*(6), 451–461. doi:10.1016/j.bushor.2006.02.006

Thomke, S., & von Hippel, E. (2002). Customers as innovators: a new way to create value. *Harvard Business Review*, *80*(4), 74–81.

Tidd, J., & Bodley, K. (2002). The influence of project novelty on the new product development process. *R & D Management*, *32*(2), 127–138. doi:10.1111/1467-9310.00245

Tietz, R., Morrison, P. D., Luthje, C., & Herstatt, C. (2005). The process of user-innovation: a case study in a consumer goods setting. *International Journal of Product Development*, *2*(4), 321–338. doi:10.1504/IJPD.2005.008005

Toffler, A. (1980). *The third wave*. New York: Bantam Books.

Tomes, A., Armstrong, P., & Clark, M. (1996). User groups in action: The management of user inputs in the NPD process. *Technovation*, *16*(10), 541–551. doi:10.1016/S0166-4972(97)82894-5

Un, C. A., Cuervo-Cazurra, A., & Asakawa, K. (2008). *R&D Collaborations and Product Innovation*. (Unpublished Paper accepted for publication in Journal of Product Innovation Management). University of South Carolina, Moore School of Business.

Urban, G. L., Hauser, J. R., Qualls, W. J., & Weinberg, B. D. (1997). Information acceleration: Validation and lessons from the field. *JMR, Journal of Marketing Research*, *34*(1), 143–153. doi:10.2307/3152071

Urban, G. L., & von Hippel, E. (1988). Lead User Analyses for the Development of New Industrial Products. *Management Science*, *34*(5), 569–582. doi:10.1287/mnsc.34.5.569

van Daelen, H. (2005). *Tools for Innovation through Customer Co-creation* (Unpublished Master Thesis). Tiburg University, Tilburg.

van Kleef, E., van Trijp, H. C. M., & Luning, P. (2005). Consumer research in the early stages of new product development: a critical review of methods and techniques. *Food Quality and Preference*, *16*(3), 181–201. doi:10.1016/j.foodqual.2004.05.012

van Zanten, W. P. C. (1998). *Technische hulpmiddelen voor groepsbesluitvorming*. Baarn/Heerlen: Nelissen/Open Universiteit NL (in Dutch).

VanderWerf, P. A. (1992). Explaining downstream innovation by commodity suppliers with expected innovation benefit. *Research Policy*, *21*(4), 315–333. doi:10.1016/0048-7333(92)90031-X

Vargo, S. L., & Lusch, R. F. (2004). Evolving to a New Dominant Logic for Marketing. *Journal of Marketing*, *68*(1), 1–17. doi:10.1509/jmkg.68.1.1.24036

von Hippel, E. (1986). Lead Users: A Source of Novel Product Concepts. *Management Science*, *32*(7), 791–806. doi:10.1287/mnsc.32.7.791

von Hippel, E. (1988). *The Sources of Innovation*. New York: Oxford University Press.

von Hippel, E. (1994). 'Sticky information' and the locus of problem solving: Implications of Innovation. *Management Science*, *40*(4), 429–439. doi:10.1287/mnsc.40.4.429

von Hippel, E. (2005). *Democratizing Innovation*. Cambridge, MA: MIT Press.

von Hippel, E., & Katz, R. (2002). Shifting innovation to users via toolkits. *Management Science*, *48*(7), 821–834. doi:10.1287/mnsc.48.7.821.2817

von Hippel, E., Thomke, S., & Sonnack, M. (1999). Creating breakthroughs at 3M. *Harvard Business Review*, *77*(5), 47–56.

von Krogh, G. (2006). Breakthrough Ideas for 2006: Customers Demand Their Slice of IP. *Harvard Business Review*, *84*(2), 45–46.

Voss, C. A. (1985). Determinants of success in the development of applications software. *Journal of Product Innovation Management*, *2*(2), 122–129. doi:10.1016/0737-6782(85)90008-6

Weber, M. (2008, June 19-20). *Customer Co-creation in the Dutch Financial Services*. Paper presented at the IMCM-PETO 2008, Copenhagen, Denmark.

Weber, M. (2009, October, 5-6). *Customer Involved Open Innovation. Involving customers and end users in product and service innovation: the case of a coffee manufacturer.* Paper presented at the Mass Customization and Personalization Conference 2009 (MCPC), Helsinki, Finland.

Wendel de Joode, R. (2005). *Understanding Open Source Communities. An organizational perspective*. Unpublished Dissertation, Technische Universiteit Delft, Delft.

West, J., & Gallagher, S. (2006). Challenges of open innovation: the paradox of firm investment in open-source software. *R & D Management*, *36*(3), 319–331. doi:10.1111/j.1467-9310.2006.00436.x

Wind, Y. J., & Mahajan, V. J. (2002). *Convergence Marketing. Strategies for Reaching the New Hybrid Consumer*. New York: Prentice Hall.

Wunsch-Vincent, S., & Vickery, G. (2007). *Participative Web: User-Created Content* (No. JT03225396). Paris: Organisation for Economic Co-operation and Development.

Yoon, E., & Lilien, G. L. (1988). Characteristics of the industrial distributor's innovation activities: An exploratory study. *Journal of Product Innovation Management*, *5*(3), 227–240. doi:10.1016/0737-6782(88)90025-2

Zaltman, G. (2003). *How Customers Think. Essential Insights into the Mind of the Market*. Boston, MA: Harvard Business School Press.

KEY TERMS AND DEFINITIONS

Co-Creation: The phenomenon where customers and firms join together in the creation of goods and services, for instance where customers on invitation by firms to supply ideas for new products and services, sometimes to even co-design, co-develop or co-produce the goods and services.

Crowdsourcing: The act of taking a job traditionally performed by a designated agent (usually an employee) and outsourcing it to an undefined, generally large group of people in the form of an open call' (Howe, 2006). So crowdsourcing stands literally for outsourcing to a crowd.

Customer Involvement in Innovation: Is defined as the process where product manufacturers and/or service providers engage with their end users or customers in (parts or phases of) innovation projects with the aim of increasing effectiveness and efficiency of the innovation process. Effectiveness refers to (1) the result of meeting users' and customers' needs and demands in a better way; and (2) increasing customer loyalty. Efficiency refers to (1) the reduction of research and development costs; and (2) the reduction of development time.

Mass Customization: Refers to a customer co-design process of products and services that meet the needs of each individual customer with regard to certain product features and where all operations are performed within a fixed solution space, characterized by stable but still flexible and responsive processes.

Online (User) Community: An internet based collaboration between users of a certain product or product category, where they exchange experience, support, and even products among each other. They exist because the users are motivated by needs, empowerment, enjoyment, reputation, and official appreciation, and work well enough to create and sustain complex innovations. By having a loosely-affiliated structure individuals with common interests or values are able to create valuable knowledge, feedback and innovations.

Open Source: Open source is a set of principles and practices that promote access to the design and production of goods and knowledge. The term is most commonly applied to the source code of software that is available to the general public with relaxed or non-existent intellectual property restrictions. This allows users to create software content through incremental individual effort or through collaboration.

Prosumerism: A composed term from the words producer and consumer, entailing the tendency towards a market where the distinction between producers and consumers decreases, where consumers are producing their own goods and services. A tendency that takes place because of individual, societal, technological and economic changes, particularly caused by the rapid advances in the information technology.

User Innovation: The phenomenon of new products and innovations being developed by customers and end users, for their own benefit, rather than by manufacturers (von Hippel, 1988).

APPENDIX A

Summary of Phases, Activities and Tools for Involvement

Table 1 gives an overview of the ways to involve customers in the innovation process. It identifies the phase or stage for involvement (1st column), the activities in which customers can participate (2nd column), the method used for interacting wit the customer(s) (3rd column) and the tool(s) suited to support that activity (4th column). Take note of the fact that methods, tools and techniques which the company uses for its own activities are not incorporated in this overview. All activities, can be conducted both online and offline, provided that this is possible (e.g. testing physical prototypes in use should be done offline).

Table 1. Ways to involve customers in the innovation process

Innovation stage/ phase	Activity with customer participation	Interaction method	Tool or technique
Ideation	Needs assessment	Customer interviews (face-to-face or group)	Long interview, Zaltman Metaphor Elicitation Technique (ZMET),
	Idea generation	Group brainstorming	Brainstorming (all variants), group creativity techniques, Lead User Approach
	Idea screening	Group discussion	Nominal Grouping, lead User Approach
Concept development	Defining requirements	Customer interviews, group discussions, workshops	Analogous or metaphorical reasoning, Lead User Approach
	Co-Design	Group discussion, workshops	Empathic or participatory design, Consumer Idealized Design
Concept testing	Concept test Prototype testing	Customer interviews, focus groups, customer observation, Self reporting (e.g. diaries)	Long Interview, ZMET, focus groups; Beta-testing
Commercial-ization	Market plan development	Brainstorm on marketing activities	Group creativity techniques
	Advertising tests	Group interviews Observation	Focus groups, Customer observation, Netnography
	Marketing of innovation	Product demonstrations, reviews, testimonials, community or social networking	Word of Mouth, Samples, Product/Service Review
Use/Re-innovation	Improvement suggestions, Complaints	Customer dialogues, User-to-user support in communities	Online forums Complaints and Suggestion Systems, Netnography

Chapter 15
Stimulating Creativity and Innovation in and Around Organisations:
Co-Creation Experiments from Ongoing Research in a Bank

Kirsten Bonde Sørensen
Kolding School of Design, Denmark

ABSTRACT

Companies are moving from an industrial strategic paradigm into a new paradigm where value co-creation has become a key issue. There are different levels of value co-creation but until now co-creation has often been operating on the outer edges of a company's value chain. We are, however, moving towards a veritable revolution where value creation will be the core activity of the organisation and customers will act as co-producers, co-creating values (Normann, 2001). This chapter outlines different approaches and developments related to co-creation, but emphasizes the capacity of a design approach. The chapter also defines the challenges of co-creation and reports experiences from ongoing research in a bank. In this example the designer is the facilitator of this organisational process – arranging conversations and debates about the values of the community (Buchanan, 2006). Co-creation is also seen as a kind of rhetorical design communication about values, and rhetorical analysis of values (Perelman, 1969), and motives (Burke, 1969) are applied. The chapter outlines experiences, possibilities and challenges from an ongoing co-creation process in a bank: design of artifacts for co-creation, the creative session with people (inside and outside the company), and analysis and transformation of the responses into a creative and useful tool for the bank. Finally future challenges and possibilities are discussed.

INTRODUCTION

Although co-creation is not a brand-new phenomenon it still represents a radically new definition of what constitutes value to consumers. From a business and innovation angle, co-creation is interesting not because everything has to be or will be co-created in the future, but because co-creation is tapping into the collective experiences, skills and ingenuity of consumers around the world. As such

DOI: 10.4018/978-1-61520-617-9.ch015

co-creation seems to be a complete departure from the inward looking, company-versus-consumer innovation model so common in companies all over the world.

From a customer perspective co-creation started out several years ago with enthusiastic amateurs who wanted a bike that could be ridden off-road. The amateurs created the mountain bike themselves, and later on it was put into production. Enthusiastic users also developed equipment for kite surfing. As the existing kites did not meet the needs of the super users, they themselves developed new and better kites. On the Internet these super qualified users exchanged drawings and models and software for rapid prototyping and developed new super kites (Von Hippel, 2005). If kite manufacturers would think democratically they would ask these super users to help innovate their products, and the result would be better kites for the company, the super users and other users. This is an example of how co-creation started in real life (Von Hippel, 2005), and in this example the Internet had an enormous influence. The Internet enables knowledge sharing and provides access to sophisticated computer programs, which is pushing forward the democratisation of innovation, co-creation activities, and open source innovation; an obvious example is Wikipedia (www.wikipedia.com).

From a business perspective co-creation may look like an easy and accessible way towards innovation, unlocking the creativity of the customers and letting their creativity be part of the development of the future business. The possibilities inherent in co-creation are immense, but co-creation also brings enormous challenges into the organisations.

The intention of this chapter is to give a short overview of different approaches to and understandings of co-creation and the development of the concept. Focus will be on a rhetorical-design approach, and on experiences from an ongoing co-creation process in a bank.

BACKGROUND

In recent years companies have moved from an industrial paradigm with focus on the production of goods for the market, with customers as passive receivers, to an intermediate stage where focus has shifted to service and customer relations. Customers are now seen as a source of information about ideas and needs. However, a further development can now be observed, a veritable revolution, where customers are seen as co-producers, involved in value creation, and the core activity of a company will be the organisation of value creation (Normann, 2001).

Different definitions and forms of co-creation have evolved. An intermediate stage is the 'service-dominant logic of marketing', which focuses on service and customer relations. In 2004 the marketing researchers Vargo & Lusch presented this new marketing perspective also called the S-D logic (2004), which redefines the relationship between the company and the customer. Goods and services are no longer viewed in the same conventional sense (G-D logic), rather the customer is promoted to being a co-producer of value and is constantly communicating with the firm to improve the quality of the offering (Vargo & Lusch, 2004). Businesses are regarded as learning organisations that do not create value by themselves, but rather are only able to propose 'offerings' to the customer. This is referred to as a 'sense-and-respond' strategy as opposed to a 'make-and-sell' strategy (Haeckel, 1999). Here the firm's objective has moved far away from 'making the sale' to maintaining a healthy and ongoing relationship with the customer.

According to the S-D logic co-creation is focused on the involvement of customers, asking them to deliver their experience. Co-creation in *the delivery of customer experience* is used most frequently in marketing these days and is referred to as viral marketing, community marketing or buzz marketing. Another form of co-creation is the *development of products or services* either in collaboration with the company or in open innovation – crowd sourcing, wikinomics or mass collaboration. In the latter case the firm is placing more emphasis on finding new services on the open market and reducing its internal resources. An example of a co-created product or service design is the open software Linux (http://www.linux.org). This type of customer involvement starts with the assumption that there are knowledgeable users, also termed lead-users, who have the competency to innovate new services (Von Hippel et al., 1999).

It is important to underline that the types of co-creation mentioned above: – the co-created delivery of customer experience and co-creation of products and services – are types of co-creation practiced at the outer edges of the company's value chain. Concurrently with customers evolving from passive customers to active citizens they are becoming increasingly engaged in ethical issues such as social responsibility and the protection of the environment. Consequently higher demands and expectations are placed on the companies, not only in relation to specific services or products but also in relation to the overall values governing the companies. Undoubtedly, this movement will continue, and co-creation will evolve and embrace the values of the organisation, not only on the outer edges of the company's value chain, but also at its very heart.

Stated values are already common in many companies. Influenced by Collins & Porras´s book "Built to Last" (1994) a tendency emerged that all managers had to formulate a specific set of guidelines, as this book claimed the best companies were the ones that adhered to a set of principles, called core values. Unfortunately this led to numerous companies formulating grandiose core values, which, however, were difficult to perceive, let alone implement in the day-to-day life of the organisation (Lencioni, 2002). But as customers are becoming more active, engaged and empowered, they find it increasingly important that the stated organisational values and the actual day-to-day practices and procedures of the organisation concur – that the lofty values are also visible and part of the reality – and furthermore that these values are accepted by the outside world. The increasing blurring of the traditional demarcation between the inside and the outside of the company is central in the understanding of co-creation as an organisational approach. Future organisations have to conduct their day-to-day operations in accordance with their core values but also in accordance with the outside world and the empowered individuals. Corporate Social Responsibility (CSR) is only the beginning. There are pros and cons about CSR, but essentially CSR is the deliberate inclusion of public interests into corporate decision-making and the honouring of the triple bottom line: People, Planet and Profit (Elkington, 1994). CSR is a self-regulating mechanism whereby businesses can monitor and ensure their adherence to law, ethical standards and international norms. In the future empowered people inside and outside the organisation will be involved in the co-creation of values – in a dynamic ongoing process within and around the company (Sanders, 2008).

In a discussion of the actual trends both in relation to product, service and organisational development, co-creation definitely *is* a way to stimulate innovation, but - as mentioned - there are different levels, and approaches to co-creation, some involving the outer edges of the company's value chain, others involving the whole organisation. How do we implement co-creation as an organisational approach?

Critics of Vargo & Lusch's S-D logic claim that although it attempts to shift focus from products to services, the vocabulary being used does not truly represent the shift, and they suggest that it still holds on to a product-centred concept (Rust, 2006). Schembri claims that a focus on the product as either goods or services eliminates the customers' experiences and thus obscures the needs of the customers (Schembri, 2006). Normann suggests a more radical redefinition of organisations as 'rule makers with a mission' or 'prime movers' (Normann, 2001). These new types of organisations are not limited to a specific product or service but see themselves as value-creating systems. In these systems co-creation is claimed to be a practical tool to involve the customers both in the value chain as a whole and also in the business system of the company (Normann, 2001). Kim & Mauborgne (2005) explicitly introduced the idea of innovating values in the book "Blue Ocean Strategy". The point in this popular book is to find or create new (blue) value communities instead of competing in the existing (red) value communities crowded with all the competitors. Later Prahalad & Krishnan emphasize the 'emphatic dialogue' (Prahalad, 2008) to be crucial and a prerequisite for co-creating values with customers. Unfortunately they never explain *how* to manage this 'empathic dialogue'.

In a summary of these new definitions of organisations including consumers, goods and services the most radical definition is Normann's characterisation of organisations as value creating systems. Normann's perspective borders on a more holistic perspective that is characteristic of design thinking. A traditional view of design is limited to form, function and product design, but parallel to the developments within the field of business, consumers and co-creation, design and the role of design has also evolved intensely. Several researchers e.g. Simon (1996), Buchanan (2001, 2006), Boland (2008), Collopy (2008), and Verganti (2009) have pointed to the relationship between managing and designing. Simon was the first to perceive design as an activity for everyone. He considers the manager to be a form-giver who shapes social organisations and economic processes to create value – just like a designer (Simon, 1996). He further claims that *"... organisational leaders of today are more responders to situations presented to them, as opposed to active makers of a future worthy of us as human beings"* (Simon, 1996). An essential point in design thinking is not to look at 'how things *are*' but 'how things *could be*'. That means design thinking is prescriptive rather than descriptive and one of the essential reasons why researcher see design thinking as a contribution to management (Buchanan, 2007). According to Buchanan (2001) designers can be mapped within four areas: The first and second areas focus on 'symbols and images' and later on 'things' and gave rise to graphic design and later on to industrial design. The third area changed its focus to 'action' and gave rise to interaction design. Here the designer emerged as a designer of processes and services, since the consumer is no longer a passive receiver. This area is also the beginning of strategic planning, information design, human-centred design and participatory design. The fourth area is related to action, but focuses on the 'environment and systems' within which action takes place. This is the area of 'thought', since it is fundamentally concerned with the organising idea of principles that operate behind environment and systems, i.e. human systems. Designers in this field can be seen as facilitators of organizational processes. *"They organise conversations and debates about the values of a community and how those values may be implemented with productive results"* (Buchanan, 2001). This latter description is very relevant to this research, including design activities and rhetorical theory in conversations and debates about values in a community.

MAIN FOCUS OF THE CHAPTER

This research regards co-creation as a way of organising conversations and debates about the values in and around a community (Buchanan, 2001). The intention of this chapter is to suggest a design-rhetorical approach to co-creation processes. This claim is supported by an illustration of how designactivities can be devised with a wider appeal that goes beyond what can be expressed in words and leaves room for the participants to express their deeper values (Sanders,

1999). Furthermore rhetorical analysis is applied to the responses, outlining the dominant logic and structures of values (Perelman, 1969) and motives (Burke, 1969). These analyses are then transformed and designed into an inspiring and dynamic tool - a tool that reflects the different participants and their preferred values and motives, without losing the presence of the people. From here the chapter outlines how co-creation can contribute to the *development of products or services* eventually in collaboration with the company – the form of co-creation related to the outer edges of the company's value chain. The chapter also goes beyond this form of co-creation and takes the first steps towards *co-creation as an organisational approach* by analyzing the dominant logic and values in the company and making comparisons between the inside and the outside of the organisation. As such this research gives a clear picture of the typical challenges in an organisation but also gives an example of how this rhetorical design approach can create an overview of the existing and maybe conflicting values in a community.

The chapter includes the following sections: Different approaches in the field of participatory design, How to design the artifacts for co-creation sessions, How to make these artifacts work in practice, How to analyze the responses, How to develop a dynamic tool, Co-Creation as an Organisational Approach, Future Trends, and finally Conclusion.

DIFFERENT APPROACHES IN THE FIELD OF PARTICIPATORY DESIGN

From a design perspective a distinctive point in the different approaches to co-creation is whether you regard the user as 'a subject' or as 'a partner'. In the business field there seems to be a tendency to regard the user as 'a partner' - Prahalad & Ramaswarmy (2006) represent an approach considering the user as a subject and von Hippel's

Figure 1. The current landscape of human-centered design research as practiced in the design and development of products and services (Sanders & Stappers, 2008)

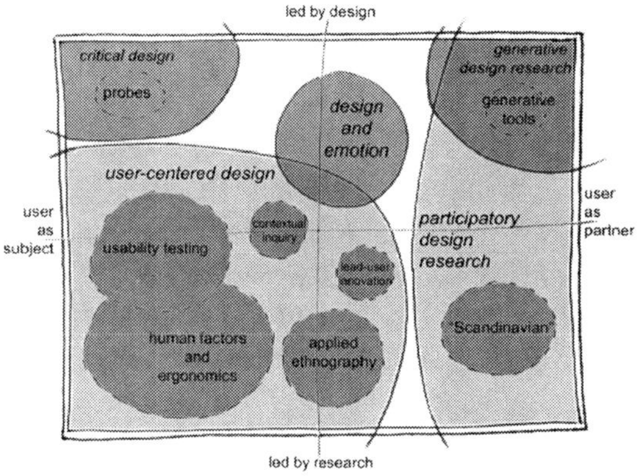

notion about 'lead user' (2005) also represents this approach. On the other hand research in the field of participatory design often represents the user as a partner. In Figure 1 below different approaches are mapped.

Central in the field of generative design research are the 'generative tools', the thinking tools non-designers can use to express their dreams (or fears) for the future. Generative design makes us see how things *could be* and empowers everyday people to generate and promote alternatives to the current situation. In this field you find researchers such as Sanders, Stappers, Visser et all. A central concept in the field of critical design is 'probes'. Cultural Probes originate from Gaver et all (1999), who designed them *"to provoke inspirational responses from elderly people in diverse communities."* Later on other types of probes were developed, e.g. Empathy Probes (Mattelmäki et Barttarbe, 2002), Technology Probes, (Hutchinson et al., 2003), Domestic Probes, (Gaver et al., 2004) and Urban Probes (Paulos and Jenkins, 2005). Originally probes are packages containing maps, postcards, a camera and a diary. These packages are sent to the private homes of the participants, who respond to them and return them to the designer. Probes are primarily used in places or circumstances too private for researchers to show up in person. Their aim is to "*make the participants think*", to "*provoke existing values*" and aim at "*generating and promoting alternatives to the current situation*" (Sanders, 2006).

In my research conducted at a small savings bank headquartered in the Danish town of Middelfart, I included experiments that mixed probes from the critical design approach with generative tools from the generative design research. People's private economy is often considered a private and intimidating issue, which I experienced in my first creative sessions. This is the reason for experimenting with a mixture of probes and generative tools. I call these 'design artifacts', in order to distinguish them both from the probes in the critical design research and from the generative tools in the generative design research. My artifacts is a box filled with different creative tasks just like the probe, but instead of sending it to the participants I experimented with inviting people in groups to a 'neutral' room or conducting the creative session in the participants' private homes. In my approach I consider the user as a dialogue partner.

HOW TO DESIGN ARTIFACTS FOR CO-CREATION SESSIONS

The material for the creative sessions was created based on Sanders (2001) notions about 'say, do, make tools'. Sanders suggests a converging perspectives approach that combines different methods, tools and artifacts that furthermore combine what people 'say', 'do' and 'make'. A combination of approaches from marketing research *('what people say')*, anthropology *('what people do')* and participatory design *('what people make')* is applied. Following the converging perspective, the participants were interviewed, asked to do things and asked to make their own expressions. Central to this way of 'asking' is the idea that the designer leads the participants through a 'guided

Figure 2. The box with all the creative tasks

discovery' appealing to their creativity and inner feelings and hereby sets a stage for them to express their thoughts and ideas.

As per Sanders there are four levels of everyday creativity, ranging from 'doing' to 'adapting', to 'making' to 'creating' (Figure 2). Where 'doing' represents an ordinary productive activity, 'adapting' means to make something one's own by changing it in some way. The motivation behind the third level of creativity, 'making', is to use one's hands and mind to make or build something that did not exist before. And finally the most advanced level of creativity is the fourth level, 'creating'. The motivation for 'creating' is to express oneself or to innovate. Truly creative efforts are fuelled by passion and guided by a high level of experience. 'Creating' differs from 'making' in that creating relies on the use of raw materials and the absence of predetermined patterns. In the food preparation domain, for example, 'making' is cooking with a recipe, whereas 'creating' is making up the recipe as you go and having to improvise along the way when you discover for example that you have run out of a key ingredient (Sanders, 2005).

When designing artifacts for the different co-creation session at the bank, I designed a box - a mixture of the generative and the critical research method. The intention was to make a private 'room' for the individual to respond in, but still having the possibility to observe and start plenary discussions. The participants were invited into the same room, and each received a box filled with material and tasks. I designed the box as a private 'room', leaving space for reflection, memories and ideas when responding to the questions, and also for the provocative statements and the creative tasks. This 'reflective room' was designed with a happy artificial long green grass carpet in the bottom, topped by the material: pictures, pieces of paper, scissor, glue, coloured pencils. The box had an appealing and accommodating look, almost like a gift with long green ribbons attached to small notes, telling people what to do.

The workshop started out the in the easy 'making' level of the creative tasks and ended up in the 'creating' level. At the 'making' level, the participants were asked to make comments to statements printed on postcards, asked to underline the preferred sentences, for example describing their relationship to their private economy. At the end, at the 'creating' level they were asked to make a collage (see 'Two Examples of How to Analyse Responses' later in the chapter) illustrating their current relationship to their economy by adding: 1) a picture illustrating themselves as customers, 2) a picture illustrating their bank/financial advisor, 3) a picture depicting the relationship between the two, 4) a picture illustrating the surroundings and 5) words underscoring the chosen pictures, all in their own words. Afterwards they were asked to make a new collage (see 'Two Examples of How to Analyse Responses' later in the chapter), in a similar way, but illustrating an ideal future situation. Finally they were asked how to get from Collage A to Collage B. These collages and answers were primarily used as material for analysis, as they were 'the final creations' presumably incorporating all the reflections. An important point is that the participants were asked to present their collages in front of a camera and these verbal presentations formed the primary material for analysis supported by the visual expressions.

HOW TO MAKE THE ARTIFACTS WORK IN PRACTICE

In rhetorical theory the function of the visual artifact is related to *the action* it creates (Foss, 2004). Functions of visual artifacts, for example, might range from memorializing individuals to creating feelings of warmth and cosiness, to encouraging viewers to explore self-imposed limitations. The different pictures, postcards, and words within the design box - called polysemic pictures (Kjeldsen, 2004) - are carefully chosen in order to be open for interpretation. The recipients see the picture and consider it´s meaning, and in this action the recipient is actively co-creating the meaning. The rhetorical function of the polysemic pictures is letting the recipients co-create the visual argument.

In the field of participatory design research the psychical artifacts are working and used as thinking tools. The very intention of using these artifacts and tools is to obtain deeper expressions and information that can inform and inspire. In accordance with the rhetorical approach the artifacts are used because of the action they create. They make room for the participants to express their experiences, feelings, values, and dreams in relation to the researched issue. When designing and using artifacts and tools for creative sessions we take advantage of the ways we use to sense, know, remember and express ourselves. The visual nature of the artifacts has a wider appeal that goes beyond what we can state in words (Sanders, 2001). Statements that fit the experiences and responses from the creative sessions:

In the completed creative sessions one of the participant, Anne Marie said:

...performing the tasks in the box was one long process, where I got deeper and deeper into the concept of 'economy': first filling out the postcards, choosing statements and so on. I think these tasks were necessary in order to make the final collages. In these collages I felt I was able

to express my reflections and final statement; my frustration about my personal finances got released.

The process performing the creative tasks made her reflect, and finally she ends up being more aware of her specific needs and wishes. She is capable of telling exactly what she wants and how she wants it. In accordance with the applied theory, she adds value to her motives, in other words, she expresses that she wants to get in control of her economy and she adds value to how this is supposed to be done. This creative session results in a lot of information about her as a customer, but it also has an effect on Anne Marie as she now is more aware of her financial situation and wants to act and change this perspective.

The participants also responded that the box looked *"welcoming", "happy", or "creative"*. Some participants uttered, the box *"was piquing my curiosity and appealing to my creativity"*. I also met remarks such as, *"I am excited about this box", "It looks really nice and a really novel and interesting way of asking about my opinion, - I like it."* What is more interesting is that the majority also expressed that they clearly liked the acknowledgement, being regarded as individual 'experts', delivering useful information to a bank. They had numerous and very different ideas about what the future bank should be like, e.g. *"I wish my financial advisors would do this with me."* The majority also said they became more aware of their relationship with their personal finances, their bank and their financial advisors and also their needs and expectations for the future. They were excited and wanted to know more about the research, its purpose, and whether this is going to be implemented some time in the future. In other words, several of the participants already had expectations of being part of a new development in the bank. This is interesting and has led to an analysis of the organisation and the dominant logic in the bank.

HOW TO ANALYSE THE RESPONSES

In design research the transformation of the participants´ responses into useable information for the designer/company constitutes an area of special attention. The reason is that it can be difficult to transform the analysis without 'killing' the presence of the customers. The most interesting part of co-creation is observing and talking to customers during these creative sessions. This is a true look into the life and thinking of the participant, into how the participant is performing the creative tasks, speaking to him or her and listening to the frustration, the irritation or the satisfaction with his or her personal finances and current banking situation. It is a challenge to translate these responses without reducing the living participants to dead analysis void of any presence or intensity. On the other hand there is also a clear demand in design research for ways to treat these responses scientifically. But there are different views on this:

Originally the cultural probes were not designed to be analysed as Gaver states:

Unlike much research, we don't emphasize precise analyses or carefully controlled methodologies; instead, we concentrate on aesthetic control, the cultural implications of our designs, and ways to open new spaces for design" (Gaver, 1999). Stappers & Sanders (2003) apply statistical methods, such as counting the co-occurrence of images and words using multidimensional scaling to reveal the patterns in chosen images and words. Others claim they need theory for analysing the responses (Graham & Rouncefild, 2008). Crabtree et al. demonstrate that if Cultural Probes are analyzed in particular ways they can serve as Informational Probes, in their case to "gather information into how the members of our user groups live their lives, their everyday circumstances, routines and rhythms, their practical concerns, everyday aesthetics and so on (Crabtree et al., 2002).

To do this, one need not alter the probes packages as such – they still contain cameras, postcards, diaries, maps, etc. – but we need to engage in analysis typical of ethnographic inquiry.

As the project can be seen as a process of organising conversation and debates about values in the community, it is obvious to apply rhetorical theory, analysing the responses and then transforming this knowledge into information that is useable in the community. In the field of co-creation there are several other approaches and foci. Mattelmaki is looking for *"an understanding of the use of situation"* or *"new eyes to look at experiences"* (Mattelmaki, 2007). Krippendorff goes further claiming that when designing artifacts for use by others, *"you need to understand how different user groups understand their worlds."* That means you need an insight into understanding, a second-order understanding (Krippendorff, 2006). Contrary to the first-order understanding this second-order understanding consists in a recursive structure. Krippendorff´s here distinguishes between first- and second-order understandings as he claims two different, but interconnected understandings are present when designing artifacts for use by others: the designer's understanding of the artifact, as well as the designer's understanding of the user's understanding of the artifact.

As human-centred design and co-creation involve design for others, it necessarily has to be grounded in second-order understanding (Krippendorff, 2007). Krippendorff has created a semantics for artifacts in the designer's second-order understanding (Figure 3). The figure illustrates how the designer's first-order understanding of the world as well as all the designer's second-order understandings makes him or her create artifacts that embrace the different stakeholders' diverse understandings of the world. The designers' second-order understanding makes room for - and in this case invites - other stakeholders' different understandings, here in relation to the subjects 'money', 'value' 'economy' etc. The responses represent new knowledge which goes back to the designer /organisation for evaluation. The responses may also generate new ideas or new questions for new co-creation processes. Evaluating the experiments, I designed artifacts with the purpose of obtaining a second-order understanding of different stakeholders in the organisation.

Figure 3. Design semantics as second-order understanding (Krippendorff, 2007)

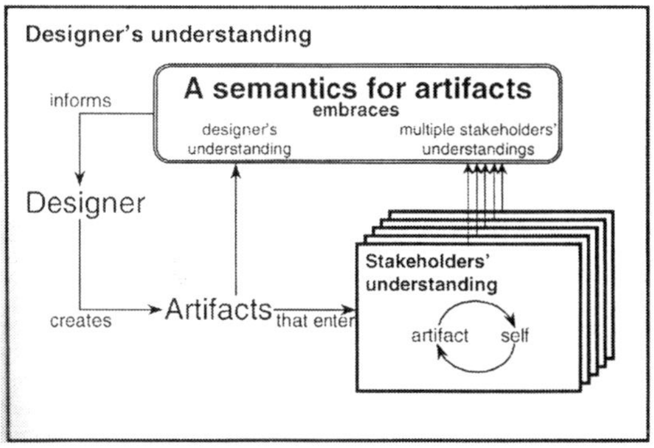

In continuation of Krippendorff´s point about securing a second-order understanding of the different stakeholders' understanding of the world, I intend to apply rhetorical theory in the analysis of the responses. This gives the possibility of outlining not only what Krippendorff names a second-order understanding; it also opens the possibility of outlining the participants' preferred values and motives.

Perelman & Olbrecths-Tyteca's theory serves as an outline of the underlying sets of values. Furthermore I find the theory of Burke (1969) which also describes the underlying motives in the participants, useful. Methodologically that means: within Krippendorff´s framework, I accentuate the rhetorical analysis of 'values' and 'motives' in the material gained from the experiments resulting in a second-order understanding of the participants, specifically outlining 'values' and 'motives'.

A parallel to both second-order understanding and structures of values and motives is Prahalad´s term (Perelman, 1969) 'the dominant logic', which describes a predictable way of thinking about opportunities, competitiveness, consumers and performance. It includes managerial behaviour and determinants such as organisational structure, performance metrics, reward systems, career management, training, beliefs and values. As per Prahalad's suggestion about analysing the dominant logic of the company (social architecture) I use rhetorical analysis to determine the dominant logic within the company.

In addition to the rhetorical analysis it also is very important to present the customers – the people working with the material – as individuals, as living and breathing human beings. This research seeks to meet the request for a scientific method without 'killing' the presence of the customer.

The Experiments

Five different workshops (approx. 25 persons altogether) were completed for the research. The participants formed a very broad group: customers, potential customers and financial advisors. These groups represent structures of values and motives and dominant logic with a broad range, including all their inputs and ideas. In the following I intend to make a short presentation of the applied rhetorical theories and examples of how to apply these to the analysis. First the applied theory will be outlined. The two participants, used as illustration, will be presented.

Within Krippendorff's framework the theory of Perelman and Burke will be applied, which will accentuate the rhetorical analysis of 'values' and 'motives' in the material gained from the workshops. This results in a second-order understanding of the participants, specifically outlining *value structures, motive structures* and *dominant logic*.

The central idea behind Burke´s theory about 'dramatism' is that human beings develop and present messages in much the same way that a play is presented. Burke claims every communication contains information about the motives of the author and how he understands and positions himself within a situation. We use rhetoric to define and present a particular view of our situation. How we describe a situation indicates our perception, the choices we see available, and the actions we are likely to take in that situation. Our language, then, provides a clue to our motives or why we do what we do. (Foss, 1996). When analysing customers' relationship to their personal finances and their banks, an interesting issue is whether the customers are in control of their money or not, and how they describe their way of being in control or not. This is the reason for applying Burke's theory in this research.

Together with his notion of 'dramatism', Burke developed a model for analysing situations, the Pentad, containing five basic elements of a drama: *act, agent, agency, scene* and *purpose*, a parallel to the questions who? (agent), what? (act), why? (purpose), when? and where? (scene), how? (agency).

Identification of the five terms results in an overview of the participant's perception of a particular situation. Furthermore it is possible to identify the dominant element by using what Burke calls *'ratio'*. A ratio is the pairing of two of the elements in the Pentad in order to discover the relationship between them and the effect that each has on the other. Working with ratios one has to continue to pair terms in ratios, to discover if one term seems to affect the nature and character of another. Review of several of the ratios will produce a pattern in which the critic discovers that one term (or sometimes more than one) is the central, controlling term and defines the other terms in the Pentad (Foss, 1996), (how a particular term influences the other elements of the rhetor´s description of the situation).

In the following I will exemplify how to apply rhetorical analysis on the responses from the workshops. But first I will introduce Perelman & Olbrechts-Tyteca.

Perelman and Olbrecths- Tyteca and Their View on Values

Perelman and Olbrechts-Tyteca (1969) claim every effective communication is based on an acceptance of equal values, "*a sense of communion centred around particular values recognized by the audience"*. Perelman's view of values and 'the sense of communion' seem to be analogous to Prahalad´s ideas about co-creation, as he considers 'dialogue' as one of the building blocks of co-creation. Prahalad defines 'dialogue' as

...interactivity, deep engagement, and a propensity to act - to both sides. Dialogue is more than listening to customers. It entails empathic understanding built around consumers' experience, and recognizing the emotional, social, and cultural context of experience. It implies shared learning and communication between two equal problem solvers (Prahalad, 2008).

When analyzing the responses from the creative sessions, I intend to focus on the implicit value structures and value hierarchies included in Roer´s value analysis (Roer, 2006).

Perelman & Olbrechts-Tyteca claim every argumentation is based on either 'the real' or 'the preferable'. 'The preferable' is represented by values, hierarchies and common places. For example the statement: "*the sun will come out tomorrow*" is a fact, whereas the statement "*days with sunshine are beautiful days*" is a statement including a preference of values.

Perelman & Olbrechts-Tyteca distinguish between abstract and concrete values. Values are abstract when they are not connected to a certain person or institution. Justice, beauty or freedom are examples of abstract values. According to Perelman & Olbrecths-Tyteca individuals who wish to maintain a certain social order use arguments with concrete values, as these are more persuasive when the aim is to maintain the present circumstances. Abstract values are suitable when you intend to argue for change whereas concrete values are suitable when you intend to maintain the existing situation.

Another distinction is between qualitative and quantitative values. Quantitative values confirm that some things are worth more than other things for quantitative reasons. Examples of quantitative values are efficiency, duration, objectivity and stability. Quantitative values are related to reason. Qualitative values include subjective concepts such as uniqueness, exceptionality and originality, which express individuality and the unique nature of a specific thing (Roar, 2006).

Hierarchies are values organized according to importance, and it is necessary to put values in hierarchies when they collide. In other words when values are inconsistent it is necessary to prioritise them by arranging them in hierarchies. Hierarchies can also be classified as 'concrete' or 'abstract' or as 'homogenous' or 'heterogeneous'. In a homogeneous hierarchy values are differentiated according to degree, for example it is preferable to have a cold rather than to have a severe illness. Conflicts involving homogeneous values are relatively easy to solve, as there is agreement on the goal, in this case good health. On the other hand, conflicts in a heterogeneous value hierarchy are more complex as they contain inter-conflicting values. For example the values 'feeling' and 'reason' can easily come into conflict, for example when you are on a diet and you are offered a piece of chocolate. Heterogeneous values contain inconsistent values that force people to choose between them.

Applying Perelman & Olbrecths-Tyteca when analysing the value structures of the responses I looked for what type of values are represented in the collages – primarily in the verbal presentations of the participants supported by the visual expressions in the collages (Figures 4 and 5).

Figure 4. Collage A

Figure 5. Collage B

Two Examples of How to Analyse Responses

Example 1, Anne Marie, 22 Years Old

Anne Marie, a 22-year-old student said (my shortened version):

I just moved to Copenhagen to study. I am the type of girl that likes to go out, I like to buy clothes, like to party…it is an expensive way of living…I know I am spending too much money and I have not done anything about it…I have kind of buried my head in the sand. I really have to do something about it NOW…I am very dissatisfied with my bank and my financial advisor – I want to change to someone who can help me make a budget…a financial advisor who can help me gain control of my personal finances NOW… Right now I feel like a number, and I feel I have to do something about my finances NOW. I would like to switch to another bank, a place where I feel welcome, where people give me a smile, a place where I feel that people see me as the person I am, not a number. I feel like a number and do not know who to speak to…. No two individuals are alike, they have to understand this and practice this, show it. I would like to experience joy and harmony, feel the 'soul' of the bank and my financial advisor. I would like or expect my financial advisor to help me set up budgets and help me get in control of my finances. I want service and I want them to be accommodating towards me. I want empathy.

During the session, when doing the collages, the young woman, Anne Marie, gets emotionally affected by the situation and repeats the word NOW several times.

Analyzing the scenes in accordance with the Pentad, I get the following results: In the current situation, the Act is: spending too much money, not being in control of her finances; the Agent is Anne Marie; the Agency is: I kind of buried my head in the sand; the Purpose is: I am the type of girl who has an expensive life style. In this example the Ratio, the two most dominant elements, are the Agency over the Agent, meaning the way she buried her head in the sand was dominating her as Agent.

In the future situation, the Act is: I am in control of my finances. The Agent is Anne Marie, and the Agency is: setting up and sticking to my budgets. The Scene is 'bad finances' and the Purpose 'I want to be in control and want to save up for e.g. new boots, a bag, a holiday etc. The Ratio is the Purpose, 'I want to be in control and save up' over the Agency, 'setting up and sticking to my budgets'.

The most interesting thing when outlining the included motives is *'the ratio'*, the two strongest elements in her presentation of her collages and the determination of which one will be in control.

When Anne Marie tells her current situation she presents the leading element to be her way of (not) controlling her finances - *'I kind of buried my head in the sand'* – and stronger than herself –'agent'. In other words, the Ratio is Agency over Agent – her over-spending is actually ruling, and she has lost control.

She says she wants to get in control of her finances: *"I want to have a new financial advisor who can help me set up budgets and help me gain control of my finances."* Her future situation, the Act, is 'to be in control of my finances', the Agent is Anne Marie, the Agency is 'setting up and sticking to her budgets' and the Scene is bad personal finances. The interesting thing is, the strongest elements have changed into Purpose and Agency', with the Purpose 'in control' and 'harmony' as the strongest element.

The Values

In Anne Marie´s collages as well as in her verbal presentation of her collages, she adds different words to the different elements. As per Perelman these words and expressions illustrate her preferences of values. As the collages are in Danish, I have translated the responses here:

In the current situation the values she has added are:

- **Me as a customer:** It is a mess, I am not in control of my expenditures and I have a headache (AM explains the pictures in the collage, here highlighted with red circles)
- **My financial advisor:** I feel there is a hidden agenda in the counselling
- **The relationship between myself and my financial advisor:** He does not care and I do not trust him. Fees and deficit.
- **Further comments:** "Who can I talk to" and handwritten; "I feel like a number". "The décor of the bank is boring."

In the current situation a hierarchy of heterogeneous values is present but in the future situation this changes into a hierarchy of homogenous values between system and individual.

In the future situation the values are:

- **Me as a customer:** I feel relaxed and satisfied, in control of my economy
- **My financial advisor:** Harmony
- **The relationship between myself and my financial advisor:** Filled with soul and empathy. Pleasant atmosphere and harmony between myself and my financial advisor
- **Further comments:** I feel welcome, people make me feel welcome and give me a smile. Counselling, focus on economy, working on how to make the right decisions. I feel special and treated well because of who I am. Services and kindness.

Example 2, Erik 45 Years Old

Erik is 45 years old. He has a higher education and works as a strategic developer in a large company. In his presentation he said (my shortened version):

I am living with my wife in a suburb of Aarhus. We live in our own house in a nice part of Aarhus close to the sea... Both of us have jobs and our combined income is relatively high. We do not spend a lot of money and we do not have a budget, but I buy more or less the same things every month. Now and then I need to buy something like shoes, clothes, or furniture. Then I go to the shop and before entering I decide how much money I want to spend... I prefer to buy good quality.

During the session, when doing the collages there are no emotional reactions; Erik thinks, and then he writes down (Figure 6).

According to the Pentad analysis, the current situation can be described as follows: the Act is balance, being in control of money, personal finances. The Agent is Erik, and the Agency is "*we do not have a budget, but I buy more or less the same things every month.*" The Scene is having

Figure 6. Collage C and Collage D

a good economy and the Purpose is: I like to be in control, to have money for future purchases, house, holiday etc. The Ratio is Purpose over Agent, which means the Purpose, to be in control, is to be in control of himself.

In the future situation, the Act is balance, to be in control of money and finances. The Agent is Erik, and the Agency is: *"we do not have a budget, but I buy more or less the same things every month"*. The Scene is having a good economy and the Purpose is to be in control, to have (more) money for future purchases, house, holiday etc. Finally the Ratio is: the Purpose, to be in control, is to be in control of the Agent, himself.

Erik's situation, according to Burke, is the following: Erik is on the Scene, in a good financial situation. The Act is being in control of his finances. In the first now-scene, Erik has illustrated himself with a picture of a relaxed man being in control also in the relation to his financial advisor. The Agency (the way the Act is done) is not illustrated, but is implied in the way he has illustrated himself as a customer. In the future situation he does not present any changes in his personal finances, but he has a lot of expectations in relation to his bank. According to Erik's presentation the purpose, to be in control of his money, is dominant in his life.

When creating these collages, he reflects on his expectations and is very precise in expressing his needs and also ideas for the future bank.

The Values

From the collages and the transcription of the individual presentation, Erik has expressed the following values:

- **Me as a customer:** Relaxed
- **My financial advisor:** Relaxed, a business partner
- **The relationship between myself and my financial advisor:** I expect he gives me service, advice etc.
- **Further comments:** Service, treats me as a human being, new ways of doing things, important knowledge.

Both in the current and in the future situation a hierarchy of homogenous values is present.

In the future situation the values are:

- **Me as a customer:** Relaxed, but more expectations
- **My financial advisor:** Proactive, ideas, finds the best for me, balance, transparency in business, more bottom lines, responsibility
- **The relationship between myself and my financial advisor:** I expect him to give me more service
- **Further comments:** A lot of ideas and expectations:

In addition to his presentation he expresses several views and ideas in relation to the banking business and his expectations as a customer:

The bank's philosophy should be: 'your dreams are our future'. The bank is a business. It has to make money. It also has to help me as a customer, help me realize my dreams and fulfil my need for cash and cash equivalents. We are all busy; the bank has to be available 24-7. I want to be able to surf the bank's website, tag my financial statements, collect my own channels of information and collect it in my virtual banking box. There has to be a tool in the net bank to create overviews and spreadsheets where I can display my budget and have permission to wholesale societies where I get discount because I am a customer in the bank. There has to be a scenario planner that gives me an overview – what if I want to take a sabbatical year and go to university? What does it cost? How much money do I have at my disposal in the bank? Is this realistic or not? I expect my bank to keep me updated about today's social conditions. Does the financial crisis have any influence on my personal finances? I expect the answer will be presented on the bank's Web TV, in an easy and understandable way - not in a technical letter from the bank filled with finger-pointing, and in a understandable language...

In both examples the value hierarchies are connected to an underlying assumption about the bank as a system and themselves as individuals. But there is a big difference in the values surrounding this relationship.

HOW TO DEVELOP A DYNAMIC AND USEFUL TOOL

Having completed the analysis of the responses there are several ways of exploring the material and getting deeper into the users' mindset, their experiences and perspectives. But traditional analysis material is often far from the inspiring and lively meeting with the participants. This is the reason for adding design in the presentation of the analysis results. The result is this sketch for an interactive tool (Figure 7):

Figure 7. Sketch for interactive tool

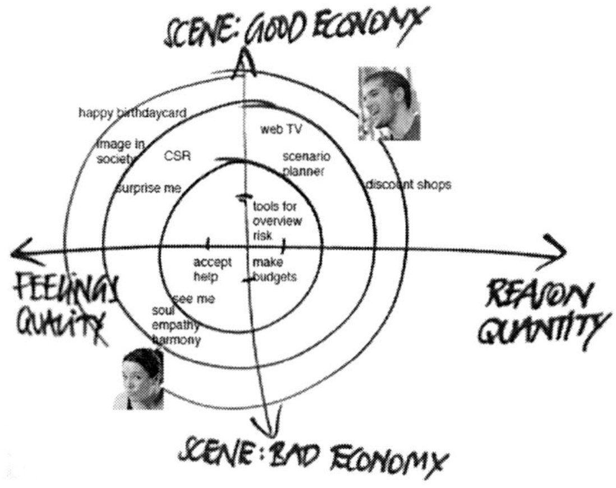

The Pentad analysis is represented vertically. Here Scene represents people having either bad or good personal finances – several banks already divide their customers into different groups related to their financial situation. Horizontally the value analysis according to Perelman is shown. The quantitative and the qualitative values, related to feelings and reason (Perelman, 1969) are presented. The idea is to plot in the participants' values in the target, the most important ones in the middle. This gives a clear indication of what is important to the participant and also how important it its. The sketch also indicates whether the values are qualitative or quantitative. The last parameter is relevant to the banking business as there has to be a certain balance between the products/services and the prices. Scene also entails the possibility of change due to people's changing situation, and maybe changing values.

The tool can work as a still picture and gives a quick overview of the participants, their values and their actual financial situation. Clicking on the picture, it turns into an interactive tool. It is possible to have a one-minute presentation of the participant and his or her ideas, dreams etc. The participants can elaborate on the different values and dreams. Other things may be included in the mapping – for example concrete ideas for CSR - if the company wants to develop more CSR activities, it is possible to choose from the value cards and find different expressions about this issue.

The overall idea is to make a mapping of people, but also to 'see' and 'feel' them. This tool still contains the presence of the people. It gives an idea of what people want, ideas, and values;

and it raises a lot of questions that can lead to more co-creation. The two customers are people in different situations, who have different foci and different ideas. They also represent two different customer groups, one currently in a good financial situation and the other currently having problems, and besides they also have different ways of controlling their finances.

The intention with the tool is to give a quick overview of different customers and their different values and motives. On the basis of these two customers, the picture gives a clear overview of the very different needs, values and expectations the two customers represent. This picture raises questions such as, what if Erik's financial situation was bad, how would his values change? And what if AM's finances were good, would her values and expectations change? Will AM's values change if she gets in control of her finances? Will she ever be in control or does she expect to have a helper by her side constantly? Do people have underlying assumptions of money and how do they affect their relationship with their bank and financial advisor?

It is also possible to go into more detail about a specific group of people. As we are in the middle of a financial crisis, it is interesting to research if customers have new and different needs for services. Part of this research it also a specific creative session with couples (4x2 people) with bad personal finances. They provided plenty of input and also defined new needs: One woman was fond of blogging and suggested the bank create a site for people who would like to get ideas for saving money, positive and motivating stories, and examples of people who had changed their way of spending money. Another person suggested making a coaching service for people who would like to be better at being in control of their money. Like a dietician helping people who eat the wrong things and ingest too many calories, this 'financial dietician' could help people get new routines buying the right things at the right places, save money for a vacation etc. There were many other ideas, but the most interesting thing was that the participants really enjoyed it, and the majority would like to spend time developing new ideas. In order to get an overview of all the inputs and different needs, the idea about a creative tool evolved.

There are different future possibilities in this tool. Within the organisation it can be used as an inspiration for future dialogue about new needs. It can also be used to invite people to engage in further co-creation, developing new services with customers etc. Expanding the tool to include both people inside and outside the organisation will open possibilities for co-creation as an organisational approach, developing the future value system in an ongoing dialogue with communication flowing both inside and outside the organisation. This requires an analysis of the existing dominant logic and values within the organisation.

CO-CREATION AS AN ORGANISATIONAL APPROACH

Before going to the analysis, here is a short description of the bank: According to the presentation on their web page (www.midspar.dk) and interviews with the board of directors, the bank has put a lot of effort into developing the organisation. The leadership is very modern and visionary in its thinking. Several years ago the bank implemented value-based leadership and later on it developed a process of self-leadership. The bank has achieved several acknowledgements of its efforts and has been nominated among the best workplaces in Denmark and Europe since 2002. The organisation is guided by 'the 6 commandments' that constitute the organisational set of values and are all addressed to the employees. Each employee is assigned a high level of responsibility, and the organisation has a lot of confidence in its employees. The bank has healthy finances, also in this current financial crisis, as it has been more prudent in its investments. The bank is founded as a kind of co-operative bank, called a 'garant bank', which means there are no shareholders, no one owns the bank, and some of the profit has to be invested in the local community.

Outlining the ruling company perspective, (the dominant logic of the company), I conducted rhetorical analyses of values/value structures and motives/dominance of three different items: the Social-Ethical Accounting Statement (which measures intangible (non-monetary) values and outlines the bank's social responsibility), the presentation on the homepage, and the creative workshops. Here is a short summary of the analyses:

Regarding the dominant logic in the Social-Ethical Accounting Statement, the company claims it is listening to its customers as well as to the surrounding community. This Statement is presented *"like the serpent in the Garden of Eden, which provides new fresh eyes to look at things."*(www.midspar.dk, *05.25.09)* The Statement emphasizes that it is not switching from an internal to an external focus, but keeps its focus by asking customers and the surrounding community, whether and *to what extent the bank lives up to its own values*. The bank has sent out a questionnaire to the local community based on an internal perspective, and in analysing motives and dominance, the bank regards itself as the dominant factor in relation to its customers and the community.

When analysing the dominant logic of the company as presented on the homepage you find this internal perspective. The headline reads: "*Excess energy for the customers*", followed by the text: *"It is important to us to be a good workplace both for the sake of our employees and for the sake of our customers. We believe a good workplace has excess energy to serve the customers; our livelihoods depend on them, you know."* (www.midspar.dk)

The argument is as follows:

The claim: *It is important to be a good workplace for the sake of our employees and for the sake of our customers.*

The support: *because a good workplace has excess energy to serve our customers, and (implicitly) our livelihoods depend on them.*

The warrant: *our livelihoods are sustained* by a good workplace.

Studying this argument highlights the ruling company-centric perspective in the organisation and a strong belief that its livelihood is sustained by a good workplace. The role of the customers is clearly uttered as *"our livelihoods depend on them"*. (This observation, however, does not indicate that the customers *feel* 'they are the ones that sustain our livelihood'.)

This strong belief is also reflected in the attitude of the employees. It is evident, from my earlier studies at the bank and just visiting the bank that the employees are very satisfied, and they are proud, almost grateful to be working in the bank. They seem very certain of themselves, and the majority of employees undoubtedly identifies strongly with the company. The employees are also considered *the most important resource in the organisation*. This strong dominant logic explains the answers and reactions in the workshop. In the creative workshop the employees were presented with the idea of letting customers do these kinds of creative workshop, expressing their dreams, needs, experiences etc. in order to have them co-create the financial consultancy. One of the financial advisors reflected on this question and posed a counter question: "*should we design our bank with competencies, sets of values and personal qualities, constructed from our world and our way of thinking and acting, in order to attract customers?"* or *"should we design our bank based on the customers' values*?"

In the last decade organisations have focused on employees, the development of the individual employee, coaching, self-leadership etc. The employees are thus often considered the most important resource in the company. If a redefinition of organisations is implemented, this perspective will have to change. Again Normann´s definition of future organisations as value creating systems is relevant, since he talks about organisations that identify with a value creating system rather than with a specific product or service. This is a more holistic view where the organisational perspective does not fit in.

FUTURE TRENDS

The co-creation trend is interesting both from a business innovation angle but also from a design and rhetorical angle.

From a design angle co-creation as an organisational approach is exciting because it outlines the future of the designer, as both designer and facilitator of organisational conversations and interactions about values. Co-creation requires an active change in perspective within the organisations. A challenge for companies is to first accept that they tend to take their traditional assumptions for granted, or, as Otto Scharmer says in his "Theory U" (Scharmer, 2007) they are 'downloading'. In order to see new possibilities it is necessary to 'suspend' (let go of) old notions and reach a level of 'seeing' (with new eyes), for example identifying the organisation with a value-creating system and not being limited to a specific product or service. Moving from 'downloading' (through 'suspending') to 'seeing' is still the biggest barrier for succeeding in creativity, innovation and co-creation (the right side of Scharmer's U figure).

Another challenge is the increasing expectations of the customers. By introducing co-creation with customers, the companies define the customers as 'experts' and 'co-creators' of values. Rhetorically co-creation can be seen as 'constitutive rhetoric' (Charland, 1982). According to Charland rhetoric is capable of 'constituting' an audience in a new subject matter – like this example of customers as 'experts', 'co-creators' or 'creative individuals'. Charland´s point in the constitutive rhetoric is that the audience will claim their rights on behalf of this new subject matter. In other words, the customers will claim their rights as 'experts', 'co-creators' or 'creative individuals'. Included in this perspective is the fact that this 'constitution' also *dismantles the role of the financial advisors* – who used to be the experts – as expressed in the dominant logic of the company. Empowering the customers, by delegating them to the role of expert, therefore immediately creates a need for a redefinition of the role of financial advisor. And this may be a challenge, not only for this bank.

In relation to everyday people given access to the design process in co-creation and customers being constituted as 'co-creators' or 'creative human beings', the question also arise - how much will design thinking and creativity diffuse into the future of everybody's curriculum for living, learning and working?

CONCLUSION

As Normann states, co-creation is a veritable revolution. It requires a redefinition of many roles and processes within the organisations. They are no longer producing goods for a market with passive consumers. These consumers are changing into active empowered citizens, occupied with ethical issues such as the environment, human rights etc. They are not satisfied with being the target of consumption; they want to participate actively but also to co-create values (Sanders, 2008). Several business researchers have stated more or less radical redefinitions of the organisation. Normann represents one of the most radical perspectives of the future organisation, as a value creating system (Normann, 2001). His view of organisations borders on a holistic perspective characteristic of design thinking.

Speaking about future organisations several researchers acknowledge the relationship between design and management. The traditional and limited understanding of design related to form and aesthetics is expanding into an increasing interest in and research of design, as design thinking and designer skills. An essential issue in the characterisation of design thinking is its focus on not how things *are*, but on how things *could* be. This prescriptive perspective is interesting from a management perspective as organisational leaders are often criticised for being reactive to situations presented to them, as opposed to being proactive creators of a desired future.

This chapter has described how a designer can design human systems and be the facilitator of conversations and debates about the values of a community (Buchanan, 2001), and I have outlined the advantages of a rhetorical design approach. First of all the chapter illustrates how the design activities for co-creation sessions had a wider appeal, which goes beyond what can be stated in words (Sanders, 2001). In accordance with conducted research in this field (Sandes, 2001) the participants expressed that they became more aware of specific issues related to the topics of money and finances during the workshop. As such this approach has its advantages and actually meets the demand of *how to* co-create values – as opposed to several researchers e.g. Prahalad & Ramaswarmy (2004) and Normann, (2001) – who claim that co-creation and 'emphatic dialogue' (Prahalad 2004) are crucial but refrain from explaining how to implement them.

The chapter also sheds light on the challenges inherent in co-creation, especially when co-creation is moving from the outer edges of the company's value chain to the centre of the organisation. Considering co-creation a type of organisational communication it is also appropriate to perform analyses of the organisation. This was the reason for including financial advisors in the co-creation sessions. Furthermore presentations of the company were analysed and the dominant logic and values were outlined. The conclusion was that the bank has a strong organisation-centred perspective with employees as the central and most important resource in their business. This view leaves not only this organisation but also the majority of other organisations with the challenge of changing their traditional views. Here Scharmer's notion about moving from 'downloading' (patterns of the past) via 'suspending' (letting go of habitual patterns) to 'seeing' (with fresh eyes) is interesting and relevant (Scharmer, 2007).

The application of rhetorical theory made it possible to analyse the conversations and debates about values in a community. As such the chapter meets the demand of design research for a scientifically based method of processing these responses. As traditional analysis is often far removed from the inspiring and spirited meetings with people, a sketch for a dynamic interactive tool was developed. In the bank this research is continuously evolving and the different debates and conversations in and around the bank is changing perspectives and opening up for new possibilities: inviting customers for the development of new services, changing the presentation on the web site, discussing SCR as a new core value etc.

The outlined rhetorical design approach and the included tool present different future options within the organisation, either as a tool for further co-creation experiments – at the outer edges of the company's value chain – or the possibility of evolving into an organisational approach. In the bank the research is continuously evolving and different debates and conversations in and around the bank is changing perspectives and opening up for new possibilities which can be seen in different actions: inviting customers for the development of new services, changing the presentation on the web site, discussing SCR as a new core value etc. As such this rhetorical design approach to co-creation, organizing debates and communication about values in and around the company, is regarded as a productive process.

REFERENCES

Buchanan, R. (2001). Design and the New Rhetoric: Productive Arts in the Philosophy of Culture in Philosophy and Rhetoric: *Vol. 34. No.3). State College*. PA: The Pennsylvania State University.

Buchanan, R., & Margolin, V. (1995). *Discovering Design. Explorations in Design Studies*. Chicago: The University of Chicago Press.

Burke, K. (1969). *A Grammar of Motives*. Berkley, CA: University of California Press.

Foss, S. K. (1996). *Rhetorical Criticism – Exploration & Practice*. Long Grove, IL: Waveland Press Inc.

Gaver, B., Dunne, T., & Pacenti, E. (1999). Cultural Probes. *Interaction*, *6*(1), 21–29. doi:10.1145/291224.291235

Graham & Rouncefild. (2008). Probes and Participation. In *Proceedings of the Participatory Design Conference* (PDC08).

Joost, G. (2007). Design as Rhetoric. In *Proceedings of the Symposium of Swiss Design Network.*

Jørgensen, C., & Onsberg, M. (1999). *Praktisk Argumentation*. København: Teknisk Forlag.

Kim, R.W., & Mauborgne (2005). *Blue Ocean Strategy*. Boston: Harvard Business School Press

Kjeldsen, J. E. (2004). *Retorik i vår tid*. Spartacus: Forlag

Klujeff, M. L. (2006). Retorisk publikum In Klujeff M.L. & Roer H. Hans (edt), Retorikkens aktualitet. Reitzels: Forlag.

Krippendorff, K. (2006). *The Semantic Turn – a New Foundation for Design*. Boca Raton, FL: Taylor & Francis Group.

Lencioni, P. (2002). *Make Your Values Mean Something*. Boston: Harvard Business School Publishing Corporation.

Mattelmäki, T. (2007). *Probes*. Helsinki, Finland: University of Art and Design Helsinki.

Matthews, B., & Horst, W. (n.d.). *What can we learn from the probes? The role of interpretation in contribution to knowledge*. Demark: Mads Clausen Institute. *University of Southern Denmark*.

Normann, R. (2001). *Reframing Business: When the Map Changes the Landscape*. New York: Wiley.

Perelman, Ch., & Olbrechts-Tyteca, L. (1969). *The New Rhetoric – A Treatise on Argumentation*. South Bend, IN: University of Notre Dame Press.

Prahalad, C. K., & Krishnan, M. S. (2008). *The New Age of Innovation, Driving Cocreated Value Through Global Networks*. New York: McGraw Hill Books.

Prahalad, C. K., & Ramaswamy, V. (2004). *The Future of Competition, Co-creating Unique Value with Customers*. Boston, MA: HBS Press.

Sanders, E. B. N. (1999). *Postdesign and Participatory Culture. Useful and Critical: The Position of Research in Design*. Tuusula, Finland: University of Art and Design Helsinki.

Sanders, E. B. N. (2006). Scaffolds for Building Everyday Creativity. In Frascara, J. (Ed.), *Design for Effective Communications: Creating Contexts for Clarity and Meaning*. New York: Allworth Press.

Sanders, E. B. N. (2006). *Design Research in 2006. Design Research Quarterly*. Design Research Society.

Sanders, E. B. N., & Stappers, P. (2003) Generative tools for context mapping: tuning the tools from the Third International Conference on Design & Emotion. Boca Raton, FL: Taylor & Francis

Sanders, E. B. N., & Stappers, P. (2008). Co-creation and the new landscape of design.

Scharmer, C. O. (2007). *Theory U – Leading from the Future as it Emerges*. Cambridge, Massachusetts: SoL.

Schön, D. A. (1982). *The Reflective Practitioner: How Professionals Think in Action*. New York: Basic Books.

Tapscott, D., & Williams, A. D. (2006). *Wikinomics: How Mass Collaboration Changes Everything*. Portfolio Hardcover.

Vargo, S., & Lusch R. (2004). Evolving to a New Dominant Logic for Marketing. *Journal of Marketing*.

Von Hippel, E. (2005). *Democratizing Innovation*. Cambridge, MA: MIT Press.

KEY TERMS AND DEFINITIONS

Values: There are several and different types of values: A personal and cultural value is a relative ethic value and an assumption upon which implementation can be extrapolated. A value system is a set of consistent values and measures of integrity are based. Values are considered subjective, vary across people and cultures and are in many ways aligned with belief and belief systems. Types of values include ethic/moral values, political/idelogical values, social values and aesthetic values. Personal values developed very early in life may be resistant to change. They may be derived from those of particular groups or systems, such as culture, religion and political party associations. However, personal values are not universal: One´s family, nation, generation and historical environment help determine one´s personal values. Cultural values exist in groups, societies, or cultures have values that are largely shared by their members. The values identify those objects, conditions or characteristics that members of the society consider important; that is, valuable. (en.wikipedia.org) Perelman and Olbrechts-Tyteca (1969) claim every effective communication is based on an acceptance of equal values: *"a sense of communion centred around particular values recognized by the audience"*. Perelman's view of values and 'the sense of communion' seem to be analogous to Prahalad´s ideas about co-creation, as he considers 'dialogue' as one of the building blocks of co-creation.

Co-Creation: Co-creation means any act of collective creativity, i.e., creativity that is shared by two or more people (Sanders, 2008). There are very different opinions on who should be involved in these collective acts of creativity, when, and in what role vary widely, as we have seen in the chapter. *Co-creation* and *co-design* are today often confused and/or treated synonymously with one another. The term *co-design* includes a collective creativity as it is applied across the whole span of a design process: for example when customers are doing the final design of a shoe. Thus, co-design is a specific instance of co-creation (Sanders, 2008).

Design: Design is the basis for the making of every tangible or intangible object or system. Design can be used as both a noun (a design) and as a verb (to design). To design refers to the very process of originating a product, a structure, a system etc. A design refers to either the final solution or plan (a sketch, a proposal, a model etc) or the result of implementing this plan in the form of the final product or a design process. Design and the understanding of design is evolving radically these days, which also is outlined in the chapter.

Designer: The person designing is called a designer, which is also a term used for people who work professionally in one of the various design areas, usually also specifying which area is being dealt with (such as a *fashion designer*, *concept designer* or *web designer*). Designing often requires a designer to consider the aesthetic, functional and many other aspects of an object or a process, which usually requires considerable research, thought, modelling, interactive adjustment and re-design. With such a broad definition, there is no universal language or unifying institution for designers of all disciplines. This allows for many differing philosophies and approaches toward the subject. However, serious study of design demands increased focus on the design process and the skills of the designer (en.wikipedia.org), (Buchanan (2001, 2006), Collopy (2008), Boland (2008), Lawson (1980)

Strategic Design: Strategic design is the application of future-orientated design principles in order to increase the innovative and competitive qualities of an organization. Its foundations lie in the analysis of external and internal trends and data, which enables design decisions to be made on the basis of facts rather than aesthetics or intuition. As such it is regarded as an effective way to bridge innovation, research, management and design. (en.wikipedia.org). There are different approaches to strategic design some focusing on the dynamic and iterative processes Collopy (2006), Buchanan (2001, 2006), Boland (2008) some on 'design landscapes' (Lønne, 2008) some on different design methods (Friis, 2007)

Creativity: Creativity is a mental and social process involving the generation of new ideas or concepts, or new associations of the creative mind between existing ideas or concepts. An alternative conception of creativeness is the act of making something new. Creativity is a complex phenomenon and has been studied from various perspectives: psychology, cognitive science, philosophy, history, economics, design research, business and management, among others. The studies have covered everyday creativity, exceptional creativity and even artificial creativity. Unlike many phenomena in science, there is no single, authoritative perspective or definition of creativity and there is no standardized measurement technique. (en.wikipedia.org). In the field of design creativity is studied either in relation to the person (the designer, artist, architect etc.), or the product (a specific design, a house, a system etc.), or the process (the creative process drawing a house etc).

Innovation: Innovation typically involves creativity, but is not identical to it: innovation involves acting on the creative ideas to make some specific and tangible difference in the domain in which the innovation occurs. For example, Amabile et al. (1996) propose: *"All innovation begins with creative ideas... We define innovation as the successful implementation of creative ideas within an organization. In this view, creativity by individuals and teams is a starting point for innovation; the first is necessary but not sufficient condition for the second"*. For innovation to occur, something more than the generation of a creative idea or insight is required: the insight must be put into action to make a genuine difference, resulting for example in new or altered business processes within the organization, or changes in the products and services provided. (en.wikipedia.org)

Rhetoric: Rhetoric is one of the arts of using language as a means to persuade. The popular view of rhetoric often is limited to the mere styling of verbal expression. For rhetoric – as for design - the deeper work lies in the invention and disposition of form and content. In approaching design from a rhetorical perspective, the hypothesis should be that all products—digital and analog, tangible and intangible - are vivid arguments about how we should lead our lives. (Buchanan, 2001)

Section 5
Social Sciences and Environment

Chapter 16
Research Project "Future of the Present:"
The Process and the Importance of Signs Observation into Fashion

Sandra Regina Rech
University of the State of Santa Catarina, Brazil

Gabrielle Stockey Chinchilha
University of the State of Santa Catarina, Brazil

ABSTRACT

This article has the purpose to describe and to report the methodology used by the research project "Future of the Present: an area to analysis and observation of signs". It will clear up in detail the process of observation, analysis, and interpretation of signs and how new trends are possible to comprehend. It's necessary to analyze and research different sections, at the same time that the human behavior is studied and also its lifestyle into the most various cultures. Understood as a language, the exploration of the signals issued by the society is a cross-behavioral study guide which enables tangible recommendations for all levels of the market supported by the irrational and emotional trend phenomenon and synthesized in images. It is a very detailed process that requires a lot of work and sensibility in order to see the evidences that the zeitgeist sends and to change them into fashion trends.

INTRODUCTION

Fashion, due to its ephemeral quality, characterizes a competitive section, pushed by globalization and by the possibility of getting new information faster and, as a result, this fast process stimulates the creation of new products that increase the competition between brands. According to Lipovetsky (1997), "fashion is the society's mirror", so taking it for granted it is possible to be studied everywhere such as in the streets, on the internet, in an art exhibit, in several news articles or in the habits and practices of a specific culture. Caldas (2006) asserts that from the 1970's the concept of the word "fashion" was identified by other authors as a total social fact, that is, its system includes all the spheres of a society and a culture.

The society sends signs that need to be perceived, analyzed and interpreted by professionals

DOI: 10.4018/978-1-61520-617-9.ch016

that are capable and qualified to understand them, so that they can get to their target in a synthetic and organized way. The present article aims at explaining how these signs are captured from the others and how they are transformed into fashion concepts and trends.

Due to the access to the large amount of information we have today, trends transform themselves into an immense variety of options. The task of a fashion trend researcher expert is to analyze the research, compare it with the times, both culturally and socio-economically speaking, in which we are living and decide what may and what may not be considered in a thoughtful line, reducing the amount of mistake regarding the determination of new trends.

This way, the research on trends and explorations[1] must be understood as a process, as a language, in a constant dialogue with the consumer guiding designers in the creation of fashion products tuned with the desires and needs of the market.

In order to get to the concept of a new product it is also necessary to find out what the consumers lack in products that already exist or create new needs. Caldas (2006) claims that, in any way, it is in the direct contact with the contemporary cultural scene in big cities besides, of course, surveys done in fairs, stores, department stores etc. where necessary references are obtained in order to prepare an updated and commercially located collection.

The use of devices as fashion trends, which echoes the spirit of its time, appears as a tactic used by companies that are trying to figure out and predict the desires of consumers and the qualities of the commercial market in a given future. "The monitoring of fashion, or fashion research, analyze information from different areas such as economics, politics, sociology, science and technology to arrive at the formulation of predictions in the form of trend" (Feghali, 2001, p.13).

Increasingly, companies in the area of fashion through the search for signs emerging in its various manifestations, grant the knowledge of information relevant to the sectors of marketing and product development, investing profits in a kind of insurance against failure. The collection and interpretation of evidence of the so-called spirit of the times must be grasped as a translation of a cultural language in a specific time, an assessment of the researcher about the times to come.

Gimeno (2000) reports that the production, promotion, distribution and setting up a set of elements that form and must be managed to develop the competitiveness of the productive chain of fashion[2] in the face of a changing market. Currently, competition is taking place based on differentiation (price-quality-creativity) and centered on the incorporation of elements such as style, design and fashion products (Rech, 2006). Therefore, in the production chain of fashion, the ranking is occurring downstream of the upstream of the end for the production of raw material. In developed countries, currently, the trend is rising in level of importance, and the activities of marketing, sales and production flexibility are significant. "The production of textiles and clothing is now driven by end customers who look for product standards and production in upstream sectors" (Prochnik, 2002, p.67).

This new format of the chain is related to the gradual division between the tangible and intangible, material and immaterial, the productive functions and the recovery (Lupatini, 2005). The author notes that although the modernization of the textile and clothing through the material assets (investment in machinery, equipment and raw materials) remains important, increasingly it is not enough. This shortage is accentuated when competitiveness of the productive chain of fashion as a whole and not as only as a few companies and / or links, since the active materials form only one dimension to sustainable gains in competitiveness of the chain.

Intangible assets are increasingly essential to the competitiveness of the textile and clothing. In this sphere, intangible assets largely include active before and after production, such as: design, product development, engineering, marketing, sales channels, brands (preferably global), logistics, maintenance and service suppliers, ability of administration and coordination of the chain. A constitutive feature of the same process that seems essential, in terms of recent changes and strategies of textile and clothing industry in Brazil, is associated with efforts to intangible assets, including product development, brand and design (Lupatini, 2005). Gimeno (2000, p. 229) asserts that fashion is an effective tool for increasing business competitiveness, as it can provide product differentiation and therefore the company itself. As a tool for competitiveness, employment is dependent on the "company's strategic plan, the market you want to occupy and the technical and artistic capacity for its application".

On this basis, it can be stated that the prospective study and its application in products of fashion design[3] is a theme of academic relevance, since the subject can produce inputs to analyze the development of the fashion system and its main actors (industry, designers and consumers) and their inclusion in the current global economic. In the academic environment, examination of the system and surveys of trends can also work for working on references from other fields important to industrial development of countries in the developing world such as Brazil. Characterized by ephemerality, fashion is presented as a competitive industry augmented by globalization, encouraging the rapid creation of new products and stimulating competition.

METHOD

The method of the present article consists in an analysis of the steps of trend research process from the methodology used by the research project "Future of the Present: an area to analysis and observation of signs", accomplishes at the University of the State of Santa Catarina, Brazil.

Initially the article analyses the concept of fashion and how it is placed in the current context as well as it considers the trend perspectives and in what way a trend is defined, established and how it contributes to the moment that we are living in.

After that the article evaluates and ponders how the project participants do their research concerning human behavior, reference sectors, technology and macrotrends, and how the results are discussed and later obtained. Along with the research process, the article also discusses which way the method was developed and what difficulties were found. Right after that the text considers informing the results and by what means they get spread. The article also highlights the difficulty of knowing and comprehending what is a fashion trend and how it takes place by the textile sector in Brazil.

Afterwards the article looks into the importance, the relevance and the multiplicity of the "blogs" into the trend research area and how "blogs" are linked with current events in fashion. Finally the article analyzes the results obtained by the "Future of the Present: an area to analysis and observation of signs" and links them with its objective

OBJECTIVES

The objectives of the present article consist in reporting, explaining and spreading the method and the results that the project of trends research "Future of the Present: an area to analysis and observation of signs" accomplishes at the University of the State of Santa Catarina. Along with that the text explains the importance of fashion as a total social fact as well as the trend research inside the worldwide context we are living in. It also tries to define the trend research as a fact of great importance into the fashion system and why its existence is essential for it.

THE FASHION SYSTEM

It is important to highlight that the word "fashion" does not have only one meaning. It is the result of many concepts developed throughout the history of a specific society, considering a lot of facts not only economical but also cultural and natural. Besides, being a long-term phenomenon its definition is very complex and complicated to be clear (BRAGA, 2005).

Fashion represents a total social fact of ephemeral dynamics, not only in garments and apparel, but also, generally speaking, in a society lifestyle, and it along with itself a lot of different meanings and significances that if studied and analyzed can be predicted.

Including everything around it, fashion acts like a dominating phenomenon of trends that can be of any kind, either behavior-like or garment-like. One of the most important and obvious participations of fashion into society is in the industrial section. It is one of the sections that generates more income and employments to a country like Brazil.

Besides the practical meanings that fashion has, in theory it represents a complex and system. It involves sociological and anthropological significances that claim that the meaning of fashion is in constant change. The permanent search for what is new has cycles and fluxes that may vary, due to the time that it is analyzed. One of the most acceptable and reasonable definitions of fashion is that it is ephemeral and represents the spirit of a time, of a certain period (CASTILHO, 2008).

Fashion has not always been what it is today, but due to the present economic system and a series of historical and cultural events it had to be adapted and improved. One can clearly see that in comparing the fashion that occurred in the end of the Middle Ages it was only useful as a factor of differentiation of sex and social position, and thanks to capitalism and its cycle of consumerism, it became dynamic, individualist and ephemeral, and still today is present in everything we live, are, experiment, study and search.

Fashion today is present in merchandising, television, internet, architecture, technology, arts. It gets and assimilates everything and transforms what has been captured into different ways of fashion, in another section or in another segment. According to Erner (2005), the fashion system bombards information, images, ideas, trends every day, and becomes an integral part of our process of identity construction and method of communication. It has its base anchored in the production of a concept of fashion and not in the production of a fashion product[4]. When the production of the fashion product and the concept of fashion are isolated it is evident why they are so different from each other (Kawamura, 2005). The concept of fashion is guided by the knowledge produced, since the production of the fashion products involves industry and is usually free from concept. However, the "concept" of fashion needs to be sustained so that consumers want to purchase a product considered beautiful and that meets the current trends of fashion.

Through technology it is possible to see the changes that happen in the world and into the society we live in and we can be affected by the latest thing in the market through the excessive consumerism of fashion. The internet allows us to see in real time the latest fashion-show or the latest events of fashion collection. We have free access to interviews, pictures, photos, that is, all the new materials just in.

This easy to access to information, causes a faster dissemination of ideas, and with it, comes the copy and the imitation not only on the product, but on the character that is using it and it is not only restricted to one social level, but to all of them, it is a general phenomenon. Most of the times the consumer is not able to escape from this system, because, apart from social level, the access to the ephemerally fashion is common to everyone and the fact of it being present in every sections of the modern world, makes the task harder to be different inside of a group where everyone is similar and at the same time wants to be different.

The logic of consumption, which is anchored to the structural pillars of the ethos[5] Fashion is becoming more accepted, that is, as Lipovetsky (1997), the domination of the individual by society is infinitely greater than ever. The needs and desires of human beings within the capitalist culture are heightened by the appeal of the fickle media and cultural industry. Bliss (1916), cited in Lynch (2007), states that one of the main reasons for the use of clothes is to address a basic sense of incompleteness and dissatisfaction. From this premise industries seek to predict the aspirations and desires of consumers, always seeking the hegemony of the fashion system.

The fashion cycle, which is efficient and persuasive, makes people want to consume items such as clothes and accessories this way encouraging the consumer to want to transform, change, evolve and to develop their own identity, but at the same time not knowing that he/she may be imitating concepts that were created to other people.

In the fashion system these contrasts are common, constant and spread by the means of communications. What happens is that someone needs and wants to be different, but at the same time, he/she needs to match the ideals imposed by the system and by the media.

According to Lipovetsky (1997), "fashion is followed by ambiguous effects" and he also claims that it is necessary to be just like the others, but not perfectly like them, it is correct to go with the flow and still have a particular taste that makes you different from everyone. Just like that the person will be always a part of the system, but will not be alienated from it; he or she will have a unique opinion and will be himself/herself. He/she will have a true and exact identity. Thus, the ideas of seduction and innovation took forehead of modern society, the opinion has become more important than being. Fashion connects the pleasures of seeing and being seen, turning the stage for the perfect reproduction of narcissism. "Fashion aestheticized and individualized human vanity, managed to make the surface a mean of salvation, one purpose of existence" (Lipovetsky, 1997, p. 39).

The fashion system makes people feel the need of feeling unique, special, different, but not excluded from the group they come from or fit in; they don't want to lose their social identity and characteristics and sometimes, due to these internal conflict, they end up being someone just to please the others around them.

Fashion multiplicity is connected, mainly, to the means of communications, built by the marketing and propaganda. Fashion stimulates people to experience, and mainly buy, new, or most of the times not so new, things, but due to a good way of seduction, they end up making up the consumers' mind, that many times, did not need the product that they are buying.

Kawamura (2005) believes that fashion is not just a piece of clothing, but it is invisible elements that it represents and that humans have added to that piece of fabric. Consequently, the fashion system, which consists of complex gears that connect social, anthropological, sociological, historical and economic logics, is far from a field of frivolous or superficial study.

From an economic explanation for the phenomenon of these changes, there are those who believe that fashion is the result of a conspiracy of the creators with the business sector, thereby making consumers be always affected by new desires, then satisfy those absences through consumption. Influences observed and analyzed by trend researchers work as a kind of mirror of the future of modernity, so expressions of zeitgeist[6] announcing its air to come. Through the study and analysis of social phenomena from the streets and disseminated by the media in general, the exploration of social trends becomes possible, which will later be translated into fashion trends for the industry.

Fashion has a cycle and is seductive. It happens in everything we live, like in the news we read, in the clothes we wear, in the music and videos we watch, in what we eat, in the billboards, highways and cities, and in a lot of things that are present in our routine where only people who are attentive to events and sensitive to what is relevant and important notice these and later transform them in a trend that will become a part of the fashion system (Kawamura, 2005).

PERSPECTIVE RESEARCH

The human being has a fascination for the future. Since ancient civilizations the future was a factor of a great importance, widely studied, however unsure. Today that is not different. We still have that fascination and desire to know what is going to happen next. But due to the high technology we have today we can predict the weather, for example, and what will be consumed and wanted in the next five years or even more or less as well as what will be desired and coveted by a diversity of social classes.

It is inside this area that the trend research is put into action. Although the future is uncertain, it is possible to get closer to a hypothesis of what tomorrow will be. In order to achieve this, it is required to have the sensibility to capture the zeitgeist that is the signs that everything around society and routines emits and that capable professionals developed the certain ability to identify these clues (Feghali, 2001). They can be on a television commercial, in what people are dressing or eating, how different groups behave to certain situation, that is, day-by-day situations where sensibility to see these trends and identify them is essential.

Since studying the future, so as to minimize the risks versus the time unknown, is not new in our society, it is worth noting that despite the prospect of emerging signs indicate generally defined direction, nothing can ensure that such directions are achieved. Therefore, the unpredictability of fashion is only the case of a linear system, where the order exists, but is not apparent when the system is seen in two dimensions. If the system of fashion could be seen in a multidimensional way, the researcher could observe an ordered pattern associated with a strong attraction. In this system, a simple action can start a complicated and extraordinary dynamics. For example a royal wedding, a movie, a war, a sporting event, a rock band can have implications that reduce the lifetime of any part of this system. Therefore, it is a complex set of elements, where many independent agents interact with each other in various ways. The richness of these interactions allows the fashion system to support a spontaneous self-organization. The system adapts itself to the events actively attempting to bring some advantages.

The trends research escapes from the frivolousness and vanity and gets deep into determination and projection of the zeitgeist, that is, what will be wanted, how people's life will be, how they will probably feel about everything that surrounds them. So it is not about an imposition process that is canalized to a result, but the recognition of a context and projection of a perspective and of many possibilities. Another point of the trends research is to understand all the influences to minimize the uncertainties we have about the future.

PROSPECTS AND TRENDS OF FASHION

It is noted currently that the main trends in fashion followed today were explored from a first observation of society and the world around us. Popcorn and Marigold (1997) cited in Oliveira (2006 p.63) note that:

[...] Seeds are everywhere – just open your eyes and look: restaurants, bars and clubs, the streets, in music we hear, in magazines we read, on television we see. Note what the government tells us and what the trade sells us.

The society "states" signs that need to be perceived, analyzed and interpreted by trained professionals that are able to understand them, coming this way to the consumer in a concise and organized manner. These influences, observed and analyzed by trend researchers act as a kind of mirror of modernity, and thus expressions of the zeitgeist advertising posthumous features.

Because one can read a ritual or a city, just as one can read a folk tale or a philosophical text. The method of exegesis may vary, but in each case the reading is done in search of meaning – the meaning inscribed by contemporaries who want to survive in a world of sense (Darnton, 2006, pg. XVI).

Currently, due to access to large amounts of information, trends turn into a multitude of options. The role of researchers is to analyze these data, compare with cultural, economic and social moments in which we live, and decide what belongs to a particular line of thought, thus reducing the degree of error in determining new trends.

Back (2008) asserts that research trend is to perceive influences exerted on a context and read their evolution trying to understand its future consequences. Nowadays the internet appears to be an effective aid in the readings of socio-economic trends, since it appears as an away of potentiating speed and democratizing trends and information from the world of fashion. With the emergence of the blogosphere[7], more and more trendy that emerges from the streets and is distributed via the internet serves as a powerful energy inspiration for researchers, creators, even consumers themselves.

Prospecting embarked on sources requires an approach that goes beyond description, enumeration and recognition of the importance of appearance in the chosen context. It requires an inventory of images that circulated. The appearances were made up, but mainly, the reception that the subjects gave it (Sant'Anna, 2003).

Therefore it is an important acute sensitivity in order to study the exploration of trends. The variability of information and the ability of its mutability are admirable, the point of confusing, what is not yet well defined. The research trend appears as a facilitator of dense and wordy concepts.

Thus, the expected outcome of the study looking for trends is not the only sanction for direction to be the correct one, but from the collection of signals in the present, architect stories "pregnant with meaning". Therefore, the filtering, the interpretation, the narrative construction and the production of meaning are the differential elements for the professional who works with the anticipation of strategic information. For this, we try to research and analyze techniques that help in the exploration and implementation of monitoring trends to guide organizations in building a process of research and innovation.

The observation of behavior and interpretation of signs is the basis of the methodology of trends, which is determined by perception of events in society. Trend, *tendentia* in Latin, means to strive, to lean or to be attracted to (CALDAS, 2006). In the field of fashion trends are basic guidelines that reflect the desires of the consumer market, that is, the application of events, feelings and sensations that settle in the human community from developments in industrial and consumer products.

The trends in fashion, or so-called short-term trends (less than two years) are ephemeral phenomena of fashion; they are considered reducing uncertainty for the productive chain of fashion and for projecting requirements for the professional practice of the designer. They are signs that foreshadow events of a time, interfering with consumer behavior, directing desires and tastes. They are given in cycles, which means they come up, become popular then decline. Trends identify colors, themes, shapes, fabrics, equipment, principles and elements of fashion designs in a particular season, pointing out possible directions for stylistic that evolve to or get close to.

Li Edelkoort (Edelkoort, 2009) a researcher in the international exploration of trends confirms that trends are not merely temporary, but long-term socio-cultural phenomena; habits ingrained in our DNA that influence how one lives, what one wears, what one eats and, as for the aesthetics changes in the conception that define a moment in time. Through a holistic approach it reflects the current multi-faceted dimension of fashion, not forgetting the economic reality, the political, social developments and technological advances.

Search for fashion trends involves a kind of perception and construction of the look in order to understand the social context in which it occurs, but also to stand back and reflect on this connection about the values that distinguish it, the wishes that are handled and generated by those who live in it. This means that the search of fashion trends is a kind of mutual action indicated by the method, the training and the construction of the researcher's perspective. Many variants of the same story intersect and intertwine, becoming part of the same plot covering an ethnography of the area and having a thread integration of the vectors in a big gear.

A considerable effort is spent to design that cuts across the rapid enlargement of the field of fashion, however most methods of identification are still concentrated on the notes of statistical signs that appear on the surface of events. Trends in the lifestyles, attitudes and culture, with particular emphasis on music, sports, television and film are increasingly used to predict and influence consumer demand for fashion products. Combining traditional methods with new forecasting methods to react more quickly to consumer desires is now a challenge for most brands and designers.

It is clear that a space for observation, analysis and interpretation of signs do not need to filter only the information of large cities and not be prejudiced and conventional. Therefore, before an industry increasingly competitive, the ability to respond quickly to market demands has undermined the traditional instruments of trend forecasting. The research methodology should focus on the consumer, noting that the "new" can emerge from anywhere. The fashion that is "born" from the streets and its urban tribes is increasingly a powerful inspirational energy and very fast as it can appear and disappear in periods of less than one season, which require on the part of designers and fashion brands from a high agility and responsiveness. The use and choice that real people do to represent their identities as well as their own life styles are reputable information predicted, far more than the models that come out of the studios of fashion design in Paris and Milan.

Currently, the Internet appears to be an effective way in the reading of socio-economic trends, since it is a way of potentiating speed and democratizing information and trends in the fashion world. Since the popular sites specializing in fashion to the pages of the different brands and designers, there is a lot of material that allows analyzing trends. They are means almost commonplace and widely used.

However, with the rise of the blogosphere, other forecasting tools and analysis based on the internet have emerged and are a wonderful source for those who want or need to be updated with the latest trends. Dozens of websites cataloging virtually the choices and styles for real people in the streets, allowing a snapshot look on the fashion trends experienced in cities like San Francisco or Rio de Janeiro. Today, this "fashion" spread and there are numerous sites with this type of position, often serving as inspiration for industry professionals. Product managers and buyers of major brands and retail chains also use these online tools to decide correctly on the styles that will be more in vogue in the coming seasons.

Fashion has become a phenomenon of expanded scope and meaning and thus to address trends one should consider the democratization of the symbolic value that the clothes and accessories have and that sophisticated understanding, which is not only the object, but the network of meaning around it was incorporated instead of the consumer. Nevertheless, the identity of the consumer is floating.

In short, it can be stated that if the trends of fashion in the past put their focus on technique, form and material as strategic information and dealt with the abstract meanings as additional elements, today the situation was reversed. These are questions that haunt contemporary life – ecology, coexistence, subjectivity etc. – grouped under the term behavior, which offers the high basic research. Finally, the fashion trends have encouraged the researcher/designer to go to other fields of knowledge.

BRAZILIAN FASHION INDUSTRY

It is important to note that the Brazilian fashion industry is made today by about 30 thousand companies, employing about 1.6 million people. The textile and clothing are becoming more modern. It is the 6th largest textile park in the world with production of 7.2 million pieces per year, in which 92% of this total goes to the internal market (Rech, 2006). Quite interesting and significant statistical data, the market of fashion, nevertheless has many shortcomings, difficulties and needs.

The constant search for product differentiation as a way to avoid the standard types, commodities, which only generates profitability in enterprises with large-scale production, remains a key to the survival of smaller industries. Therefore the productive chain of fashion should be restructured starting with a direct attack on the fragility and fragmentation of small businesses, the prevailing feature in the main productive centers in Brazil. This feature influences, inevitably, the level of productivity and competitive insertion of these organizations in different market niches, since a sustainable competitive advantage is the foundation of industrial performance over the long term average.

Thus, the constant and transverse (all sectors) market trends – permanently attached to macro-cultural trends in which they feed – is transformed into packets of information and strategic tool for development of products and services for the management of innovation processes of branding.

Thus, from these ideas, the research project "Future of the Present: a space for observation, analysis and interpretation of signs" attempts at proposing a new method of research trends in which there is a concern about systemic and complex relation and also that there can be exploring opportunities, capturing and analyzing the signals properly embedded in social behaviors, technological influences, social, economic and even historical factors.

STANDARD REFERENCE

The version used as a reference of study and example in this present article is the trend research project "Future of the Present: an area to analysis and observation of signs" that takes place in Florianopolis at the University of the State of Santa Catarina, in Brazil. The participants of the project get together once a week to report the facts and the happenings researched and analyzed in order to discuss what, in a global and also social context of a certain moment, has the potential to become a trend or not.

The main methodological steps are:

- **Collection of empirical data:** The methodology of the premises of the research participant observation, a method traditionally used by social anthropology, and particularly suitable for the exploration of new markets or targets;
- **Procedures for coding or data analysis:** In a second step, the resulting signals will be subject to socio-semiotic analysis, synthesis of the sociological methods of interpreting data and of semiology, which makes it possible to decipher meanings and representations. The point of this anal-

ysis is a concise overview of trends, styles and sensibilities emerging. It comprises three steps:

- Open coding;
- Axial coding or training and development of the concept;
- Selective coding or modification and integration of the concept;

- **Distribution of the theory:** It is the application of the methodology of the Critical Genetic analysis of the documents to the research process by which you can intuit, confirm and systematize concepts.

METHODS AND MATERIALS

Influences observed and analyzed by trend researchers work as a sort of mirror of the future of modernity, so expressions of the "spirit of the time" advertising features posthumously. Through the study and analysis of social phenomena from the streets and disseminated by the media in general, it makes the exploration of social trends possible, which will later be translated into fashion trends for the industry.

We can see that the observation, analysis and interpretation of signs is what drives the major sectors of the wishes and desires of consumers. However, to analyze them it is necessary to create a certain distance to see such practices in the way they actually present themselves. To do this it is imperative the use of a structured method aiming at organization, validity and alleged impartiality of the information collected through the study of three categories (the category Technology seems implicit in the other categories):

a. Influence of macro-trends;
b. Behavior (male and female);
c. Sectors of reference (male and female).

The method developed by the trend research project "Future of the Present: an area to analysis and observation of signs" to prospection of trends consists of three parts:

1. Preparing for the search (by category);
2. Collection and analysis of data (blogosphere);
3. Analysis of these data or coding (open coding, axial and selective);
4. Delimitation of theories (exploration of fashion trends and their dissemination via the website of the research project - http://www.ceart.udesc.br/futurodopresente).

The research is the longest and most intuitive phase of the process. It is necessary to analyze the most quantity of information as possible, like internet, books, magazines, television, art, cinema and even the streets. The everyday life is one of the most crucial and emphasized sources in the project, since it deals directly with the consumers and their needs, that is, the biggest and most important center of obtainment of information for the behavior section. In this line of thought, one of the most important sources to research are the blogs, where the behavior observation, especially "tribes" like the "emos" or the "indies", are essential to a definition of behavior trends through the internet. The research in the most various areas is fundamental in the determination of new habits and trends, because it enables distinct points of view about the same subject, or even about completely different topics. It is in this phase that the human behavior is also analyzed, because to a more open study that relates life style with fashion trends, there is a need to examine a system that involves everything, from what is consumed to how life is lived or even what is the opinion about certain subjects. During the process of observation and investigation the results obtained from this research are written down in a sketchbook[8], that consists of a notebook with clean pages free of lines or anything that can keep your mind from writing, drawing or figuring out your point of view (Gomes, 2001). It also makes easier for the participants to express their research an analyses of these subjects later on.

In the research area, during the first stage of the project that occurred from August 2007 to August 2008, each researcher should observe and investigate every area that contained any valued information, not taking any section as a basis. After that, it was time to figure out some subjects that got more defined than the others and give them a shape and connect it to a large amount of ideas and considerations crossed by each other. This method of research (Back, 2008), later, got surpassed, due to the accessibility to the huge amount of information and due to the difficulty of grasping everything that was valid into different sectors, taken as a basis.

After noticing this complexity, already in the second stage of the project, from August 2008 to August 2009, the chosen method as a research reference was improved. This improvement occurred the moment that four main and general research areas were defined. They are: Behavior, Reference Sector, Macrotrends and Technology.

In the area of Behavior, the conduct of the human being was studied and analyzed, whether it was a man or a woman, or any social class it was researched in the media, internet, television, magazines, researches or reports that discussed the performance of a person before any conflicting situation or even in his/her everyday life.

As for the Reference Sectors, sources were considered in order to be taken as identifiers of what may become a trend, like fairs, fashion weeks, conferences that have as a subject not necessarily fashion or trends, but themes that are connected or that can be related with fashion subjects or even topics linked to the society performance before the world context.

In the Macrotrends area, sources which take as theme subjects related to what will happen in a far future. It could be about how people will react about consumerism, in what way they will act into the society, how their lifestyle will be, what the weather will be like as well as the environment we live in, whether it is the cities or remote areas of the Planet Earth. How technology will be and how all these areas will interact among one another and if they will be in harmony or not. It is in this area that the Macrotrends act, among other several situations, always having in mind a not so close future.

At last in the Technology area what is studied and considered are the innovations and improvements of products that already exist in the market which, one way or another, can be related to fashion. It can be in a new technological fabric, or in an advanced product or method that makes people's life easier.

As for the second half of the second stage of the project, the need of a sub division of these four sections into two more was essential to a better research. So, the sectors were divided into: Masculine Behavior, Feminine Behavior, Masculine Reference Sector, Feminine Reference Sector and Macrotrends. This happened due to the facts that women and men are a very different public and they cannot be fitted into the same trend or the same line of thinking, by the fact that they can not be studied and compared in the same way. This way the line of research of trends follows subdivided by the masculine and feminine sectors of behavior, masculine and feminine sectors of reference and Macrotrends. The latter was not divided because it is a much wider and subjective area than the others and also possible to be analyzed under the eye of a collective society. The Technology section was extinct, but not abolished from the research. Each researcher must study it into his/her section of his/her target, aiming at making a much more direct and dynamic research intending to make technology more included into a perceived and predetermined context.

Following this information analyses method, the next step is the analysis of the data obtained which consists in the most accurate and complex stage of a trend research. In order to have cohesion in a connected group of trends, a general appreciation of all obtained ideas is needed to define a concept that makes sense and logic in each month. For this cohesion and harmony to occur between every sector a meeting is organized specifically with the purpose to expose all the defined themes, the ones that had stood out the most from the others, attempting to check which concept the team will go for, intending to make all the trends work together in a balanced discourse.

After the conclusion of the chosen themes, a picture that contains images that make references to these themes, in each sector, is developed. The picture tells and sets out the story in a very subjective way what has been deliberated during that month of research and later will be divulged through the website http://www.ceart.udesc.br/futurodopresente. With the picture, an article is elaborated, telling and explaining each trend, also in a very subjective way, but easy enough to get what is being told. The divulgation consists in the availability of the synthetic and studied content, as well the picture and the article, into the project's website. As an example of the aforementioned article, here follows a text about Macrotrends, from June, 2008.

Macrotrends: June 2008

The consumers will tell their own stories, objectifying their own status. Brands will help them and give them the ingredient not only to tell the stories, but experience them. That makes the client loyal to only one brand, formatting a new consumption behavior, creating a new niche into the market for brands, where they should proportionate that the experience dominates the world of prestige symbols.

The experiences proportionate instantaneous gratification, once they are exclusive by their nature: they are cheaper and are more numerous than in the world of status and past symbols. Therefore to be much more attractive, the status symbol needs to have the approval from the others: friends, work colleagues, family.

Slowly, societies are beginning to recognize obtained abilities, "eco questions", and activities without lucrative ways or the number of on-line visitors. Due to its main part not physic (not visible) and with a dark nature, these new ways of prestige run over status stories to proportionate a permanent status to their followers.

The new brands, not well known, capture the audience, the interest and the respect from their public with products that should give the necessary details or the ingredients for a real status story to their costumers: the origin of these products, their singularity, their eco responsibility.

How was the product made? By whom? How did it get to the store? What are the effects on the environment after buying it? Status stories and eco concerning are equally important. Due the consumer's desire to know (and tell others) about a product's origin in increasing, companies are obligated to innovate on the way that they are answering these questions that construct any buying process nowadays.

Participation, creation and the sentence "make it yourself", more than a way of consuming, means even 1001 ways for brands involve their clients (even the participation inside the creation process), including the non avoiding divulgation of the rights and stories about status. If telling and sharing stories means experiences (status), so the trend life caching (the day-by-day stories, told on internet, blogs) fits perfectly. It is at this moment that people pick and store their memories.

As trends are a manifestation of new existent ways to cease to blockade human necessities, we can consider that life caching frees the permanent need to "approach", "have", "achieve". The human beings (moved by the necessity of self-esteem, validation, control, vanity and even immortality) like to collect and save possessions, memories, experiences, in order to create a personal album or just to follow the normal way of life. And, in a system moved by the market, which a lot of consumers are favorable to intangible than tangible, "collect", "save", "show" and "share" experiences are exciting, and, to an extent, necessary.

We hope that consumers (especially the youngest ones) become happy to save and share everything they do, create or experiment, obtaining different status levels during the process. This will give an avalanche of innovating ideas about ways of helping the consumers to experiment this process.

In short:

- *It's not about brands telling its stories to the crowd, but helping individuals telling their own;*
- *It's not about something momentary;*
- *We are talking about consumers, real people, telling stories about their own experiences or acquisitions, in order to achieve status by impulse.*

The consumers do not recommend or defame a brand. The divulgation "mouth to mouth" will have a devastating effect, but it can not be the mainly objective form brands when helping the costumers with their stories.

Therefore, status stores will not substitute status symbols, but will become more and more important, and can compete with some well known and old fashioned status symbols on the crowd. And maybe someday it will turn out to be one of them.

THE IMPORTANCE OF THE BLOGS INTO FASHION

The blogs are more and more frequently being used by the internet users, as well to self promotion or to tell more about economic or fashion subjects. The blogs own the on-line pages. The main reason for its success must be the ephemeral, individualistic and multiple characteristics that capture the users, due to the possibility of self exaltation and self affirmation into an environment that he or she can be transformed in whatever they want. Campos (2008) says that, with the sprouting of blogosphere, each time more trends that emerge from streets and divulged by Internet serve as a powerful inspiring energy for researchers, creators and even personal consumers.

Through the blogs, people from different places in the world can communicate and relate, increasing the number of chances of finding people that think and act the same way as the user and as a matter of fact, also helping to disseminate new styles, thoughts or ideas. This is why blogs are so rich in new information, and so important to the behavior research.

The main reason of the importance of the blogs into the trend research, apart from its easiness and amount of information is due to the fact that in studying a blog people react like who they really are, there are no external interferences in what the person is posting. The poster is determined to expose his/her point of view, or ideas without any social retaliation, that comes from people who think differently. It is important to the trend research because it is possible to analyze the people's behavior more easily and with more freedom to ask questions and make observations, not invading his/her environment and avoiding a distortion of information by the one being studied. The importance of the blogs into the research project "Future of the Present: an area to analysis and observation of signs" is so big that 2/3 of the researches of Behavior, Reference Sectors or Macrotrends come from these blogs.

THE IMPORTANCE OF THE PROJECT INSIDE THE UNIVERSITY OF THE STATE OF SANTA CATARINA

The research project "Future of the Present: an area to analysis and observation of signs" seeks the idea that the words "fashion trends" do not only mean the materialization of the ideas built only into the clothes area, but a study that comes from an observation of social or environmental elements and facts that get the construction of a concept based on a sensibility to capture what we call "spirit of a time". Vincent-Ricard (1989) says that the person who is good radar has instinct, is receptive to all around her, feels in the air the zeitgeist of the time and gives importance to her own ideas without establishing any linking with the product to be manufactured. Also, the project aims the understanding of the importance of the research into the fashion area, therefore existing into a collection of clothes a point of view according to the environment where the clothes will be in, consequently becoming a group of objects with a base of ideas superior from the others that do not speak with a concept.

As already stated in the article, fashion is a total social fact that, in one of its segments, is always seeking what is new and what is valid for what is relevant in a certain time. In order for this to happen the trend research aims at discovering and evaluating what is this "new" or what is more appropriate to a defined period of time (Lynch, 2007).

The project aims at reaching the studious public of the fashion system that searches for the comprehension of the trends through the website. The reason that only the public that understands the concept of fashion is the project's goal, is due to the Brazilian reality that most of the people who work into the fashion area, are not aware of the importance of a fashion trend research specifically to their clients, due to the fast and easier information that comes from abroad and that is ready to go to the runway and windows. Most of the time, the entrepreneur prefers the technological improvement, not giving up of the importance of this fact, than a more elaborated research, that would give more and better results to their business (Norman, 2008).

The right thing would be to teach these people the fundamental importance of the fashion trend research for a collection, but, due to some difficulties found, this point, can not, yet, be discussed and concretized by the Project.

The website of the trend research project "Future of the Present: an area to analysis and observation of signs" (http://www.ceart.udesc.br/futurodopresente) reports trends each month, and is the materialization of ideas and concepts discussed throughout the months and the meetings. Inside the website there is a blog, where each day things are posted of what is researched and already considered valid and suitable for the discourse of the present moment.

FINAL CONSIDERATIONS

As Freitas (2003) asserts, the fashion is an area that claims to be multidisciplinary. This occurs due to external facts, like the sectors researched discussed in this article, that influence and exercise a bigger concept with a trend that is going to be applied into an object that will be manufactured and produced.

Sant'Anna (2003) claims that the prospection over the sources requires an approach that goes over the description, the enumeration and the recognition of the importance of a chosen context. It requires a base of images that used to flow, of appearances that composed, but mainly, of the reception that people gave to it. This way, it is possible to emphasize the importance of sensibility in order to be more sensitive to a trend prospection. The variability of information nowadays and the easy way that it changes is huge in order to make confuse what is not clear yet. The trends research comes as a divisor between all these ideas and as an opener of these compact and heavy concepts.

The aim of this article is to tell the method of work and study into fashion trends, from how a trend begins to exist and the degree of its importance intending to increase the development of the Brazilian textile sector through more capable future professionals that will be aware of the complex theme of such a trend.

It is still valid to consider the importance of the understanding of the concept of fashion and the process of how its trends are developed by the professionals in the market, so there are responsible brands wise enough so that products with quality are launched and the creative skills of stylists and designers is even more sharpened and inspired by the events.

REFERENCES

Back, S. (2008). *Pesquisa de Tendências – um modelo de referência para pesquisa prospectiva* (Unpublished master's thesis). Florianópolis/SC, Brazil: Universidade Federal de Santa Catarina.

Braga, J. (2005). *Reflexões Sobre Moda. São Paulo/SP*. Brazil: Anhembi Morumbi.

Caldas, D. (2006). *Observatório de Sinais: teoria e prática da pesquisa de tendências. Rio de Janeiro/RJ*. Brazil: Senac Rio.

Campos, A. Q., & Rech, S. R. (2008, october). *Futuro do Presente: porque e como pesquisar tendências*. Paper present at 8º. Congresso Brasileiro de Pesquisa & Desenvolvimento em Design – P&D Design, São Paulo/SP, Brazil.

Castilho, K. (2008). *Corpo e Moda: por uma compreensão do contemporâneo. Barueri/SP*. Brazil: Estação das Letras e Cores.

Darnton, R. (2006). *O Grande Massacre de Gatos e Outros Episódios da História Cultural Francesa. Rio de Janeiro/RJ*. Brazil: Graal.

de Oliveira, A. S. C. (2006). *O Estudo das Tendências para o Processo de Design*. (Unpublished graduate's thesis). Programa de Graduação em Design Industrial, UDESC, Florianópolis/SC, Brazil.

Edelkoort (2009). *Lidewij Edelkoort*. Retrieved July 13, 2009, from http:www.edelkoort.com

Erner, G. (2005). *Vítimas da moda? como a criamos e por que a seguimos? São Paulo/SP*. Brazil: Editora Senac.

Feghali, M. (2001). *Senac* (Brazil, R. J., Ed.). Rio de Janeiro: As Engrenagens da Moda.

Freitas, N. K. (2003). Visões Paralelas de Moda e Psicologia do Consumidor. In: QUIRINO, S. F. S. (Ed.), *Modapalavra*, (vol. 2, nº. 2, pp. 112-117). Florianópolis/SC, Brazil: UDESC/CEART, 2003.

Gimeno, J. M. I. (2000). *La Gestión del Diseño en la Empresa*. Madrid, Spain: McGraw-Hill/Interamericana de España.

Gomes, L. V. N. (2001). *Criatividade: Projeto, desenho, produto*. Santa Maria/RS, Brazil: sCHDs.

Kawamura, Y. (2005). *Fashion-ology: an introduction to fashion studies*. New York: Berg.

Lipovetsky, G. (1997). *O Império do Efêmero - a moda e seu destino na sociedade moderna. São Paulo/SP*. Brazil: Companhia das Letras.

Lupatini, M. (2005). *Relatório Setorial Preliminar: setor têxtil e vestuário. FINEP - Rede DPP*. Retrieved May 14, 2005, from www.finep.gov.br/portalDPP/relatório_setorial

Lynch, A., & Strauss, M. (2007). *Changing Fashion: a critical introduction to trend analysis and meaning*. New York: Berg.

Norman, D. A. (2008). *Design Emocional: por que adoramos (ou detestamos) os objetos do dia-a-dia. Rio de Janeiro/RJ*. Brazil: Rocco.

Prochnik, V. (2002). *Estudo da Competitividade de Cadeias Produtivas no Brasil: impactos das zonas de livre comércio*. Cadeia Têxtil e Confecções (Nota técnica final), dez.

Rech, S. R. (2002). *Moda: por um fio de qualidade. Florianópolis/SC*. Brazil: Editora da UDESC.

Rech, S. R. (2006). *Cadeia Produtiva da Moda: um modelo conceitual de análise da competitividade no elo confecção* (Unpublished PhD's thesis). Programa de Pós-Graduação em Engenharia de Produção, UFSC: Florianópolis/SC, Brazil.

Sant'anna, M. R. (2003). A Moda – o desafio de pensá-la além do produto. In: QUIRINO, S. F. S. (Ed.), *Modapalavra* (vol. 2, nº. 2, pp. 72-86). Florianópolis/SC, Brazil: UDESC/CEART.

Sant'anna, M. R. (2007). *Teoria da Moda: sociedade, imagem e consumo. Barueri/SP*. Brazil: Estação das Letras Editora.

Vincent-Ricard, F. (1989). *As Espirais da Moda. Rio de Janeiro/RJ*. Brazil: Paz e Terra.

KEY TERMS AND DEFINITIONS

Trends and Prospects: Basic guidelines that reflect the desire of the consumer market, that is, the application of events, feelings and sensations that settle in the human community from developments in industrial and consumer products.

Production Chain: Can also be termed as Filière a term of French origin and has the meaning of row, that is, a sequence of business activities that lead to a subsequent processing of goods, from raw to finished or consumer designated. It consists of several stages of production interrelated, each with its special characteristics that contribute to the development of the next phase. Globally, the production chain of fashion is a system that has certain singularities covering five links: (a) wiring, (b) weaving, (c) processing, (d) elaboration, (e) market. [...] Across, support services operate at the same time such as mechanical and textile and intermediary services such as equipment suppliers, the production of software, specialized publishing, fashion fairs, advertising agencies and communication studios creation in fashion design and corporate functions (marketing, finance, trade, among others). These phases are interdependent and have numerous links among themselves and with other industrial sectors. The process of textile production is, however, relatively straightforward and independent: the outcome of each stage of production can feed the next step apart from factors such as scale and production technology" (Rech, 2006, p.68).

Product from the Fashion Design: Any "item or service that combines the properties of creation (design and fashion trends), quality (conceptual and physical), wearability, appearance and prices from the wishes and desires of the market segment to which the product is intented" (Rech, 2002, p.37).

Fashion Product: Any goods, material or immaterial which is the aggregate concept of fashion, that is, that introduces itself as the new carrier from the place of its diffusion and consumption until its "packaging". It expresses that differential; the claim that was produced from a creative act (Sant'anna, 2007).

The Fashion and Ethos of Modern, Individualistic Society, Formed In Significant: Articulates the relations between social actors through the appearance and introducing the new category as a hierarchy of meanings (Sant'anna, 2007).

Zeitgeist: Considered like "spirit of time".

Blogosphere: Refers to the collective term that includes blogs and their connections. It is important to emphasize the idea that the blogosphere works as a social phenomenon, while the pages of blogs are essentially web pages. The internet is a network of people, not computers.

Sketchbook: An unlined notebook – preferably ivory paper, hardcover, sewn. "This contract which may also serve as a calendar should be used as a notebook of phrases, sayings, thoughts and ideas for schemes that emerge suddenly. In the case of the student, this book should serve as a booklet in scribbles, doodles, sketches" (GOMES, 2001, p.48).

Chapter 17
Design for Desirability: A Collaborative Innovation–Initiative between New Zealand Design Academia and Industry

Mark Goellner
Massey University, New Zealand

Anders Warell
Lund University, Sweden

Rodney Adank
Massey University, New Zealand

Lyn Garrett
Massey University, New Zealand

Tony Parker
Massey University, New Zealand

ABSTRACT

This chapter outlines an innovative and collaborative design research project that connects New Zealand SME manufacturers with advanced design thinking about affective design. This project was developed and implemented by the centre for affective design research (Affect) at Massey University. Design researchers and five NZ companies have collaborated to foster 'design-for-desirability' thinking and develop capabilities by means of knowledge sharing, enterprise training and individualised projects. This created visionary product concepts utilizing the perceptual product experience framework (Warell, 2008). The chapter provides a novel model for collaboration between industry and academia that focuses on implementing 'design-for-desirability' thinking in SME companies with the aim of improving their international competitiveness. Well-designed, functional products are expected in today's competitive global markets. Gaining success in global markets requires a step beyond this level of usability in order to develop products that are desirable and appeal to the users on emotional, social and intuitive levels.

DOI: 10.4018/978-1-61520-617-9.ch017

INTRODUCTION: ECONOMY AND DESIGN IN NEW ZEALAND

New Zealand (NZ) is a relatively small, industrialized free-market country with 4.3 million people (NationMaster.com, 2009). It is it an entrepreneurial society that is known for its ingenuity and innovation in industry and research. Frederick describes New Zealand as one of the world's most entrepreneurial countries, with more than 17% of all employment in New Zealand being in entrepreneurial companies (Frederick, 2002). In the design sector many New Zealand companies have portrayed this entrepreneurial thinking by producing highly innovative design solutions. This is either done by internal design teams (for example the highly successful company Icebreaker[1]) or by independent design consultancies working alongside manufacturers. Regardless of type these product design teams are generally small in size: one to three designers and rarely more than four designers (Goellner, 2005).

Currently New Zealand manufacturing and design industries are facing issues related to: global competition from the internationalisation of design and engineering services; the manufacturing capabilities from Asian markets offering low-cost products; an increasingly well- informed consumers that expect exceptional products (which they can access globally through the internet); and the increasing customer's awareness and demand for sustainable products and solutions.

In New Zealand the design community and the government is seeing the integration of sustainable solutions and the certification of sustainability as some of the most important aspects for New Zealand industry in order to maintain their "100% pure and green" New Zealand image. Conferences with focus on sustainability have been organised by the Designers Institute of New Zealand (DINZ 2007, 2008). The "Sustainable Business Network" was formed to promote sustainable practice in New Zealand and support businesses on the path to becoming sustainable, and in March 2009, the "Greenlist.co.nz" website was launched as is the world's first online directory of green products and services that are compared against basic principles of sustainability. In addition, the Ministry of Economic Development is working on a sustainability assessment and certification for New Zealand products.

With all these current challenges, one needs to keep in mind that most New Zealand manufacturing companies are small to medium sized by world standards (Darroch, 2002; New Zealand Trade and Enterprise, 2004) and despite their entrepreneurial thinking, exports of products and services to the global market have increased only in recent years (Frederick, 2002; NationMaster.com, 2009). The main issues for NZ manufacturing companies are their geographical remoteness to major markets such as America, Europe or Asia, and their relatively small size, which, through resource scarcity, makes it difficult to compete internationally. At the same time, the absence of domestic markets has also led the small to medium companies in New Zealand to draw knowledge from international networks and "to provide high levels of customisation for their international market customers which in turn drove R&D and innovation practices" (Davenport, 2005).

With the need for sustained economic growth forcing international market orientation, the New Zealand Design Industry Taskforce (Design Industry Taskforce, 2002, 2003) found that NZ businesses do not often recognise the required design input in R&D necessary to succeed internationally. Similar to New Zealand, small-medium enterprise (SME) companies in Europe (France, Germany, Sweden, etc.) also rely on innovation as one of the factors for high growth, but only few of them have significant investment in specific R&D departments (OECD (Organisation for Economic Co-Operation and Development), 2002). Previous studies have acknowledged the need for more sophisticated understanding of business improvement and the pursuit of international competitiveness (Knuckey, 2002) as well as the need for increased utilisation of industrial design (Slappendel, 1996). A key strategy to support design-led innovation thinking is the New Zealand Trade and Enterprise's initiative called "Better by Design", a programme aimed to support NZ companies to integrate design principles right across their business (Better by Design, 2009). This programme helps New Zealand business understand the need to generate innovative solutions and the role of Design in achieving internationally successful and competitive product solutions.

THE "GROWTH INNOVATION PILOT INITIATIVE" PROJECT

In alignment with the "Better by Design" program of advancing New Zealand manufacturing industry's competence in utilising design to gain a competitive advantage in the global marketplace, a government-funded Growth Innovation Pilot Initiative (GIPI) "Industrial Competitiveness through Design for Desirability" was initiated by Massey University's industrial design researchers in 2006.

As globalisation has led to a change in traditional ways of achieving competitive advantages a new foundation of high growth through innovation has been created: intangible assets, which include knowledge, competence and intellectual property. Knowledge acquisition becomes an important field in relation to innovation and high growth for SMEs (Teece, 2006).

In alignment with this new shift, the aim of the GIPI project (2007 to 2009) is to support New Zealand manufacturing companies to advance their design understanding and capabilities to generate products and services that are not only functional, reliable and easy-to-use, but moreover: are innovative, exciting and inspirational; inviting, pleasing and intuitive to use; and address users' socio-cultural and emotional aspirations and needs. The aim is to create products that meet and can surpass worldwide competitors in functionality, usability, appeal, quality and experience. To achieve this the project encompasses activities primarily in the fields of "Knowledge Sharing" and "Enterprise Training".

The project brought together industrial design academics (research associates) and representatives from five New Zealand design and manufacturing companies, from the following industries: radio communication solutions, outdoor apparel / equipment, injection moulding development, animal management systems and kitchen appliances. The GIPI project has been conducted through a newly founded research centre: Affect – the Centre for Affective Design Research *(www.affectdesignresearch.com)* at the College of Creative Arts (Massey University). This research centre focuses on research on affective issues in industrial/ product design and is lead by several experienced industrial design academics and supported by multi-disciplinary team of research colleagues, external associates and administrative staff. Affect currently runs several projects exploring the synergies of cross-collaboration and affective design aspects by means of industry-projects, academic research and graduate and under-graduate study projects. The GIPI project is designed to be able to draw upon the support of a multi-disciplinary team, including experts in small-medium enterprise (SME) research, engineering and ergonomics research, as well as marketing, ideation and usability-testing consultancies in order to support the participating companies.

In order to met the aim of the project and the needs of the companies the researchers focused on meeting the needs of these companies through focusing on three of the most critical areas for product success in today's marketplace:

- **Innovation:** The project will help companies identify weak areas in products, activities and strategies. With the objective to establish processes and mindsets in companies that are flexible and responsive to

ever changing market conditions in order to foster innovation.
- **Aspiration:** People increasingly buy products because of their expressive virtues, such as desires, personal values, aspirations and emotions. This project will develop and implement methods and processes that help companies define who they are designing for, capture the desires of users and support the design for the intended market.
- **Experience:** A major shift from the product being the focus of a development, to the experience (i.e. use, interaction and enjoyment with the product) being the point of interest has occurred in the recent time. This project aims to facilitate the creation of enjoyable processes of use and product interaction by leveraging the companies' competency in design for increased "use-worthiness" in products.

THE GIPI PROJECT IN CONTEXT

Collaborative design projects between academia and industry are certainly not a new phenomenon. On the contrary, it has been extensively utilised for a long time internationally for example in design and engineering programmes. The value of such collaborations in applied subjects such as design has long been noted (Evans, 2005). Benefits of such collaborations include well-prepared graduates that can articulate requirements and constraints, prioritise opportunities and justify their decisions.

However, as noted by Nelkin *et al.* (Nelkin, 1987) collaborative projects focussing on research in design is not as established as it is in other fields such as technology, and has not received much scholarly attention. According to Almquist *et al.* (Almquist J., 2007), "few models of successful collaboration exist and there is little knowledge about how such collaboration can work, what could go wrong and what safeguards might be implemented to increase the chances of success". Almquist *et al.* discuss a number of aspects of importance for successful industry-academia collaboration:

- It is beneficial if neither group, on its own, is equipped to answer the research challenge
- Previous history of mutual interests between the involved parties is beneficial
- Good relationships on individual, social and organisational levels is crucial to support communication and the development of mutual valuation, respect, and even trust
- Both groups needs to share fundamental goals

Reporting from a design research collaboration project focussing on the development of a mobile clinical collaboration tool, Battarbee *et al.* (Battarbee, 2005) argue that academic researchers "can bring new experiences and insights to a company, but academics also benefit from applying their knowledge and responding to the challenges of the corporate design world". The project centred around three key ideas:

- To enable design collaboration and proceed to concept creation
- To link with research interests as well as corporate interests and projects
- To approach the study topic with a broad, open focus to allow new development

The creation of organisational mechanisms helps the project to succeed, why Almquist *et al.* (2007) stress the importance of giving sufficient thought to how the collaboration could happen. Therefore, the establishment of an organisational (sub) unit designed to mediate the collaboration is suggested.

In the collaboration described in this paper, this entity had been previously created in the form of the Affect research centre. The existence of the centre proved to be key in establishing, focussing and pushing the collaboration forward. Part of reason is that it provided the project a defined 'home base', rather than living a life as, yet another, task amongst the numerous teaching and administrative duties of everyday academic work. Furthermore, the formation of Affect created a brand name, which could be effectively used in external as well as internal networking and marketing of the project. This was helpful in focussing the participants' attention and efforts as everyday tasks at the academic and industrial organisations competed for their attention.

A particular characteristic of the GIPI project described here, is that it established an academic-industry collaboration involving not only one partner of each kind, but is rather a multi-corporate collaboration initiative. As such, it is envisioned that this will allow the industrial manufacturing companies to meet and interact on neutral ground and to freely participate in recent and relevant research in affective design, and to engage in product development and strategic design projects of direct benefit for the companies at a high level of gain but at a comparatively low level of effort. Thus, industrial networks can be created, knowledge sharing will be established, and research opportunities will arise for the mutual benefit of industry and the Affect research centre.

DESIGN FOR DESIRABILITY

The central proposition of this paper is that the concept of 'desirability' is fundamental for creating a product/service that will subsequently be positively perceived and become successful in relevant markets. Nowadays, it's all about offering products and services that are desirable. They need to be appealing - emotionally, socially and intuitively. Function and usability are still key issues of good industrial design but unless the product is instantly appealing and continues to appeal, chances for success are slim.

The term 'desirability' is used in this research to bring the focus to the perceived benefits of the product or service. At the core of the concept of desirability, are perceptions of the stakeholders, who may be internal or external. Internal may be senior managers, designers, engineers, marketers, purchasers, sales or production. External stakeholders may be wholesalers, retailers, customers and end users. It is recognised that given such a wide range of stakeholders, the interpretation of desirability will be wide ranging. Hence, the study focuses on desirability aspects related to factors that may be influenced by industrial design in whole or in part.

The term desirability has been used in product design related research, though not extensively. Benedek and Trish (Benedek, 2002) describe 'desirable' as "worth having or seeking, as by being useful, advantageous, or pleasing". Furthermore, they define 'desirability' as: "fun, engaging, natural, enjoyable, valuable, essential, suitable, worthwhile, beneficial, likeable, in demand, amusing, and appealing." Thus, the standpoint taken in this chapter is that a higher level of desirability in products must be achieved in order to lead to a more desirable product offer that is preferred by customers, and hence, a higher level of success in the marketplace. Inherently, the role of industrial design is thus not to focus only on functionality and usability, but instead to address the entire experience of the product, including the use, attitudes, perception, emotion, ideological and social aspects, that contribute to desirability. This leads us to the adoption of an alternative definition of design in this context, as: "The creation of desirable, meaningful and relevant experiences for people" (A. Warell, Adank R., Garrett, L., Goellner, M., Parker, T., 2008).

It is well known that design contributes to the experience of products. For example, Givechi and Velasquez (Givechi, 2004) showed that product/industrial design is capable of eliciting positive reactions such as joy, inspiration and achievement. Desmet (Desmet, 2007) constituted nine sources of product-evoked emotions into a framework for designers and researchers to discover and define concerns and stimuli for those emotions. In addition, Mano and Oliver (Mano, 1993) emphasize the connection between product satisfaction and product-elicited emotions, highlighting the importance of the subjective response for a positive product satisfaction. They report that the dimension of 'hedonic' or 'aesthetic' performance is one of the two major dimensions of product relevance; the other being the utilitarian performance.

In the field of affective ergonomics, Jordan (Jordan, 1998) argued for the importance of creating products that are not only useful and usable, but also pleasurable to use. He categorised these product-related pleasure aspects into four categories: physiological, psychological, sociological and ideological pleasure. He found that the properties of products that are salient in terms of influencing the level of pleasure/displeasure with a product include aesthetics, apart from traditional aspects such as features, usability, performance and reliability. Seva *et al.* (Seva, Duh & Helander, 2007) found that affect created by industrial design strongly influence purchase decision. Authors in a number of other research areas, from psychology (Norman, 2004) to design have offered a variety of viewpoints which provide insights into the complex nature of product experience and its relation to, e.g., meaning (Vihma, 1995), formal aesthetics (Mueller, 2001; Warell, 2001), and emotions (Desmet, 2002). Clearly, desirability is created through a range of aspects related to the user's experience of the product.

THE NOTION OF 'DESIRABILITY' WITHIN COMPANIES

In order to explore the notion of 'design for desirability' of the companies within this collaborative innovation-initiative, initial interviews, followed by workshops, have been conducted, as described in the following section.

The Initial Interviews

At the first stage in the GIPI project, each company was initially interviewed to explore the company's internal perceptions of desirability, to assess their current design level and consequently to identify the necessary and individualised projects that would provide the companies with the support to expand their competence in "designing for desirability" (A. Warell, Adank R., Garrett, L., Goellner, M., Parker, T., 2008).

The companies participating in the study represent significant NZ exporters to the global market, operating in the business-to-business as well as in the consumer markets. The method used in the study was the qualitative research interview (Kvale, 1996). A semi-structured interview was conducted at each of the companies' premises, which took approx. 2 hrs. The interviews were carried out together with 2-3 researchers and 2-3 representatives from the company, consisting mainly of group and marketing managers, design directors, industrial designers and design engineers. The interviews were recorded and transcribed for later analysis.

For the interviews no definition of 'desirability' was presented, instead the views and experiences of the respondents formed the basis for the qualitative data collected. The reason for this is that it was essential to allow the participants to explore the concept of desirability in their own terms with regards to the particular context of their own company. Thus, taking a grounded theory approach, all interpretations of what 'desirability' may be perceived to be were seen as relevant and valuable.

The Company Profiles

The findings about the companies' understanding of desirability were summarised as follows (see also Warell, A., Adank R., Garrett, L., Goellner, M., Parker, T., 2008):

Company 1 operates in a conservative market of professional radio communication. The company's idea of desirability focuses on meeting customer requirements with respect to development and manufacturing processes. A challenge in understating their user's requirements is that the customer is not usually the end user. The need for improved focus on end user preferences and requirements is recognised, and despite that industrial design is still a small activity, it is recognised and currently progressed within the company.

Company 2 develops products for animal management. They have a good understanding of their market situation and what design can do to improve the perception of their products. They understand the desirability factors from the viewpoints of internal stakeholders, retailers and customers, with each of them having equal importance to them. Historically, the added value of design to the product is seen as minimal apart from improved appearance. However, they increasingly engage in a deeper and more frequent way with design thinking.

Company 3 operates on the business-to-business market offering plastic materials, tooling and manufacturing solutions. They see desirability of their business model as being one that offers manufacturing and knowledge brokering in their market, with the competence to provide their customers with unique solutions. Desirability focuses on cost, function and durability.

Company 4 primarily develops consumer products in the outdoor apparel and equipment market. They have a thorough understanding of the end user and well-developed processes for elicitation of end user perceptions and requirements, using e.g. lead users for testing. They operate on a well-understood market and recognise that many customers have a strong emotional element in their purchase decision.

Company 5 is a leading developer of kitchen appliances, which can be called a design-led company that introduced several innovative new products that pushed the envelope of the existing appliance solutions on the market. They have a thorough understanding of their design process and end-users. They employ multidisciplinary processes to identify end-user perception and requirements and are open-minded for new design-driven approaches to discover new territory. They also recognise that their customers have a strong emotional element in their purchase decision.

Based on these findings, several projects and workshops were developed to achieve the goal of supporting the companies in advancing their design understanding and capability to generate desirable products and services. A series of workshops were conducted and in addition several research, postgraduate, as well as undergraduate design projects have been run to explore the opportunities of design for desirability thinking for each of the companies. In the following, a short overview of the foci of the workshops will be provided and two of the master of design projects will be presented in more detail.

THE COLLABORATIVE WORKSHOPS

Three workshops were conducted throughout the project to introduce, discuss and develop approaches and goals for the project, in collaboration with industrial partners.

The Initial Workshop

Following the interviews, an initial workshop with Affect researchers and representatives of each company was held at Massey University in Wellington at the end of 2007. The workshop focused on presenting identified project opportunities for each company, which were then discussed and the level and scope refined for each company. As hoped for, at this workshop the GIPI partner companies actively engaged in an open knowledge exchange about their product development experiences. As envisioned this positive side affect likely occurred due to the neutral ground provided by the University for the

Figure 1. The perceptual product experience (PPE) framework (based on Warell, 2008)

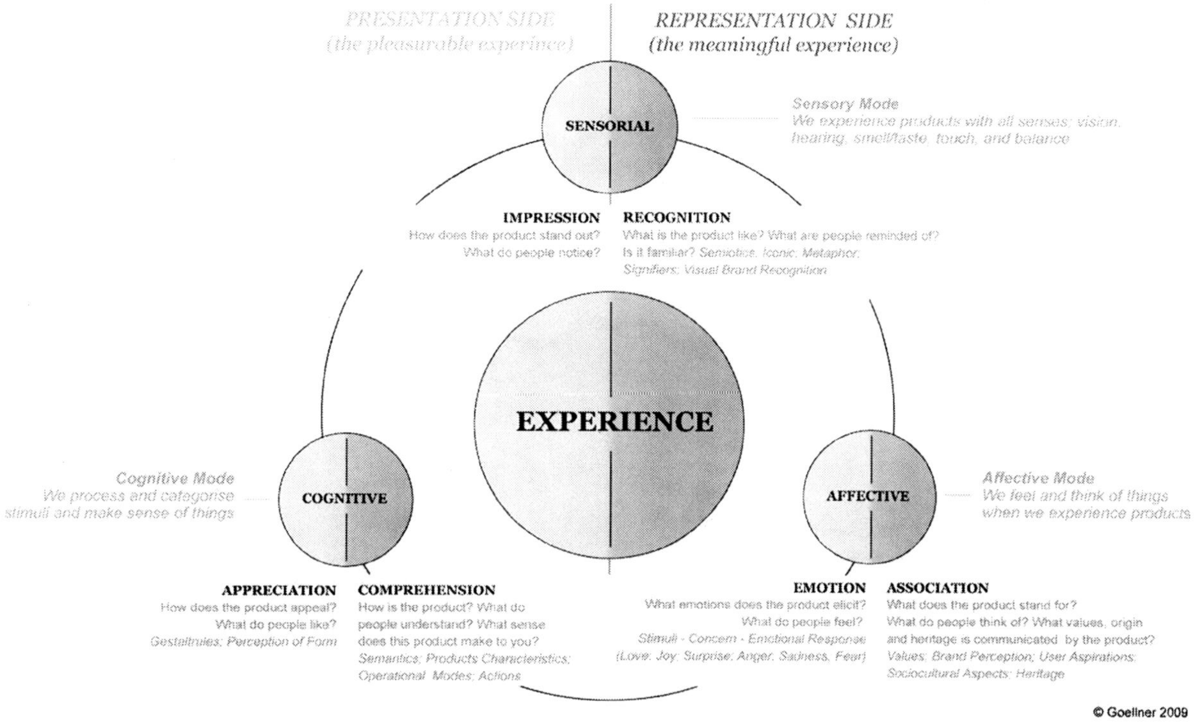

GIPI project. The participating companies, while being protected by mutual confidentiality agreements, met in a neutral environment that created the opportunity for the companies to interact in a non-competitive environment.

The Perceptual Product Experience (PPE) Workshop

The second workshop was conducted within the premises of one of the company partners in May 2008, as requested and agreed by all other companies after trust had been established in the first workshop. The idea was to provide the participating companies with the opportunity to view the R&D and production facilities of the hosting company and share insights and knowledge in a relaxed way. The focus of the second workshop was on product experience, introducing the Perceptual Product Experience (PPE) framework (A. Warell, 2008), which is based on two major elements:

a. **Perception:** the process of becoming aware of physical objects / phenomena through our senses.
b. **Experience:** the entire set of effects that is elicited by the interaction between the user and a product, including experiences related to aesthetics, meaning and emotions.

The PPE framework is designed to allow users to focus on perceived, non-instrumental aspects of product interaction in order to understand, map, organise and analyse possible user experiences. The PPE framework (see Figure 1) has three core modes that help understanding how people experience products:

Figure 2. The perceptual product experience (PPE) worksheets

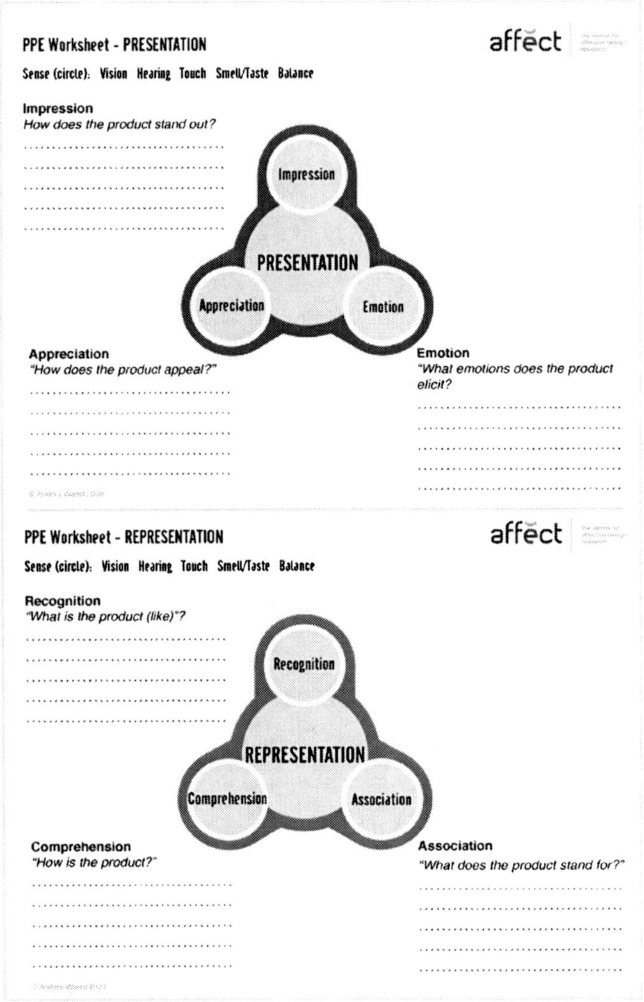

1. Sensory mode (i.e. how we experience products with all senses)
2. Cognitive mode (i.e. how we process stimuli)
3. Affective mode (i.e. and how we feel and think of things)

These three modes can be analysed from two different perspectives:

a. The presentation side (consisting of the experience modes of impression, appreciation and emotion)
b. The representation side (consisting of the experience modes of recognition, comprehension and association)

In the workshop this PPE framework was demonstrated and then tested by the companies. Using worksheets (see Figure 2), participants analysed one of each other's current products, followed by a presentation to all participants.

Figure 3. Products reviewed in the 3rd workshop: Phil & Teds Vibe Stroller; Fidlock magnetic closures; Fiskars pruner; APPLE IPod Touch and Nano; VAUDE Versametric Backpack

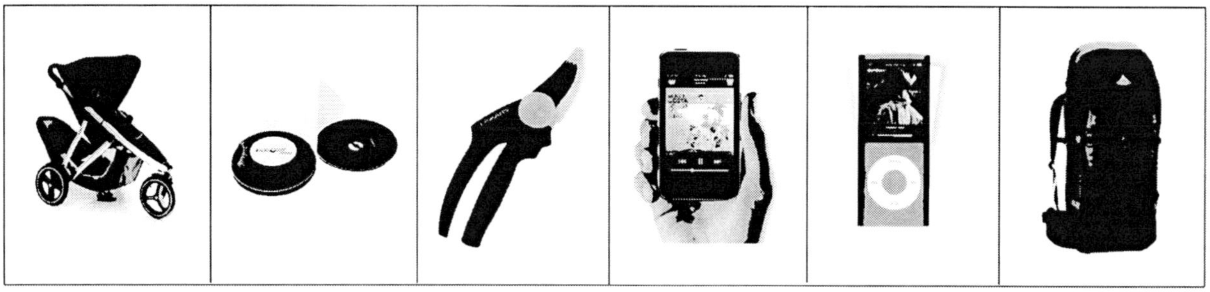

The Emotional Design Workshop

The third workshop was conducted at the premises of Affect in November 2008. The focus of this workshop was the introduction of several design tools/methods to explore the user-product experiences in more depth with focus on emotions and included: the "Nine Sources of Product-Evoked Emotions" (Desmet, 2007) and the "Product Personality Assignment" (Jordan, 2002). In addition, the "Sensory Experience Assessment" (Adank, 2008) was demonstrated as an applied exercise to analyse a variety of products recently released to the market (see Figure 3). This technique focuses a design team on the sensory experience and interaction through consideration of the intimacy of sensation. The Sensory Experience Assessment technique focuses a design team on the sensory experience and interaction through consideration of the intimacy of the sensation. The purpose of this technique is to provide an immersion experience for members of a development team (e.g. designers, engineers and marketers) focussing their attention on the intimacy of sensory experience through subjective proximal interactions with an existing product. This technique consists of the following four stages:

1. Pre-use assessment (product examination prior to use)
2. In-use assessment (product examination during use)
3. Analysis of the outcomes
4. Follow-up idea generation

This method was applied in the exploration of a variety of products: Phil & Teds Vibe Stroller; FIDLOCK magnetic closures; Fiskars pruner; APPLE IPod Touch and Nano; VAUDE Versametric Backpack (see Figure 3). The investigated products were selected on the basis of providing a high level of product novelty, such as: a new aesthetic style; innovative magnetic locking mechanism; ergonomic force reduction; multi-touch screen; shake-to-shuffle technology and material fusions.

The workshop was conducted with support of several industrial design students of Massey University and provided a rich discussion around each of the products with regards to its sensual perception and emotional values. The workshop provided the companies with "enterprise training" in utilizing new tools and methods that they can apply for their design development, and thereby opening new avenues for integrating design for desirability in their product development.

The fourth, and final workshop is currently being planned and will be held during November 2009. The workshop aims to broaden the companies understanding about design for experience by collaborating with cross-disciplinary fields that, to a great extent, deal with design for experience, such as the movie industry.

Figure 4. Scenario sketches of current installation practices of permanent electric fences in New Zealand (McKinley, 2009)

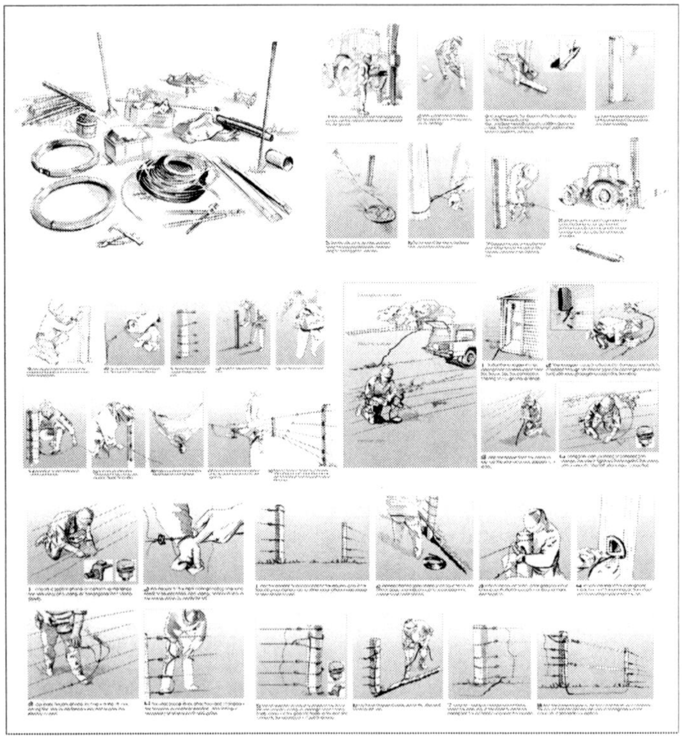

THE MASTER OF DESIGN PROJECTS

Following the initial interview and the first workshop, two Master of Design (MDes) projects were initiated to explore how "Design for Desirability" thinking can be applied to the product range of two of the companies, as part of the GIPI project. Both MDes projects utilized the PPE framework (A. Warell, 2008) to analyse the existing products of the respective companies and to direct their product development towards an experientially rich and desirable design outcome, beyond the pragmatic level of pure functionality and usability.

Desirable Electric Fencing (MDes Project 1)

The first MDes project dealt with the development of a desirable electric fencing system for livestock focusing primarily on the European lifestyle (recreational) farmer market. The new fence concept had to be desirable for the customer not only from the point of purchase, but more importantly through its installation and operation. An innovative product was envisioned, that would give the customer the traditional satisfaction of constructing their fence themselves, but without the necessity of costly machinery. Research interviews with end-users, professional fencing contractors and observations of fence installations were conducted and visualised in scenarios (see Figure 4). This research revealed that the new system had to be intuitive and physically less arduous to install, easy to operate and aesthetically pleasing within a lifestyle farm environment.

The research findings were categorised into detailed 'performance' and 'experience' criteria that were used to drive design development. Subsequent to initial research and generation of

Figure 5. Design development sketches (McKinley, 2009)

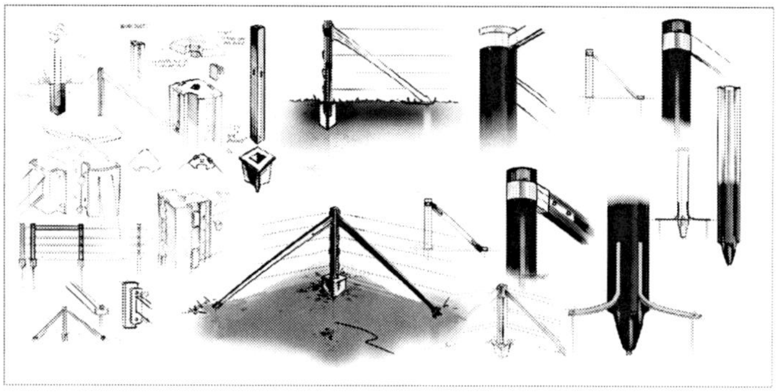

design criteria, a structured concept generation, prototype development and iterative design development phase followed. This was done in close correspondence with company representative being involved in the GIPI project and several focus groups with the company's engineering designers to explain and discuss the performance and experience aspects related to the affectively rich design. Following an initial prototype testing to assure that functionality (i.e. stability in different soils) and the desired experience (i.e. easy to install without heavy machinery) were met, the design development focused on creating a visual expression that elicited the high-quality value of traditional fences, while also being visually pleasing within the lifestyle farm environment. Iterative design development using sketching (see Figure 5) and prototyping was utilised, while the use of non-traditional materials (polymers rather than wood) was discussed with the company regarding manufacturing, but also regarding the increasing awareness of sustainable products in New Zealand and overseas markets.

The design development in conjunction with the researchers of the GIPI project, who supervised the student, lead to a final design solution that was presented to the company's global sales representatives at their annual meeting in New Zealand. The overwhelmingly positive feedback from this meeting has prompted the company to develop the product concept to market over the next years, leading to patent all viable design ideas, employing the master graduate for this process.

Future New Zealand Police Radio Communication (MDes Project 2)

The second MDes project focused on developing a future-forecasted concept for a police radio communication tool for the New Zealand police (Ellis, 2009). Using the PPE framework (Warell, 2008) to analyse current radio products, speculative scenario planning (Schoemaker, 1995) forecasting future requirements of police radios, as well as end-user investigations (i.e. interviews, observations and concept evaluation with New Zealand police officers), the study was provided with a comprehensive research background leading to a realistic design concept that pushed the envelope of the existing conservative market of professional radio communication in which the company operates.

Based on the initial police interviews (in-office) and "ride-along" police observations the master student identified future police radio requirements, revealed relevant desirability aspects of police officers, and developed several possible future scenarios. Subsequently, the scenarios were developed into an explorative approach utilising four 'product streams', consisting of an iconic vehicle, a cell phone and one concept matching

Figure 6. Four product streams representing the associative vehicle and cell-phones and the design concepts (Ellis, 2009)

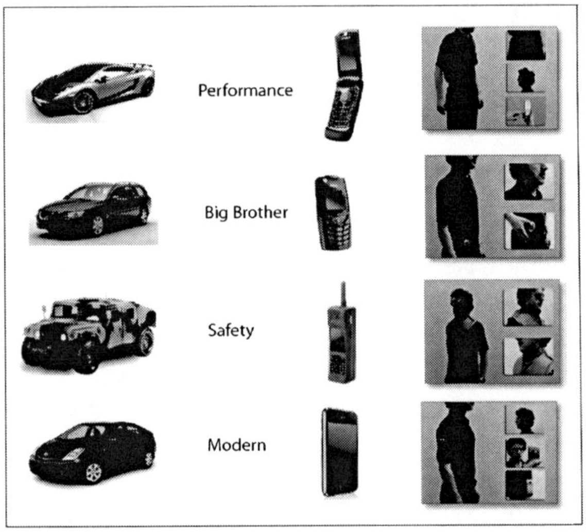

a future scenario of the NZ police work in 2018 (see Figure 6). The purpose of this work was to develop new opportunities for police radios and to capture the basic 'nature' of each product scenario with familiar products (i.e. cell-phone and cars). This method supported the understanding of the associative design requirements for the end-user: "what the product stands for" (Warell, 2008). Each product in a stream was selected based on the similarities in the product messages being communicated. Cars were chosen because they are heavily designed to embody iconic signs and branding (for example, the "Hummer" embodies ideas of strength and robustness), while the cell-phones were chosen due to similarity in design and functionality to radios.

Four scenario-based concepts were developed:

1. PERFORMANCE: A performance driven multi-functional hand-held device (*performance driven; consumer product aesthetic; sleek; advanced technology*)
2. BIG BROTHER: A voice-driven device providing permanent contact to a central unit *(high usability; practical; safe; reliable; conservative)*
3. SAFETY: A bullet-proof vest incorporating all radio functions *(tough; robust; powerful; rugged; reliable)*
4. MODERN: A face-recognition device for quick suspect identification *(dynamic; innovative; state-of-the-art; intuitive; modern)*

These four product streams were subsequently used in the user investigation involving interviews with twelve police officers. The officers were asked to elaborate on which of the vehicles and cell-phones they would choose to use for their job (product association exercise). The officers were then asked to highlight their positive and negative associations of each one in context to their job. The following qualities were identified as desirable for the NZ Police:

- Vehicle: High level of handling and performance; Durability; Robustness; Reliability; Utility; Responsible use of resources
- Cell-Phone: Easy to understand interface; Robustness; Reliability; Utility; Cost effectiveness

Following this product association exploration, officers were shown a simplified physical model of the concept (i.e. size, form, weight and texture) and a short video clip that explained the scenario-based concepts including their context and functionality. Consistently created video representations were chosen in order to present each scenario concept in the same way in order to avoid any subjective comments by the researcher presenting the concepts himself.

The findings indicated that the favourite concept was the "performance" driven multi-functional hand-held device. The police officers were very receptive to the weight, shape and form of the concept. Ellis (2009) discovered that officers enjoyed the additional functionality of the touch-screen interface and the thumbprint identification

Figure 7. Final design concept of the future-forecasted radio communication device for the police (Ellis, 2009)

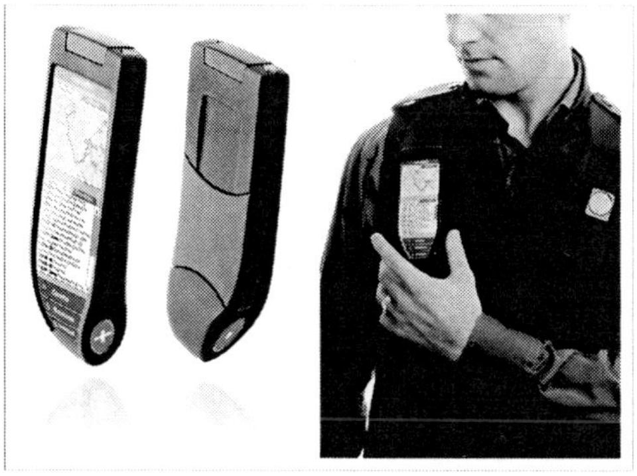

tool, however there were doubts regarding the perceived robustness and appropriateness of the touch-screen interface. The officers suggested the thumbprint identification could not only be used to identify the radio user, but also to identify potential suspects. The inclusion of a camera (still and video) was discussed as being favourable for collection and storage of evidential information. These findings, as well as the results from the ride-along observations, led to performance and experience criteria which informed the development of the final design concept (see Figure 7). Following iterative evaluation and consultation, the final design was presented to the R&D and marketing team of the company and resulted in an overwhelming interest in the design outcome and the 'design for desirability' driven approach employed in the project. The product concept is currently being followed up through further discussion between the company and the MDes graduate.

Both Master of Design projects provided a thorough insight and deep understanding of the study area for the companies. The design concepts developed, one being more applied and one being more conceptual, provided the companies with a demonstration of how a 'desirability' driven project, based on in-depth understanding of user perceptions by employing the PPE framework, can provide an innovative and affectively attractive product vision.

In addition to these research projects interaction with these companies has resulted in further meaningful engagement with design. Both companies have gained a deeper understanding of the subtleties of affective design and aspects of developing desirability in their products. This has resulted in a request for further research and assistance in establishing strategic company positions in relation to ideological and strategic positions on design with regards to company design philosophy. In addition, an annual design scholarship for the industrial design programme at Massey University has been established by one of the companies to support research between industrial design and them.

DISCUSSION: FUTURE TRENDS

In-depth, applied research with focus on the often intangible, yet highly important aspects of design for desirability, as demonstrated in these two Master projects, is often neglected during design and development in small and medium-sized companies prevalent in New Zealand. This is likely to be attributed to the demands of everyday projects that are challenged by an increasing product complexity, multidisciplinary teamwork and time pressure (Earl, Eckert, & Johnson, 2004; Freudenthal, 1999; Roozenburg & Eekels, 1995; von Stamm, 2003).

Such user-focused research, exploring the perceptions of desirability of the product, is often referred to as a 'blue-sky' design project, which is often difficult to conduct in SME environments due to time and cost constraints. However, these desirability-focused and in-depth user-focused design developments are highly important for companies to create products that are desirable and increase their chance of success in the competitive international market. As discussed by Warell *et al.* (A. Warell, Adank R., Garrett, L., Goellner,

M., Parker, T, 2008), the concept of desirability is multi-facetted and elusive in nature. This is also a reason why companies struggle to define and to achieve specific objectives with respect to desirability. The level of understanding of desirability, as well as the ability to engage with designing for desirability, varies across the companies portrayed in this chapter. It is also important for managers to realise that design for desirability is complex and interactive in nature and requires a holistic approach, which spans across company activities in order to be effective. For this reason, it is specifically important with respect to industrial design to ensure effective interaction with other disciplines involved in new product development to assure this holistic approach. From a management point-of-view, the development of desirability should be driven by strategic goals and be informed by a thorough understanding of target markets and end users.

As demonstrated in this chapter, the collaborative innovation-driven initiative between academia and industry employed shows great potential as an appropriate method to foster advanced thinking and implementation of design for desirability in companies. As indicated, SMEs in New Zealand and overseas largely lack the knowledge and resources to advance this type of thinking and to conduct desirability-focused design developments that provide a holistic understanding of user aspirations and desires. Academia has the potential to offer the advanced knowledge required to create innovative products/services in today's global market (Teece, 2006). Moreover, academia also has the capacity to conduct such projects in collaboration with industry in order to foster the companies' potential to create innovative, inspirational and desirable products for people. As indicated in this chapter, multiple avenues need to be pursued to achieve a holistic knowledge transfer and enterprise training with regards to design for desirability. These initiatives include: initial analysis of the companies' knowledge level, individual project definitions, collaborative workshops and conducting company specific design projects that present exemplary design thinking to advance the understanding of 'Design for Desirability' in relation to the companies product area.

The time frame for the project has allowed relationships and trust to build between the participants and generated understanding of what academic design research can offer industry. This has lead to further projects initiated at the request of individual companies. In addition new projects that have introduced other design disciplines to the companies have been proposed and undertaken. This provides a rich texture for design research and innovative possibilities for the companies. Projects that now include Fashion, Spatial, Business and Industrial design collaboration are in progress. From the position of the Centre for Affective Design Research this has stimulated research activity within the College of Creative Arts and been instrumental in achieving cross discipline and cross institute interaction and engagement with New Zealand industry.

CONCLUSION

The collaborative innovation-driven initiative described in this chapter shows great potential as a successful collaboration model between academia and the manufacturing industry. While SMEs are pressured by the need to create innovative products to succeed on international markets, academic environments possess the knowledge and capabilities of supporting SMEs in achieving this by knowledge sharing, enterprise training and individualised research projects. Moreover, by establishing a neutral ground, in this case in the form of the Affect research centre, this industry-academia collaboration becomes a multi-corporate collaboration initiative, allowing industrial networks to be created, knowledge sharing to be established, and research opportunities to rise for the mutual benefit of industry and "Affect – The Centre for Affective Design Research" at Massey University.

This chapter describes how the advancement of desirability through individual design projects is pursued, but also how training of companies' understanding of new approaches and methods to focus on the whole product experience in their product development processes can be achieved in a mutual collaboration between academia and industry. It outlines a novel approach consisting of "knowledge sharing", "enterprise training" and "individualized research projects" to foster collaboration between manufacturing industry in order to advance companies' thinking and capabilities in designing for desirability. As presented, multiple avenues were pursued to assure a holistic knowledge transfer and enterprise training with regards to design for desirability, such as:

a. Initial analysis of the companies' knowledge level regarding "Design for Desirability" thinking and mindsets
b. Identification of company specific projects
c. Collaborative workshops involving different manufacturing companies and academic researchers
d. Conduction of the specific design projects that foster and advance design for desirability thinking and approaches together with the companies

Initial outcomes indicate that newly developed knowledge and capabilities with regards to design for desirability thinking that goes beyond functionality and usability could be established within the companies. Particularly, the individual projects (MDes projects) demonstrated the companies' commitment to advance their thinking to foster product success in international markets by focussing on emotion, aspiration and experience in their product design processes, which represents a radically different approach compared to current practice.

As this collaboration is still ongoing and the workshops and most individual projects have just recently been completed, it would be too early to draw final conclusions about the hard-to-measure long-term consequences of this initiative for the companies (e.g. increased brand recognition; growing success in international markets; move to design-led product development and innovation processes, etc.). However, the appreciation of the companies for the opportunity to engage in this advanced design for desirability thinking, their commitment to develop the individual masters projects further and their desire for further collaboration with academia in the future, indicate that this collaborative innovation-initiative is likely to contribute to change and advancement in the companies' design for desirability thinking with regards to the development of new products. The future will show to which extent the products that are currently in development, and hopefully have been influenced by this 'Design for Desirability' initiative, will actually be successful in the global market.

With regards to any issues of such collaborative innovation-initiatives, the authors would like to highlight that such large and multi-faceted collaborative projects are naturally complex. They do not only demand the dedication and commitment of the research team to sustain the project, but also require the resources of administration, time and funding, as well as an open-mindedness of everyone involved, in order to achieve the goal of advancing design understanding and capabilities to generate desirable products. However, the authors are hopeful that more of these collaborative innovation-initiatives will occur in the future, not only to support industry in future-focussed design thinking, but also to advance the design research by applied projects that are relevant to both industry and academia.

ACKNOWLEDGMENT

The authors would like to thank the New Zealand Tertiary Education Commission for the funding that made this Growth Innovation Pilot Initiative possible, and Matt McKinley and Frazer Ellis for their support to publish their master projects in this chapter. Furthermore, the support provided by Massey University and Affect - the Centre for Affective Design Research, is gratefully acknowledged.

REFERENCES

Adank, R., & Warell, A. (2008). Five Senses Testing: Assessing and Predicting Sensory Experience of Product Design. In K. M. Desmet P., van Erp J. (Ed.), *Design & Emotion Moves*. Cambridge, UK: Cambridge Scholars Publishing.

Almquist, J. M. S., Kelly C., Bromberg J., & Montori V. M. (2007). *Academia-industry collaboration as an emerging model of design research: An examination of the Steelcase–Mayo Clinic case.* Paper presented at the The International Conference of International Association of Societies of Design Research, Hong Kong.

Battarbee, K., Mattelmäki, T., Ylirisku, S., Koskel, H., Soosalu, M., Salo, H., & Allén, M. (2005, September 22-24). *Looking beyond the product: Design research in industrial and aceademin collaboration.* Paper presented at the Joining Forces, University of Art and Design Helsinki.

Benedek, J., & Miner, T. (2002). *Measuring Desirability: New methods for evaluating desirability in a usability lab setting. A paper on the Desirability Toolkit.* Paper presented at the Proceedings of the Usability Professionals Association Conference.

Better by Design (Producer). (2009, March). *Better by Design Title Page* (www.betterbydesign.org.nz) *Podcast*. Retrieved from www.betterbydesign.org.nz.

Darroch, J. (2002). *Innovation and Knowledge Management.* (Unpublished PhD Thesis). University of Otago, Dunedin.

Davenport, S. (2005). Exploring the role of proximity in SME knowledge-acquisition. *Research Policy*, *35*(5), 189–216.

Design Industry Taskforce. (2002). *integrate! A critical look at the interface between business and design.* Wellington, New Zealand: New Zealand Trade and Enterprise.

Design Industry Taskforce. (2003). *Success by Design – A Report and Strategic Plan.* Wellington, New Zealand: New Zealand Trade and Enterprise.

Designers Institute of New Zealand (DINZ). (2007). The Business of Sustainable Design. In *Proceedings of the DINZ National Conference*. Auckland.

Designers Institute of New Zealand (DINZ). (2008). Sustainability and Imminent Change. In *Proceedings of the DINZ National Conference*. Auckland.

Desmet, P. M. A. (2002). *Designing Emotions*. Delft (NL): Technical University of Delft.

Desmet, P. M. A. (2007). *Nine Sources of Product Emotion.* Paper presented at the International Association of Societies of Design Research IASDR07: Emerging Trends in Design Research, The Hong Kong Polytechnic University.

Earl, C., Eckert, C., & Johnson, J. (2004). *Proceedings of the 8th International Design Conference: Complexity models in design.* Paper presented at the International Design Conference - Design 2004, Dubrovnik (Croatia).

Ellis, F. (2009). *How will Robocop communicate?: The design of a conceptual portable radio communication product for NZ Police in 2018.* Wellington: Massey University.

Evans, M., & Spruce, J. (2005, September 15-16). *Knowledge Networks: Collaboration between Industry and Academia in Design.* Paper presented at the Engineering and Product Design Education Conference, Napier University, Edinburgh.

Frederick, H. H. (2002). *The Unitec Global Entrepreneurship Monitor 2003/2004: Toward High growth Enterprise in New Zealand* (World Wide Web (Internet)): New Zealand Centre for Innovation & Entrepreneurship.

Freudenthal, A. (1999). *The design of home appliances for young and old consumers.* Delft: Delft University Press.

Givechi, R., & Velasquez, V. (2004). Positive Space. In Mcdonaugh, D., Hekkert, P., van Erp, J., & Gyi, D. (Eds.), *Design and Emotion: The Experience of Everyday Things* (pp. 12–17). London: Taylor and Francis.

Goellner, M. (2005). *Addressing Complexity in Product Design: Guidelines for Product Designers* (Unpublished PhD Thesis). University of Otago, Dunedin (New Zealand).

Hekkert, P. (2006). Design aesthetics: Principles of pleasure in product design. *Psychological Science, 48*(2), 157–172.

Industrial Designers Society of America (IDSA). (2009). ID Defined (Publication no. http://www.idsa.org/absolutenm/templates/?a=89&z=23). Retrieved March 2009, from Industrial Designers Society of America. Retrieved from http://www.idsa.org/absolutenm/templates/?a=89&z=23

Jordan, P. (1998). Human factors for pleasure in product use. *Applied Ergonomics, 29*(1), 25–33. doi:10.1016/S0003-6870(97)00022-7

Jordan, P. (2002). The Personalities of Products. In Green, W., & Jordan, P. (Eds.), *Pleasure with Products: Beyond Usability.* London: Taylor & Francis.

Knuckey, S., Johnston, H., Campbell-Hunt, C., Carlaw, K., Corbett, L., & Massey, C. (2002). *Firm Foundations: A Study of New Zealand Business Practices & Performance.* Wellington: New Zealand Ministry of Economic Development.

Kvale, S. (1996). *InterViews. An Introduction to Qualitative Research Interviewing.* Thousand Oaks, CA: SAGE Publications.

Mano, H., & Oliver, R. L. (1993). Assessing the Dimensionality and Structure of the Consumption Experience: Evaluation, Feeling, and Satisfaction. *The Journal of Consumer Research, 20*(3), 451–466. doi:10.1086/209361

McKinley, M. (2009). *Re-Energizing the Fence: Designing a Desirable Electric Fencing System* (Unpublished Master Thesis). Massey University, Wellington.

Muller, W. (2001). *Order and Meaning in Design.* Utrecht: Lemma Publishers.

NationMaster.com. (2009). *NationMaster.* New Zealand: Economy Retrieved March 2009, from http://www.nationmaster.com/country/nz-new-zealand/eco-economy

Nelkin, D., Nelson, R., & Kiernan, C. (1987). Commentary: University-Industry alliances. *Science, Technology & Human Values, 12*(1), 65–74.

New Zealand Trade and Enterprise. (2004). *Creative Sector Engagement Strategy.* Retrieved February 2005, from http://www.nzte.govt.nz/common/files/ses-creative.pdf

Norman, D. (2004). *Emotional Design: Why we love (or hate) everyday things.* New York: Basic Books.

OCED (Organisation for Economic Co-Operation and Development). (2002). *High-Growth SMEs and Employment.* Paris: Organisation for Economic Co-Operation and Development.

Roozenburg, N. F. M., & Eekels, J. (1995). *Product Design: Fundamentals and Methods*. New York: John Wiley and Sons.

Schoemaker, P. J. H. (1995). Scenario planning: A tool for strategic thinking. *Sloan Management Review*, *36*(2), 25–40.

Seva, R. R., Duh, H. B.-L., & Helander, M. G. (2007). The marketing implications of affective product design. *Applied Ergonomics*, *38*, 723–731. doi:10.1016/j.apergo.2006.12.001

Slappendel, C. (1996). Industrial design utilization in New Zealand firms. *Design Studies*, *17*, 3–18. doi:10.1016/0142-694X(95)00004-B

Teece, D. J. (2006). *Managing intellectual capital: Organizational, strategic, and policy dimension.* Oxford, UK: Oxford University Press.

Vihma, S. (1995). *Products as Representations: A Semiotic and Aesthetic Study of Design Products*. Helsinki: University of Art and Design.

von Stamm, B. (2003). *Managing Innovation, Design and Creativity*. Chichester, UK: John Wiley & Sons.

Warell, A. (2001). *Design Syntactics – A Functional Approach to Visual Product Form. Theory, Models, and Methods*. Gothenburg: Chalmers University of Technology.

Warell, A. (2008). *Modeling perceptual product experience – Towards a cohesive framework of presentation and representation in design"*. Paper presented at the 6th Design and Emotion Conference: Dare to Desire, Hong Kong Polytechnic University; School of Design, 6-9 October 2008.

Warell, A., Adank, R., Garrett, L., Goellner, M., & Parker, T. (2008). *Design for Desirability': Perceptions and Implications in New Zealand Export Industry.* Paper presented at the 15th International Product Development Management Conference, Hamburg: Helmut Schmidt University. June 29 – July 1.

KEY TERMS AND DEFINITIONS

Industrial (Product) Design: "Industrial design (ID) is the professional service of creating and developing concepts and specifications that optimise the function, value and appearance of products and systems for the mutual benefit of both user and manufacturer." (Industrial Designers Society of America (IDSA), 2009)

Design for Desirability: Design for desirability is the creation of relevant, desirable and meaningful experiences in products and user-product interactions. It addresses the user's aspiration of owning and using the product. This is related to personal identity, cultural and social values, and self-representation. In order to be successful, the product must be more desirable than the competition.

Product Experience: Product experience is the entire set of effects that is elicited by the interaction between a user and a product, including: (1) the degree to which all our senses are gratified (aesthetic experience); (2) the meanings we attach to the product (experience of meaning); (3) the feelings and emotions that are elicited (emotional experience) (Hekkert, 2006)

Affective Design: Affective design describes a design (of a product) that is capable of eliciting certain emotional experiences from users. Affective design aims to identify the subjective emotional relationships between users and products and to create products that are capable of eliciting pleasures for the users.

Emotional Design: Emotional design is concerned with the discovery and understanding of human emotions elicited by user-product interactions. Emotions are the results of appraisal process that is based on our experiences and concerns regarding the activity / interaction. They can be positive, negative or absent (Desmet, 2002).

Growth Innovation Pilot Initiative: A New Zealand governmentally funded tertiary education commission initiative (2007 - 2009) that aims to foster industrial competitiveness through design for desirability. The intention of this project is to assist New Zealand manufacturers and designers to achieve increased levels of success in international markets through focusing on Design for Desirability by establishing collaboration between manufacturing industries and academia.

Perceptual Product Experience (PPE): The Perceptual Product Experience framework is a model to understand, map, organise and analyse possible user experiences, by means of the sensory, cognitive and affective mode of user-product interactions. The PPE framework focuses on how we experience products with all senses; how we process and categorise stimuli and make sense of things; and how we feel and think of things when we experience products (Warell, 2008)

ENDNOTE

1 Icebreaker is a New Zealand company that produces outdoor-focused clothing made of 100% merino wool, a natural product that is comfortable to wear, breathable, naturally temperature regulating and durable.

Chapter 18
The Influence of Ageing on User Experience

Ana Cristina Medeiros
University of Cambridge, UK

Nathan Crilly
University of Cambridge, UK

P. John Clarkson
University of Cambridge, UK

ABSTRACT

The world population is ageing. At both the individual and collective levels, ageing causes several changes in people's lives that influence their needs and the way in which they interact with products. This chapter aims first, to provide a literature review on different aspects of ageing in order to better understand how people will respond to products as they grow older. Second, it describes how the physical and cognitive decline of older users increases rather than decreases the need to focus on all aspects of their experience, including their physical, sensory, cognitive and affective responses. Having considered these issues, suggestions are made in order that designers might best focus their attention as they design for a population that is rapidly ageing, and where members of that population are encouraged to maintain full participation in society.

INTRODUCTION

The world population is ageing at an unprecedented rate. The number of people aged 60 years and above is growing considerably faster than the population as a whole, and is projected to reach nearly 2 billion in 2050 (United Nations, 2001). Of the world's major areas, Europe is currently that with the highest proportion of older adults, and is expected to remain so for the next 40 years. However, with the exception of the United States of America, by 2050 the older population will be increasingly concentrated in the less developed countries, including China, India, Indonesia and Brazil. Irrespective of regional variations, the trend of population ageing is pervasive – a global phenomenon affecting every man, woman and child; it is profound – with major implications for all facets of human life; and it is enduring – as this trend is expected to continue through the

DOI: 10.4018/978-1-61520-617-9.ch018

twenty-first century. It is also said to be largely irreversible, with the young populations of the past unlikely to reoccur. Collectively, these changes will demand economic and social adjustments to be made in most countries (United Nations, 2001).

The rapid ageing of the world population has numerous implications for product development. At the biological level, ageing leads to gradual functional losses and behavioural adaptations that influence people's interaction with products. At the collective level, ageing leads to changes in people's social life and to shifts in the availability of financial resources. Such changes threaten older people's social integration and independent life, but technology is one way of assuring their continued well-being. A mobile telephone, for example, is a device that can bring confidence, security and independence to older adults, as long as they are able to use the product and are motivated to do so. However, for many older people, mobile phones are simply experienced as frustrating and unattractive because this range of products is designed for the average, young, healthy user. If such products are to be successfully developed to truly meet the needs of this expanding market then it is essential to develop an appreciation of people's capabilities and needs during different life stages.

In considering the relationship between population ageing and product development, this chapter seeks to improve understanding of how ageing influences user experience. The objectives of this chapter are first, to provide an overview of the ageing process. However, the intention here is not to give a comprehensive description of biological and sociological ageing, but to focus on those aspects of ageing that influence the user experience. The second objective of the chapter is to give a brief overview on how technology products have been developed in the last few decades, and how this also influences user experience and the adoption of technology. In particular, these issues are considered with respect to how people with different levels of capability may or may not approach, use and keep using such products. Having considered matters of ageing and user experience separately, the chapter closes by addressing the need to design for the ageing experience – i.e. design in ways that consider users in different life stages and circumstances.

AGEING

The ageing of the population is a result of a long-run decline in fertility, and a reduction in mortality at younger ages (Lee & Tuljapurkar, 1998). These factors, in turn, result from ongoing developments in medicine, infrastructure and education. However, whilst population ageing is a consequence of various social achievements, at the individual level, the ageing process involves not only elements of continued fulfilment, but also of progressive degeneration (Burton et al., 2005). Since the ageing process is complex and varied, researchers have focused on different aspects of ageing separately. For ease of understanding, in this chapter these aspects are clustered in five different groups – physical, sensory, cognitive, affective and social ageing. As we shall see later, when designing for the broader population, it is important to consider each of these aspects, how they may influence one another, and how they may affect people's responses and well-being. Therefore, each group of aspects is discussed in turn below so that their implications for design might be better understood.

Physical Ageing

In biological terms, around a quarter to one third of what determines life span is genetic. Therefore, ageing takes place at different rates within a population depending on the individual's genetic constitution. For example, as people reach extreme old age, the increase in mortality rate slows down. The most likely explanation for this is not that the ageing process itself decelerates, but that the slowing reflects population diversity – i.e. the frailer individuals die first, whereas the more robust individuals live longer (Kirkwood, 2002). Ageing also takes place at different rates within an individual, and decline in different organs are not necessarily related to each other. For instance, the main factors in cognitive decline may make little contribution to cardiovascular degeneration (Kipling & Faragher, 1999).

In evolutionary terms, living into old age is mostly an artefact of developed societies, where natural selection is less relevant. In this sense, improvements in nutrition, housing, sanitation and health care have led to reduced mortality at younger ages, resulting in an increased fraction of older people in the world (Kirkwood, 2002). The downside of living longer is that age-related diseases affect an ever-growing number of adults (Faragher & Kipling, 1998). Such diseases include diabetes, cardiovascular disorders, osteoporosis, arthritis and cancer among others. Diseases of this kind constitute the main cause of disability, morbidity and death in later life. They compromise older adults' well-being not only physically, but also emotionally and socially.

Physical ageing translates directly into loss of motor capabilities, which include locomotion (walking, stair climbing, bending and balancing), reach-and-stretch (upper-body strength and control), and dexterity (picking things up, carrying, holding and twisting using hands) (Clarkson, 2008). Findings from the analysis of a longitudinal study in Great Britain (Seidel et al., 2009) show that the loss of motor capabilities follows a hierarchical pattern, with locomotion being the first to be lost by half the sample (around age 77 for women and 78 for men), then reaching (around age 78 for women and 84 for men), and then dexterity (around age 89 for women and 91 for men).

Sensory Ageing

In order to interact with the world around us we use our senses. The five classical sensory functions of our bodies are vision, audition, touch, taste and smell, but there are many other senses commonly accepted, such as the sense of balance and body awareness. As we age, changes in the nervous system contribute to slower processing of sensory stimuli and also to the slowing of motor and sensory conduction speed (Wickremaratchi & Llewelyn, 2005). Moreover, ageing of the sensory system's tissues results in a gradual decrease in the intensity of sensation. Sensory ageing takes place at different rates but it is usually after the age of sixty that it presents more drastic implications, such as loss of functional abilities and loss of independence. In the sections below, we briefly outline the physiological changes for each sense and also the implications these changes have for interaction.

Vision

Visual limitations or blindness are directly related to ageing. Research has also indicated that ageing is responsible for a slowing in the speed with which visual information is processed (Fisk et al., 2004). Difficulties with physical abilities, such as activities of daily living, and specifically with mobility, are directly related to increased incidence of visual impairments (West et al., 1997) resulting in higher incidence of tripping and falling as well. Risk of car accidents among older adults is also more frequent due to reduction in the useful field of view (Owsley et al., 1998).

Audition

Hearing loss is one of the most common self-reported chronic conditions among older adults. The prevalence of hearing loss increases greatly with age, with men four times more likely to show hearing deficits than women (Cruickshanks et al., 1998). Consequences are difficulties with communication, which translate into frustration and a reduced quality of emotional and social life. Although hearing loss affects the individual, it is likely that family members and caregivers experience as much, or even more, frustration as a result of inefficient communication (Dalton et al., 2003).

Touch

Age-related changes in the skin lead to a decrease in spatial acuity in light touch, while a decrease in the sensation of pain and temperature is caused by changes in the nerves. Loss of vibration sense has also been observed in older adults (Wickremaratchi & Llewelyn, 2005). The main consequences are on hand grip, articulation of speech, due to reduced oral sensitivity, and postural stability, due to reduced foot tactile sense, leading to an increase in the risk of falls.

Taste and Smell

Also known as the chemical senses, taste and smell are used to detect and identify stimuli such as foods, aromas and gases in the environment (Hoffman et al., 1998). These senses deteriorate with age and this causes a higher risk of nutritional problems and fire hazards, especially for older adults living alone, who are generally unaware of the impairment (Murphy et al., 2002).

Cognitive Ageing

Cognitive development is a lifelong process and may involve changes that do not originate at birth, but later, during the life span (Baltes et al., 1980). Besides age, changes in the level of cognitive functioning are codetermined by multiple systems of influence, such as improvements in education, health and work. It is assumed that any developmental progression includes at the same time both growth and decline in adaptative capacity. This inherent dynamic between gains and losses is often described as *multidirectionality* of changes, and involves compensation mechanisms such as plasticity (Baltes, 1987).

An example of multidirectionality (and multidimensionality) is the *Theory of Fluid-Crystallized Intelligences*. According to Horn and Cattell (1967), fluid intelligence involves concept formation and attainment, reasoning and abstracting. It is marked by a turning point in early adulthood and gradually declines throughout later life. Crystallized intelligence, on the other hand, indicates the extent to which people have appropriated the collective intelligence of their culture for their own use. Because crystallized intelligence is largely based on experience, its increase will tend to be associated with ageing. Building on this theory, Baltes (1987) explains that intelligence consists of several subcomponents (multidimensionality) clustered in two main groups – the content-free fluid intelligence and the pragmatic crystallized intelligence. These groups differ in the direction of their development showing multidirectionality of gains and losses. In a theoretical framework of cognitive ageing, Baltes and Baltes (1989) outline a process called *Selective Optimization with Compensation*. According to this framework, for this process to occur people must be able to actively select what tasks they believe they are capable of accomplishing depending on their motivational and cognitive resources (Staundiger et al., 1989).

In cognitive ageing research, one form of compensation is known as plasticity of development or within-person variability. It refers to learning gains and designates the potential that people have for development after instruction and practice (Singer et al., 2003). In a more recent study, Boron and colleagues (2007) found that although those individuals at high risk of dementia are not likely to experience the same magnitude of gain as individuals who are not at risk, they are able to experience minimal improvements in fluid-type abilities. This provides evidence that some plasticity is still present, even in the early stages of cognitive impairment. Another kind of compensation is that general knowledge and specific knowledge in areas of expertise increase across adulthood. Such expertise can even offset cognitive decline (Carstensen et al., 2006).

In summary, although cognitive ageing is generally characterized by a process of decline, it is possible to compensate for this through training and through the optimization of intellectual resources by selecting appropriate activities. Moreover, as we will see in the next section, with ageing there is an increase in emotion control. Consequently, performance on emotionally charged tasks shows greater improvement than performance on non-emotionally charged tasks (Carstensen et al., 2006).

Affective Ageing

The terms affect, feeling and emotion relate to different aspects of the experience. Affect refers to the unconscious way in which the body prepares itself for action in a given circumstance, and plays an important role in determining person-to-person and person-to-environment relationships. These relationships are based on communication and translate into experiences. Emotion refers to the display of a feeling and can vary according to the social context a person is in. Whilst feelings are genuine sensations that have been checked against previous experiences and labelled, emotions can be genuine or not (Shouse, 2005).

The basic components of emotion, namely subjective experience, expression and physiological responsivity, change little with age. In their study on emotion, physiology and expression in later life, Levenson et al. (1991) found only a few aspects of emotional experience in which there is decline – the magnitude of physiological reactions, such as heart rate and finger temperature cooling. In compensation, there appear to be gains in emotional control (Carstensen et al, 2006), because the cognitive processes that regulate emotionality emerge only gradually as people mature (Panksepp, 2003). The ability to control emotion is important for human adaptative functioning and largely determines the impact negative events will have on our mental and physical well-being. Moreover, greater focus on emotional satisfaction among older adults, lead them to favour positive and avoid negative information in their attention (Mather and Carstensen, 2005). A possible explanation is that as people approach the end of life, goals associated with immediate well-being become more urgent because of the perception that "time is running out". In such a context, selecting one's experiences operates as a mechanism for affect regulation and allows people to conserve physical energy, a task that becomes more important with age (Fredrikson & Carstensen, 1990). This model of actively selecting one's experiences is advocated in a theory called *Socioemotional Selectivity*. This theory is in line with the previously mentioned framework of *Selective Optimization with Compensation* and can be considered one application of Baltes' framework to the social realm (Carstensen, 1992).

Interactions between emotion and cognition might be particularly relevant for understanding and improving cognitive performance in older adults. People in later life, at a subconscious or conscious level, may not attend to negative information simply because they are not motivated to do so (Carstensen & Mikels 2005). Like attention, memory is selective. Mather and Carstensen (2005) suggest that older adults are led by their focus on emotional goals to store information and subsequently retrieve it in ways that enhance their well-being. They found no differences due to gender, race or socioeconomic status. Likewise, age-related deficits were eliminated in a study on working memory maintenance tasks when the elements to be remembered were emotional, especially when positively valenced (Mikels et al., 2005).

In summary, in contrast with cognition, subjective experience in later life is enhanced. Older adults not only show better emotional control, they also actively select positive experiences over negative ones in order to enhance mood and well-being (Carstensen et al., 2006). This has consequences on the uptake of technology, as frustrated users are more likely to store negative experiences with products in their memories and to develop a pattern of avoidance towards these products.

Social Ageing

The preceding sections have discussed the physiological and psychological components of the ageing process. Let us now consider the social aspects of ageing. As people grow older there is a narrowing in their social life, which is related to but not necessarily caused by retirement. Different theories have tried to explain this phenomenon. *Disengagement Theory* (Cumming et al., 1960) posits that interaction lessens from the sixth decade of life as the individual disengages him/herself from society and as society withdraws from the individual. Hochschild (1975) assumes that ageing itself is not a determinant of disengagement, but that disengagement results from a combination of personal factors associated with ageing (e.g. poor health, widowhood) and other factors related to the nature of the social group a person is in. *Activity Theory* suggests that the reduced involvement of older adults in society is imposed largely by younger segments of the society (Phillips & Sternthal, 1977). *Socioemotional Selectivity Theory* (discussed previously in the section *Affective Ageing*) attributes this process of lessening one's social interaction to a change in emotional goals in later life.

Another aspect of social ageing is that gains in life expectancy are projected to substantially extend the number of years spent in retirement, a period typically characterised by high levels of consumption and expenditure on health care (Lee & Carter, 1992; Lee & Tuljapurkar, 1997). If no changes in policies are applied regarding the retirement age, there will be three retired citizens for each working-age person in Europe, in the middle of the twenty-first century (United Nations, 2001). With fewer adults available to support the elderly, retirement in the USA and Europe is likely to be soon delayed from age 65 to 67 (Wise, 1997; Berkel & Boersch-Supan, 2004). Later retirement is unlikely to be prevented by poor health; not only due to improvements in medicine, but also because people now age more healthily than in the past – the incidence of diseases followed by disability and dependence is declining (Cutler * Scheiner, 1998) as more people adhere earlier in life to healthy habits such as eating well, exercising, abstaining from smoking and controlling their alcohol intake. In general, the more educated and health-conscious people are, the longer they live (Vaillant & Mukamal, 2001). It is expected that some, if not most, people may react negatively to the prospect of working longer. However, on the plus side, staying professionally active can bring benefits to ageing because people then have the opportunity to more easily maintain social contacts. In a study on social relationships, it was found that keeping in contact can have positive effects on people's physical and psychological functioning (Lang & Carstensen, 2002).

In summary, as people grow older they gradually participate less frequently in social life. However, it is not the quantity of social interactions, but their quality, that matters to older adults. As life is being extended and people now age more healthily, there is a tendency for older adults to spend more years in work. This can help people maintain a more active social life, which is in turn beneficial for their health. With respect to product development, there are a number of areas – sport and recreational equipments, food packaging, kitchen appliances, information communication technology, work environment and devices, to name a few – where design could provide people with products that contribute towards healthy productive years in later life.

This section described ageing from both the individual and the social point of view, and the changes in capabilities and behaviour caused by the ageing process have been discussed. The next sections take these changes into the context of design and experience, and translate them into the user needs which designers must address.

DESIGN

There are many different definitions of design, but according to the Industrial Designers Society of America, industrial design is the professional service of creating and developing concepts and specifications that optimize the function, value and appearance of products and systems for the mutual benefit of both user and manufacturer (see www.idsa.org). In this sense, understanding users needs, their capabilities and behaviours is essential for the success of the product and the satisfaction of the user. With population ageing and the extension of working life, an increasing number of older users are interacting with products, systems and services whether or not such users were explicitly considered during the design process. For design to successfully meet the needs of users and successfully exploit their capabilities it is important to consider not only the product but the user experience as well. Therefore this section discusses first, how products have been developed and the consequences this has had on the adoption of technology. Second, it outlines the shifts that have taken place in design – from the product to the experience, and defines the aspects and responses involved in the user experience.

Users, Product Development and Technology Adoption

Products have often been designed as if everyone were a 25-30 year old successful professional (Benktzon, 1993). This is because most designers still focus on satisfying users who match their own characteristics – young, male and able-bodied (Keates and Clarkson, 2003). Therefore early adopters are generally identified as educated, economically active adults with good information on new technologies, willing to invest in the latest commercially available innovation (Wozniak, 1987). Such people represent an attractive market opportunity, and so receive much attention from industry.

Young people (usually economically dependent students), whose fluid intelligence has not yet reached decline, happen to learn very quickly and adopt new technologies easily. When a product that was originally developed for adults is later adapted for young people, this adaptation often involves very few functional changes, instead simply focusing on changing the product's outer form. Because of that, this age group is shaped by technology in a process described by Linch (2004) as techno-determinism (also see Mumford, 1946). Whilst young people may be considered 'natives' to the digital era, older adults in contrast, might be considered as 'immigrants', who have had to adapt to technologies introduced during their lifetime (Prensky, 2001). In addition to this lack of familiarity, older adults are more selective about their experiences, and have sensory impairments and slower cognitive processes. In order to improve usability for this group, designers must therefore lower the product demands (Gupta et al., 2003). Only after many modifications and simplifications have been made, are products likely to be taken up by older adults. In this case, technology is shaped by the elderly in a social-deterministic process (Pinch & Bijker, 1984). This process, however, takes a long time as, in most cases, only later in the development of mainstream products are the needs of older adults considered. As a result, although older adults might be interested in new technology (Fisk et al., 2004), they rely on their crystallized intelligence and need more time to adapt to novelties. Another reason for this delay in technology adoption is the older adults' resistance to change (Gilly & Zeithaml, 1985). This is caused not just by inadequate design (Forlizzi et al., 2001), but also by older people favouring the routine over the unknown and suffering from lack of information.

Routine Over the Unknown

Older adults prefer to maintain their routine instead of trying something new. According to Berger and Luckman (1966), all human activity is subject to habitualization. Any action that is repeated frequently becomes cast into a pattern, which can then be reproduced with economy of effort. Hence, following a routine is energy saving, especially for older adults.

Lack of Information

Due to less exposure to the market, older adults are usually less informed about new technologies and their benefits. As a result, older adults tend to avoid new products. The main cause of this lessening in market exposure is apparently a decrease in social interaction (discussed in the section *Social Ageing*).

From Product Design to the User Experience

In different areas of research, such as anthropology, sociology, psychology and history, a concern to understand the role of emotion in individual and social life has grown. This is in response to the dissatisfaction with the dominant cognitive view of people as mechanical "information processors" (Lutz & White, 1986). The same concern seems to be now shared by researchers in the design field.

According to Redström (2005), design has been shifting its focus from the object towards the user, advancing from form to function, from function to communication, and from communication to experience. This interest in user-centred approaches has stimulated a shift from the users' behavioural and cognitive experience to the users' affective experience of product interaction as well (Desmet & Hekkert, 2007). In this context, product experience specifically, has been defined as "the awareness of the psychological effects elicited by the interaction with a product, including the degree to which our senses are stimulated, the meanings and values we attach to the product, and the feelings and emotions that are elicited" (Schifferstein & Hekkert, 2008: p2). Because physical, sensory, cognitive and affective responses change with age, we can expect product experience to vary with age also. To tackle this issue, a number of design methodologies propose that mainstream products, and not just assistive products, should provide for the needs of less capable and/or older users in order not to exclude these individuals from full participation in society. In doing so, design may help users with limited capabilities to have a dignified, fulfilling and productive life (Fisk, 1993).

The opportunities for designing for the ageing population are many, but care must be taken to avoid the design of stigmatizing products, which just worsen older adults' performance on cognitive tasks (Carstensen et al., 2006) and discourage adoption. Instead, designers should address the diversity in older adults' changing capabilities that accompanies the ageing process (Gregor et al., 2002). This should be done in such a way that design promotes both physical and social inclusion, helping older adults stay healthy, occupied and integrated (Woudhuysen, 1993). As seen in the *Social Ageing* section, people will not only live longer, they will work for more years as well. In this sense, Coughlin (1999) states that an older work force introduces new challenges to the work place. For example, issues on older adults' interaction with technology will have to be addressed and for this reason changes in the design of workspaces will have to include more than features that simply enable improved physical movement and safety. Therefore, design for older adults should provide independence and social interaction, in domestic, public and professional spheres. Moreover, it is believed that products designed to address emotional needs contribute positively to the user's self-image and might be more rapidly adopted among the elderly (Hirsch et al., 2000).

Better Understanding User Experience

In recent years many attempts have been made to model user experience and to classify the different elements that constitute that experience. The resulting models often decompose experience into a number of aspects that relate more to the product (such as utility, accessibility, usability and desirability), and also into aspects that relate more to the responses of the user (such as physical, sensory, cognitive and affective). Although not entirely possible, separating the aspects in these two groups can be useful to conceptually understand the elements of the user experience. The following sections describe both groups of aspects.

Aspects of the Product

Different aspects of the product, which include usability, utility and accessibility have been studied individually or partially combined. Nielsen (1993: p25) has referred to usability and utility as aspects of an acceptability system based on cognition, while Keates and Clarkson (2003: p51) refer to accessibility, utility and usability as aspects of a design approach for inclusivity based on functional capabilities. The use of these terms across the literature is still confusing, as researchers may use different words to describe the same thing and vice-versa, and there are many open issues regarding affective reactions (Mahlke, 2005). Desirability, for example, is not just influenced by factors such as aesthetics, but also by practical concerns. To avoid any further confusion, in this chapter these terms will be defined as follows:

Utility refers to the functionality of a product or system that is deemed useful to the user. According to Jordan (2000), a product cannot be useful if it does not contain the functions necessary to perform the tasks for which it is intended. If a product does not have the right functionality it will cause dissatisfaction. He continues saying that in order to fulfil people's needs, designers must have an understanding of what the product will be used for and in which context and environment it will be used.

Accessibility refers to the potential the user has to interact with a product or system. It has become a major topic of debate in the Design community and different accessibility design approaches, such as Transgenerational Design (Woudhuysen, 1993), Universal Design (Preiser & Ostroff, 2001) and Inclusive Design (Keates & Clarkson, 2003) have emerged. Researchers in these areas argue that designers should take into account the capabilities of a broader spectrum of the population instead of considering just the average user.

Usability refers to the potential for the user to benefit from a product or system's utility through successful interaction. This includes ease of learning and remembering, ease of use, and pleasantness of use (Gould & Lewis, 1985). Nielsen (1993: p26) refers to these attributes of usability as learnability, memorability, efficiency and satisfaction, and also adds a fifth attribute: errors. He explains this by asking how many errors users make, how severe these errors are, and how easily users can recover from the errors. From Jordan's (2000) point of view usable products will not necessarily be pleasurable, although products that are not usable are unlikely to be pleasurable; and he concludes that in this way, usability should be seen as a key component of pleasurability.

Desirability refers to a product or system's potential to motivate a user to approach, obtain or interact with it. Traktinsky et al. (2000), state that emphasis on functionality stresses the importance of the product's usability and utility, while accentuating the product's form serves more the aesthetic and social needs of the users. In contrast, this chapter considers desirability to be based not only on aesthetic experience, but also on issues on utility, usability and accessibility. Desirability may thus be considered to be partly influenced by personality, age, gender, social context and previous experiences; and in part results from the user's measurable functional capabilities.

Aspects of the User

In order to develop user-centred methodologies to help designers design better interfaces, human factors researchers have traditionally tried to understand people's functional capabilities and responses. More specifically researchers have proposed different methods to assess user capability levels, which include physical, sensory and cognitive abilities. Assessing the affective dimension has generally been discussed separately and a variety of different approaches have emerged. These approaches are often used according to geographic and disciplinary boundaries. For ease of understanding, in this chapter these responses are divided into four groups (similar to the aspects of ageing described earlier), which refer to the following:

Physical Interaction is mainly based on the user's motor capabilities, which include locomotion, reach-and-stretch and dexterity. In order to interact with most mainstream products the user must be able to physically access the product's functionality. It was seen in the section on ageing that as we grow older, our abilities change. These changes take place at different rates across the population. Therefore, older adults encompass an incredibly diverse group of users, and even small subsets of this group tend to have a greater diversity of capabilities than is found in groups of younger people (Gregor et al., 2002).

Sensory Perception is our translation system of the features of the environment. It guides us through lights, sounds, colours, textures, aromas and tastes sending information to our brains, which will then decode it into meanings. Separating perception from interpretation in this way is not necessarily entirely accurate, but it can be conceptually useful when considering that sensory responses tend to deteriorate with ageing, which results in different levels of capabilities that must be designed for. In addition, in the context of user experience, products perceived as beautiful, are likely to be desirable and potentially pleasurable (Hekkert, 2006).

Cognitive Processing is the management of all information in the brain, and the commander of many of our responses. It involves associations, interpretations, and considerations, but is highly influenced by emotion. Cognitive responses are conscious and controlled activities (Rapaille, 2006). In the context of consumption, cognitive response refers to the judgments that the user makes about the product based on the information perceived by the senses. These judgements result in a behavioural response of approaching or avoiding the product. Similar behaviour can be observed in the context of product use (Bloch, 1995). Because ageing influences cognition, it is expected that cognitive responses will change along life.

Affective Selection is the expression of personality, sociocultural aspects, previous experiences and age. As discussed, as people age they tend to focus on emotional satisfaction which leads them to become steadily more selective regarding their experiences. Affect governs the quality of interaction with a product in the user's environment; it influences and mediates specific aspects of interaction before, during and after the use of a product. It also acts as a cognitive artifact in task achievement and is central to how other artefacts are interpreted and how pleasure is perceived (Spillers, 1994). For example, assistive devices have often caused embarrassment to the users because of their appearance. This negative impression of the product may even prevent users from adopting these products (Benktzon, 1993).

DESIGNING FOR THE AGEING EXPERIENCE

So far in this chapter we have seen how ageing-related changes take place in people's lives and how design has shifted from a functional to an experiential view. Having done so, we can now bring these two issues together and consider design for the ageing experience.

As discussed, ageing is a continuous irreversible process. Maintaining independence and social integration in daily life is important for well-being. Because of that, people generally favour technology products that are used for communication of some sort, especially information communication technology (ICT) products (Medeiros et al., 2008). In the context of product experience, taking ICT products as an example, ageing leads users to develop different needs and expectations with respect to product functionality. The degree to which these needs and expectations are met generates feelings such as satisfaction or frustration, excitement or fear, motivation or avoidance. The sum of these feelings makes up the so called affective response. According to its positive or negative nature, this response will influence the acceptance and use of a product and, beyond that, will influence the acceptance and use of the technology provided by the product. In other words, depending on the life stage people are in, they will require products that satisfy the specific needs and expectations of that life stage. For instance, while young students may be thrilled by mobile phones with big touch screens, media players and easy access to the Internet, economically active adults may prefer handsets which can send and receive emails, and be synchronised with personal computers for professional reasons. In contrast, retirees might want a mobile phone which presents an easily readable keypad, and exhibits restricted functionality, perhaps only making phone calls and sending text messages. What this age group is usually unaware of is that they could compensate for cognitive decline by making use of other features such as task reminders, voice memos, notes, alarms, address book and organiser if they just knew how to access and use these functions. Or better, if these functions were designed with these users in mind.

Young people often adapt to new products quickly and buy new technology more frequently. This means that products are expected to last only a couple of years, because within that period they will be replaced by a new model. Although not an efficient use of resources, this rate of consumption helps these users to keep up to date with the ever-changing and evolving technology, and makes the ongoing adaptation process less noticeable. This is not true in the case of older users – first, because in most cases problems with usability are a real issue; second, because this age group tends to resist technological novelties; third, because of the lack of information on new products (as seen in the section *Users, Product Development and Technology Adoption*). As a result, when older users finally adapt to a new product, it is expected that the product will last more than a couple of years and that obsolescence will not impede its use prematurely. In general, this pattern of consumption is far more environmentally friendly. The drawback here is that after a decade of use (should it last that long), when the product has to be finally disposed of and there are endless new choices on the market, the user risks having become technologically illiterate again.

Although preferable in many respects, there is yet to be designed a single mobile phone (to keep with the example mentioned above) that is attractive to and usable by the vast majority of society, and performs well for a reasonable length of time. Segmentation of the market relieves some of the issues, but can lead to the perception that some products are associated with disability which results in stigmatizing products that are often avoided by all consumers, even those that they are designed for. Bearing this in mind, we are left with two basic rules when designing a product: design it according to the users' capabilities, and design it to satisfy their needs.

Design it According to the Users' Capabilities

Whatever the product (or service) is, it is important to know if its demands will meet the users' capabilities. When demands are too high, interaction is compromised; users become easily frustrated and simply avoid using the product. It is also worth noticing the hierarchy of physical, sensory and cognitive capabilities. Loss of sensory capabilities, for example, may unnecessarily exclude users from operating a simple product in spite of their high cognitive capabilities. In the same way, full vision and audition are not enough to compensate for low cognition, when demands on thinking exceed the users' capabilities (Clarkson, 2008). Capability loss due to ageing appears on average in the seventh decade of life, with women losing capabilities slightly earlier than men. The pattern of loss among the six abilities most commonly associated with product interaction was found to be – first locomotion, followed by reaching, thinking, audition, vision, and dexterity (Seidel et al., 2009).

Design it to Satisfy the Users' Needs

Generally, people's needs and the resulting expectations towards a given product vary depending on their life stage as seen in the above example of the mobile phone. Often products designed for young people or adults will not necessarily satisfy older adults' needs. The amount of functionality in the product has to be balanced according to the users' real needs. However, if a product is intended to fulfil the needs of a large and probably diverse group, functionality could be designed in ways that enable users to access more advanced features gradually, after they have mastered the basic functions. This could be especially helpful to older adults, who usually believe they do not need extra functions and do not feel encouraged to explore them.

In the face of the phenomenon of world population ageing, this chapter has given emphasis to designing for the ageing experience; more specifically it has focused on the later life stages. However, it is important to remember that although life stages are frequently defined by different age groups, chronological age alone is not the single determinant of a life stage. Other factors that co-determine life stages are health, professional and family statuses. For example, let us consider two different people living alone – the first is a young adult retired early due to cognitive impairment caused by an accident; the second is an older adult with dementia, recently widowed. They may have, in some respects, the same needs and capabilities – both spend probably most of their time alone, need easy access to communication in order to facilitate social interaction and to ask for help, and would benefit from easy-to-use technology which could help compensate for memory loss and cognitive decline. These are obviously extreme examples, but if a mainstream product or service were designed considering older users' capabilities and needs, then it would automatically include a much broader group of the population (including young adults such as the one described above) who also need easy access to communication and would benefit from technology that augments their cognitive abilities.

It is also important to anticipate that users will not stop ageing at the moment of product purchase. The use of upgrades to existing ICT products seems to be an interesting option in this case. Instead of having to learn how to operate a completely new product, the user is given the opportunity to add new features gradually onto the product. This has the benefit of increasing the product useful life (postponing disposal, which is good for the environment) and offering users a more adaptative choice.

Finally, products are not experienced in isolation, and it is important to create a suitable context for that experience (Overbeek et al., 2002). In this sense, designers must ensure that it is not just products that are designed in accordance to the users' needs, but that services are too. If we consider services as an extended part of the product and therefore part of what is experienced, then it should also address the same needs, for how useful is a well designed bus if it does not serve a useful route according to a reasonable timetable?

CONCLUSION

This chapter has focused on the ageing process, the consequences it brings to people's lives and how designers should accommodate those consequences. Specifically, four main aspects of individual ageing – physical, sensory, cognitive and affective – have been described. Social ageing has also been discussed and design opportunities to develop or improve products were indicated. Moreover, this chapter has addressed the need to better understand the user experience amongst older adults in order to improve their interaction with products and services.

Products designed for adults generally will not satisfy older adults' needs, as the ageing process causes functional capability losses and results in several behavioural adaptations. These changes directly impact how people respond to products and also how they experience them. The resulting experience will ultimately influence people's quality of life as they will not adopt technology they cannot use, or are not motivated to use, even if that technology could potentially be of benefit to them. It is noteworthy that, although technology is rapidly developing, it takes a long time for the entire population to adapt to innovations. Therefore, designers should not just take into account user's past experiences, present physical capabilities and contexts of use; it is equally important to look ahead and design for the future, integrating technology products and systems into people's different life stages in a way that is less frustrating and less frightening. After all, positive experiences, including those with products, are more likely to be remembered and repeated, and this makes the adoption of technology a more natural and pleasurable process.

REFERENCES

Baltes, P. B. (1987). Theoretical Propositions of Life-Span Developmental Psychology: On the Dynamics between Growth and Decline. *Developmental Psychology*, *23*(5), 611–626. doi:10.1037/0012-1649.23.5.611

Baltes, P. B., & Baltes, M. M. (1989). Psychological perspectives on successful aging: The model of selective optimization with compensation. In Baltes, P. B., & Baltes, M. M. (Eds.), *Successful Aging: Perspectives from the Behavioral Sciences*. New York: Cambridge University Press.

Baltes, P. B., Reese, H. W., & Lipsitt, L. P. (1980). Life-Span Developmental Psychology. *Annual Review of Psychology*, *31*, 65–110. doi:10.1146/annurev.ps.31.020180.000433

Benktzon, M. (1993). Designing for our Future Selves: the Swedish Experience. *Applied Ergonomics*, *24*(1), 19–27. doi:10.1016/0003-6870(93)90155-3

Berger, P. L., & Luckman, T. (1966). *The Social Construction of Reality: A Treatise in the Sociology of Knowledge*. New York: Anchor Books.

Berkel, B., & Boersch-Supan, A. (2004). *Pension Reform in Germany: The Impact on Retirement Decisions*. Mannheimer Research Institute for Economics and Demographics.

Bloch, P. H. (1995). Seeking the Ideal Form: Product Design and Consumer Response. *Journal of Marketing*, *59*(3), 16–29. doi:10.2307/1252116

Boron, J. B., Willis, S. L., & Schaie, K. W. (2007). Cognitive Training Gain as a Predictor of Mental Status. *Journal of Gerontology: Psychological Sciences*, *62B*(1), 45–52.

Burton, D. G. A., Allen, M. C., Bird, J. L. E., & Faragher, R. G. A. (2005). Bridging the Gap: Ageing, Pharmacokinetics and Pharmacodynamics. *The Journal of Pharmacy and Pharmacology*, *57*, 671–679. doi:10.1211/0022357056163

Carstensen, L. L. (1992). Social and Emotional Patterns in Adulthood: Support for Socioemotional Selectivity Theory. *Psychology and Aging*, *7*(3), 331–338. doi:10.1037/0882-7974.7.3.331

Carstensen, L. L., & Mikels, J. A. (2005). At the Intersection of Emotion and Cognition: Aging and the Positivity Effect. *American Psychology Society*, *14*(3), 117–121.

Carstensen, L. L., Mikels, J. A., & Mather, M. (2006). Aging and the Intersection of Cognition, Motivation, and Emotion. In Birren, J., & Schaie, K. W. (Eds.), *Handbook of the Psychology of Aging* (pp. 343–362). New York: Academic Press. doi:10.1016/B978-012101264-9/50018-5

Clarkson, P. J. (2008). Human Capability and Product Design. In Schifferstein, H. N. J., & Hekkert, P. (Eds.), *Product Experience* (pp. 165–198). San Diego: Elsevier. doi:10.1016/B978-008045089-6.50009-5

Coughlin, J. F. (1999). Technology Needs of Aging Boomers. *Issues in Science and Technology* (Fall).

Cruickshanks, K. J., Wiley, T. L., Tweed, T. S., Klein, B. E. K., Klein, R., Mares-Perlman, J. A., & Nondahl, D. M. (1998). Prevalence of Hearing Loss in Older Adults in Beaver Dam, Wisconsin. *American Journal of Epidemiology*, *148*(9), 879–886.

Cumming, E., Dean, L. R., Newell, D. S., & McCaffrey, I. (1960). Disengagement: A Tentative Theory of Aging. *Sociometry*, *23*(1), 23–35. doi:10.2307/2786135

Cutler, D. M., & Sheiner, L. (1998). Standard and Non-standard Effects. In *Proceeding of Demographic Change and Fiscal Policy*. Demographics and Medical Care Spending.

Dalton, D. S., Cruickshanks, K. J., Klein, B. E. K., Klein, R., Wiley, T. L., & Nondahl, D. M. (2003). The Impact of Hearing Loss on Quality of Life in Older Adults. *The Gerontologist*, *43*(5), 661–668.

Desmet, P. M. A., & Hekkert, P. (2007). Framework of Product Experience. *International Journal of Design*, *1*(1). http://www.ijdesign.org/ojs/index.php/IJDesign/article/view/66/15.

Faragher, R. G. A., & Kipling, D. (1998). How might Replicative Senescence Contribute to Human Ageing? *BioEssays*, *20*, 985–991. doi:10.1002/(SICI)1521-1878(199812)20:12<985::AID-BIES4>3.0.CO;2-A

Fisk, A., Rogers, W. A., Charness, N., Czaja, S. J., & Sharit, J. (2004). *Designing for Older Adults: Principles and Creative Human Factors Approaches*. Boca Raton, FL: CRC Press.

Fisk, J. (1993). Design for the Elderly: a Biological Perspective. *Applied Ergonomics*, *24*(1), 47–50. doi:10.1016/0003-6870(93)90161-2

Forlizzi, J., Hirsch, T., Hyder, E., & Goetz, J. (2001). Designing Pleasurable Technology for Elders. In *Proceedings of INCLUDE 01 - International Conference on Inclusive Design and Communication*.

Fredrikson, B. L., & Carstensen, L. L. (1990). Choosing Social Partners: How Old Age and Anticipated Endings Make People More Selective. *Psychology and Aging*, *5*(3), 335–347. doi:10.1037/0882-7974.5.3.335

Gilly, M. C., & Zeithaml, V. A. (1985). The Elderly Consumer and Adoption of Technologies. *The Journal of Consumer Research*, *12*(3), 353–357. doi:10.1086/208521

Gould, J. D., & Lewis, C. (1985). Designing for Usability: Key Principles and What Designers Think. *Communications of the ACM*, *28*(3), 300–311. doi:10.1145/3166.3170

Gregor, P., Newell, A. F., & Zajicek, M. (2002). Designing for Dynamic Diversity: Interfaces for Older People. *Assets*, *2002*, 151–156.

Gupta, S. P., Keates, S., & Clarkson, P. J. (2003). Usability Considerations in the Design of Handheld Electronics. In *Proceedings of the 14th International Conference on Engineering Design ICED*, Stockholm, Sweeden.

Hekkert, P. (2006). Design aesthetics: principles of pleasure in design. *Psychological Science*, *48*(2), 157–172.

Hirsch, T., Forlizzi, J., Hyder, E., Goetz, J., Stroback, J., & Kurtz, C. (2000). The ELDer Project: Social, Emotional, and Environmental Factors in the Design of Eldercare Technologies. In *Proceedings of the Conference on Universal Usability CUU.* Virginia.

Hochschild, A. R. (1975). Disengagement Theory: A Critique and Proposal. *American Sociological Review*, *40*(5), 553–569. doi:10.2307/2094195

Hoffman, H. J., Ishii, E. K., & Macturk, R. H. (1998). Age-related Changes in the Prevalence of Smell/Taste Problems among the United States Adult Population. *Annals of the New York Academy of Sciences*, *855*, 716–722. doi:10.1111/j.1749-6632.1998.tb10650.x

Horn, J. I., & Cattel, R. B. (1967). Age Differences in Fluid and Crystallized Intelligence. *Acta Psychologica*, *26*, 107–129. doi:10.1016/0001-6918(67)90011-X

IDSA. (n.d.). *IDSA*. Retrieved from http://www.idsa.org

Jordan, P. W. (2000). *Designing Pleasurable Things*. London: Taylor & Francis.

Keates, S., & Clarkson, P. J. (2003). *Countering Design Exclusion: An Introduction to Inclusive Design*. London: Springer Verlag.

Kipling, D., & Faragher, R. G. A. (1999). Ageing Hard or Hardly Ageing? *Nature*, *398*, 191–193. doi:10.1038/18306

Kirkwood, T. B. L. (2002). Evolution of Ageing. *Mechanisms of Ageing and Development*, *123*, 737–745. doi:10.1016/S0047-6374(01)00419-5

Lang, F. R., & Carstensen, L. L. (2002). Time Counts: Future Time Perspective, Goals, and Social Relationships. *Psychology and Aging*, *17*(1), 125–139. doi:10.1037/0882-7974.17.1.125

Lee, R., & Tuljapurkar, S. (1997). Death and Taxes: Longer Life, Consumption, and Social Security. *Demography. The Demography of Aging*, *34*(1), 67–81. doi:10.2307/2061660

Lee, R., & Tuljapurkar, S. (1998). *Uncertain Demographic Futures and Social Security Finances*. The American Economic Review, Papers and Proceedings of the Hundred and Tenth Annual Meeting of the American Economic Association.

Lee, R. D., & Carter, L. R. (1992). Modeling and Forecasting U. S. Mortality. *Journal of the American Statistical Association*, *87*(419), 659–671. doi:10.2307/2290201

Levenson, R. W., Carstensen, L. L., Friesen, W. V., & Ekman, P. (1991). Emotion, Physiology, and Expression in Old Age. *Psychology and Aging*, *6*(1), 28–35. doi:10.1037/0882-7974.6.1.28

Linch, R. (2004). *The Mobile Connection: The Cell Phone's Impact on Society*. San Francisco: Morgan Kaufmann.

Lutz, C., & White, G. M. (1986). The Anthropology of Emotions. *Annual Review of Anthropology*, *15*, 405–436. doi:10.1146/annurev.an.15.100186.002201

Mahlke, S. (2005). Understanding Users' Experience of Interaction. Source ACM International Conference Proceeding Series. In *Proceedings of the 2005 Annual Conference on European Association of Cognitive Ergonomics.*

Mather, M., & Carstensen, L. L. (2005). Aging and Motivated Cognition: The Positivity Effect in Attention and Memory. *Trends in Cognitive Sciences*, *9*(10), 496–502. doi:10.1016/j.tics.2005.08.005

Medeiros, A. C. B., Crilly, N., & Clarkson, P. J. (2008). The Influence of Ageing on Product Experience. In *Proceedings of the 6th Design & Emotion Conference*. Hong Kong.

Mikels, J. A., Larkin, G. R., Reuter-Lorenz, P. A., & Carstensen, L. L. (2005). Divergent Trajectories in the Aging Mind: Changes in Working Memory for Affective Versus Visual Information with Age. *Psychology and Aging*, *20*(4), 542–533. doi:10.1037/0882-7974.20.4.542

Mumford, L. (1946). *Technics and Civilization*. London: George Routledge & Sons.

Murphy, C., Schubert, C. R., Cruickshanks, K. J., Klein, B. E. K., Klein, R., & Nondahl, D. M. (2002). Prevalence of Olfactory Impairment in Older Adults. *Journal of the American Medical Association*, *288*(18), 2307–2312. doi:10.1001/jama.288.18.2307

Nielsen, J. (1993). *Usability Engineering*. San Francisco: Morgan Kaufmann.

Overbeek, K., Djajadiningrat, T., Hummels, C., & Wensveen, S. (2002). Beauty in Usability: Forget about Ease of Use! In Green, W. S., & Jordan, P. W. (Eds.), *Pleasure with products: Beyond Usability*. London: Taylor & Francis.

Owsley, C., Ball, K., McGwin, G., Sloane, M. E., Roenker, D. L., White, M. F., & Overley, E. T. (1998). Visual Processing Impairment and Risk of Motor Vehicle Crash among Older Adults. *Journal of the American Medical Association*, *279*(14), 1083–1088. doi:10.1001/jama.279.14.1083

Panksepp, J. (2003). At the Interface of Affective, Behavioral, and Cognitive Neurosciences: Decoding the Emotional Feelings of the Brain. *Brain and Cognition*, *52*, 4–14. doi:10.1016/S0278-2626(03)00003-4

Phillips, L. W., & Sternthal, B. (1977). Age Differences in Information Processing: A Perspective on the Aged Consumer. *JMR, Journal of Marketing Research*, *14*(4), 444–457. doi:10.2307/3151185

Pinch, T. J., & Bijker, W. E. (1984). The Social Construction of Facts and Artefacts: or How the Sociology of Science and the Sociology of Technology Might Benefit Each Other. *Social Studies of Science*, *14*(3), 399–441. doi:10.1177/030631284014003004

Preiser, W. F. E., & Ostroff, E. (2001). *Universal Design Handbook*. New York: McGraw-Hill Professional.

Prensky, M. (2001). Digital Natives, Digital Immigrants. *On the Horizon. MCB University Press*, *9*(5), 1–6.

Rapaille, C. (2006). *The Culture Code: An Ingenious Way to Understand Why People around the World Live and Buy as They Do*. New York: Broadway.

Redström, J. (2006). Towards User Design? On the Shift from Object to User as the Subject of Design. *Design Studies*, *27*(2), 123–139. doi:10.1016/j.destud.2005.06.001

Schifferstein, H. N. J., & Hekkert, P. (2008). Introducing Product Experience. In Schifferstein, H. N. J., & Hekkert, P. (Eds.), *Product Experience*. San Diego: Elsevier.

Seidel, D., Crilly, N., Langdon, P. M., Matthews, F. E., Jagger, C., Brayne, C., Clarkson, P. J. & MRC CFAS (2009). Understanding Patterns of Capability Loss among Elderly Users. In *Proceedings of the 17th International Conference on Engineering Design*. ICED, Stanford.

Shouse, E. (2005). Feeling, Emotion, Affect. *M/C Journal*, *8*(6). Retrieved September 2, 2009, from http://journal.media-culture.org.au/0512/03-shouse.php

Singer, T., Lindenberger, U., & Baltes, P. B. (2003). Plasticity of Memory for New Learning in Very Old Age: A Story of Major Loss? *Psychology and Aging*, *18*(2). doi:10.1037/0882-7974.18.2.306

Spillers, F. (2004). Emotion as a Cognitive Artifact and the Design Implications for Products that are Perceived as Pleasurable. In *Proceedings of the 4th Conference on Design and Emotion*. Ankara Turkey.

Staundiger, U. M., Cornelius, S. W., & Baltes, P. B. (1989). The Aging of Intelligence: Potential and Limits. *The Annals of the American Academy of Political and Social Science*, *503*, 43–59. doi:10.1177/0002716289503001004

Traktinsky, N., Katz, A. S., & Ikar, D. (2000). What is Beautiful is Usable. *Interacting with Computers*, *13*, 127–145. doi:10.1016/S0953-5438(00)00031-X

United Nations. (2001). *World Population Ageing: 1950-2050*. Washington, DC: Department of Economics and Social Affairs, Population Division.

Vaillant, G. E., & Mukamal, K. (2001). Successful Aging. *The American Journal of Psychiatry*, *158*, 839–847.

West, S. K., Munoz, B., Rubin, G. S., Schein, O. D., Bandeen-Roche, K., & Zeger, S. (1997). Function and Visual Impairment in a Population-Based Study of Older Adults. *Investigative Ophthalmology & Visual Science*, *38*(1), 72–81.

Wickremaratchi, M. M., & LLlewelyn, J. G. (2006). Effects of Ageing on Touch. *Postgraduate Medical Journal*, *82*, 301–304. doi:10.1136/pgmj.2005.039651

Wise, D. A. (1997). Retirement against the Demographic Trend: More Older People Living Longer, Working Less, and Saving Less. *Demography. The Demography of Aging*, *34*(1), 83–95. doi:10.2307/2061661

Woudhuysen, J. (1993). A Call for Transgenerational Design. *Applied Ergonomics*, *24*(1), 44–46. doi:10.1016/0003-6870(93)90160-B

Wozniak, G. D. (1987). Human Capital, Information, and the Early Adoption of New Technology. *The Journal of Human Resources*, *22*(1), 101–112. doi:10.2307/145869

KEY TERMS AND DEFINITONS

(Industrial) Design: Is defined as: "the professional service of creating and developing concepts and specifications that optimize the function, value and appearance of products and systems for the mutual benefit of both user and manufacturer" (www.idsa.org)

Product Experience: Is defined as: "the awareness of the psychological effects elicited by the interaction with a product, including the degree to which our senses are stimulated, the meanings and values we attach to the product, and the feelings and emotions that are elicited" (Schifferstein and Hekkert, 2008: p2)

Utility: Is defined as: the functionality of a product or system that is deemed useful to the user.

Accessibility: Is defined as: the potential the user has to interact with a product or system.

Usability: Is defined as: the potential for the user to benefit from a product or system's utility through successful interaction.

Desirability: Is defined as: a product or system's potential to motivate a user to approach, obtain or interact with it.

User Capabilities: Are defined as: the user's motion, sensory and cognitive resources. It includes locomotion, reach-and-stretch, dexterity, vision, hearing, communication and intellectual functions.

Chapter 19
The Contribution of Ergonomic Analysis in the Product Design for Recycling

Eduardo Romeiro Filho
Federal University of Minas Gerais, Brazil

Rose Mary Rosa de Lima
Pitagoras Faculty of Belo Horizonte, Brazil

ABSTRACT

This chapter presents a case study about approaches dealing with Design for "X" (DFX), with emphasis on Design for Environment (DFE), specifically Design for Recycling (DFR) and Design for Disassembly (DFD), in a particular social and economic situation, in Belo Horizonte, Minas Gerais, Brazil. This approach is justified by the necessity of considering the environmental demands in the production process as well as the information about social demands needed by the designers in the product development process. This research shows that the products are designed only for adaptation on the production process and the demands of consumers. Having the environmental questions in mind, the designers need also consider how to disassembly these products to realize a possibility of recycling.

INTRODUCTION

Different methods of analyze of the product design deficiencies for recycling process have been described in the literature. The most of these methods are quantitative or based in computer simulation, and do not take into consideration the real activity of the "recycling workers". This is a very important approach in the developing countries like Brazil and others in Latin America, Africa and Asia, where people employment is a crucial social problem. In these situations, the recycling processes are in most times a manual activity that could improve the familiar income. Therefore, the opportunity of a simple and (if possible) manual disassembly is a very important characteristic of recyclable products. In this case, the point of view of the recycling workers is fundamental to an adequate product design for recycling. The Ergonomic Analysis of Work on the recyclers' workers emerges as a helpful methodology for the DFE and DFD to evaluate the difficulties of products disassembly that interfere in the recycling process.

DOI: 10.4018/978-1-61520-617-9.ch019

This study was executed at the Association of Collectors of Paper, Cardboard and Reusable Materials (Asmare), has its focus in the analysis of the disassembly of liquid containers made by Polyethylene Terepthalate (PET). It is demonstrated in the analysis that most of the difficulties of manual disassembly process derived from the product characteristics defined during its design development that interferes with the recycling process. Therefore, the strategies of disassembly and the work conditions of recycling workers, which are involved in these processes, should be brought into consideration at the initial phase of the product design. This will have benefits for all workers, the recycling companies, plastic industries and to the environment.

At the end of the analysis, investigations were made in factories that use PET containers, to obtain information about the design process and the designers' point of view about the addition of some DFR characteristics in the PET containers. The investigations demonstrate that the factories did not withhold knowledge of the difficulties that the products give to the "recycling workers". Finally, this approach is proposed as a strategy to improve the environmental characteristics in the products, in an adequate relationship with the social aspects.

BACKGROUND

The interest for the environment questions has increased in the latest years. The environment demand in the production process looks for alternative ways to design products, which are "environmentally correct". The pattern of consumption in the current society increasingly requires the manufacture of new products, which at the end of its useful life, results in millions of tons of discarded materials, most of which ends in "junk". Only in Brazil are produced, daily, about 240.000 tons of garbage (Lima e Romeiro, 2001). Several approaches have been suggested towards solving problems related to this subject, with some pointing to changes in the stage of design of the products, trying to incorporate some characteristics in those, being able to make them less harmful to the environment, or more easily recyclable. Techniques are being designed to be inserted in the project development of the product. Those techniques, considered specialties of the project, known as DFX's ("Design for Anything," when "X" represents the main objective in the design process, or simple "Design for Excellence" when the goal is the improvement of product design) and can be attributed to the assembly, manufacturing, environment etc. For the design back to the environment, the approaches are directed primarily to the products, suggesting that even in the design phase, are considered matters of the final stages of the life cycle (from the disposal) or specific to facilitate early a new life cycle For the project back to the environment, the approaches are directed primarily to the products, suggesting that even in the design phase, are considered matters of the final stages of the life cycle (from the disposal) or specificities to facilitate early a new life cycle. According to Bitencourt (2001), the Design for Environment (DFE) consists in developing environmentally sound products, without compromising the technical and economical viability of them. The DFE covers all the stages of the product life cycle, from design to final destination.

The main part of the project proposals within the context of the DFE is: design for reuse of materials and components, design for manufacturing, design for greater energy efficiency, design for recycling, design for disassembly etc. This latter approach can provide significant results in the phase of the project, therefore anticipates the consideration of the product "recycling". For example, if the initial phase of the product project considers only the type of material (recycled or non toxic) and not the non-destructive disassembly, it can may cause an increase in the complexity of the task and the workload of those involved, thus preventing the recycling process. The main components are originated from the DFE materials, processing, use and disposal. In the case of disposal, the directions will be differed depending on the type of product, materials used and the methods available to them can be specified as a desired goal in the project: incineration, landfill and recycling. In the DFE, when the recycling target is specified as the desired disposal, it is important to establish that the environmental damage, resulting from the use of recycling a material, is lower than that one resulting from the use of virgin material. In case of using of non-renewable resources, typically, the recycling is environmentally the most sensible option (Huang, 1996, p. 386). The differential has been being a new environment so that companies are adopting to the competitiveness of its products. Implementation of DFE in product development involves organizational and technological changes on the enterprise. However, in some companies, where the DFE is deployed, some difficulties have been identified and can be seen below (Huang, Op Cit., P. 85).

- **Lack of vision:** The managers is not aware of the influence of decisions made in the development of product;
- **Lack of motivation:** Neither the managers nor the research and development area or marketing express an interest for the DFE, once they do not see the benefits of environmental consideration, although they are aware about the environmental impact;
- **Insecurity:** The management feels unsafe on the regulatory initiatives and the trade effects of the environment demand in the development of product;
- **Complexity in implementation:**
    - The company does not have a systematic approach to product development, thus they do not know how to integrate the DFE in a structured way;
    - There are no resources that help the company to set the stage for the implementation of the DFE (organization of pre-conditions and organization of multidisciplinary teams), which can result in a lack of necessary knowledge;
    - The company is discouraged by the cost of acquiring necessary information for the implementation of the DFE (advice of experts, guides, databases, and others);
- **Other priorities:**
    - The company prioritizes investments in other activities, or
    - The company gives priority to consideration of other environmental activities;
- **Lack of environmental conscience:** the company never thought about its effect on the environmental issue.

Considerations of Recycling in the Project for the Environment

Designing a product for the environment means designing an eco-efficient product: non-polluting, non-toxic, recyclable and particularly beneficial to the environment and human health, within a sustainable development model. The environmental component in the development of an eco-efficient product should not be restricted to the production and marketing. Also should be taken into account the life cycle of the product, mainly the inclusion of post-consumer stage (discharge). Besides considering the characteristics of the product for recycling, must be considered also the working conditions of those involved in the activity of disassembly, so that recycling is becoming easier. Thus, it is essential to the proper procedure for recycling to the product be easily dismounted (by, wherever possible, cases of non-destructive) or can be fully recycled in the same process. Typical example of projects oriented to the environment are plastic products which, although they have separately recyclable components, they are constructed so that disassembly is impossible or economically costly, which in practice makes the reuse of material be almost impossible.

THE DISASSEMBLY AND RECYCLING IN PRODUCT DESIGN

Disassembly is the process of separation of components or materials of a product with the purpose of recovery or recycling of them. From an economic perspective, the dismantling of the product must be implemented with minimal cost, to facilitate recycling. In developed countries, research on the dismantling are being developed for electronic products, electronic appliances and automobiles, because of the life cycle of these products is becoming increasingly short and due to environmental laws are becoming stricter. Most studies (Gupta and Maclean, 1996; Mok et al, 1997; Ron and Penev, 1995 and others) shows that the products manufactured by the 1980s, currently discarded as garbage, were not designed to be dismantled easily, causing increasing problems as the depletion of landfill. The separation of components from discarded products is a process that can greatly reduce the problem, and extend the life of certain raw materials, reducing energy consumption for its production and reducing the demand of the environment.

Hanft and Kroll (1996) conducted a study using a simple procedure for evaluation of disassembly for recycling the product. The procedure is demonstrated through the evaluation of a computer keyboard and consists of a graphical assessment of disassembly corresponding of difficult tasks. The hard results from the analysis to measure the work of standard dismantling provided means to identify limitations in the product design. To maximize the return to the designers, the method analyzes the weaknesses of the product, capturing the sources of difficulties in the performance of each task. Kroll and Carver (1999) present an analysis of the disassembly of the product using the estimated time of disassembly in order to help designers to identify the weaknesses of the project, in terms of dismantling. The assessment based on the time of disassembly is applied to an electric drill and the difficulties associated with the creation of other metrics disassembly are discussed. The authors emphasize that the method provides important measures for an ease disassembly, when used to compare alternative projects of the same product. Kriwet (1995) propose a study, which analyzes the disassembly of a washing machine. The study demonstrated the need for change in the project, through a sub-assembly. The authors pointed out two suggestions: groups of electrical components / electronics and the use of special clamps to connect the lid. Both suggestions were given to enable easy disassembly. After the completion of the project modification, it was concluded that there was a better reuse of parts for an ease disassembly, with views in the recycling of materials. The authors present an approach of recycling through a systematic integration, where is introduced by a network of recycling defining different types of messages between the "server" (designer) and "customers" (consumers, recyclers and suppliers) and between the system recycling. The recycling system works with the life cycle of the system and provides relevant information to each phase. From the conceptual stage of the project, the network allows designers to represent and communicate relevant information to the project. The information exchange that happens in the network is on environmental and financial issues, e.g. in the database are provided information on environmental legislation, recycling processes, materials and history of the product. The authors present simple guidelines, easy to apply and evaluate the work structure of the network of the project for recycling. There was a review of the manual disassembly of a household refrigerator to see the difficulties of the process. The manual process was compared to a process that uses electric or pneumatic tools. After disassembling the product, each material was pooled, weighed and valued

Figure 1. Redesign throughout the product life cycle (BITENCOURT, Op. Cit., p. 35)

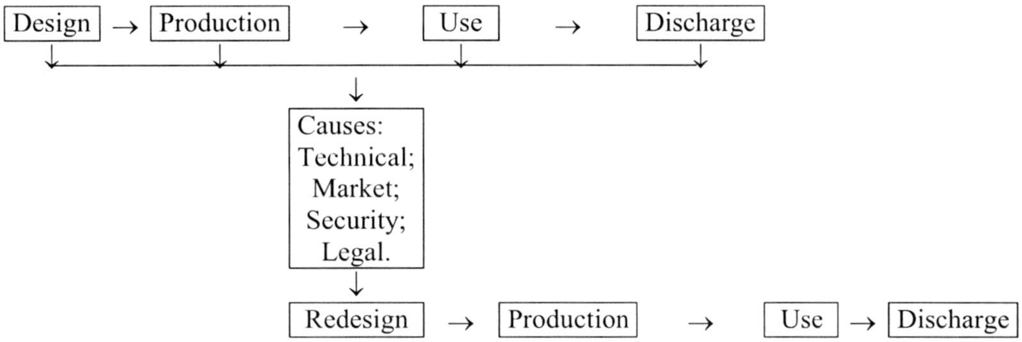

according to market value. The study showed that the economic feasibility of the systematic dismantling of the product leads the new designs of products to facilitate disassembly.

According to Kriwet et al (1995), the designer has a key role in improving the recyclability of the product because it has to take into consideration the requirements of the recycling process. It is difficult for the designer, design for recycling thinking, for example, the activity of the worker who will deal with this process. Duarte (1997) emphasizes that this type of activity, being more distant from the manufacturer, rarely receives attention of designers and the consequences are already visible through the saturation of the sites of final disposal of products - the landfill. Accordingly, the Ergonomics Analysis of Work of recyclers, through the survey of problems in real labor with the disassembly of products for everyday use in Brazil, can help providing guidelines to designers for the development of new products or the redesign of others, to facilitate recycling.

THE PRODUCT REDESIGN

The redesign is a process of changes and / or modifications that can be made throughout the life cycle of the existing product. According to Maribondo (2000), these changes are designed to meet new requirements or improve the performance of an existing project. According to Bitencourt (2001), changes can be made to meet demands of various features or characteristics, may be due of technical reasons, market, security or legal, as shown in Figure 1.

An example of redesign to meet demands related to legal causes of the program is redesign packages of Hewlett-Packard Company (cited by Bitencourt, Op. Cit.). The company made changes, replacing the material used in the packaging of their products by other with higher recycled material and replacing ink by another of lower toxicity. According to Otto and Wood (1998), the process of redesign is divided into four phases: information, conceptual, preliminary and detailed.

- **Redesign informational:** the first phase of the redesign. This phase is the analysis of existing product information and it is raised information about the demands (market, safety, environmental etc.).
- **Redesign conceptual:** it is the stage that involves more complex changes in the product. At this stage, changes can be made from the product design.
- **Redesign preliminary:** in this phase alterations can be done, as changes in the configuration of the product.

Table 1. Levels of redesigns products

Level of Redesign	Main feature
Original	Includes the highest level of changes in the existing product. Is linked to functional changes and / or the principles of solution. For example: change in organizational structure and selection of principles of solution that corresponds to greater energy efficiency. Includes the phases of conceptual, preliminary and detailed redesign.
Adaptive	Changes occur in the layout or in the configuration of the existing product. For example: need for easily assembly, recycling, disassembly, improvement in the separation of components etc. It covers the stages of preliminary and detailed redesign.
Parametric	The necessary changes are related only to engineering parameters. For example: change of material, increase or reduction of a certain size, change in power required etc... Is related to the stage of detailed redesign.

- **Redesign detailed:** it is the last stage, when you run up the details of the selected layout, set up the materials and processes and identify the gains to be obtained by redesign.

The process of redesign can be used to make either complex or simple changes in the product. These changes will depend on the level of change to better fit on the requirements of redesign. The levels of changes are called levels of redesign. According to the authors, the levels of redesign can be classified into: original, adaptive and parametric and the main features can be seen through Table 1.

Following, it is a presented a summary of an analysis of the activity of a product disassembly which from the results, some guidelines can be drawn to assist the designers in the Project for Environment. Problems are demonstrated by recyclers that were ultimately caused by design decisions that overlooked the real activity performed in the process of dismantling.

The Ergonomic Analysis of Work on Liquid Containers Made by Polyethylene Terepthalate (PET)

This study aimed to examine the disassembly of a product of high consumption, conducted by an association dedicated to recycling. It was considered a situation of working with the Association of Collectors of Paper, Cardboard & Materials Reused (ASMARE) located in Belo Horizonte (MG), through the dismantling of PET containers for recycling. The PET (Polyethylene Terepthalate) is a polyester, thermoplastic polymer that was introduced in Brazil in 1988, used primarily in the textile industry, since 1993, it had a strong expression in the packaging, mainly for soft drinks (ABEPET, 2009). The choice for this product was determined from its insertion in Brazil and because it presents a strong growth in consumption of plastics in Brazil, as shown in Figure 2.

The packaging used for soft drinks, to be composed of different types of resin, as shown in Table 2, for it is recycling requires the separation of these components. Then, you receive a simplified description of the analysis of the study, the difficulties found in the work situation related,

Figure 2. Changes in the use of PET packaging in Brazil. Source: ABEPET (2009)

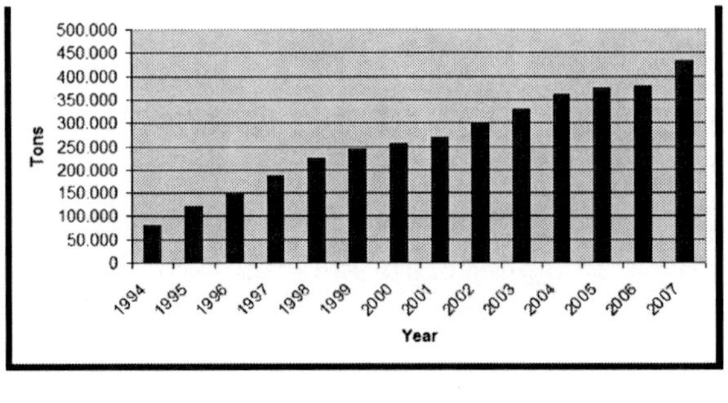

Table 2. Composition of the material of the PET packaging

Component	Material
Lid	PEAD (high density polyethylene)
Seal (ring)	PEAD (high-density polyethylene)
Label	PEBD (low-density polyethylene)
Bottle	PET (polyethylene ethylene)

specifically to problems of design and guidelines directed the designers.

The recovery of different types of resin that comprise the range of plastic packaging for recycling can be obtained by two types of process: manual and automated. The manual process is that one used in associations and sorting plants to serve a market that absorbs the material previously separated. But the automated process does not require a prior separation, id est., it is channeled all the mixed resin and its separation occurs in a process later. Table 3 shows the advantages and disadvantages of each of the processes:

This study arose as a way to identify the difficulties found in the process of recycling, particularly the packaging of household use. For this, we used principles of ergonomics, which can be defined as the study of relations between man and his work, to improve the conditions for implementing the proposed task. It is understood as the science of work, it seeks to understand the relationship between the behavior of man and his work. The ergonomics is also widely used in product development, especially in defining an appropriate size of the product to the user and the development of solutions that facilitate the use of the product in their daily lives.

In this case, however, the application of ergonomics has been devoted to analysis of the work of the recycler and his difficulties in the process with PET bottles to the separation of its components and recycling of plastic. The objective was to establish relationships between the difficulty of disassembly in a real situation and issues defined in the design of the product. It was used the methodology of Ergonomics Analysis of Work by Guérin et al, (1991) and Wisner (1976), focusing on ergonomics located. According to Echternacht and Veríssimo (2009), "The Ergonomics located has a methodology known as Ergonomics Analysis of Work and techniques of observation and interviews whose objectives are the collection of empirical material sufficient to meet the dynamic organization of the activity and variability of the situation and detailing the procedure for workers." In this research, various

Table 3. Advantages and disadvantages of the types of processes of resins separation Source: Data extracted from IPT (2000th p. 150-154)

KIND OF PROCESS	*ADVANTAGES*	*DISADVANTAGES*
Separation of resins by hand	- Requires low investment; - Provides greater generation of jobs; - Get a pure material; - Less chance of contamination in the process; - Production of goods with better quality.	- Slower process; - It is more difficult to separate and identify the various types of resins.
Automated separation of resins	- Faster process; - Presents an easier separation and identification of various types of resins.	- Requires high investment in special equipment; - Increased possibility of contamination during the process; - Restriction on production of artifacts (allows the manufacture of products with large thickness, for example: picket of fences, concrete forms); - Production of low quality goods; - Reduction of jobs possibility. .

tools were used for data collection: (a) meetings with groups; (b) interviews; (c) non-participating observations; (d) participants comments and (e) photographs of the work process.

The focus of this research activity was in the sorting of recyclable materials, specifically in the area of sorting of plastic products, local considered critical due to a large volume, variety and accumulation of materials, in addition to difficulties in performing the tasks.

Analysis of the Dismantling of PET Packaging

Asmare, institution where the examination was held, is formed by a central committee, divided into seven committees to work: religion, infrastructure, publicity and media, health, finances, education and culture, environment, and one member representing each one. It has 356 members, of which 250 are considered permanent and most has been collector of paper. The unit has seen about 150 members organized in an office, a carpentry, a workshop for handicrafts, a kitchen and three warehouses, which operate as follows:

1. Shed the sorting of material from the selective collection and donations from companies;
2. Shed of screening material of informal collectors;
3. Operational shed (packing and marketing of the material sorted).

The Association has three units in Belo Horizonte and works with the materials resulting from selective collection of donations and materials collected in the streets. Only one of the units works with the material of the selective collection of donations and the informal collection. The other two only work with the material of the informal gathering. 1 balance, 2 assistants and 23 sorters form the working population of the screening area. The group of sorters consists of 20 women and 3 men aged 16 to 66 years, mainly the range of 18 to 26 years, with less than eight years of formal study. The time of service varies from 2 months to 12 years, most of which have 2 to 3 years of experience. The main complaints of the workers are: back pain and spine, scratching and burning eyes, pain in legs and feet, very mixed material, loading weight, cut the hands and fingers, presence of animals such as rats.

Disassembly of the PET packaging requires essentially (1) take off the lid, (2) remove the seal (ring) and (3) remove the label. The task is described according to the current requirement of the buyers of recycled material, which only buy lots of the same material. Thus, those components produced from different types of plastic must be separated from the bottle. After dismantling, the parts must be separated and stored in bundles according to the type of PET material and packaging must be separated according to color. For this, the sector sorters remove the seal and remove the label using a serrated knife, using a disassembly considered destructive. In the example described in this study, two critical points were raised for the dismantling of PET packaging:

- **Seal:** The seal withdrawal of the lid of the pack was considered the most critical point of dismantling, because besides the failure to be removed without the use of a tool, there is not an appropriate tool to implement its removal.
- **Label:** Removing the label was also considered another critical point, mainly because of the use of glue on them. In many cases, excess of glue make it difficult to remove the label, contaminating the material to be recycled, because it is incompatible with the plastic packaging.

Despite the criticisms, some positive aspects were raised during the analysis of the disassembly of the product, such as:

Figure 3. Work of dismantling the PET

- Existence, so little representative of another type of seal that breaks with the lid, or does not remain fixed to the container after opening it.
- Existence of a different kind of label that does not use the glue to fix on the packaging.

Figure 3 shows the process of work involved in dismantling the PET bottle.

From the issues raised during the observation of the disassembly activity of the packaging, technical visits were carried out in two soft drink companies that use PET in their packaging, for a research about the different types of seal and label. The objective in this case was the intersection between information gathered during the AET and the criteria for the design of packaging.

RESULTS OF TECHNICAL VISITS

From the visit in the company "A", a plant of medium-sized soft drinks, it was found that there are two types of lid to seal the packaging PET (Figure 4) used in soft drinks (which was already observed in the market). One is called *double lock*, in which the ring (seal) is forced to be completely broken when the lid is removed and is fixed to the container (which brings problems to disassemble). In another type, called *win lock*, the seal is not completely separate from the lid and so it is removed from the container (with the lid). On the label, that is used for the plant, it is fixed in the package with glue, a task performed by a firm outsourced. It was found from the analysis of the sorters that the packages from this company contain excess glue on the labels of their products, which again brings considerable problems in the activity of dismantling.

Figure 4. Different covers for sealing of PET packaging: win lock(left) and double lock (right)

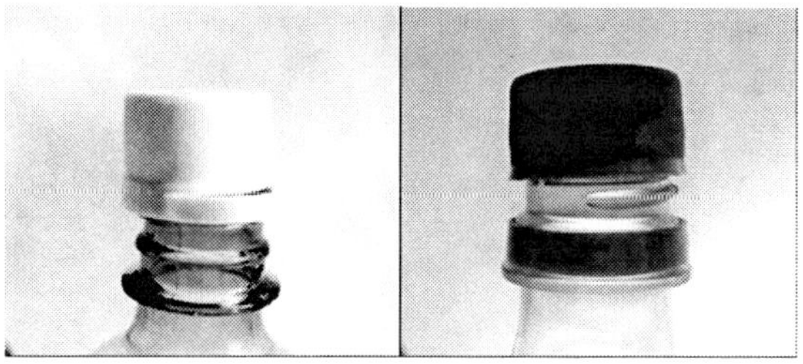

In the visit to the company "B", a multinational drink company, it was found that the use of *win lock* in their products was eliminated for security reasons, or not to allow the violation of their products, a problem reported by designer interviewed. According to this, complaints from consumers (in which the packages were violated) indicated that the lid could be removed and replaced without the seal was broken, which would not be possible with the adoption of the *double lock* type lids. The *win lock* type is still used in packs of glass that, according the interviewee does not allow the content to be violated without the removal of the seal.

About the labels, until now, the plant uses two different types in its product: (Figure 5): (left) BOPP, which is a type of label placed on the container using glue and (right) "Sleeve": type of label affixed on the container without glue. This label is "closed" with a special solder before them.

It is emphasized that the label kind "sleeve" is no longer used by the company, due to the higher cost of raw material and operational, as well the image of the product because the quality of printing that label is also lower. Table 4 shows

Figure 5. Kinds of labels: BOPP(left) and "Sleeve"(right)

the advantages and disadvantages of types of labels.

From the situation examined in relation to the product made, recommendations were established on two points of view to improving the working conditions of sorters and consequently the conditions for recycling of packaging: Focus on the product and focus on the job. Table 5 shows a comparison between the two types of intervention.

All things considered, different alterations can be suggested to bottle design improvement in the recycling aspects. Some examples are presented below:

1. Different forms to security warranty of bottle seal, with use of different models of bottle cap, for easy disassembly. In fact, a most adequate model (win-lock) is already used in the traditional glass bottle (returnable).
2. Use of label without glue, like the "sleeve" type. In this case, new label models, in different types of recyclable plastics, bring some advantage in relation at traditional "sleeve" label. For example, reduction of the requirement of control of production process.
3. The adoption of components (bottle, label, lid) made of the same material, that permit a common recycle process.

FUTURE TRENDS

From the example it was possible to see several opportunities for new approaches in design of products. In particular, develop studies that use the ergonomic methodology applied in real situations of activities involving the disassembly of complex products, for recycling. Propose a practical method of the Ergonomics Analysis of Work to industries (automobile, appliances, electronics, packaging, etc.) with the same aim of this research to obtain information about the product so that they can be incorporated into new projects or redesign.

In the analysis performed in this case, although not part of the scope of this research were raised issues related to the job with regard to positions taken by sorters, the displacements and environmental aspects. Thus, further possible applications, to propose improvements in the jobs available in developing countries, where automation is not always possible, are considered relevant. In these situations, technical solutions originally developed

Table 4. Advantages and disadvantages of types of labels

Kind of label	Advantages	Disadvantages
BOPP	- Lower cost of raw material and operational. - Increased capacity by labeling equipment (up to 18 thousand containers per hour). - Better print quality.	- Difficult to remove the label in the process of dismantling. - Uses glue on them.
"Sleeve"	- Facilitates the removal of the label in the process of dismantling. - Non-destructive process of separation (because of the absence of glue).	- Increased cost of raw material and operational (10-15% more expensive than the BOPP). - Requires greater control over the process. - It can drop undermining the image of the product. - Smaller capacity by labeling equipment (up to 10,500 containers per hour).

Table 5. Comparison of the possibilities of intervention related to the product

Focus	Possibility of intervention	Positives Points	Negative Points
Product	Product redesign (level of redesign: adaptive)	- Ease of disassembly of the product; - Efficiency in the activity of recycling; - Reducing the risk of perforating injuries; - Increased productivity in the business of recycling; - Greater coverage (service to other sites that pull the seal).	- Need for more investment from industry;
Workstation	Developing a tool to work properly	- Need to lower investment; - Reducing the risk of perforating injuries.	- Requires greater adaptation by the workers; - Less scope (local service).

for industrialized countries are not necessarily appropriate for other countries that do not have the same technical and human reality.

In this chapter it was shown need to redesign components of the package. In this sense they also become relevant studies to investigate alternatives in terms of material (use of lid and container of the same material) or of solutions of design (a mixed solution between win lock and double lock), which incorporates safety and removal of the seal, change this to allow easy disassembly and recycling.

CONCLUSION

The case study showed that the difficulties identified by the analysis of work activity of the dismantling of the product provide ways to identify problems and to extract design guidelines that can be directed to projects where the approach of "dismantling" can be incorporated to facilitate the design for Recycling (DFR). The ergonomic approach centered on the manual disassembly of a product rather simple, but can be applied to procedures for disassembly of more complex products such as cars, the appliances, computers, among others. The ergonomics has an important role in the analysis of problems to disassemble the product to recycling, to be performed directly from the worker (in real situation). Thus designers can be helped in decision making from elements consistent through guidelines to be adopted in the design of a new product or even in redesigns of an existing product.

It was not the main objective of this case study to investigate the companies using the PET packaging in their products. But the data collected in the two companies investigated, showed that there is a demand for information to designers on environmental requirements for its products, regarding the ease of disassembly for recycling. Through interviews with professionals involved with the production process, it was concluded that companies do not have knowledge of the difficulties that the sorters have for a manual disassembly of their products. And even those changes in design can really bring benefits to recycling processes. The need for the use of cutting tools, in the case, serrated knife, in the labor situation analyzed stems from features defined in the design of the product, id est., the ring seal attached to the bottle. Thus, considerations from the designers are necessary, of their responsibility to avoid the possibility of risk of perforating injuries in the work situation. It will demands changes in the design of the product, to a greater ease of disassembly, making effective improvements in the process of recycling and working conditions for the sorters. It seems clear that in this way, the working conditions of those involved in the recycling process must be considered from the design of the product, which will facilitate the process of dismantling striking.

REFERENCES

ABEPET – Brazilian Association of PET Packages Producers. *Associação Brasileira dos Fabricantes de Embalagens de PET.* (2009). *ABEPET.* Retrieved February 15, 2009 from http://www.abepet.com.br.

Bitencourt, A. C. P. (2001). *The development of a Re-Design Methodology for Environment*. (Thesis (Mechanical Engineering)). Federal University of Santa Catarina, Brazil. (in Portuguese)

Duarte, M. D. (1997). Characterization of Ambiental Label of Products. (Thesis (Production Engineering)). Federal University of Santa Catarina, Brazil. (in Portuguese)

Echternacht, E. H. O., & Verissimo, M. (2009). Cheminenement et perspectives de l'ergologie au Brésil: La santé au travail et les sciences de l'éducation en dialogue. In: Yves Schwartz; Louis Durrive. (Org.), *L'Activité en Dialogues: Entretiens sur l'activité humaine* (II). I ed. Toulouse - France: OCTARÈS Éditions, 2009, v. II, p. 191-214. (in French)

Guérin, F., Laville, A.; Daniellou, F.; Durrafourg, J., & e Kerguelen, A. (1991). Comprendre le travel pour le transformer. Paris, ANACT. (in French)

Gupta, S. M., & Mclean, C. R. (1996). Disassembly of products. In *Proceedings of the 19th International Conference on Computers and Industrial Engineering*, (v. 31, n. 1-2, p. 225-228).

Hanft, T. A., & Kroll, E. (1996). Ease-of-disassembly evaluation in design for recycling. In Huang, G. Q. (Ed.), *Design for X: Concurrent Engineering Imperatives*. London: Chapman & Hail.

Huang, G. Q. (1996). *Design for X: Concurrent Engineering Imperatives*. London: Chapman & Hall.

IPT. Instituto de Pesquisas Tecnológicas. (2000). *Lixo municipal: manual de gerenciamento de lixo*. (2. ed.), São Paulo: Imprensa Oficial. (in Portuguese)

Kriwet, A., Zussman, E., & Seliger, G. (1995). Systematic integration of design-for-recycling into product design. *International Journal of Production Economics*, *38*(1), 15–22. doi:10.1016/0925-5273(95)99062-A

Kroll, E., & Carver, B. S. (1999). Disassembly analysis through time estimation and metrics. *Robotics and Computer-integrated Manufacturing*, *15*, 191–200. doi:10.1016/S0736-5845(99)00026-5

Lima, R. M. R., & Romeiro, F. E. (2001). *The Material Recycling and Application in New Product Development: A Case Study*. (3rd Brazilian Congress of Product Development Management). Florianópolis, Santa Catarina: IGDP. (in Portuguese)

Maribondo, J. F. (2000). *Development of a Design Methodology for Modular Systems, Applied on Units Processing of Home Solid Residues*. (Thesis (Mechanical Engineering)). Federal University of Santa Catarina, Brazil. (in Portuguese)

Mok, H. S., Kim, H. J., & Moon, K. S. (1997). Disassemblability of mechanical parts in automobile for recycling. *Computers & Industrial Engineering*, *33*(3-4), 621–624. doi:10.1016/S0360-8352(97)00207-6

Otto, K. N., & Wood, K. L. (1998). *Product evolution: a reverse engineering and redesign methodology. Research in Engineering Design – Theory, Applications and Concurrent Engineering*. London: Springer.

Ron, A., & Penev, K. (1995). Disassembly and recycling of electronic consumer products: an overview. *Technovation*, *15*(6), 163–374.

Wisner, A. *(1976)*. Analyse de la Situation de Travail, Méthodes et Critères. *Paris: CNAN Ed. (in French)*

KEY TERMS AND DEFINITIONS

Design for "X": Techniques are considered specialties of the design process, known as "Design for Anything," when "X" represents the main objective in the design process, or simple "Design for Excellence" when the goal is the improvement of product design. The "X" approach and can be attributed to the assembly, manufacturing, environment, quality, testing etc.

Design for Environment: An specific approach in design, directed primarily to the products, suggesting that even in the design phase, are considered matters of the final stages of the life cycle (from the disposal) or specific to facilitate early a new life cycle. The main part of the project proposals within the context of the DFE is: design for reuse of materials and components, design for manufacturing, design for greater energy efficiency, design for recycling, design for disassembly etc.

Disassembly: The process of separation of components or materials of a product with the purpose of recovery or recycling of them. From an economic perspective, the dismantling of the product must be implemented with minimal cost, to facilitate recycling.

Redesign: A process that can be used to make either complex or simple changes in the product, along it lifecycle. These changes will depend on the level of change to better fit on the requirements of redesign. The levels of redesign can be classified into: original, adaptive and parametric.

Polyethylene Terepthalate (PET): A polyester, thermoplastic polymer, used primarily in the textile industry. It had a progressive and strong expression in the packaging industry, mainly for soft drinks bottles.

Ergonomics: Defined as the science related to man and his work, embodying the anatomic, physiologic, and mechanical principles affecting the efficient use of human energy. As examples of ergonomics applications: Safe lifting techniques, proper posture, appropriate seating position, and adaptive equipment are only a few of the many examples of ergonomics in the workplace.

Ergonomics Analysis of Work: is a specific methodology in ergonomics, which has techniques of observation and interviews whose objectives are the collection of empirical material sufficient to meet the dynamic organization of the activity and variability of the situation and detailing the procedure for workers.

Section 6
Systems Integration

Chapter 20
Understand Complex Design Problems Using Systems Thinking

Tao Huang
Columbia College Chicago, USA

Eric E. Anderson
Independent Consultant, USA

ABSTRACT

This chapter provides a brief overview of systems theory and suggests that product designers could use systems theory and systems dynamics models to improve our understanding of complex Product Design research problems, to anticipate how and where changes in these dynamically evolving systems might occur and how they might interact with the current system to produce a new system with new behaviors, and to identify leverage points within the system where potential policy or design process changes might be introduced to produce effective solutions to these problems with minimum policy resistance. By investigating the current and future trends of the application of systems theory in Product Design research, this chapter invites multidisciplinary discussions of these topics.

INTRODUCTION

Systems theory has been widely used in business decision making process and policy analysis and has been proven to be effective in understanding the dynamic complexity that characterizes these domains. Product designers also need to adopt a systems approach to thinking about the world to comprehend the complexity of the environmental, social, and technical issues facing humanity and our profession. Solving design problems requires an understanding of "hard problems" such as process in materials and manufacturing, however it cannot be simplified as mere mechanical systems. At its core, design is an interaction of human activities. It is complex, inherently non-linear, and often unpredictable. As such, problems typically encountered during the design process can be better conceptualized as "soft problems" with fuzzy boundaries that are not easily defined that are characterized by complex dynamic interaction and feedback among the designer, the objects of the design, and other participants of design in the society who affect and are affected by the design.

DOI: 10.4018/978-1-61520-617-9.ch020

This chapter will introduce product designers and researchers to soft systems theory and provide them with the basic tools to utilize it effectively as a tool to facilitate product design research and analysis. We begin with an overview of the basic concepts of systems thinking and systems dynamics theory. We revisit the classic "wicked problem", and show how soft systems thinking can improve our understanding of the innate non-linearity of Product Design and facilitate the discovery of new knowledge. We summarize and discuss the main characteristics of complex systems and illustrate these distilled concepts with a few examples of the behavior of feedback loops and introduce the use of systems dynamics modeling as a method to understand complex systems. To conclude, this chapter discusses systems theory's manifestation in product design, particularly in sustainable design and product development, service design, and design management.

BACKGROUND: GENERAL SYSTEMS THEORY

The Definition of Systems

According to general systems theory, a system is "an entity that maintains its existence and functions as a whole through the interaction of its parts" (O'Connor & McDermott, 1997). One does not need to look far for examples of complex systems. Humans and their activities constitute a subsystem of a system of the Earth, the interconnected dynamic geophysical, geochemical, and biological processes that collectively serve to make our planet a living world.

Systems thinkers must consider three "inters": the interactions of components within a process, the inter-relationships of processes within a system, and the inter-connections between systems and across time (Waldman 2007). Two types of systems are generally recognized, categorized in terms of their domain boundaries: hard systems and soft systems. Hard systems are those with "hard" problems where the primary concern is with objects, which can be objectively defined (Hutchinson, 2008). Hard systems are also called "mechanical systems" whose behaviors and boundaries can be successfully described and predicted in mechanical terms. On the other hand, soft systems have fuzzy boundaries with parts or elements that are difficult to quantify. Closed systems and open systems are two other common terms used to describe systems with distinct or fuzzy boundaries. Hard systems thinking has been used to maximize efficiency of a system, whilst soft systems thinking attempts to develop effective solutions by being inclusive of different viewpoints (Hutchinson, 2008).

Systems Thinking as Epistemology

General systems theory uses systems thinking as the epistemological foundation as well as an approach to problem solving. As a poet might say, systems thinking is a point of view. Historically, the "systematic view" has been referred to as a "holistic worldview." Systems thinking makes the assumption that a system has emergent properties and is therefore greater than the sum of its combined parts (O'Connor & McDermott, 1997).

On the contrary, the reductionist worldview slices the world into small pieces then analyzes these pieces to understand the world based on the assumption that the whole of a system is equal to the sum of the parts. This approach proved very effective at discovering new knowledge in closed systems at the micro scale where external environmental conditions could be controlled by the experimenter. However, this technique has its limitations and could not be applied to open systems and large scale systems because the external conditions were beyond control of the researchers. Furthermore, it cannot predict the emergent properties in the systems that are not found in their parts. The founder of modern systems theory, Ludwig Von Bertalanffy, concludes that studying the parts in isolation will not provide an accurate picture of the systems (cited in Waldman 2007, p.278).

Systems thinking is generally applicable to holistic concepts of organizations and systems, and it has evolved to be the foundation of complexity theory, chaos theory, systems dynamics theory, and game theory, etc. Systems theory also contributes to our understanding of how and where changes in various systems might occur and how these changes might interact with the current system to produce a new system with new behaviors.

Product Design Process as Systems

Product designers work with both hard and soft systems. Most tangible products can be conceptualized as "hard" systems because their materials and mechanical features can be defined and quantified. As individual products, their systems are closed. However, products are constantly interacting with their environment and users. Sociology, psychology, economics, as well as culture and aesthetics all play a part in these complicated interactions. In contrast, although it typically includes physical objects, due to the inseparability of products and their environment and users, the design process *cannot* be simplified to mere mechanical systems. Design is an interaction of human activities. It is complex, highly non-linear and often produces unpredictable results. Likewise, design problems in Product Design have rich context. Design process and problems can however be conceptualized as "soft systems" with fuzzy boundaries that are not easily defined that are characterized by complex dynamic interaction and feedback among the designer, the objects of the design, and other members of the society who affect and are affected by the design. Bela H. Banathy (Banathy, 1996) explains the problem situation in design systems as below:

In designing social systems we are confronted with problem situations that compose a system of problems rather than a collection of problems. They are embedded in uncertainty and require subjective interpretations. Above all, design problems are ill structured and defy straightforward analysis. In design there is a continuous interaction between problems and solutions.

Soft systems thinking is more appropriate in fuzzy and ill-defined situations involving both human beings and cultural considerations for a lateral design process (Checkland, 1999). As designers can attest from experience, the design process cannot always be described in rational, linear and mechanical terms. In fact, architect Botund Bognar (Seamon & Mugerauer, 2000) believes that the tendency of "rational" design reduces architecture "to measurable effects and results…which seriously limits architecture's primary grounding in human experience." Peter Green's "Design Process" diagram (Green, 1974) vividly demonstrates that the design process is multidisciplinary and non-linear, fundamentally a system of large amount of interacting elements.

The systematic nature of design process determines that there are generally two dimensions in research conducted in Product Design: micro-research and macro-research. A design process normally begins with research on existing or competing products, market, user groups, and technology that might provide the necessary support to help solve the design problem, or revolutionize the process to present a new intervention. The objects of this type of research are typically specific design problems or products, with limited resources or time to allow in-depth investigation of the macro environment that the products exist in or are intended to dwell within.

The macro-research studies the entire "eco-system" of artifacts: the interactions of objects, relationships between physical objects and humans, the social structure and natural environment that effect the acceptance or demand of certain types of products, the evolution of organizations that involve in product development, etc. This type of research is multidisciplinary in nature and broad in its problem interest, demanding research to be conducted beyond physical objects. Hence, macro-research in Product Design cannot progress by employing a reductionist worldview; it must adopt systems thinking.

Tackling a complex system might seem a daunting task to many designers. However, as John Gribbin points out: a complex system is really just a system that is made up of several simpler components interacting with one another. A complex system is not necessarily complicated (Gribbin, 2004). As the research project "Embracing Complexity in Design" (ECiD for short, funded by Engineering and Physical Sciences Research Council and Arts and Humanities Research Council under the Designing for the 21st Century initiative in the United Kingdom) concludes, there are four areas in which design and complexity interact (J. Johnson, 2006):

1. Many designed products and systems are inherently complex, e.g. aeroplanes, buildings, cities, microchips, information systems, manufacturers, organizations.
2. Designers need to understand the often complex dynamic processes used to fabricate and manufacture products and systems: design, products and processes co-evolve.
3. The social and economic context of design is complex, embracing market economics, legal regulation, social trends, mass culture, fashion, and much more.
4. The process of designing can involve complex social dynamics, with many people processing and exchanging complex heterogeneous information over complex human and communication networks, in the context of many changing constraints.

The objectives of linking systems theory and design are to enhance design by avoiding unwanted results and to increase the potential for design to play a more important role in the society.

Wicked Problem Revisit

Before systems theory became more available to disciplines other than mathematics, Horst Rittel transferred the knowledge of systems theory from the science and engineering profession to the design profession by coining the term "wicked problem". Rittel states that our infrastructure is essentially developed. The easy problems have been solved. Designing systems today is difficult because there is no consensus on what the problems are, let alone how to resolve them. He and Melvin Webber defined ten significant characteristics of "wicked problem" as below (Rittel & Webber, 1973):

1. There is no definitive formulation of a wicked problem.
2. Wicked problems have no stopping rule.
3. Solutions to wicked problems are not true-or-false but good-or-bad.
4. There is no immediate and no ultimate test of a solution to a wicked problem.
5. Every implemented solution to a wicked problem has consequences.
6. Wicked problems do not have a well-described set of potential solutions.
7. Every wicked problem is essentially unique.
8. Every wicked problem can be considered a symptom of another problem.
9. The causes of a wicked problem can be explained in numerous ways.
10. The planner (designer) has no right to be wrong.

More recently, the Australian Public Service Commission (APSC, 2007) described "wicked problems" as:

...problems that go beyond the capacity of any one organisation to understand and respond to, and [where] there is often disagreement about the causes of the problems and the best way to tackle them. .....Usually, part of the solution to wicked

problems involves changing the behaviour of groups of citizens or all citizens. Other key ingredients in solving or at least managing complex policy problems include successfully working across both internal and external organisational boundaries and engaging citizens and stakeholders in policy making and implementation... wicked problems require innovative, comprehensive solutions that can be modified in the light of experience and on-the-ground feedback ...all of the above can pose challenges to traditional approaches to policy making and programme implementation.

As the above authors and organization point out, design problems cannot be defined as "tamed problems" and systems thinking should be applied in solving these problems.

MAIN FOCUS: DYNAMIC COMPLEXITY OF SYSTEMS

When attempting to modify systems such as design systems, dynamic complexity always arises because systems have these characteristics (Sterman, 2000):

Dynamic

The most important character of a system is that it is dynamic, which means that changes and interventions in systems occur at many *time* and spatial scales. The micro-research in Product Design is only concerned with three dimensional objects, whereas the macro-research involves the fourth dimension: time. Examples can be easily found in the history of modern Product Design that over time, as the context of the design changes, an originally well-planned product or idea might be proven to be inappropriate, inadequate or even dangerous to the human society. One example stands out is the "Planned Obsolescence" concept popularized by American product designer Brooks Stevens since 1954. This practice artificially shortens product lifecycles by designing features that will fail shortly after the warranty period or by making services to the older products inconvenient in order to stimulate new demands. While originally this product development strategy was considered as revolutionary and beneficial to manufacturers, over the years it has attracted more and more criticism because it is based on dishonesty to consumers. While initially accepted as beneficial or at least benign, this practice, widely adopted by manufacturers, stands in direct opposition to the now prevalent idea of sustainability, which demands product designers eliminate unnecessary waste instead of generating more waste by simply changing the styling of products and making them less durable.

Tightly Coupled

"Tightly coupled" is a term that is used in computer science to describe components in a system that share workloads and which must interact to perform a given functional task. The system will fail if any of these collaborating components fails. Parts in a system are connected and work together (O'Connor & McDermott, 1997). One change in a part will have a ripple effect on the entire system. Social systems in which products interact with users have human participants whom are called stakeholders. In the research of Product Design, stakeholders include designers, users, market researchers, manufacturers, and engineers who influence decision making in the process of design. These participants in the system interact strongly with one another and with the design objects and the natural world. Because of the unpredictable nature of these interactions, design can have far reaching and unanticipated consequences and responses.

Governed by Feedback

All dynamic behavior arises from the interaction of just two types of feedback loops, positive (or self-reinforcing) and negative (or self-correcting) loops. Feedback from all the elements in the system constantly reinforces or weakens the changes. Systems dynamics modeling focuses on understanding the factors that drive these feedback loops. In systems, the dynamics of cause and effect is nonlinear. There are no linear chains of events initiated by a root cause that ultimately give rise to an associated effect. Instead, due to the nature of the looping structure, all elements in a loop represent both causes and effects of other elements in the loop. A given element can comprise more than one loop. The emergent systemic effects result from the dynamic interaction of the interconnected positive and negative loops. A simplified method to identify meaningful feedback loops is discussed further in this chapter.

Nonlinear

Causal relationships can be difficult to identify in a system. Confirmation of the existence of suspected relationships requires the collection and analysis of data. Time and resource constraints may preclude the collection of the quantitative data required to definitively establish and quantify such relationships. Furthermore, since effect is rarely proportional to cause, significant effects of reform might take generations to emerge. Despite these limitations, causal loop models remain useful as the existence and significance of proposed relationships can be interpreted, discussed and debated in a qualitative sense. Often, this level of analysis is all that is required to improve understanding of the problem. Well-positioned interventions in the system with high leverage can be identified with the potential to affect large and rapid changes. Similarly, interventions, which initially sound attractive, might be illuminated as being ineffective or as having unintended side-effects. The design process, also a nonlinear process that takes iterative steps, lends itself to this kind of qualitative "causal loop" analysis.

History-Dependent (Path-Dependence)

John Gribbin claims that this business of chaos and complexity is based on two simple ideas – the sensitivity of a system to its starting conditions, and feedback (Gribbin, 2004). The starting conditions of a system build the history of the system that cannot be ignored as the system continues to evolve. Though systems are dynamic and interact with other systems constantly, they still are greatly influenced by their starting conditions. One example is the long life of the QWERTY keyboard. The path-dependence of any institution determines that changing a system must consider the historical evolution and the current state of the system.

Emergence/Self-Organizing

The movement from low-level rules to higher-level sophistication is what we call emergence (S. Johnson, 2001). The dynamics of systems arises spontaneously from their internal structure. There are constant renegotiations within the system. Systems react to interventions and adjust themselves internally. Many grass-root social movements emerge as self-organized efforts: they have no defined founder, no defined starting point, and no pacemaker. Emergence is also often used to describe social phenomena that have simple causes but sprawl to unexpected areas and altitude. A powerful example is the rise of blog in recent years. Enabled by information technology, blog has become not only a platform to express personal opinions but also a powerful political tool to influence public policy.

Adaptive

Emergence and adaptive are two closely related characteristics of systems. The capabilities and decision rules of the objects and participants in complex systems change over time. The participants in the system learn new ways to achieve their goals in the face of interventions over time. These stakeholder actions result in the creation of new elements and new relationships between elements that might serve to weaken or reinforce the intervention. In the practice and research of Product Design, we often discover that users devise creative ways to use products beyond their intended designed functions. Many examples can be also drawn from online computer games, arts, and music. Especially in the era of open-source codes and internet, people adapt to their systems much more quickly and innovatively. This is the movement we observe as "Everyone is a designer."

Counterintuitive

Journalist Matt Powers claims that pound for pound, making a Prius contributes more carbon to the atmosphere than making a Hummer, largely due to the environmental cost of the 30 pounds of nickel in the hybrid's battery (Powers, 2008). Viewed in isolation, this comparison seems to suggest that a Hummer is more environmentally friendly than a Prius. Using systems thinking, one can easily point out that the hybrid quickly erases that carbon deficit on the road, thanks to its vastly superior fuel economy. However, Powers continues to say that because a used car has already paid off the carbon debt associated with its manufacturing process, a new Prius can never catch up with a highly efficient used car. This counterintuitive example demonstrates that when analyzing a particular product, product designers must adopt the holistic view, be proactive, and think beyond the current events; the consequences of doing so are sometimes counter-intuitive and can lead to important discoveries. Causes and effects might be distant in time and space in a complex system. In some system configurations, a short-run gain leads to a long-term decline. Similarly, a short-run loss can lead to a long-term gain.

Policy Resistant

Policies may create unanticipated side effects. It has long been acknowledged that people seeking to solve a problem often make it worse (Sterman, 2000). Policy resistance is the tendency for interventions to be delayed, diluted, or defeated by the response of the system to the intervention itself (Meadows, Richardson, & Bruckmann, 1982). It arises because we often do not understand the full range of negative feedbacks operating in the system. Product designers usually do not take part in public policy making directly. However, the results of our services enter the public realm and sometimes fundamentally change people's behaviors (cases in point are the walkman and Ipod). When introducing policy intervention, we should expect it to experience resistance. We must be able to foresee some of the possible outcomes and design strategy to counter the policy resistance. Policy resistance is always caused by balancing (negative) loops. Elements in these loops that will affect the changes in the system must be identified and their relationships with the intended changes have to be understood to ensure the policy resistance can be minimized.

Trade-Offs

Product designers are all too familiar with trade-offs: compromises one has to make to complete certain design tasks as we are constantly interacting with people from different disciplines and designing for users of all ages and needs. These trade-offs, and stakeholders who demand them, have to be taken into consideration. Using systems thinking, we also understand that time delays in feedback channels mean the long-run response of a system to an intervention is often different from its short-run response. There will be trade-offs when implementing interventions.

Figure 1. An example of feedback loops in the product design educational system

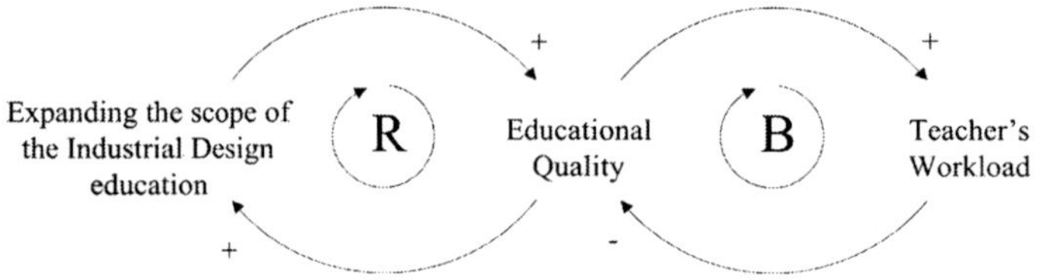

Understanding Complex Design Problem Using Systems Dynamics Models

Based on the characteristics of a dynamic system discussed above, many systems theorists use systems dynamics models to simulate and analyze complex systems. Systems dynamics modeling is a powerful method to gain useful insight into situations of dynamic complexity and policy resistance (Sterman, 2000). Researchers studying soft systems often find that numerical data are unavailable. In such cases, systems dynamics models can be constructed in the form of "causal loop" diagrams that utilize qualitative data to predict trends and explain relationships.

The steps of modeling process are as follows (Sterman, 2000):

- Problem articulation;
- Formulation of dynamic hypothesis;
- Formulation of a simulation model;
- Testing policy design and evaluation.

Systems dynamics models mainly use feedback loops and stocks and flows models to imitate systems. These models define the elements and participants of systems and construct their relationship based on time and space. The act of building these models will improve the understanding of a system and facilitate formulation of policy interventions to change the current system. An example of a simple feedback loop in the Product Design educational system is shown in Figure 1.

Symbols: "R/B" circles are loop identifiers; "R" circle means reinforcing loop; "B" circle means balancing loop (negative loop); Lines with arrows represent causal links; "+/-" are link polarities; "+" means that one part of the system has positive influence on the other part of the system and "-" means that one part of the system has negative influence on the other part of the system

Figure 1 suggests that by expanding the scope of the Product Design education (i.e. discussing more social and cultural issues in the curriculum) will presumably enhance the educational quality, but also increases the teachers' workload. Therefore, educators might resist the expansion and the educational quality might in fact decline and act to resist the demand for expansion. This resistance is always caused by balancing (negative) loops. A system with a strong balancing loop seeks its emergent goal. Changes to the system are not permanent because once the pressure is off, unless the balancing loop has been eliminated, weakened, or counter-balanced by an additional reinforcing (positive) loop, the system simply returns to its original goal. Hence, to change a system, one has to create or strengthen reinforcing loops and remove or weaken balancing loops. In the above example, educational systems, like all established institutions, are complex and have numerous negative feedback loops associated with established processes and practices. Predictably, when the goal of an educational system is long-term and spans multiple institutions, it is difficult to sustain support for new policy interventions at the departmental level. Other stakeholders in the system such as students and administrators will add to the complexity of the model.

Another example is the product attractiveness model illustrated as Figure 2.

Figure 2. Product attractiveness model

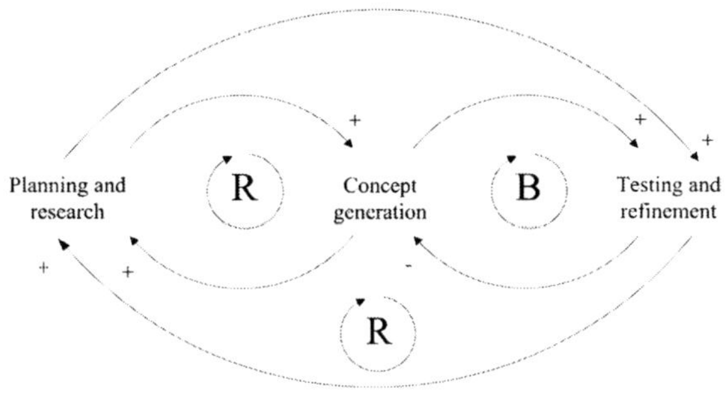

Figure 2 shows two balance loops, that when product attractiveness increases, the price of the product will most likely increase, thus reduces the customer acceptance, and ultimately reduces the product attractiveness.

Adaptation of the Systems Dynamics Model in Product Development

Define Product Development Process Using Systems Dynamics Model

Karl T. Ulrich and Steven D. Eppinger articulate a generic product development process in six phases (Ulrich & Eppinger, 2004):

1. Planning – identify opportunities; evaluate and prioritize projects; allocate resources and plan timing; complete pre-project planning;
2. Concept Development
3. System-level Design – define the product architecture; decompose the product into components.
4. Detail Design
5. Testing and Refinement
6. Production Ramp-up – test production and training of work force.

Although there are many renditions of this process, this version is comprehensive and widely accepted. Based on these six phases, Figure 3 is a proposed simplified systems dynamics model for this process.

The model proposed above articulates the dynamics of the three main phases in the product development process. In this model, the planning and research phase supports the concept generation phase, which is a combination of phase 2 to 4 of the process described by Karl T. Ulrich and Steven D. Eppinger. One might predict that the more planning and research that the design team conducts, the more concepts might be generated. The testing and refinement phase (combination of phase 5 and 6) modifies the design and reduces the feasible options. Thus, they form a balance loop. The reinforcement loop of the planning and research phase and testing and refinement phase is based on the consideration that as the new products are being tested and refined, more research will be needed. Planning and research usually also determine the methods of testing and refinement. These phases provide feedbacks to

Figure 3. A simplified systems dynamics model for the product development process

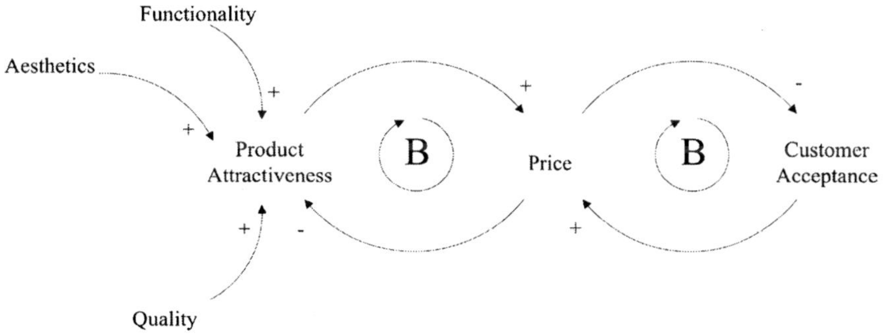

each other and are in constant interaction. This model illustrates the product development process as an iterative and non-linear process.

Principles of Developing Systems Dynamics Models for Product Designers

Introducing this modeling method not only expands the modeling capability of product designers and researchers, but also creates new graphic rhetoric for demonstrating research results to clients. In many cases, it articulates rationales for product development more clearly than numbers or qualitative research narratives. Clearly, the goal of building systems dynamics models is not to draw mathematically precise diagrams, but to more accurately identify the associations of the parts in a system, thus to understand trade-offs, predict possible policy resistance, and design products and strategy accordingly. Using these models, designers and researchers can focus on eliminating negative feedbacks and increase positive feedbacks in product development. The causal relationships in the systems dynamics models are not linear; instead they form a "network" of reasoning. Thus, the design process can be more efficient and sensitive to the interactions of products and their environment.

To use systems dynamics modeling successfully, product designers and researchers should follow these principles (Forrester, 1961; Morecroft & Sterman, 1994; Sterman, 2000):

- *Develop a model to solve a particular problem, not to model the entire system.* System dynamics modeling is no silver bullet. One must understand the importance of purpose of modeling and define the perimeter of any problem before adopting the techniques. Product designers should orient themselves to the problems within their capacities instead of re-designing the entire system.
- *Modeling should be integrated into a project from the beginning.* Systems theory must serve as the fundamental epistemology for the research conducted for the project.
- *Be skeptical about the value of modeling and force the "why do we need it" discussion at the start of the project.* Systems dynamics modeling includes a set of useful tools to understand systems but it has its limitations. These tools might be overly time-consuming or the results of modeling might be too broad to be used in the design project. In addition, the subjectivity of the designer and researcher should be closely examined. For product designers and researchers, we should focus on implementation from the start of the project.
- *Systems dynamics model does not stand alone; use other tools and methods as appropriate.*
- *Modeling works best as an iterative process of joint inquiry between client and consultant.* Like the process of design, the process of modeling is also non-linear and requires inputs from all stakeholders involved in the system. Product Design researchers must avoid black box modeling and work with their clients.
- *Validation is a continuous process of testing and building confidence in the model.* Validation should be done either as part of the contract with the client or as individual investigation after the design services end. For the further understanding of the system, implementation of a model does not end with a single project.
- *Get a preliminary model working as soon as possible, add detail only as necessary.* A broad model boundary is more important than a great deal of detail.
- *Use expert modelers.* Designers and researchers with expertise in solving the particular problem under investigation will

provide insights for building a comprehensive systems dynamics model.

Limitations of Systems Dynamics Modeling

Although graphically systems dynamics models are easy to understand for product designers, building an insightful and realistic model is an iterative process that requires in-depth knowledge of a system. Figures shown in this chapter are simplified example of systems dynamics models, with no elements such as "stocks and flows" that are usually employed in more complicated models, or mathematical formula to precisely measure the causal relationship. As the number of parts in a system grows, the model will grow exponentially more complicated. Causal relationships will be difficult to identify or validate. It is therefore advisable to keep the model as simple as possible by focusing on identifying the most important drivers of system behavior. However, when the model is overly simplified, it cannot simulate the real world sufficiently. What elements are important and what inter-relationships are significant should be the focus of multidisciplinary discussion and debate. Developing an insightful model is difficult; however using modeling to help change organizations and implement new policies is even harder (Sterman, 2000).

FUTURE TRENDS

Systems Theory in Sustainable Design and Product Development

In recent decades, systems theory has been applied to many aspects of Product Design research, most noticeably in sustainable design and product development. Systems thinking is a holistic approach to the solution of complex problems and is well-suited to analyze the complexity associated with sustainability. Using this approach, The Club of Rome, which was formed in 1968 at the instigation of Italian industrialist Dr. Aurelio Peccei, undertook an ambitious project "to examine the complex of problems troubling men of all nations: poverty in the midst of plenty; degradation of the environment; loss of faith in institutions; uncontrolled urban spread; insecurity of employment; alienation of youth; rejection of traditional values; and inflation and other monetary and economic disruption" (Meadows, Club of Rome, & Potomac Associates, 1974). The report by the World Commission on Environment and Development (World Commission on Environment and Development., 1987) concluded that, "[Sustainability is to be attained by] more rapid economic growth in both industrial and developing countries, freer market access for the products of developing countries, lower interest rates, greater technology transfer, and significantly larger capital flows". While the inclusion and exclusion of specific elements and the wisdom of the policies recommended by the organizations are and should be a matter of debate, it is clear that sustainability encompasses all aspects of human society and requires systematic analysis for its scale.

In product development, Life Cycle Assessment (LCA) is an example of adopting systems theory to analyze the environmental impacts of products throughout their lifecycles. In LCA's data process, it starts with listing a bill of materials and their manufacturing/abstraction processes, energy used, transportation cost and impacts, end-of-life impacts, and toxicity; it then lists the inventory cost of these materials, including land use, emission, and resource depletion. This information is usually converted into "scores" that can be conveniently used to compare the environmental impacts of materials and processes.

While not every aspect of a system can be reduced to a number in Life Cycle Assessment, it is important to use a systems dynamics model to include ecology, sociology, economics, and politics in the analysis of the system. These aspects of the system have to be carefully weighted to determine the trade-offs and suggest optimization.

One of the other example of adopting systems theory in sustainable design is the "Cradle-to-Cradle" model developed by two of the most important advocates for sustainable design in North America, architect William McDonough and chemist Michael Braungart (McDonough & Braungart, 2002). They suggest a new model for industry that will imitate the natural cycle of materials flow. This model eliminates the concept of waste by considering waste as "food". It discards the old linear model in industrial growth where waste does not enter the system. This model also demonstrates systems thinking by developing products with *time* in mind.

Systems Theory in Service Design

In an interview, Ezio Manzini points out that past definitions of service described a "form of interaction where one person generates a benefit for somebody else". A product, on the other hand, was a "physical form of material in space" ("Weaving designs in time and space to create a "sustainable service society": Ezio Manzini, " 2004). While the term "service design" is not new, it has not appeared in the curriculum of design schools until recently. Though some researchers might not consider service design as a branch of Product Design, product designers and researchers have been involved in designing services for public or commercial use for decades as services are combinations of products, recourses, organization, local and global economy, and human relations. Service design is multidisciplinary in nature but it is in its core a problem solving process similar to Product Design. Ezio Manzini also points out in many of his publications that service design is a hybrid of materials and immaterials that contribute to a more sustainable community because it reduces the individual consumption of resources (Manzini & Jégou, 2003; Manzini, Vezzoli, United Nations Environment Programme. Division of Technology Industry and Economics. Production and Consumption Unit., Politecnico di Milano. Interdepartmental Research Centre Innovation for the Environmental Sustainability., & Lunds universitet. International Institute for Industrial Environmental Economics., 2003). For example, instead of focusing on designing washers and dryers for every household, product designers could design laundry mat services to service a local community. Needless to say, the products for a public service would have a different list of design considerations and design specs.

Service design must adopt systems theory because it is usually much more complex than any physical product since it involves a great deal of human to human and human to machine interactions that are beyond the control of the designers. More importantly, service design involves user experiences through time and space. These user experiences are usually quite different from the user experiences of a physical object because the users are actively providing feedback to the designers and the service will adjust accordingly at a much faster and more interactive manner. Thus, system dynamics models should be used to constantly monitor these feedbacks.

Systems Theory in Design Management

Design management is another domain of Product Design that could adopt systems theory. Pioneer of systems dynamics theory, Jay W. Forrester, states that "[Systems dynamics] is an approach that should help in important top-management problems…The goal should be to find management policies and organizational structures that lead to greater success" (Forrester, 1961). As the interface of design and management, design management involves both the organization making design-related decisions for the services and products it provides and managing its own design team.

One successful example of adaptation of systems theory into design management is the famous "14 Principles of the Toyota Ways" developed by the Toyota Corporation (Liker, 2004), which many believe was based on the production management methodology developed by W. Edwards Deming (Deming, 1986). The first principle states: base your management decisions on a long-term philosophy, even at the expense of short-term financial goals. This counter-intuitive method recognizes the delay of responses in a system. The 11th principle states: respect your extended network of partners and suppliers by challenging them and helping them improve. This is also an indicator that Toyota recognizes that its partners in production are closely-coupled stakeholders in the system and Toyota's success depends on the collaborations of all the stakeholders.

CONCLUSION

This chapter briefly discusses systems theory and the use of systems dynamics models to understand complex Product Design problems. Systems theory and its applications in Product Design research remains a controversial research topic due to the inconsistency of definitions of design, complexity, and other terms as used in different disciplines. For instance, "service design" has different meanings for managers, designers, and operation design specialists. The parameters of design problems remain unclear when product designers try to apply systems theory, thus they might be accused of being overly ambitious.

The most important purpose of developing systems dynamics models is to help researchers understand the correlations of different components in a system over time. Therefore it can be utilized in analyzing any system and process when constant interactions of parts are involved. However, this chapter does not suggest that systems theory represents a panacea that will provide solutions for all the complex design problems product designers are facing today. Rather, we intend for it to act as a catalyst to invite broader conversations on the potential of systems theory, systems dynamics modeling, and its applications to facilitate Product Design research and product development. The design community understands that as designers we need to learn to speak languages other than our "design language". We need to communicate with professionals in other domains to learn their paradigms and languages. This chapter is an attempt to breach one of those gaps between natural science, social science, and Product Design. By having this conversation, designers and scientists will learn to appreciate each other's perspectives, and thus learn to work collaboratively together.

REFERENCES

APSC. (2007). *Tackling Wicked Problems. A Public Policy Perspective.* Canberra: Australian Government/Australian Public Service Commission.

Banathy, B. H. (1996). *Designing social systems in a changing world.* New York: Plenum Press.

Checkland, P. (1999). *Soft systems methodology: a 30-year retrospective.* New York: John Wiley.

Deming, W. E. (1986). *Out of the crisis.* Cambridge, MA: Massachusetts Institute of Technology, Center for Advanced Engineering Study.

Forrester, J. W. (1961). *Industrial dynamics.* Cambridge, MA: M.I.T. Press.

Green, P. A. (1974). *Design education; problem solving and visual experience.* London: Batsford.

Gribbin, J. R. (2004). *Deep simplicity: bringing order to chaos and complexity* (1st U.S. ed.). New York: Random House.

Hutchinson, B. (2008). *Systems Thinking and Associated Methodologies.* Perth, Australia: Praxis Education.

Johnson, J. (2006). Embracing Complexity in Design. Retrieved April 10, 2009, from http://www.complexityanddesign.net/

Johnson, S. (2001). *Emergence: the connected lives of ants, brains, cities, and software*. New York: Scribner.

Liker, J. K. (2004). *The Toyota way: 14 management principles from the world's greatest manufacturer*. New York: McGraw-Hill.

Manzini, E., & Jégou, F. (2003). *Sustainable everyday: scenarios of urban life*. Milano: Ambiente.

Manzini, E., & Vezzoli, C. (2003). *Product-service systems and sustainability: opportunities for sustainable solutions*. Paris: UNEP.

McDonough, W., & Braungart, M. (2002). *Cradle to cradle: remaking the way we make things* (1st ed.). New York: North Point Press.

Meadows, D. H. Club of Rome, & Potomac Associates. (1974). *The Limits to growth: a report for the Club of Rome's Project on the Predicament of Mankind* (2d ed.). New York: Universe Books.

Meadows, D. H., Richardson, J. M., & Bruckmann, G. (1982). *Groping in the dark: the first decade of global modelling*. New York: Wiley.

Morecroft, J. D. W., & Sterman, J. (1994). *Modeling for learning organizations*. Portland, OR: Productivity Press.

O'Connor, J., & McDermott, I. (1997). *The Art of Systems Thinking*. San Francisco: Thorsons.

Pierson, P. (2004). *Politics in time: history, institutions, and social analysis*. Princeton, NJ: Princeton University Press.

Powers, M. (2008, May 19). Get Ready to Rethink What It Means to Be Green. *Wired, 16.06*.

Rittel, H., & Webber, M. (1973). Dilemmas in a General Theory of Planning. *Policy Sciences, 4*, 155–169. doi:10.1007/BF01405730

Seamon, D., & Mugerauer, R. (2000). *Dwelling, place, and environment: towards a phenomenology of person and world*. Malabar, FL: Krieger Pub. Co.

Sterman, J. (2000). *Business dynamics: systems thinking and modeling for a complex world*. Boston: Irwin/McGraw-Hill.

Ulrich, K. T., & Eppinger, S. D. (2004). *Product design and development* (3rd ed.). Boston: McGraw-Hill/Irwin.

Weaving designs in time and space to create a "sustainable service society": Ezio Manzini. (2004). *Future News*. Retrieved from www.changedesign.org/Students/Changes/Enabling/Enabling%20Docs/ffmanzini.pdf

World Commission on Environment and Development. (1987). *Our common future*. New York: Oxford University Press.

KEY TERMS AND DEFINITIONS

Closed System (Isolated Systems): Systems that do not interact with their surroundings.

Emergence: The macro-level result experienced as a consequence of the collective behavior of individual system components.

Hard Systems: Also called "mechanical systems", are systems whose behaviors and boundaries can be successfully described and predicted in mechanical terms, where the primary concern is with objects, which can be objectively defined (Hutchinson, 2008).

Leverage: In systems theory, leverage points are places where policy resistance is lower and small changes will be amplified.

Open Systems: Systems that can be affected by events outside of the actual or conceptual boundaries of the systems.

Path-Dependence: Means the self-reinforcement mechanism of institutions or systems

(Pierson, 2004). In a broader conception, it also means "history matters" in the development of institutions and systems.

Soft Systems: Systems that have fuzzy boundaries that are not easily defined.

Systems Dynamics Model: Systems dynamics model is a modeling approach to simulating system and understanding the behavior of complex systems over time.

Chapter 21
Integrated Approach to Product and Process Design Based on Life Cycle Engineering

Paulo Peças
Instituto Superior Técnico, Portugal

Elsa Henriques
Instituto Superior Técnico, Portugal

Inês Ribeiro
Instituto Superior Técnico, Portugal

ABSTRACT

An integrated approach to product and process design and development based on Life Cycle Engineering principles is proposed. The aim is to make available a structured framework to drive the design team in their discussions and analysis towards the creation of purposeful design concepts and process alternatives and in their global evaluations towards informed decisions on a life cycle perspective. A case study is presented based on a cloths peg already in the market, which is intended to be redesigned in order to reduce its cost and increase its attractiveness. The simplicity of the product allows giving emphasis to the models and its methods. Considering several design and process alternatives, it is revealed how the proposed approach is applied and the type of conclusions and findings of such type of methodology.

INTRODUCTION

Product innovation through continuous improvement of existent products and processes is a key factor for companies to sustain competitiveness in nowadays markets. To deal with this challenge marketing, design and manufacturing areas should be combined, bringing together a multidisciplinary team with different expertises and perspectives (Ulrich, 2003). Product innovation is achieved by implementing the product development as a process, starting to capture information from the exterior, identifying the customer's needs, and following the best way to achieve products with more value. Value can be perceived as a ratio between the customer needs and the product cost. Pursuing high Value means to conceive products with the functions that customers are willing to pay at the

DOI: 10.4018/978-1-61520-617-9.ch021

lowest cost. Value Analysis is an organized and creative approach to efficiently identify unnecessary costs that don't add value to the perceived quality characteristics. When applied to products, this approach assists in the methodical utilization of better approaches, alternative materials, newer processes, and abilities of specialized suppliers. It focuses engineering, manufacturing, and purchasing attention on one objective-equivalent performance for lower production cost, providing step-by-step procedures for its accomplishment in an efficient and assertive manner (Miles, 1989).

Although Value Analysis is an effective tool in product development, nowadays products need address not only the production costs, but also all the costs occurred throughout their entire life cycle (Life Cycle Cost - LCC). Moreover, the LCC approaches by themselves, without additional assessments, are not sufficient to provide drivers and indicators for a sustainable practice (Rebitzer, 2003). It is advised to evaluate the product on an environmental basis following also a life cycle approach, namely involving Life Cycle Assessment (LCA) methods. The main reason for including a life cycle perspective since the early stages of product development is that decisions taken at product design stage largely influence the product's costs and environmental impacts for its entire life (Kurk, 2008). Therefore, to design better products for their entire life Design-for-X (DfX) strategies, supported by the corresponding DfX tools, have been increasingly and successfully applied. These strategies drive the design team in the creation of products and services that achieve a specific target or that maximize the performance in a wide range of engineering fields (cost, environment, assembly, etc.). Despite the wide spectrum of design accomplishments, the DfX strategies restrict the analysis to trade off among several specific alternatives in each engineering field. In fact, the general objective of design for cost is to minimize the production cost; design for assembly aims the definition of the product's features to minimize the effort of the assembly tasks; and design for environment targets a lower environmental impact focusing essentially on product material issues. The integration of the decisions considering multiple design objectives and perspectives is difficult and sometimes impossible. The addition most individual DfX strategies are focused on a product life window and do not tackle conveniently the overall impact of a design decision. The consideration of the all life cycle stages of a product in the early design phase allows a more complete perception of the product's importance and differentiation in the market and in the society. This way of design and develop a product can be called as Design for the Life Cycle. To differentiate it from the regular DfX strategies, several authors prefer the denomination: Life Cycle Engineering (LCE), which is defined as a decision-making methodology that considers technical performance, environmental, and cost dimensions throughout the duration of a product, guiding design engineers towards informed decisions (Wanyama, 2003; Betz, 1998).

LIFE CYCLE ENGINEERING IN PRODUCT DESIGN & DEVELOPMENT

LCE stands for the need to consider from the early product concept its complete projected life, including market research, design phases, manufacturing processes, qualification, reliability aspects and customer service-maintainability-supportability issues (Wanyama, 2003; Keys, 1990; Alting, 1995). To develop increasingly sophisticated products (systems or facilities) in shorter time-frames is a tough challenge that can be better achieved by a holistic understanding of products and processes life cycle (Alting, 1995; Ishii, 1995). Additionally, a change in the whole concept of industrial production and in the behaviour of the consumer society has been mandated by profound demographic and economic evolutions, and by a growing awareness of environmental issues

(Alting, 1995; Fava, 2000). So, to support LCE philosophy in early design phases, implementation guidelines were published, considering LCE an ongoing process to develop specifications to meet a set of requirements and goals that span the product life cycle (Cooper, 2001). In addition to those guidelines tools and methodologies must be developed to foster the implementation of the LCE approach to product design as "engineering activities involving the application of technological and scientific principles to the design and manufacture of products with the goal of environment protection and resources conserving, while encouraging economic progress, keeping in mind the need for sustainability, and at the same time optimizing the product life cycle and minimizing pollution and waste" (Keys, 1990; Jeswiet, 2003). A LCE tool integrates in a single decision framework all the relevant information that must be available within the design phase, regarding the technical, economical and environmental dimensions, and considers all the life cycle stages of the product to avoid tradeoffs (Tseng, 2008; Aurich, 2006).

The analysis of the technical, economic and environmental performance provides a solid foundation for designers to understand the trade-offs and implications of product design alternatives (Wanyama, 2003; Fava, 2000). The reduction to three dimensions can be considered as a drawback as compared with DfX, which foster the consideration of a wide spectrum of disciplines. The answer to this relies in the way the three dimensions are analyzed. For the economic performance the use of methodologies like Life Cycle Cost is recommended. The estimation of all the costs associated with a product throughout the product's life, from "cradle" to "grave" integrates the analysis of the impact of design for cost, design for maintainability, design for assembly, etc. Accordingly, the use of methodologies like Life Cycle Assessment to estimate the environmental performance includes the disciplines of design for environment, design for recycling, design for standards, etc. Finally the designs for reliability, for service, for use, among others are surely integrated in the technical performance assessment. For the integrated analysis of the three dimensions comparison methods can be used, e.g. graphic theory and matrix approach (GTMA) and fuzzy multiple attribute decision-making methods (MADM).

The research in LCE challenges the academic world because it is based on a multidisciplinary approach in dealing with a problem solving framework (Keys, 1990). In fact the development of LCE tools and its implementation in product design and development requires the collaboration of different expertises during several phases of such project. Therefore, the incorporation of concurrent engineering practices is recommended if not mandatory (Bento, 2008; Ribeiro, 2008).

Although it is convenient to develop the LCE approach in the design early phase, it can involve a large set of alternative materials and other design options which means an enormous effort in terms of life cycle analyses (LCC and LCA). Providentially, there are several factors that limit the number of alternatives to study on the LCE approach. In fact, the company technological and strategic framework restricts the number of alternatives to the ones possible to handle, to have access, to process, among others. Additionally, the product logic and specific requirements limit the alternatives domain to a few options (Peças, 2009; Pousa, 2009). Despite this focus on specific alternatives range, the spectrum of possible design variables is still large. So, to efficiently and effectively apply the LCE approach, the design problem must be well established based on the definition of a few set of design alternatives for further analysis and comparison. The use of Product Design and Development (PDD) related methods to compare the initial set of alternatives, without the need a huge set o data, permits to eliminate a significant number of possibilities and focus the LCE analysis in the most promising ones.

This chapter presents an integrated approach of PDD, derived from Value Analysis and LCE, aiming the comparison and identification of best design alternatives. The comparison is the output of the quantitative analysis deriving from the LCE approach for several production (or market) alternative scenarios, which are previously defined by a sequential product development procedure. This procedure begins with the product planning, continues with concepts development and ends with the design concepts selection based on Value Analysis. The design concepts selected are the ones further developed under the LCE analysis.

The method used for the final comparison is based on ternary diagrams, allowing mapping the best alternatives in the universe of the three performance dimensions: cost, environmental and technical. In this way the selection of weighting criteria for the performance dimensions is avoided, inhibiting unfounded favouring (or disfavouring) of any alternatives. The use of ternary diagrams for mapping the LCE solutions has been applied to different types of products and process in the last years (Ribeiro, 2008; Peças, 2009; Betz, 1998; Pousa, 2009).

After presenting the general integrated approach proposed for PDD, it is applied to a specific product in a case study basis. The studied product - a cloths peg – is simple enough to clearly demonstrate the type of outputs and comparative analysis made possible with the proposed approach. The data and results presented are real since they were obtained on an industrial project between a University (IST/IDMEC, Lisboa, Portugal) and an industrial company (Fapil S.A., Portugal). From the industrial point of view the objective was to redesign the cloths peg, already in the market, in order to reduce its cost and increase its attractiveness. In fact, some positive features of the available cloths peg (double-sided hang cloths feature) were intended to be kept in the future design. As the design of the resulting product was patented, the real solutions achieved are presented in this document. Nevertheless, as regards to costs, there are confidentiality restriction involved so the economic aspects are referred in this document to a fictitious monetary unity (MU).

AN INTEGRATED APPROACH TO PRODUCT DESIGN & DEVELOPMENT

General Methodology

The proposed approach integrates the Development and Selection of Product Design Concepts (Phase 1) with Process Development and Comparison (Phase 2). In phase 1 the typical steps of PDD are followed (Ulrich, 2003) to achieve the Design Concepts of the product (Figure 1). The first step consists in Product Planning where the target market and the project main objectives are stated. In the case of a product redesign, this step also includes the current product diagnostics. The second step is the Design Concepts (DC) development, characterisation and selection. Through the identification of physical, mechanical, chemical and economic constraints, together with customer requirements and project assumptions, it is possible to define the target specifications of the product. After concepts generation, their screening is performed by Value Analysis, resulting in a limit number of DC alternatives to be further analysed in Phase 2. Although some product costs estimations are included in Phase 1, these are rough approaches based on the what seems to be the most relevant cost factors and don't envisage the whole life cycle of the product. Moreover, no environmental analysis is performed in this early phase.

In Phase 2 a life cycle assessment methodology is applied to the alternative design concepts identified and selected in phase 1 for further exploration. Besides the environmental and economical evaluation, through LCA and LCC methods respectively, a functional/technical evaluation of the product is also performed. Finally, these three dimensions of analysis are aggregated in a single framework in order to better understand which decisions to take regarding the alternatives studied. A more detailed description

Figure 1. General description of the Integrated Approach

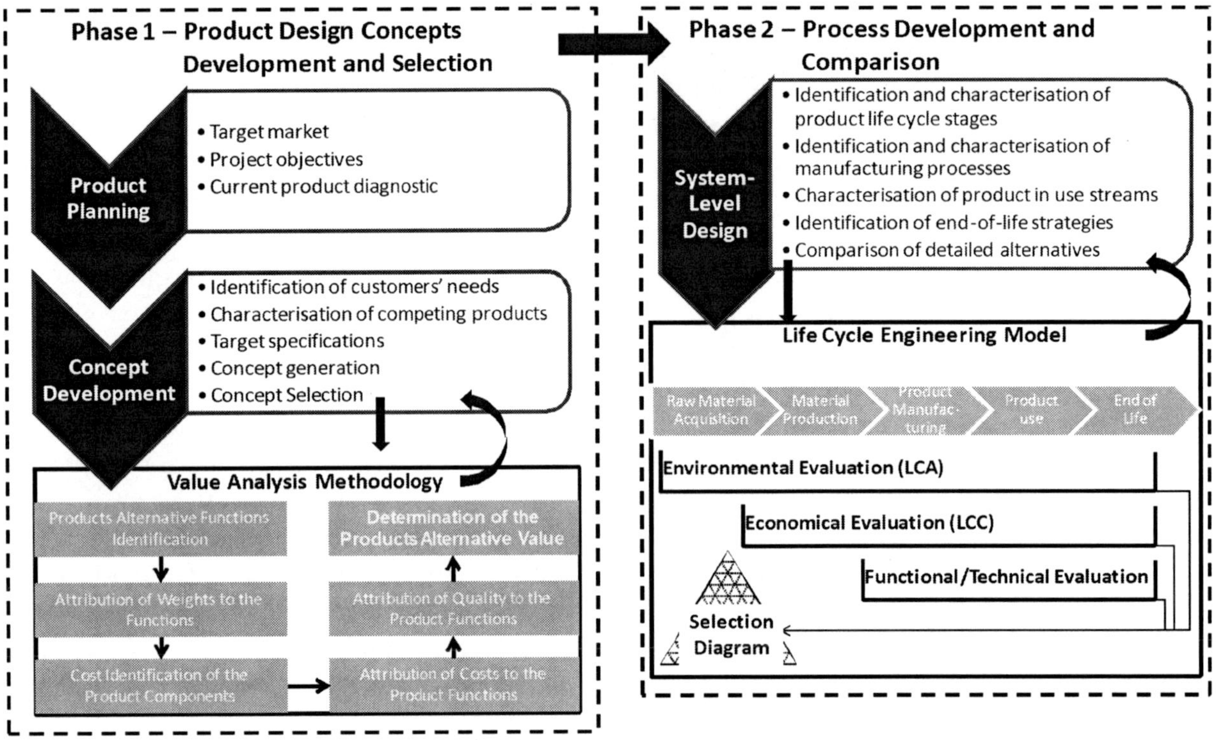

of the models developed and used is presented in the following section.

Models used on Life Cycle Engineering

As mentioned, through the LCE methodology the life cycle stages of different design alternatives are analysed in terms of economical, environmental and technical/functional performances. These three dimensions are aggregated in a single framework analysis, in which depending on the importance given to each one, an alternative appears as the best choice. The proposed LCE model is composed by a Life Cycle Cost (LCC) model built-up to include and correlate the cost factors, allowing the economic performance assessment. The environmental impact is quantified by a Life Cycle Assessment (LCA) model that allows the corresponding evaluation of the environmental performance of the alternatives. Finally, it includes a model based on Multiple Attributes Decision-Making (MADM) method to assess the alternatives as regards the fulfilment of functional requirements and their contribution for the technical performance over the life cycle.

Life Cycle Cost Model

The Life Cycle Cost analysis generally refers to the assessment of all costs associated with a product throughout its life, including the costs that are not normally expressed in the product market price (Krozer, 2006), such as costs incurred during the usage and disposal phases. LCC is essentially an evaluation and comparison tool in the sense that it uses a metric for choosing the most cost-effective solution from a set of alternatives.

In Figure2, the LCC model developed to be integrated in the proposed LCE general model is presented. The LCC model must be tuned to the specific characteristics of the product under

Figure 2. Life Cycle Cost model

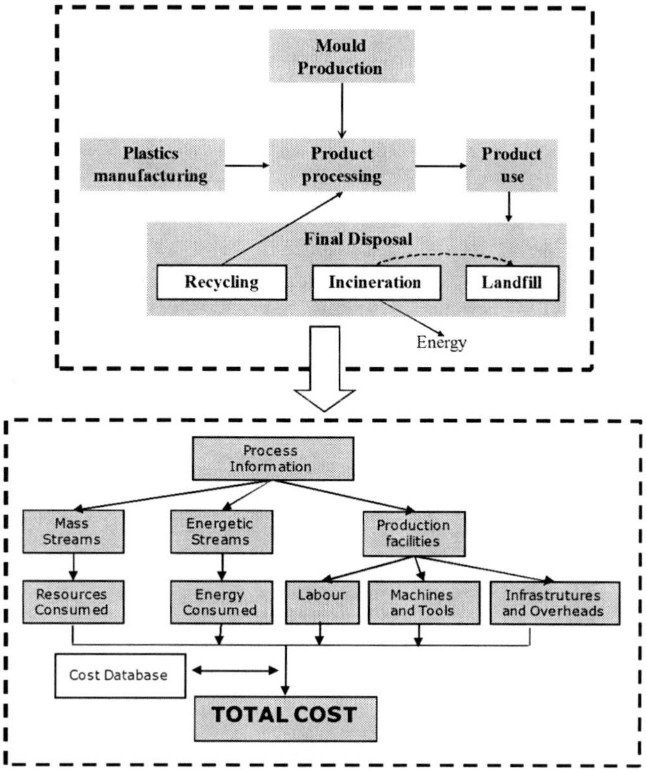

study or development. Some products may require several manufacturing processes, others just one. Some may incur in complex use phases involving intricate cost issues, others have cost free use phase. For the case of the cloths peg, the production phase involves a dedicated injection mould, so its design and manufacturing costs must be included in the product LCC.

Life Cycle Assessment Model

To perform the environmental analysis the Life Cycle Assessment is proposed. This is a structured method to quantify potential environmental impacts of products or services over their full life cycle (Johansson, 2007; Udo de Haes, 2007). Although LCA had emerged in the early 1970s, its methodological basis at that time was chaotic and conflicting results were produced. As an attempt to organize and harmonize LCA developments, SETAC (1980s), ISO (1990s) and UNEP/SETAC (1990s) established a terminology, a standard methodology (ISO, 1997, 1998, 2000) and, as LCA is still a field of innovation and discovery, became privileged spaces to submit developments for discussion (Goedkoop, 2007). Presently, LCA consists of four steps: definition of the goal and scope of the study, construction of the product life cycle model with all environmental inflows and outflows (life cycle inventory stage – LCI), evaluation of the environmental relevance of all the inflows and outflows (life cycle impacts assessment stage – LCIA) and, finally, the interpretation of the results (ISO, 2000#1). There are several methods for LCIA stage compatible with ISO requirements (ISO, 1997) and therefore, most experts prefer to select a published method instead of developing a new one (Goedkoop, 2007).

The LCA model developed and applied to the proposed LCE general model is present in

Figure 3. LCA model and environmental streams considered throughout the product life cycle

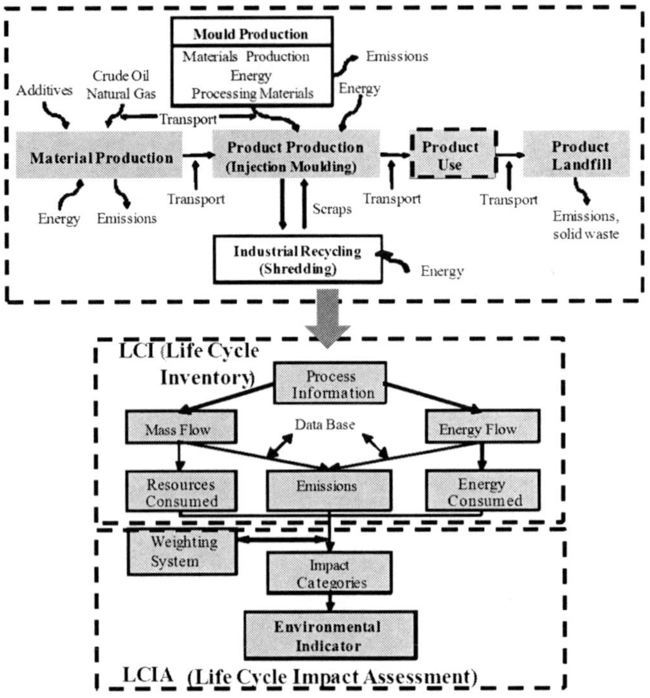

Figure3, together with the environmental streams considered throughout the cloths peg life cycle. As mentioned for the LCC model, the LCA model must be tailored to the type of product under development. The methodology proposed for the impact assessment considers 11 environmental impact categories, aggregating all the emissions and resources consumption all over the life cycle into three main impact blocks: Human Health (HH), Ecosystem Quality (EQ) and Resources (R). Then, the scores achieved for the impact blocks are weighted into a single value, called the "eco-indicator 99" (EI'99) (Goedkoop, 2000). The weighting coefficients are applied according to the hierarchic/average (H/A) perspective, which is a moderate view generally accepted by the scientific community, attributing 40–40–20% of weight to the three considered impact blocks, HH–EQ–R, respectively (Ermolaeva, 2004). One should note that these blocks have specific separated units: Disability-Adjusted Life Years (DALY) is the health indicator; Potentially Disappeared Fraction multiplied by an area unit, m^2, and a time unit, year (PDF $\times$ m^2 $\times$ year) translates the damage to ecosystems; and Mega Joule (MJ) surplus is the measure of resource depletion.

Functional/Technical Evaluation

The evaluation of the functional/technical (F/T) performance of a product, tool or other equipment relies on the know-how of professionals (and users) to identify the set of relevant technical attributes and quantify their importance, translating in engineering data the relevant quality requirements of the product from the user (the most relevant stakeholder for the identification) point of view, in order to support evaluation and comparison. Several decision making methods have been applied on this kind of comparisons, such as graphic theory and matrix approach (GTMA) and fuzzy multiple attribute decision-making methods (MADM) (Rao, 2007). However, the expertise

Table 1. Decision matrix in MADM methods

Alternatives	Attributes					
	B_1	B_2	B_3	-	-	B_M
	(w_1)	(w_2)	(w_3)	(-)	(-)	(w_M)
A_1	m_{11}	m_{12}	m_{13}	-	-	m_{1M}
A_2	m_{21}	m_{22}	m_{23}	-	-	m_{2M}
A_3	m_{31}	m_{32}	m_{33}	-	-	m_{3M}
-	-	-	-	-	-	-
A_N	m_{N1}	m_{N2}	m_{N3}	-	-	m_{NM}

of professionals remains crucial for this type of evaluations.

In the LCE model proposed in this research, the F/T evaluation refers to technical attributes and their contribution to the technical performance of the products over its life cycle. It can include not only the product F/T performance in the use phase, but also material and processes technical behaviour, productivity, capability, etc. during the product and tools (if required) manufacturing phase. The evaluation of the F/T performance of the alternatives is based on Multiple Attributes Decision-making (MADM) methods, in particular the Simple Additive Weighting (SAW) one, which is the simplest and still the widest used MADM method. MADM is an approach employed to measure the performance of a finite number of alternatives, involving a decision matrix and a decision-making method. The decision matrix, outlined in Table 1, has four main parts: alternatives (A_i, for i=1,2,…,N), attributes (B_j, for j=1,2,…,M) and their respective importance weights (w_j, for j=1,2,…,M), and the performance measure of the alternatives in each attribute (m_{ij}, for i=1,2,…,N; j=1,2,…,M). The overall performance score of the alternative i, P_i, is given by Equation 1:

$$P_i = \sum_{j=1}^{M} w_j m_{ij} \tag{1}$$

The measures of performances, m_{ij}, can be real numbers (e.g. cycle time, duration) or fuzzy and/or linguistic information (e.g. technology availability and complexity, product aesthetical). In the first case, if they are not expressed in identical units of measure, all the elements of the decision table must be normalized. In this case, the overall performance score of an alternative is given by Equation 2:

$$P_i = \sum_{j=1}^{M} w_j m_{ij\,normal} \tag{2}$$

If the measures of performance are fuzzy and/or linguistic, fuzzy MADM methods should be addressed. Rao (2007) propose a method for converting linguistic terms into fuzzy number and then the fuzzy numbers into crisp scores. Different scales can be used and Table 2 illustrates the method for a 5-point scale.

Table 2. Conversion of linguistic terms into crisp scores (5-point scale) (Rao, 2007)

Linguistic term	Crisp Score
Low	0.115
Bellow average	0.295
Average	0.495
Above average	0.695
High	0.895

Global Evaluation/Selection Diagram

The final step of the LCE model is the analysis of the results of the three dimensions aiming to compare and select the "best" DC together with the "best" manufacturing process alternative. The first action consists in the adimentionalization of the final numerical results obtained in the LCC, LCA and F/T performance for each alternative by the best score (minimum in LCC and LCA, maximum in F/T performance), and in their translation into a numerical value in a scale of 10 points. As in common LCE analysis, the following action could be to attribute importance weights to each dimension, with the sum of the three dimensions weights to be 100%. Nevertheless, if this approach is followed different combinations of weights might result in a different "best alternative". The difficulty to attribute importance weights to each dimension of analysis, which closely reflect a corporation strategy, and the sensibility of the achieved results to such weights, are the major drawbacks normally pointed to a global evaluation based on weights attribution (Ribeiro, 2008; Peças, 2009). To overcome this disadvantage and have a clear view of the possible "best alternative" correlated to its domain of weights, a global evaluation is performed through a ternary diagram, where each axis represents one dimension of analysis. The diagram illustrates not only the "best alternative" for a particular set of importance weights but also the domain of weights for each one. This approach fosters a global comparison of the alternatives according to different business scenarios and corporate strategies, supporting more informed decision-making processes.

THE CLOTHS PEG CASE STUDY

The application of the integrated approach of product design and development to a simple product - a cloths peg – is following described. The main objective of the study is to redesign an existent cloths peg already in the market, in order to reduce the product cost and yet keep or even surpass the functional and technical characteristics.

Development and Selection of Product Design Concepts

After determining the target market and project objectives, the current product diagnosis was the main issue in order to understand and determine the technical characteristics of the current product. This phase consisted mainly in data gathering from several sources, namely the type of materials used in the current clothes peg and the main characteristics of these materials, geometric data, manufacturing process and production costs. In order to mechanically characterize the functional behaviour of the current product an experimental apparatus was assembled (Figure4). The data generated was considered confidential by the company, and so it is not completely exhibited.

The second step consisted in the concepts development. The several required tasks happened simultaneously, overlapping in time. Identifying customer's needs was the first of these tasks, mainly to understand what the customer wants, and how much he/she is willing to pay for a clothes peg. Several clothes pegs users were interviewed, to better understand what was at stake. The new product's specification was done simultaneously, by integrating the customer needs and expectations with the competing products characteristics and with the diagnosis previously done. After this, new concepts were generated, changing particular aspects of the existing peg body and clamping scheme. The main drivers for redesign were the weight reduction (fostering price reduction), and the increasing of stiffness, strength and user comfort.

To drive the generation of concepts, Value Analysis was used in the comparison of product functions and in the identification of the ones that should be improved in cost or quality to increase product value. The functions were identified

Figure 4. Current clothes peg. a) Experimental apparatus; b) Mechanical forces diagram; c) Maximum Von Mises stresses

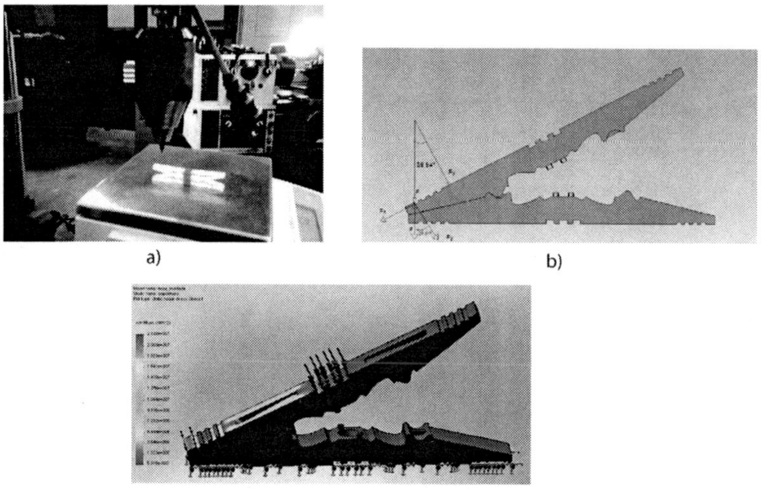
a) b) c)

and classified in terms of importance. Then the materials and manufacturing costs of the product were identified, computed and distributed by the product components. Based on the product architecture (extremely simple in this case) a correlation between product components and functions can be established and the cost of each function can be estimated. The cost-function histogram was built allowing identifying the relevance of each function in terms of cost and importance to the customer/user (Figure5). Additionally, the level of relative importance of each design feature was derived in brainstorming sessions involving elements from the company and clothes pegs users. The features related to usability and ergonomics were, in general, more important to the customers, while the company was more preoccupied with features related to cost and manufacturing.

Three new Design Concepts emerged. The Design Concept 1 and 2 involve material reduction, satisfying the intentions of the company of cost reduction (Figure6). The DC 1 involves a more dramatic reduction of material, with a slight reduction on the clamping force to hold the cloths. As it can be observed in Figure5, the function Hold Cloths has a high influence on cost, significantly higher than its relative importance, pointing to a potential slightly decrease in the performance of this function. Since the high material reduction done in DC 1 can reduce significantly the perceived Value of the cloths peg, the DC 2 is a compromise amid material reduction and clamping force. It must be noticed that using the same material of the current cloths peg, both DC 1 and DC 2 fulfil the technical requirements determined in the product specifications. Also for both concepts the ergonomics of the product was improved as can be seen in Figure6. The Design Concept 3, corresponds to the one that more

Figure 5. Cost-Function histogram for the cloth peg

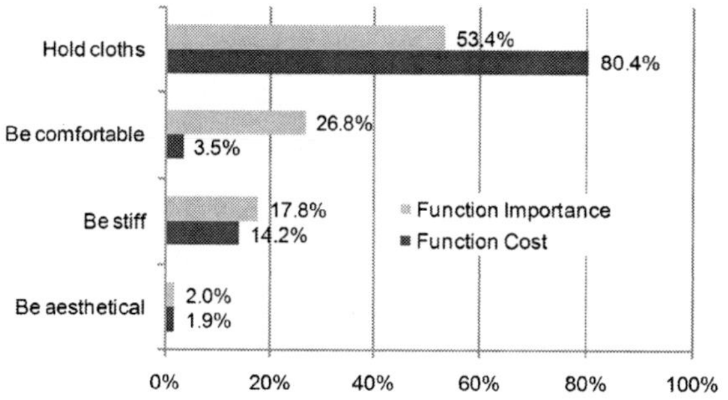

Figure 6. Design Concepts: a) Current Product; b) Design Concept 1(Patent PT103874, PT1012); c) Design Concept 2 (Patent PT103874, PT1013) and d) Design Concept 3

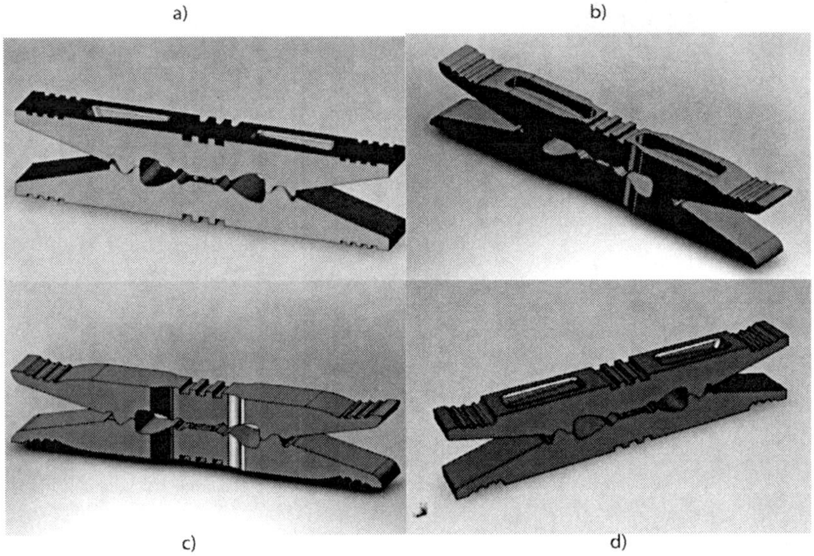

emphasizes the customer's needs. So, the current product design was changed mainly to increase usability (comfort), bearing in mind that stiffness and the clamping force should be less damaged than in previous concepts. The DC 3 considers that material reduction is not an objective from the users' point of view (Figure6).

The current cloths peg is made of a polypropylene homopolymer (HMU 216). Other materials were evaluated to find the "best" combination with the geometrical concepts. In order to avoid long descriptions related with material properties it is only presented the reference of the involved materials. The selection of 2 new materials (a lower grade of a polypropylene homopolymer, HMU 210, and a polypropylene copolymer, PP 1063 L1) was based on Ashby diagrams and on information supplied by the material manufacturer related with resistance to moister and ultraviolet radiance. For each material and each geometry finite element analysis were run to confirm and slightly adapt the cloths pegs design to each material. Together with the shape characteristics, the numerical analysis allows determining the performance of each design concept in each critical-to-quality function (Table 3). To determine a cost comparative measure of each design concept associated to a material, the material quantity multiplied by its specific cost was calculated. A measure of the Value of each design solution is then the ratio between the level of fulfilment of the function (functions total score) and the material cost.

From Table 3 is possible to conclude that all the new Design Concepts achieve a higher Value than the actual product. As regards to the material the copolymer is the one that permits higher Values due to its lower cost. As regards to the design concepts, the DC 1 and DC 2 present the best results, which is strictly related with the lower materials cost resulting from a lower consumption of materials. The application of Value Analysis highlights the two design concepts that, lowering quality characteristics not crucial to the product performance or to the client perception of quality, allow a significant cost reduction. A higher Value was achieved by reducing cost in attributes that are not essential. The DC 1 and DC

Table 3. Value determination for the combinations of Design Concepts and materials

		Functions						
	Material	**Hold Cloths**	**Be Comfortable**	**Be Stiff**	**Be Aesthetical**	**Total Score**	**Material Cost**	**VALUE**
Current	HMU 216	4.27	0.80	1.60	0.16	6.83	0.022	310.5
Design Concept 1	PP 1063 L1	2.32	2.00	1.07	0.30	5.69	0.011	**517.3**
	HMU 216			1.42		6.04	0.015	402.7
	HMU 210			1.42		6.04	0.014	431.4
Design Concept 2	PP 1063 L1	2.67	1.88	1.25	0.30	6.10	0.012	**508.3**
	HMU 216			1.42		6.27	0.016	391.9
	HMU 210			1.42		6.27	0.014	447.9
Design Concept 3	PP 1063 L1	3.20	2.14	1.28	0.16	6.78	0.015	452.0
	HMU 216			1.50		7.00	0.020	350.0
	HMU 210			1.50		7.00	0.019	368.4

2 made of PP 1063 L1 were the design solutions selected for the second phase of development, in which a deeper analysis was carried out on both concepts as regards the evaluation of their manufacturing processes.

Process Development and Comparison

This section is dedicated to the analysis and discussion of the results obtained by applying the LCE model (Phase 2 of the integrated approach of Product Design and Development) to compare alternative manufacturing processes of the Design Concepts previously selected. The aim is to evaluate on a life cycle perspective different process designs and also the effect of the DC on the processes life cycle performance. As mentioned before, the process design is an integrative phase of the product development and it can influence significantly the economic and environmental impact of the product. Even if the design concept determines largely the global manufacturing process, a deeper breakdown into more refined process alternatives provides valuable inputs for the selection of the best design.

Accordingly, in this case study, two types of moulds were considered differentiated by their plastic feeding system: mould with cold-runners and mould with hot-runners. The former are less expensive and easier to manufacture; the later allow lower cycle times and lower labour input after injection moulding.

Beside the type of mould, the number of cavities was also allowed to differ between 16, 32 and 96 cavities per mould. The number of cloths pegs (2 body parts per peg) obtained in each injection cycle is half of the number of cavities (parts).

From the above, the number of process design alternatives is 6 for each design solution, making 12 alternatives in total. For each design / process combination an application for injection moulding simulation was used to estimate the cycle time, the mass of material required, the number and position of feeding points (Figure7), the appropriated design of the cooling system to avoid shrinkage problem (Figure8), according with appropriate process parameters.

Table 4 presents the main results obtained in the injection moulding simulation of the 2 design solutions. It's clear that DC 1 requests less plastic material and permits a lower cycle time. When

Figure 7. Feeding channels designed for the a) 32 cavities / Cold runners / DC 1 and b) 32 cavities / Hot runners / DC 1

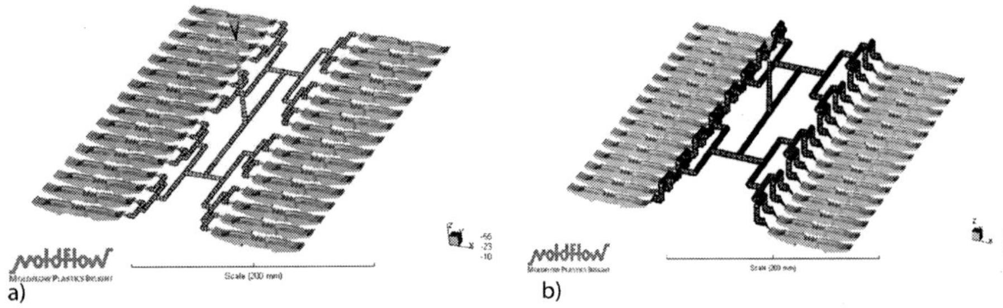

compared with cold runners, the use of hot runners reduces the consumption of plastic material per cycle and the cycle time. As the number of cavities increases the cycle time per part naturally reduces. Nevertheless, for the cold runners moulds the increase of the mould size causes the need for more channels and longer feedings, so the plastic mass required is higher.

It can be observed that the mould cost has an inverse trend with mould productivity (cycle time/part and mass/part), as hot runners and a large number of cavities endorse the increase of the mould cost.

The different influence of each design variable and process parameter will be included in the following analysis, in which LCC and LCA models are applied to the 12 different alternatives.

Results and Discussion of the Life Cycle Cost

The LCC of the cloths peg was assessed through a process-based cost model in order to allow posterior sensitivity analysis related with the impacts of changes in design, materials, manufacturing processes, production volume, etc. It should be noted that the use phase of the cloths peg was considered similar for all the alternative concepts under analysis and so it was not included for comparative purposes. The material cost is understanding from a manufacturing point of view meaning that it takes in all the material mass required for the injection process multiplied by the its commercial specific price. The plastic material required for the injection process but not included in the final product (plastic waste in the mould feeding system) can be reused in the injec-

Figure 8. Cooling systems designed for the a) 16 cavities / Cold runners / DC 1 and b) 16 cavities / Hot runners / DC 1

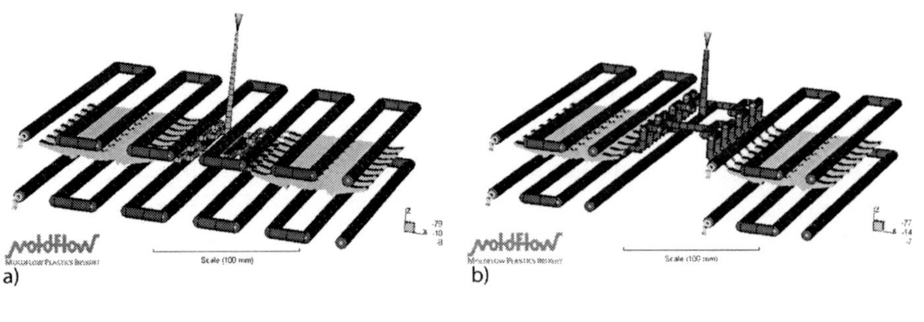

Table 4. Results from the injection moulding simulations that are inputs for the LCE model

	Mould Type	Cavities	Reference	Cycle Time (s)	Cycle Time/part (s/part)	Mass/cycle (g/cycle)	Mass/part (g/part)
Design Concept 1 DC 1 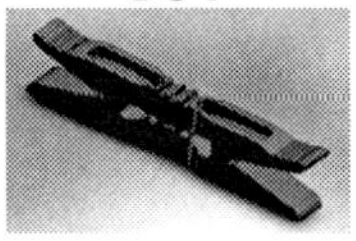	Hot-Runners	16	16H-DC1	16.3	1.019	47.9	3.00
	Cold-Runners	16	16C-DC1	20.9	1.306	56.4	3.53
	Hot-Runners	32	32H-DC1	17.1	0.534	95.9	3.00
	Cold-Runners	32	32C-DC1	22.8	0.713	114.4	3.57
	Hot-Runners	96	96H-DC1	19.5	0.203	287.6	3.00
	Cold-Runners	96	96C-DC1	25.4	0.265	414.0	4.31
Design Concept 2 DC 2 	Hot-Runners	16	16H-DC2	27.9	1.744	52.1	3.26
	Cold-Runners	16	16C-DC2	31.5	1.969	60.6	3.79
	Hot-Runners	32	32H-DC2	29.0	0.906	104.3	3.26
	Cold-Runners	32	32C-DC2	35.4	1.106	122.9	3.84
	Hot-Runners	96	96H-DC2	34.5	0.359	312.8	3.26
	Cold-Runners	96	96C-DC2	38.0	0.396	439.3	4.58

tion process. Nevertheless, for quality reasons a limit of 10% of material recycling was allowed for each injection. The remaining material was considered to be sold or incorporated in other products as recycled plastic. The moulds cost was estimated by matching quotations of several moulds manufacturers. The injection moulding costs were calculated based on three direct and time dependent parcels: labour, injection equipment and energy. The indirect production costs, fixed or variable, were neglected given that they are similar for all cases and do not affect the relative life cycle results of the alternatives. Regarding the end of life, the mould was considered to be sent to recycling and so sold as scrap alloy steel.

Finally all costs were gathered and the total LCC of each alternative was computed (Figure 9) for a production volume of $4x10^6$ cloths pegs (4 Mpegs) involving the injection of $8x10^6$ body parts. The results evidence that DC 1 based alternatives tend to exhibit a lower life cycle cost. The main reason is associated with lower injection moulding costs that are a consequence of the lower cycle time required for the DC 1 alternatives. The importance of the injection moulding costs, for this production volume, is also revealed by the lower total costs of the moulds with higher number of cavities (32 and 96).

Despite the influence of the injection moulding costs, the mould cost is also essential mainly for the DC1 alternatives. In fact, the lower mould cost is responsible for the best cost result of 32C-DC1 (design concept 1, 32 cavities, cold runners) alternative, compensating for example an higher cost on injection moulding and plastic material than on 96x-DC1 alternatives. It must be noticed that with the increasing of the production volume, the importance of the mould cost will vanish in opposition to importance of the injection moulding and material costs, as will be discussed later.

The consumption of material constitutes an important percentage of the total LCC. However, its variability is not as high as the other cost factors. The main differences in material consumption are observed between the hot and cold runners alternative moulds. Since the cold runners moulds consume (waste) more plastic material per cycle its cost is higher than in the ones using hot runners. Once again, for this production volume, the influence of the mould cost determines that alternatives with hot runners (more expensive

Figure 9. Life Cycle Cost of the Design Concepts/Process alternatives for 4 Mpegs

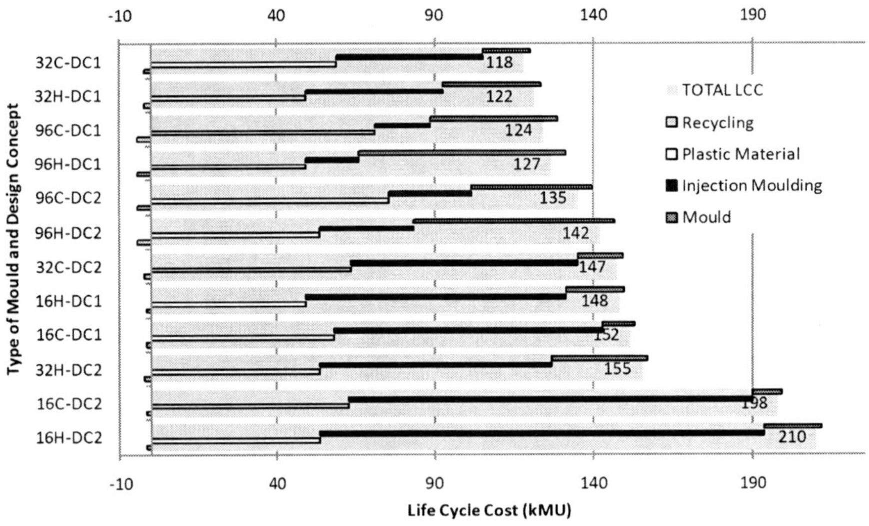

moulds) have always a higher LCC than the ones with cold runners.

For a further understanding it is interesting to look at the cost parcels of the injection moulding phase for the production volume of 4 MPegs (Figure 10). There is a trend for the higher mould cost to be balanced with a lower value of the other injection moulding cost parcels (material, labour, equipments, and energy). In fact, the 32C-DC1 alternative, the one with lower LCC for 4Mpegs (Figure9), is in the middle of the graph with an average injection moulding cost balanced with one of the lowest mould cost. As stated in Figure9, the alternatives with lower injection moulding cost are the ones involving moulds with a larger number of cavities and hot-runners. These two

Figure 10. Cost parcels of injection moulding for 4 Mpegs

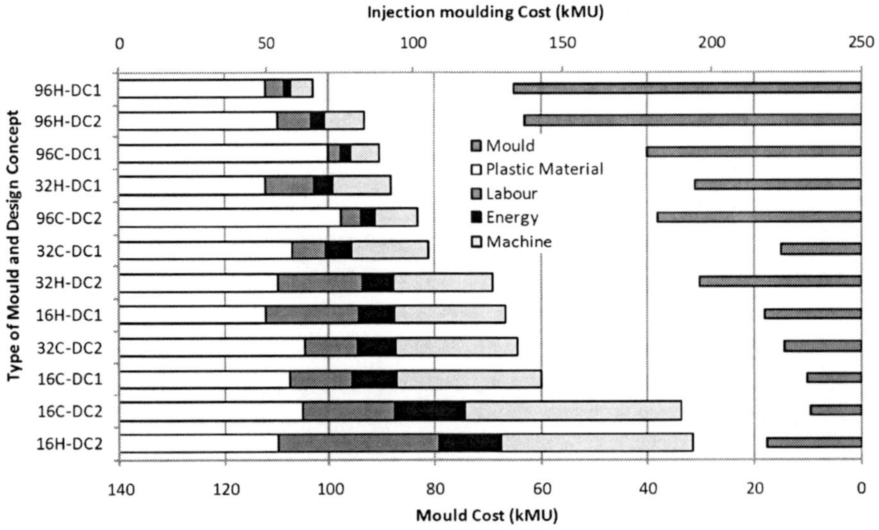

Figure 11. Best (lowest LCC) alternatives for different production volumes

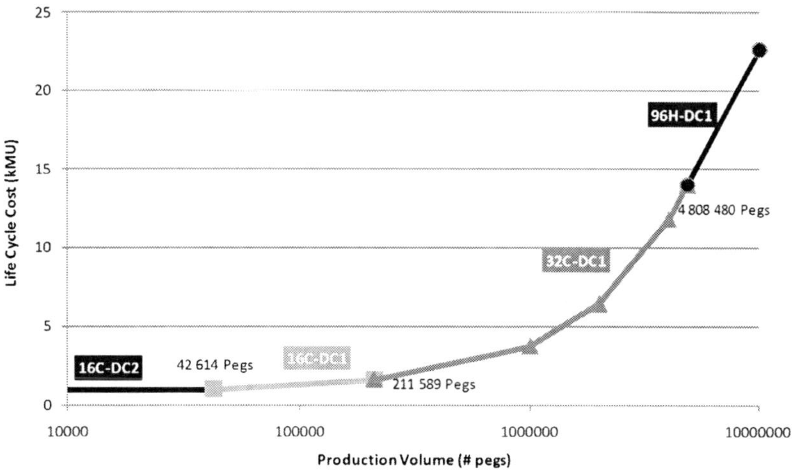

characteristics contribute for the mould higher productivity due to the lower number of injections cycles and lower cycle time (process time reduction with impact on associated time driven costs). Additionally, the use of hot-runner eliminates the need of post-injection operations related with the removing of feeding channels and the running of materials for recycling. So, it can be said that the lower cycle time of the DC1 alternatives together with the lower cycle time per part of the moulds with a large number of cavities contribute for the low injection moulding costs of those alternatives.

Until now the analysis has been done based on a certain production volume, which being dependent of the market demand is a delicate figure to estimate. Therefore, it is important to understand what is the sensitivity of the best alternative to different production volumes (Figure11). For the sake of interpretation Figure 11 represents only the alternatives with the lowest LCC for each production volume. For the highest production volumes (more than 4.81 MPegs) the DC1 injected in 96 cavities moulds with hot-ruuners achieves the lower cost. It means that the lower cycle times allow lower injection moulding costs that together with lower material consumption (DC type and hot ruuners) compensates the higher cost of the mould, which does not vary on the analyzed range of production volumes. For lower volumes, until around 211.6 kPegs, the 32C-DC1 alternative is the best one, due mainly to the increasing importance of the mould cost for the cloths peg life cycle cost. Finally, it should be remarked that the higher quantity of required material and mainly the higher injection costs (cycle time) of the DC2 type alternatives represent a drawback for this Design Concept when compared to the DC1 geometry. Only for extremely small series, below 42.6 kPegs (calculated only for analysis purpose since it's not a reasonable production volume), a DC2 alternative (16 cavities and cold runners) achieves the best choice position. In that situation, the mould productivity largely reduces its impact and the LCC of the cloths peg becomes essentially driven by the mould cost.

Results and Discussion of the Life Cycle Assessment

Goal and Scope Definition

In order to evaluate the environmental performance of the different alternatives, a Life Cycle Assessment was performed, using all data previously collected for the product's specific conditions. A cradle to grave approach was used, according to LCA standards (ISO, 2006, 2006).

The analysed system was already described in the life cycle cost analysis and it refers to the processes involved in all the life cycle stages of the clothes pegs. In order to maintain coherence in the whole methodology, the functional unit is the same used in the LCC analysis and is the annual cloths pegs foreseen production: 4Mpegs.

The first step was to define the system boundaries, as illustrated in Figure3. The use phase of the cloths peg was disregarded since it is neither responsible for significant impacts nor for significant differences among the alternatives. Regarding the materials impact all the stages of the plastic material from the raw materials extraction to the plastic production were considered. The injection moulding and the mould production were analysed separately, as more accurate and parcelled results can be obtained. Finally, the end of life of the plastic parts, as already mentioned in LCC analysis, were divided in two main streams: the used cloths pegs were considered to be deposited in landfill since there is no recycling framework; the plastic material required for the injection process but not included in the product (waste) was considered to be recycled, in the company or by any recycling framework. The environmental streams of the clothes peg life cycle were weighed using EI'99 method permitting to compute the life cycle environmental impact.

Inventory Analysis and Impact Assessment

To compute all the streams with an environmentally burden aspect, the model used for the cost analysis previously presented was used with the required adjustments to the LCA analysis. In particular some environmental fluxes related to material consumption during the production process were gathered using SimaPro7 database [SimaPro7]. The use of the same model for cost and environmental analyses allowed maintaining the consistency and further aggregations of results over the two dimensions of analysis. All the data obtained through the process-based model referred to the same functional unit.

The impact assessment of the environmental streams gathered in the previous stage of LCA was performed using the environmental indicator Eco-Indicator 99 (EI'99) and the weighting coefficients were applied according to the hierarchic/average perspective (H/A), as mentioned in point 2.2 of the section "An Integrated Approach to Product Design & Development".

Figure12 presents the results of LCI for 4 Mpegs for the different alternatives under analysis. The first conclusion is the extremely reduced environmental impact of the mould. The plastic material has the major slice of impact mainly in alternatives with 96 and 32 cavities. In fact, as observed for the costs, the environmental impact of plastic material varies in a relative strict range.

Interpretation

The alternatives with lower performance are the ones that use cold runners, due to the extra material required for the feeding channels and the limit of 10% of recycled material in the injection process. It must be noticed that since this wasted material was considered for recycling, there is a recovering of impacts for the alternatives using cold runners (negative impact figures). This fact motivates a relative balance of the overall impact of the consumed materials, which results in a higher importance of the impact of the injection moulding process (energy consumption) in the alternatives ranking. Therefore, alternatives with lower cycle time per part (moulds with high number of cavities and hot runners) have a lower environmental impact. The alternatives with cold runner moulds present always higher impact than the similar ones with hot runners, due to higher impact of injection moulding and material consumption. For the some type of runners and cavities the DC2 based alternatives result in a lower environmental performance.

Figure 12. Life Cycle Impact of the Design Concept/Process alternatives for 4 Mpegs

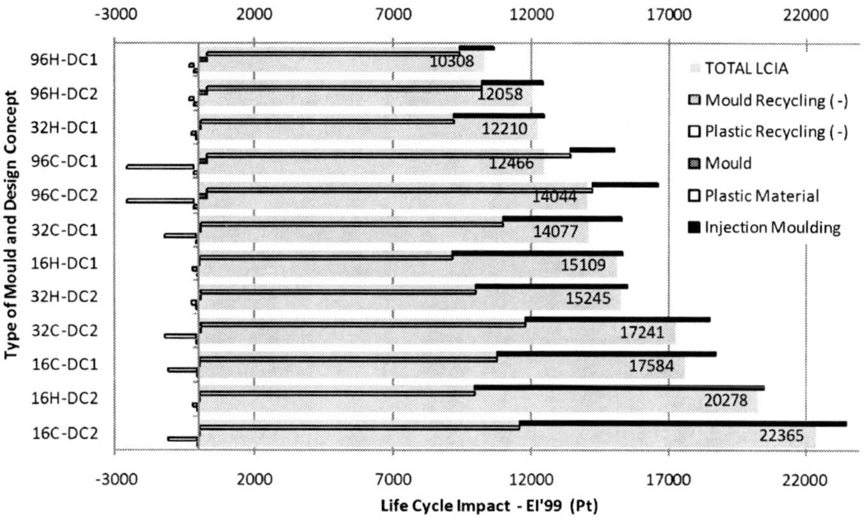

Again, understanding the behaviour of the alternatives with different production volumes is important to accommodate the uncertainty of market demand (Figure13). The 96H-DC1 alternative has the lowest environmental impact for production volumes higher than 190.9 kPegs. The reference in the graph for 190.9 and 32.6 kPegs is only to complete the analysis, since these production volumes are too low to be economically applied in industrial environment. Nevertheless, for such volumes DC1 with hot runners alternatives have a better environmental performance, in accordance with the analysis above.

Results and Discussion of the Functional/Technical Analysis

The F/T evaluation of the different design / process alternatives was applied only to the mould performance on injection moulding phase. The

Figure 13. Best (lowest environmental impacts) alternatives for different production volumes

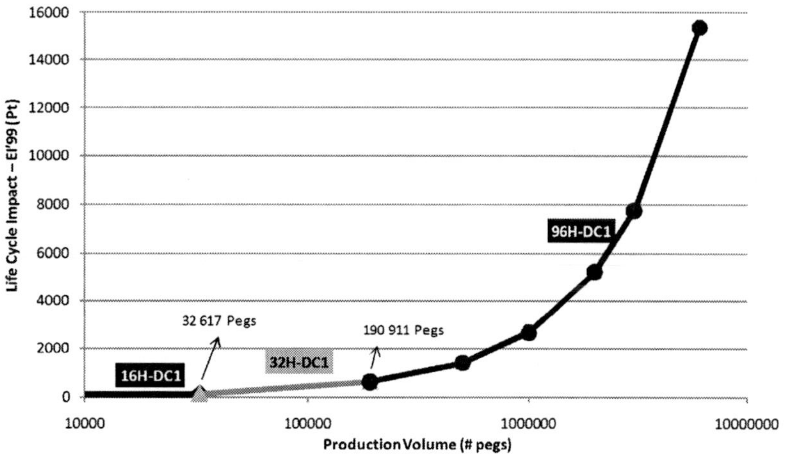

Table 5. General attributes for the process and their weights for this case study

Attributes	Stage	Attributes weights (w_j)
Technology Availability (B_1)	Mould Production 13%	6.4%
Technology Reliability (B_2)		4.8%
Number of Production steps (B3)		0.2%
Time-to-plastic part production (B_4)		1.6%
Mould Capability (B_5)	Use of the Mould 87%	32.7%
Mould Robustness (B_6)		10.9%
Mould Reliability (B_7)		38.1%
Production Time (B_8)		5.2%

design alternatives were considered similar enough to not be a differentiation issue in the process performance.

Following the model previously explained regarding the F/T evaluation, the most important requirements for the life cycle performance of each mould are modelled as attributes. The weights (w_j) of the attributes (B_j) were defined trough pairwise comparisons, discussed among specialists (mould manufacturer and mould user), giving a score (0 - as important; 1 - weakly more important; 2 – moderately more important; 3 - strongly more important) to the more important attribute in each comparison. At the end, the given total scores are summed up to each attribute and the absolute scores are converted into relative ones. Having defined the attributes and their weights (Table 5), the decision matrix can be built (Table 6). As the measures of the performance of the alternatives (m_{ij}) are fuzzy, the Chen and Hwang (1992) method was used for converting these measures into crisp scores, using a 5-point scale. Having the decision table completed with the values regarding the alternatives for the particular case under analysis, the final technical performance score, P_i, can be determined for each alternative. Notice that the product alternative designs were considered to have the same F/T performance, so only the mould/injection moulding alternative technologies were evaluated regarding this dimension of analysis. As results show, the best alternative in terms of F/T analysis is the 96H type, mainly because of its best performance during injection moulding phase.

Table 6. Decision matrix for this case study

Alternatives		Attributes								
		Mould Production Phase				Injection Phase				
	B_j	B_1	B_2	B_3	B_4	B_5	B_6	B_7	B_8	
	w_j	6.4%	4.8%	0.2%	1.6%	32.7%	10.9%	38.1%	5.2%	P_i
16H		0.495	0.495	0.495	0.695	0.295	0.295	0.495	0.495	0.355
16C		0.495	0.495	0.495	0.495	0.115	0.295	0.695	0.295	0.359
32H		0.495	0.495	0.495	0.495	0.495	0.495	0.495	0.695	0.449
32C		0.495	0.495	0.495	0.295	0.295	0.495	0.695	0.495	0.447
96H		0.495	0.495	0.495	0.295	0.695	0.695	0.495	0.895	**0.544**
96C		0.495	0.495	0.495	0.115	0.495	0.695	0.695	0.695	0.541

Figure 14. Adimensionalised scores for the design concepts alternatives for 4 Mpegs

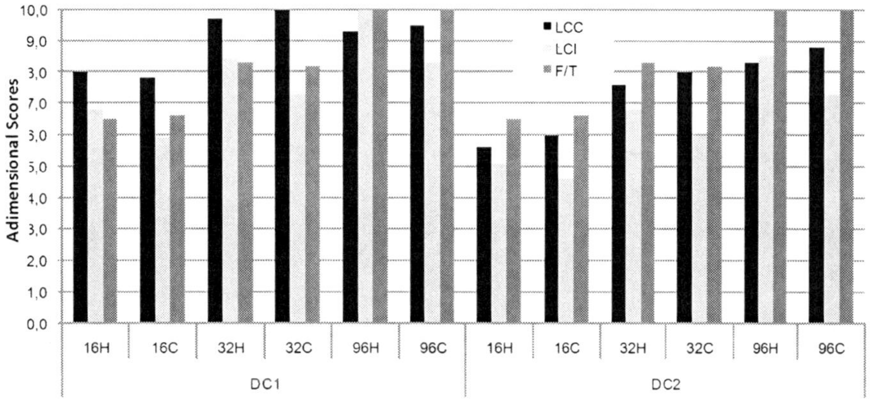

Global Evaluation

The assessment of the results achieved allows the development of a global evaluation of the alternatives. The outcomes of each individual dimension of analysis (economical, environmental performance and functional/technical performances) were adimensionalized by the best score (minimum in LCC and LCA, maximum in F/T performance), and translated into a numerical value in a scale of 10 points. The results achieved for a production volume of 4Mpegs (Figure14) show that the higher scores are obtained for DC1 alternatives type and for the larger number of cavities. The 16 cavities alternative types are confirmed as the lowest competitive ones.

However, it is difficult to select the "best" alternative from this analysis. Actually, in Figure 14 all the three dimensions of analysis are scored as they would have the same importance for the decision. In industrial contexts, typical design and process decisions are known by including several criteria tacitly pondered according to their strategic importance for the business competitiveness. So, importance weights could have been attributed to each of the three dimensions of analysis to select the "best" alternative. Nevertheless, this approach only selects the "best" one for a specific weights combination. To overcome this drawback ternary diagrams are proposed taking advantage of their capacity to represent the "best" solutions map.

Figure15 presents the ternary diagram for 4 different production volumes: 0.5, 2, 4 and 6 MPegs. Independently of the production volume, the DC2 based alternatives are never among the "best" ones. This was an expected result since the DC2 demands for more plastic material and so for a higher cycle time than DC1. The alternative 32C-DC1 has a balance between the mould cost and injection moulding (cost and energy consumption), so in the range of 0.5 to 4 MPegs is the solution with lower LCC. Consequently, for a high importance given to economical performance this is the "best" solution. The 96C-DC1 alternative is the one with the best scores in F/T performance. Additionally, its economical and environmental performance is high due to the cost of the mould (lower than 96H alternatives) and the cycle time per part (high number of cavities with lower cost per part, than the 32 cavities alternatives). The 96H-DC1 alternative is present in the region of high importance of environmental performance as well as high importance of technical performance. In fact, it is one of the alternatives with the best score on F/T performance and is the alternative with the lowest cycle time per part, due to the use of hot runners and the high number of cavities.

Figure 15. Ternary selection diagrams representing the "best" solutions map for different production volumes: a) 0.5 MPegs; b) 2 MPegs; 4 MPegs and d) 6 MPegs

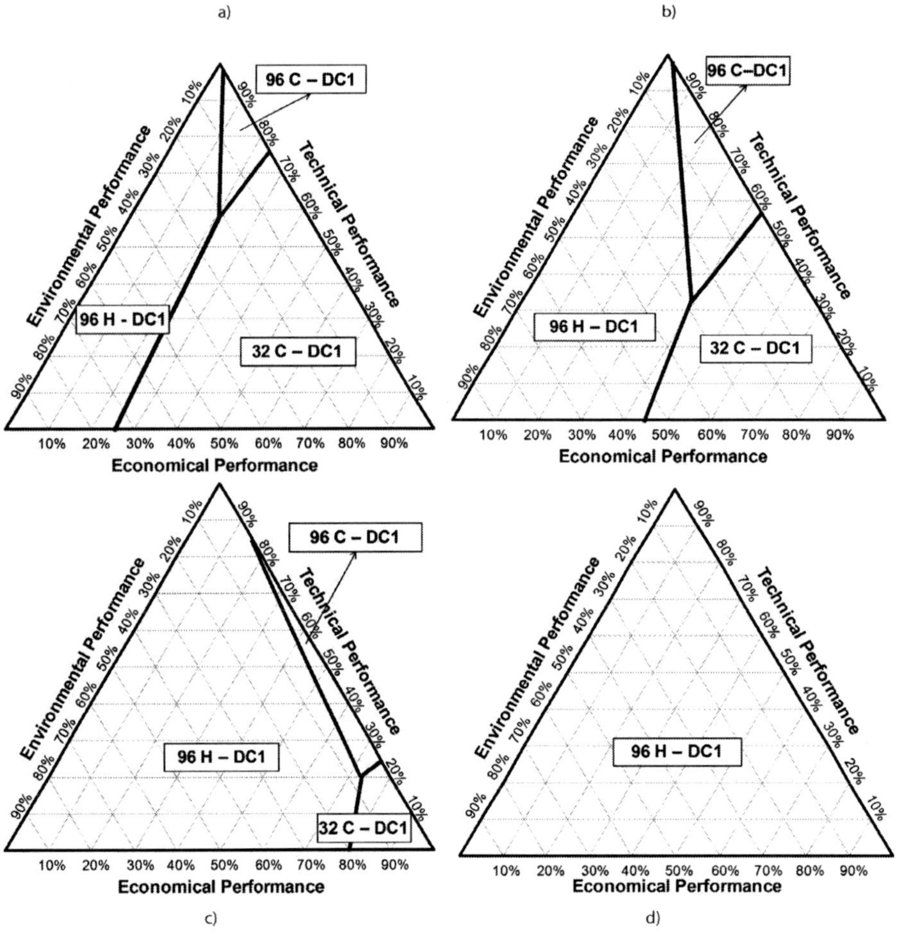

This characteristic allows having the best environmental performance in a large range of production volumes. Its main disadvantage is the mould cost, which influence decreases for larger production volumes. Accordingly, with the increasing of the volume the 96H-DC1 alternative tends to dominate the "best" solutions map, with its hegemony stated for productions larger than 4.8 MPegs.

The ternary diagrams developed for this case study can support the product designing team in the selection of the most appropriated design concept and manufacturing process. As an example of its application, consider a market demand of 2 MPegs (Figure15 b). Consider also that, after some interaction and discussion among the design team, it was stated 70% importance for the economical performance, 15% for the environmental and 15% for the technical performance. For this scenario, the 32C-DC1 alternative should clearly be the selected one. For the same scenario, but assuming an uncertain market demand between 2 MPegs and 4 Mpegs, two "best" solutions arose: 32C-DC1 for the lowest production volume and 96H-DC1 for the highest. So, a deeper knowledge on the market demand estimative might be recommended to support a more accurate decision-making process. However, even the perception of the consequences of a particular choice under uncertain conditions is a step forward towards more informed decisions.

CONCLUSION

This chapter proposed an integrated approach to product design and development based on Life Cycle Engineering principles. The model to implement this approach comprises two main phases. In the first phase the product design concepts are developed following the product design and development typical steps. The Value Analysis method is used to promote and drive discussions among the design team, making available a structured framework to compare and select design concepts with higher Value and contributing to identify the potential best options and consciously reduce the number of design concepts for deeper development. The selected design concepts are the focus of the second phase, which consists in the development of the manufacturing processes and in the evaluation of technological options on a life cycle basis. The methodology used for evaluation and comparison is based on Life Cycle Engineering principles, involving three dimensions of analysis (technical, economical and environmental performance) of the design concepts and manufacturing process alternatives. The global assessment is based on ternary diagrams, which permit the visualization of the "best" solutions map and so foster a robust decision-making process.

The proposed integrated approach was applied to a simple case study to allow its straightforward demonstration: a cloths peg. The simplicity of the product allow revealing the analysis potential of the proposed approach. After defining the design project aim, the design concepts developed are presented and compared in terms of Value Analysis. The Life Cycle Engineering model is applied to the two design concepts selected in the first development phase and the life cycle cost, environmental impacts and functional/technical performance were calculated including the process design options (several injection moulds types as well as for different production volumes), having in mind a comparison framework for decision making purposed. Finally, the ternary diagrams were exhibited allowing understanding the importance and usefulness of the proposed approach as well as how it can be applied in real industrial environment.

REFERENCES

Alting, L., & Legarth, J. B. (1995). Life Cycle Engineering and Design. *Annals of CIRP*, *44*(2), 569–580. doi:10.1016/S0007-8506(07)60504-6

Aurich, J. C., Fuchs, C., & Wagenknecht, C. (2006). Life cycle oriented design of technical Product-Service Systems. *Journal of Cleaner Production*, *14*, 1480–1494. doi:10.1016/j.jclepro.2006.01.019

Bento, C., Peças, P., Henriques, E., & Silva, A. (2008). Product Development guided by Value Analysis through Collaborative R&D. In Gonçalves, R. (Ed.), *14th International Conference on Concurrent Enterprising – ICE 2008* (pp. 751-758). Lisboa: ESoCE-NET.

Betz, M., Schuckert, M., & Herrmann, C. (1998). Life Cycle Engineering as Decision Making Support in the Electronics Industry. In *Proceedings of the 1998 IEEE International Symposium on Electronics and the Environment* (pp. 231-236). Oak Brook, IL: IEEE.

Chen, S. J., & Hwang, C. L. (1992). *Fuzzy multiple attribute decision making methods and applications. (Lecture Notes in Economics and Mathematical Systems)*. New York: Springer.

Cooper, J. S., & Vigon, B. (2001). *Life Cycle Engineering Guidelines. (Report 600/R-01/101)*. Washington, DC: Environmental Protection Agency.

Ermolaeva, S., Castro, G., & Kandachar, V. (2004). Materials selection for an automotive structure by integrating structural optimization with environmental impact assessment. *Materials & Design*, *25*(8), 689–698. doi:10.1016/j.matdes.2004.02.021

Fava, J. A., Brady, K., Young, S. B., & Saur, K. (2000). Sustainable Strategies Using Life Cycle Approaches. *Environment and Progress*, *19*(2), 61–64. doi:10.1002/ep.670190203

Goedkoop, M., Effting, S., & Collignon, M. (2000). *The Eco-indicator 99 - A damage oriented method for Life Cycle Impact Assessment Manual for Designers*. Amersfoort, The Netherlands: PRé Consultants.

Goedkoop, M., Schryver, A., & Oele, M. (2007). *Introduction to LCA with SimaPro 7*. Amersfoort, The Netherlands: PRé Consultants.

Ishii, K. (1995). Life-Cycle Engineering Design. *ASME Journal of Vibration and Acoustics, 117*(B), 42-47.

Jeswiet, J. A. (2003). Definition of life cycle engineering. In Bley H. (Ed.), *Proceedings of the 36th CIRP international seminar on manufacturing systems* (pp. 17-20). Saarbrücken: CIRP.

Johansson, K., Perzon, M., Fröling, M., & Mossakowska, A. (2007). Sewage sludge handling with phosphorus utilization – [REMOVED HYPERLINK FIELD]life cycle[REMOVED HYPERLINK FIELD] assessment of four alternatives. *Journal of Cleaner Production*, *16*(1), 135–151. doi:10.1016/j.jclepro.2006.12.004

Keys, L. K. (1990). System Life Cycle Engineering and DF"X". *IEEE Transactions on Components. Hybrids, and Manufacturing Technology*, *13*, 83–93. doi:10.1109/33.52854

Krozer, Y., (2006). Life cycle costing for innovations in product chains. *Journal of Cleaner Production, 16*(3), 310–321.

Kurk, F., & Eagan, P. (2008). The value of adding design-for-the-environment to pollution prevention assistance options. *Journal of Cleaner Production*, *16*(6), 722–726. doi:10.1016/j.jclepro.2007.02.022

Miles, L. D. (1989). *Techniques of Value Analysis and Engineering*. London: Eleanor Miles Walker.

Peças, P., Ribeiro, I., Folgado, R., & Henriques, E. (2009). A Life Cycle Engineering Approach for the Selection of Technology: a Case Study on Plastic Injection Moulds for Low Production Volumes. *International Journal of Cleaner Production*, *17*(9), 846–856. doi:10.1016/j.jclepro.2009.01.001

Pousa, C., Ribeiro, I., Folgado, R., Peças, P., & Henriques, E. (2009). LCC and LCA Simplified Models to Foster the Design of Sustainable Plastic Injection Moulds. In Waguih ElMaraghy (Ed.), *16th CIRP International Conference on Life Cycle Engineering – LCE 2009* (pp. 99-104). Cairo, Eqypt: University of Windsor.

Rao, R. V. (2007). *Decision Making in the Manufacturing Environment: using graph theory and fuzzy multiple attribute decision making methods*. New York: Springer.

Rebitzer, G., & Hunkeler, D. (2003). Life Cycle Costing in LCM: Ambitions, Opportunities, and Limitations. *International Journal of Life Cycle Assessment*, *8*(5), 253–256. doi:10.1007/BF02978913

Ribeiro, I., Peças, P., Silva, A., & Henriques, E. (2008). Life cycle engineering methodology applied to material selection, a fender case study. *Journal of Cleaner Production*, *16*(17), 1887–1899. doi:10.1016/j.jclepro.2008.01.002

SimaPro7 software. (n.d.). *PRe´ Consultants*. Retrieved from: http://www.pre.nl

Tseng, H. E., Chang, C. C., & Li, J. D. (2008). Modular design to support green life-cycle engineering. *Expert Systems with Applications*, *34*, 2524–2537. doi:10.1016/j.eswa.2007.04.018

Udo de Haes, H. A., & Heijungs, R. (2007). Life-cycle assessment for energy analysis and management. *Applied Energy*, *84*(7–8), 817–827.

Ulrich, K. T., & Eppinger, S. D. (2003). *Product Design and Development*. Boston: McGraw Hill-Irvin.

Wanyama, W., Ertas, A., Zhang, H. C., & Ekwaro-Osire, S. (2003). Life-cycle engineering: issues, tools and research. *International Journal of Computer Integrated Manufacturing*, *16*(4), 307–316. doi:10.1080/0951192031000089255

KEY TERMS AND DEFINITIONS

Life Cycle Engineering: A decision-making methodology that considers performance, environmental, and cost dimensions throughout the duration of a product, guiding design engineers towards informed decisions.

Life Cycle Cost: All the costs associated with a product throughout the product's life.

Life Cycle Assessment: A structured process to evaluate the environmental burdens associated with a product, process or activity by identifying and quantifying energy, used materials and wastes released to the environment, to assess their impact on the environment, and to evaluate opportunities and implement initiatives aiming environmental improvements. The assessment includes the entire life cycle of the product, process or activity, encompassing raw materials extracting and processing; manufacturing, transportation and distribution; use, reuse and maintenance; recycling and final disposal.

Life Cycle Impact Analysis: Technical, quantitative, and/or qualitative methods to characterize and assess the effects of the environmental burdens identified in the inventory of the energy and materials flows. Impact Assessment is presently under development and has yet no fixed methodologies.

Value Analysis: A systematic method of improving the "value" of goods and services through the examination of functions and costs.

Multiple Attribute Decision-Making Methods: An approach employed to measure the performance of a finite number of alternatives, involving a decision matrix and a decision-making method.

Injection Moulding: A manufacturing process for producing parts from both thermoplastic and thermosetting polymeric materials. Material is fed into a heated barrel, mixed, and forced into a mold cavity where it cools and hardens to the configuration of the mould cavity.

Section 7
Case Studies

Chapter 22
The "Madame Butterfly" Robot: A Case Study in Product Design and Development

Luisa Fontana
FONTANAtelier, Italy

Davide Fornari
SUPSI - University of Applied Sciences and Arts, Switzerland

ABSTRACT

"Madame Butterfly" is an industrial robot produced by Euroimpianti Spa. It is the result of the unusual cooperation between a cutting-edge robot factory and Fontanatelier, an unconventional architecture studio. Although it had all started as a common machinery project, and it was almost at the end of its development phase, the need for an iconic design led the designer to involve the architect Luisa Fontana in the project. The contrasting opinions between the engineers and the architect forced everybody to change points of view, in such a complicated and challenging field as robot production. The final design conveys a strong image of the product, very uncommon for the automation industry, maintaining the functionality of a standard robot, with improvements for some specific usage.

INTRODUCTION

Nicknamed "Madame Butterfly", the robot "Skilled 604" was first presented to the public in April 2008 at the Düsseldorf INTERPACK, the world's most important fair for the processing, packaging and line automation industries.

It was produced by Euroimpianti Spa of Schio (Vicenza), which has been a leading company in this sector for 35 years. Combining high quality design and performance, the articulated robot arm is used for stacking pallets.

It has four interpolated axes and has a work radius of 360°. Weighing 2.140 K, it can move up to 750 Kg. The result is a robot which houses cables, working mechanisms and a core structure in a flowing shape, inspired by the wings of a butterfly.

This innovative design seems to be taken out of a science-fiction comic strip: its shape, however, underlines the strength and flexibility of the aluminum structure shaped around the operating system of the machine. "Madame Butterfly" is

DOI: 10.4018/978-1-61520-617-9.ch022

Figure 1. M.me Butterfly at the Düsseldorf INTERPACK, April 2008

unique in modern day robot construction. A true synthesis of function and beauty, it is certainly destined to influence the future development of robot design and production.» (Luca M.F. Fabris, 2008, p. 16)

BACKGROUND

Machines are simulations of the organs of human body. The lever, as an example, is an extension of the arm. The lever has an amplified ability of raising weights, while all the other features belonging to an arm are neglected. (Vilém Flusser, 2003, p. 43)

The palletizer Skilled 604 can be considered as an extension of the human arm, that allows the stacking of product layers on a pallet for logistics or transportation, for the purpose of boosting human abilities. It is the ultimate evolution of the humble lever mechanism and has many other functions to help the worker. The top performances and the great design of the robot increase both productivity and labour standards in workplaces.

Machinery Design

Since the first assembly line developed by Ford Motor Company in 1913, machinery design has focused on the efficiency of tools to optimize the production process. Mechanical engineers focused on the increase of manufacturing speed and reduction of costs to boost productivity on the assembly line. This is the main reason why it is difficult to find examples of user-centered design applied to machinery. During the development of robotics for industrial productions, the machines were coupled to the human worker's body in order to copy the repetitive movements For example during the car painting process, the automatic painter arm could learn and memorize the movements of the worker that it was going to be replace. However this "anthropomorphic approach" was never used for the proper machine design.

For this reason we must go back to previous historical examples to discuss the background of anthropomorphic machines.

Automata

Early examples of anthropomorphic machines can be found in the history of architecture and technology. Since the Classical Greek period and in the Arab culture, automata were mechanical devices based on hydraulics or pneumatics, used for ritual purposes, such as measuring time through a regulated flow of liquids, or for entertainment. In the Western culture since the Middle Ages until the French Revolution, anthropomorphic machines played an important role in the experimentation of technological innovations, such as the spring mechanism (Losano, 1990).

Despite the Catholic Church's opposition, the production of anthropomorphic robots became a testing ground for many scientists who were later involved in industrial production. Jacques de Vaucanson, the author of the *Anatomies mouvantes* mentioned in the *Encyclopédie* by Diderot

Table 1.

General information	
Project name	M. me Butterfly
Object	Skilled 604, an articulated arm robot
Client	EUROIMPIANTI spa, Schio (VI)
Projects credits	
Design	Luisa Fontana architect (FONTANAtelier)
Mechanical engineering	G. Pierantoni
Chronology	
Design	2006-2007
Manufacture	2008
Technical features	
The SKILLED 604 is the best of European Technology. It has a flexible aluminum structure, the electronics is based on the highest international standards while the software is designed by EUROIMPIANTI. The Skilled 604 is capable of stacking product layers on standard pallets up to max 2,9 m (114") net height; with special pallets can reach up to a max 2,5 m (98,4") net height. It is ideal for: - picking-up complete rows of products - up to 4-5 different products lines - large operative areas - multiple lines low speed	
Maximum load	400 Kg (881 lb) incl. the gripper
Maximum reach	3200 mm (126")
Number of axes	4
Cycles/hour	800 max
Repeatability	+/- 0,25 mm (0,0098")
Weight	2140 Kg (4718 lb)
Operating noise level	less than 75 Db
Protection rating	IP 54
Mechanical Robot	EUROIMPIANTI
Electronic System	EUROIMPIANTI
Software	EUROIMPIANTI
Structure	anthropomorphic

and D'Alembert, was then chosen as supervisor of the royal silk industry in Lyon.

Since the Jesuits took the basic mechanical knowledge to Japan in the 17th century, robotics was developed until it reached the current level of excellence. During the sixties the first experimentations of the Tokyo based team of professor Ichiro Kato constructed a prosthetic hand, which duplicated limbs and their motion,and the first robot "Wabot" (Hornyak, 2006, pp. 73-76). Robotics found applications in all fields of industry and replaced the workers in the most heaviest tasks.

The relevant differences between the West and Japan in the acceptance of these robots is connected to their social and cultural perception. Since the twenties, robots have been considered as grandchildren of Nature and sons of Man in the Japanese culture. They were welcomed and seen as blessings by Buddhist monks when installed in factories, and they are fully integrated in the society, contributing to the "mechanosphere", a blending of human and mechanical functions. Gomarasca, 2001, pp. 220-262). In the Western culture robots were viewed with suspicion by the Catholic Church and had a negative reputation in the collective mind after the film "Metropolis" by Fritz Lang was released. (Fornari, 2008; Fortunati, 2003). Generally, a "mechanized" or "automated" work place was seen as something inhumane.

The Design of Machines in Italy

The most important Italian industries employed the Ford assembly line much later. Adriano Olivetti was one of the first Italian entrepreneurs to understand its importance. But even during the seventies, a large car industry such as Fiat was still employing human workers for very repetitive tasks: for a detailed description of the state of the art on working in Italian industries, we recommend looking at the documents on the working environment, workers conditions and society's reaction, collected in the book by Deaglio, 2009. The application of robotic systems in the industrial production in Italy came very late, due to the need to import machinery (Gini & Gini, 1983).

Despite this delay, an Italian firm stressed the importance of quality of working machines: the Ivrea based Olivetti company became famous for its product design. Camillo, Adriano and Roberto Olivetti turned the importance of the aesthetic of machines into a work ethic: «Machines are not pleasurable objects, neither for leisure nor for short or occasional use. People [...] should be able to establish an acceptable relationship with them,» (Giuntella & Zucconi, 1984, p. 103). The level of excellence reached by Olivetti's typing machines is well-known: designed to be not only functional and portable, but also with a particular stress on their aesthetic features, in order to reach visual, graphic and ergonomic comfort. This excellence was the result of the collaboration between the company and a number of famous designers: among them Marcello Nizzoli, Ettore Sottsass and Tomás Maldonado. Despite this success, the further developments of Olivetti in the field of computers are less known. The first large calculators needed enormous specific spaces to be hosted in, but the company managed to shrink them to human scale in order to avoid psychological discomfort to the users. They commissioned King & Miranda to design a specific syaesthetic system of interaction and interface for the control panels (Barbacetto, 1987). The philosophy of Olivetti was a precursor of modern day research and development patterns, but it never developed fully since it did not reach a level of thin the international context. Combining design and industry in the production of machine tools is the challenge of many other companies, such as IBM and Apple in the field of ICT, including those with a lower media exposure (Chiapponi & Riccini, 2008, pp. 168-169).

The Semantic Approach

One of the most remarkable features of Italian design that Olivetti helped to was the product identity (Branzi, 2008). In many cases this identity consists not only in the immediate recognition of formal characteristics, but in the relationship between product shapes and metaphorical factors. These semantic features of the products help to establish an interaction between user and product, called "semantic turn" (Krippendorff, 2006). However, the semantic and semiotic value of the objects was first highlighted by two Italian theorists of industrial design: Tomás Maldonado and Gillo Dorfles (Maldonado, 1976; Dorfles, 1963). Tomás Maldonado who was a director of the Hochschule für Gestaltung (School of Design) in Ulm (where Krippendorff also studied) first introduced semiotics in an industrial design teaching program. Maldonaldo was later hired as a consultant by Ettore Sottsass, art director at Olivetti (Anceschi, 2009).

The semantic approach to objects was always an important characteristic of Italian design Referring to typing machines, the "Valentina" designed by Ettore Sottsass for Olivetti in 1969 is a case study. It is one of the first example of portable typing machines with a case that is part of the design: a work tool to carry on trips. It was produced in a bright red color and named after a comic character, "Valentina" by Guido Crepax. In this case the product design is transformed into something else by its name, in order to make the interaction between user and machine pleasurable, relating not only to human factors but mostly to the semantic approach and formal innovation.

Thus, an anthropomorphic robot, reproducing the function of a human arm and called after the main character of a lyrical opera, alludes in a synesthetic way to other features, and its shape is a metaphor (Riccò, 2008), but it is also a result of the same "semantic turn". "Madame Butterfly" refers to the lightness of the machine in the undertaking of its specific tasks. Such visual lightness is due to the unexpected void in the bulk matter of the arm and to the sinuosity of shapes, recalling the motion of a butterfly while in flight. This camouflage, both anthropomorphic and zoomorphic, fulfills the need to establish an emotional relationship with the users, beyond the simple visual pleasure. It is a different setting from typing machines, large calculators and personal computers, although a loading robot is a machine that interacts with human beings and it has a strong influence in workplaces.

A Scenario of Anthropomorphic Machines

In this sense, the semantic approach to the design of machines features, including tool machines such as this palletizer, is a strategic area for industrial design. According to future trends, robotics will be ubiquitous in many fields of application (Ichbiah, 2005; Somalvico, 2003), and the management of the man-machine interaction will be of main concern, not only for the sake of pleasure. In a period of strong typological and morphological innovation, the ability of giving identity to new typologies of artifacts is indispensable for their social acceptance and for the technological transfer to mass markets, as happened during the boom of industrial design domestic products and furniture in Italy (Annichiarico, 2002; Chiapponi, 2005). It is a matter of making technology acceptable by designing features which are at the boundaries of interaction with man. In the case of Madame Butterfly, the synesthetic features refer to lightness and grace, cultural factors, as well making the interaction with an anthropomorphic robot for logistics much richer in terms of intangible feedback (Montefusco, 1992).

DEVELOPMENT PROCESS

The Producer: Euroimpianti S.p.A.

Euroimpianti was founded by Gastone Trecco, the current Managing Director, in 1973 in Zanè (Vicenza, Italy); with its SKILLED® brand the company has grown and made a name for itself among the producers of end line automated systems and achieved recognition as an industrial success story thanks to its openness to innovation and its readiness in taking up any challenge made by new technology and by the market. In 2008 it acquired all the assets of C&D Skilled Robotics, Inc., Texas to expand and to consolidate its presence in North America. The history of palletizing systems has gone hand in hand with the growth and development of the Company, which started with traditional palletizers, passed on to polar machines and then to anthropomorphic robots. In 1992 it added a new product line for Automatic Guided Vehicles. The company's actual strength lies in the added value of its products and services: design solutions based on simple concepts, designed, supplied and guaranteed by a sole supplier. The decision to produce not only standard robots, but complete integrated automated systems has proved to be a winning philosophy for Euroimpianti, and a great advantage for its Customers who can rely on a single source for design, product, commissioning and customer service.

A New Robot

There was no specific need for the project; everything started because Euroimpianti wanted a new machine in stock. They wanted this machine to have a higher lifting power than all the other products on the market, to be ahead of all other producers. Eventually they decided that a diverse design could add more value, and thus have more market appeal, that is why the managing director, Gastone Trecco, called Luisa Fontana. He focused on the design because the geometry of the machines for the packaging sector is quite flat, since technical requirements and movements of these machines are well- known. From the technical viewpoint there was nothing to be designed, so he turned to real market analysis, where he saw all these articulated arm robots, all resembling one another. For this reason he said "Why don't we try to be different from all the other rival business firms?" He realized immediately that the best way to deliver a different image was by developing a different, possibly unique, design.

The Architect

«When the director in Euroimpianti of Schio asked me to design the robot I was surprised, as I had no experience in industrial design. But I accepted the challenge. I went to the factory to observe the robots at work. The movements were like those of the human arm, but without the smooth harmony. The mechanical parts, the motors and the plant equipments all seemed to have been designed by different people and then assembled. I understood that for this project every detail had to be studied. We explored the degrees of freedom within which we could act to find the limitations of the design. And it was thanks to this "scientifically creative" approach adopted by the atelier that we managed to come up with a substantially innovative robot... and give him/it a soul.» (Luca M.F. Fabris, 2008, p. 16).

It was the first time that Luisa Fontana had tackled a "mechanical" design project, bringing to the task both creativity and self-confidence. Revealing great talent toward experimental design, she managed to combine both creativity and technology.

Concept

The machine replaces man in lifting and moving things, and it tries to replicate human movements, in complete harmony given a planned sequence. The robot is a set up as a big and powerful human arm. Human arm means perfection in movement and completeness. Every component... artery skeleton, muscles, veins, skin. The brain sends the impulse, the breathing apparatus and the feeding system produce the movement. Man reproduces its species, assuring continuity of its presence in time, in a production process which adapts to changes in the surrounding environment, evolving. The whole process is fluid: the human body is a beautiful and perfect "machine". The robot is not that different: through its simulation of the movement of the human body, it aspires to achieve its perfection by forcing its performances (L. Fontana, personal press release, May 22, 2009)

The Market Production

The architect decided to explore the market to understand the state of the art in robot production through a careful evaluation of other brands, like Kuka, Fanuc, ABB, and so on. The main issue is that the functionality of the machines does not lead to a formal unity. Every company implements some variations and optimization in detailing, without getting to a real result from an aesthetic viewpoint. As the beauty of a system comes from the harmony between its components, what was missing in those machines was integration. This approach is dear to the designer of "M.me Butterfly", who is a professional working daily to implement sustainability, bioclimatic and "green" architecture in her architecture. Such an approach aims to create buildings which behave like living creatures, to improve relationships between men and the environment. She defines such buildings as "human building/machinery". Products in the robot market lacked "soul" and beauty which comes from form and function

Two Solutions for the New Robot

The architect sketched two drawings, for the customer to choose from.

Strep Tease

The first idea is to cover the mechanical arm and its systems (cables, activators) in a "skin". Such a surface could cover the inner mechanics without the need for a joint design, to optimize functions and costs.

This is a really simplified solution, where the innovation is represented by the cover. Such a surface should adapt to every movement and underlining shape. Easily removable, for easy maintenance, it protects the internal machinery. Such a solution is feasible especially for the food and drug industry, where cleanliness of any component is a key factor. The cover can be removed and washed as an "apron". The skin becomes like a canvas, and can be made transparent or matt, in colorful spots or stripes. The model can help build a corporate identity or adapt marketing strategies: the possibilities are infinite.

Madam Butterfly

The second idea implies a complete re-design of the robot, because of the integration between mechanical structure, electro-mechanical systems, shape and movement. The architect developed her design by studying the pantograph, which is the key movement for the mechanical arm. From

Figure 2. The first idea: "Strip tease" Robot

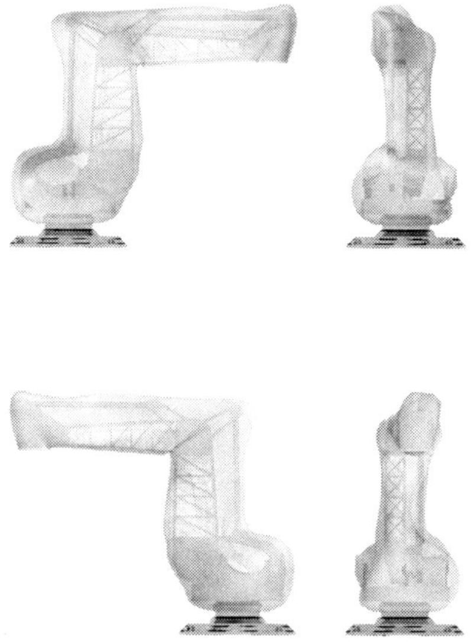

Figure 3. The skin like a canvas in the Strip tease robot

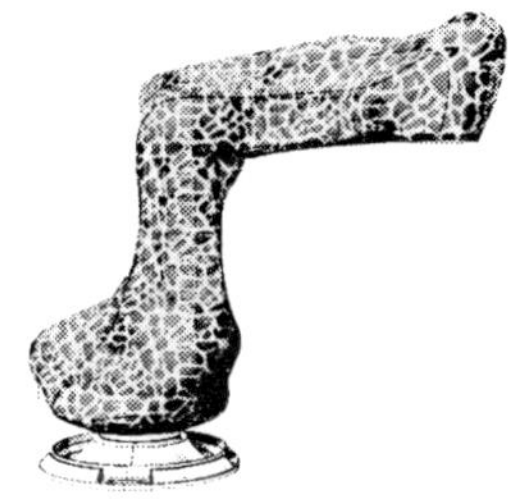

Figure 4. The Second idea: The Madame Butterfly Robot. Sketch by Luisa Fontana

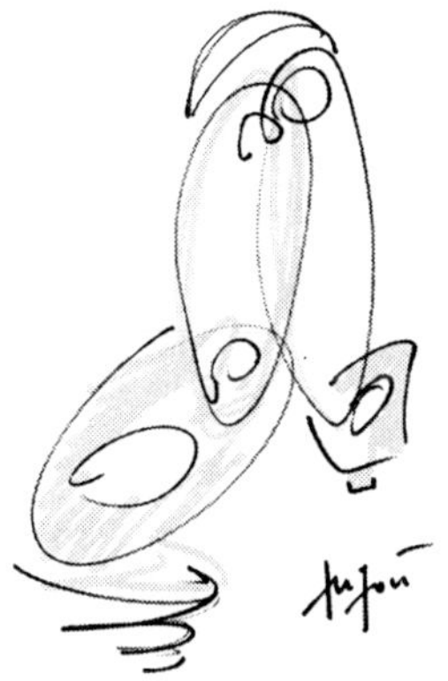

Figure 5. Technical design

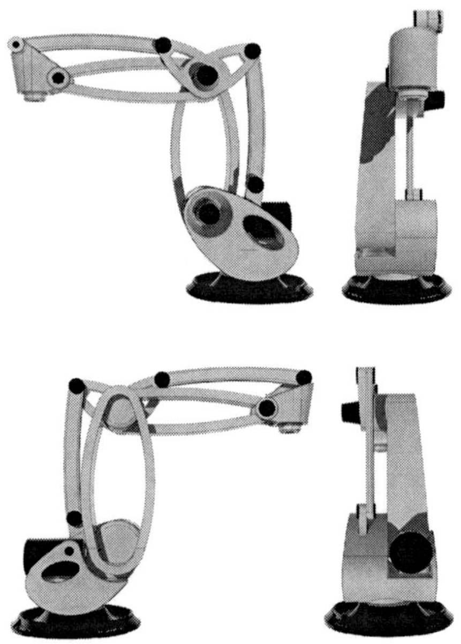

this starting point, the structure and the system of the machine was planned from scratch. The client chose the second suggestion, which led to intense work due to the strong link between design choices and performance needs.

The Team Work

The new robot "M.me Butterfly" is the result of a joint design between the architect, a mechanical engineer, and the production manager. In such a product, function and cost need to be properly evaluated. The designer needs to know the limits and constraints to be really able to innovate. At the beginning, though, the customer had a different requirement: he called for a redesign of the robot, claiming that this could be done without any change in the structure and functionality of the product. The architect's opinion was the opposite: only a profound knowledge of the machine, the engineering behind it, the performances and functionality could produce a new design. This second solution was eventually chosen, and a special team was set up in the company to implement it.

The aim of Euroimpianti was to control all commercial aspects and service requirements of the machine, to be able to sell the product to a larger market. An external technical office, specialized in mechanical engineering, took care of the mechanical aspect, while the electronics, like machine power, cables, electrical motors and so on were developed by technical staff members inside Euroimpianti. As far as the general shape is concerned, Luisa Fontana designed all the innovative aspects of the machine. The synergy between all members of the professional team was the key in delivering the final result.

Development Approach

First of all Euroimpianti needed to stress the complete change in the usual development of such a machine. Usually there is a primary phase of data acquisition: they set up technical specifications and analyze market trends to get as much information as possible. This is followed by an actual design phase where Euroimpianti produces a shape and a working model based on the original drawing.

In this specific project this process was turned upside down: the design came first, then they had to fit all the technical requirements of the machine in this design At the same time, they were trying to set a market trend, rather than following emerging needs in the industry. This choice was for the technical side of the project: specific structural and mechanical solutions had to be found to adhere to the original sketch.

The mechanical engineer explains: "The approach was different too because while generally we could compare our work with something preexisting, in this case we were doing something that had never been tried before in mechanics. We were basically turning away from usual methods, thus we struggled to find a solution structurally and mechanically compatible with the harmony of the design." (G. Pierantoni, personal communication, April 9, 2009)

Figure 6. Test room

Figure 7. Details

The engineer used a 3D modeling software and then verified the final result with a FEM (Finite Element Method) program. He analyzed all the pressure on the structure to verify the overall resistance of the structure. They worked harder compared to a standard robot project because they were moving away from the usual development path. In this situation all cross sections, all elements had to be verified properly, they could not rely on previous experience.

Mechanical Structure and Movement

During the development of the new robot the designer pushed for slenderness and lightness of the arm. Total weight could be reduced by using titanium, but this solution was too expensive. The planning team worked only on the shape, refining the drawing by hollowing out the mechanical arm. The hard work was finding the right balance between lifting needs and final shape, to assure the correct flow of forces in the structure. The drawing sketches were converted into a 3D model and then processed by the finite element software to optimize cross sections.

The biggest problem was modeling this new shape and being consistent with the structural needs. The arm is not made by a single and continuous element there aren't load bearing elements where we usually find them, and everything had to be designed upside down. Thus we had to reinforce some cross sections, and develop an in-depth mechanical analysis for the elements to have the desired shape.

By using "Rhinoceros" modeling software a first model based on the architects sketches was drawn up. The work model was transferred to "Mechanical desktop" mechanical design software. In this way it was possible to optimize the shape and mechanical sections.

Vantages of the New Design

The innovative design turned out to have positive effects on the functionality of the machine,

In fact the outstanding design was not the only result of the consultation with the architect. For example, there are no cables outside the elements, because she wanted the machine to look cleaner and lighter. As a result the machine can be used safely in areas of welding operations, lowering the risk of accidental fire or explosions. Moreover, without all the cables in between, it can be easily cleansed. This is important where health procedures are strict, as in the food industry. Furthermore that there are fewer elements assembled in the robot, which means fewer components which can break, thus lowering the usage cost for the customer. Basically the design forced the team to work harder, but delivered also unforeseen practical advantages for the product, adding value to the overall result.

The key issue of the project is that the technical qualities are the same, or even superior to other products, but this design gives a new shape to the robot. This design conveys an innovative image which makes Euroimpianti, and consequently their clients, more competitive in the world market. It gave everybody the chance to choose a product taking into account not only the practical aspects, but also the external qualities of the machine they need.

Costs and Results

This work was a little bit more expensive for Euroimpianti because the design weighed with an increase of 10%. on the global cost. Due to difficulties in production, it took a year and a half to complete the machine, which for Euroimpianti means 30% more than the usual time. Nevertheless Euroimpianti thinks that the positive reaction of their customers is a good payoff, not to mention the significant know-how they gained.

The reactions were all positive, it was considered a breakthrough in the industry; especially at the fair in Dusseldorf, where all the other robots looked exactly the same. From a commercial viewpoint it was a real success, because people were attracted by the soft lines and sleek design. Thus Euroimpianti could present not only Skilled 604, but all the product line. The reaction of their competitors was very interesting as well, because everybody wanted to know more about the robot. A firm which sells components asked Euroimpianti for the technical project because they were so impressed when they saw it. They could not make all the information public, of course, but this is to show what an impact they made on the market.

CONCLUSION

It was really an exciting project which highlighted Fontanatelier unconventional approach to design. Luisa Fontana's theory could be verified on field: Architecture can function on many themes, with the right approach. The most important thing is the way we look at challenges: being "scientificamente creativo" means being creative in a scientific way. This is what I try to do in my studio, in my daily work and with special projects like this one". (L. Fontana, personal press release. Luca M.F. Fabris, 2008, p. 24).

Euroimpianti believes that what is beautiful is also functional. The client is usually unaware of all the mechanical issues of the product: he wants to buy it because it solves a practical problem, and maybe because it looks nice. By evaluating the harmony of the external components, the client realizes that the robot is well made. It is clear for the customer that such a shape cannot spring up from a sudden idea, but is the result of a long process of study in all its components. Basically we applied the same approach to design which is used in other sectors like cars, or mobile phones.

REFERENCES

Anceschi, G. (2009). Maldonado, semioticist of knowledge. In *Annicchiarico, S. (2009). Tomás Maldonado* (pp. 136–155). Milan, Italy: Skira.

Annicchiarico, S. (Ed.). (2002). *Animal House. When Objects Have Animals' Names*. Milan, Italy: Charta.

Barbacetto, G. (Ed.). (1987). *Design Interface. How Man and Machine Communicate. Olivetti Design Research by King & Miranda*. Milan, Italy: Arcadia.

Branzi, A. (2008). *Introduzione al design italiano. Una modernità incompleta*. Milan, Italy: Baldini Castoldi Dalai.

Chiapponi, M. (2005). *Le forme degli oggetti*, in *Il Verri*, 27, (pp. 17-34).

Chiapponi, M., & Riccini, R. (Eds.). (2008). *Made in Iuav. L'università del design fra ricerca e progetto*, Tavagnacco: Dindi.

Deaglio, E. (2009). *Patria 1978-2008*, Milan, Italy: il Saggiatore.

Dorfles, G. (1963). *Introduzione al disegno industriale*. Bologna: Cappelli.

Fabris, L. (Ed.). (2008). *LuisaFONTANAtelier. Global architecture*, Dogana, San Marino: Litografia Titanlito s.a.

Flusser, V. (2003). *Filosofia del design*. Milan, Italy: Bruno Mondadori.

Fornari, D. (2008). *Perception of Humanoid Artifacts: Cultural Differences*, in Fujita, H. (Ed.), *Proceedings of the 6th ICDHS Conference*, (pp. 312-315). Osaka, Japan: KSI.

Fortunati, L. (2003). Real People, Artificial Bodies. In Fortunati, L., Katz, J. E., & Riccini, R. (Eds.), *Mediating the Human Body: Technology, Communication, and Fashion* (pp. 61–70). Mahwah, NJ: Lawrence Erlbaum.

Gini, G., & Gini, M. (1983). *Robot: Controllo, programmazione, interazione con l'ambiente*. Milan, Italy: Clup.

Giuntella, F., & Zucconi, A. (1984). *Fabbrica, comunità, democrazia: testimonianze su Adriano Olivetti e il Movimento Comunità*. Rome: Fondazione Adriano Olivetti.

Gomarasca, A. (Ed.). (2001). *La bambola e il robottone. Cultura pop nel Giappone contemporaneo*. Turin, Italy: Einaudi.

Hornyak, T. N. (2006). *Loving the Machine. The Art and Science of Japanese Robots*. Tokyo: Kodansha.

Ichbiah, D. (2005). *Robots. From Science Fiction to Technological Revolution*. New York: Harry N. Abrams.

Krippendorff, K. (2006). *The Semantic Turn*. Boca Raton, FL: CRC press.

Losano, M. G. (1990). *Storie di automi. Dalla Grecia classica alla Belle Epoque*. Turin, Italy: Einaudi.

Maldonado, T. (1976). *Disegno industriale: un riesame*. Milan, Italy: Feltrinelli.

Montefusco, P. (1992). Interazione, non interfacce. In Anceschi, G. (Ed.), *Il progetto delle interfacce. Oggetti colloquiali e protesi virtuali* (pp. 97–131). Milan, Italy: Domus Academy.

Riccò, D. (2008). *Sentire il design. Sinestesie del progetto di comunicazione*. Rome: Carocci.

Somalvico, M. (2003). Bodies and Robots. In Fortunati, L., Katz, J. E., & Riccini, R. (Eds.), *Mediating the Human Body: Technology, Communication, and Fashion* (pp. 31–38). Mahwah, NJ: Lawrence Erlbaum.

KEY TERMS AND DEFINITIONS

Comprehensive Approach: The multidisciplinary approach may be hard work - and unusual - but it enables one to draw on expertise in various sectors in order to achieve the best possible solution to a problem.

Creativity: Only when based on a knowledge of the limits and regulations that it can use and overcome, creativity becomes synonymous with freedom. It enables us to establish truly new relations between individuals and the environment, experimenting with the expressive possibilities offered by contemporary architecture and design. Without creativity there is no innovation.

Design: Design means giving harmony to the all project, means accepting the importance of the aesthetic of machines into the work area, means finding new solutions for common problems, it adds value to the overall result, sometimes it means a major cost but also a bigger market appeal.

Research: Life is continually presenting us with new problems for which there are no ready-made answers. Whether we like it or not, if we are to avoid repeating the mistakes of the past then we must commit ourselves to the avant-garde – that is, we must be continually open to the exploration of new horizons. Such research is the only way in which progress is possible.

Robot: Robot represents a big help for the workers and must be functional but it can establish an emotional relationship with the users too if it is not only made by parts, motors and plant equipments assembled together. It can be anthropomorphic, zoomorphic and it can have a soul too.

Innovative Approach: Any design project requires the effort to rethink things, to go back to zero. This means rejecting any easy resort to past models, form or styles. It means a willingness to make a tabula rasa of all preconceived ideas, to look at reality from a new point of view. This is the only way one can come up with something new.

Chapter 23
Transformal Role of Product Design in Singapore's Transition to a Service Economy

Peer M. Sathikh
Nanyang Technological University, Singapore

ABSTRACT

Singapore, a city state of 4.8 million people, located at the tip of the Malaysian Peninsula, was founded in 1819, when Sir Stamford Raffles of the East India Company established a trading settlement in Singapore. The meeting point for Chinese, Malays, Indians, Arabs, Europeans and others on their journey through the southern seas, Singapore achieved its initial economic success through international trade as a free port and free market. Given the status of an independent country in 1965, Singapore suddenly found itself in a struggle to survive. It's small population and scarce resources meant that regional and world markets were larger than the domestic market, presenting the government and its policymakers with distinctive economic challenges and opportunities. This chapter tries to recount the policies and subsequent actions put in place in Singapore from the 1960s till the present, promoting the creative industry, including product design, in order to transform a market dependent economy into a service centered economy. This chapter also discusses if and how such a 'planned intervention' played an important role in building up the resources and infrastructure within Singapore and in attracting multi-national companies to locate their R&D and design facilities in Singapore, pointing to where it has succeeded and where it has not.

INTRODUCTION

The modern history of Singapore starts on the 9th August 1965, when Singapore was 'expelled' from the federation of Malaysia due to differences between the political establishments within Singapore and the Malaysian government of that time. Overnight, the economic prospects of Singapore looked bleak. Looking at it from a purely Western point of view, Singapore had all the numbers stacked up against them. The literacy rate was just 57% of the small population of around 2.7 million

DOI: 10.4018/978-1-61520-617-9.ch023

with an unemployment rate at double-digit figure. With most of the population involved in mercantile activities, it is surprising to learn that the number of graduates in the country was very low and that most of them studied arts or science, with engineering training almost unheard of by the locals. Almost all products, save some furniture, were being imported for local consumption. Singapore had been used to, up until this time, predominantly British, and to a lesser extent American goods. How did this country, which had only 51,595 (1.9%) persons employed in the manufacturing sector in 1965, grow to attract companies such as Philips, Motorola, Apple, Seagate, Siemens, Matsushita and other global names to set up their operations here? How did it further convince these companies to move up the value of their presence in Singapore from mere manufacturing assembly units to establishing research and development (R&D) and eventually to having product design capabilities as well?

The answer lies in the various carefully planned initiatives taken up by the Government of Singapore from as early as 1961. As a country with a 'mixed' multi-ethnic culture without any root to the practice of manufacturing, Singapore was faced with the challenge of reaching 'instant' industrialisation.

Being small, policies could be evolved, debated by the various lawmakers and stakeholders and ratified in the Parliament quite rapidly with the implementation almost immediate. These initiatives, again, were not the action of individuals, but were from the various government bodies and boards that the government established on high priority as soon as the country gained its independence or even earlier.

To sustain the initiatives taken earlier, new initiatives are introduced regularly, as the global situation warrants. New agencies and boards are formed to sustain these initiatives keeping Singapore 'always on the edge'. Associate Professor Suresh Sethi, the Acting Chair of the School of Art, Design and Media at the Nanyang Technological University, Singapore paraphrases this phenomenon in today's terms as, "fast forward by design". As the initiatives started getting more and more sophisticated, Singapore came to realise in the late 1980s that design, in general, and product design in particular has to play a key role in sustaining the relevance of Singapore way past the new millennium.

FROM 1965: SETTING THE SCENE

Being a popular air and seaport, it was always easy for Singapore to import goods and products for local consumption. Perishables such as fruits and vegetables were much easier with Malaysia being just across a causeway. However, the products for everyday use such as radio, television, stove, oven, fan, etc., were imported. While Ford Motors and Mercedes Benz had factories assembling cars, the parts used for assembling these cars were completely imported.

At what level was the local expertise in product design and development at this point in Singapore then? Since there was not any manufacture of complete products for mass consumption, the 'first generation' engineers had their first encounter of product design in the form of finished technical drawings and plans for parts manufacture and assembly that essentially came from established companies in the United States, Europe and, to a lesser extent, Japan, which were rapidly establishing their manufacturing facilities in Singapore. How was this possible in such a short time?

On the advice of a United Nations expert named Dr. Albert Winsemius, the provincial government of Singapore (still a part of a united Malaysia) established the Singapore Economic Development Board (EDB) in 1961, to convince foreign investors that Singapore was a good place for business. Upon independence, the first Finance Minister of Singapore, Dr. Goh Keng Swee decided to support and hasten EDB's effort by converting 3000 hectares of swampy land into the Jurong Industrial Estate. During a time when industrial estates in general were considered polluted and not a healthy place to work, the first Prime Minister Mr. Lee Kwan Yew declared, "We are determined that Jurong shall be clean and green. Anti-pollution measures will be relentlessly carried out". This helped in two ways. Firstly the investors were convinced that the area in which the industrial estate would be located will have first world environment conditions and secondly, Singaporeans were willing to move and live at the new Jurong townships that were being built around the industrial estate. The success of this was followed up with industrial estates in Tanglin and Red Hill areas of Singapore.

Though the first capital and technology intensive industries attracted were Shell Petroleum and the National Iron and Steel mills, the first product centric factories produced garments, textiles, toys, wood products and wigs, which were essentially low skill, low knowledge and low value products.

Even after the EDB had opened offices in New York and Hong Kong to spread their nets for potential investors, there was this realisation that in order to expand the industrial base and its economic sustainability, Singapore needs to cultivate technical manpower at different levels, right form operations and supervision till management. A foundation was then laid to build up this manpower, as usual, in a 'fast forward' mode.

To fulfill the immediate requirement Singapore's Ministry of Education started concentrating on 'technical and vocational schools'. Several such schools such as the Singapore Vocational Institute (SVI) graduated workers with certificates in Mechanical Engineering Practice, Electrical Fitting, Electrical Installation, Radio Servicing, Motor Vehicle Mechanics, Refrigeration and Air-conditioning, Plumbing, Wood Craft and Construction, Building Drawing, Ship Building, Sheet Metal and Welding, etc. Even the Teachers Training College was pressed into introducing a two year course in Certificate of Teaching (Technical) in order to fill the vacancies for qualified 'technical' teachers.

Looking at the challenges in attracting investors and industries, the EDB soon realised that Singapore needed sufficient technical manpower beyond that offered by vocational schools. Its response was to seek technical and financial assistantship through the United Nations Development Programme (UNDP). Under this programme, Singapore was able to receive technical expertise and machinery from countries such as Japan, Britain and France, and by 1968, it established six training centres, namely, Metal Industries Development Centre, Prototype Production and Training Centre, Electro-mechanical Training Centre, Electro-chemical Engineering Centre, Woodworking Industries Training Centre and Precision Engineering Development Centre. In one way or the other, these six centres played a role in promoting the emergence of new product design and development capabilities in Singapore at a later date.

Looking beyond technical skills towards supervisory level manpower, Singapore established the first two polytechnics, the Singapore Polytechnic and Ngee Ann Polytechnic (initially named Ngee Ann Technical College), which paved the way for post secondary school (post high school) education in areas such as Mechanical Engineering, Electrical and Electronic Engineering, Civil Engineering, Architectural Studies, etc. Graduates from these polytechnics were awarded a diploma after three years of study.

In anticipation of the knowledge based technical manpower requirements, the University of Singapore (now named National University of Singapore) started a bachelor level degree course in engineering in the areas of mechanical, electrical and electronics and civil engineering by the late 1960s.

It is important to note here that none of the courses referred to 'product design or development' *per se*. The graduating workforce had sufficient knowledge to read 'blue prints' (engineering drawings) and execute the requirements of the drawing supplied to them. Considering the stage of development that Singapore was at, it becomes apparent that Singapore was focused on industrialisation and was not concentrating on the peripheral but important aspects of art, culture and heritage that makes up the environment (together with industry) for new product design and development. What was perhaps lacking in the work force of Singapore, if one captures a moment in time, at the end of the 1960s, was this ability to question, in order to understand better, the process of manufacturing that could lead to knowledge of product design and development. It is apparent from the subsequent events in Singapore that the people involved at the government and policy level were aware of this and much more. Here again is an ingredient of success of Singapore; the policy makers are always thinking beyond 'what is next' to 'what could be after next'. In order to prepare for what is beyond next, bright Singaporeans were being sent to study at the best of the universities in the West, mostly in the areas of engineering, finance and administration, but not in design!

THE 1970S: LAYING THE FOUNDATION

The momentum created by the EDB in the late 1960s set Singapore on a path, which continued through the 1970s, by end of which, manufacturing became the largest economy, surpassing trade. Anticipating this, Singapore set about expanding its ports and building a new airport to handle the heavy movement of goods out of Singapore, and the equally heavy influx of raw materials from elsewhere. Road transportations from the factories to the port and airport was modernised and telecommunication infrastructure was expanded to handle the heavy information traffic.

Due to EDB's presence in Europe, USA and Asia, companies involved in sophisticated component level products such as computer parts, computer peripherals and silicon wafers set up their operations in Singapore. The major *coup de tat* for Singapore happened when Texas Instruments of USA rolled out its production line within fifty days of committing to invest $6 million. Texas Instruments was also manufacturing semiconductors and integrated circuits. At this point Singapore did not have to worry about unemployment. In fact lower skilled workers were being brought in from Johor Bahru, across the causeway in Malaysia, to fill the requirements. While EDB was successfully marketing Singapore as an ideal place to start up with minimum bureaucracy and tax break, with highly skilled work force, the companies setting up were only prepared to operate their part level manufacturing and assembly in Singapore. The creation and development of the products, whose parts were being manufactured or which were being assembled, was being done nearer to their headquarters.

While this was not a cause for concern, it was puzzling that companies were not starting to conceive their products in Singapore. Eventual analysis started to show that there were more gaps to cover before this could be achieved. First of all, there was continuing need for specialist training, which could not be provided by the present technical institutions. Secondly there was a need for concerted effort in promoting productivity through standardisation and quality control. Thirdly, with the first two in place, to build manpower, with both skill and knowledge, that will embrace quality control and productivity in order to convince the companies to set up their research and development (R&D) activities.

The need for specialist training was filled by the establishment of Joint Government Training Centres, with Tata of India, Philips of Holland and Rollei of Germany. These were started on the pretext that specialist technical training could be imparted only by specialist industries. The idea, then, was to attract and help large industrial companies to set up facilities to train people for their own needs and the trick was to make them train double their requirements and let the excess to help meet the manpower requirements of other companies. Usually the land and buildings were provided for by the government together with financial assistance to cover up to 70 percent of the operating cost of the training centre. The trainees were also given a stipend and signed a bond to serve the EDB or any company as directed by the government, for a period of five years.

To improve the standards and quality control in Singapore, the National Productivity Board (NPB – which was renamed initially to Singapore Institute of Standards and Industrial Research (SISIR) and later as Productivity and Standards Board (PSB) before settling to its present name of Standards, Productivity and Innovation Board (SPRINGSingapore)) was established in 1972. The NPB was tasked with imparting knowledge and best practices through courses and seminars on productivity techniques, in-plant training in productivity and standards and was mandated to give consulting services. Eventually PSB became a certified body for accreditation under the ISO certification system.

Full fledged operations by 'multi-national corporations' (MNCs) in Singapore called for support industries and 'vendors' to whom parts level manufacturing can be outsourced. This trend saw the founding of local component manufacturers such as Amtek Engineering Ltd., in Singapore. By now, Singapore had several industrial estates spread across the island. The EDB and NPB supported these local industries in terms of manpower development, business and quality control development. Banks and financing institutions stepped in to help in financing the equipment and machinery requirements.

Going by the developments during this decade, in terms of investments, infrastructure development and manpower availability, one would think that by the end of the 1970s Singapore would have gained enough momentum to have companies doing new product design and development within Singapore. While the high volume manufacture of parts and product assembly brought in high revenue boosting Singapore's Gross Domestic Product (GDP) beyond $13.5 billion, Singapore realised that sustainability of this growth required the industries located within to Singapore to design and develop new products in order to multiply the revenue several folds.

At first, Singapore policy makers seemed to have concluded that establishing the 'research and development '(R&D) functions by established companies would be the way forward. With encouragement and incentives from EDB and other government bodies, MNCs began R&D activities as an extension of their manufacturing operations, which seemed like a demonstration of long-term confidence by these companies in Singapore. Establishing R&D alone did not lead to new product design and development activities. Instead, it was leading to new processes, better parts, better efficiency and cost savings, which by itself was good but not enough. Studying established companies in USA, Europe and Japan, it was apparent that, whenever new product design and development was involved, these companies were supported by a unique type of professionals called 'industrial designers' and that industrial design as a profession had matured in these countries over five decades or so.

Singapore was not graduating anyone with Industrial Design qualifications, nor were there any local designers available with such qualifications. However, Singapore did have an art school named the Nanyang Academy of Fine Arts (NAFA- established in 1938) and the Baharuddin Vocational Institute (BVI- established in 1969) which were sending out students with diplomas in 'pure arts' and 'applied arts'. This meant that Singapore had to put its fingers on the 'fast forward' button again.

THE 1980S: DESIGN AWARENESS AND PROMOTION

The next phase of the accelerated development of Singapore was termed the 'Second Industrial Revolution' by the government, which decided to move into 'knowledge-intensive activities such as R&D, engineering design and computer software services' according to the EDB. Once again emphasis was placed on 'R&D' and 'engineering design' without any mention of industrial design. As in many other countries in a haste to catch up, very few people at the policy making level seem to have understood that new design, especially new product design, is not the result of pure market pull and pure technology push, but a balanced combination of both with elements of art/culture and enterprise.

While EDB continued incessantly to find new companies for investment from the West and Japan, it was quite clear that local companies and enterprises had to play a greater role in the next phase of development of Singapore. The government decide to form the Singapore Trade Development Board (TDB) in 1983 to develop 'Singapore as a premier international trade hub and to promote the nation's goods and services '. Now renamed as International Enterprise Singapore (IESingapore), the TDB more or less 'discovered' a secret ingredient in the success of developed countries, which were known for introducing new products. This secret ingredient turned out to be 'industrial design'. At once TDB decided to press itself on to the 'fast forward' mode to bring this ingredient of industrial design into Singapore.

The National Productivity Board (NPB), which was by then known as SISIR, also discovered industrial design as the ingredient needed in Singapore, if it had to help the local companies and enterprises to move up the value chain to offer their own products and services. It was SISIR's turn at the 'fast-forward' process.

Both the agencies swung into action by encouraging a pioneer batch of talented students, giving

Figure 1. The author (left) with Michael Gan (centre) and Sannie Abdul in 2005 (from author's own collection)

scholarship where necessary, to study industrial design in established design schools and colleges around the world. This early batch included Mr. Sannie Abdul, an architect by training who went to do his masters in industrial design in the USA. Today, Mr.Abdul, who retired as the Director of the Singapore Design Centre (ofTDB) and teaches space design at the Nanyang Polytechnic.

The second batch of students included Ms. Lim Lay Ngoh (a director of Design 3 consultants) and Mr. Michael Gan (Manager of Industrial Design Course at the Nanyang Polytechnic) followed by Mr. Tan Khee Soon (a lecturer at Nanyang Polytechnic). All three of them are well known and respected within the design community in Singapore today. A few Singaporeans went on their own to study industrial design abroad. One of them was Mr. David Chen, son of a well known local artist, who studied industrial design in the United Kingdom and worked briefly in Italy. Upon returning to Singapore Mr. Chen started, probably, the first independent industrial design consultancy known as Studio Industrial Design in Singapore, which he still runs today. Figure 1 shows the author with two of the pioneers in industrial design in Singapore.

When the first few of the pioneering designers returned back, both TDB and SISIR had important tasks waiting for them. TDB went on to start a

Design Division within TDB and appointed Mr.Abdul to head this division. Mr. Abdul's mandate was to formulate a design promotion policy. This design policy was to have three main thrusts. Firstly, it had to evolve programs to familiarise the local companies on the importance of introducing design (both industrial design and graphics/packaging design) early in their product development process. Secondly, it had to build up the pool of designers in Singapore and, thirdly, it had to come up way and means to encourage the local companies and enterprise to use the services of designers in their new product design ventures.

SISIR decided that, in order to help the local enterprise into new product design and development activities, it had to get directly involved by starting a 'quasi-commercial' design consultancy of its own. Named the Product Design and Engineering Centre (PDEC) the new consultancy needed to be staffed with, not only talented designers, but also with experienced ones. Mr. Gan, Ms. Ng and Mr.Tan came back to Singapore to find themselves as pioneering designers at PDEC. SISIR also invited a management consultant with product design capabilities from the United Kingdom to partner PDEC, to bring in the level of industry experience needed.

While these developments were going on at TDB and SISIR, MNCs were starting to introduce industrial design personnel (mostly from their existing design centres) into their Singapore operations. Amongst them were Philips (the Netherlands), Thomson Consumer Electronics (France) and Matsushita Electronics (Japan). Local engineers were now entrusted with some part of the development stages of product design and development. These companies trained the local skilled manpower in model making and prototyping, for their in-house faculties. To this day, the local prototyping industry in Singapore is dominated by those trained by Philips and Thomson Consumer Electronics in the 1980s. Due to the presence of industrial designers in the MNCs, awareness to aesthetic elements, human factors (then known as ergonomics) started to rub onto local enterprises as well.

As this development of 'in-house' product design and development was taking place within large corporation, Singapore was getting ready for a very different phenomenon that was shaping up around the world in the 1980s. High technology companies were 'starting up' around famous universities such as Stanford University in California, MIT in Massachusetts and Cambridge University in the UK. Following this trend, the EDB set up the Science Park near to the National University of Singapore (NUS). The Science Park phenomenon became such a huge success with the internet boom of the 1990s, triggering so many start ups that Singapore eventually built three Science Parks and several 'Technology Parks'.

Even though the Science Park was up and running, SISIR's PDEC was in business and a small pool of local companies were ready to do new product design and development, one vital ingredient seemed still missing in the ecosystem. There was a need for a 'business proposition' to make design services affordable for the local industries, enterprises and start-ups. There was also need to enlarge the designer pool to offer variety and variations in design and design thinking.

It is at this juncture that the EDB, TDB and SISIR had to come together to formulate a strategy. As a first step they offered two incentive schemes:

1. The Design Venture Programme - DVP
2. Local Enterprise Technical Assistance Scheme – LETAS (which was already available for other areas of engineering and technology development)

Under these schemes a local company or enterprise (small and medium scale industries) were reimbursed up to 70% of cost of engaging the services of a design and development consultancy/company. This scheme extended up to the cost for the fabrication of mock-ups and prototypes. Attractive as it seemed, these two schemes required a lot of effort to 'sell' to the potential recipients. The next step was to look at designers themselves for ideas.

Sometime in the mid 1980s, a group of 11 designers, a mix of product and graphics designers, got together to form the Designers Association Singapore (DAS) as a non-profit organisation to promote professional attitudes and conduct amongst designers. The TDB then asked DAS to come out with a white paper to propose how design in Singapore can be established and promoted in a sustainable manner.

DAS presented TDB with a plan that included:

1. Establishing a 'Singapore Design Award' (SDA): To encourage industry and the designer to showcase their experience in new product development
2. Establish a 'Young Designers Award' (YDA): To encourage youngsters at high schools to become designers and to identify talents
3. Establish a bi-annual international event on design with a conference and an international showcase with pavilions from nations known for design such as USA, UK, (West) Germany and Japan

The idea that design needs to be embraced in its wider context, beyond the limited efforts of the government bodies, as means to take the nation to the next level, was brought to the attention of the Prime Minister's Office by the ministries in charge of finance, industry, art and education. The white paper presented by the DAS to TDB caught the attention of the policy makers there. The TDB was given the necessary funds to establish all the three ideas put forth by DAS. The Design Division of TDB was promoted into the Design Centre under the TDB with an expanded team. The Singapore Design Council was established, with a prominent Singapore businessman as its chairman, to develop strategies and policies on design, which would be executed through the Design Centre. Mr. Sannie Abdul was appointed as the Director of the Design Centre. The Design Centre was also given space in a building right in the centre of Singapore, to establish itself. The Design Centre had a design library, exhibition spaces and a souvenir shop, besides office area and meeting rooms. The Design Centre was mandated to:

- Promote design awareness
- Encourage the development of good design and high quality products in the industry
- Be an information source for design disciplines, including industrial design, graphic design and interior design
- Support the development of design talent
- Be a focal point of interaction among manufacturers, designers and design students
- Link up with the international design community by representing Singapore in world bodies like the International Council of Societies of Industrial Design (ICSID)

In 1988, the TDB, together with DAS and supported by other keys agencies, organised, the First International Design Forum, incorporating the Singapore Design Award (SDA) and the Young Designers Award (YDA), together with an exhibition featuring good product designs and graphic design from countries known for design such as the Great Britain, the United States, Germany and Italy. With very high level of interest within the government, the Prime Minister of Great Britain, Mrs. Margaret Thatcher, the Prime Minister of Australia, Mr. Bob Hawk and Singapore's Prime Minister Mr. Lee Kwan Yew were amongst the well-wishers for this forum. Famous designers, such as England's Kenneth Grange, were invited to speak at this high profile forum, which was opened by the President of Singapore, Mr. Wee Kim Wee himself. The International Forum and the two design awards became regular events in the international and national design calendar, every two years, till 2004. Figures 2 and 3 show banners from two such forums.

Figure 2. Banners for two International Design Forums (from author's own collection)

Figure 3. Banners for two International Design Forums (from author's own collection)

The success of the first International Design Forum in 1988 saw a spurt in the involvement of design in Singapore. Companies such as Motorola and HP started their own design studios in Singapore. Local design consultancies such as Design 3, Studio Industrial Design and Lawton& Yeo Associates established themselves as a good private alternative to PDEC at SISIR. Individual industrial designers, including the author, were encouraged to migrate to Singapore and find their professional roots during the late 1980s.

To match this rapid growth in design in Singapore, several home-grown companies found success in the regional and international market place. Newly public listed companies such as Aztech Systems Ltd., IPC Corporations Ltd., and Creative Technology Ltd., with high level of investments in R&D, engineering and marketing, created new products in the fast growing area of electronics and communications. To add to the equation 'Silicon Valley' style venture capital groups were starting to make their presence attracting the new tenants of the Science Centre.

The result of this was that local designers and in-house designers located in Singapore were starting to design personal computers, electronic calculators, televisions, and monitors, audio and video systems, which were marketed worldwide.

The ministry of education, sensing a demand for 'creative' talents, established a third polytechnic in 1989. Led by its dynamic Principal and 'CEO', Dr. Varaprasad, Temasek Polytechnic started Temasek Design School, which began offering a three year diploma programme in product design along with diplomas in graphics Design and interior Design. With the Temasek Design School in place, Singapore moved towards establishing itself as 'producers' of design talents in Singapore (and the region). With the first batch coming out in 1991, Temasek Design School's most distinguished of Temasek's alumni could be Mr.Low Cheaw Hwei, a first batch student who has risen to be a Director at the Philips Design Center at global level.

The effect of all this excitement, at the end of the 1980s, was an euphoric feeling that industrial design was going to be 'the' next money making venture and that many of the world's innovative products would be coming out of Singapore. The author's opinion is that all the buzz and excitement and the first level of success in using product design was based on a conventional product development model which emerged out of the post world war industrialisation efforts. As per this model, the industrial designers finish their job of creating 'clever' design solutions with modern aesthetics and pass the 'dirty' job of engineering, prototyping, troubleshooting and first (pilot) production to others. The industrial designers were available for discussion and usually gave criticisms of 'what could be done better' and leave the rest to the engineers. While this could work in a Western environment, the home-grown companies needed more than a 'consultant' just for industrial design alone. They needed product design 'innovators' who could join them right from their ideation stage through product design, engineering, prototyping and troubleshooting till the first production of the new product(s). The feeling amongst the Singapore client base that industrial designers were artists and aestheticians who were needed only at the end stage to beautify the 'housing' needed to be addressed if product design was to play a significant role in Singapore's economy.

THE 1990S: INFORMATION TECHNOLOGY, INTERNET AND DESIGN RENAISSANCE

With the Design Centre (TDB) operating out of its new office in the heart of the Singapore, product design enjoyed a period of renaissance during the 1990s. By this time the Singapore Manufacturers Association (SMA - now known as Singapore Manufacturer's Federation), founded in 1932, also got involved in exploring how product design can bring benefit their member companies to move to the level of manufacturing their own products. The Designers Association Singapore (DAS), which had attracted close to hundred members, started working closely with the Design Centre (TDB) and SMA. During the 1990s, the Design Centre (TDB) and DAS, successfully organised, five International Design Forums (IDF) in 1990,92,94,96 and 98. All the five IDFs were well received, with design pioneers like Kenji Ekuan of Japan, famous designers like Robert Brunner of Apple and renowned design thinkers like Victor Papanek, sharing the podium. Singapore Design Award (SDA) and the Young Designers Award (YDA) were also successfully organised during the five IDFs and became well known amongst the industry and the educational institutes respectively. Figure 4 shows one such IDF, where Mr.Kenji Ekuan of Japan was a key speaker.

Figure 4. Designer Kenji Ekuan of Japan (centre) at one of the International Design Forums (from author's own collection)

While designers were enjoying their limelight, EDB had identified 'service industries' as a second pillar (manufacturing being the first) and engine for growth of the Singapore economy. Their eyes were cast on two important emerging sectors, information technology (IT) and medical science industry that included pharmaceuticals, biotechnology and medical technology.

With 'internet' and 'world wide web' being the new buzz words, the IT sector started blossoming with local and western high-technology start-ups vying for the space available at the new Science Parks and Technology Parks. Software, with its ability to be 'stand alone' as well as be 'embedded' within circuits, became an important aspect of the technology industries. Well-known software companies such as Tata Consultancy of India and Convergence of the USA set up their operations in Singapore.

Early 1990s saw two industrial design consultancies establishing operations in Singapore. They were (the famous) Frog Design from USA and NeoTechnics from Australia. The third, CFI Design of Japan joined hands with Design 3 of Singapore to open CFI Design Singapore in 1992. These three consultancies, together with Lawton&Yeo and PDEC of SISIR, were servicing

Figure 5. Hearing aid for Siemens (Inovasia Design) (from author's own collection)

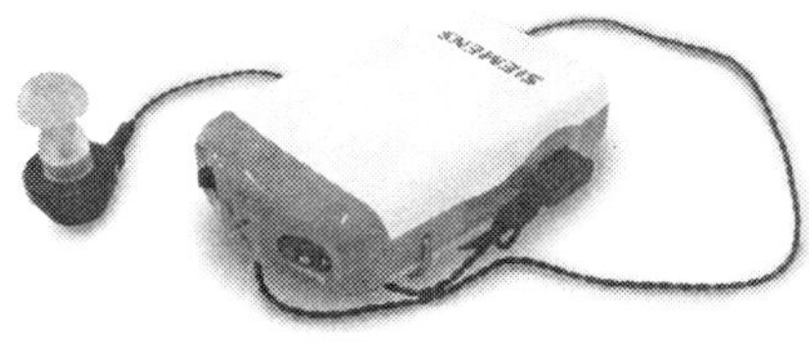

the large number of product design projects from local clients under the two incentive programmes offered by TDB and SISIR. Fitch, an American design and branding consultants joined hands with GE of USA to set up GE Fitch in Singapore and began servicing multinationals such as HP in their new product design efforts. While many of the multinationals were using Singapore based design consultancies to augment their in-house design capabilities, some, like Black&Decker of USA decided to use only external product design resources together with their own in house engineering team. To add to the list of new clients, Government linked companies such as Singapore Technologies Electronics, which traditionally takes large infrastructure projects, started using the services of product designers in mass rapid transport projects that they had won for the North Eastern Line in Singapore during this period.

Unfortunately, Frog Design and NeoTechnics could not sustain their operations beyond 1995 due to the high overheads attributed to expatriate design staff and low revenues generated. CFI Design Singapore survived for another decade since it was predominantly staffed with local designers. With the exit of Frog Design and NeoTechnics leaving a large void in the design service available locally, and with the first wave of internet boom hitting Singapore in 1995, two notable design consultancies were formed which started doing a large chunk of the new product development projects requiring product design services. The first of the two, Design Exchange, was founded by Mr. Song Kee Hong, who had studied industrial design in the USA. The second, Inovasia Design

Figure 6. Bluetooth pendant for add value technologies, Singapore (Inovasia Design) (from author's own collection)

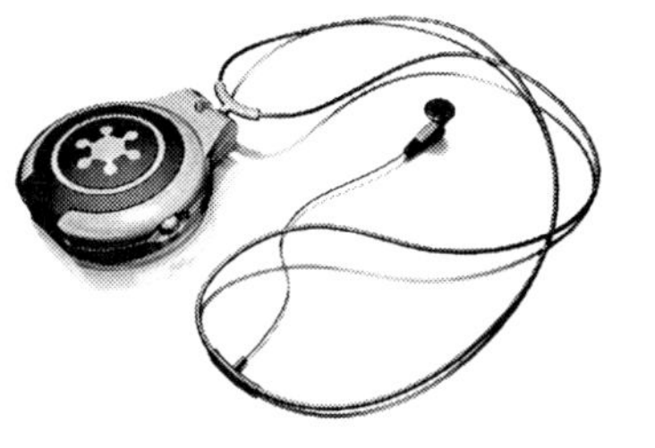

was co-founded by the author and Mr. Derek Lim, who studied industrial design in Australia. While Design Exchange concentrated on servicing purely the high tech electronic industry, Inovasia took on projects ranging from electronics to industrial to products for mass rapid systems. Orca Design, a Malaysia based design consultancy opened an office in Singapore and enjoyed the resurgent product design projects available in Singapore. Joining these two, later, was 'aNd Studio', founded by Mr. Alex Lau, another industrial design graduate from Australia, who included 'branding' also into the services his studio offered. Figures 5, 6, 7 and 8 products designed by Inovasia during this period.

The medical science industry, on the other hand, needed larger areas for clean room operations, and hence needed space in larger industrial estates. Becton Dickinson Medical from the USA, for example is located at a large industrial area in the west side of Singapore. Hospitals in

Figure 7. DSL router for add value technologies, Singapore (Inovasia Design) (from author's own collection)

Figure 8. Integrated electronic safes for Chubb (Inovasia Design) (from owner's own collection)

Figure 9. Portable high-tech urimeter for a Singapore hospital (Inovasia Design) (from author's private collection)

Singapore, such as Tan Tock Hospital, set up their own biomedical research team which were tasked with developing unique products, such as the high-tech portable urimeter (Figure 9) for patients in intensive care, for which it received the Local Enterprise Technical Assistance Scheme from SPRINGSingapore (former SISIR). In projects such as these, local product design expertise was in high demand, providing the design consultancies continued work.

Product design had picked up enough momentum to warrant the fourth polytechnic, Nanyang Polytechnic, to start a diploma programme in Industrial design in the mid 1990s. Interestingly, the erstwhile German-Singapore Institute, a specialist training centre under EDB, was amalgamated into the Nanyang Polytechnic to form a part of the design school there. In 1999, The National University of Singapore (NUS), which had become the premier teaching and research university in Singapore, started a four year Bachelor of Art (Honours) programme in Industrial Design under the Faculty of Architecture supported by Faculty of Engineering and the School of Business.

NUS's industrial design programme, for the first time, mentioned about bringing the 3D design culture together with advanced technologies of engineering and the entrepreneurial spirit of business. This was an indication of what EDB, TDB, SISIR, DAS, SMA and all the education institutes were trying to do, i.e., bring together *design, technology and enterprise* in order to create viable economic environment within Singapore for sustainable growth.

The fairy tale of product design being an 'important factor' in Singapore's economic development did take a huge hit in 1996, when TDB decided to downgrade the Design Centre back into a division within TDB. This came as a surprise to the product design community as awhole, and to the Designers Association Singapore (DAS) in particular. Having taken the lead in organising the International Design Forum, Singapore Design Awards and the Young Designers Award for the last eight years or so and having worked closely with the Design Centre to promote design throughout the region and beyond, DAS felt that it was not the right time to let the momentum falter. While it looked like everything was going well, with the local companies developing new products, and the design consultancies and its supporting prototyping industry thriving, Singapore's touch with the design world outside, and the goodwill that had been created, was starting to wane. With the Director of the Design Centre deciding to leave, the staff at the Design Division lost the momentum that was created in the 1980s.

DAS and the Design Division at the newly named International Enterprise Singapore (IESingapore) carried on with the 1998 International

Design Forum, the Singapore Design Award and the Young Designers Award, even though the industry sponsors and the international support seemed hesitant. With the result, the public at large and the local companies and MNCs sensed a lack of priority for product design in Singapore.

This did not stop the design industry from moving forward, even after a blistering economic down turn in 1997. While many of the internet companies collapsed, those with a strong technology behind their business, such as Innomedia (a spin off from Creative Technology) and Aztech Systems, survived, providing projects for the local product design community.

In the end, 1990s was truly the decade of design renaissance when one could acknowledge that product design played a significant 'transformal' role in Singapore's transition into a service economy. Other sectors that were contributing to this transition were the finance industry and management consultancy. To those in the product design industry, the 1990s may have been the last chance to experience the classic product design process, for the winds of change that the new millennium was going to bring was already in the air by 1999.

THE 2000S TILL 2009—NEW PARADIGM: WHERE DOES PRODUCT DESIGN FIT IN THIS?

Entering a new millennium, the product design industry seemed to be cruising on autopilot with all the design groups having enough work in their hands. However, products being developed by Singapore companies were getting to be more and more about user interface (GUI) and interaction design. The ubiquitous screen became the 'sanctum sanctorum' around which the physical shapes were being designed. The new buzzword emerging was 'innovation'. Aesthetics in products started moving towards 'relevant form to fit the product function' rather than beautiful or smart looking. Material thickness was shrinking below 1mm, injection molded plastics were being replaced by die cast magnesium in some cases. With more 'global products' being developed locally, compliance to statutory requirements of multiple countries was being transferred as a design constraint to industrial designers upfront. While design of mobile information/communication products became a major 'bread and butter' job for the design groups, there were also a few projects from health and lifestyle products companies to design 'massage chairs' or 'cyclonic air purifiers'.

The traditional manufacturing sector had slowly moved out of Singapore to China, Indonesia, Malaysia and Vietnam due to the high cost of production locally. Only the technology intense, 'high value-add' manufacturing industries remained. Singapore designers had to agree to the slogan 'Designed in Singapore, manufactures elsewhere'. Based on the traditional 'design promotion' policies, IESingapore (former TDB) had done their best to put Singapore design in the world map. The time was ripe for a new entity to propel design into new territories to come.

Sensing clearly that a new paradigm is required to make Singapore into a 'dynamic global city, thriving in a changed world', Singapore government set up the Economic Review Committee (ERC) bringing together local and foreign experts from each sector of the economy to fundamentally review Singapore's development strategy, and formulate strategies to upgrade, transform and revitalise the economy.

Relevant to design is the report from one of the Subcommittee on Service Industries to evolve a 'Creative Industries Development Strategy'. After several rounds of intense discussions amongst all the stakeholders in the creative industry, the subcommittee submitted its report to the government. Where did product design figure in this? A foot note in the second page of the media release to notify the submission of the strategy combines the following under 'Design': Advertising, architecture, web and software, graphics *industrial product*, fashion, communications, interior and environmental.

The same report highlights that, even though Singapore's product design did not score favourably in a Global Competitiveness Report in 2001-02, the economic spin-offs from 'Product and Visual Design' was much higher than that for any other industries within Singapore (1.76 as an output multiplier vs. 1.44 for banking services) in the result of an independent study commissioned by IESingapore in 2000. This finding is the clearest evidence to date of the transformative role played by product design in the transition of Singapore into a service economy during the 'renaissance' decade of the 1990s.

More important development from the subcommittee report was the formation of the 'DesignSingapore Council' within the Ministry of Communication and the Arts (MICA). Dr. Milton Tan, an Associate Professor and former Dean of the Faculty of Architecture and Planning in NUS, became the Director of the DesignSingapore Council in 2003 with a mandate to implement the following strategies as per the ERC report:

- Strategy 1: Integrate design in enterprise
- Strategy 2: Develop a vibrant & professional design community
- Strategy 3: Position Singapore as a global design hub
- Strategy 4: Foster a design culture & awareness

The ERC Subcommittee report on Creative Industries Development, which can be easily located in the internet, is an important read for all those interested in government led strategies for the creative industry. Going back to the question, 'Where is product design in all this?, one notices that in the same report, product design is being associated with 'iconic' products, placing more emphasis on 'distinctive design identity and its qualities' from a branding angle. The report is also looking 'for iconic "killer applications" that would help establish the standards for design excellence in Singapore'.

The enthusiasm for developing high technology products, that was prevalent during the 1990s, has yet to be repeated. The number of design consultancies that are involved in designing high technology products seems to be dropping, even though many of the younger designers graduating from polytechnics and NUS have opened their own design consultancies in the last five years or so. With the last of the International Design Forum (IDF), together with the Singapore Design Award (SDA) and the Young Designers Award (YDA) being held in 2004, the international design community and the Singaporean public are not fully aware of the product design activities in Singapore.

DesignSingapore Council has several councils and committees to advise, support and ratify the plans and actions of the Council. Some of DesignSingapore's initiatives include the bi-annual Singapore Design Week, participation in the annual Milan Furniture Fair, participation in London Design Week, participation in the Vienna Biennale, introduction of the annual President's Award, providing financial incentive for companies using design through the Design Pioneer Programme, overseas scholarships for talented students to study design, to mention a few. Besides these, DesignSingapore invited the well known BMW DesignWorks, a multi-product design consultancy to set up their studio in Singapore.

Alongside DesignSingapore's initiatives, Singapore's second university, Nanyang Technological University (NTU) started a four year Bachelor of Fine Arts (Honours) programme in Product Design at the School of Art, Design and Media. The first batch from this school graduated in July 2009. Ngee Ann Polytechnic's Mechanical Engineering Department also started a three year diploma programme in Product Design and Innovation (PDI), combining aspects of industrial design and creative engineering design to bring out, what the author would call, product development designers. The second batch from the PDI course graduated in 2009.

It must be said that DesignSingapore has done well in its effort to bring in a whole 'design ambiance' within the creative industry in Singapore. It is staffed with young and enthusiastic people, who have taken the ERC Subcommittee mandate to their heart. Since product design is only one of the many areas in their purview within the creative industry, it is perhaps best for the product design community to help themselves by proposing plans and initiatives that fall within their mandate to DesignSingapore, rather than wait for intiatives from them.

With the present economic climate restricting many major initiatives in R&D in Singapore, the product design community seems to have put a high level of expectation on the ICSID 2009 Conference to be held in Singapore. The product design community, one can say is looking forward to a new path, a new direction and new energy, in its efforts to offer globally competent design services when the world comes out of it present economic condition.

DISCUSSION

What will be the future of product design in Singapore then? As a small community in a small city-state, it is apparent that product design is dependant on the events and initiatives that the government of Singapore puts in place, if it is to part take at the highest level of economic development. At present, the focus seems to be more on media and media related design, game design, digital animation and digital film making, which is being spearheaded by the Media Development Authority (MDA) of Singapore. This has put many in the product design community back in the 'survival mode' of the 1980s.

Like a capable Formula One racecar, caught in traffic jam during the race, product design is engulfed by all other areas within the creative industries in Singapore, at this present time. Like a successful Formula One racing team, product design needs a good driver, a good car with a good engine, adequate fuel and an efficient pit stop crew, in order to get a podium finish again.

This calls for several 'far sighted' initiatives from within the product design community in Singapore itself.

First, would be to realign product design back with the R&D activities happening in the startup companies and enterprises in the Science Parks, Technology Parks, Universities and Research Centres of Singapore.

Second, would be to come to a realisation that product design has to go back to the roots of its success, i.e., to bring together *design, technology and enterprise* to create viable economic environment within Singapore for sustainable growth in new product development. This would mean reestablishing close links with EDB, IESingapore (TDB) and the Singapore Manufacturers Federation, while continuing to have strong links with DesignSingapore Council

Third, would be re-evaluate the curriculum of product design/industrial design in the polytechnics and universities in order to graduate product designers who understand the languages of design, technology and business, while being able to explore and experiment like artists and scientists.

Fourth, would be to educate the stakeholders and users of design that sustainable product design is a 'serious' profession that looks beyond 'image and branding' into the heart and soul of the company, the products and the technologies that it holds. There is a need today, more than ever, to take design to the Board Room of successful companies, where it gets the right voice, as it happens in Apple Incorporated in USA, where Mr. Jonathan Ive, the Vice President for Design, reports directly to Mr. Steve Jobs, the CEO of Apple.

Perhaps a good first initiative would be to give immediate attention to the environmental sustainability concerns of today, and to design products and systems for alternate energy sources. With

EDB and other government agencies taking a large interest in bringing in and developing high technology companies involved in 'green initiatives', alternate fuels and sustainability, product design could earn the 'first mover' advantage by being the first to work in these areas. This could re-ignite the spirit of the renaissance era again.

REFERENCES

Chang, M. (1998). *From Economic Debacle to Economic Miracle*. Singapore: Ministry of Education.

Chua, R. (Ed.). (1994). *Design Works*. Singapore: Designers Association Singapore.http://www.edb.gov.sg/edb/sg/en_uk/index/about_edb/our_history.html

IdN Magazine (Ed.). (2005). *SINGAPOREdge: Design Culture in Singapore*. Singapore: Sysems Design Limited.

International Enterprise Singapore. (2009). *About us; Milestones*. Retrieved from www.iesingapore.gov.sg

Japan Industrial Design Promotion Organization (Ed.). (2002). *Design Awards Asia*. Tokyo: Japan Industrial Design Promotion Organization (JIDPO).

Sathikh, P. M. (1998). Singapore: A Healthy Environment for Innovation and Design. [). Tokyo, Japan]. *Axis Magazine*, *73*, 143.

Singapore Economic Development Board. (2009). *Our History*. Retrieved from Singapore Economic Development Board. (2009). *Singapore Facts & Figures*. Retrieved from www.singstat.gov.sg/stats/keyind.html

Tan, B. S. (Ed.). (2000). *Singapore Design Award 2000*. Singapore: Trade Development Board & Designers Association Singapore.

Tan, B. S. (Ed.). (2002). *Singapore Design Award 2002*. Singapore: Designers Association Singapore.

The Ministry of Trade and Industry. (2009). *Design Singapore Initiative: Summary Table from Preliminary Strategies and Recommendations*. Retrieved from http://app.mti.gov.sg/data/pages/507/doc/ERC_SVS_CRE_Annex3.2.pdf

The Ministry of Trade and Industry. (2009). *Media Release by Economic Review Committee, Subcommittee on Service Industries*. Retrieved from http://app.mti.gov.sg/data/pages/507/doc/ERC_SVS_CRE_PressStatement.pdf

KEY TERMS AND DEFINITIONS

Singapore Economic Development Board (EDB): The lead government agency in Singapore that plans and executes economic strategies that enhance Singapore's position as a global hub for business, investment and talent.

International Enterprise Singapore (IESingapore): The lead agency under the Ministry of Trade and Industry to promote the overseas growth of Singapore-based enterprises and international trade.

Singapore Productivity and Innovation Board (SPRINGSingapore): Is the enterprise development agency for growing innovative companies and fostering a competitive Small and Medium Enterprises (SME) sector. It also develops and promotes internationally-recognised standards and quality assurance to enhance competitiveness and facilitate trade.

DesignSingapore Council: Is the national agency to promote design in business and education, and to develop Singapore as one of the leading international centres for design creativity.

Designers Association Singapore (DAS): A non-profit association of practicing designers and design educators from all fields of design.

Chapter 24
The Value of Storytelling in Product Design

Rina Bernabei
University of New South Wales, Australia

Kelly Freeman
University of New South Wales, Australia

Jacqueline Power
University of New South Wales, Australia

ABSTRACT

Contemporary product designers are increasingly attempting to utilize the latent potential of the product narrative and the impact it can have on the end-user and their relationship with products. Storytelling in product design allows a dialogue and conceptual exchange to be established between products and end-users. This chapter will consider the end-user's relationship with products and the designer's role in this emotional and conceptual exchange of storytelling in product design. This will be outlined through a framework defined by the authors as Narrative; Manufacture; History; and Interaction and will consider the work of contemporary designers Philippe Starke, Hella Jongerius, Dick Van Hoff, Ron Arad, Tokujin Yoshioka, Peter Jagt and Frank Tjepkema. The paper will be supported by a main case study of the work of Bernabeifreeman, the award winning industrial design practice of the authors.

INTRODUCTION

Why do you need stories?
Everyone needs stories
Adults watch the telly
When children go to bed
And fathers read the papers
And everyone likes movies
And mothers buy the magazines
To see what people said
-The Land where Stories End, David Foster

This whimsical quotation focuses upon fulfilling a desire for stories from the printed word and flicking screen image- instant story gratification obtainable from mass media. Yet stories need not just be found in ephemera or as expressions of mass culture. Products themselves can be designed to encourage reminiscence to occur and allow stories to be attached to them over time by

DOI: 10.4018/978-1-61520-617-9.ch024

end-users. Products are well suited to this collection of stories- they are tangible, alterable by the effects of time and are silent witnesses to the life of end-users. "Past memories, present experiences, and future dreams of each person are inextricably linked to the objects that comprise his or her environment" (Csikszentmihalyi & Rochberg-Halton, 1981, p. ix). Or similarly as Marcus Kwint (1999) explains, within Western cultures objects may "furnish recollection…stimulate remembering… form records: analogues to living memory (p. 2)."

Contemporary product designers are increasingly utilising the latent potential of the *product narrative* recognising, as highlighted by Donald Norman, that emotional responses to products are as valid as other reactive responses (Zachry, 2005). For as Norman has said, "People understand things that have a personal connection, which is where stories play a very important role in human behaviour. In some sense, that is what you want a good design to do--which is to guide you through the story" (Norman in Zachry, 2005).

The response generated by a product is contextually and culturally influenced. Adam Richardson (1993) has suggested in *The Death of the Designer* that "how users and cultures respond to the products which designers help create is not well understood" (p. 34). Responses triggered may tap into the reservoir of memory and design knowledge of the end-user. Jonathan Chapman (2005) suggests that product meaning is actually comprised of three elements: polysemy (the multivalent nature of objects), contextual sensitivity, and consensus (the need to share information about the product in order to successfully communicate with others about it) (pp. 39-40). Product designers must consider this trinity and its impact upon the product narrative. A successful product designer or *storyteller* may embed triggers in the hope that these will act as a catalyst for end-users to initiate a story. The triggers are as literary techniques of metaphor, simile, and personification: poetic and rhythmic. The design elements or triggers at the disposal of the designer are such aspects as form, material, finish, colour, repetition, scale, manufacture methods and design reference.

To assess how end-users respond to their possessions and environments, an extensive series of interviews were conducted by Mihaly Csikszentmihalyi that formed the basis for the text *The Meaning of Things: Domestic symbols and the self* and an associated article. Csikszentmihalyi (1981) explains in the text that the objects cited as being special to end-users "often lacked any discernable esthetic value, but they were charged with meanings that conveyed a sense of integrity and purpose to the lives of the owners" (p. 27). Similarly in the article *The Gift,* author Clive Dilnot discusses how products are infused with much more than their immediately perceivable purpose. Dilnot considers the gift-object and the exchange that takes place when a gift is proffered. He highlights that in the exchange of gifts more than just the product is exchanged. It is for such a reason that a product may then be recognized as special by an end-user because of significant memories associated with it. In particular it is domestic products that Csikszentmihalyi (1981) argues as being the most special because "household objects are chosen and could be freely discarded if they produced too much conflict within the self" (p. 17). Csikszentmihalyi (1981) suggests that products within the domestic interior "represent, at least potentially, the endogenous being of the owner" (p. 17).

How the product narrative reveals itself in the work of contemporary designers Philippe Starke, Hella Jongerius, Dick Van Hoff, Ron Arad, Tokujin Yoshioka, Peter Jagt and Frank Tjepkema will be considered in respect to 4 categories: Narrative; Manufacture; History and Interaction. These four storytelling categories have been identified by product designers' Bernabeifreeman in both their work and that of other designers.

It is the Narrative and Manufacture methods of storytelling over which the product designer has the most control. Narratives are the stories or historical references the designer seeks to refer;

the precedence that the product speaks of. The Manufacture techniques are the stories that can be told from the physical making of the product. History on the other hand is the personal stories that the product may conjure in the minds of end-users, and Interaction speaks more specifically of the physical changes that may occur to the product over its lifetime in its contact with end-users.

HOW STORYTELLING REVEALS ITSELF IN THE WORK OF CONTEMPORARY INDUSTRIAL DESIGNERS

The storytelling framework identified by Bernabeifreeman of Narrative, Manufacture, History and Interaction, presents itself as a valuable tool through which to consider the stories told by the work of other contemporary product designers. Whether these stories have been intentionally designed into the products described here or not, the products have enabled a story to be associated with them, that importantly may heighten the 'value' of the product to an end-user.

Narrative

The work of French product designer Philippe Starck, sometimes referred to as a "superstar" (Conran & Fraser, 2004, p. 20) designer, is identified here as one of the most successful product designers/storytellers who conveys rich narratives through his products. Starck has often been quoted as not being interested in design itself, by this he means he is not interested in the mere physicality of the object, he is interested in the emotion of the product and the voyage which the product takes the end-user on. Perhaps this idea is best illustrated by his *Louis Ghost* chair which he designed for Kartell in 2002. The *Louis Ghost* chair is a plastic mass produced product reasonably priced as a result of its manufacture method. It is also a product rich in narrative that engages the end-user in its playful modern interpretation of the 18th century Louis XV armchair that populated the Palace of Versailles. There is an amusing irony in the juxtaposition of the chair's precedence and manufacture technique, as is alluded by its name; the product is designed to reach a broad demographic as a result of its manufacture methods, yet the product references chairs once far removed from the financial means of many people. Underlying this is a reminder in the chair's ghostly nature that it is symbolic of a courtly world that was eventually to give way to a rising middle-class.

The barely-there tinted one piece injected molded clear polycarbonate design, is free from any joints or bindings and creates a piece that is both delicate and indestructible. It is the ghost of the chair past, with a highly recognizable iconic form interpreted into transparent manufacturing. It speaks of wealth, royalty, elegance and good taste, yet also of modern technological manufacturing and digitalized design. Starck sees his role of designer as that of a film director, in a recent interview he was quoted as saying "I'm not into the product. I'm into the effect the product is going to create. I'm only interested in the scenario which people are going to live out" (Martin, 2009, p. 108). He sees his product's role to transport people out of the ordinariness of their daily lives, to be moved by the product "by putting them in places which is supposed to make them sexier, more beautiful..."(Martin, 2009, p. 108). Starke goes on to explain that it is the repetition of encounters with products "the daily encounter with....an invisible chair, which is going to little by little take people to the places where I want to take them" (Martin, 2009, p. 232).

The work of product designer Hella Jongerius also conveys narratives to their end-user. She believes that objects are "saturated with references" (Jongerius & Schouwenberg, 2003) as is evident in her *Animal Bowls* produced for the German porcelain manufacturer *Nymphenburg*. Jongerius interpreted designs and shapes from the repository of the Porzellan Manufaktur Nymphenburg collections. She selected five animal forms historically used by the manufacturer and placed them in bowls (snail, bird, hippopotamus, fawn and hare), and applied hand-painted patterns over the creatures drawn from the *Nymphenburg* painting archives (Jackson, 2005). The snail seems to extrude a trail of golden flowers, the hippopotamus is draped in a cloak of flowers, the ears and underbelly of the hare is decorated with a delicate textile pattern and the fawn seems to drip in a sauce of guinea hen feathers. Not only do the completed pieces reference the *Nymphenburg* repository of past work but they also seem to possess an additional layer of narrative. The various porcelain animals such as the fawn, snail and rabbit rest unsuspectingly in the centre of bowls. It seems no mistake that the fawn in particular is not just a highly sculptural creature but often finds its way onto the dinner plate and as such seems to drip in a sauce-like pattern. Jongerius' narrative seems to be a tongue-in-cheek comment about fine dining, attitudes towards food and traditional display 'china' and figurines that once decorated the domestic interior. The *Animal Bowls* have been described as "luxurious, but subtly seditious" (Jackson, 2005). The refined forms are contrasted by the "factory stamps and colour trials" (Jackson, 2005) evident on the final product, revealing to end-users the often concealed manufacture processes associated with porcelain.

Manufacture

Hella Jongerius also conveys narratives to end-users through the utilization of manufacturing techniques. She believes that products are "saturated with references" (Jongerius & Schouwenberg, 2003) and that stories can be told through the manufacture of a product, as is evident in her *Soft Urn* vases (Figure 1). Constructed from polyurethane, the vases "expose the story of their making, showing scratches, bubbles and moulding joints" (Conran & Fraser, 2004, p. 140). Thus

Figure 1. Urn vase for Droog by Hella Jongerius. (©1996, Droog B.V. Used with permission)

the vases celebrate their manufacture with join lines and imperfections retained on the finished product. "The vase hides nothing, the traces of its making are still there. The vase is a mass of scratches and bubbles, the edge is frayed, even the moulding joins are visible" (Ramakers & Bakker, 1998, p. 19). There is a sense of honesty and integrity expressed by this approach to the manufacture of a product.

In some cases modification of the product's form may occur in a less controlled manner, allowing for a sense of individualization to result from the Manufacturing narrative of the product. Such an approach is well illustrated by Dick Van Hoff's *Extrusion Plates* (Figure 2) for the *Droog Design for Rosenthal* collection. The porcelain plates transform the traditional blue and white porcelain into a highly individual collection of

Figure 2. Dick Van Hoff's Extrusion Plates., (© 1997, Droog B.V. Used with permission)

products dictated entirely by the manufacturing process. The use of colour references 'blue-and-white porcelain' yet subverts the traditional manufacture technique of "painting in cobalt blue under the glaze" (Hillier, 1975, p. 482) a technique that only requires a single firing. In this instance two colours of clay are used, blue and white, and the two colours are mixed in the extrusion machine and then cast into traditional plate forms (Ramakers & Bakker, 1998, p. 135). Unpredictable individual effects result, each swirl and mix of colour dependent upon the extrusion machine. The end-user can see the liquid-like folds and swirls of colours as they would have emerged from the machine, the plates thus capturing the moment of production. The pattern produced is no less relevant a pattern than the formalized figures and landscapes that decorate blue-and-white porcelain or Delftware. As each plate is individual and there is no matching set, the product is made more exclusive as a result of the Manufacturing narrative. The end-user is provided with a visual expression of the product's manufacture as a result of its patterning, a patterning that goes deeper than the surface, but is an integral part of the product.

History

The History of the product also instills narratives into the product. This method is illustrated by a traditional Chinese and Japanese approach to dealing with what would typically be viewed as the end-life of pottery within other cultures. As is explained by Louise Schouwenberg and Hella Jongerius (2003) in the book entitled *Hella Jongerius*:

You're presumably familiar with the traditional Chinese and Japanese practice of repairing broken pottery with gold lacquer. Throwing things away isn't an option, and the defect becomes a thing of beauty in its own right.

This example speaks of both the history of the product and interaction (with an unintended outcome) which occurs at the hands of its end-user. The lacquer joint creates a physical record of the life events of the pottery. The golden join is not viewed as a blemish but instead is framed as a beauty-mark adding character to the piece. This is an event generally out of the control of the designer yet no less valid in the shaping of a valuable product in the eyes of its end-user.

Yet the History of the product may sometimes fall into the purview of the designer as exhibited by the *Rover* chair. The designer can enhance the value of an existing product by modification or re-interpretation of products for a new context, as seen in Ron Arad's *Rover* chair. Produced in 1981 the *Rover* chair unites two ready-made products salvaged from a junkyard; a seat from a Rover 200 car mounted on a scaffold frame (Conran & Fraser, 2004, p. 34). Through the re-interpretation and modification of these two seemingly disjoined objects the designer has created a new context and new product from the ready-made and pre-used products. The *Rover* chair is framed as a new product yet on closer inspection the end-user sees the signs of age and decay of the highly sought after and now almost vanished Rover car. It is the context and narrative of the ready-made parts that provide an established history to this new product, yet over time as the object further decays it will bring with it an additional history, a history of use in its new context; a narrative as an armchair. Much like the highly prized Chesterfield lounges, age and wear can create strong sensorial links to memory and past lives. Sometimes this can be acquired after a life-time of use of a product but other times, as in this instance, it can be designed into a new product.

Interaction

Specific interaction with a product can also be designed to occur. Thus a story is also communicated effectively when some degree of contact takes place with the end-user. The value of ageing in products is full of sentiment – sometimes contradictory and dependent on the emotions of the end-user (Chapman, 2005, p. 131). In many products the 'aged surface' can add character giving a sense of history, thus leading the designer to choose materials that determine the speed at which this aged surface appears, and to what extent it will capture the history of the product. The *Honey Pop* chair (Figure 3) by Tokujin Yoshioka, is an example of the product graphically capturing interaction with the end-user over time. The poetic storytelling potential of design is recognized by Yoshioka who has said that, "The essence of design lies not in form but in something invisible or intangible" (Conran & Fraser, 2004, p. 248). Yoshioka has cited such natural elements as clouds and snow crystals as poetic and inspirational, leading him to seek the "intrinsic beauty of materials" rather than spearheading the design process by form-making (personal communication, August 27, 2009). In respect to his rationale towards design Yoshioka has outlined in a *Design Theory*:

My design has never existed before, but for some reason, they are well integrated with our everyday lives and remain...For a chair, I might be trying to design just the sense of seating. I cannot ask for anything more than a chair, which realized the comfort in seating and the sense of [the] sitter floating in the air released from...gravity (personal communication, August 27, 2009).

The *Honey Pop* chair interacts with the end-user to individualize the seating experience. The chair is created from a two dimensional build up of one hundred and twenty pieces of glassine paper which are glued together and precisely cut. The structure is then opened forming a strong three-dimensional honeycomb structure. The chair folds out from its spine much like a book, with the web of paper connections giving the chair stability and its ethereal quality. When the end-user sits on the chair, numerous layers of thin paper 'register' this

Figure 3. Honey Pop Chair. (© 2004 Tokujin Yoshioka. Used with permission)

interaction forming an impression of their form. Over time this impression is made more obvious as the chair literally moulds to the end-user's form. Yoshioka has said that "honeycombs, the human body (bottom)" (Conran & Fraser, 2004, p. 248) influenced the work. The many folds and distortion of the paper edges are descriptive metaphors of the weight and time spent in the chair by the end-user. The designer has constructed a three-dimensional 'patina' in a seemingly throw away product yet it is its delicateness which also gives this product its value (Chapman, 2005, p. 131).

The Interaction method of storytelling is also well illustrated by the Peter Jagt and Frank Tjepkema's *Do Break* (Figure 4) designed as one of eight products in the *Do Create* collection (Louie, 2001, p. 3). This product is in the same vein as the description of the broken pottery by Louise Schouwenberg, although this idea has been re-interpreted and designed into the product as an intended interactive narrative rather than an

Figure 4. Peter Jagt and Frank Tjepkema's Do Break vase. (©2000, Droog B.V]. Used with permission)

incidental occurrence. The concept behind the vase, which has a porcelain exterior and rubber inner shell, was to create a product that could be smashed to form unique cracks in the porcelain and still remain intact, held together by the rubber interior. Frank Tjepkema has been quoted as saying that "'You can 'break it a thousand times…But jumping on it is another story'" (Louie, 2001, p. 3). The memory of the event or moment when the vase is 'broken' no doubt becomes inextricably linked to the cracks or craquelure of the outer shell of the vase. A strong connection between the event and product is facilitated through the careful design of the vase. It is through interaction with the product that a bond may form with the end-user, and allow the completion of the product with its intended state realized (Louie, 2001, p. 3).

CASE STUDY OF BERNABEIFREEMAN

The incandescent lamp is a product described by Elaine Scarry (1985) in *The Body in Pain* as one that;

Transforms the human being from a creature who would spend approximately a third of each day groping in the dark, to one who sees simply by wishing to see: its impossibly fragile, milky-white globe curved protectively around an even more fragile, upright-then-folding filament of wire is the materialization of neither retina, nor pupil, nor day-seeing, nor night-seeing; it is the materialization of a counter-factual perception about the dependence of human sight on the rhythm of the earth's rotation; no wonder it is in its form so beautiful (p. 292).

This poetic description of the incandescent lamp reminds us that it is "not merely a physical object" (Dilnot, 1993, p. 57). It is latent with potential and has the potential to transform the lives of those who come into contact with it. The metaphoric nature of light and its ability to be utilized in both domestic and exhibition applications has been recognized by award winning Australian product and lighting designers' Bernabeifreeman. The Sydney-based practice of Bernabeifreeman specializes in lighting design and the designers seek to engage with the product narrative in their work. Storytelling in product and lighting design generally allows a dialogue and conceptual exchange to occur between products, end-users, and the product designer. In many instances when the product passes from the hands of the designer to its end-user, the relationship of the designer with the product in many respects is dissolved. Bernabeifreeman have desired to re-engage this relationship and understand how end-users relate to the stories embedded in their products and how end-users attach their own. As explained by Rina Bernabei, *"For us to develop a new product, we have to be able to justify it. It has to have a reason for being there, it had to say something new, have its own story, have its own place in the world" (Gallagher, 2006, p. 10).* Stemming from this desire, Bernabeifreeman organised a small sample of interviews with end-users in order to further understand storytelling in their work. The Bernabeifreeman designers were keen to further understand the product narrative and in what ways it is successfully communicated to end-users through their products. This process was also deemed important by Bernabeifreeman to ensure that future products might be tailored to successfully communicate with and be precious to their end-users.

THE INTERVIEW PROCEDURE AND METHODOLOGY

The selection criteria for involvement in the interview process were those who have dealt with Bernabeifreeman products in a variety of capacities: those who own the products and have them in their homes or those who have dealt with the

Figure 5. Leaf light (left), Lace light (centre) and Brodie Table (right). (©2006, Bernabeifreeman. Used with permission.)

products as exhibition pieces. Bernabeifreeman collated a list of clients from which an interviewer was able to telephone and determine their willingness to participate in the study. The format the interviews followed was to outline the general purpose of the study and present to the interviewee the questionnaire accompanied with a sheet of images of the products. The questionnaire was divided into sections pertaining to the four storytelling methods identified by Bernabeifreeman: Narrative, Manufacture, History and Interaction. Three products formed the focus of the interviews: *Leaf light*, *Lace light* and *Brodie table* (Figure 5). The interviewer and interviewee discussed each question and the interviewer made notes on the questionnaire, wherever possible noting down as quotes key phrases and terms used. The total sample was small and was therefore not a conclusive study, yet the exploration of storytelling and how it is perceived by end-users made the study highly valuable to the designers.[1] It is hoped that these interviews will be the start of an ongoing investigation of storytelling in Bernabeifreeman products and that the data gathered might influence development of future products.

LEAF LIGHT AND LACE LIGHT

Narrative

The Bernabeifreeman designers draw precedence from historic interiors, traditional textiles and textile-making techniques, particularly lace-making and embroidery. Drawing inspiration from needlework techniques traditionally a female-craft, presents something of a contrast between the masculinity of industrial product design and the feminine world of handicrafts. The use of 'hard' materials such as sheet metal and the overlay of softening patterns presents a gentle irony in the form, materials, manufacture methods and decorative nature of the products. Bernabeifreeman's work has been described as pushing "the machine and modern manufacturing techniques to create something more organic and feminine" (Gallagher, 2006, p. 11).

The storytelling method of Narrative is characterized by the role of precedence and how that influences the final design outcome. The *Leaf light* and *Lace light* were inspired by an Arts and Craft print and lace-making techniques respectively. During the interview process, interviewees were asked to identify the underlying inspiration for the two luminaires. Their responses have been summarized in Table 1 revealing especially in respect to the *Lace light* a strong appreciation of the Narrative embedded in the products.[2] One interviewee perceptively explained the precedence serving as inspiration for the two products as involving, "textile design and women's domestic craft" as well as "motifs from textiles."

Interviewees were asked 'What words would you associate with the *Lace light* and *Leaf light*? E.g. Feminine, nostalgia, textile.' Concerning the *Leaf light* the descriptive terms included: "warmth," "curviness," "organic," "pattern" and "feminine." In response to the *Lace light* descrip-

Table 1. Compilation of responses from 'What do you think was the designers' source of inspiration?' (Adapted from table contained in: Bernabei, Freeman and Power, Designers as Storytellers: the weaving of narratives in the work of Bernabeifreeman)

Interviewee	Leaf Light	Lace Light
01	Leaf, nature	Handicrafts, needlework
02	Asian/Japanese influence	Lace work
03	'Technology' of perforation	Pattern of lace-holes/overlap/shadow
04	Nature/textile patterns	Nature/textile patterns
05	William Morris leaf patterns	The characteristics of traditional lace

tions included: "retro," "textile," "screen," "busy" and "shadow." As a number of the responses that of "shadow" was further elaborated, the interviewee explaining that "the shadow evokes the memories not the light." The shadow is "an intrinsic metaphorical narrative element."

Manufacture

During the manufacturing process stories may also be embedded into the product. Stories may actually be embedded during the manufacturing process and be discernable to end-users due to: the material/s used, the fixings, finishes, the industrial or handmade quality instilled in the product and any resulting imperfections from this process.

In discussing the manufacture processes involved one of the interviewees said that "adherence to sheet metal and exploring its qualities by rolling, bending, punching is very interesting… the material is a 'shape shifter.'" This response indicates an interviewee appreciative of 'what can be done with a material' and the transformative qualities of the material. He further went on to say that "in all cases the process of manufacture seems very simple this is part of what makes it [the products] so elegant." This suggests that the manufacture process subtly influences the perceptions formed about the products.

When asked *'What other activity or techniques would you relate these processes to, if any?'* two images were provided to spur responses, as it was thought the manufacture process may be somewhat distant and seem conceptual to interviewees. Responses related the processes to other techniques and included a mixture of other industrial processes and embroidery techniques. Responses included; "fine detailing on cars," "art processes," "embroidery meets panel beating...a marrying of conflicting sensibilities" and "needlework techniques that use a combination of stitch/knot with hole or negative space" (Bernabei, Freeman, & Power, 2008).

History

The History narrative allows end-users to embed their own personal stories thus heightening their emotional connection to the product. As Csikszentmihalyi (1981) has explained,

When a thing 'means something' to someone, it is interpreted in the context of past experiences, either consciously or unconsciously in the form of habit. The emotion that things evoke is also an interpretation or inference, a sign or symbol of one's attitude (p. 21).

Bernabeifreeman hope that the personal stories end-users attach to their products will result in the greater valuing of the product by the end-user. In the interviews carried out the interviewees had questions put to them about feeling and fondness in relation to the products. Table 2 provides

Table 2. When asked 'What memories do these objects conjure for you?' the responses were wide ranging and at times seemingly the most personal proffered during the interview (Adapted from table contained in: Bernabei, Freeman and Power, Designers as Storytellers: the weaving of narratives in the work of Bernabeifreeman)

Interviewee	Leaf light	Lace light
01	Outdoor experiences. Autumn. Nature- referred specifically to sea-shell mobile positioned near the light.	'Retro.' Feminine.
02	William Morris prints. Art Nouveau, prints from Japan.	Islamic screens. Arab Institute in Paris designed by Jean Nouvel.
03	*Sydney Style Exhibition* (interviewee was curator).	Grandmother and other older peoples' homes.
04	Image of late 19th century textiles and wallpapers- stylized natural patterns of Victorian excess.	Turn of the century Wiener Werkstätte design reform movement.
05	Dappled night light in my daughter's nursery.	It hangs over my bed and my daughter loved staring at it as a baby.

a summary of their responses. Responses ranged from reference to historic design styles, family members, and personal experiences. One interviewee explained that the *Leaf light* reminded her of outdoor experiences and Autumn. When discussing this memory she specifically referred to a sea-shell mobile that hung in close proximity to the *Leaf light*. Thus other objects in the domestic interior may potentially become intertwined in the History narrative of a product.

Interaction

Interaction with a product may take place with end-users in a subtle manner perhaps altering the domestic interior, for example when a light switch is flicked on. Both the *Leaf light* and *Lace light* have been designed to cast an intricate pattern of shadows onto surrounding surfaces. The shadows are as much a part of the product design as the more tangible, physical object itself. Interviewees were questioned about the shadows produced by the two luminaires. In their responses interviewees stated that the shadows "imply and soften the home" and "provide a secondary pattern that puts up a dialogue with the light…a wallpaper effect." Interestingly the wallpaper effect of the shadows seems even more accentuated when the lights are placed within complementary architectural settings such as Federation Style homes. Bernabeifreeman believe that the act of illumination is essentially one of spectacle and is greatly utilized in their exhibition products.

Brodie Table

The *Brodie Table* (one of the three products which formed the focus of the interview) was originally designed for an exhibition but due of its industrial nature was well suited for and began to be produced as a commercial piece. Bernabeifreeman also sought to convey a story through this product. Within this product the Narrative and Manufacture storytelling methods are inextricably linked. For the design of the *Brodie Table* (Figure 6) Bernabeifreeman drew precedence from the technique and aesthetic of an embroidery technique called *broderie anglaise*. Figure 6 shows the sample of broderie anglaise held in the *Collection of the Embroiders Guild Victoria* from which the designers drew their inspiration. This "whitework" embroidery technique dating from 1820 was used for designs of geometric patterns or "of conventionalized flowers and leaves" (Hillier, 1975, p. 400). The most delicate examples of broderie anglaise were primarily used for baby

Figure 6. Brodie Table (left) and Broderie anglaise (right). ©2006, Bernabeifreeman. Used with permission.)

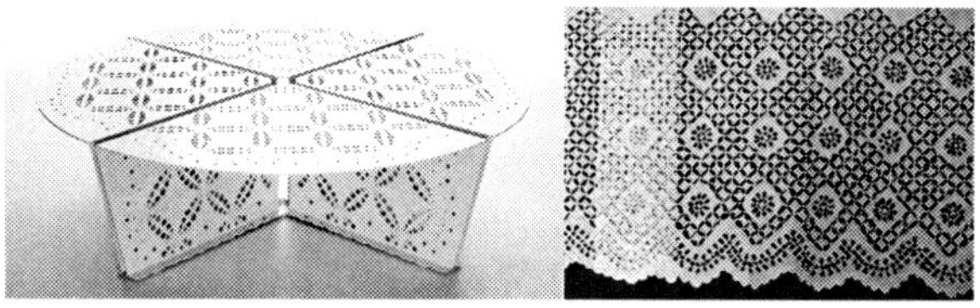

garments and undergarments, thus ornamentation of the most intimate and precious items of clothing (Hillier, 1975, p. 388). The design of the *Brodie Table* alludes to past interiors and the intimacy of the 'tête-à-tête,' which makes the reference to broderie anglaise seem particularly appropriate in its association with intimacy (Bernabei et al., 2008).

During the interviews questions relating to the *Brodie Table* were asked. Table Three outlines responses to questions relating to the *Brodie Table* and their relevance to the storytelling method of Narrative (Question One) and History (Question Two). Responses to Question One clearly indicate that the precedence and conceptual underpinning of the *Brodie Table* was being communicated. Question Two reveals the personal stories of the interviewees beginning to be attached to the product. Interviewee Two for example was reminded of "origami" and "Islamic screens" indicating an architectural background and attachment of these associated ideas to the product.

Images (Figure 7) were shown to interviewees who were asked to select the one which most reminded them of the *Brodie Table*. Also provided with the images was a short descriptive sentence explaining the embroidery technique used for each example shown. 60% selected the correct image (Figure 7, image on far left) supporting their decision with such explanations as "because of the strong geometric pattern" and "play of positive and negative pattern." The other 40% selected the image of the lace doily (Figure 7, image on far right) because of the "scalloped edges" and "up-scaled [the doily]

Table 3. Sample of collated responses from interviews pertaining to Narrative (Question 1) and History (Question 2) storytelling methods

Interviewee	Question 1: Do you think the *Brodie Table* conjures any memories of the past, or of particular dining rituals?	Question 2: What memories do these objects [*Brodie Table*] conjure for you? They may be personal or of time past, such as activities or spaces.
01	It is reminiscent of traditional textile crafts. The decorative pattern created is reminiscent of a tablecloth.	Afternoon Tea.
02	"Table cloth," "Old-fashioned lace napkins," "Stiffly starched cloths," "Starched cloth and folds."	Origami- folding an integral part. Perforations of Islamic screens.
03	"Evocative of lace and lace reminds me of my grandparents." "A lace overlay or lace-edged things." A "historical look- given that lace doesn't appear in the modern home."	Looks more 'modern.' A 50s look. Regulated and ordered patterns. Dots arranged in abstract configurations. Early 20th century German and Austrian modernism.
04	"Afternoon tea/morning tea."	N/A
05	"Sense of importance of table in family ritual- European traditions." "Lace/doily evokes a sense of "preciousness, homeliness and femininity." "Fond memories of those people and environments- mothers/grandmothers." Matrilineal.	Grandmother and other older people's homes.

Figure 7. Broderie anglaise (left), Surface embroidery (centre) and Lace doily (right). (©2006, Bernabeifreeman. Used with permission.)

becomes the object itself." The descriptive explanations of the embroidery techniques did not seem to influence the responses, as the responses in fact seemed to be reactions to the visual qualities of each piece rather than considered responses to the embroidery techniques employed.

Interviewees were questioned in regard to the Manufacture narrative and were asked, 'What elements of your Bernabeifreeman *Brodie Table* manufacture are you particularly drawn to?' One interviewee suggested that the "repeat of single components four times" has an "obvious elegance in how it's constructed." This observation was elaborated further, in which sustainable implications as a result of the storing and shipping of the *Brodie Table* as "the components 'cradle' into one another" was mentioned. The use of repetition has been recognized by this end-user in the overall construction of the product in its division into four segments. This highlights that a product might be engaged with by an end-users to varying degrees, as is the case with all four of the identified storytelling methods.

Curator Stephen Gallagher's (2006) descriptive explanation of products generally captures the notion being tapped into by the *Brodie Table*, as he explains "Objects allude to the people who owned them and the social context in which they were used; evoking moments that once were- a particular place at a particular time" (p. 8). It is through the four storytelling methods, particularly Narrative and Manufacture that provides the *Brodie Table* the conceptual techniques to achieve this aim.

CONCLUSION

The four storytelling categories identified by product designers' Bernabeifreeman in both their work and that of other designers, the Narrative and Manufacture methods and History and Interaction storytelling methods, provides potential methods for designers to use while conceptualizing designs. These methods offer possible avenues for creating emotionally rich industrially designed products which may be then further personalized by end-users, enabling the product to become special and integrated into an end-user's life. The Narratives are the precedence the product speaks of, the Manufacture techniques are stories told through its making, the History of the product are the memories and associations the product stirs in the mind of end-users and Interaction is particularly concerned with the physical alterations that may occur to the product in its contact with end-users. Each of the four storytelling methods provide different opportunities and level of designer involvement.

To assess the success of storytelling in their own work Bernabeifreeman instigated interviews concerning three of their products: the *Leaf light*, the *Lace light* and *Brodie Table*. The interview process generated some useful qualitative feedback concerning the three products and future role of storytelling in Bernabeifreeman's work. The Narrative, Manufacture, History and Interaction storytelling methods were recognized and engaged with by end-users to differing degrees. The case study by no means forms conclusive evidence that embedding stories into products results in a stronger end-user attachment. Whether or not the end-users interviewed could 'read' the stories as intended by the designers is to a certain extent immaterial, although satisfying when indeed interpreted. The 'triggers' embedded in the products importantly provided opportunities for memory and sensorial engagement and attachment to take place. The end-users were in full control of the stories they perceived and of the future stories the product collected over time. All the interviewees were much attached to the Bernabeifreeman products whether they completely embraced the conceptual stories or not. The products communicated to some level their Narrative, Manufacture, History and Interaction.

Perhaps for stories to be ongoing and enduring between end-user and product, designers will need to go beyond the methods illustrated above and consider more profound or multilayered sensorial dimensions in the designing of products. The life-cycle of the product may need to unfold new stories, triggers, and provide insight and stimulation. There is still far to go to achieve these aims, and as yet we can only,

Imagine a world in which products are designed to support the investment of emotion- cherishable products, appliances and tools that not only sustain but also amplify sensations of attachment- a world where technological modernity is not the sole value indicator... (Chapman, 2005, p. 163).

REFERENCES

Bernabei, R., Freeman, K., & Power, J. (2008). Designers as Storytellers: The Weaving of Narratives in the Work of Bernabeifreeman. In Desmet, P.M.A, Tzvetanova, S.A., Hekkert, P. & Justice, L. (Eds.), *Proceedings from the 6th conference on design and emotion 2008*. Hong Kong.

Chapman, J. (2005). *Emotionally durable design: Objects, experiences and empathy*. London: Earthscan.

Conran, T., & Fraser, M. (2004). *Designers on design*. London: Conran Octopus.

Csikszentmihalyi, M., & Rochberg-Halton, E. (1981). *The meaning of things: Domestic symbols and the self*. Cambridge, UK: Cambridge University Press.

Dilnot, C. (1993). The gift. *Design Issues*, *9*(2), 51–63. doi:10.2307/1511674

Foster, D. (2001). *The land where stories end: As narrated by the angel depicted in Madonna Con Bambino E Due Angeli by Felippo Lippi. Potts Point, N.S.W.* Duffy and Snellgrove.

Gallagher, S. (Ed.). (2006). *The presence of things: Sense, veneer and guise*. Victoria: The Embroiders Guild Victoria.

Hillier, B. (Ed.). (1975). *The connoisseur complete encyclopedia of antiques*. London: Peerage Books.

Jackson, L. (2005, August). Hella Jongerius. *Icon., 26*. Retrieved from www.iconeye.com

Jongerius, H., & Schouwenberg, L. (2003). *Hella Jongerius*. London: Phaidon.

Kwint, M., Breward, C., & Aynsley, J. (Eds.). (1999). *Material memories*. Oxford, UK: Berg.

Louie, E. (2001, May 24). Objects designed to foster a little havoc in the home. *New York Times, 3*.

Martin, C. (2009). *Philippe Starck interview with Catherine Martin*. Vogue Living Australia.

Ramakers, R., & Bakker, G. (Eds.). (1998). *Droog design: Spirit of the nineties*. Rotterdam, The Netherlands: 010 Publishers.

Richardson, A. (1993). The death of the designer. *Design Issues*, *9*(2), 34–43. doi:10.2307/1511672

Scarry, E. (1985). *The body in pain: The making and unmaking of the world.* New York: Oxford University Press.

Zachry, M. (2005). An interview with Donald A. Norman. *Technical Communication Quarterly*, *14*(4), 469–487. doi:10.1207/s15427625tcq1404_5

KEY TERMS AND DEFINITIONS

Product Narrative: The potential for a product to communicate or initiate the telling of a story.

Narrative: The stories or historical references to which the designer seeks to refer; the influence of precedence on the finished product.

Manufacture: The traces of stories that remain from the physical processes of making the product.

History: The personal stories that the product may conjure in the minds of end-users.

Interaction: The physical changes that may occur to the product over its lifetime as a result of its contact with end-users.

Triggers: The design elements (such as form, material, finish, colour, repetition, scale, manufacture methods and design reference) at the disposal of the designer that may contribute to the product narrative and therefore act as a storytelling catalyst.

ENDNOTES

1 Any future interviews that take place will do so after an assessment of how the interview process and questionnaire might be further improved.

2 It is important to note that the exact recognition of the designers' Narrative instilled in the product was deemed desirable but not crucial, as the designers feel that elements of the Narrative may act as a storytelling catalyst for end-users to initiate their own stories.

Chapter 25
Deploying and Adapting an Indoor Positioning System in the Clinical Setting

James Stahl
Massachusetts General Hospital, USA

Julie Holt
Massachusetts General Hospital, USA

Michael Lye
Rhode Island School of Design, USA

ABSTRACT

We are living in an era where the demands on our healthcare system are relentlessly rising while at the same time key resources, such as, the number of physicians and time available to see patients, are declining. In order to diagnose what is wrong and treat it appropriately we need to be able to objectively measure and describe how our healthcare system behaves. At Massachusetts General Hospital, an innovative project weaves together industrial design, operations research, outcomes research with emerging technologies to provide a means for objectively and reliably measuring time in the primary care setting. The RFID in Clinical Workflow Project aims to provide a tool with which to understand resource allocation and to shape appropriate and effective policy. In order to successfully incorporate the use of an emerging technology that enables accurate and reliable measurement into the demanding and critical clinical setting, the multidisciplinary team used a hybrid of design techniques sourced from the different disciplines represented on the team.

INTRODUCTION

The United States healthcare system has reached a critical point. The demand for healthcare services is rising, driving up costs and expenditures. The US currently spends more per capita on healthcare than any other industrialized nation(1) yet at the same time healthcare resources – including the supply of physicians– are getting scarcer and more difficult to access.

Policymakers, healthcare providers and researchers have long struggled to improve the

DOI: 10.4018/978-1-61520-617-9.ch025

cost, quality and efficiency of our healthcare system. However, the lack of key quantifiable data has often inhibited or distorted effective decision-making. Decision makers are constrained to work with what information is available, for example, from easily measured areas such as in the procedural or imaging specialties. Key information related to time, resource utilization system behavior and organization and face-to-face time between patient and physician is often missing.

In order to effectively address this policy-information gap we need to develop better methods for quantifying, measuring and collecting data in healthcare beyond the simple counts of surgeries done, images taken or vaccines given that currently characterize our healthcare system for decision-makers. In medicine, time is perhaps the most important resource – yet measuring time spent on core clinical activities, such as the time patients and clinicians spend face-to-face, has been difficult. This has resulted in inadequate weighting and reimbursement of primary care activities (informed decision making, problem solving, resource coordination, counseling) which are typically time intensive but otherwise difficult to characterize. Without comprehensive data quantifying these activities understanding and planning optimal use of our resources is difficult if not impossible.

At Massachusetts General Hospital, an innovative project weaves together industrial design, operations research, outcomes research with emerging technologies to provide a means for objectively and reliably measuring time in the primary care setting. The RFID in Clinical Workflow Project aims to provide a tool with which to understand resource allocation and to shape appropriate and effective policy. Using such information in conjunction with other clinical data and simulation models will allow policy makers to identify optimal current and future configurations of our healthcare resources. In order to successfully incorporate the use of an emerging technology that enables accurate and reliable measurement into the demanding and critical clinical setting, the multidisciplinary team used a hybrid of design techniques sourced from the different disciplines represented on the team.

What This Chapter Will Cover

This case study presents an opportunity to explore a wider role for industrial design in the healthcare system. It also highlights ways that other complex problems in non-healthcare systems can be best approached by a multidisciplinary team that includes industrial design as a crucial component alongside other, more quantitative disciplines, such as operations research and outcomes research. In the increasingly common situations involving technology adoption or adaptation, the skills and techniques of industrial design complement the research and technology capabilities in other disciplines, providing a far greater chance of success integrating technologies into new settings and cultures. Though this case study looks specifically at the role of design applied to a clinical healthcare setting, the lessons learned can be applied to a wide variety of other circumstances. Industrial design can play an essential role as a part of a multidisciplinary team in finding new approaches to complex and intractable problems.

In the first section of the chapter, we introduce the problems currently faced by healthcare professionals with regard to the shifting demographics and needs of patients in the primary care setting, as well as the difficulty of measuring work product in these settings utilizing traditional research methods.

We will then introduce Indoor Positioning Systems as a tool for measuring time in the clinical setting and discuss the design challenges of both technology adoption and adapting a technology commonly used in other industries to the specific needs of healthcare systems. Because the technology is mature and well-developed for other uses, the success of the technology adoption in the healthcare setting depends heavily upon utilizing design to make the technology accepted by the clinical culture. As we will discuss, this requires significant effort with regard to integrating design principles into the end-user experience.

Because this project drew upon many industries and fields of study, maximizing the odds of success required being able to draw and act upon the broadest possible option set of ideas and methods. We discuss the ways in which good design comes out of good collaboration and integrates the technology, the needs of the end users, and the needs of all of the other stakeholders.

We then explore how industrial design was crucial in three main components of the project implementation: the physical interface, the information interface, and the "face" or branding of the project.

Finally, we offer a discussion of the future of such projects in healthcare and other areas in which industrial designers can provide much needed expertise.

BACKGROUND

The State of Primary Care

Healthcare in the United States is currently in crisis. We have entered an age of increasing demand and decreasing resources. Our population is aging and the burden of chronic disease is rising. At present just 5% of our population generates over 50% of our healthcare costs and resource use(2). Total healthcare expenditures are set to rise rapidly as our population ages and our population becomes ever more obese. It is predicted that demand for physicians will rise 13 - 29% while physician supply will only rise 2-7%(3) over the next decade. At the same time, the burden of care is being increasingly shifted to the outpatient setting. Primary care physicians (PCP) are at the interface between the public and the healthcare system. It falls to them to identify and manage disease early before becoming critical, organize resources and navigate patients through a complex high tech healthcare system. Unfortunately, primary care clinicians are becoming increasingly scarce as fewer people are entering primary care and fewer are staying in the field. All of these conditions conspire to put ever more demands on the PCP resulting in burnout, decreased patient and staff satisfaction and rising medical error rates, driving more and more of those remaining out of the business.

A Crisis in Measurement

A major cause of the healthcare crisis is the current unbalanced incentive structure in medicine. There is an overemphasis on domains with easily measured output, such as procedural specialties surgery and radiology. In these areas of medicine, procedures are easy to count and thus reimburse – for example, surgeries completed per day or x-rays read per day. It is not as easy to measure and reimburse the principal outputs of primary care: early detection, decision making, counseling and planning. While these activities are difficult to quantify as individual units, they are all valuable and highly correlated with both better outcomes and time spent with patients.

Measuring time consistently, objectively and unobtrusively in a live clinical setting is a challenge. The gold standard of time measurement is the time-motion observation method first developed by Frederick Taylor (1911). Although time-motion studies are thought to generate accurate and reliable data and have in the past often been used in healthcare(4-6), they can be quite costly and require highly skilled observers that can quickly interpret staff and patient activities. They also can suffer from observation bias - providers know that their activities are being monitored and may alter their normal behavior. Some recent alternatives are: the work sampling technique(7) (subjects are called randomly through the day and asked what they are doing), patient flow analysis(8) and provider recall(9, 10). In patient flow

Figure 1. Patient tag workflow

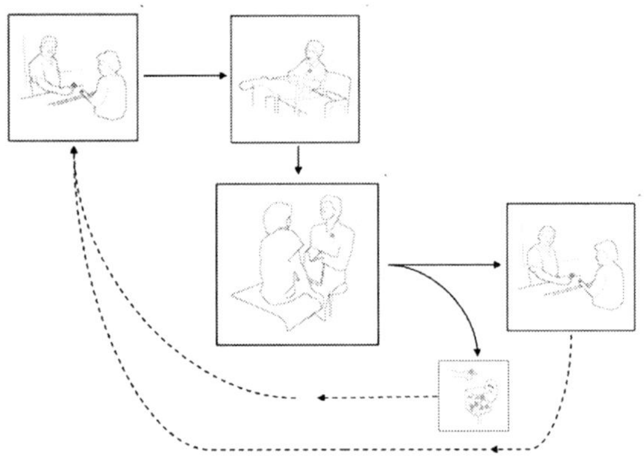

analysis a patient is asked to carry a pre-coded form through their visit (Figure 1). The patient is then instructed to present the form to clinic staff with whom the patient interacts. The staff uses synchronized-watches to record the time at the beginning and end of each contact. This has the problem of requiring multiple parties to remember to present and stamp their forms and to consistently identify events whose definitions maybe ambiguous to participants. Provider recall is gathered in the form of structured interviews and self-administered timesheets.

All these techniques have significant problems primarily with regard to cost or accuracy. Bratt et al(8) found statistical differences across all three alternate methods in their study. According to Bratt et al. "If a less complicated method can be shown to produce similar results to the time-motion study it would make future studies of time use and cost easier to conduct." Bratt presents a clear opportunity to innovate.

Outpatient clinical practices, the primary point of access and main interface between the healthcare system and the population as a whole, are far more complex than is generally acknowledged and may be described as "tightly coupled" systems(11). This means that there are few buffers in the system and any perturbation in the flow and process can cause major disruptions.

The first step in understanding how to optimize these clinical delivery systems is obtaining good objective information about how these systems behave, particularly at the human level. If it were possible to replace costly or inaccurate methods of data collection with new tools that allow for rapid and accurate quantification of presently poorly understood clinical delivery systems, policymakers, researchers and clinicians would have a better understanding how these systems actually behave – critically important information that is currently missing.

A NEW OPTION: INDOOR POSITIONING SYSTEMS

Just as the Global Positioning System has revolutionized navigation and changed how we locate ourselves in the outdoors, Indoor Positioning Systems (IPS) can offer similar and just as revolutionary benefits inside buildings. There are several developing IPS technologies that are becoming practical and cost-effective in measuring time and location in the healthcare setting. These include technologies based on ultrasound, infrared, and radiofrequency identification (RFID). Using any of them requires that the same essential issues be understood: (1) how to use this technology as a replacement for traditional time-motion studies, (2) how to create meaning from time-location data streams they all produce, and (3) how to scale and integrate this information to make our system smarter and more responsive. Indoor positioning systems hold the promise of being able to track the time and location of individual objects and people objectively, unobtrusively and comprehensively within the work environment.

Like many new technologies, RFID's proliferation across various industries has evolved as people have gained experience. Initially, RFID replaced barcode and other scanner-based technologies. The advantage of RFID is that it

Table 1. Stages of technology adoption

	Business		Healthcare	
	Opportunity	*Value*	*Opportunity*	*Value*
Substitution Effect	Inventory Audit, data capture	Lower cost for data capture; Timely accurate data at back end operations	Time-motion data, Inventory management	Lower cost for data capture; Timely accurate data at back end operations
Scale Effect	Item level, multisite visibility	Planning with visibility in inventory accuracy, product origin tracking	Person and item level, multisite visibility	Planning with visibility, staffing accuracy. Patient and data flow, patient safety
Structural Effect	Intelligent supply chain and services	Supply chain security, process redesign, smart product/ service	Intelligent access to care and services, HIPAA, Guidelines	Process redesign, linkage to other knowledge, security, smart products/ services, reduce medical errors
After Hau L Lee, Thomas Professor of Operations, Information & Technology, Graduate School of Business, Stanford University				

is readable when in range of a receiver, even in the middle of a packed palette of goods, whereas barcodes cannot be read if the code is damaged or obscured. Following this initial technology replacement, RFID was able to scale up rapidly. As a result large companies such as Wal-Mart and Amazon now have near instantaneous visibility up and down their supply chains. As the technology has matured, RFID has been able to generate new ideas in addition to extending existing functions. RFID is now frequently used to provide structural insights in systems or for linking other data sets for new security paradigms or smart product service. The parallels in medicine are outlined in Table 1.

RFID-based IPS has been used in industry, most notably in logistics and retail, for many years. However, until this project no IPS has been applied in the high speed, constant motion, physically tight and complex clinic setting. For this project an RFID-based IPS was selected as the most practical option for use in the live clinical setting.

We now have a measurement tool which can quantify and thus empower primary care providers. Understanding what people need to provide high quality service while at the same time allowing maximum access will help determine what human resources we need on a national level.

Project Goals

This project has had several goals: (1) to develop a tool to measure and describe the behavior of human level clinical systems, (2) to provide both pre- and post-intervention effectiveness data for process interventions, and (3) to create a framework to explore system utilization, improve care and to develop intelligent services for the delivery of healthcare.

In addition, the project team was tasked with integrating the RFID-based IPS into the clinical workspace in a seamless way, making it part of everyday workflow so that it receded into the background after implementation. This task meant looking at the technology from the perspective of all stakeholders, from the perspective of the individual end-user to the perspective of the larger organization, and address issues of buy-in, technologic function, human-technology interaction and measurement. From the outset, expectations were that this implementation would require great thought and care with regard to fostering successful technology adoption.

Design and Technology Adoption

Every technology introduction is also by necessity a change in culture and a change in system dynamics. New technology, whether a device or policy, perturbs the system into which it is introduced and changes how the end-users are able to achieve their goals. In order to be successful, every new technology must satisfy both the needs of the system and the needs of the people using it. It is often the human component to which success is most sensitive. This is particularly true in cases where a relatively well developed or mature technology is imported into a new setting for the first time.

Perturbing a system with a new technology, even slightly, can often have large effects on the people and culture in which it is introduced. It is often the many small details of design and technology implementation that determine if it will be adopted and used to its fullest by end-users.

Satisfying both the technical requirements of a problem and the human requirements are necessary conditions for a successful technology adoption. Neither is a condition sufficient by itself. There has been considerable work in the area of technology adoption in the realm of information technology and industrial design. These range from the Theory of Reasoned Action(12), the Technology Acceptance Model(13), the Theory of Planned Behavior(14), Innovation Diffusion Theory(15), Social Cognitive Theory(16, 17) and the Unified Theory of Acceptance and the Use of Technology (UTAUT) described by Venkatesh et al(18) one of the more generally accepted theories attempting to explain user intentions and subsequent end-user behavior, in this case the patients and the clinicians. This theory has four core ideas: performance expectancy, effort expectancy, social influence and facilitating conditions. Venkatesh et al showed that these factors are strong predictors of ultimate technology usage and end-user behavior. Many factors influence how these four core ideas play out. For example, the end-user's age and gender can influence expectancy of how the technology will or will not work. Past experience with the same or similar technology can influence effort expectancy and the end-user's voluntariness or adopter mind set (early, middle, or late) can influence the facilitating conditions in a technology roll out.

Another more concrete approach has been Technology Readiness Levels (TRL)(19). This is a method developed by NASA and the Department of Defense and used by many large organizations. TRL categorizes a technology into nine stages from the first observation of the scientific principles underlying a technology to the actual successful implementation of the technology in the field, such as a successful space flight. Technology Readiness Level 7 describes an entire system of technology, not just the individual components, that is being prototyped and tested in the actual environment for which it is intended. Our RFID-based IPS is at TRL 7. Though the technology is generally well developed, this particular implementation in a live clinic had not yet been developed or tested.

It is at this level and higher, the system level, that technological requirements and constraints need to be balanced with the constraints and expectations of the stakeholders and end-users. If the technology does not meet expectations the introduction will likely fail. A properly functioning technology is merely a necessary prerequisite. It also needs to work in the environment and microculture of the end-users and stakeholders. For a technology introduction to be successful it has to allow the end-users to achieve their goals.

The many dysfunctional products in the market – unusable mobile phones, confusing digital cameras, over-featured microwave ovens, VCRs relentlessly flashing 12:00– are clear evidence that robust technology alone is not enough. In these cases and many others, the technology is ready but the design does not meet the needs or expectations of the user. One central aspect of good design is integrating the disparate constraints and goals of the end-users with the available technologies into one unified whole.

Integrating an RFID IPS into a live clinical system was one of our core design challenges. In this project the RFID technology, which is relatively well established in some markets, had not been tried in a complex clinical setting like the one attempted here. There still remained several technical barriers to be overcome, specifically those elements linking the technology to the end-user. Any one element of the system, even seemingly minor ones far removed from the core technology, from the Graphical User Interface (GUI) to the physical design of tags and wristbands to the information brochures, branding and signage needed to be considered as potential failure points impeding the adoption of the integrated system. In fact, as discussed later, the GUI upon which the end-users relied to interact with the system proved to be a significant stumbling block in some clinics. Even though some of the potential design solutions were clear, implementing those solutions often took longer than expected.

Making technology work for the people that need to interact with it is a major and often overlooked aspect of the design of any system. It is something that needs to be considered early in the design process as well as in the technology development process.

Project Stakeholders

There were many stakeholders in this project: the hospital administration, the primary care clinics where the project would be implemented, the patients and the project team. Each had differing expectations.

The primary stakeholder was the hospital itself - or more broadly the hospital network – since it supplied the financial resources to make the project possible. The hospital network is comprised a tertiary care centers and broad grouping of closely and loosely associated facilities, including approximately a dozen primary care practices, reporting to one centralized administration. The participating primary care clinics supplied the venues for the project. The patients and staff were both stakeholders and the end-users. The project staff were also stakeholders and supplied manpower and expertise.

The stated aims of the project from the hospital's perspective were to use the IPS to (1) measure baseline clinical system behavior (2) measure changes over time or after other process interventions, and (3) ensure that the technology would become seamlessly integrated into the end users' workflow. Once the project was approved, the project team was responsible for managing the various expectations and goals of the other stakeholders.

Getting "Buy–In"

Developing buy-in with the lead stakeholders took time. The project leader had significant experience with the RFID technology through several prior pilot projects. This created a level of credibility with the hospital decision makers in the project leadership but was insufficient by itself. Prior to roll-out the technology had to be re-piloted in the specific clinical circumstance in which it would be eventually applied. Researchers had to demonstrate that the IPS system could not only work in operating room environment (its previous use) but also in the outpatient clinic. In addition, the project had to present a convincing business plan to the administration and compete with other projects on the budget agenda.

After the administration was convinced that the project was feasible and that it would provide valuable information for the hospital and staff, clinic sites had to be identified and convinced to participate. The affiliated clinics in the hospital network have both a great deal of autonomy as well as a wide range of micro-cultures. Therefore, even though the hospital administration was the primary stakeholder, clinics could not be compelled to participate. As a result, the project leader needed to work with the leadership of each clinic to gain acceptance for the project and potential clinic participation.

Clinic Cultures

Each primary care clinic considered for participation in the project could be characterized by where it fell in a wide multidimensional matrix, describing its culture, organization and customer base. While all the clinics provided high quality care, they were as a group rather heterogeneous. They each served different populations with varying socioeconomic backgrounds, ethnicities, genders and ages. On the practitioner side, clinicians could have small or large panel sizes (patients under the direct care of an individual clinician). The practices could be comprised of mostly full-time or mostly part-time clinicians, utilize a mixed-practitioner model (MD, NP, and RN) or be physician centered. In terms of organizational structure, these clinics ranged from the hierarchical to team oriented, decision making style ranged from being centered on individual leaders to consensual decision-making. In terms of communication, some practices felt very comfortable sharing performance feedback; others were very reluctant to do so. Finally, the clinics varied on their prior experience with process improvement initiatives and where they as a group fell on the adoption/voluntariness curve as defined by Rogers, et al (15).

Where a clinic fell on the adoption curve was a strong predictor of which ones opted to participate (as "early adopters") and which decided to take a "wait and see" position ("early majority" or "late majority"). This in turn was strongly influenced by a combination of clinic leadership style, past experience with process improvement initiatives and how much they felt they were being judged by the administration and their peers. All these factors needed to be aligned for clinics to participate. For example, in one case, while clinic leadership was in favor of the project, resistance to the project due to innovation fatigue and fear of being judged prevailed. In other cases much of the staff was keen to join the project but the leadership was reluctant. Past negative feelings about process improvement initiatives work against adoption. Process improvement initiatives going on at the same time or around the same time also worked against adoption. These clinics felt they didn't have the bandwidth to manage more than one intervention at a time.

Framing

How a project is framed can profoundly influence whether or not a project is adopted. It was very important from the beginning to make stakeholders aware of the aspects of the project of most interest and benefit to them and to make sure their worries or fears were addressed explicitly and directly. For the primary care clinics, the team first demonstrated how this system of technology could be used as a tool for self management. Second, stakeholders were shown how the IPS could become a framework for building future tools to improve the care they delivered - for example, wait time alarms, allowing support staff to find clinicians quickly and being able to locate where patients are during the course of their visits. This information was disseminated through frequent meetings with clinic staff and leadership, having project staff available for questions anytime and making a point to elicit clinic staff concerns in focus group-style meetings.

From the patients' perspective, it was emphasized that their interaction with the system would be brief, anonymous, only for the duration of the patient encounter, and that it would be used to help improve the delivery of services and provide optimal time with their clinicians. This information was disseminated to patients though brochures and signage as well as making project staff available for questions. Patients and clinic staff were given option to opt out at any time.

Sites Identified

After several months and many meetings with clinics, two initial sites were identified and agreed to participate. These clinics differed in several important ways. The first clinic was an urgent care clinic. This clinic followed a first-in, first-out scheduling process for patients. It had four exam rooms that were also used as offices in a relatively small architectural footprint. It had no recent experience with process improvement interventions. A member of the project team worked there clinically and could act as a local champion. It also has a relatively strong team culture among the core staff. The second clinic was a more traditional primary care clinic with scheduled appointments. Clinicians used separate exam rooms and offices - eight of each. It had very recent experience in process improvement interventions and its leaders were considered early adopters. Organizationally, it had a traditional hierarchical organizational structure.

Multi-Disciplinary Implementation Team

Creating change in the healthcare system may be considered to be identifying a problem, understanding the problem space surrounding the issue, crafting a solution and striking the right balance between the perceived pain of the crisis and the perceived pain of adopting the solution(20). The project leadership felt that maximizing the odds of success requires being able to draw and act upon the broadest possible option set of ideas and methods.

To address this we organized a multidisciplinary team of clinicians, outcomes researchers, industrial designers, operations researchers and others. The multidisciplinary approach gave us flexibility to (1) identify critical points in the system, (2) generate end-user and use-centric solutions and interventions, (3) prototype solutions, (4) generate evidence for decision makers, and (5) cultivate systems-level thinking and innovative problem solving among the stakeholders.

This multidisciplinary team approach also offered a very robust approach to design of the system. All members of the team contributed unique perspectives and skills, expanding the pool of ideas and potential solutions, and each member offered essential input throughout the entire process of the design of the system. The resulting design was developed more thoroughly and comprehensively than any one discipline could manage and ultimately was greater than the sum of its individual parts. In this sense, good design comes out of good collaboration and integrates the technology, the needs of the end users, and the needs of all of the other stakeholders.

An essential piece of this project was the design of the team itself and the communication protocols used by the team. The project leader made significant efforts to assemble a team that included a wide variety of perspectives and skills but made certain that the team was conscious of the value of the differing approaches. To avoid having the team descend into chaos, each member was made conscious of the other team member's skills and abilities. A shared a common desire to listen and learn from the other disciplines was assiduously cultivated.

As in the clinics, some of the more intangible aspects of the team became crucial deciding factors in the success of the project. The right mix of personalities was needed and required a willingness to work together, as well as a high level of curiosity in general and about the subject and the disciplines represented on the team. While initially challenging, learning to speak the language of other disciplines was an essential element in the design of the team and the project. By assembling a team that shared the goals for the project but had vastly different ways of defining problems and opportunities, approaching solutions, and seeing the people and environments, the team was able to utilize a wide array of skill sets to expand the conceptual outlines and landscape of the project.

In this case the greater the diversity represented on the team the greater the possible insights and valuable solutions. One important example of this approach was the inclusion of the end users into the team, including the clinic physicians, front desk and other clinic staff. This participatory design approach had many valuable benefits including a better understanding of the user needs but it also opened up the idea space to include inputs that might have otherwise been ignored until later, or potentially too late, in the process. By working with the end users the project was also able to better communicate the evolving nature of the iterative process used to refine the systems and better manage the users' expectations.

One of the shortcomings of the team was that the RFID vendor was not as integrated into the process as was desired which would have been beneficial. Though the PI had worked with the company on an earlier project, over the intervening time the company though still small had evolved from a start up to a more mature business. This resulted in a change in the company's culture and more demands on the time of their still relatively small staff, limiting their staff's availability to be part of the team process and lengthening the feedback loop between the project team and the vendor.

Without the vendor playing a central role on the team, some potential solutions took longer to discover and implement than expected. Though there were not many options to choose from in this area, had we been more aware of the changes in the vendor relationship early in the project, the makeup of the team could have been adjusted earlier to accommodate some of the missing skills and perspectives. This only underlines the value of a close working relationship between all the team members and all the key contributors to any system design.

Each team member represents a different but complementary perspective on problem solving with unique ways of generating ideas and achieving solutions. Used in concert, each compensates for potential weaknesses (in promoting change in healthcare) and reinforces strengths.

Outcomes Researchers: Health outcomes link clinical decisions to their consequences. Outcomes researchers try to promote change by accumulating evidence through prospective and retrospective studies. The underlying premise is that the weight of evidence will promote change in a Kuhnian(21) fashion. However, though diagnoses and prognoses may be accurate, and indicate change is needed, they rarely indicate "how" best to proceed.

Operations Research: Operations research (OR)/ industrial engineering is concerned with coordinating and executing operations within an organization or system. OR uses tools such as statistics, optimization, queuing theory, game theory, and simulation to improve and evaluate systems of people, money, information, processes, etc. Systems analysis and engineering is needed to create and implement sustainable solutions and identify their strong and weak points. While OR often address the "how" of solutions, rarely do engineering solutions expressly take into account end-user wants and needs(20), reducing the activation energy needed to adopt change or long term health consequences. This deficit can sabotage even the most innovative solutions.

Industrial Design: Industrial designers (ID) are trained to effectively identify end-users' needs and requirements, the opportunities for change, and to create solutions to multivariate problems while articulating solutions in a way that excites and motivates people to adopt changes. This includes understanding the market (in this case the clinical environment and its stakeholders) and the technological and other constraints. Designers use a process of rapid iteration of prototypes to refine and evaluate the ideas and concepts. This iterative process of refining, improving and evaluating is an essential part of the design process and one that translated well into the entire team.

Two industrial designers were enlisted to assess the social/cultural factors affecting technology adoption and to develop and prototype adaptations/solutions for barriers to adoption. All process change or technical intervention by necessity is also culture change. Here industrial design's focus on the end-user and their needs is critical. Designers are accustomed to researching, generating and iterating ideas rapidly and are trained to act as intellectual catalysts in interdisciplinary teams. Formal design is an iterative process that follows a planned trajectory: problem and scope definition often using ethnography, focus groups and brainstorming; inquiry and experimentation involving generative workshops; sketch model testing and physical mock ups; design development, testing and evaluation. Designers are also advocates for people, e.g., patients. However, IDs are typically more involved with creation and evaluation of potential solutions than the rigorous analysis and development of the quantitative evidence healthcare decision makers usually need. Healthcare decision-makers, need not only good ideas but evidence about safety, costs and effectiveness if they are to assume the risk of change.

To help this disparate group work together, the project group purposely chose a flat team-oriented organizational structure. This included frequent formal and informal communication, a consciously cultivated atmosphere of mutual respect and regular mutual education about each others' fields.

Though they speak different languages (clinical, visual, systems-based, outcomes), all diagnose problems, provide solutions and have many parallels in how they solve problems. Each group however works through a system of progressive inquiry as they problem solve. This may be described as discovery, definition, and evaluation phases, or more broadly described as formative and summative evaluation phases. Good design plus good engineering plus good science equals good problem solving. When used in concert, as seen below, they generate powerful results- targeted innovations supported by evidence.

Designs for a Sustainable System

We considered the design focus of the project to include three main elements: the physical interface between the people and the system, the information interface between the people and the system, and the face of the project – how the project communicated with the end-users and stakeholders.

The technological infrastructure for this project consisted of an IPS that the principal investigator helped beta test in an earlier workflow project. The system was known to be reliable and was being installed in several locations in the hospital. It was intended that the IPS be installed at two to three outpatient primary care clinics during Year 1 of the project. The system is capable of using both IR and RF to determine location and is comprised of five main components- a transponder, a receiving unit, a wired or wireless network, end-user software interface, and database where location and timestamp data are stored. The RFID system uses industry standards, e.g., Ethernet, MySQL, which makes it relatively easy to integrate with existing information systems.

An RFID tag is an integrated circuit connected to an antenna, with or without a battery. RFID comes in two general varieties, *passive* and *active,* with several permutations in between. Passive tags require no internal power source. They are only active when a reader is nearby to power them. In passive tags, the incoming RF signal from the transmitter induces an electric current in the tag which powers the integrated circuit in the tag to transmit a response. Active tags require a power source, usually a small battery. Communications from active tags to readers is typically much more reliable (i.e. fewer errors) than from passive tags. In addition, active tags may transmit at higher power levels, allowing them to be detected in technologically noisy environments or at longer distances.

The *physical interface* was comprised primarily of the transponders and the receivers. The receivers did not pose a problem as they were primarily located innocuously high up on the walls and of neutral color. To the end-users it was just one more piece of passive innocuous looking equipment similar in form if somewhat bigger than a smoke detector, which people have already learned to ignore. The tags on the other hand required some thought. The version of these devices we used was a rectangular prism approximately half the size of a pager. Active tags, though continually getting smaller, do not have the range in size and form that passive RFID tags do. This is because active tags require a battery supply and transmitter. We needed to find a form that was wearable for both short and long duration. We also had to develop form factors that minimized tag loss while keeping them comfortable enough for end-users to be willing to carry them. In the end we settled on three basic forms: one for the patients, one for clinicians and one for administrative staff. Depending on the clinic patients wore the tags on a brightly colored wrist coil which was comfortable and relatively obvious which helped minimize loss due to patients accidentally taking tags home. To prevent tag loss in clinics where a high rate of "walk out" was anticipated, the "bathroom key" solution of attaching a large colorful object was applied (in this instance, wooden fish hand-painted by clinic staff – an activity that also helped create buy in and ownership in the clinic). Administrative staff were offered wrist coil and lanyard options. Most administrative staff chose the lanyard, as having the tag attached to the wrist impeded the ability to type. Finally, clinical staff were offered lanyards, alligator clips, wrist coils and belt clips. Most chose to either put it in their lab coat or wear it as a pager. Staff frequently walked off with the tags still attached but almost invariably returned with them. This was either due to taking advantage of the habit of clinicians putting on pagers or the fact that many clinicians rarely empty their pockets.

The *information interface* between the system and end users consisted primarily of the Graphical User Interface (GUI) used by the front desk staff, placing the greatest work burden on these end users. There were two activities that they were asked to do: check in and check out patients. The check in process involved scanning a bar code on the tag, scanning a barcode on the patient chart and tabbing through several screens. This started a new unique session for the tag and associated the tag with an individual. The check in process was similar but several steps shorter. The challenge here was to figure out how to integrate a few steps of additional work the registration process using a system that was originally designed for inventory management rather than people tracking. The GUI provided by the vendor was designed and expected to be used only intermittently to track expensive equipment that rarely leaves or moves more than a few times per day. Patients, on the other hand, can arrive and leave every few minutes. Expecting that this might be an issue, front desk staff were offered a per-patient incentive during the development phase and were asked how we might improve the system.

The *project "face"* was considered a critical part of the project from the beginning. The face is what the end-users see. It is what helps them develop an explanatory model of how the system works, why it is important and why they would be interested in using it. Many technology roll outs fail because insufficient attention is paid to this aspect of a project. The face is comprised of branding, the physical form of the system being deployed and the people executing the deployment. In the case of branding, two messages were provided - one for the patients and one for clinical staff end-users. For the patients, the message had to be delivered in a brief format with minimal person to person interaction since the patients were only encountering the system for brief and infrequent periods. The message was "Help us help you have more time with your clinician." A good brand message emphasizes the truth and is concordant with the end-users' concerns. This message was delivered via graphic posters and a brochure explaining the project. A research assistant was available to answer any immediate concerns. For the clinical staff, the message was, "Help us develop and provide you with a tool to improve how you do your job." Because staff was in constant contact with the system, this message was delivered through several mechanisms before and after the technology was deployed. The mechanisms used were training sessions which included time for open ended questions, focus groups, identifying local champions and super users, and a website where people could post ideas, suggestions or complaints. In addition, project staff was made available to meet with anyone who had questions or concerns one-on-one.

Because the basic nature of the IPS system is one of measurement, several clinics felt that measuring their behavior could only result in a negative outcome. It was better from their perspective not to have objective metrics. This "Big Brother" problem must be addressed with any system that introduces a new form of performance-related measurement. People generally have one of two reactions to performance measurement. They either feel it empowers them to take control of their lives and circumstances, or that it gives power over them to someone else. Like all tools, RFID is morally neutral but the emotional reaction is real and reflects underlying attitudes towards change. Since this technology tracks movement these attitudes are very close to the surface and need to be addressed.

This same issue was raised in the first project to use this technology, the Operating Room of the Future project(5, 6, 22). Concerns were raised that the investigators would use the technology to track people going to the bathroom or going outside for a cigarette. Other expressed relief that surgeries would no longer be delayed or put at risk because people or equipment couldn't be found. In the ORF project, and subsequently in this project, the fears tended to disappear with use - particularly as it became clear that no one used the technology in the way feared. However, there were individuals who never overcame this fear. The issue is who controls the information and consequently how will the information be used, and some continued to fear "yet another tool to punish them." Our solution was to allow individuals to opt out, and to keep all information anonymous to personal identifiers until the clinic asked for detail to be added. Almost all of the clinics eventually asked to include more detail.

Implementation

Introducing the system and preparing the clinic staff took time. The cultures of each clinic determined how the project was introduced. Several preparatory meetings were held with as much lead time as possible. The purpose of each meeting was to listen openly to concerns and address as much as possible before implementation. The hierarchical clinic asked us to meet with classes of staff separately – Registered Nurses as a group, Medical Assistants as a group, and physicians as a group. The urgent care clinic asked us to meet mixed ad hoc groups based solely on availability. This led to different types of questions and concerns raised. In general the younger or more junior staff had the least concerns over the technology whereas the older staff had more concerns over privacy and being judged. Initial resistance in all setting centered around four issues: (1) where the clinic or individual sat on the adoption curve, (2) potential innovation fatigue, (3) worries about additional work load, and (4) fear of being judged. In addition to addressing these concerns to the extent possible in group meetings, all staff were offered one-on-one and group training on using the GUI and the feedback website.

Because every intervention perturbs the system in which it is introduced, making it easier or harder for end users to perform their work, we wanted to measure the effect of the system before and after its introduction. To that end we attempted to survey the staff using a version of the

Maslach Burnout inventory(23, 24). This tool has been validated in the healthcare setting and has the useful property of capturing a range of work perception from the propensity towards burnout to high job satisfaction.

Shakedown Phase/Initial Rollout & Assessment (~1 Month)

The first month in the clinics was considered the rollout or shake-down phase. During this period project staff was present almost constantly for training, troubleshooting and tuning the system. It was during this period the team learned how well it anticipated potential problems.

In any technology implementation there is always the trade-off between getting it perfect and getting it implemented quickly (quality vs. time). Therefore, one must plan for imperfection and a continued dialogue with the end-users. The areas where we anticipated problems were mostly related to the technology system itself. The unexpected issues and barriers we discovered were mostly related to unanticipated political or communication issues.

Expected Issues

The basic commercial IPS system was designed for low traffic inventory management as opposed to the fast-paced, high volume patient environment in tight spaces. Therefore as expected considerable early effort was spent refining the time and spatial resolution of the system both before and after system go-live. This was particularly true in the smaller clinic; the spatial resolution required was near the system's limits. The next technical problem related to the choice of bar code scanners. The initial scanners, while compatible with the system specifications, turned out to be slow and had difficulty reading the bar codes. The replacement scanners were much better at reading the bar codes but turned out to be initially incompatible with the older computers at the front desks which had to then be replaced with newer machines. This was a result of an interesting unexpected aspect of clinic culture. In these clinics, it is typical for the clinicians to receive the newest computers and for their computers to be handed down to the rest of the clinic. However, the clinicians placed the least demand on their computers as they mostly used them for word processing and linking into the hospital EMR where over 99% of the charting took place. On the other hand, front desk staff placed the highest demands on the system, having multiple applications open simultaneously in order to efficiently do their work.

The graphical user interface also was expected to pose some minor problems as it required 18 steps to register counting tabbing fields and drop down menus per patient. The GUI was designed by the engineers at the IPS vendor company and reflected a valid but generic approach to interacting with the IPS system. This approach did not take into account the specific ways in which the clinical setting differs from other uses of the technology, how clinics might differ from each other, or the pressures the staff in a clinical setting might face. It was initially designed with many more options than needed for the clinic environment as it was initially designed to register a wide diversity of equipment to be tracked. While this wasn't substantially different than the number of steps required to register patient contact, insurance and other routine registration procedures it was new and unfamiliar. The whole process took between 15 – 30 seconds per patient which can seem like a very long time in a busy clinic. We provided a per person registered incentive and in the initial phase allowed staff to skip patients if there were any problems. We anticipated the need to simplify the interface. What was unexpected was the delay in getting the vendor, a small and fairly young company, to make the necessary changes. The vendor's staff were spread very thin and service was slower than expected.

Unexpected Issues and Barriers

Unexpected issues and barriers were mostly confined to problems with end-user feedback and politics. Anticipating that we wanted close feedback with the end users, we expended significant effort setting up a website with all the training material, background literature, a forum for comments, complaints and suggestions, and with direct access to the project team and trained super-users in each clinic in its use. The clinic staff did not use it. During focus groups and one-on-one discussions, the staff explained that they were too busy to sit down and log into it. In addition, the staff did not feel the need to use the website when people from the project team were always around.

The next feedback failure was the per person register incentive for the front desk staff. The incentive was set up to yield between $10 – $30 per day depending on the number of people successfully registered and unregistered. Front desk staff were only asked to keep count of the number of people the personally registered so we could reward most those who did most. The staff did not keep track of how many patients they registered each day. During focus groups they explained that they were too busy to record this number. Several became angry when the incentives were not initially paid out because there were no numbers per person with which to calculate their incentive. Eventually we gave the front desk a flat incentive at the end of the initial study period – a situation that was far from ideal.

Finally, we ran into unexpected politics surrounding the pre-/post survey in one of the clinics. Specifically, middle management in this clinic felt the survey created dissatisfaction that wasn't there before the survey. Our initial hypothesis was that any new technology adoption must influence how the staffers get their jobs done. If the technology worsens baseline job satisfaction then there is a problem with the system and it should not be adopted. This baseline assessment unexpectedly and significantly alienated middle management. From the perspective of the relevant middle managers, asking job satisfaction questions gave staff the opportunity to consider problems in their jobs, creating dissatisfaction rather than measuring the spectrum of satisfaction to dissatisfaction found in their work environment. In addition, the managers felt they would be penalized for any negative information coming out of the clinic regardless of context. Thus the pre-implementation survey alienated the management. This was unexpected as a very similar pre- and post survey was conducted in the surgical environment by the project leader without significant problems or complaints(5).

How System was Changed to Meet Users' Needs

Once these problems were identified we needed to address the end-users' concerns. This was done first through a focused assessment by the industrial design team; several focus groups with the staff led by the design team, and follow up observations by the project team. The team met regularly during this process and sets of solutions were proposed and vetted for each of the problems. Solutions were then fleshed out and prototyped and the best solutions were advanced to implementation.

Ultimately solutions were ranked based on need and feasibility. These were the implemented with the expectation of improved experience and expectation that team would continue to address ongoing needs.

The items addressed to date have been a simplified GUI, replacement and upgrading of scanner and computer equipment, changes to the physical form of the tag and improvement in data analysis, providing time-flow data to the stakeholders. Changes still in development are a hands free registration process and a new incentive structure.

Tempo

A critical issue that arose, for the multidisciplinary team and diverse group of stakeholders involved in the system roll-out, was tempo. Each set of stakeholders had a different set of expectations on the timing of feedback and solution fulfillment. The project team had varying internal expectations on solution formulation and turn-around time. The sense of an appropriate tempo influences the perception of success. Balancing these competing expectations can be tricky but is essential. Balancing expectations within the project team was more successful than balancing those of the stakeholders. As a general rule for the stakeholders, the closer they were to the front line the greater their sense of urgency. For the project team, the sense of tempo was driven in part by what was considered adequate evidence upon which to base a decision and the scope of the problem. In general, industrial designers had the fastest sense of tempo while the outcomes researchers had the slowest.

Later Resistance to Change

Despite addressing the early issues in the roll-out phase, later issues evolved. These centered around workflow, the perception of being judged, innovation fatigue, cultural issues, tempo and transparency of explanatory models.

Despite simplifying the GUI, cutting the steps in half, and upgrading the scanners, the staff in one of the clinics remained very sensitive to what appeared to the management and the project team to be minor changes in workflow. The staff in the same clinic also developed a sense of being judged, particularly by onsite project observers (who represented the face of the team and face of the project). We feel this was in part due to the mixed research and process improvement message of the project. We asked our research assistant to observe workflow issues and act as a troubleshooter. We feel this created ambiguity as to this person's role and created a focus for anxiety in the staff. This may in part have been due to cultural issues. In the hierarchical organization staff didn't appear to feel empowered to raise issues or adopt changes independent of clinic leadership. They felt they needed approval from supervisors and managers. On the other hand the more team-oriented clinic seemed more ready to support each other with new processes and making project work. These cultural issues may have influenced later confusion over explanatory models, which are needed to orient people on how to interact with systems and how to predict system behavior. In this case an explanatory model helps end users understand why wearing the tags is important and what kind of information might be expected from the system. The explanatory model in this group seemed unstable. After the initial introduction and feedback groups most appeared to understand the purpose of the project and roughly the nature of the technology. This understanding seemed to fade with time.

Innovation fatigue also appeared do become an issue in this clinic. Specifically introducing too many things at once can create burnout – emotional exhaustion, depersonalization and feelings of lack of control or personal accomplishment. Too many new systems at once can run the risk of cognitive and emotional buffer overload.

The cycle time of the feedback loop in this same clinic also became a regular issue. Although clinic staff complimented the project team on providing an unprecedented level of receptivity to the end-users (who had prior experience with projects that provided no feedback), the project team did not fulfill their hopes with regard to timing of feedback. This may be due in part to setting expectations too high early in the project. It was also due to limited team resources and limited vendor support.

Lessons Learned

What didn't work? Managing feedback and expectations was only partially successful. In addition the stability of the explanatory model which in turn influences expectations was not as stable as expected. The lesson learned here is that greater effort must be spent on establishing clear communication channels and on constantly educating and reinforcing explanatory models with the end-users. It should also be communicated to stakeholders from the beginning that the project will not be perfect immediately and that procedures can be expected to evolve as we learn how to do things better. In addition, it is evident that everyone who comes in contact with the system is both a potential stakeholder and a potential point of failure. Individuals can act as facilitators or champions, but can also block a successful implementation. "Blockers" may have disproportionate influence on the failure of the project or technology adoption by slowing its progress and providing a focus for discontent among peers. However, their concerns may be legitimate and should be addressed. Addressing these concerns will make any project stronger. Therefore, it is important to identify blockers early and develop a strategy for managing them. In retrospect, had we had a strategy to identify and manage "blockers" from the start, we believe the fulfillment of project goals would have been faster and the adoption process smoother.

With any technology project, one must balance getting it right the first time versus getting it out into the field. This implies that a failsafe for project success is high quality and frequent feedback from the end users.

What did work? This project had many successes. The interdisciplinary approach to the project was a significant success. Incorporating ID into the team, a relatively uncommon approach in the quantitatively driven healthcare policy and research area, helped to identify and overcome key hurdles for a successful and sustainable project and technology rollout. The ID focus on the interaction of the technology with human elements of the system, including all the user and stakeholders, along with specific techniques for design research and observation, helped to avoid many of the potential points of failure and ensured ease of use for an already stressed and overworked clinical staff. ID helped to identify problems others may not have seen and acted as interlocutors among different groups by visualizing and communicating solutions in ways that many different audiences could understand. By closely examining the human element and clinical cultures, ID helped the project team understand how people were going to interact with the physical, social and organizational aspects of the project; for example, how the RFID transponder tags were to be worn or carried by patients and staff and optimizing the check-in process. This helped the project team to be flexible, adaptable, and responsive.

The face and framing of the project was also a success. Developing a brand and framing the project carefully to the stakeholders helped facilitate all of the project's endeavors. The project was also successful in the empowerment of the end-users particularly in the urgent care clinic. They felt free to provide critique and feedback and also became active participants in the design process. The project personnel including the local champions also contributed to a positive face to the project. The local champions helped provide at least in some case a very short feedback circuit. The feedback sessions and focus groups, while perhaps not as frequent as some of the end users desired, were clearly exceptional from their perspective and promoted a sense of team between project staff and end-users.

Another sign of success is that many other clinics in the hospital system now want to adopt the IPS system for their own use. The project has produced quantitative data (such as patient wait time, flow time, and face time with physician) that has demonstrated enough value to decision-makers be rolled out in other clinical settings (25). In addition, several patents have come out of the project and each of the current groups and soon to join groups have helped generate new ideas on

Figure 2. Example primary care dwell-flow map (Each color represents an individual's path through clinic. Each circle is proportional to the duration of the patient at the location)

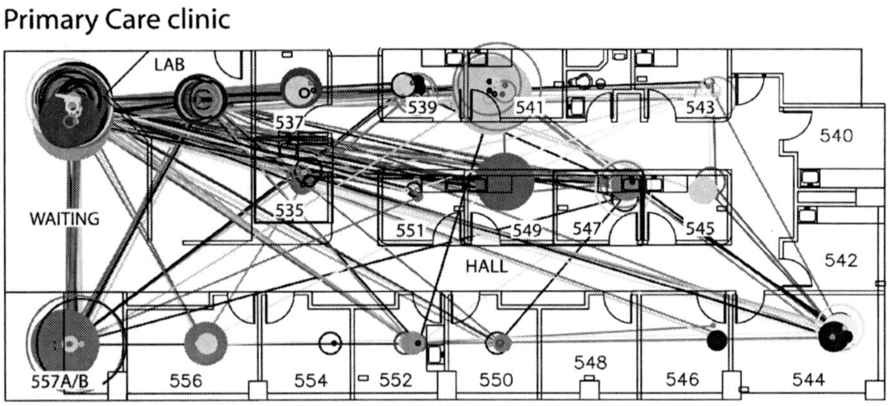

how this system maybe applied for future projects and collaborations (Figure 2).

FUTURE TRENDS

All technologies are means for the end user to extend his or her reach and/or reduce his or her burden of work. Introducing a new technology is not free. The cost, in terms of time, money, time and overcoming technological, organizational and cultural barriers, of a new technology must be substantially less than its benefit.

The long term success of any project depends on whether or not the end users will continue to use the new system once the original urgency for change is over. A successful technology implementation cannot solely rely on the presence of a champion or advocate who may move on; rather, a durable technology change needs deeper change. This kind of change requires insight into the system and culture as a whole.

This is especially true for medical technologies, where a technology with a new functionality rarely influences just one task. Changes in the medical environment is typically have widely distributed knock-on effects on diagnosis and information and knowledge acquisition, communication between clinicians, staff and patients, and the effectiveness with which care is delivered.

Therefore, successfully introducing a new technology in the healthcare environment requires attention to a wide variety of barriers and design issues. Not doing so can result in new technologies stalling or failing.

For example, adoption must also be driven by the technology's ability to improve workflow and increase end users' satisfaction (whether the end user is the patient or staff).

Multidisciplinary teams are critical in this respect as they are able to address all aspects of the issue instead of focusing on a single driver. Similarly, such teams are equipped to handle multiple types of users and stakeholders due to their varying backgrounds. Multidisciplinary teams force individuals to externalize knowledge and thought processes - explaining opinions allows others to contribute. Having multiple perspectives in a complex implementation is essential to success.

Some future areas requiring a multidisciplinary approach in medicine include:

- Outpatient Electronic Medical Record Systems
- Point of care testing

- Information/Communication Technologies
- Online medical record systems and mobile medical records

As our system becomes bigger and more complex, our clinical and cognitive resources become more precious and less clear to manage. In order to effectively manage these resources and integrate new knowledge and technology we have to look at the system as a whole and be prepared to learn from everyone around us.

CONCLUSION

Policymakers, healthcare providers and researchers have long struggled to improve the cost, quality and efficiency of our healthcare system. Indoor Position Systems (IPS) offer a means to help solve the problem of measurement and perverse incentives. The goal of this project was to roll out an IPS in a live complex clinical environment and to seamlessly integrate the technology into the clinical workflow.

The professional team was comprised of Industrial Designers, Operations Researchers, and Outcomes Researchers, though clearly the team as a whole also included the system end users: clinicians, staff, and leadership. The team needed to face and solve the challenges of a complex technology, multiple stakeholders with varying motivations, and diverse local cultures and organizational structures. This effort centered on three main components: the physical interface, the information interface, and the face of the project or branding. Because technology adoption disrupts the existing system the framing and face of the project for the individuals involved becomes crucial. This was especially true as the nature of IPS is measurement; the project team had to ensure end users that data would be used to their benefit and not punitively. As is often in the case of complex projects, it took more time than anticipated to solve even the expected barriers to implementation such as simplifying the computer interface. Unexpected barriers were related primarily to local politics, culture and the previous unknown historical experience of the end-users with change and had little to do with technology itself.

Deploying a new system of technology is intrinsically a multidisciplinary effort. Technical solutions cannot succeed independent of cultural and organizational solutions. At a minimum, to be successful, an ongoing conversation between those introducing the technology and those who are using it is essential. Promoting change requires listening, persuasion and evidence. Therefore, one needs to pay close attention to culture and systemic measures. Having a broad arrange of skills at your disposal allows the ability to respond flexibly and rapidly, increasing the likelihood of success. The challenge, however, is being able to listen to what all the stakeholders, including the implementation team, are saying.

REFERENCES

Ajzen, I. (1991). The Theory of Planned Behavior. *Organizational Behavior and Human Decision Processes*, *50*, 179–211. doi:10.1016/0749-5978(91)90020-T

Ajzen, I., & Fishbein, M. (1980). *Understanding attitudes and predicting social behavior*. Englewood Cliffs, NJ: Prentice-Hall.

Bandura, A. (1988). Organizational Application of Social Cognitive Theory. *Australian Journal of Management*, *13*(2), 275–302. doi:10.1177/031289628801300210

Bandura, A. (1989). Human Agency in Social Cognitive Theory. *The American Psychologist*, *44*, 1175–1184. doi:10.1037/0003-066X.44.9.1175

Bratt, J., Foreit, J., & Chen, P. (1999). A comparison of four approaches for measuring clinician time use. *Health Policy and Planning*, *14*, 374–381. doi:10.1093/heapol/14.4.374

Chernichovsky, D., & Anson, J. (1989). (*The World Bank*). Cost Recovery and True Cost-Effectiveness Ratios of Contraceptive Provision.

Coburn, P. (2006). *The Change Function: Why Some Technologies Take Off and Others Crash and Burn*. New York: Penguin Group.

Colwill, J., Cultice, J., & Kruse, R. (2008). Will Generalist Physician Supply Meet Demands Of An Increasing And Aging Population? *Health Affairs (Project Hope)*, *27*(3), w232–w241. doi:10.1377/hlthaff.27.3.w232

Conwell, L., & Cohen, J. (2002). *Characteristics of people with high medical expenses in the U.S. civilian noninstitutionalized population*. Washington, DC: Agency for Healthcare Research and Quality.

Davis, F. D. (1989). Perceived Usefulness, Perceived Ease of Use, and User Acceptance of Information Technology. *Management Information Systems Quarterly*, *13*(3), 319–340. doi:10.2307/249008

Finkler, S., Kinckman, J., Hendrickson, G., Lipkin, M., & Thompson, W. (1993). A comparison of work sampling and time-and motion techniques for studies in health services research. *Health Services Research*, *28*, 578–597.

Fisher, J., & Monahan, T. (2008). Tracking the social dimensions of RFID systems in hospitals. *International Journal of Medical Informatics*, *77*(3), 176–183. doi:10.1016/j.ijmedinf.2007.04.010

Health Data, O. E. C. D. (2006). *How does the United States Compare*. http://www.oecd.org/dataoecd/29/52/36960035.pdf

Kuhn, T. (1996). *The Structure of Scientific Revolutions* (3rd ed.). Chicago: University Of Chicago Press.

Lurie, N., Rank, B., Parenti, C., Wooley, T., & Snoke, W. (1989). How Do House Officers Spend Their Nights? A Time Study of Internal Medicine House Staff on Call. *The New England Journal of Medicine*, *320*, 1673–1677.

Mankins, J. C. (1995). Technology Readiness Levels: A White Paper. In Technology ACO-OoSAa, (ed.), *Advanced Concepts Office of Space Access and Technology*. Washington DC: NASA

Maslach, C., & Jackson, S. (1989). Burnout in health professions: a social psychological analysis. In Sanders, G. S. J. (Ed.), *Social psychology of health and illness*. Hillsdale, NJ: Lawrence Erlbaum Associates.

Maslach, C., Schaufeli, W., & Leiter, M. (2001). Job Burnout. *Annual Review of Psychology*, *52*, 397–422. doi:10.1146/annurev.psych.52.1.397

Perrow, C. B. (1999). *Normal Accidents: Living With High Risk Technologie*. Princeton, NJ: Princeton University Press.

Rodgers, E. (1995). *Diffusion of Innovations* (4th ed.). New York: The Free Press, Simon & Schuster.

Stahl, J., Egan, M., & Goldman, J. (2005). *Introducing new technology into the OR: Measuring the impact on job performance and satisfaction*. Surgery.

Stahl, J., Goldman, J., Rattner, D., & Gazelle, G. (2007). Adapting to a new system of surgical technologies and perioperative processes among clinicians. *The Journal of Surgical Research*, *139*(1), 61–67. doi:10.1016/j.jss.2006.08.030

Stahl, J., Holt, J., & Gagliano, N. (n.d.). Understanding Performance and Behavior of Tightly Coupled Outpatient Systems Using RFID: Initial Experience. *Journal of Medical Systems*.

Valadez, J., Varga,s W., & DiPrete, L. (1990). Supervision of primary care in Costa Rica: time well spent? *Health Policy and Planning*, *5*, 118–125. doi:10.1093/heapol/5.2.118

Venkatesh, V., Morris, M. G., Davis, G. B., & Davis, F. D. (2003). User Acceptance of Information Technology: Toward a Unified View. *Management Information Systems Quarterly*, *27*(3), 425–478.

KEY TERMS AND DEFINITIONS

Operations Research: An interdisciplinary branch of applied mathematics that uses methods such as mathematical modeling, queue theory, statistics, and simulation to understand real and theoretical systems.

Outcomes Research: A field of study that seeks to understand the end results of particular health care practices and interventions. End results include effects that people experience and care about, such as change in a patient's ability to function.

Indoor Position System: A system of technology used to locate individuals or items in time and space within closed structures. Many are based on radiofrequency, infrared, and ultrasound based technologies.

Outpatient Care: Clinical services delivered to non-hospitalized patients, often in primary care practices, urgent care centers or emergency rooms.

Healthcare System: The system of human and capital resources used to deliver healthcare related services to a population.

Radiofrequency Identification (RFID): A class of technology used for the purpose of identification and tracking using radio waves.

Tightly Coupled System: A system in which the resources within the system are highly dependent on each other and typically have very little spare buffering capacity between resources.

Chapter 26
Designing Toys, Gifts and Games:
Learning through Knowledge Transfer Partnerships

Friedemann Schaber
The University of Northampton, UK

Vicki Thomas
The University of Northampton, UK

Randle Turner
The University of Northampton, UK

ABSTRACT

Knowledge Transfer Partnerships (KTP) is a program in the United Kingdom (UK), administrated on behalf of the Technology Strategy Board, a Government Agency, where academics give their expertise to provide a solution for a particular problem, helping organisations to improve their competitiveness and productivity. The academic partners at universities, colleges or research institutes bring their knowledge, skill and technology. In return, KTP enhances the business relevance of their research and teaching, in particular, deepening their expertise of 'real life' projects. This investigation exemplifies some of the strengths and pitfalls of this type of knowledge transfer in the context of the global toy, games and gift markets, through case studies of the design and prototype development process for a range of toys, giftware and seasonal products.

INTRODUCTION

Part funded by the government and led by staff at The University of Northampton, UK, design and design management competences are being encouraged and developed through knowledge transfer, in a range of businesses in the toy and giftware sector, such as the retail division of a charity and several producers and importers of giftware and leisure games. Design graduates are placed within the companies to research, plan and implement a new product design methodology by designing new products in-house. These so-called KTP Associates will gain business-based experi-

DOI: 10.4018/978-1-61520-617-9.ch026

ence and personal and professional development opportunities. Ideally, they are offered, and accept a post with the company on completion of the project, to head a new design department.

Toys and gifts are produced globally and most of the ranges sold by the organizations detailed in case study one, two and three are produced in the Far East. By large, through the involvement in KTP, the learning and teaching at the university has been enhanced, through exposure to highly current solutions of global production, new markets and regulatory requirements and by amalgamating the KTP experience into 'live' teaching. The Associate also introduces undergraduates at the university to an operation involving manufacturing overseas, with some parts made by automated process, shipping from the Far East and UK distribution. This process augments, in significant ways, the practice of sharing knowledge by inviting industry into the university to run live projects, employing outside designers on a part-time basis and entering international competition, as stated by Wilkinson and Sale (2000).

The objective of this chapter is to share the knowledge gained from ten years of working on KTP with a wider audience, and to raise questions as to whether the process has set trends that should be extended and followed by others working in different circumstances.

BACKGROUND

Theoretically, this chapter is about the concept of knowledge transfer and how it has been used as a way to encourage firms to invest in product design. Knowledge transfer is an idea that has been developed in management theory. Nonaka & Takeuchi (1995), Blackler (1995), Collins (1996), Davenport & Prusak (1998) discuss types of knowledge. For the past three decades the KTP initiative, along with its predecessor, the Teaching Company Scheme (T.C.S.), established in the mid-1970s by the Science & Engineering Research Council (T.C.S., 2009), has given British firms access to new methodologies, via the "knowledge bases" (the term used in the initiative for universities). KTP assumes the notion of knowledge at its core, but is it just management trend? Do our case studies confirm the continuing relevance of the approach?

The transfers of knowledge are between different networks; some are structured organizations like the education system, or complex social ones like, Chinese culture in the twenty-first century and others are short-lived networks just "alive" for one project. Actor network theory outlined by Latour and others (Law & Hassard, 1999) has been found to be useful by many, when exploring the relationship between technology and society. In this chapter's case studies, the Associates are the key actors working as a catalyst and transferring knowledge between many of the networks. But there are other actors; the traders, supervisors, retail buyers and most importantly, the models, drawings and products themselves. Does seeing the experience of KTP in network terms help us understand better how the knowledge is being transferred?

Gift, toys and games are not necessarily engineered. Many would not consider them as functional products or necessities. Rather, they would consider them as luxuries and indulgences, or, even material resources that could be put to better use. Mauss (1954) analyzed what he called gift exchange in the 1920s, and argued that it is a social process found in all societies generating positive social relationships and trade. Gifts generate obligations and are designed for specific occasions and people. (Thomas, 1984) They often contain a message. Toys and games are often given between parents and children, bringing people together whilst providing an opportunity for them to develop social skills. The ability to play is regularly used in teaching management; understanding how people play and encouraging play is seen as a way of generating creativity and as a form of education. In short, when designing these products, should the designer have an understanding of their own social and cultural background as well as to have the ability to communicate these to others whose cultural traditions may differ from their own?

Academics such as geographers, social anthropologists, design and cultural historians have been researching and acting as advisors to firms working in the global market place. (Atkins & Bowler, 2001; Hofstede, 2001) One approach this chapter could have taken is to compare how different countries have facilitated their knowledge base for the advantage of the nation as a whole in order to promote trade and international collaboration. The scope of this study, however, is limited to the UK experience. Evolved from the Teaching Company Scheme, the Technology Strategy Board oversees and part-funds the KTP program where academics provide solutions for UK businesses while putting research into practice. (T.C.S., 2009) At the KTP Business Leader of Tomorrow award ceremony, Jonathan Kestenbaum, chief executive at the National Endowment for Science, Technology and the Arts NESTA, said that:

If business and universities do not work together on innovation, the UK will face a very uncertain future. Our competitiveness depends on them sharing the rich base of knowledge they produce." (IET Newsletter, 2008)

The Lambert Review of Business-University Collaboration (2003) made a series of recommendations regarding Britain's strong science base and the business community. The Lambert Review concluded that:

The British dual support system acts as a disincentive to business-university collaboration and that the biggest single challenge for knowledge transfer is in boosting the demand for research from non-academic communities, rather in increasing the supply of ideas and services from universities. (p.12)

As we shall see, the findings in our case studies do echo the following recommendations by the report:

- "The best form of knowledge transfer comes when a talented researcher moves out of the university and into business, or vice versa." In a KTP, young 1st class honor graduates are appointed Associates working within the company.
- "The most exciting collaborations arise as a result of like-minded people getting together – sometimes by chance – to address a problem." In these case studies, there is a skills mix at the academic support base and also within the industrial partner.
- "Encouraging academics and business people to spend more time together should be a high priority for knowledge transfer professionals." At Northampton, there are weekly meetings throughout the two-year project lifespan.
- "The innovation process is non-linear, so knowledge transfer is not simply a question of channeling clever ideas from researchers down a production line into commercialization … [It is] critical to build dynamic networks between academic researchers and their business counterparts." KTP benefits from the multiple networks that the partners are involved.
- "Diversity is good, both in mission and in funding. The type of business collaboration that would make sense for one kind of university might be either impossible or irrelevant for another." Toys, games and gift design seem to have worked within School of the Arts skills mix, enhanced by consultancy background and engineering knowhow.
- "Proximity matters when it comes to business collaboration." As we shall see, distance is perceived a pitfall in case study three (pp.12-13).

The KTP approach should be seen in the context of current UK research policy, which places great emphasis on knowledge transfer – evidence that public investment in academic research ultimately results in economic benefit from the sale of products resulting from that research. Blackwell (2008), however questions, by referring to the Lambert report, the commercial success of university industry partnership models by exemplifying ventures at Cambridge. He states that

They would appear to be the most dramatic evidence of this economic opportunity, for reasons closely related to disciplines and design. ... Instead, the process of knowledge transfer is often presented as if it were a pipeline, with scientific research results entering at one end, being converted into patents and license agreements, and emerging from the other as products and commercial enterprises. This is almost the reverse of any reasonable design process, in which it is the requirement or market opportunity that forms the starting point for design. (p.5)

Cambridge hosts many corporate research laboratories. However, these labs also struggle to contribute to the products of their companies. It is argued that:

The corporate labs of technology companies are structured and staffed according to academic disciplines and scientific endeavors, not according to market opportunities. ... Furthermore, the research policy fiction of 'knowledge transfer' – that economic benefit can result from pursuing the best standards of academic research – has no evidence to support it. Those academic fields that are most engaged in responding to outside problems, including architecture, education, product design, are those that struggle most with the regimes of academic quality assessment. (Blackwell, 2008: 6)

This chapter is informed by the authors' decade long engagement, with the such funding organizations as academic partners of knowledge seeking companies, but at a newly established university that has grown out of a technical college and art school. The authors have been researching and contributing to the debates on the role of knowledge transfer in the global marketplace. (Schaber & Thomas, 2008) Furthermore, they have also learned from the practical experience of acting as supervisors and working with industry on KTP, thus have gained the understanding of the benefits and negative aspects for all the partners involved. (Betts et al, 2007) Conference participation and publications are the routine ways that academics share knowledge, allowing the authors to identify and to map out the best practice and, at the same time, to develop networks within the academic system. For example, Griffith and Bamford (2007) provide the international context of industry-academy collaboration in engineering and design.

Universities promote and generate opportunities for KTP within regional business communities through specialist business development offices; in Northampton, it is known as the Knowledge Exchange. The government also provides networking opportunities for fellow supervisors to exchange information with the advisors and executives at the KTP agency. In short, knowledge is shared between Universities involved with KTP.

Other disciplines are following the KTP model and a trend began to emerge. For example, the Centre for Research in the Arts, Social Sciences and Humanities, Cambridge (CRASSH, 2009) states that:

Knowledge is produced in a wide range of contexts within and beyond universities, including practice-based research. It often involves organizations or individuals linked by networks rather than institutions.... Universities of the future will have to create platforms for such networks while staking out new space for themselves in the marketplace of ideas.... But the transfer of knowledge between disciplines does not necessarily guarantee innovation. (p.1)

So, what models are needed to facilitate disciplinary innovation? How does disciplinary innovation relate to institutional changes in Arts and Humanities research? What knowledge have we learnt from the KTP experience that we would share with other disciplines? By writing this chapter maybe a discussion can be started with other countries on whether this approach resonates or works within their own networks.

Sir Ian Needham (2008), Director at Dyson, addressed the unique challenges of innovation alongside creative design. He explained: "Innovation needs to be carried out against a clear design framework to manage what you're doing to a successful conclusion"(p.1). In the global context, he considers the strength of British businesses as solution providers, succeeding in areas with high added value. Let us better understand the KTP role in this process.

MAIN FOCUS OF THE CHAPTER

All the three case studies presented below explore these educational benefits in some detail. The businesses' benefits are varied: access to innovation and expertise, computer aided design (CAD) file exchange, the Associate is working on the premises, embedding design into the organization, visiting manufacturing sites outside the UK, dealing with communication and cultural issues, handling intellectual property rights, using visual and digital skills to communicate to buyers, internal boards and off-site manufacturing. The Associate is guiding the key design decisions of the industrial partner, using a variety of research and presentation techniques; thus setting up and managing the design process. The Associate has the opportunity to take a vocational management qualification and pursue a higher degree. Although based at the industrial partner, they are treated as a research student at the University.

Whilst the role of CAD on product development varies and as each company's objectives change, the Associate shares and develops crucial skills, such as visual communication, an ability to respond to trend forecasting and market research, to communicate effectively within teams, and to interpret and create viable products from minimal briefs. Prototyping and rapid fabrication as product development enhancement tools can be observed and practiced. The Associate learns about the process of designing in a global market place, dealing with the constraints of manufacturing overseas and responding to market demands and economic cycles.

There have been some concerns raised that Universities could be seen as working beyond their remit by acting as a design consultancy. The scheme, however, originated from the apprenticeship culture and the internships for engineering graduates, facilitated by (what was then) the Teaching Company Scheme. So, at its core it is still about learning on the job, but the transfers of knowledge occur over a network between the different cultures and/or systems of managing and generating knowledge, the Associate being one of the key actors. As we shall see from the case studies, everyone in the network has the opportunity to share and learn, but the lessons are not always the same.

Funded under the Science and Technology Act 1965, the Teaching Company Scheme (TCS) was established in 1975 by the Science and Engineering Research Council, based upon the teaching hospital idea - 'learning by doing'. Originally aimed at engineering projects, KTP today covers a wide business spectrum to meet the social, technological and economic priorities of the UK. (Ktpoline, 2009) So what can we learn from the authors' hands on experience?

CASE STUDY ONE: TOYS

The first case study of new product design and development through Knowledge Transfer between design education and industry describes a partnership of developing a range of toys and gifts for the UK market. The aim of this KTP project, between the retail arm of a Charity, the University of Northampton and the Associate, is to create a design capability and thus limit purchasing of non-bespoke items sourced from China. This will enable the Charity to design and sell their own bespoke products through a chain of 400 shops on high street locations throughout the UK. In this case, the partner company is a charity whose generated profits would provide compassionate care for people living with serious illnesses, while expanding and securing the charity's future. The University interpreted the requirements of the knowledge seeking company as follows:

The area Sue Ryder Care Direct was looking to improve specifically was its offerings of bespoke and individual goods ... designed by Sue Ryder Care Direct in accordance with market demands. An in-house 3D design capability, which they were looking for the KTP to provide, would allow the charity to better specify its existing Chinese product manufacturers, resulting in unique end-products less exposed to negative price comparisons. It was also envisaged that through the KTP, Sue Ryder Care Direct would be able to concentrate its new 3D design capability on its own range of toys, giftware, ornaments and hobby items which could be charged at a competitive retail price and have their own brand identity (The University of Northampton, 2006)

Intellectual Property Issues

The initial contact with the University was made as the Charity experienced infringements of copyright by sourcing non-bespoke items from China and not being able to control local manufacturers copying and supplying the competition with look-a-like products (Figure 1). Since developing its own design expertise though the KTP activities, the branding and packaging of new product ranges were designed by the Associate and he collaborated with the London based team publishing the catalogue and developing the internet presence. So the Associate acted as supervisor of the operation; again this supports the aim of embedding design management experience. Copyright remains an issue, as the contracted manufacturers overseas continue to copy and sell on the contracted designs. As with many western companies, the Charity is (through the KTP) developing its brand identity and increasing customer loyalty and emotional engagement, through collector's clubs, reuse bags and loyalty schemes. (Schaber and Thomas, 2008)

Communicating Design and the Network with China

In the process of idea generation, designing and prototyping, marker visuals on layout paper have been used to generate a range of images describing features of the fairy castle design. These are communicated to the charity's project manager in a presentation of proposals prior to developing three-dimensional models or CAD work. Based upon the CAD file and associated renderings received, the manufacturer in China interprets and generates a first model. This prototype shows features key to exporting such flat pack assembly. For example, the sections of turret nest in order to compact the packaging. There is communication with China specifying color ways and materials; here alterations took place in accordance with local knowledge and available materials such as plywood.

Each KTP has a main design project and in his case it was the design of dollhouses and castles, which was extended to include fantasy figures, giving the Associate the opportunity to oversee the design of 40 pieces and learning about mould

Figure 1. Retail catalogue. (© 200], The Sue Ryder Care Company. Used with permission.)

charges. In support, computer visualization training courses were undertaken. There was a visit to China to develop his understanding of the wood manufacturing capabilities of suppliers, allowing him to learn from the Chinese knowledge base. Together with his academic supervisors, he was able to present the KTP case at regional knowledge exchanges and business shows and also at an international conference (Figure 2). He investigated UK manufacturing, developing contacts with British toy and model making companies that are staying competitive by producing to a higher quality, able to customize and adapt ranges swiftly, with easier communication and transport.

Figure 2. KTP associate. (© 2008, The University of Northampton. Used with permission.)

It was shown, that use of CNC and laser-cutting machinery allows for short production runs and high dimensional accuracy, and can partially offset the labour cost issue. (Betts, et al 2007) This KTP has been successful, with an increase in sales and market share, and also by offering new larger and more diverse product range, including dolls houses, fantasy figurines, musical instruments and furniture. (Sue Ryder Care Final Report, 2008)

Flexibility, Short-Term Networks and Future Development

Although the overall project plan structure has stayed the same, the tasks within the project changed due in part to the unpredictably of the commercial environment. So the needs of the project changed accordingly with these commercial opportunities and threats. A promotional partnership with a UK Newspaper (The Daily Mail) is a good example of this, exclusively offering its readers a miniature fantasy castle and accessories, which were then partially developed with rapid prototyping technology at the University. A number of new projects have arisen through the liaison and these opportunities usually have very short and demanding delivery deadlines.

The design processes and systems embedded in the charity through the KTP were able to deal with the increased demands. The new larger and diverse product range will eventually lead to a more diverse range of buyers. Storage, furniture and is being considered as future product ranges, after monitoring the UK charities shop market and changing social patterns. The Charity is conscious to widen their customer base, described as "tend to be of an age of 40 and upwards, female and mothers who are socially disadvantaged or have commitments at home" and catered for with miniatures, items "that provoke nostalgia and encourage the customer base to make a purchase". (Betts et al 2007) The concurrent global financial difficulties have attracted new customers, so the Charity and Associate have had to respond swiftly with new product development and their retail offer.

Impact on Teaching

In support of learning and teaching at the University, the associate composed a student project brief based on one of his current projects. Design students were invited to begin the concept stages of an Art Deco dolls house project. This supported the curriculum on all academic aspects, including teaching of social and design history, business and marketing studies, as well as practical studio based work, developing a range of 2D concept ideas, 3D sketch models and CAD skills.

The KTP Associate acts as client and is instrumental in devising the project's objectives, communicated in a brief. Students gained knowledge from the live project about the importance of adhering to deadlines and identifying deliverables, teamwork, problem solving through investigation, experimentation and making. Undergraduates thus developed problem-solving skills early on, and used them in future assignments. Further, they could add a project to their portfolios with a brand name, potentially seeing the winning design being further developed for manufacturing and marketable products.

CASE STUDY TWO: GIFTS AND SEASONAL PRODUCTS

The second case study explores two companies producing ranges for outdoor eating – picnic ware and melamine. Both firms, like the first case study, have very close collaborations with manufacturing facilities in China. In both examples they are selling to major buyers in Europe who are increasingly demanding "own-brand" goods and seeking to import products from the Far East either directly or through key suppliers. The KTP allowed them to introduce a design capability into their own organization rather than being led by the buyer or the supplier. In one case, they needed three-dimensional design skills in the UK and in the other visualization and presentation skills. In both cases they were keen to innovate and diversify and change the product fundamentally and not just the surface decoration or finish.

The first company selected an Associate, with the necessary three-dimensional skills, and they worked closely with the University and the company to set up CAD within the firm. The company had already started product design using outside a consultant, but the graduate was able to develop this further using the University's rapid prototyping facilities. Each KTP is focused around a major and subsidiary design project and by undertaking these tasks the process of product development is embedded within the firm. The management is rigorous and the Associate had to report back not only to the company but also to the university and the funding body over a period of two years.

In this case the process of knowledge transfer in the KTP, the supervision and implementation process, was seen as being exceedingly slow to respond by the company. (Fleck, 1999) The market was demanding quicker responses and the suppliers in the Far East were offering new shapes "off the shelf". It was simply easier and cheaper to work closer with manufacturers in China, amending existing forms, rather than developing designs themselves that may requiring expensive tooling.

The ceramic import regulations changed at the same time allowing ceramics to be imported without restrictions. This meant that the instead of buying in melamine tableware they could replace it with ceramic alternatives where the tooling is cheaper. The result as far as the KTP was concerned is that it was ended prematurely and the Associate was forced to find another design post.

There was a "live" project about outdoor eating and the garden, investigating ways of encouraging the consumers to eat out more on patios and terraces, at the same time the garden was being promoted as "the fifth room". But with only the company director visiting to brief and see the resultant presentations, the knowledge exchanged was far less than had been in case study one (Figure 3).

The firm now concentrates on responding to different buyers needs by changing the surface pattern and developing gift products by combining ceramics, melamine with distinctive pattern, for example creating mug and coasters sets as a gift. They have extended their graphic based studio staff and have concentrated on licensing imagery and surface pattern design. For example they have recently won an industry prize for reproducing historic pattern work from a museum for a collection of boxed gift mugs and have sourced images from freelance designers, including graduates from Northampton, for kitchen products requested by major retail groups.

The second firm had invested directly in a plant in China and had direct control of a design development team there as well as technical knowledge of vacuum containers and cool bags. Nevertheless, the firm lacked the necessary design knowledge and visualization skills, to be able to communicate confidently the briefs outlining the buyers' requirements to the factory and make the factory's specific capabilities relevant and exciting to the buyers. Required was market intelligence such as color and design trend information as well as the graphic and modeling skills to be able to visualize and mock-up the products for the UK buyer and the design team in the factory

Figure 3. Surface pattern design. (© 2008, S Felton. Used with permission.)

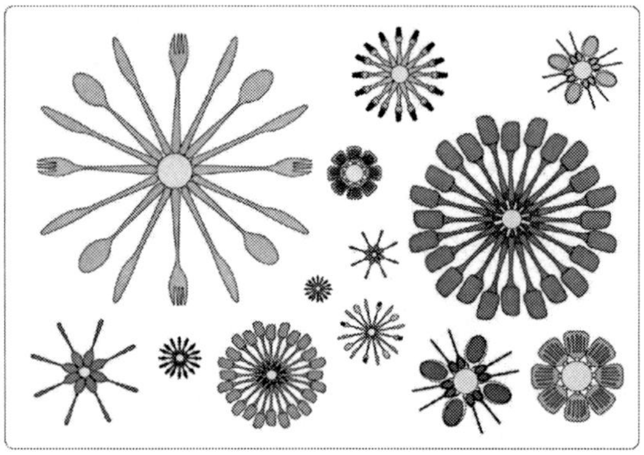

in China. The Associate's key role would be as a communicator and catalyst between the two distinct networks.

Just before an Associate was due to be appointed the UK suffered a very poor summer and the key buyers scaled back their buying. The market became competitive and they decided not to invest in full-time design capabilities in the UK, instead responding to buyer requests by using sales staff's expertise to brief their factory in thc China, relying on following the design expertise of the sales staff and following market trends rather than being innovators, as they had planned to do by having product and graphic design on hand in the UK.

In these cases the University learnt a great deal about training product designers to work in a global market, passing on the knowledge of the manufacturing and the market, but also the key element was being able to visualize and present a distinctive product range to the buyer and the Far East manufacturer. The market is seasonal and so are the trends, turnaround has to be in weeks in order to win the orders and insure timely delivery. For both these firms the KTP was seen as too slow a process. The industry partners learnt to appreciate the expertise they had built up as buyers and sellers in the global market. They had acquired a tacit knowledge of how to source surface pattern and buy-in existing products and present them in the appropriate way to suit each of their retail clients. The ranges are linked together, not primarily by their packaging but by their appropriateness for the season and gift occasion, through form, color and surface design. Product developers for these firms must have a good knowledge of the design trends in the market and knowledge of manufacturing techniques so that they can communicate with all involved in the global market. Knowledgeable designers in China can be key to the network and in some industries such as ceramics they are coming to the UK to train (Figure 4).

In the final case study presented, here too, the KTP introduces a design capability into an organization, in order not to be led, by the retail buyer or the supplier.

CASE STUDY THREE: GAMES

The third case study is based on a KTP with BCE (Distribution) Ltd.; "BCE" are a major player in the toys and entertainment sector, providing tabletop and also garden games to the UK and European markets. The company has a diversified portfolio of products, ranging from air hockey; foldaway pool tables to fully professional slate bed snooker tables. BCE traditionally sourced its equipment through buyers who sourced products from large international trade fairs hosted in China, reflecting Chinese manufactured goods.

The challenge with this KTP was to appoint an Associate who could provide an in-house design capability, the strategy being twofold:

1. To eliminate current design costs in terms of visualization of ideas, by not having to outsource 2D graphical / printing origination,
2. To develop a new product design methodology, and be proactive in generating new 3D ideas based on research, including feedback obtained from the company's client base.

Figure 4. Prototpe production at the university. (© 2008, The University of Northampton. Used with permission.)

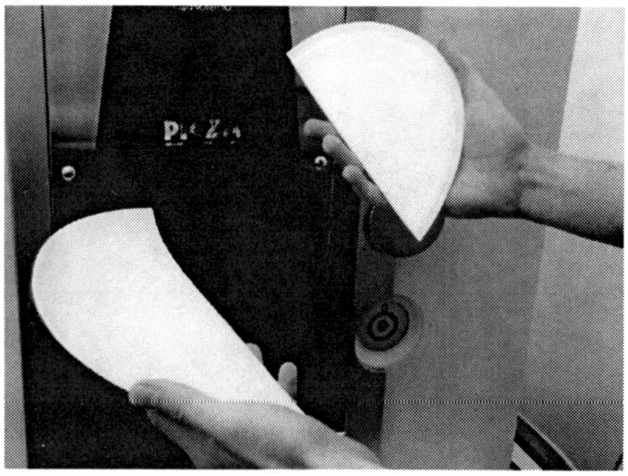

BCE's UK client base include the retailer/department stores John Lewis Ltd, Argos Ltd and several high street sports stores, each client desires a bespoke design that reflects its unique brand identity, hence, the job of the Associate was to provide a variation from the original generic design that meets the aspirations of the purchaser.

The recruitment of the Associate involved an extensive short-listing process, conducted by the University, based on application and portfolio assessment, leading to interviews, the final selection being made by the company. The company had identified a person specification, with a skill set that crossed between a graphic designer and a product designer, so a hybrid 2D/3D designer was being sought (Figure 5).

The first time through the selection process ended unexpectedly, when the strongest candidate, having been offered the position, declined to accept it, he had been offered a more lucrative opportunity in a design house in Australia. The second time through, however, was more awkward as the Associate was originally intended to work out of the company office situated relatively close to the University, however, with the recent changes in the financial climate an economic decision was made to close down that site and to

Figure 5. Pool billard table. (© 2008, BCE Ltd. Used with permission.)

locate the designer in BCE's main distribution centre, which is 120 miles from the University campus. All the three partners within the KTP scheme discussed the impact of the change in location, and an operational strategy was developed which could allow the KTP to be managed despite all the changes identified.

The Associate having joined the company was immediately presented with a large workload. BCE operationally had peaks and troughs, and the Associate had taken up post at a very busy time in the calendar. The nature of the work was largely graphical in nature, and whilst he had some understanding of Adobe Creative Suite TM he had little understanding of the requirements of print. It was therefore very clear that a fast track bespoke Adobe training course was essential to help productivity. The course was identified and completed, just in time to migrate the work from paper / catalogue to exhibition vinyl banner design and also 3D structural packaging. The university had already been very successful in KTP management and delivery within a product design environment, this however, posed issues outside the normal 3D disciplines and required additional specialist technical advice from the graphic communication lecturers within the division. In the UK educational system the training of product and graphic designers is split but the KTP experience is encouraging closer collaboration and may in turn lead to more shared modules and collaborative projects.

In keeping with previous KTP, a decision to purchase a laptop capable of running multiple operating systems was taken, and this proved to be most useful in allowing files to be transferred natively between designer and print house. The minor design project, a key ingredient within any KTP was identified to be an in house production of a monthly special offer flyer, which would be distributed to the client data base with a view to increase sales. In order to produce these flyers, an archive of all the company's product images had to be created, with an associated process / methodology. The by-product was therefore the creation of a visual database as a spin-off from the minor design project.

BCE also has a presence at a number of highly important trade fairs, held within Europe, it fell as part of the responsibility of the Associate to design the interior space and exhibition banners for these shows, this identified yet again the need for further specialist graphical training to be provided, as the required large format prints presented several unexpected complications in terms of their production.

At the time of writing, this KTP has completed six months of a twenty-four month cycle, it is interesting to note the comments made by the Managing Director at the last local management committee (LMC) meeting, that the reduction in costs by having an in-house designer has amounted to approximately £80,000 to date, but this does not consider any increase in profit due to new design generation (Figure 6).

Running concurrently with all the 2D design origination has been the development of new 3D games tables, these have been more evolution of existing designs rather than starting from a blank sheet of paper. The tables are then prototyped in China and returned to BCE for approval and modification, they are also used to inform BCE clients about new product ranges, which hope-

Figure 6. Board game. (© 2009, BCE Ltd. Used with permission.)

fully will receive orders for the next season. The role of the Associate in this context is classical product design, typically modification of form and refinement of mechanisms. The success of this aspect of new product design is difficult to quantify, so early in the design process, but should improve the bottom line in the same way that taking costs out of the production of 2D flatwork have already evidently done. As with one of the other KTP, it was easier to start by adapting designs already in production in the Far East.

The development of new games tables has also inspired one of the final year product design undergraduate students, who having met with the BCE Associate, seen the work undertaken to date, has focused on how to improve family engagement in a modern society, through families playing and eating together. This windfall learning opportunity is consistent with KTP enhancing the academic curriculum, usually though through student project or case studies. In this particular KTP the University has learned that classical barriers between different design disciplines: graphics, illustration, product design and interior design need to be reviewed, as highlighted by a relatively small company which has only a single designer, and limited skill sets. The challenge is to provide a curriculum which provides breadth but not at the cost of depth.

The use of 3D Solid modeling has had less emphasis within the first six months of operation than was expected, but due to operational necessity graphic, packaging and exhibition design has been key to the success of the work undertaken. Due to the large distance between company and university, the regular meetings have had to be carefully structured, typically less frequently held than previous KTP, but to compensate also lasting longer. In addition, the use of video conferencing once or twice a week to supplement the face-to-face meetings has proved invaluable.

FUTURE TRENDS

After ten years experience of using the KTP and Teaching Company Schemes to encourage Product design and development (PDD) what trends have we seen emerging and what do they mean for the individual partners in the process.

For the Associate

For the Associates the opportunity to get a post with real responsibility early in a career is vital, will continue to be valuable especially in a competitive market. The opportunity to get additional academic qualifications at the same time will provide a competitive advantage. The University is there to be called on to provide the graduate with training and design support, they may have lacked at undergraduate level. The routes open to postgraduate research (MPhil) and management studies leading to a MBA qualification. In these case studies, many of the product or industrial design students required support with graphics knowledge. There were also opportunities to set up networks with other Associates, through the funding body and this has resulted in new on-line forums and networks disseminating knowledge.

For the University

For the University partners, the process of supervising such cross-cultural KTP projects is to encourage them to develop new courses and teaching approaches, to help prepare design undergraduates to operate in the new global market place. The University of Northampton, for example, has started specialist surface pattern courses. There are a series of linked courses with marketing, and there is continuing emphasis on visualization throughout the undergraduate product courses, through ongoing sketchbook projects to the teaching of specialist graphics packages. At the end of the twentieth century, design and marketing firms became more multi-disciplinary and developed integrated solutions. The teaching experience is growing to compliment these trends.

The knowledge gained feeds directly into the academic research network in Northampton. Here, the focus is on the future use of leather based fibrous materials in different cultures and the role of the diasporas in design history and the international markets for illustration. As well as developing new teaching approaches, there is a new emphasis on live projects with the Associates and transferring the knowledge they have gained to the next generation. Many of the universities have been primarily teaching institutions. The new emphasis on research may mean changes in the balance between teaching and research activities. There are plans to extend KTP into the Further Education colleges, where the Associates could offer specialist knowledge that is valuable and not found in the University sector. So the trend is to tap into knowledge at all levels of the education system.

For the Government

For the funding body, the government in effect, the KTP has been a success. The formula or pattern has been extended to other disciplines and areas of academic research. In some cases, lessons have been learned: the projects have been shortened and they were then called Knowledge Catalysts. As they last for a few months, they tend to need less of an investment in time and money by the business or community organization. Bringing the University's expertise for a smaller specific project, which is outside their internal remit or expertise, would allow the institutions to offer a new service.

In a global market place where a majority of goods continues to be manufactured overseas, the KTP is seen as a key tool in the growth of the creative industries. Valuable export earnings continue to be gained through intellectual property, international licensing agreements and technology transfer.

The KTP advisors, linked to the funding bodies and advising on the applications are gradually building up knowledge base of their own. They too are sharing knowledge between themselves and networks of expertise are developing. The trend is towards professionalization of knowledge transfer practitioners.

For Knowledge Seeking Firms

For the partner firm, design is an area of increasing importance to industrial competitiveness. In a global market, the competitive advantage of companies lies not only on the mastering of existing processes and methodologies and developing their own expertise, but most of all on their ability to pursue different avenues, with added value to their customers. We have seen that the conditions need to be right for them to make the investment and rely on internal product development. Most firms in the markets described in these case studies need to change their offer at least twice a year and the product cycles are quite quick. In all cases, a great deal has been learnt by simply "trying" to implement an in-house design facility and the two year, major and minor project model does not tend to suit all. There is evidence of new shorter schemes that are more appropriate to smaller organizations.

In all cases, they came to appreciate the tacit knowledge they had built through their existing trade relationships with the Far East. They did not need to learn, as far as our case studies are concerned, how to operate in a global market and it was this knowledge that has been most valuable to the other partners. They were keen to improve and add value through design to their partners in China and the UK. It maybe that some of the knowledge about cross cultural issues does not reside within the Universities in the UK but needs to be gained through wider international collaboration at an academic level. The University team at Northampton is seeking collaborations through the European Union to undertake cross-cultural design research, particularly linked to visual imagery and communication. Sue Ryder is a charity; so, this KTP is likely to be used as an example to encourage other types of organizations, such as other non-profit making institutions to apply.

For the Global Market

Knowledge transfers that result in new product innovation can only be achieved with an up-to-date knowledge of the field. Technology in the form of CAD, graphics software and the Internet can make the exchanges easier and swifter but also with the inherent risks associated to innovative product development processes and methodologies. The KTP model will have to adapt to the current economic circumstances. Will firms be designing more for local and regional markets? (Iwabuchi, 1988) A UK based consultancy operating in these markets report that they are being asked to supply surface design for tableware to be sold in China but produced in Indonesia and also an increased demand for souvenirs, both product types rely on ability to learn and apply cultural knowledge. The pattern design can be transferred over the web, but the products are moved shorter distances or carried by individuals rather than containers.

CONCLUSION

KTP in Product Design has provided opportunities to learn for all those involved. Some may argue that as for embedding or implementing a product design and development capability, the challenges are now reduced. As far as the academics are concerned, they tend to repeat the same exercise with a limited amount that can be learnt. These case studies all have similarities: all are small to medium sized companies, all trade with China, and in each case the vital role played by the two-dimensional graphic skills was highlighted. They also have different challenges; a charity seeking not to buy standard products but to develop its own intellectual property is different from a firm with design skills in China seeking to use British design skills and cultural knowledge to help them import and sell ranges in the UK.

It is clear from all the case studies that there is still a great deal to be learnt about other markets and cultures, and design education is called to develop the knowledge and skills at undergraduate level. The KTP model we have seen extending to other areas of academia. It maybe that more transfers of knowledge need to occur between design and art based courses with other departments, such as history, languages, and media studies. Departments such as law, business, social anthropology and geography are already building up cross-cultural knowledge valuable to the designer. At Northampton, the product course started with a European focus, teaching a foreign language and encouraging students to look to markets for their skills within the European Union. It may be that the expertise will be called on in a localized global market. (Iwabuchi, 1988)

Learning by doing, by trying, by participating in networks by exchanging ideas and information are all key strengths of the KTP approach. As a department we have started to encourage students to analyze the process in which they are involved in order to take note of their decision-making and where they get the knowledge. This is done not just through formal referencing but also through visual diaries showing their decision/product development trail through the networks. It is seen as important to know how transfers of knowledge are occurring, through seeing, doing and networking.

A key factor in the successful KTP projects appears to be the ability to be flexible. Like the spider constructing a web, they may have to start again should wind or a successful catch alter the web. The networks of knowledge are dynamic ones that can be built up and strengthened over time. The case studies also indicate that they need to be allowed to fail and through that process knowledge is gained. (Fleck, 1999) KTP may not be about adapting and adding value through design of goods imported from China; it is rather about selling the expertise to help them design for their home market whose consumers are seeking innovation.

It is our experiences that design and cultural knowledge are also transferred in the knowledge transfer process, in a variety of different ways and directions and through very different types of networks. The KTP have proved a valuable learning experience for all those involved. For the university, it will continue to change design and design management teaching as it to meet the needs of the changing global market places.

REFERENCES

Atkins, P., & Bowler, I. (2001). *Food in Society - Economy, Culture, Geography*. London: Arnold Headline.

Betts, S., Schaber, F., & Turner, R. (2007). *From Ivory Tower to Fantasy Castle: A Design Case Study of Industrial Collaboration.* International Conference on Engineering Education ICEE. Coimbra: The University of Coimbra.

Blackler, F. (1995). Knowledge, Knowledge Work and Organizations: An Overview and Interpretation. *Organization Studies*, *16*(6), 1021–1046. doi:10.1177/017084069501600605

Blackwell, A. (2008). Design Research and Academic Principles. *Design Research Quarterly*, *3/4*, 5–6.

Braziel, J. E., & Mannur, A. (2003). *Theorizing Diaspora.* Oxford, UK: Blackwell Publishing.

Centre for Research in the Arts. (n.d.). Social Sciences and Humanities, Cambridge *CRASSH2009*, Retrieved from http://www.crassh.cam.ac.uk/events/783

Collins, H. M. (1996). Embedded or Embodied: Hubert Dreyfus's What Computers Still Can't Do. *Artificial Intelligence*, *80*(1), 99–117. doi:10.1016/0004-3702(96)00083-6

Cox, G. (2005). *The Cox Review of Creativity in Business*. London: HM Treasury.

Davenport, T. H., & Prusak, L. (1998). *Working Knowledge: How Organizations Manage What They Know*. Boston: Harvard Business School.

Dixon, N. (2000). *Common Knowledge: How Companies Thrive by Sharing what They Know*. Boston: Harvard Business School Press.

Fleck, J. (1999). Learning by trying: the implementation of configurational technology. In MacKenzie, D., & Wajcman, J. (Eds.), *The Social Shaping of Technology* (2nd ed.). Buckingham, UK: Open University Press.

Griffith, S. (2007, May 15-17). Designing Undergraduate and Postgraduate Design Management and Practice Courses, Design—Challenge. In *Proceedings of the Design Management Conference.*

Griffith, S., & Bamford, R. (2007, 9-12). Embedding the Principles of Responsible Design into Student Practice from 1st year, Connected 2007. In *Proceedings of International Conference on Design Education*. Sidney, Australia: University of New South Wales.

Highmore, B. (2009). *The Design Culture Reader*. London: Routledge.

Hofstede, G. (2001). *Culture's Consequences: Comparing Values, Behaviors, Institutions and Organizations across Nations*. Thousand Oaks, CA: Sage.

Iwabuchi, K. (1988). Marketing "Japan": Japanese Cultural Presence under a Global Gaze. In Highmore, B. (Ed.), *The Design Culture Reader*. London: Routledge.

Lambert Review of Business-University Collaboration. (2003). *Lambert Review of Business-University Collaboration*. London: HM Treasury. Retrieved April 13, 2009, from http://www.hm-treasury.gov.uk/d/lambert_review_final_450.pdf

Law, J., & Hassard, J. (1999). *Actor Network Theory and after*. Oxford, UK: Blackwell Publishers/The Sociological Review.

Machlup, F. (1984). *Knowledge: Its Creation, Distribution and Economic Significance (3 Volumes)*. Princeton, NJ: Princeton University Press.

MacKenzie, D., & Wajcman, J. (Eds.). (1999). *The Social Shaping of Technology* (2nd ed.). Buckingham, UK: Open University Press.

Mauss, M. (1954). *The Gift Forms and Functions of Exchange in Archaic Societies*. London: Routledge & Kegan Paul.

Nonaka, I., & Takeuchi, H. (1995). *The Knowledge Creating Company*. New York: The Oxford University Press.

Schaber, F., & Thomas, V. (2007). Knowledge Transfer: Industry, Academia and the Global Gift Market. *Design Management Conference*. Seoul: Korea Institute of Design Promotion.

Schaber, F., & Thomas, V. (2008). Knowledge Transfer: Industry, Academia and the Global Gift Market. *Design Management Journal.*, *1*(3), 1–14.

Smith. N. (2009). From Mortarboard to Drawingboards. *The Institution of Engineering and Technology Knowledge Network*. Retrieved from http://kn.theiet.org/magazine/issues/0808/boards.cfm

Teaching Company Scheme. (2009). *History*. Retrieved from http://www.ktponline.org.uk/strategy/history.aspx

The University of Northampton. (2008). *News Release 6 May 2008*. Retrieved from http://www.northampton.ac.uk/news/release/?id=UON8199

The University of Northampton Sue Ryder Care Final Report. (2008). *Sue Ryder Care Final Report* (Unpublished). Northampton, UK: The University of Northampton.

Thomas, V. (1984). *Gifts: The Designer's Role in the Commercialization of Gift Exchange* (Unpublished). London: Royal College of Art/Victoria & Albert Museum.

Wilkinson, M., & Sale, R. (2000). *Pressures for Change in the Education of Product Designers. Design Pedagogy: Basic Design and Academic Experiences*. Birmingham, UK: University of Birmingham.

KEY TERMS AND DEFINITIONS

Added Value: Refers in this study to the additional value design brings of a product over the cost of materials and production.

Emotional Engagement: By the user is considered as important as having them enthusiastic about the functional benefits of a product: "*It is becoming increasingly apparent that a changing social climate is now demanding brands be assessed by their integrity and genuine commitment to communicating honestly about who they are, what they do, and what they represent.*" (Simmons, 2007)

Cross-Cultural Design: Is growing in importance because of the global market place. Products are designed in one country, manufactured in another and sold worldwide. Designers have to have the skills to make them understood and have a wide range of cultural knowledge as well as design knowledge.

Knowledge Transfer Partnerships: Is a UK program where academics provide business solutions while putting research into practice. A recent graduate, called Associate, is placed within the knowledge seeking company to research, plan and implement new methodologies, supported throughout the project time by an academic and his university, the knowledge base.

Live Projects: Are student assignments that are not textbook exercises, but projects linked to real clients from business or the community. The results may or not go into production and the students gain valuable experience in dealing with real design problems, access to production processes and feedback from marketing teams and sales staff.

Rapid Prototyping: Allows for computer-generated geometries of models and prototypes, with realization through a 3D printer, such as a fused deposition modeler, producing accurate forms in a polymer.

Trends: A way of describing how things (designs, fashions, colors and ideas) tend to move together. They are used to predict changes in the society and particularly markets. Trends can be set, manipulated and followed. They are used by designers to lead us, by retailers to be back up their buying decisions and could be said to narrow the consumers' choice at point of sale.

Chapter 27
Product Form Evolution

Andrew Muir Wood
University of Cambridge, UK

James Moultrie
University of Cambridge, UK

Claudia Eckert
Open University, UK

ABSTRACT

Companies are coming round to the idea that function and form are complimentary factors in improving the user's experience of a product and competing in today's saturated consumer goods markets. However, consumer perception of form is constantly changing, and this manifests itself in the evolving forms of the products that they adopt. From clothes to cameras to cars, change in form is inevitable, and design teams must account for these trends in their product design and development strategies. Through literature, semi-structured interviews with design and trend practitioners, and an archival case study of mobile phone evolution, the authors have developed theories about the continuities that occur in product forms over time, and the forces that can disrupt this behaviour. They then go on to suggest how this view of form as evolving trajectories can benefit future product design strategies.

INTRODUCTION

In the design and development of functional artefacts, non-functional aspects such as aesthetics and social symbolism are key differentiators for success in increasingly saturated markets (Bloch, 1995). But societies and markets never stand still, they are constantly changing. Many consumer product categories are cycling faster and faster each year. For example, it has been demonstrated that the development cycles and market life of high tech consumer goods (such as video recorders and DVD players) has reduced rapidly in recent years (Minderhoud & Fraser, 2005). For a product or brand to succeed in this environment, it is vital that managers and product-planners are aware of the dynamic context in which their product or brand is perceived and consumed. The appearance or *form* of a product is what links its technology and functional content with its context of use and the emotions of the consumer. Thus the challenge is to ensure that a product's form is attractive and

DOI: 10.4018/978-1-61520-617-9.ch027

relevant to its target consumer segment at its time of release, while at the same time following and evolving the strategy and values of the brand.

The form of a product can objectively be described as a combination of characteristics such as colour, geometry, material, texture, details, dimensions and graphics (Crilly, Moultrie, & Clarkson, 2004). When a collection of products share similar forms or characteristic components of form, this is called a *style*. For example, a minimalist style is not necessarily typified by a single object, but rather by a group of objects that share a similar lack of ornamentation (Cappetta, Cillo, & Ponti, 2006). A consumer's cognitive response when viewing a product has several layers, Norman (2004) proposes that we process product form on three levels: *aesthetic*, *semantic*, and *symbolic*. In brief, aesthetic impression is the perception of how attractive or unattractive we find the object; semantic interpretation is the understanding of a product's function and practical qualities; and symbolic association is the social significance of owning or using the product (Crilly, Moultrie & Clarkson, 2004).

Products don't just have a functional lifespan; they also have a finite period during which their form is attractive and relevant to its context, which can often be shorter. But some products have a longer duration of aesthetic appeal than others. According to Coates (2003), a *classic* is a product which has a *seminal* form that "establishes a new visual standard for a class of products and seeds a new trend that competitors are obliged to follow" (p. 151). When this is combined with a product with *epochal* innovation that "fills a practical need or solves a problem not yet met or does so in a markedly better way than previous products" (p. 151), then competitors have no choice other than to imitate and thus reinforce the authenticity of the original product and brand.

Not all new products can be classics, and imitation is one potential strategy for responsive companies who quickly identify the potential of riding the wave of a competitor's success. This strategy has a reduced risk associated with it, but also a finite duration of impact as the innovation or form becomes more widespread. Even in highly functional scientific equipment, form can enable products to outsell technically superior competitors and command a higher price tag (Yamamoto & Lambert, 1994). At the other end of the spectrum, the form of a product can make it date quickly – which can actually be desirable to manufacturers in rapidly cycling markets such as mobile phones or consumer electronics, where visual obsolescence might encourage consumers to replace their devices more regularly. This chapter will explore the dynamic nature of form, and provide ideas and insights that will help design managers and product planners to understand how this view of form can enhance their product strategies.

Objectives

The key objectives of this chapter are as follows:

- To present the theoretical foundations of taste change in consumers and society that lead to change and trends in product form.
- To display evidence from previous studies and the authors' own research that there are identifiable patterns of behaviour in product form evolution.
- To help design managers and product planners to identify where their products fit into an evolving context and take advantage of this position.
- To challenge companies to develop products that respond effectively to trends and cycles, or even to create their own new ones.

Methods/Information Sources

The background for this chapter has come from an extensive literature review of books, journals, magazines and online newsgroups and blogs in a diverse range of disciplines, from consumer psychology to technology foresight. Further insight comes from semi-structured interviews with 17 designers, product managers, trend researchers and strategists who have a specific role in envisioning how product forms might evolve. The interviews were conducted between 2007-09 in 3 continents and a variety of product categories; they were transcribed and coded in order to understand the usage of trends across products and companies and to compare real commercial practice with existing theory in literature (Muir Wood, Moultrie, & Eckert, 2008).

Contextual examples will be provided from products that have ultimately succeeded or failed to resonate with the trends of the times, thus elaborating on the challenges and hazards of particular aspects of form evolution. Further detail will be taken from an in-depth study on mobile phone design. The study explores the evolution of mobile phones as they have become increasingly accepted as fashionable products, in contrast to when they were first introduced, targeted primarily at the business sector. The form of mobile phones has evolved substantially, based in part on technological, societal, cultural and material factors. Archival content analysis of gadget magazines made it possible to track mobile phone forms over a 13 year time period. Magazines in general provide an interesting and reliable source of data – products are selected and articles written by industry experts with a level of neutrality and consistency that earns them the respect and loyalty of their readership (Cappetta, Cillo, & Ponti, 2006; Kroeber, 1919; Lowe & Lowe, 1985; Robinson, 1975). Gadget magazines are ahead of the curve because they tend to highlight premium or flagship products that feature new forms, technology and functionality which will eventually trickle down into the mainstream categories.

WHY CAN'T PRODUCT FORMS STAND STILL?

This section will provide some theoretical background to explain why taking a dynamic view of product form is necessary. First to be examined are the *internal* cognitive reasons why people like to consume (and design) new forms and the *external* factors that influence their preferences. Next, the evolution of a product category will be investigated to see how and when form contributes most to differentiation. The final subsection will provide a literature review of the current practice of brands, designers and trend researchers and how they respond to these changing tastes and demand.

Novelty Never Lasts

Once a society has matured to a level of economic stability, patterns of consumption change from fulfilling physiological and security based needs toward the indulgence of *taste*: "the acts and arts of appreciation, discrimination, critical judgement and the pleasurable savouring of man-made objects" (Lloyd Jones, 1991). Over the years philosophers, designers and researchers have tried to establish what it is that makes a form that is universally attractive to all tastes. Objective rules of proportion such as the Golden Section were widely used in historical art and architecture. In the 20th century the design principles behind the highly rational products of the Bauhaus school reflected the rules of Gestalt psychologists, which emphasise the importance of symmetry, regularity and harmony in creating attractive designs (Crilly, Moultrie, & Clarkson, 2004). However, it is now widely accepted that the visual appeal of an artefact is also influenced by the social, cultural and historical experiences of a consumer or social group (Bloch, 1995; Crozier, 1994). What's more, consumers experience the most pleasure from an object that offers them a balance of boredom and contrast (Berlyne, in Lloyd Jones, 1991). In other words, we are attracted by new or unusual objects, but pleasurable products are not "frighteningly unfamiliar", which is the basis of the MAYA principle: "most advanced yet acceptable", proposed by the great American industrial designer Raymond Loewy (Schifferstein & Hekkert, 2007).

This explains the internal desire for novelty, which of course can never be quenched as new product forms eventually become familiar themselves (Cappetta, Cillo, & Ponti, 2006). The level of novelty that is desirable varies from consumer to consumer and relates to their individual *innovativeness* (Midgley & Dowling, 1978); or their propensity to adopt an innovation early in its *diffusion*. Diffusion is the process by which new products or ideas dissipate across consumer groups and society (Rogers, 2003), and communication between people is key to this. The key catalysts of the diffusion process are the *innovators* or *opinion leaders*, who can exist in all areas of society – from within peer groups or specific *subcultures* to international celebrities, sports stars or other elite thought leaders in wider society (Sproles & Burns, 1994). Fashion conscious consumers aim to differentiate themselves from the mainstream by drawing inspiration from the styles of their opinion leaders.(Cappetta, Cillo, & Ponti, 2006). The social component of taste is perhaps most telling in apparel fashion, but imitation and differentiation also occur in the adoption of other types of product. For example, cars and watches might be purchased as a symbolic demonstration of wealth or aspiration to be a member of a particular social group (Baxter, 1995).

In addition to social forces, there are other external factors that cause changes in the mood, symbolism and cultural imagery of a social group. Coates (2003) describes this as *zeitgeist* or "the spirit of the times" and explains that major events, iconic products, and cultural activity can shape the tastes and hence the product forms of an era. He gives the example of car colours during the Great Depression of the 1930s, when consumers preferred black and other sombre colours, while bright pastels were desired when optimism returned in the 1950s. By definition the zeitgeist is constantly shifting, and hence products and styles must evolve to stay relevant to their context of consumption. For a product to successfully sustain attraction, it needs to be durable in both function and form. This is particularly difficult to achieve in an evolving product *class* or *category*.

Synthesising this literature understanding, Figure 1 is the authors' visualisation of the external factors acting on consumers to drive taste change. The three sections correspond to the types of stimulus socio-cultural *influences*; the consumption behaviour of *influencers* – prominent individuals and groups across society; and the artefacts that provide the consumer with comparison and *reference*. Starting from the inside ring, which is specific to the consumer's immediate *peers* and the product's *class*, each ring represents a layer of factors with increasing duration and scope of impact on products and social groups. In the consumption behaviour and artefact sections, the groups of people or objects are aligned with an

Figure 1. The layers of external factors that shape tastes and trends

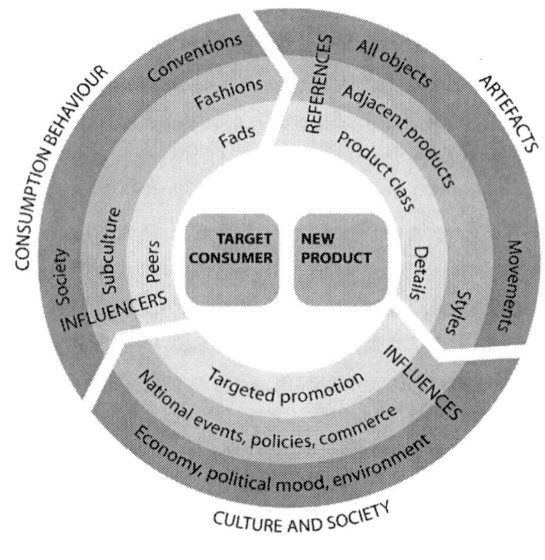

Table 1. Stages in category maturity: bikes (from Roy 1994) and mobile phones

	Bicycles	*Mobile Phones*
Exploration	**1818** Hobby horse initiated diverse innovation which included penny farthings and 3-wheelers	**1983** Motorola DynaTAC – mobile phone category established, early phones resemble cordless fixed line phones or CB radios
Consolidation	**1890** Humber safety – form settles around chain driven, 2 same-sized wheels	**1992** Nokia 1011 – mass produced GSM phones become feasible
Mature	**1900**s Incremental manufacturing and material improvement centred around a standard form	**1995** Price reduction, some segmentation, personalisation, standard forms: candy bar and clamshell
Re-innovation	**1962** Moulton bicycle catalyses functional divergence: fashion, lifestyle, racing, street, mountain, commuter and other categories emerge. Advanced design and materials continues	**2000** Divergent form factors, price groupings, functional classes **2005** Homogenising with other product categories: MP3, digital cameras, PDA, GPS.
Decline/ Displacement	**2000**s Powered personal mobility devices are the nearest competitor but show no sign of displacing bicycles as transportation or a leisure pursuit.	Mobile functionality has the potential to displace the above categories **2007** Apple iPhone and other touch screen smart phones set to trigger convergence and exploration in software and interface design

equivalent trend description. For example, products with different functions can share the same *style* which has more longevity than *details* found within a single product class. The directionality of each section signifies continuous change and interrelationship between types of factors.

Category Maturity

Product categories can't necessarily be compared directly, because they are often in different stages of maturity. During each stage of the life of a product class, different factors are in flux, or are seen as a key selling point of the product. Also, it takes time for both designers and consumers to understand the functionality of a completely new product category before it can develop an identity of its own (Karjalainen, 2006).

Roy (1994, pp. 364-366) proposes a 5-phase evolutionary cycle of a successful, innovative product category. The cycle starts with *exploration* – when prototypes and experimental products emerge. Next is *consolidation* – when a small selection of dominant designs begins to prevail. This is followed by the *mature* stage, when the production processes are refined and the product becomes widely available to consumers. At this initial point of maturity, differentiation becomes important: variety emerges through price points, surface aesthetics and minor functional improvements. However, it is only when (disruptive) *re-innovation* occurs that new product families are created to capture new markets. A final stage for the product category is its *decline*, where its functionality is displaced by a new concept or technology. Table 1 gives two examples of products following this evolution – bicycles from Roy's study and mobile phones from the authors' research.

Current Application of Form in Design and Development Practice in Literature

This study has been greatly informed by existing literature in a multitude of adjacent areas. However, little direct academic coverage has been give to the application of form evolution in the context of product design and development. In this section we will briefly review some of the related academic work which this chapter aims to compliment and build upon.

Consumer research is a well documented source of insight into the practical and social needs of a target market segment, but Crilly and Clarkson (2006) find that in practice, attitudes of designers are very mixed with respect to using consumer research to inform or justify the appearance of their product concepts. In design and marketing literature, systematic approaches such as Cagan and Vogel's (2001) SET factors (social, economic, technology) are used to establish and fulfil product opportunities. Innovation literature details companies using foresight techniques such as scenario planning, economic forecasting, and product scanning also to predict future strategic opportunities, but limited reference is made to product form (Andriopoulos & Gotsi, 2006; Costanzo, 2004). Rutter and Agne (1998) explored the possibility of a Darwinian evolutionary comparison in design, but could find no equivalent to natural selection amongst products.

In the apparel fashion industry Cillo and Verona (2008) describe two (very successful) types of "style searching": *market driven* and *designer driven*. The former approach involves design teams constantly adapting to the market and positioning the brand relative to its competitors, while researching emerging styles abroad and from the street (e.g. Diesel clothing). The latter relies on a signature designer who is the primary interface between the brand and the market (e.g. Escada). Eckert and Stacey (2000) investigated sources of inspiration for knitwear designers and found that new products must fit into an evolving envelope of acceptable design – echoing the MAYA principle introduced earlier.

Whether tacit, or explicit, designers have a "contextual radar" that allows them to receive socio-cultural information and translate it into product strategies. In fashion, authors such as Blumer (1969) noticed the apparent convergence of brands and designers in the clothing lines that they released each season.

The designers were attuned to an impressive degree to modern developments and were seeking to capture and express in dress design the spirit of such developments. I think that this explains why the dress designers – again a competitive and secretive group, working apart from each other in a large number of different fashion houses – create independently of each other such remarkably similar designs. (Blumer, 1969, pp. 279-280)

In their fashion study, Eckert and Stacey describe the use of *mood boards*: a collection of images that represent a particular fashion or style. Similar approaches are used in product design – an established user group can be visually explored through *lifestyle* or *trend boards* which can help to guide the development of aesthetic principles for the product (Baxter, 1995; Karjalainen, 2006). But these techniques produce static snapshots of context, with no historical perspective or future predictions. No methods have been identified in the literature that link these snapshots together in order to create an evolving view of products or context over time.

This is not an exhaustive review of literature or commercial practice, but it demonstrates that in terms of literature understanding, the fashion industry has prescribed methods of systematically identifying, characterising and translating trends into product forms. Product design has a good prescriptive coverage of practice in textbooks, papers and case studies, yet the authors have not identified any source that systematically examines the trajectories of product form trends and relates them to the design and development of real products. From articles, interviews and personal experience, we are aware that these practices are occurring in industry, both tacitly and through structured methodologies, but the evidence is anecdotal, there is no empirical theory behind it. We are not denying the importance of designer's individual tastes, training, cultural background, ingenuity and art, but feel that better understanding of form evolution will benefit the product design and development process.

WHERE ARE PRODUCT FORMS GOING?

As we have established in the previous sections, product forms must change in order to both inspire and respond to new directions of taste. But in what tangible ways does form change? Are the changes merely in superficial decoration, or are there underlying patterns or trends that emerge? Do different product types follow different paths? In order to answer these questions, it makes sense to reverse the situation. We will identify the actual trajectories that occur in aspects of product form, and then relate them back to the contextual factors surrounding them.

This reverse approach is inspired by a study conducted almost 100 years ago by an influential American anthropologist named Alfred Kroeber. He conducted a content analysis of dimensional changes in women's dress, from pictures in Parisian fashion journals between 1844 and 1919. At first he tried to follow surface decoration and accessories such as collars or ruffles, but these changed too much or disappeared completely, preventing any extended comparison. He realised a more abstract approach was necessary and selected 8 measurements to be taken from solely eveningwear pictures. The measurements, such as waist and skirt diameters, formed the basis of an underlying form that existed throughout the entire duration of the study. The results from his study were visualised as dimensionless ratios – such as that of skirt width at its hem to the height of its wearer (from tip of toe to mouth) in Figure 2. This and the other figures in his paper (Kroeber, 1919) showed a remarkable continuity of proportion, in long cycles that only became apparent over the long time period being studied. For Kroeber, the results were evidence of trends in collective social behaviour whose "magnitude dwarfs the influence which any individual can possibly have exerted in an alteration of costume" (Kroeber, 1919, p. 260).

Figure 2. Ratio of skirt hem width to height of wearer. (Adapted from Kroeber, 1919)

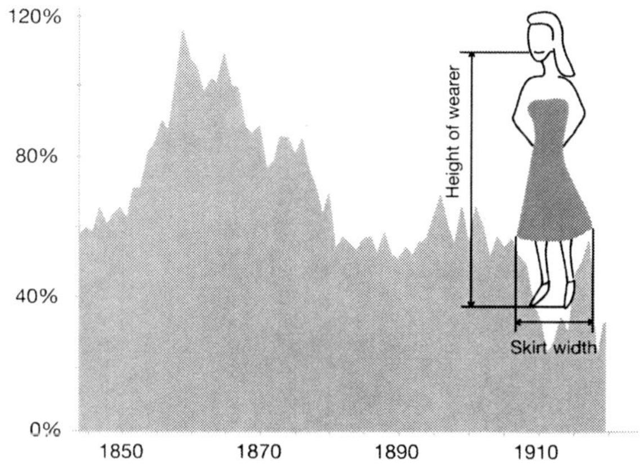

Following a similar approach, the authors of this chapter conducted a case study of mobile phones in gadget magazines between 1996 and 2009. Kroeber's concept of tracking the underlying dimensions and characteristics of the devices was adopted. The study aimed to find whether it is possible that consumer products follow equivalent patterns of behaviour to women's dress? What are the limiting factors of this behaviour? What does it take to disrupt the trends that may occur? This chapter will use examples from this phone study and prior studies of other products to examine form behaviour over time.

Continuity and Cycles in Form

The long term form trends that Kroeber identified were later to be described in fashion literature as historical or cultural *continuity*. Continuity in fashion theory holds that successful new styles are rarely revolutionary, instead they are an "evolutionary outgrowth and elaboration of previously existing fashions" (Sproles & Burns, 1994, p. 32). Lowe and Lowe (1985) argue that "most fashion flops have flaunted the principle of cultural continuity" (p. 205), and also that there

Figure 3. Mobile phone colour trends between 1996 and early 2009

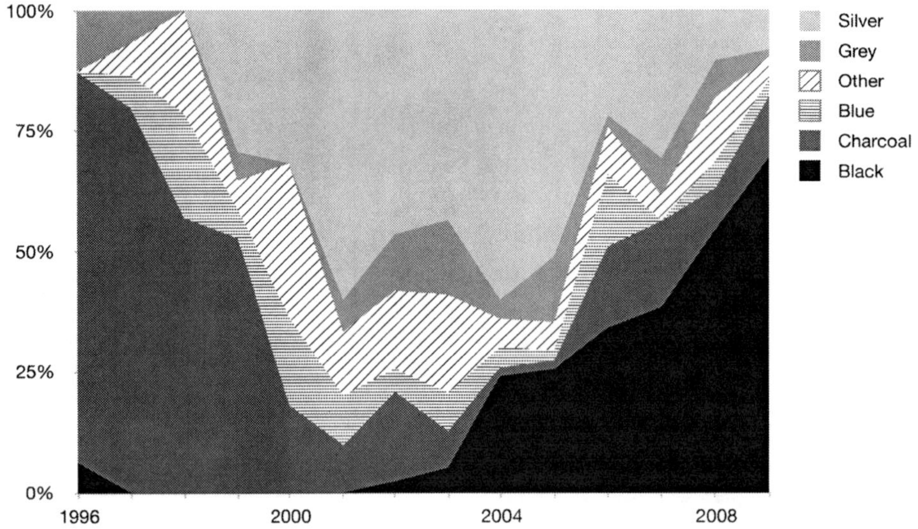

are "inherent brakes on the rate at which fashion changes may occur". This understanding makes perfect sense when aligned with the consumer's internal equilibrium between a desire for familiarity with past product forms and the need for novelty through novel, innovative forms (Schifferstein & Hekkert, 2007).

Several authors have identified a decrease in the period of design cycles. Lowe and Lowe (1985) claim that Kroeber and Carman's 45 year fashion cycle theory is no longer valid, and that since the 1960s the speed of communication and manufacture has enabled more rapid change in the underlying forms of apparel. Many years earlier, architectural historian Pevsner predicted an "unhaltable trend to constantly accelerating change" (in Jencks, 1971 (p. 23)) – and given more recent evidence, this certainly seems to have come true in all areas of design.

But can continuous cycles be seen in consumer products? Figure 3 shows the relative frequency of colours of mobile phones displayed in gadget magazines between 1996 and early 2009. The colours charcoal, silver and black consecutively outperform any other colour for several years. Note how silver slowly emerged as charcoal declined, and now black has gradually grown to become the dominant colour. Silver in particular may have been popular as a result of optimism and a focus on the future around the millennium. But now many of the matte silver devices from 5-10 years ago appear tacky or dated – despite seeming futuristic at the time. What's more, the current trend in the colour black can be seen across several product categories from computing, personal and household electronics to interior design. Perhaps this is a sombre reflection of the declining economic situation. Like Kroeber's dress width, the continuity shown by each of these colours is surprising for such a superficial, external component of form.

These colour cycles show convergence between mobile handset developers in the decision to make phones a certain colour. This must reflect consumer demand to some degree – although consumers are limited to the selection that is made available by the networks and retailers. The slow change-over period also indicates inertia in the trends – there are no big jumps in a single year. Also of interest is the lack of white mobiles around the 2001/2 introduction and mass popularity of Apple's iPod – whose polished white plastic was greatly imitated in other product categories (white is included in the "other" colour segment, because it had no real impact). Cappetta, Cillo & Ponti (2006) studied similar cycles of convergence of styles in fine fashion, and found that after a peak of convergence and incremental change, the style would decline and a "period of ferment" would occur, from which new style innovations would be born.

Turning Points

Cappetta et al.'s study showed that while there is inertia in form trends, there are limiting factors that constrain the trajectory that forms may take. Take Kroeber's dress study in Figure 2, the curve reaches its peak around 1860 when the *average* diameter of the base of the skirt exceeded the height of its wearer. The ratio then decreased almost continuously over the following 40-50 years before some shorter, shallower cycles were seen. The continuous, gradual nature of the decrease indicated a cycle that was unlikely to happen by chance. But what initiated this reduction? Was it just that it would be impractical to exceed such an already decadent size of dress without widening doorways? A number of authors (Carman, 1966; Jencks, 1971; Robinson, 1975; Young, 1937) have investigated the cycles that occur in apparel fashion, architecture and automotive design and established some of the possible limiting factors. We have grouped these into categories of *objective* and *subjective* limits.

Objective limits are the economic, technological and anatomical constraints that exist for any product form – beyond which the functionality of the product would be compromised or infeasible. For example, in post-World War 2 America, cars such as the Cadillac became increasingly longer and lower, until they were so low that drivers struggled to fit in them, and so long that they required bigger engines to shift them. According to Robinson (1975) this is why, in the 50s and 60s, shorter taller European cars began to proliferate. The width of mobile phones appears to have reached an objective limit at an average of 52mm, and very few phones less than 40mm wide are available. This may be due to the minimum comfortable size of the numerical keypad to be used for typing numbers or text messages.

Subjective limits relate to and are specific to the context and tastes of consumer groups and society as described earlier in this chapter. For example, *over-abundance* of a product or style in one consumer's environment can lead to boredom or *aesthetic fatigue* (Göller, 1994) or *cliché* (Coates, 2003), which leads to rejection, and hence a reduction in the quantity of that form consumed by that group. But this may happen at different speeds, after a time-lag, or not at all for other consumer groups. *Scarcity*, the opposite of abundance, is why some iconic objects have retained their appeal:

...super-cars from the 1980s are still cool-looking because we don't see them on an everyday basis, but [the] Honda Civic, after 4 years it gets really boring, because we see it on an everyday basis... (Industrial Designer, Canada)

Discontinuity: Changing the Boundaries of Form

It has been established that underlying form trends exist that will follow smooth trajectories, cycling between limits that are either tangible or socially constructed. But what does it take to disrupt these patterns? Is it possible to break the inertia? From quantitative studies of men's beards, women's dress and car design, Robinson (1975) concluded that no external event or single innovation can break the long cycle of a form trend. However, the products and styles that he selected were functionally stable during the time period of his study. Cars were an established product, with settled consumer segments, long lead times, and high demand which meant gradual technological improvement (van Duijn, 1981). Consequently

Table 2. Examples of different types of disruption

Disruption	*Impact on consumers*	*Form implications*
Recession	Reduced conspicuous consumption	Luxury goods become less "bling", charity shop chic styles emerge
Apple iPod + iTunes	Personal music systems become highly portable, easy to use, and cool	iPod styling pervades other product categories, compatible products emerge
Kyocera Camera Phone	Camera phone category emerges	Bigger screen needed to see/take photos, larger phones become acceptable after a period of miniaturisation
Fuel price increase	Fuel consumption becomes major driver of vehicle choice	Large rugged SUVs replaced by smaller, sleeker, fuel efficient or hybrid vehicle forms

aesthetics were a disproportionate driver of consumer attraction.

However, in less stable time periods, new technological and functional innovations can disrupt form trends. Bower and Christensen (1995) explored disruptive innovations in various industries, and found that new technologies or technological applications could enable previously unrealised usage behaviours. After adoption by an initially small consumer group, the innovation would often evolve quickly until it could outperform mainstream products and technologies and even render them obsolete. However, despite rapid growth, adoption of disruptive innovations still follows a smooth s-curve of adoption.

It is not always technology shifts that lead to changes in form – it can be the other way round. Often when a designer has an ideal of how a product should look, engineers and technologists are required to create a new device, or develop new material or manufacturing solutions to make the ideal possible and economically feasible. Motorola's ultra-thin RAZR mobile phone was the result of this difficult process, and was immensely popular – showing growth in parallel with Apple's iPod, sparking an array of imitations from other brands and saving the handset manufacturer from a period of decline (Lashinsky, 2006).

For the context of product form evolution, the source of disruption cannot be limited to products or technologies, but rather to include external socio-cultural and political innovations and influential events – anything that shifts the zeitgeist, redefining product categories and consumer behaviour in an unexpected way (Coates, 2003). For example, suit design in Japan is linked to the economic climate:

...our company produces business suits... when the economic situation is quite good, we usually produce many colours. If the economic situation is [in] decline, we usually abandon other colours [and] produce basic colours like navy, grey and black... (Textile Designer, Japan)

Disruptions have direct implications such as the new geometries, configurations, scales and finishes that have been made possible by technology. But at the same time indirect consequences such as behaviours, lifestyles, and symbolism may also be enabled or enhanced, which will lead to the generation and adoption of new product forms and styles (Ulrich & Eppinger, 2004). Table 2 provides some examples of disruptions that have occurred in the past.

WHAT DOES THIS MEAN TO COMPANIES AND DESIGN TEAMS?

The previous sections have explained why product form evolves over time and illustrated the direction and disruption of the paths that underlying forms may follow. This is a way of viewing products and product categories as evolving trajectories in a dynamic context. The authors believe that adopting this way of thinking can enhance the visual strategy of any company that develops products for consumers, to compliment and contextualise the technological and functional components of the product and to provide differentiation in a saturated category.

The way that a design team should work with trends is unique to the product and the strategy of the company or brand. The theories developed in this chapter are intended to provide the reader with the foundations to develop their own methodology for identifying, interpreting and responding to form trends. To achieve this, the following factors should be considered:

The brand: In terms of form, where on the innovation curve does the brand sit – does it lead or follow its competitors? What are the brand's policies and values? Is it willing to take risks? Also what is the existing and desired consumer perception of the brand?

As with new functionalities or technologies, it is not always the best strategy to be the first to bring a new form innovation to the market. Unless the consumer expectation is for new and unique styles, then piggy-backing established forms is a safe tactic – as long as the reference form has the necessary longevity. This is the approach used in many "own brand" products on the market today.

... [a large supermarket chain] might do a very successful own brand shampoo range or something, if you're thinking of packaging, that kind of looks like something you'd see in kind of high end boutique. It doesn't damage them at all to create something that's already on the market, because the price points are much lower and you know you're getting a good product and they look really up-to-date and modern. (Product Strategist, London)

Other brands may prefer to follow different strategies, depending on the product category and target consumer.

The product category: How mature is the category – is it changing rapidly? What is the typical lifespan of a product in this category? Are there many form factors or segments within the category? Is there evidence of convergence or divergence of forms? What is the main differentiating factor (e.g. a certain functionality, some aspect of form)? What are the key competitor and predecessor products? What are the adjacent categories?

There is currently a convergence of forms in the UK mobile phone category. Handset manufacturers have moved away from the tapered sides and characteristic shapes that used to define their brand identities. The majority of phones on the market at the moment have a simple rectangular form with slightly rounded corners. Content delivery, touchscreen interfaces and the high profile, disruptive introduction of Apple's iPhone in 2007 has contributed to this convergence, which will potentially lead the form to appear clichéd and create an opportunity for differentiation.

Target consumers: where does the user group sit on the adoption curve (early adopters, late adopters, laggards etc.)? What subcultural groups does the target segment belong to, or aspire to belong to and what artefacts or imagery are associated with these groups? Who are their key opinion leaders? What are the user group's moods, problems, aspirations etc. at the moment, and how might these change over time?

These are essentially standard consumer research questions, but asked with a time component in mind. Returns will be limited for a product's visual imagery that is associated with a particular group or individual thought leader if their popularity is in decline. For example, past youth culture movements in the UK such as "garage" or "rave" have had finite lifecycles, so the imagery and behaviour associated with them have lost their impact, or evolved into something else.

Disruptions: what recent, emerging or future disruptions have the potential to change the current zeitgeist? What is the duration and scale of impact of each disruption? In what way might these changes affect consumer tastes and product forms?

It is easy to identify a disruption after it has happened. In addition to those in table 2, there are many examples of past products that appear to have captured, thrived on, or even catalysed a new pattern of behaviour. To predict, or rapidly identify disruptions and their influence on product form is an interpretive process that requires the experience and intuition of designers, trend researchers, and product strategists:

...in a month's time maybe you buy one or two magazines and I buy forty. So that's the difference, I can look at them all, rip everything out and put them in piles and you have your trends. It's not rocket science; it's just intuition or gathering the right information. (Freelance Art Director, Holland)

In today's rapidly changing, saturated markets, it is insight such as this that can directly benefit the design and development of new products that stand out from the competition.

CONCLUSION

This chapter has established that consumer taste changes over time by initially looking at the psychological, social and artefactual drivers of an individual's attraction to a product form. Next it was recognised that in the creative industries there are processes which successfully track and respond to such changes, especially in the field of apparel fashion. However, there is little formal theory regarding the evolution of form in consumer products, beyond anecdotal reports of commercial activity. Thus, to fill this gap, the authors investigated trends in the forms of products as a tangible manifestation of evolving taste.

It was found that the direction of change in product forms is not entirely random, that there are patterns of behaviour such as continuity and cycles that have predictable characteristics. Sources of disruption of trends were examined and their effect was seen as reshaping the boundaries of what is possible or acceptable in form. Finally, it was concluded that this theory development, and way of thinking about evolving forms, can benefit the design and development of products by providing a means to ensure that the appearance of the product is relevant and attractive, and compliments its functional and technological components.

ACKNOWLEDGMENT

The authors would like to thank the trend managers and design practitioners around the world who participated in this study. This work would not have been possible without their insights, valued knowledge and willingness to participate. Thanks also to Megan Strickfaden of the University of Alberta, Edmonton for providing help and support with the original proposal. Finally we would like to thank Cambridge University's Innovative Manufacturing Research Centre (IMRC) for funding this work.

REFERENCES

Andriopoulos, C., & Gotsi, M. (2006). Probing the future: Mobilising foresight in multiple product innovation firms. *Futures*, *38*(1), 50–66. doi:10.1016/j.futures.2005.04.003

Baxter, M. (1995). *Product Design: Practical methods for the systematic development of new products*. Cheltenham, UK: Stanley Thornes.

Bloch, P. H. (1995). Seeking the Ideal Form: Product Design and Consumer Response. *Journal of Marketing*, *59*(3), 16–29. doi:10.2307/1252116

Blumer, H. (1969). Fashion: From Class Differentiation to Collective Selection. *The Sociological Quarterly, 10*(3), 275–291. doi:10.1111/j.1533-8525.1969.tb01292.x

Bower, J. L., & Christensen, C. M. (1995). Disruptive Technologies: Catching the Wave. *Harvard Business Review, 73*(1), 43–53.

Cagan, J., & Vogel, C. M. (2001). *Creating Breakthrough Products: Innovation from Product Planning to Program Approval*. Upper Saddle River, NJ: FT Press.

Cappetta, R., Cillo, P., & Ponti, A. (2006). Convergent designs in fine fashion: An evolutionary model for stylistic innovation. *Research Policy, 35*(9), 1273–1290. doi:10.1016/j.respol.2006.02.009

Carman, J. M. (1966). The fate of fashion cycles in our modern society. In *Proceedings of Science and Technology*. Chicago: American Marketing Association.

Cillo, P., & Verona, G. (2008). Search Styles in Style Searching: Exploring Innovation Strategies in Fashion Firms. *Long Range Planning, 41*(6), 650–671. doi:10.1016/j.lrp.2008.08.001

Coates, D. (2003). *Watches Tell More Than Time*. New York: McGraw-Hill.

Costanzo, L. A. (2004). Strategic foresight in a high-speed environment. *Futures, 36*(2), 219–235. doi:10.1016/S0016-3287(03)00145-9

Crilly, N., & Clarkson, P. J. (2006). *The influence of consumer research on product aesthetics*. Paper presented at the 9th International Design Conference - DESIGN 2006.

Crilly, N., Moultrie, J., & Clarkson, J. (2004). Seeing things: consumer response to the visual domain in product design. *Design Studies, 25*(6), 547–577. doi:10.1016/j.destud.2004.03.001

Crozier, R. (1994). *Manufactured Pleasures: Psychological Responses to Design*. Manchester, UK: Manchester University Press.

Eckert, C., & Stacey, M. (2000). Sources of inspiration: a language of design. *Design Studies, 21*(5), 523–538. doi:10.1016/S0142-694X(00)00022-3

Göller, A. (1994). What is the Cause of Perpetual Style Change in Architecture? In Mallgrave, H. F., & Ikonomou, E. (Eds.), *Empathy, Form, and Space: Problems in German Aesthetics* (pp. 1873–1893). Los Angeles: The Getty Center For The History Of Art.

Jencks, C. (1971). [London: Studio Vista Ltd.]. *Architecture (Washington, D.C.)*, 2000.

Karjalainen, T.-M. (2006). *Semantic Transformation in Design*. Helsinki, Finland: University of Art and Design Helsinki.

Kroeber, A. L. (1919). On the principle of Order in Civilization as Exemplified by Changes of Fashion. *American Anthropologist, 21*(3), 235–263. doi:10.1525/aa.1919.21.3.02a00010

Lashinsky, A. (2006, June 1). The team that saved Motorola. *Fortune Online*.

Lloyd Jones, P. (1991). *Taste Today*. Oxford, UK: Pergamon Press.

Lowe, E. D., & Lowe, J. W. G. (1985). Quantitative Analysis of Women's Dress. In Solomon, M. R. (Ed.), *The Psychology of Fashion*. New York: Institute of Retail Management.

Midgley, D. F., & Dowling, G. R. (1978). Innovativeness: The Concept and Its Measurement. *The Journal of Consumer Research, 4*(4), 229–242. doi:10.1086/208701

Minderhoud, S., & Fraser, P. (2005). Shifting paradigms of product development in fast and dynamic markets. *Reliability Engineering & System Safety, 88*(2), 127–135. doi:10.1016/j.ress.2004.07.002

Muir Wood, A., Moultrie, J., & Eckert, C. (2008). *Trends and Design: Relating literature to industry practice*. Paper presented at the Design Research Society Conference.

Norman, D. (2004). *Emotional Design: why we love (or hate) everyday things*. New York: Basic Books.

Robinson, D. E. (1975). Style changes: cyclical, inexorable, and foreseeable. *Harvard Business Review*, *53*(6), 121–131.

Rogers, E. M. (2003). *Diffusion of Innovation* (5th ed., *Vol. 5*). New York: Free Press.

Roy, R. (1994). The evolution of ecodesign. *Technovation*, *14*(6), 363–380. doi:10.1016/0166-4972(94)90016-7

Rutter, B., & Agne, J. (1998). A Darwinian Theory of Good Design. *DMI Review*, *9*(4), 36–41.

Schifferstein, H. N. J., & Hekkert, P. (2007). *Product Experience*. London: Elsevier Science.

Sproles, G. B., & Burns, L. D. (1994). *Changing Appearances*. New York: Fairchild Publications.

Ulrich, K. T., & Eppinger, S. D. (2004). *Product Design and Development*. New York: McGraw Hill.

van Duijn, J. J. (1981). Fluctuations in innovations over time. *Futures*, *13*(4), 264–275. doi:10.1016/0016-3287(81)90143-9

Yamamoto, M., & Lambert, D. R. (1994). The Impact of Product Aesthetics on the Evaluation of Industrial Products. *Journal of Product Innovation Management*, *11*(4), 309–324. doi:10.1016/0737-6782(94)90086-8

Young, A. (1937). *Recurring Cycles of Fashion*. New York: Harper and Row.

KEY TERMS AND DEFINITIONS

Product Form: The appearance of a product, including both its surface properties and its semantic and symbolic qualities.

Underlying Form: The basic silhouette of an object, not including detailing and ornamentation.

Seminal Form: A new product whose form revolutionises its category – forcing competitors to imitate.

Continuity: A theory that in order to be successful, the underlying form of new products cannot be too different from those that precede them, resulting in smooth, evolving trends.

Disruption: A product or event that changes consumer behaviour and interrupts the continuity of form evolution.

Category Maturity: The stage of development of a product category which defines the relative importance of form or function as a differentiating factor.

Compilation of References

Aaker, D. A. (1996). *Building strong brands*. New York: Free Press.

Aaker, J. (1997). Dimensions of Brand Personality. *Journal of Marketing Research, Los Angeles, August*, (pp. 347-356).

Abbott, A. (2004). *Methods of Discovery: Heuristics for the Social Sciences*. New York: W.W. Norton & Company.

ABEPET – Brazilian Association of PET Packages Producers. *Associação Brasileira dos Fabricantes de Embalagens de PET.* (2009). *ABEPET*. Retrieved February 15, 2009 from http://www.abepet.com.br.

About Cerebral Palsy. A resource for Parents and Families (n.d.). *About Cerebral Palsy*. Retrieved February 21, 2009, from http://www.about-cerebral-palsy.org/definition/athetoid-cerebral-palsy.html

Adams, C., & Hartley, R. (2000). *The Chipping of goods initiative. Property crime reduction through the use of electronic tagging systems. (Police Scientific Development Branch, Publication N° 23/00)*. London: Home Office.

Adamson, R. E. (1952). Functional Fixedness as Related to Problem Solving: A Repetition of Three Experiments. *Journal of Experimental Psychology*, *44*, 288–921. doi:10.1037/h0062487

Adank, R., & Warell, A. (2008). Five Senses Testing: Assessing and Predicting Sensory Experience of Product Design. In K. M. Desmet P., van Erp J. (Ed.), *Design & Emotion Moves*. Cambridge, UK: Cambridge Scholars Publishing.

Ahmed, S. (2007). An Industrial Case Study: Identification of Competencies of Design Engineers. *Journal of Mechanical Design*, 709–716. doi:10.1115/1.2723807

Ajzen, I. (1991). The Theory of Planned Behavior. *Organizational Behavior and Human Decision Processes*, *50*, 179–211. doi:10.1016/0749-5978(91)90020-T

Ajzen, I., & Fishbein, M. (1980). *Understanding attitudes and predicting social behavior*. Englewood Cliffs, NJ: Prentice-Hall.

Akamavi, R. K. (2005). A research agenda for investigation of product innovation in the financial services sector. *Journal of Services Marketing*, *19*(6), 359–378. doi:10.1108/08876040510620148

Akay, A. (2003). The renaissance engineer: educating engineers in a post-9/11 world. *European Journal of Engineering Education*, *28*(2), 145–150. doi:10.1080/0304379031000078979

Alam, I. (2002). An Exploratory Investigation of User Involvement in New Service Development. *Journal of the Academy of Marketing Science*, *30*(3), 250–261.

Alam, I. (2006). Process of Customer Interaction in New Service Development. In Edvardsson, B., Gustafsson, A., Kristensson, P., Magnusson, P., & Matthing, J. (Eds.), *Involving Customers in New Service Development* (pp. 15–32). London: Imperial College Press. doi:10.1142/9781860948893_0002

Alam, I., & Perry, C. (2002). A customer-oriented new service development process. *Journal of Services Marketing*, *16*(6), 515–534. doi:10.1108/08876040210443391

Albinsson, L., & Forsgren, O. (2004, Feb. 23-24). *Ideal Oriented Co-Design of e-Services*. Paper presented at the First Scandinavian eGovernment Conference, Örebro, Sweden.

Albinsson, L., & Forsgren, O. (2005, June 19-20). *Co-Design Metaphors and Scenarios – Two Elements in a Design Language for Co-Design.* Paper presented at the Language Action Perspective (LAP) 2005, Kiruna, Sweden.

Aldersey-Williams, H., Bound, J., & Coleman, R. (Eds.). (1999). The Methods Lab: user research for design. In Hofmeester K & de Charon de Saint Germain E. (Eds.), *PRESENCE: new media for older people*. Amsterdam: Netherlands Design Institute.

Almquist, J. M. S., Kelly C., Bromberg J., & Montori V. M. (2007). *Academia-industry collaboration as an emerging model of design research: An examination of the Steelcase–Mayo Clinic case.* Paper presented at the The International Conference of International Association of Societies of Design Research, Hong Kong.

Alting, L., & Legarth, J. B. (1995). Life Cycle Engineering and Design. *Annals of CIRP*, *44*(2), 569–580. doi:10.1016/S0007-8506(07)60504-6

Altshuller, G. (2005). *40 Principles Extended Edition: TRIZ keys to Technical Innovation*. Worcester, MA: Technical Innovation Center.

Amabile, T. M. (1983). *The Social Psychology of Creativity*. New York: Springer-Verlag.

Amabile, T. M. (1996). *Creativity in Context: Update to the social psychology of creativity*. Boulder, CO: Westview Press.

Amant, K. S., & Still, B. (2007). *Handbook of research on open source software: Technological, economic, and social perspectives*. London: Information Science Reference.

Ancarani, F., & Shankar, V. (2003). Symbian: Customer interaction through collaboration and competition in a convergent industry. *Journal of Interactive Marketing*, *17*(1), 56–76. doi:10.1002/dir.10048

Anceschi, G. (2009). Maldonado, semioticist of knowledge. In *Annicchiarico, S. (2009). Tomás Maldonado* (pp. 136–155). Milan, Italy: Skira.

Anderson, W. L., & Crocca, W. T. (1993). Engineering practice and codevelopment of product prototypes. *Communications of the ACM*, *36*(6), 49–56. doi:10.1145/153571.256015

Andriopoulos, C., & Gotsi, M. (2006). Probing the future: Mobilising foresight in multiple product innovation firms. *Futures*, *38*(1), 50–66. doi:10.1016/j.futures.2005.04.003

Annicchiarico, S. (Ed.). (2002). *Animal House. When Objects Have Animals' Names*. Milan, Italy: Charta.

Anogianakis, G., Bühler, C., & Soede, M. (Eds.). (1997). *Advancement of assistive technology*. Amsterdam: IOS Press.

Antikainen, M., & Väätäjä, H. (2008). *Rewarding in open innovation communities – How to motivate members?* (Unpublished Paper Accepted to be published in Proceedings of the 2008 ISPIM conference.). Tampere University of Technology.

Antikainen, M., Ahonen, M., & Mäkipää, M. (2006). *Customers as Designers, Sellers and Buyers: Utilizing Web 2.0 Business Models in Mass Customization*. (Unpublished Working Paper). University of Tampere, Finland.

Antorini, Y. M., & Schultz, M. (2007, May 31 - June 2). *Brand Revitalization through User-Innovation: The Adult Fan of LEGO Community.* Paper presented at the 11th International Conference on Reputation, Brand, Identity & Competitiveness Bergen, Norway.

APSC. (2007). *Tackling Wicked Problems. A Public Policy Perspective*. Canberra: Australian Government/Australian Public Service Commission.

Ariyo, O. O., Eckert, C. M., & Clarkson, P. J. (2008). Hierarchical decompositions for complex product representation. In *Proceedings of International Design Conference – DESIGN 2008* (pp. 737-744). Dubrovnik.

Assmann, E., & Ramezani, H. (2000). *The use of virtuality in car development*. San Diego: Houghton Mifflin Harcourt.

Atkins, P., & Bowler, I. (2001). *Food in Society - Economy, Culture, Geography*. London: Arnold Headline.

Atkinson, R. (1999). Project Management: Cost, Time, and Quality; Two Best Guesses and a Phenomenon, Its Time to Accept Other Success Criteria. *International Journal of Project Management*, 337–342. doi:10.1016/S0263-7863(98)00069-6

Atman, C. J., Chimka, J. R., Bursic, K. M., & Nachtmann, H. L. (1999). A comparison of Freshman and Senior

Engineering Design Processes. *Design Studies*, *20*(2), 131–152. doi:10.1016/S0142-694X(98)00031-3

Atuahene-Gima, K. (1996). Differential potency of factors affecting innovation performance in manufacturing and services firms in Australia. *Journal of Product Innovation Management*, *13*(1), 35–52. doi:10.1016/0737-6782(95)00090-9

Aurich, J. C., Fuchs, C., & Wagenknecht, C. (2006). Life cycle oriented design of technical Product-Service Systems. *Journal of Cleaner Production*, *14*, 1480–1494. doi:10.1016/j.jclepro.2006.01.019

Austin, S., Steele, J., Macmillan, S., Kirby, P., & Spence, R. (2001). Mapping the conceptual design activity of interdisciplinary teams. *Design Studies*, *22*(3), 211–232. doi:10.1016/S0142-694X(00)00026-0

Back, S. (2008). *Pesquisa de Tendências □ um modelo de referência para pesquisa prospectiva* (Unpublished master's thesis). Florianópolis/SC, Brazil: Universidade Federal de Santa Catarina.

Baird, F., Moore, C. J., & Jagodzinski, A. P. (2000). An ethnographic study of engineering design teams at Rolls-Royce Aerospace. *Design Studies*, *21*, 333–355. doi:10.1016/S0142-694X(00)00006-5

Ball, L. J., & Christensen, B. T. (2009). Analogical Reasoning and Mental Simulation in Design: Two Strategies linked to Uncertainty Resolution. *Design Studies*, *30*(2), 169–186. doi:10.1016/j.destud.2008.12.005

Baltes, P. B. (1987). Theoretical Propositions of Life-Span Developmental Psychology: On the Dynamics between Growth and Decline. *Developmental Psychology*, *23*(5), 611–626. doi:10.1037/0012-1649.23.5.611

Baltes, P. B., & Baltes, M. M. (1989). Psychological perspectives on successful aging: The model of selective optimization with compensation. In Baltes, P. B., & Baltes, M. M. (Eds.), *Successful Aging: Perspectives from the Behavioral Sciences*. New York: Cambridge University Press.

Baltes, P. B., Reese, H. W., & Lipsitt, L. P. (1980). Life-Span Developmental Psychology. *Annual Review of Psychology*, *31*, 65–110. doi:10.1146/annurev.ps.31.020180.000433

Banathy, B. H. (1996). *Designing social systems in a changing world*. New York: Plenum Press.

Bandura, A. (1988). Organizational Application of Social Cognitive Theory. *Australian Journal of Management*, *13*(2), 275–302. doi:10.1177/031289628801300210

Bandura, A. (1989). Human Agency in Social Cognitive Theory. *The American Psychologist*, *44*, 1175–1184. doi:10.1037/0003-066X.44.9.1175

Bannister, D., & Mair, J. M. M. (1968). *The Evaluation of Personal Constructs*. London: Academic Press.

Barbacetto, G. (Ed.). (1987). *Design Interface. How Man and Machine Communicate. Olivetti Design Research by King & Miranda*. Milan, Italy: Arcadia.

Battarbee, K., Mattelmäki, T., Ylirisku, S., Koskel, H., Soosalu, M., Salo, H., & Allén, M. (2005, September 22-24). *Looking beyond the product: Design research in industrial and aceademin collaboration*. Paper presented at the Joining Forces, University of Art and Design Helsinki.

Battista, G. D., Pietrosanti, E., Tamassia, R., & Tollis, I. G. (1989). *Automatic Layout of PERT diagrams with X-PERT* (pp. 171–176). Washington, DC: IEEE.

Baxter, M. (1995). *Product Design: Practical methods for the systematic development of new products*. Cheltenham, UK: Stanley Thornes.

Beaudouin-Lafon, M. (2004). Designing interaction, not interfaces. In *AVI'04: Proceedings of the working conference on Advanced visual interfaces*. New York: ACM Press.

Benedek, J., & Miner, T. (2002). *Measuring Desirability: New methods for evaluating desirability in a usability lab setting. A paper on the Desirability Toolkit.* Paper presented at the Proceedings of the Usability Professionals Association Conference.

Benktzon, M. (1993). Designing for our Future Selves: the Swedish Experience. *Applied Ergonomics*, *24*(1), 19–27. doi:10.1016/0003-6870(93)90155-3

Bento, C., Peças, P., Henriques, E., & Silva, A. (2008). Product Development guided by Value Analysis through Collaborative R&D. In Gonçalves, R. (Ed.), *14th International Conference on Concurrent Enterprising – ICE 2008* (pp. 751-758). Lisboa: ESoCE-NET.

Berger, A. (2004). *Ads, fads and consumer culture: advertising's impact on American character and society. Lanham, MD*. Oxford: Rowman & Littlefield Pub.

Berger, C., Möslein, K., Piller, F. T., & Reichwald, R. (2005). Co-designing modes of cooperation at the customer interface: learning from exploratory research. *European Management Review*, *2*(1), 70–87. doi:10.1057/palgrave.emr.1500030

Berger, P. L., & Luckman, T. (1966). *The Social Construction of Reality: A Treatise in the Sociology of Knowledge*. New York: Anchor Books.

Berkel, B., & Boersch-Supan, A. (2004). *Pension Reform in Germany: The Impact on Retirement Decisions*. Mannheimer Research Institute for Economics and Demographics.

Bernabei, R., Freeman, K., & Power, J. (2008). Designers as Storytellers: The Weaving of Narratives in the Work of Bernabeifreeman. In Desmet, P.M.A, Tzvetanova, S.A., Hekkert, P. & Justice, L. (Eds.), *Proceedings from the 6th conference on design and emotion 2008*. Hong Kong.

Berta, J. (1999). Integrating VR and CAD [Los Alamitos, CA: IEEE Computer Society.]. *IEEE Computer Graphics and Applications*, 19.

Berthon, P., Hulbert, J. M., & Pitt, L. (2004). Innovation or customer orientation? An empirical investigation. *European Journal of Marketing*, *38*(9/10), 1065–1090. doi:10.1108/03090560410548870

Best, K. (2006). *Design Management: Managing Design Strategy, Process and Implementation*. Lausanne: AVA Publishing.

Better by Design (Producer). (2009, March). *Better by Design Title Page* (www.betterbydesign.org.nz) *Podcast*. Retrieved from www.betterbydesign.org.nz.

Betts, S., Schaber, F., & Turner, R. (2007). *From Ivory Tower to Fantasy Castle: A Design Case Study of Industrial Collaboration*. International Conference on Engineering Education ICEE. Coimbra: The University of Coimbra.

Betz, M., Schuckert, M., & Herrmann, C. (1998). Life Cycle Engineering as Decision Making Support in the Electronics Industry. In *Proceedings of the 1998 IEEE International Symposium on Electronics and the Environment* (pp. 231-236). Oak Brook, IL: IEEE.

Beyer, H., & Holtzblatt, K. (1998). *Contextual Design: Defining Customer-Centered Systems*. San Francisco: Morgan Kaufmann.

Biemans, W. G. (1991). User and third-party involvement in developing medical equipment innovations. *Technovation*, *11*(3), 163–182. doi:10.1016/0166-4972(91)90032-Y

Bikeoff. (n.d.). *Bikeoff*. Retrieved April 9, 2009, from http://www.bikeoff.org

Birkhofer, H. (2006). The consolidation of design science – a critical review of the status and some proposals to improve it. In Vanek, V., & Hosnedl, S. (Eds.), *AEDS 2006 Workshop* (pp. 13–22). Pilzen, Czech Republic: University of West Bohemia.

Bitencourt, A. C. P. (2001). *The development of a Re-Design Methodology for Environment*. (Thesis (Mechanical Engineering)). Federal University of Santa Catarina, Brazil. (in Portuguese)

Bjelland, O. M., & Wood, R. C. (2008). An Inside View of IBM's 'Innovation Jam'. *MIT Sloan Management Review*, *50*(1), 32–46.

Blackler, F. (1995). Knowledge, Knowledge Work and Organizations: An Overview and Interpretation. *Organization Studies*, *16*(6), 1021–1046. doi:10.1177/017084069501600605

Blackwell, A. (2008). Design Research and Academic Principles. *Design Research Quarterly*, *3/4*, 5–6.

Bloch, P. H. (1995). Seeking the Ideal Form: Product Design and Consumer Response. *Journal of Marketing*, *59*(3), 16–29. doi:10.2307/1252116

Bloem, J., & van Doorn, M. (2007). *Open voor business: Open-source-inspiratie voor innovatie*. Groningen: Uitgeverij Kleine Uil. (in Dutch)

Blumer, H. (1969). Fashion: From Class Differentiation to Collective Selection. *The Sociological Quarterly*, *10*(3), 275–291. doi:10.1111/j.1533-8525.1969.tb01292.x

Boddington, A., Brown, B., Chapman, J., Cooper, R., Doordan, D., & Garl, K. (2008). Brighton 05-06-07. *Design Issues*, *24*(1), 91–93. doi:10.1162/desi.2008.24.1.91

Boden, M. A. (1990). *The Creative Mind: Myths and Mechanisms*. London: Weidenfeld and Nicolson.

Boden, M. A. (1998). Creativity and artificial intelligence. *Artificial Intelligence*, *103*, 347–356. doi:10.1016/S0004-3702(98)00055-1

Bohm, D., & Peat, F. D. (1989). *Science, Order and Creativity*. London: Routledge.

Bonabeau, E. (2009). Decision 2.0: The Power of Collective Intelligence. *MIT Sloan Management Review*, *50*(2), 45–52.

Boron, J. B., Willis, S. L., & Schaie, K. W. (2007). Cognitive Training Gain as a Predictor of Mental Status. *Journal of Gerontology: Psychological Sciences*, *62B*(1), 45–52.

Bower, J. L., & Christensen, C. M. (1995). Disruptive Technologies: Catching the Wave. *Harvard Business Review*, *73*(1), 43–53.

Bowman, D. A., & Wingrave, C. A. (2001). *Design and evaluation of menu systems for immersive virtual environments*. Institute of Electrical and Electronics Engineers Computer Society.

Bowman, D. A., Kruijff, E., LaViola, J. J. Jr, & Poupyrev, I. (2005). *3D user interfaces*. Reading, MA: Addison-Wesley.

Brabant-Zeeuwse Werkgeversvereniging. (2004). Klantenleercentrum: Van klantgericht naar klantgedreven innoveren. *InterActie, jaargang 2 nr. 7*, p. 4 (in Dutch).

Braga, J. (2005). *Reflexões Sobre Moda. São Paulo/SP*. Brazil: Anhembi Morumbi.

Braha, D., & Reich, Y. (2003). Topological structures for modeling engineering design processes. *Research in Engineering Design*, *14*, 185–199. doi:10.1007/s00163-003-0035-3

Brand, R. (2005). The citizen-innovator. *The Innovation Journal: The Public Sector Innovation Journal*, *10*(1), 1–11.

Branzi, A. (2008). *Introduzione al design italiano. Una modernità incompleta*. Milan, Italy: Baldini Castoldi Dalai.

Bratt, J., Foreit, J., & Chen, P. (1999). A comparison of four approaches for measuring clinician time use. *Health Policy and Planning*, *14*, 374–381. doi:10.1093/heapol/14.4.374

Braziel, J. E., & Mannur, A. (2003). *Theorizing Diaspora*. Oxford, UK: Blackwell Publishing.

Bretzner, L., & Lindeberg, T. (1998). Use your hand as a 3D mouse, or, relative orientation from extended sequences of sparse point and line correspondences using the affine trifocal tensor. In Proceedings of the *5th European Conf. on Computer Vision*. Freiburg, Germany.

Brown, S. L., & Eisenhardt, K. M. (1995). Product development: Past research, present findings, and future directions. *Academy of Management. Academy of Management Review*, *20*(2), 343–378. doi:10.2307/258850

Brown, T. (2008, June). Design Thinking. *Harvard Business Review*, •••, 84–92.

Brunel University. (2009). *Product Design Engineering - Course Summary*. Retrieved in 2009, from www.brunel.ac.uk/courses/ug/cdata/p/ProductDesignEngineeringBSc

Bruseberg, A., & McDonagh-Philp, D. (2001). New product development by eliciting user experience and aspirations. *International Journal of Human Computer Studies*, *55*(4) 435-452, October. ISSN 1071-5819

Buber, R., Gadner, J., & Richards, L. (Eds.). (2004). *Applying Qualitative Methods to Marketing Management Research*. New York: Palgrave MacMillan.

Buchanan, R. (2001). Design and the New Rhetoric: Productive Arts in the Philosophy of Culture in Philosophy and Rhetoric: *Vol. 34. No.3). State College*. PA: The Pennsylvania State University.

Buchanan, R., & Margolin, V. (1995). *Discovering Design. Explorations in Design Studies*. Chicago: The University of Chicago Press.

Burdea, G. C., & Coiffet, P. (2003). *Virtual reality technology*. Hoboken, NJ: Wiley-IEEE Press.

Burge, J., & Brown, D. (2002). Integrating Design Rationale with a Process Model. *Workshop on Process Modelling, AID'02: International AI in Design Conference*. Cambridge, UK.

Burke, K. (1969). *A Grammar of Motives*. Berkley, CA: University of California Press.

Burton, D. G. A., Allen, M. C., Bird, J. L. E., & Faragher, R. G. A. (2005). Bridging the Gap: Ageing, Pharmacokinetics and Pharmacodynamics. *The Journal of Pharmacy and Pharmacology*, *57*, 671–679. doi:10.1211/0022357056163

Busseri, M. A., & Palmer, J. M. (2000). Improving teamwork: the effect of self-assessment on construction design teams. *Design Studies*, *21*(3), 223–238. doi:10.1016/S0142-694X(99)00024-1

Buzan, T., & Buzan, B. (1995). *The Mind Map book*. London: BBC Books.

Cabral, M. C., Morimoto, C. H., & Zuffo, M. K. (2005). On the usability of gesture interfaces in virtual reality environments. In *CLIHC '05: Proceedings of the 2005 Latin American conference on Human-computer interaction*. Cuernavaca, Mexico: ACM Press.

Cagan, J., & Vogel, C. M. (2001). *Creating Breakthrough Products: Innovation from Product Planning to Program Approval*. Upper Saddle River, NJ: FT Press.

Calantone, R. J., Di Benedetto, C. A., & Haggblom, T. (1995). Principles of new product management: Exploring the beliefs of product practitioners. *Journal of Product Innovation Management*, *12*(3), 235–247. doi:10.1111/1540-5885.1230235

Caldas, D. (2006). *Observatório de Sinais: teoria e prática da pesquisa de tendências. Rio de Janeiro/RJ*. Brazil: Senac Rio.

Calhoun, E. F. (2009). Action research: Three approaches. In Schmuck, R. A. (Ed.), *Practical action research* (pp. 61–67). Thousand Oaks, CA: Corwin Press.

Callahan, J., & Lasry, E. (2004). The importance of customer input in the development of very new products. *R & D Management*, *34*(2), 107–120. doi:10.1111/j.1467-9310.2004.00327.x

Callon, M. (1986). Some elements of a sociology of translation: domestication of the scallops and the fishermen of St Brieuc Bay. In J. Law (ed.), *Power, action and belief: a new sociology of knowledge?* London: Routledge. Retrieved February 23, 2009, from http://www.csi.ensmp.fr/Perso/Callon/MCarticlesavant1987.html

Callon, M. (2005). Actor-network theory – the market test. In Law, J., & Hassard, J. (Eds.), *Actor Network Theory and After*. Oxford, UK: Blackwell.

Campbell, A. J., & Cooper, R. G. (1999). Do Customer Partnerships Improve New Product Success Rates? *Industrial Marketing Management*, *28*(5), 507–519. doi:10.1016/S0019-8501(99)00058-9

Campos, A. Q., & Rech, S. R. (2008, october). *Futuro do Presente: porque e como pesquisar tendências*. Paper present at 8º. Congresso Brasileiro de Pesquisa & Desenvolvimento em Design – P&D Design, São Paulo/SP, Brazil.

Canaan, D. (2003). Research to fuel the creative process. In Laurel, B. (Ed.), *Design Research: Methods and Perspectives* (p. 236). Cambridge, MA: MIT Press.

Cantanessa, M. (2005). Product portfolio management. In Clarkson, J., & Eckert, C. (Eds.), *Design Process Improvement – A review of current practice* (pp. 404–435). London: Springer.

Cappetta, R., Cillo, P., & Ponti, A. (2006). Convergent designs in fine fashion: An evolutionary model for stylistic innovation. *Research Policy*, *35*(9), 1273–1290. doi:10.1016/j.respol.2006.02.009

Carlile, P. R. (2002). A Pragmatic View of Knowledge and Boundaries: Boundary Objects in New Product Development. *Organization Science*, *13*(4), 442–455. doi:10.1287/orsc.13.4.442.2953

Carman, J. M. (1966). The fate of fashion cycles in our modern society. In *Proceedings of Science and Technology*. Chicago: American Marketing Association.

Carrillo, A. G. (2002). *Engineering design team performance: quantitative evidence that membership diversity effects are time dependent*. (Unpublished doctoral dissertation). Stanford, CA: Stanford University

Carstensen, L. L. (1992). Social and Emotional Patterns in Adulthood: Support for Socioemotional Selectivity Theory. *Psychology and Aging*, *7*(3), 331–338. doi:10.1037/0882-7974.7.3.331

Carstensen, L. L., & Mikels, J. A. (2005). At the Intersection of Emotion and Cognition: Aging and the Positivity Effect. *American Psychology Society*, *14*(3), 117–121.

Carstensen, L. L., Mikels, J. A., & Mather, M. (2006). Aging and the Intersection of Cognition, Motivation, and Emotion. In Birren, J., & Schaie, K. W. (Eds.), *Handbook of the Psychology of Aging* (pp. 343–362). New York: Academic Press. doi:10.1016/B978-012101264-9/50018-5

Castilho, K. (2008). *Corpo e Moda: por uma compreensão do contemporâneo. Barueri/SP*. Brazil: Estação das Letras e Cores.

Centre for Research in the Arts. (n.d.). Social Sciences and Humanities, Cambridge *CRASSH2009*, Retrieved from http://www.crassh.cam.ac.uk/events/783

Chan, T.-Y., & Lee, J.-F. (2004, April 3–7). *A Comparative Study of Online User Communities Involvement in Product Innovation and Development.* Paper presented at the 13th International Conference on Management of Technology (IAMOT), Washington D.C.

Chaney, D. (1996). *Lifestyles*. London, New York: Routledge.

Chang, C. M. (2008). Engaging the creative minds - the ENGAGE models. *International Journal of Innovation and Technology Management*, *5*(1), 149–165. doi:10.1142/S0219877008001254

Chang, M. (1998). *From Economic Debacle to Economic Miracle*. Singapore: Ministry of Education.

Chapman, J. (2005). *Emotionally durable design: objects, experiences and empathy*. London: Earthscan.

Checkland, P. (1999). *Soft systems methodology: a 30-year retrospective*. New York: John Wiley.

Chen, S. J., & Hwang, C. L. (1992). *Fuzzy multiple attribute decision making methods and applications. (Lecture Notes in Economics and Mathematical Systems)*. New York: Springer.

Cherbonneau, G. (2005). A framework for improving the effectiveness of distributed project teams. (Unpublished doctoral dissertation). Cambridge, MA: Massachusetts Institute of Technology

Chernichovsky, D., & Anson, J. (1989). (*The World Bank*). Cost Recovery and True Cost-Effectiveness Ratios of Contraceptive Provision.

Chesbrough, H. W. (2003). *Open Innovation. The New Imperative For Creating and Profiting from Technology*. Boston, MA: Harvard Business School Publishing.

Chiapponi, M. (2005). *Le forme degli oggetti,* in *Il Verri*, 27, (pp. 17-34).

Chiapponi, M., & Riccini, R. (Eds.). (2008). *Made in Iuav. L'università del design fra ricerca e progetto*, Tavagnacco: Dindi.

Child, M. Help and Hope for Life, Learn About Cerebral Palsy (n.d.). *4 My Child*. Retrieved February 21, 2009, from http://www.cerebralpalsy.org/what-is-cerebral-palsy/

Christensen, B. T., & Schunn, C. D. (2007). The Relationship of Analogical Distance to Analogical Function and pre-inventive Structure: The Case of Enginerring Design. *Memory & Cognition*, *35*(1), 29–38.

Christensen, C. C. (1997). *The Innovator's Dilemma: When New Technologies Cause Great Firms to Fail*. Boston, MA: Harvard Business School Press.

Christensen, C. M., & Bower, J. L. (1996). Customer power, strategic investment, and the failure of leading firms. *Strategic Management Journal*, *17*(3), 197–218. doi:10.1002/(SICI)1097-0266(199603)17:3<197::AID-SMJ804>3.0.CO;2-U

Christensen, C. M., & Raynor, M. E. (2003). *The Innovator's Solution. Creating and Sustaining Successful Growth*. Boston, MA: Harvard Business School Press.

Christensen, P. R., & Guildford, J. P. (1958). *Creativity/fluency scales*. Beverly Hills, CA: Sheridan Psychological Services.

Christiaans, H., & Dorst, C. (1992). Cognitive Models in Industrial Design Engineering: a protocol study. In Taylor, D. L., & Stauffer, D. A. (Eds.), *Design Theory and Methodology – DTM92*. New York: American Society of Mechanical Engineers.

Chua, R. (Ed.). (1994). *Design Works*. Singapore: Designers Association Singapore.http://www.edb.gov.sg/edb/sg/en_uk/index/about_edb/our_history.html

Ciborra, C. U. (1998). From Tool to Gestell: Agendas for managing the information infrastructure. *Information Technology & People*, *11*(4), 305–327. doi:10.1108/09593849810246129

Ciccantelli, S., & Magidson, J. (1993). From experience: Consumer idealized design: Involving consumers in the product development process. *Journal of Product Innovation Management, 10*(4), 341–347. doi:10.1016/0737-6782(93)90076-3

Ciger, J., Gutierrez, M., Vexo, F., & Thalmann, D. (2003). *The magic wand*. New York: ACM Press.

Cillo, P., & Verona, G. (2008). Search Styles in Style Searching: Exploring Innovation Strategies in Fashion Firms. *Long Range Planning, 41*(6), 650–671. doi:10.1016/j.lrp.2008.08.001

Clark, K., & Smith, R. (2008). Unleashing the Power of Design Thinking, *Design. Management Review, 19*(3), 8–15.

Clarke, P., & George, J. (2005). Big Box Thinking: Overcoming Barriers to Creativity in Manufacturing. *Design Management Review, 16*(2), 42–48. doi:10.1111/j.1948-7169.2005.tb00192.x

Clarkson, J., Coleman, R., Coates, S., & Lebbon, C. (Eds.). (2003). *Inclusive design: design for the whole population*. London: Springer-Verlag.

Clarkson, P. J. (2008). Human Capability and Product Design. In Schifferstein, H. N. J., & Hekkert, P. (Eds.), *Product Experience* (pp. 165–198). San Diego: Elsevier. doi:10.1016/B978-008045089-6.50009-5

Clemensen, J., Larsen, S., Kyng, M., & Kirkevold, M. (2007). Participatory Design in health sciences: Using cooperative experimental methods in developing health services and computer technology. *Qualitative Health Research, 17*, 122–130. doi:10.1177/1049732306293664

Coates, D. (2003). *Watches Tell More Than Time*. New York: McGraw-Hill.

Coburn, P. (2006). *The Change Function: Why Some Technologies Take Off and Others Crash and Burn*. New York: Penguin Group.

Collins, H. M. (1996). Embedded or Embodied: Hubert Dreyfus's What Computers Still Can't Do. *Artificial Intelligence, 80*(1), 99–117. doi:10.1016/0004-3702(96)00083-6

Colwill, J., Cultice, J., & Kruse, R. (2008). Will Generalist Physician Supply Meet Demands Of An Increasing And Aging Population? *Health Affairs (Project Hope), 27*(3), w232–w241. doi:10.1377/hlthaff.27.3.w232

Comer, J. M., & Zirger, B. J. (1997). Building a Supplier-Customer Relationship Using Joint New Product Development. *Industrial Marketing Management, 26*, 203–211. doi:10.1016/S0019-8501(96)00114-9

Condit, P. M. (1994). Focusing on the customer: How Boeing does it. *Research Technology Management, 37*(1), 33–37.

Conran, T., & Fraser, M. (2004). *Designers on design*. London: Conran Octopus.

Conwell, L., & Cohen, J. (2002). *Characteristics of people with high medical expenses in the U.S. civilian noninstitutionalized population*. Washington, DC: Agency for Healthcare Research and Quality.

Cooper, J. (Ed.). (2000). *Law, rights and disability*. London: Jessica Kingsley Publishers.

Cooper, J. S., & Vigon, B. (2001). *Life Cycle Engineering Guidelines. (Report 600/R-01/101)*. Washington, DC: Environmental Protection Agency.

Cooper, R. G. (1994). Perspective third-generation new product processes. *Journal of Product Innovation Management, 11*(1), 3–14. doi:10.1016/0737-6782(94)90115-5

Cooper, R. G. (1999). From Experience: The Invisible Success Factors in Product Innovation. *Journal of Product Innovation Management, 16*, 115–133. doi:10.1016/S0737-6782(98)00061-7

Cooper, R. G., & Edgett, S. J. (1996). Critical success factors for new financial services. *Marketing Management, 5*(3), 26–37.

Cooper, R. G., & Kleinschmidt, E. J. (1986). An investigation into the new product process: Steps, deficiencies, and impact. *Journal of Product Innovation Management, 3*(2), 71–85. doi:10.1016/0737-6782(86)90030-5

Corbolan-Berna, F. J. (1992). Creativity as a cognitive style. *The Journal of Creative Behavior, 26*(3), 163–164.

Costanzo, L. A. (2004). Strategic foresight in a high-speed environment. *Futures, 36*(2), 219–235. doi:10.1016/S0016-3287(03)00145-9

Coughlin, J. F. (1999). Technology Needs of Aging Boomers. *Issues in Science and Technology* (Fall).

Cox, G. (2005). *The Cox Review of Creativity in Business*. London: HM Treasury.

Coyne, R., & Snodgrass, A. (1995). Problem Setting within Prevalent Metaphors of Design. *Design Issues*, *11*(2), 31–61. doi:10.2307/1511758

CPH127. (n.d.). *Design and Innovation*. Retrieved April 1, 2007, from http://www.cph127.com/cph127/html

Crilly, N., & Clarkson, P. J. (2006). *The influence of consumer research on product aesthetics*. Paper presented at the 9th International Design Conference - DESIGN 2006.

Crilly, N., Moultrie, J., & Clarkson, J. (2004). Seeing things: consumer response to the visual domain in product design. *Design Studies*, *25*(6), 547–577. doi:10.1016/j.destud.2004.03.001

Cropley, D. H., & Cropley, A. J. (2000). Fostering Creativity in Engineering Undergraduates. *High Ability Studies*, *11*(2), 207–218. doi:10.1080/13598130020001223

Cross, N. (2000). *Engineering Design Methods: Strategies for Product Design* (3rd ed.). New York: Wiley.

Cross, N. (2001). Designerly ways of knowing: design discipline versus design Science. *Design Issues*, *17*(3), 49–55. doi:10.1162/074793601750357196

Cross, N. (2002, February). *Design as a discipline*. Paper presented to assist discussion at the conference "The Interdisciplinary Design Quandary", De Montfort University, Leicester, UK.

Cross, N. (2007). Forty years of design research. *Design Studies*, *28*(1), 1–4. doi:10.1016/j.destud.2006.11.004

Crozier, R. (1994). *Manufactured Pleasures: Psychological Responses to Design*. Manchester, UK: Manchester University Press.

Cruickshanks, K. J., Wiley, T. L., Tweed, T. S., Klein, B. E. K., Klein, R., Mares-Perlman, J. A., & Nondahl, D. M. (1998). Prevalence of Hearing Loss in Older Adults in Beaver Dam, Wisconsin. *American Journal of Epidemiology*, *148*(9), 879–886.

Csikszentmihalyi, M. (1996). *Creativity: Flow and the Psychology of Discovery and Invention*. New York: Harper Collins Publishers.

Csikszentmihalyi, M., & Rochberg-Halton, E. (1981). *The meaning of things: Domestic symbols and the self*. Cambridge, UK: Cambridge University Press.

Cumming, E., Dean, L. R., Newell, D. S., & McCaffrey, I. (1960). Disengagement: A Tentative Theory of Aging. *Sociometry*, *23*(1), 23–35. doi:10.2307/2786135

Cumulus Kyoto. (2008). *Cumulus Kyoto*. Retrieved on April 11, 2008, from http://www.kyoto-seika.ac.jp/cumulus/e_prospectus/index.html

Cutler, D. M., & Sheiner, L. (1998). Standard and Nonstandard Effects. In *Proceeding of Demographic Change and Fiscal Policy*. Demographics and Medical Care Spending.

Dahan, E., & Hauser, J. R. (2002). The virtual customer. *Journal of Product Innovation Management*, *19*(5), 332–353. doi:10.1016/S0737-6782(02)00151-0

Daily RFID Co, Ltd. (n.d.). *Daily RFID*. Retrieved April 9, 2009, from http://www.rfid-in-china.com

Dalton, D. S., Cruickshanks, K. J., Klein, B. E. K., Klein, R., Wiley, T. L., & Nondahl, D. M. (2003). The Impact of Hearing Loss on Quality of Life in Older Adults. *The Gerontologist*, *43*(5), 661–668.

Danneels, E. (2004). Disruptive Technology Reconsidered: A Critique and Research Agenda. *Journal of Product Innovation Management*, *21*(4), 246–258. doi:10.1111/j.0737-6782.2004.00076.x

Dant, T. (1999). *Material Culture in the Social World. Values, Activities, Lifestyles*. Buckingham, Philadelphia: Open University Press.

Darke, J. (1979). The Primary Generator and the Design Process. In Cross, N. (Ed.), *Developments in Design Methodology*.

Darmofal, D. L., Soderholm, D. H., & Brodeur, D. R. (2002). *Using Concept Maps and Concept Questions to Enhance Conceptual Understanding*. Retrieved December 5, 2008, from http://www.cdio.org/papers/papers.html

Darnton, R. (2006). *O Grande Massacre de Gatos e Outros Episódios da História Cultural Francesa. Rio de Janeiro/RJ.* Brazil: Graal.

Darroch, J. (2002). *Innovation and Knowledge Management.* (Unpublished PhD Thesis). University of Otago, Dunedin.

Das, R. (2002). *An Introduction to RFID and Tagging Technologies*, Cambridge, UK: White Paper, IDTechEX

Davenport, S. (2005). Exploring the role of proximity in SME knowledge-acquisition. *Research Policy, 35*(5), 189–216.

Davenport, T. H., & Prusak, L. (1998). *Working Knowledge: How Organizations Manage What They Know.* Boston: Harvard Business School.

Davenport, T. H., Leibold, M., & Voelpel, S. (2006). *Strategic Management in the Innovation Economy. Strategy Approaches and Tools for Dynamic Innovation Capabilities.* Berlin Heidelberg, Germany: Springer Verlag.

daVinci, L. (n.d.). Leonardo daVinci quotes. *ThinkExist.com Quotations Online.* Retrieved March 15, 2009, from http://thinkexist.com/quotation/i_have_been_impressed_with_the_urgency_of_doing/214030.html

Davis, F. D. (1989). Perceived Usefulness, Perceived Ease of Use, and User Acceptance of Information Technology. *Management Information Systems Quarterly, 13*(3), 319–340. doi:10.2307/249008

Davis, M. (2008). *Toto, I've got a feeling we're not in Kansas anymore ... AIGA Design Educators Conference Massaging Media 2*, (April 4-6, 2008, Boston). Retrieved March 29, 2009, from http://wwwmassagingmedia.org/keynote-address-meredith-davis

Dawkins, R. (1976). *The Selfish Gene. (1989 edition).* Oxford, UK: Oxford University Press.

de Bono, E. (1992). *Serious creativity.* London: Harper Collins Publishers.

de Jong, J. P. J., & von Hippel, E. (2009). *Measuring user innovation in Dutch high tech SMEs: Frequency, nature and transfer to producers.* (Unpublished Working paper). EIM Business and Policy Research, Zoetermeer, The Netherlands.

de Oliveira, A. S. C. (2006). *O Estudo das Tendências para o Processo de Design.* (Unpublished graduate's thesis). Programa de Graduação em Design Industrial, UDESC, Florianópolis/SC, Brazil.

de Ruyter, K., & Scholl, N. (2004). *Qualitative Market Research. Theory and Practice.* Utrecht: Lemma.

de Zeeuw, G. (2005). The Acquisition of High Quality Experience. *Journal of Research Practice, 1*(1).

Deaglio, E. (2009). *Patria 1978-2008*, Milan, Italy: il Saggiatore.

Deavours, D., Ramakrishnan, K. M., & Syed, A. (2005). *RFID Performance Tag Analysis (Tech. Rep.).* University of Kansas, Information and Telecommunication Technology Center.

Deci, E. L., Koestner, R., & Ryan, R. M. (1999). A meta-analytic review of experiments examining the effects of extrinsic rewards on intrinsic motivation. *Psychological Bulletin, 125*(6), 627–668. doi:10.1037/0033-2909.125.6.627

Delre, S. A. (2007). *Effects of social networks on innovation diffusion and marketing dynamics* (Unpublished Ph D Dissertation). University of Groningen, Groningen.

Dembski, W. A. (1997). *Explaining Specified Complexity.* Retrieved from http://inia.cls.org/~welsberr/evobio/evc/ae/dembski_wa/19990913_explaining_csi.html

Deming, W. E. (1986). *Out of the crisis.* Cambridge, MA: Massachusetts Institute of Technology, Center for Advanced Engineering Study.

Denton, H., & McDonagh, D. (2003). Using focus group methods to improve students' design project research in schools: drawing parallels from action research at undergraduate level. *International Journal of Technology and Design Education, 13*(2), 129–144. doi:10.1023/A:1024149703800

Denyer, D., Tranfield, D., & van Aken, J. E. (2008). Developing design propositions through research synthesis. *Organization Studies, 29*, 393–413. doi:10.1177/0170840607088020

Derbaix, C., & Vanhamme, J. (2003). Inducing Word-of-Mouth by Eliciting Surprise – a Pilot Investigation. *Journal of Economic Psychology*, *24*(1), 99–116. doi:10.1016/S0167-4870(02)00157-5

Design Council. (2002). *Evidence Pack. DAC case studies*. London: Design Council.

Design Council. User Research – Case Study: OXO Good Grips (2006). *Design Council*. Retrieved September 14, 2008, from http://www.designcouncil.org.uk/en/Case-Studies/All-Case-Studies/OXO-Good-Grips/User-research/

Design Industry Taskforce. (2002). *integrate! A critical look at the interface between business and design*. Wellington, New Zealand: New Zealand Trade and Enterprise.

Design Industry Taskforce. (2003). *Success by Design – A Report and Strategic Plan*. Wellington, New Zealand: New Zealand Trade and Enterprise.

Design PhD Conference. (2009). *Design PhD Conference*. Retrieved March 20, 2009, from http://imagination.lancaster.ac.uk/conference2009/

Design Victoria. (2008). *Five Years On: Victoria's Design Sector 2003-2008*. Melbourne: Retrieved from http://www.designvic.com/~/media/PDF/FiveYearsOn.ashx

Designers Institute of New Zealand (DINZ). (2007). The Business of Sustainable Design. In *Proceedings of the DINZ National Conference*. Auckland.

Designers Institute of New Zealand (DINZ). (2008). Sustainability and Imminent Change. In *Proceedings of the DINZ National Conference*. Auckland.

Desmet, P. M. A. (2002). *Designing Emotions*. Delft (NL): Technical University of Delft.

Desmet, P. M. A. (2007). *Nine Sources of Product Emotion*. Paper presented at the International Association of Societies of Design Research IASDR07: Emerging Trends in Design Research, The Hong Kong Polytechnic University.

Desmet, P. M. A., & Hekkert, P. (2007). Framework of Product Experience. *International Journal of Design*, *1*(1). http://www.ijdesign.org/ojs/index.php/IJDesign/article/view/66/15.

Dilnot, C. (1993). The gift. *Design Issues*, *9*(2), 51–63. doi:10.2307/1511674

Disability Rights Commission. (2009). *Disabled people*. Retrieved June 22, 2010, from http://www.direct.gov.uk/en/DisabledPeople/RightsAndObligations/DisabilityRights/DG_4001070

Dixon, J. R. (1988). On research methodology towards a scientific theory of engineering design. *Artificial Intelligence for Engineering Design, Analysis and Manufacturing*, *1*(3), 145–157. doi:10.1017/S0890060400000251

Dixon, N. (2000). *Common Knowledge: How Companies Thrive by Sharing what They Know*. Boston: Harvard Business School Press.

Dolan, R. J., & Matthews, J. M. (1993). Maximizing the utility of customer product testing: Beta test design and management. *Journal of Product Innovation Management*, *10*(4), 318–330. doi:10.1016/0737-6782(93)90074-Z

Donath, B. (1992). The Customer as Consultant. *Sales and Marketing Management*, *144*(11), 84–87.

Dorador, J. M., & Young, R. I. (2000). Application of IDEF0, IDEF3, and UML methodologies in the creation of information models. *International Journal of Computer Integrated Manufacturing*, 430–445.

Dorfles, G. (1963). *Introduzione al disegno industriale*. Bologna: Cappelli.

Drew, S. A. W. (1995). Strategic benchmarking: innovation practices in financial institutions. *International Journal of Bank Marketing*, *13*(1), 4–16. doi:10.1108/02652329510075418

Dreyfus, H. L. (1991). *Being-in-the-World. A Commentary on Heidegger's Being and Time*. Cambridge, MA: The MIT Press.

Duarte, M. D. (1997). Characterization of Ambiental Label of Products. (Thesis (Production Engineering)). Federal University of Santa Catarina, Brazil. (in Portuguese)

Dudek, S., & Cote, R. (1994). Problem finding revisited. In Runco, M. A. (Ed.), *Problem finding, problem solving, and creativity* (pp. 130–150). Norwood, NJ: Ablex.

duGay, P., Hall, S., Janes, L., Mackay, H., & Negus, K. (1997). *Doing Cultural Studies: The Story of the Sony Walkman*. London: Sage.

Dumas, J. S., & Redish, J. C. (1999). *A Practical Guide to Usability Testing*. Portland: Intellect Books.

Dym, C. A. (1999). Learning Engineering: Design, Languages and Experiences. *Journal of Engineering Education*, *88*(2), 145–148.

Dym, C. A., Agogino, A. M., Eris, O., Frey, D. D., & Leifer, L. J. (2005). Engineering Design Thinking, Teaching, and Learning. *Journal of Engineering Education, January 2005*, 103-119.

Earl, C., Eckert, C., & Johnson, J. (2004). *Proceedings of the 8th International Design Conference: Complexity models in design.* Paper presented at the International Design Conference - Design 2004, Dubrovnik (Croatia).

Eastman, C. M. (1970). On the Analysis of Intuitive Design Processes. In Moore, G. T. (Ed.), *Emerging Methods in Environmental Design and Planning*. Cambridge, MA: MIT Press.

Echternacht, E. H. O., & Verissimo, M. (2009). Chemincnement et perspectives de l'ergologie au Brésil: La santé au travail et les sciences de l'éducation en dialogue. In: Yves Schwartz; Louis Durrive. (Org.), *L□Activité en Dialogues: Entretiens sur l□activité humaine* (II). I ed. Toulouse - France: OCTARÈS Éditions, 2009, v. II, p. 191-214. (in French)

Eckert, C. M., Cross, N., & Johnson, J. H. (2000). Intelligent support for communication in design teams: garment shape specifications in the knitwear industry. *Design Studies*, *21*(1), 99–112. doi:10.1016/S0142-694X(99)00006-X

Eckert, C., & Stacey, M. (2000). Sources of inspiration: a language of design. *Design Studies*, *21*(5), 523–538. doi:10.1016/S0142-694X(00)00022-3

Edelkoort (2009). *Lidewij Edelkoort.* Retrieved July 13, 2009, from http:www.edelkoort.com

Eder, W., & Hosnedl, S. (2008). *Design Engineering, A Manual for Enhanced Creativity.*. Florida, USA: CRC Press.

Edvardsson, B., Gustafsson, A., Kristensson, P., Magnusson, P., & Matthing, J. (Eds.). (2006). *Involving Customers in New Service Development*. London: Imperial College Press.

Eekels, J. (1987). Guidelines for Engineering Teachers Concerning Educating the Engineer for Innovation and Entrepreneurial Activity. *European Journal of Engineering Education*, *12*(3), 259–270. doi:10.1080/03043798708939369

Eekels, J. (1994). The Engineer as Designer and as a Morally Responsible Individual. *Journal of Engineering Design*, *5*(1), 7–23. doi:10.1080/09544829408907868

Eilander, E. (2008). Word of Mouth rond Omo Klein & Krachtig. [in Dutch]. *Tijdschrift voor Marketing*, *42*(10), 52–55.

Eisenhardt, K. M. (1989). Building Theories from Case Study Research. *Academy of Management Review*, *14*(4), 532–550. doi:10.2307/258557

El-Haik, B. S. (2005). *Axiomatic quality. Integrating axiomatic design with six-sigma, reliability and quality engineering*. Hoboken, NJ: John Wiley & Sons.

Ellis, F. (2009). *How will Robocop communicate?: The design of a conceptual portable radio communication product for NZ Police in 2018*. Wellington: Massey University.

Ennew, C. T., & Binks, M. R. (1996). Good and bad customers: the benefits of participating in the banking relationship. *International Journal of Bank Marketing*, *14*(2), 5–13. doi:10.1108/02652329610106872

Enos, J. L. (1962). *Petroleum Progress and Profits: A History of Process Innovation*. Cambridge, MA: MIT Press.

Eppinger, S. D. (2002). *Patterns of Product Development Interactions*. (Working paper ESD-WP-2003-01.05). Cambridge, MA: MIT Engineering Systems Division.

Erickson, T. D. (1990). Working with Interface Metaphors. In Laurel, B. (Ed.), *The Art of Human-Computer Interface Design*. Reading, MA: Addison-Wesley.

Ermolaeva, S., Castro, G., & Kandachar, V. (2004). Materials selection for an automotive structure by integrating structural optimization with environmental impact assessment. *Materials & Design*, *25*(8), 689–698. doi:10.1016/j.matdes.2004.02.021

Erner, G. (2005). *Vítimas da moda? como a criamos e por que a seguimos? São Paulo/SP*. Brazil: Editora Senac.

ESD Symposium Committee (2002). *ESD terms and definitions*, (version 14). Retrieved September 1, 2009, from http://ocw.mit.edu/OcwWEB/engineering-systems-division/ESD-342Spring-2006/readings/

ESD. (2008). *ESD Releases New Strategic Plan*. Retrieved December, 2, 2008, from http://esd.mit.edu/HeadLine/esd101408.html

Evans, M., & Sommerville, S. (2006). Educating the future: embedding futures thinking in the design curriculum. In Rothbucher, Kolar, Ion, Clarke (Ed.), Educating designers for a Global Context? In *Proceedings of the 4th Engineering & Product Design Education Conference*, Salzburg, Austria, September 7-8 2006 (pp. 355-360). Basildon: Hadleys Ltd.

Evans, M., & Spruce, J. (2005, September 15-16). *Knowledge Networks: Collaboration between Industry and Academia in Design*. Paper presented at the Engineering and Product Design Education Conference, Napier University, Edinburgh.

Fabris, L. (Ed.). (2008). *LuisaFONTANAtelier. Global architecture*, Dogana, San Marino: Litografia Titanlito s.a.

Fairen (2004). MiniVR: a portable virtual reality system. *Computers graphics, 28*. New York: Pergamon Press.

Faragher, R. G. A., & Kipling, D. (1998). How might Replicative Senescence Contribute to Human Ageing? *BioEssays*, *20*, 985–991. doi:10.1002/(SICI)1521-1878(199812)20:12<985::AID-BIES4>3.0.CO;2-A

Fava, J. A., Brady, K., Young, S. B., & Saur, K. (2000). Sustainable Strategies Using Life Cycle Approaches. *Environment and Progress*, *19*(2), 61–64. doi:10.1002/ep.670190203

Feghali, M. (2001). *Senac* (Brazil, R. J., Ed.). Rio de Janeiro: As Engrenagens da Moda.

Felder, R. M. (1987). On Creating Creative Engineers. *English Education*, *77*(4), 222–227.

Ferguson, R. (2007). *The blind need not apply: A history of overcoming prejudice in the orientation and mobility profession*. Charlotte, NC: Information Age Publishing.

Fernandes, J. M., Romeiro Filho, E., Araki, L. A., Reis, L. P., Figueiredo, A. C., & Vaz, C. R. (2005). Cross-functional Teams and Concurrent Engineering: Contributions to the Development of Product Design Through Multidisciplinary Integration Using CAD Systems. *Product: Management and Development*, *3*(1), 5–9.

Findeli, A. (2001). Rethinking Design Education for the 21st Century: Theoretical, Methodological, and Ethical Discussion. *Design Issues*, *17*(1), 5–17. doi:10.1162/07479360152103796

Finke, R. A., Ward, T. B., & Smith, S. M. (1992). *Creative Cognition: Theory, research, and applications*. Cambridge, MA: MIT Press.

Finkler, S., Kinckman, J., Hendrickson, G., Lipkin, M., & Thompson, W. (1993). A comparison of work sampling and time-and motion techniques for studies in health services research. *Health Services Research*, *28*, 578–597.

Fisher, J., & Monahan, T. (2008). Tracking the social dimensions of RFID systems in hospitals. *International Journal of Medical Informatics*, *77*(3), 176–183. doi:10.1016/j.ijmedinf.2007.04.010

Fisk, A., Rogers, W. A., Charness, N., Czaja, S. J., & Sharit, J. (2004). *Designing for Older Adults: Principles and Creative Human Factors Approaches*. Boca Raton, FL: CRC Press.

Fisk, J. (1993). Design for the Elderly: a Biological Perspective. *Applied Ergonomics*, *24*(1), 47–50. doi:10.1016/0003-6870(93)90161-2

Fixson, S. K. (2001). *Three Perspectives on Modularity – A Literature Review of a Product Concept for Assembled Hardware Products*. (Working Paper ESD-WP-2001-06). Cambridge, MA: MIT Engineering Systems Division.

Fleck, J. (1999). Learning by trying: the implementation of configurational technology. In MacKenzie, D., & Wajcman, J. (Eds.), *The Social Shaping of Technology* (2nd ed.). Buckingham, UK: Open University Press.

Flores, F. (1993). Innovation by listening carefully to customers. *Long Range Planning*, *26*(3), 95–102. doi:10.1016/0024-6301(93)90011-4

Flusser, V. (2003). *Filosofia del design*. Milan, Italy: Bruno Mondadori.

FNB. (2008). *FNB's Stokvel Account, making saving more attractive*. Retrieved May 6, 2009, from http://www.fnb.co.za/news/archives/2008/20080413stokvel.html

Fong, P. S. W. (2003). Knowledge creation in multidisciplinary project teams: an empirical study of the processes and their dynamic interrelationships. *International Journal of Project Management*, *21*(7), 479–486. doi:10.1016/S0263-7863(03)00047-4

Forlizzi, J., Hirsch, T., Hyder, E., & Goetz, J. (2001). Designing Pleasurable Technology for Elders. In *Proceedings of INCLUDE 01 - International Conference on Inclusive Design and Communication.*

Formosa, D. (2006). *Social responsibility through design: Smart Design calls for a wide-angle view of Universal Design*. Retrieved September 14, 2009, from http://www.smartdesignworldwide.com/news/article.php?id=98http://www.smartdesignworldwide.com/news/article.php?id=98 (accessed 14th September 2009)

Fornari, D. (2008). *Perception of Humanoid Artifacts: Cultural Differences*, in Fujita, H. (Ed.), *Proceedings of the 6th ICDHS Conference*, (pp. 312-315). Osaka, Japan: KSI.

Forrester, J. W. (1961). *Industrial dynamics*. Cambridge, MA: M.I.T. Press.

Forsythe, D. E. (1999). It's just a matter of common sense, ethnography as invisible work. *Computer Supported Cooperative Work*, *8*, 127–145. doi:10.1023/A:1008692231284

Fortunati, L. (2003). Real People, Artificial Bodies. In Fortunati, L., Katz, J. E., & Riccini, R. (Eds.), *Mediating the Human Body: Technology, Communication, and Fashion* (pp. 61–70). Mahwah, NJ: Lawrence Erlbaum.

Foss, S. K. (1996). *Rhetorical Criticism – Exploration & Practice*. Long Grove, IL: Waveland Press Inc.

Foster, D. (2001). *The land where stories end: As narrated by the angel depicted in Madonna Con Bambino E Due Angeli by Felippo Lippi. Potts Point, N.S.W.* Duffy and Snellgrove.

Fox, R. (1981). Design based studies: an action-based 'form of knowledge' for thinking, reasoning and operating. *Design Studies*, *2*(1), 33–40. doi:10.1016/0142-694X(81)90027-2

Foxall, G. (1996). *Consumers in Context*. London: Routledge.

Foxall, G., & Tierney, J. (1984). From CAP 1 to CAP 2: User-Initiated Innovation from the User's Point of View. *Management Decision*, *22*(5), 3–15. doi:10.1108/eb001358

Frambach, R. T., Barkema, H. G., Nooteboom, B., & Wedel, M. (1998). Adoption of a Service Innovation in the Business Market An Empirical Test of Supply-Side Variables. *Journal of Business Research*, *41*, 161–174. doi:10.1016/S0148-2963(97)00005-2

Franke, N., & Piller, F. (2004). Value Creation by Toolkits for User Innovation and Design: The Case of the Watch Market. *Journal of Product Innovation Management*, *21*(6), 401–415. doi:10.1111/j.0737-6782.2004.00094.x

Franke, N., & Piller, F. T. (2003). Key research issues in user interaction with user toolkits in a mass customisation system. *International Journal of Technology Management*, *26*(5-6), 578–599. doi:10.1504/IJTM.2003.003424

Franke, N., & Shah, S. (2003). How communities support innovative activities: an exploration of assistance and sharing among end-users. *Research Policy*, *32*(1), 157–178. doi:10.1016/S0048-7333(02)00006-9

Franke, N., von Hippel, E., & Schreier, M. (2006). Finding Commercially Attractive User Innovations: A Test of Lead-User Theory. *Journal of Product Innovation Management*, *23*(4), 301–315. doi:10.1111/j.1540-5885.2006.00203.x

Frederick, H. H. (2002). *The Unitec Global Entrepreneurship Monitor 2003/2004: Toward High growth Enterprise in New Zealand* (World Wide Web (Internet)): New Zealand Centre for Innovation & Entrepreneurship.

Fredrikson, B. L., & Carstensen, L. L. (1990). Choosing Social Partners: How Old Age and Anticipated Endings Make People More Selective. *Psychology and Aging*, *5*(3), 335–347. doi:10.1037/0882-7974.5.3.335

Freeman, C. (1968). Chemical Process Plant: Innovation and the World Market. *National Institute Economic Review*, *45*(1), 29–51. doi:10.1177/002795016804500104

Freitas, N. K. (2003). Visões Paralelas de Moda e Psicologia do Consumidor. In: QUIRINO, S. F. S. (Ed.), *Modapalavra*, (vol. 2, nº. 2, pp. 112-117). Florianópolis/SC, Brazil: UDESC/CEART, 2003.

Freud, S. (1949). *An outline of psychoanalysis*. New York: Norton.

Freudenthal, A. (1999). *The design of home appliances for young and old consumers*. Delft: Delft University Press.

Fricke, G. (1993). Empirical Investigations of Successful Approaches When Dealing with Differently Precised Design Problems. *International Conference on Engineering Design ICED93*, Heurista, Zurich.

Fricke, G. (1996). Successful Individual Approaches in Engineering Design. *Research in Engineering Design*, *8*, 151–165. doi:10.1007/BF01608350

Friesen, G. B. (2001). Co-creation: When 1 and 1 make 11. *Consulting to Management*, *12*(1), 28–31.

Fruchter, R., & Townsend, A. (2003). Multi-Cultural Dimensions and Multi-Modal Communication in distributed, Cross-Disciplinary Teamwork. *International Journal of Engineering Education*, *19*(1), 53–61.

Fry, R. (2006). *Defining the Obvious: Explaining Creativity and Design Thinking to Nondesigners*. Paper presented at the 2006 IDSA National Education Conference.

Füller, J., & Bartl, M. (2007, October 7-10). *What Consumers Expect from Virtual Co-Creation.* Paper presented at the Mass Customization & Personalization Conference 2007, Cambridge, MA.

Füller, J., Bartl, M., Ernst, H., & Mühlbacher, H. (2006). Community based innovation: How to integrate members of virtual communities into new product development. *Electronic Commerce Research*, *6*(1), 57–73. doi:10.1007/s10660-006-5988-7

Fulton Suri, J. (2003). Empathic design: informed and inspired by other people's experience. In Koskinen, I., Battarbee, K., & Mattelmäki, T. (Eds.), *Empathic Design: User Experience in Product Design*. Finland: IT Press.

Fung, A., Lo, A., & Rao, M. N. (2005). *Creative Tools*. Hong Kong: The School of Design, The Hong Kong Polytechnic University.

Gad, T. (2001). *4-D Branding: Cracking the Corporate Code of the Network Economy*. New York: Financial Times Prentice Hall.

Gadon, E. W. (2006). My life's journey as researcher. *Journal of Research Practice, 2*(2), Article M1. Retrieved December 8, 2008, from http://jrp.icaap.orh/index.php/jrp/article/view/40/53

Gallagher, S. (Ed.). (2006). *The presence of things: Sense, veneer and guise*. Victoria: The Embroiders Guild Victoria.

Gardiner, P., & Rothwell, R. (1985). Tough Customers: Good Designs. *Design Studies*, *6*(1), 7–17. doi:10.1016/0142-694X(85)90036-5

Gault, F., & von Hippel, E. (2009). *The prevalence of user innovation and free innovation transfers: Implications for statistical indicators and innovation policy* (Unpublished Working Paper). Cambridge, MA: MIT Sloan School of Management.

Gaver, B., Dunne, T., & Pacenti, E. (1999). Cultural Probes. *Interaction*, *6*(1), 21–29. doi:10.1145/291224.291235

Gedenryd, H. (1998). *How Designers Work. Making Sense of Authentic Cognitive Activities. Lund University Cognitive Studies 75*. Lund, Sweden: Lund University.

Geertz, C. (1973). *Thick Description: Toward an Interpretative Theory of culture. The Interpretation of Cultures*. New York: Basic Books.

Gellersen, H. W., Schmidt, A., & Beigl, M. (2000). *Adding Some Smartness to Devices and Everyday Things*. Monterrey, USA: IEEE Workshop on Mobile Computing Systems and Applications.

Gero, J. S. (1990). Design Prototypes: A Knowledge Representation Schema for Design. *AI Magazine*, *11*(4), 26–36.

Ghosh, S. (1993). An Exercise in Inducing Creativity in Undergraduate Engineering Students Through Challenging Examinations and Open-Ended Design Problems. *IEEE Transactions on Education*, *36*(1), 113–119. doi:10.1109/13.204828

Giddens, A. (1984). *The Constitution of Society*. Berkeley, CA: University of California Press.

Giddens, A. (1991). *Modernity and Self-Identity: Self and Society in the Late Modern Age*. Stanford, CA: Stanford University Press.

Gilly, M. C., & Zeithaml, V. A. (1985). The Elderly Consumer and Adoption of Technologies. *The Journal of Consumer Research*, *12*(3), 353–357. doi:10.1086/208521

Gimeno, J. M. I. (2000). *La Gestión del Diseño en la Empresa*. Madrid, Spain: McGraw-Hill/Interamericana de España.

Gini, G., & Gini, M. (1983). *Robot: Controllo, programmazione, interazione con l'ambiente*. Milan, Italy: Clup.

Ginnerup, S. (2009). *Achieving full participation through Universal Design* (pp. 7–8). Strasbourg: Council of European Publishing.

Giuntella, F., & Zucconi, A. (1984). *Fabbrica, comunità, democrazia: testimonianze su Adriano Olivetti e il Movimento Comunità*. Rome: Fondazione Adriano Olivetti.

Givechi, R., & Velasquez, V. (2004). Positive Space. In Mcdonaugh, D., Hekkert, P., van Erp, J., & Gyi, D. (Eds.), *Design and Emotion: The Experience of Everyday Things* (pp. 12–17). London: Taylor and Francis.

Glasgow School of Art. (2009). PDE Program - Relevance to Industry and Society. Retrieved from www.gsa.ac.uk/gsa.cfm?pid=1421

Gobeli, D. H., & Brown, D. J. (1993). Improving the process of product innovation. *Research Technology Management*, *32*(2), 38–45.

Goedkoop, M., Effting, S., & Collignon, M. (2000). *The Eco-indicator 99 - A damage oriented method for Life Cycle Impact Assessment Manual for Designers*. Amersfoort, The Netherlands: PRé Consultants.

Goedkoop, M., Schryver, A., & Oele, M. (2007). *Introduction to LCA with SimaPro 7*. Amersfoort, The Netherlands: PRé Consultants.

Goel, V. (1995). *Sketches of Thought*. Cambridge, MA: MIT Press.

Goellner, M. (2005). *Addressing Complexity in Product Design: Guidelines for Product Designers* (Unpublished PhD Thesis). University of Otago, Dunedin (New Zealand).

Goldenberg, J., & Mazursky, D. (2002). *Creativity in Product Innovation*. Cambridge, UK: Cambridge University Press. doi:10.1017/CBO9780511674464

Goldschmidt, G. (1991). The Dialectics of Sketching. *Creativity Research Journal*, *4*(2), 123–143. doi:10.1080/10400419109534381

Göller, A. (1994). What is the Cause of Perpetual Style Change in Architecture? In Mallgrave, H. F., & Ikonomou, E. (Eds.), *Empathy, Form, and Space: Problems in German Aesthetics* (pp. 1873–1893). Los Angeles: The Getty Center For The History Of Art.

Gomarasca, A. (Ed.). (2001). *La bambola e il robottone. Cultura pop nel Giappone contemporaneo*. Turin, Italy: Einaudi.

Gomes, L. V. N. (2001). *Criatividade: Projeto, desenho, produto*. Santa Maria/RS, Brazil: sCHDs.

Gordon, W. J. J. (1961). *Synectics, the development of creative capacity*. New York: Harper and Row Publishers.

Gould, J. D., & Lewis, C. (1985). Designing for Usability: Key Principles and What Designers Think. *Communications of the ACM*, *28*(3), 300–311. doi:10.1145/3166.3170

Grabowski, H., Lossack, R. S., & Weis, C. (1995). A Design Process Model based on Design Working Spaces. In *T. T., M. M., & S. Finger, Knowledge Intensive CAD-1* (pp. 26–29). New York: Springer.

Grady, P. (1999). *The age of Modularity*. Iowa: Adams and Steele Publishers.

Graham & Rouncefild. (2008). Probes and Participation. In *Proceedings of the Participatory Design Conference* (PDC08).

Grasso, D., & Martinelli, D. (2007). Holistic Engineering. *The Chronical Review*, *53*(28), B8.

Gratton, L., Volgt, A., & Erickson, T. (2007). Bridging faultlines in diverse teams. *MIT Sloan Management Review*, *48*(4), 21–29.

Green, D., Stanton, N., Walker, G., & Salmon, P. (2005). *Using wireless technology to develop a virtual reality command and control centre, Virtual Reality*. New York: Springer.

Green, G., & Kennedy, P. (2001). Redefining Engineering Education: The Reflective Practice of Product Design Engineering. *International Journal of Engineering Education*, *17*(1), 3–9.

Green, P. A. (1974). *Design education; problem solving and visual experience*. London: Batsford.

Green, W., & Jordan, P. W. (Eds.). (2002). *Pleasure with Products: Beyond Usability*. London: Taylor & Francis.

Greenbaum, J., & Kyng, M. (Eds.). (1991). *Design at Work: Cooperative Design of Computer Systems*. Hillsdale, NJ: Lawrence Erlbaum Ass.

Greenberg, K. (October 20, 2003). Rock on, folks: Caddy's Back. *BrandWeek*. Retrieved November 6, 2003, from http://O-web.lexis-nexis.com.library.lib.asu.edu:80/universe/printdoc

Gregor, P., Newell, A. F., & Zajicek, M. (2002). Designing for Dynamic Diversity: Interfaces for Older People. *Assets, 2002*, 151–156.

Gribbin, J. R. (2004). *Deep simplicity: bringing order to chaos and complexity* (1st U.S. ed.). New York: Random House.

Griffin, A. (1992). Evaluating QFD's use in US firms as a process for developing products. *Journal of Product Innovation Management*, *9*(3), 171–187. doi:10.1016/0737-6782(92)90028-B

Griffin, A., & Hauser, J. R. (1993). The Voice of the Customer. *Marketing Science*, *12*(1), 1–27. doi:10.1287/mksc.12.1.1

Griffith, S. (2007, May 15-17). Designing Undergraduate and Postgraduate Design Management and Practice Courses, Design—Challenge. In *Proceedings of the Design Management Conference*.

Griffith, S., & Bamford, R. (2007, 9-12). Embedding the Principles of Responsible Design into Student Practice from 1st year, Connected 2007. In *Proceedings of International Conference on Design Education*. Sidney, Australia: University of New South Wales.

Groat, L., & Wang, D. (2002). *Architectural Research Methods*. New York: J. Wiley.

Gruner, K. E., & Homburg, C. (2000). Does Customer Interaction Enhance New Product Success? *Journal of Business Research*, *49*(1), 1–14. doi:10.1016/S0148-2963(99)00013-2

Guérin, F., Laville, A.; Daniellou, F.; Durrafourg, J., & e Kerguelen, A. (1991). Comprendre le travel pour le transformer. Paris, ANACT. (in French)

Guilford, J. P. (1950). Creativity. *The American Psychologist*, *5*, 444–454. doi:10.1037/h0063487

Gulati, R. K., & Eppinger, S. D. (1996). *The Coupling of Product Architecture and Organizational Structure Decisions*. (Working Paper #3906-96). Cambridge, MA: MIT Sloan School of Management.

Gupta, S. M., & Mclean, C. R. (1996). Disassembly of products. In *Proceedings of the 19th International Conference on Computers and Industrial Engineering*, (v. 31, n. 1-2, p. 225-228).

Gupta, S. P., Keates, S., & Clarkson, P. J. (2003). Usability Considerations in the Design of Handheld Electronics. In *Proceedings of the 14th International Conference on Engineering Design ICED*, Stockholm, Sweeden.

Hall, P. (2009). A Good Argument. *Metropolis*. Retrieved March 3, 2009, from http://www.metropolismag.com/story/20090318/a-good-argument

Halter, R. (2005). Remembering the Future: Memory and Interaction. In Buurman, G. M. (Ed.), *Total Interaction: Theory and practice of a new paradigm for the design disciplines*. Basel: Birkhäuser.

Hamel, G., & Prahalad, C. K. (1994). *Competing for the Future*. Boston: Harvard Business School Press.

Hammer, J. (2007). Sustainability - an Introduction from the National President. Retrieved November 12, 2008, from http://www.engineersaustralia.org.au/ieaust//quicklinks/sustainability.cfm

Hammersley, M., & Atkinson, P. (2003). *Ethnography: principles in practice*. London: Routledge.

Han Boon, L., Ko, M. S. H., Gay, R. K. L., Kah, F., Leong, K., & Chua, C. (1992). Using computer-based tools and technology to improve jewellery design and manufacturing. *International Journal of Computer Applications in Technology*, *5*(1), 72–80.

Hancock, T. (2005). Towards a free matter economy (Part 3). Designing the Narya Bazaar [Electronic Version]. *Free Software Magazine*. Retrieved September 10, 2006 from http://www.freesoftwaremagazine.com/free_issues/issue_09/free_matter_economy_3.

Hanft, T. A., & Kroll, E. (1996). Ease-of-disassembly evaluation in design for recycling. In Huang, G. Q. (Ed.), *Design for X: Concurrent Engineering Imperatives*. London: Chapman & Hail.

Hansen, M. T., & Birkinshaw, J. (2007). The Innovation Value Chain. *Harvard Business Review*, *85*(6), 121–130.

Hargadon, A. (2005). Leading with Vision: The Design of New Ventures. *Design Management Review*, *16*(1), 33–39. doi:10.1111/j.1948-7169.2005.tb00005.x

Harhoff, D., Henkel, J., & Hippel, E. v. (2003). Profiting from voluntary information spillovers: how users benefit by freely revealing their innovations. *Research Policy*, *32*(10), 1753–1769. doi:10.1016/S0048-7333(03)00061-1

Harmaakorpi, V., & Melkas, H. (2005). Knowledge Management in Regional Innovation Networks: The Case of Lahti, Finland. *European Planning Studies*, *13*(5), 641–659. doi:10.1080/09654310500139277

Hastings, D. (2004). *The future of engineering systems: development of engineering leaders. Engineering Systems Monograph*. Cambridge, MA: MIT Engineering Systems Division.

Hatchuel, A., & Weil, B. (2003). A new approach of innovative design: an introduction to C-K theory. In *Proceedings of the International Conference on Engineering Design – ICED'03* (pp. 109-110, executive summary). On CD-ROM. Stockholm.

Hazelrigg, G. A. (1998). A framework for decision based engineering design. *Journal of Mechanical Design*, 653–658. doi:10.1115/1.2829328

Health Data, O. E. C. D. (2006). *How does the United States Compare*. http://www.oecd.org/dataoecd/29/52/36960035.pdf

Heidegger, M. (1962). *Being and Time. John Macquarrie & Edward Robinson (trnls.)*. New York: Harper Collins.

Hekkert, P. (2006). Design aesthetics: Principles of pleasure in product design. *Psychological Science*, *48*(2), 157–172.

Helkema, R., Vugt, T. v., & Bosveld, K. (2008). Grote Marketing Enquete 2008: Marketeer bang voor mondige consument. *Tijdschrift voor Marketing, 2008*(Juli-Augustus), 16-24 (in Dutch).

Henderson, K. (1999). *On Line and on Paper: Visual Representations, Visual Culture, and Computer Graphics in Design Engineering*. Boston: MIT Press.

Henkel, J. (2006). Selective revealing in open innovation processes: The case of embedded Linux. *Research Policy*, *35*(7), 953–969. doi:10.1016/j.respol.2006.04.010

Henkel, J., & von Hippel, E. (2005). Welfare Implications of User Innovation. *The Journal of Technology Transfer*, *30*(1/2), 73–87.

Herb, E., Leslie, K., & Price, C. (2001). Teamwork at the top. *The McKinsey Quarterly*, *2*, 32–43.

Hernandez, G., Allen, J. K., Woodruff, G. W., Simpson, T. W., Bascaran, E., & Avila, L. F. (2001). Robust design of families of products with production modeling and evaluation. *Journal of Mechanical Design*, 183–190. doi:10.1115/1.1359786

Herstatt, C., & Kalogerakis, K. (2005). How to use analogies for breakthrough innovations. *International Journal of Innovation and Technology Management*, *2*(3), 331–347. doi:10.1142/S0219877005000538

Herstatt, C., & von Hippel, E. (1992). From experience: Developing new product concepts via the lead user method: A case study in a "low-tech" field. *Journal of Product Innovation Management*, *9*(3), 213–221. doi:10.1016/0737-6782(92)90031-7

Hertel, G., Niedner, S., & Herrmann, S. (2003). Motivation of software developers in Open Source projects: an Internet-based survey of contributors to the Linux kernel. *Research Policy*, *32*(7), 1159–1177. doi:10.1016/S0048-7333(03)00047-7

Hertzberger, H. (1991). *Lessons for students of architecture*. Rotterdam, UK: Uitgeverij 010.

Hienerth, C. (2006). The commercialization of user innovations: the development of the rodeo kayak industry. *R & D Management*, *36*(3), 273–294. doi:10.1111/j.1467-9310.2006.00430.x

Highmore, B. (2009). *The Design Culture Reader*. London: Routledge.

Hillier, B. (Ed.). (1975). *The connoisseur complete encyclopedia of antiques*. London: Peerage Books.

Hilton, P. D., & Jacobs, P. F. (2000). *Rapid Tooling: Technologies and Industrial Applications*. New York: Marcel Dekker, Inc.doi:10.1201/9780203908020

Hippel, E. (2005). *Democratizing Innovation*. Cambridge, MA: The MIT Press.

Hirsch, T., Forlizzi, J., Hyder, E., Goetz, J., Stroback, J., & Kurtz, C. (2000). The ELDer Project: Social, Emotional, and Environmental Factors in the Design of Eldercare Technologies. In *Proceedings of the Conference on Universal Usability CUU*. Virginia.

Hochschild, A. R. (1975). Disengagement Theory: A Critique and Proposal. *American Sociological Review*, *40*(5), 553–569. doi:10.2307/2094195

Hockenberry, J. (2009). Within the Product of No Product. *Metropolis*. Retrieved March 20m 2009 from http://www.metropolismag.com/story/20090318/within-the-product-of-no-product

Hodges, S., & McFarlane, D. (2005). *Radio frequency identification: technology, applications and impact*, AutoID Labs white paper.

Hoffman, H. J., Ishii, E. K., & Macturk, R. H. (1998). Age-related Changes in the Prevalence of Smell/Taste Problems among the United States Adult Population. *Annals of the New York Academy of Sciences*, *855*, 716–722. doi:10.1111/j.1749-6632.1998.tb10650.x

Hofstede, G. (2001). *Culture's Consequences: Comparing Values, Behaviors, Institutions and Organizations across Nations*. Thousand Oaks, CA: Sage.

Holt, K. (1987). User-oriented product innovation in theory and practice. *Technovation*, *6*(1), 37–45. doi:10.1016/0166-4972(87)90037-X

Holt, K. (1988). The role of the user in product innovation. *Technovation*, *8*, 249–258. doi:10.1016/0166-4972(88)90028-4

Holt, K., Geschka, H., & Peterlongo, G. (1984). *Need Assessment—A Key to User-Oriented Product Innovation*. Chichester, UK: Wiley and Sons.

Hong Kong Blind Union. (2007). *Ten tips to get along with visually impaired persons*. Retrieved June 22, 2010, from http://www.hkbu.org.hk/44.0.html/

Horn, J. I., & Cattel, R. B. (1967). Age Differences in Fluid and Crystallized Intelligence. *Acta Psychologica*, *26*, 107–129. doi:10.1016/0001-6918(67)90011-X

Hornyak, T. N. (2006). *Loving the Machine. The Art and Science of Japanese Robots*. Tokyo: Kodansha.

Howard-Jones, P. A., & Murray, S. (2003). Ideational productivity, focus of attention, and context. *Creativity Research Journal*, *15*(2&3), 153–166. doi:10.1207/S15326934CRJ152&3_07

Howe, J. (2006, June 2006). The Rise of Crowdsourcing. *Wired, 14.06*.

Howe, J. (2008). *Crowdsourcing: Why the power of the crowd is driving the future of business*. New York: Crown Business.

Huang, G. Q. (1996). *Design for X: Concurrent Engineering Imperatives*. London: Chapman & Hall.

Huberman, A. M., & Miles, M. B. (1994). Early Steps in Analysis. In Norman K. Denzin and Yvonna S. Lincoln (Eds.), *Handbook of Qualitative Research*. Thousand Oaks, CA: Sage. (pp.50-88, 428-444).

Hundal, M. (1990). Research in design theory and methodology in west Germany. In *Proceedings of the ASME Design Theory and Methodology Conference – DTM'90* (pp. 235-238), Chicago. New York: ASME.

Hutchinson, B. (2008). *Systems Thinking and Associated Methodologies*. Perth, Australia: Praxis Education.

Ichbiah, D. (2005). *Robots. From Science Fiction to Technological Revolution*. New York: Harry N. Abrams.

ICSID. (2001). *Design for Future Needs*. Retrieved December 10, 2008, from http://www.icsid.org/resources/library.htm

ICSID/Danish Designers. (2008). *The increasingly vital role of design*. Retrieved December 10, 2008, from http://www.icsid.org/resources/library.htm

IdN Magazine (Ed.). (2005). *SINGAPOREdge: Design Culture in Singapore*. Singapore: Sysems Design Limited.

IDSA. (n.d.). *IDSA*. Retrieved from http://www.idsa.org

IME. (1985). *The Formation of Mechanical Engineers: Present and Future Needs* (The Grant Report).

Imrie, R., & Hall, P. (2001). *Inclusive design: Designing and developing accessible environments*. London: Spon Press.

Industrial Designers Society of America (IDSA). (2009). ID Defined (Publication no. http://www.idsa.org/absolutenm/templates/?a=89&z=23). Retrieved March 2009, from Industrial Designers Society of America. Retrieved from http://www.idsa.org/absolutenm/templates/?a=89&z=23

Innosupport: Supporting Innovation in SMEs. (2005). *Education and Culture. Leonardo da Vinci. Pilot Projects.* Retrieved September 16, 2006, from www.innosupport.net

International Enterprise Singapore. (2009). *About us; Milestones.* Retrieved from www.iesingapore.gov.sg

Intrachooto, S. (2004). Lead users concept in building design: its applicability to member selection in technologically innovative projects. *The TQM Magazine, 16*(5), 359–368. doi:10.1108/09544780410551304

IPT. Instituto de Pesquisas Tecnológicas. (2000). *Lixo municipal: manual de gerenciamento de lixo.* (2. ed.), São Paulo: Imprensa Oficial. (in Portuguese)

Ishii, K. (1995). Life-Cycle Engineering Design. *ASME Journal of Vibration and Acoustics, 117*(B), 42-47.

Iwabuchi, K. (1988). Marketing "Japan": Japanese Cultural Presence under a Global Gaze. In Highmore, B. (Ed.), *The Design Culture Reader.* London: Routledge.

Jackson, L. (2005, August). Hella Jongerius. *Icon., 26.* Retrieved from www.iconeye.com

Jana, R. (2009). Innovation Trickles in a New Direction. *BusinessWeek.* Retrieved March 20, 2009, from http://www.businessweek.com/magazine/content/09_12/b4124038287365.htm

Jansson, D. G., & Smith, S. M. (1991). Design Fixation. *Design Studies, 12*(1), 3–11. doi:10.1016/0142-694X(91)90003-F

Janzik, L., & Herstatt, C. (2008, September 21-24). *Innovation Communities: Motivation and Incentives for Community Members to Contribute.* Paper presented at the International Conference on Management of Innovation and Technology (ICMIT) 2008, Bangkok, Thailand.

Japan Industrial Design Promotion Organization (Ed.). (2002). *Design Awards Asia.* Tokyo: Japan Industrial Design Promotion Organization (JIDPO).

Jarvis, C. (2008). *The Business Idea – Product and Service Concept.* Retrieved April 10, 2008, from http://www.bola.biz/businesses/busplan/product.html

Jencks, C. (1971). [London: Studio Vista Ltd.]. *Architecture (Washington, D.C.),* 2000.

Jeppesen, L. B. (2005). User Toolkits for Innovation: Consumers Support Each Other. *Journal of Product Innovation Management, 22*(4), 347–362. doi:10.1111/j.0737-6782.2005.00131.x

Jeppesen, L. B., & Frederiksen, L. (2004, January 5). *Innovative Users in a Firm-Established User Community: The characteristics of users co-developing "computer-controlled music instruments".* Paper presented at the DRUID, PhD Winter Conference, Aalborg, DK.

Jeppesen, L. B., & Molin, M. (2003). Consumers as Co-developers: Learning and Innovation Outside the Firm. *Technology Analysis and Strategic Management, 15*(3), 363–381. doi:10.1080/09537320310001601531

Jeswiet, J. A. (2003). Definition of life cycle engineering. In Bley H. (Ed.), *Proceedings of the 36th CIRP international seminar on manufacturing systems* (pp. 17-20). Saarbrücken: CIRP.

Johansson, F. (2004). *The Medici Effect: breakthrough insights at the intersection of ideas.* Boston, MA: Harvard Business School Press.

Johansson, K., Perzon, M., Fröling, M., & Mossakowska, A. (2007). Sewage sludge handling with phosphorus utilization –life cycle assessment of four alternatives. *Journal of Cleaner Production, 16*(1), 135–151. doi:10.1016/j.jclepro.2006.12.004

Johnson, B. M. (2003). The Paradox of Design research – The role of Informance. In Laurel, B. (Ed.), *Design Research Methods and Perspectives.* London: The MIT Press.

Johnson, J. (2006). Embracing Complexity in Design. Retrieved April 10, 2009, from http://www.complexityanddesign.net/

Johnson, S. (2001). *Emergence: the connected lives of ants, brains, cities, and software.* New York: Scribner.

Jones, C. R. M., & Fazio, R. H. (2008). Associative Strength and Consumer Choice Behavior. Haugtvedt C. P.,

Herr P. M., & Kardes F. R. (Eds.), *Handbook of Consumer Psychology*. New York: Taylor & Francis.

Jones, J. C. (1984). *Essays in Design*. Chichester, UK: Wiley.

Jones, J. C. (1988). Softecnica. In Thakara, J. (Ed.), *Design After Modernism: Beyond the Object*. London: Thames & Hudson.

Jongerius, H., & Schouwenberg, L. (2003). *Hella Jongerius*. London: Phaidon.

Joost, G. (2007). Design as Rhetoric. In *Proceedings of the Symposium of Swiss Design Network*.

Jordan, P. (2002). The Personalities of Products. In Green, W., & Jordan, P. (Eds.), *Pleasure with Products: Beyond Usability*. London: Taylor & Francis.

Jordan, P. W. (1999). Pleasure with products: Human factors for body, mind and soul. In Green, W. S., & Jordan, P. W. (Eds.), *Human Factors in Design: Current Practice and Future Trends*. London: Taylor & Francis.

Jordan, P. W. (2000). *Designing Pleasurable Products*. London: Taylor Francis.

Jordan, P. W., & Green, W. S. (1999). *Human factors in product design: Current practice and future trends*. London: Taylor and Francis.

Jørgensen, C., & Onsberg, M. (1999). *Praktisk Argumentation*. København: Teknisk Forlag.

Kahney, L. (2002). *Apple: It's All About the Brand*. Retrieved December 18, 2003, from http://www.wired.com/news/mac/0,2125,56677,00.html

Kailes, J. (2002). *Emergency evacuation preparedness taking responsibility for your safety: a guide for people with disabilities and other activity limitations*. Center for Disability Issues and the Health Professions.

Kalpakjian, S. (1995). *Manufacturing Engineering Technology*. Reading, MA: Addison-Wesley Publishing Company.

Kälviäinen, M. (2002). Product design for consumer taste. In Green, W., & Jordan, P. (Eds.), *Pleasure with Products. Beyond Usability* (pp. 77–96). London: Taylor & Francis.

Kälviäinen, M. (2007). User experience capturing through Make Do innovation sessions. In *Proceedings of QMOD 2007 & First European Conference on Affective Design and Kansei Engineering*. Helsingborg, Sweden, 18-20 June. Lund University electronic publications.

Kälviäinen, M., & Miller, H. (2005, September 22-24). Visual research: means of producing shared meanings. In *Proceedings of Joining Forces International Conference on Design Research*. Helsinki: UIAH.

Kan, I. P., Barsalou, L. W., Soloman, K. O., Minor, J. K., & Thompson-Shill, S. L. (2003). Role of mental imagery in a property verification task: fMRI evidence for perceptual representations of conceptual knowledge. *Cognitive Neuropsychology*, 2003, *20*(3/4/5/6), 525-540.

Kapferer, J. N. (1998). *Strategic Brand Management: Creating and Sustaining Brand Equity Long Term*. London: Kogen Page Limited.

Karagozoglu, N., & Brown, W. B. (1993). Time-based management of the new product development process. *Journal of Product Innovation Management*, *10*(3), 204–215. doi:10.1016/0737-6782(93)90026-M

Karjalainen, T.-M. (2006). *Semantic Transformation in Design*. Helsinki, Finland: University of Art and Design Helsinki.

Kauffman, S. (1995). *At Home in the Universe*. New York: Oxford University Press.

Kaulio, M. A. (1998). Customer, consumer and user involvement in product development: A framework and a review of selected methods. *Total Quality Management*, *9*(1), 141–149. doi:10.1080/0954412989333

Kawamura, Y. (2005). *Fashion-ology: an introduction to fashion studies*. New York: Berg.

Keates, S., & Clarkson, P. J. (2003). *Countering Design Exclusion: An Introduction to Inclusive Design*. London: Springer Verlag.

Keates, S., Clarkson, P. J., Harrison, L. J., & Robinson, P. (2000). *Towards a practical inclusive design approach. In 1st ACM Conference on Universal Usability* (CUU 2000), Arlington, VA, USA, 45-52. Retrieved April 13, 2009, from http://www-edc.eng.cam.ac.uk/cgi-bin/publications.cgi?740

Keen, A. (2007). *The Cult of the Amateur. How Today's Internet Is Killing Our Culture.* New York: Random House.

Kei, C. C., Meng, C. S., Ching, L. S., Tei, L. S., & Aung, S. C. (2000). Facial prosthetic model fabrication using rapid prototyping tools. *Integrated Manufacturing Systems, 11*(1), 42–53. doi:10.1108/09576060010303668

Keller, R., Eckert, C. M., & Clarkson, P. J. (2006). Matrices of node-link diagrams: which visual representation is better for visualizing connectivity models? *Information Visualization*, 62-76.

Kelley, P., Sanspree, M., & Davidson, R. (2000). Vision impairment in children and youth. In Silverstone, B., Lang, M. A., Rosenthal, B., & Faye, E. E. (Eds.), *The lighthouse handbook on vision impairment and vision rehabilitation* (*Vol. 2*, pp. 1137–1151). New York: Oxford University Press.

Kelley, T., & Littman, J. (2001). *The Art of Innovation. Lessons in Creativity from IDEO.* San Francisco: IDEO.

Kelley, T., & Littman, J. (2006). *The ten faces of innovation: strategies for heightening creativity.* London: Profile Books.

Keys, L. K. (1990). System Life Cycle Engineering and DF"X". *IEEE Transactions on Components. Hybrids, and Manufacturing Technology, 13*, 83–93. doi:10.1109/33.52854

Khuri, M. L. (2004). Working with emotion in educational intergroup dialogue. *International Journal of Intercultural Relations, 28*, 595–612. doi:10.1016/j.ijintrel.2005.01.012

Kim, H., & Fellner, D. W. (2004). *Interaction with hand gesture for a back-projection wall. Computer Graphics International.* Washington, DC: IEEE.

Kim, R.W., & Mauborgne (2005). *Blue Ocean Strategy.* Boston: Harvard Business School Press

Kim, S.-H., & Jang, K.-J. (1999). Designing Performance Analysis and IDEF0 for Enterprise Modelling in BPR. *International Journal of Production Economics*, 121–133.

Kipling, D., & Faragher, R. G. A. (1999). Ageing Hard or Hardly Ageing? *Nature, 398*, 191–193. doi:10.1038/18306

Kirkwood, T. B. L. (2002). Evolution of Ageing. *Mechanisms of Ageing and Development, 123*, 737–745. doi:10.1016/S0047-6374(01)00419-5

Kirschman, C. F., Fadel, G. M., & Jara-Almonte, C. C. (1996). Classifying Functions for Mechanical Design. *Design Engineering Technical Conference.* Las Vegas, NV: ASME.

Kjeldsen, J. E. (2004). *Retorik i vår tid.* Spartacus: Forlag

Klujeff, M. L. (2006). Retorisk publikum In Klujeff M.L. & Roer H. Hans (edt), Retorikkens aktualitet. Reitzels: Forlag.

Knapp, D. (2000). *The Brand Mindset.* New York: McGraw-Hill.

Knuckey, S., Johnston, H., Campbell-Hunt, C., Carlaw, K., Corbett, L., & Massey, C. (2002). *Firm Foundations: A Study of New Zealand Business Practices & Performance.* Wellington: New Zealand Ministry of Economic Development.

Kölsch, M., & Turk, M., (2002). Keyboards without Keyboards: A Survey of Virtual Keyboards, *UCSB Technical Report, 21.*

Kratzer, J., Leenders, R. T., & Engelen, J. M. (2004). Stimulating the potential: creative performance and communication in innovation teams. *Creativity and Innovation Management, 13*, 63–71. doi:10.1111/j.1467-8691.2004.00294.x

Kratzer, J., Leenders, R., & Engelen, J. (2006). Managing creative performance in virtual environments: an empirical study in 44 R&D teams. *Technovation, 26*, 42–49. doi:10.1016/j.technovation.2004.07.016

Kratzer, J., Leenders, R., & Engelen, J. (2008). The social structure of leadership and creativity in engineering design teams: an empirical analysis. *Journal of Engineering and Technology Management, 25*, 269–286. doi:10.1016/j.jengtecman.2008.10.004

Krippendorff, K. (2006). *The Semantic Turn – a New Foundation for Design.* Boca Raton, FL: Taylor & Francis Group.

Kristensson, P., Gustafsson, A., & Archer, T. (2004). Harnessing the Creative Potential among Users. *Jour-*

nal of Product Innovation Management, 21(1), 4–14. doi:10.1111/j.0737-6782.2004.00050.x

Kriwet, A., Zussman, E., & Seliger, G. (1995). Systematic integration of design-for-recycling into product design. *International Journal of Production Economics, 38*(1), 15–22. doi:10.1016/0925-5273(95)99062-A

Kroeber, A. L. (1919). On the principle of Order in Civilization as Exemplified by Changes of Fashion. *American Anthropologist, 21*(3), 235–263. doi:10.1525/aa.1919.21.3.02a00010

Kroll, E., & Carver, B. S. (1999). Disassembly analysis through time estimation and metrics. *Robotics and Computer-integrated Manufacturing, 15*, 191–200. doi:10.1016/S0736-5845(99)00026-5

Krozer, Y., (2006). Life cycle costing for innovations in product chains. *Journal of Cleaner Production, 16*(3), 310–321.

Kuhn, T. (1996). *The Structure of Scientific Revolutions* (3rd ed.). Chicago: University Of Chicago Press.

Kukimoto, N. (2005). Scientific visualization in collaborative virtual environment with PDA-based control and 3D annotation. [Tokyo: Japan Society of Mechanical Engineers.]. *JSME International Journal. Series B, Fluids and Thermal Engineering*, 48.

Kumar, P., & Mocko, G. M. (2007). Modeling and Analysis Of Engineering Design Activities Ontology Using the Design Structure Matrix. *Design Engineering Technical Conference.* Las Vegas, NV: ASME.

Kumar, S. (2002). *Methods for community participation: A complete guide for practitioners*. London: ITDG Publisher.

Kuniavsky, M. (2003). *Observing the user experience: A practitioner's guide to user research.* San Francisco, CA: Morgan Kaufmann.

Kurk, F., & Eagan, P. (2008). The value of adding design-for-the-environment to pollution prevention assistance options. *Journal of Cleaner Production, 16*(6), 722–726. doi:10.1016/j.jclepro.2007.02.022

Kurtzberg, T. R. (2005). Feeling creative, being creative: an empirical study of diversity and creativity in teams. *Creativity Research Journal, 17*(1), 51–65. doi:10.1207/s15326934crj1701_5

Kvale, S. (1996). *InterViews. An Introduction to Qualitative Research Interviewing*. Thousand Oaks, CA: SAGE Publications.

Kwint, M., Breward, C., & Aynsley, J. (Eds.). (1999). *Material memories*. Oxford, UK: Berg.

Kwok, Y. C. J., & Ku, H. B. B. (2001). *We are alive: Photonarrative of the everyday life of the elderly in Hong Kong* (Chinese ed.). Hong Kong: The Hong Kong Polytechnic University.

Kwok, Y. C. J., & Siu, K. W. M. (2002). *Participatory research for the designing of children and youth integrated services centres*. Hong Kong: The Hong Kong Polytechnic University.

Kyoto Design Declaration. (2008). *Cumulus Kyoto 2008 Conference*. Retrieved April 11, 2008, from http://www.cumulusassociation.org/

Lagrosen, S. (2005). Customer involvement in new product development: A relationship marketing perspective. *European Journal of Innovation Management, 8*(4), 424–436. doi:10.1108/14601060510627803

Lakhani, K. R., & von Hippel, E. (2003). How open source software works: "free" user-to-user assistance. *Research Policy, 32*(6), 923–943. doi:10.1016/S0048-7333(02)00095-1

Lambert Review of Business-University Collaboration. (2003). *Lambert Review of Business-University Collaboration.* London: HM Treasury. Retrieved April 13, 2009, from http://www.hmtreasury.gov.uk/d/lambert_review_final_450.pdf

Lang, F. R., & Carstensen, L. L. (2002). Time Counts: Future Time Perspective, Goals, and Social Relationships. *Psychology and Aging, 17*(1), 125–139. doi:10.1037/0882-7974.17.1.125

Langer, H. J. (2008). I want it now - pushes manufacturers into rapid responses. *Metal Powder Report, 63*(4), 8–9. doi:10.1016/S0026-0657(09)70035-9

LaSalle, D., & Britton, T. (2003). *Priceless: turning ordinary products into extraordinary experiences*. Boston, MA: Harvard Business Press.

Lashinsky, A. (2006, June 1). The team that saved Motorola. *Fortune Online*.

Latour, B. (2005). *Reassembling the Social*. Oxford, UK: Oxford University Press.

Laurel, B. (2003). *Design Research Methods and Perspectives*. Cambridge, MA: MIT Press.

Law, J. (2005). After ANT: complexity, naming and topology. In Law, J., & Hassard, J. (Eds.), *Actor Network Theory and After*. Oxford, UK: Blackwell.

Law, J., & Hassard, J. (1999). *Actor Network Theory and after*. Oxford, UK: Blackwell Publishers/The Sociological Review.

Lawson, B. R. (1979). Cognitive strategies in architectural design. *Ergonomics*, *22*(1), 59–68. doi:10.1080/00140137908924589

Leadbeater, C., & Miller, P. (2004). *The Pro-AmRevolution. How enthusiasts are changing our economy and society*. London: Demos.

Learmount, S., Press, M., & Cooper, R. (2000). *Design Against Crime. Report to the Design Council, Home Office and Department of Trade and Industry. ODPM (Office of the Deputy Prime Minister) (2004). Safer places. The planning system and crime prevention*. London, UK: Thomas Telford.

Lee, R. D., & Carter, L. R. (1992). Modeling and Forecasting U. S. Mortality. *Journal of the American Statistical Association*, *87*(419), 659–671. doi:10.2307/2290201

Lee, R., & Tuljapurkar, S. (1997). Death and Taxes: Longer Life, Consumption, and Social Security. *Demography. The Demography of Aging*, *34*(1), 67–81. doi:10.2307/2061660

Lee, R., & Tuljapurkar, S. (1998). *Uncertain Demographic Futures and Social Security Finances*. The American Economic Review, Papers and Proceedings of the Hundred and Tenth Annual Meeting of the American Economic Association.

Leifer, L. J., & Tang, J. C. (1988). *A Framework for Understanding the Workspace Activity of Design Teams*. New York: ACM.

Leiviskä, E. (2001). *Creative Interdisciplinarity. IDBM Programme. Department of Teacher Education. (Research Report 227)*. Helsinki: University of Helsinki.

Lenau, T., & Boelskifte, P. (October 1, 2004) Soft and Hard Product Attributes in Design. *Semantic & Aesthetic Functions in Design*. (Working Papers 28). Helsinki, Finland: University of Art and Design Helsinki.

Lencioni, P. (2002). *Make Your Values Mean Something*. Boston: Harvard Business School Publishing Corporation.

Lenss, V. G., & Pumpuni, G. A. (2001). A systems engineering approach to managing communication in globally dispersed teams. (Unpublished doctoral dissertation). Cambridge, MA: Massachusetts Institute of Technology.

Leonard, D., & Rayport, J. F. (1997). Spark Innovation Through Empathic Design. *Harvard Business Review*, *75*(6), 104–113.

Leonard-Barton, D. (1992). Core Capabilities and Core Rigidities: A Paradox in Managing New Product Development. *Strategic Management Journal*, *13*, 111–125. doi:10.1002/smj.4250131009

Leong, K. F., Chua, C. K., Chua, G. S., & Tan, C. H. (1998). Abrasive jet deburring of jewellery models built by stereolithography apparatus (SLA). *Journal of Materials Processing Technology*, *83*, 36–47. doi:10.1016/S0924-0136(98)00041-7

Leubner, C., Brockmann, C., & Muller, H. (2001). Computer-vision-based humancomputer interaction with a back projection wall using arm gestures. In *Proceedings of the Euromicro Conference*, Varsavia, Polonia.

Levenson, R. W., Carstensen, L. L., Friesen, W. V., & Ekman, P. (1991). Emotion, Physiology, and Expression in Old Age. *Psychology and Aging*, *6*(1), 28–35. doi:10.1037/0882-7974.6.1.28

Levitt, T. (n.d.). Ted Levitt quotes. *ThinkExist.com Quotations Online*. Retrieved September 20, 2009, from http://thinkexist.com/quotation/just_as_energy_is_the_basis_of_life_itself-and/161705.html

Lewis, L., & Farris, E. (1999). *An institutional perspective on students with disabilities in postsecondary education. Greene, B. project officer*. Washington, DC: U.S. Department of Education, National Center for Education Statistics.

Li, C., & Bernoff, J. (2008). *Groundswell: Winning in a World Transformed by Social Technologies*. Boston: Harvard Business Press.

Li, Z. (1998). Creativity and aesthetic sense. *Creativity Research Journal*, *11*(4), 309–313. doi:10.1207/s15326934crj1104_4

Lifchez, R., & Winslow, B. (1979). *Design for independent living: the environment and physically disabled people*. Berkeley, Los Angeles: University of California Press.

Liker, J. K. (2004). *The Toyota way: 14 management principles from the world's greatest manufacturer*. New York: McGraw-Hill.

Lilien, G. L., Morrison, P. D., Searls, K., Sonnack, M., & Hippel, E. v. (2002). Performance assessment of the lead user idea-generation process for new product development. *Management Science*, *48*(8), 1042–1059. doi:10.1287/mnsc.48.8.1042.171

Lima, R. M. R., & Romeiro, F. E. (2001). *The Material Recycling and Application in New Product Development: A Case Study*. (3rd Brazilian Congress of Product Development Management). Florianópolis, Santa Catarina: IGDP. (in Portuguese)

Linch, R. (2004). *The Mobile Connection: The Cell Phone's Impact on Society*. San Francisco: Morgan Kaufmann.

Lindholm, C., Keinonen, T., & Kiljander, H. (2003). *User interface styles. Mobile Usability: How Nokia Changed the Face of Mobile Phone*. New York: McGraw-Hill.

Linstone, H. A., & Turoff, M. (Eds.). (2002). *The Delphi Method: Techniques and Applications*. Newark, NJ: New Jersey Institute of Technology.

Lipovetsky, G. (1997). *O Império do Efêmero - a moda e seu destino na sociedade moderna. São Paulo/SP*. Brazil: Companhia das Letras.

Lipson, H., & Shpitalni, M. (2000). Conceptual design and analysis by sketching. *Artificial Intelligence for Engineering Design, Analysis and Manufacturing*, *14*, 391–401. doi:10.1017/S0890060400145044

Lloyd Jones, P. (1991). *Taste Today*. Oxford, UK: Pergamon Press.

Lockledge, J. C., & Salustri, F. A. (2001). Communication Using a Variation of the Design Structure Matrix. *International Conference on Engineering Design*. Glasgow: ICED.

Losano, M. G. (1990). *Storie di automi. Dalla Grecia classica alla Belle Epoque*. Turin, Italy: Einaudi.

Lossack, R., & Grabowski, H. (2000). The axiomatic approach in the universal design theory. First International Conference on Axiomatic Design. Cambridge, MA.

Louie, E. (2001, May 24). Objects designed to foster a little havoc in the home. *New York Times, 3*.

Lovelock, R. (1995). *Visual impairment, social support: Recent research in context*. Aldershort: Avebury.

Lowe, E. D., & Lowe, J. W. G. (1985). Quantitative Analysis of Women's Dress. In Solomon, M. R. (Ed.), *The Psychology of Fashion*. New York: Institute of Retail Management.

Lubart, T. I. (1997). Emotion, metaphor, and the creative process. *Creativity Research Journal*, *10*(4), 285–301. doi:10.1207/s15326934crj1004_1

Lubart, T. I. (2000-2001). Models of the creative process: past, present and future. *Creativity Research Journal*, *13*(3&4), 295–308.

Ludden, G. D. S., Schifferstein, H. N. J., & Hekkert, P. (2008). Surprise as a Design Strategy. *Design Issues*, *24*(2), 28–38. doi:10.1162/desi.2008.24.2.28

Ludwig, A. M. (1995). *The Price of greatness*. New York: Guilford.

Luhmann, N. (2002). *Theories of Distinction*. Stanford, CA: Stanford University Press.

Lundkvist, A., & Yakhlef, A. (2004). Customer involvement in new service development: a conversational approach. *Managing Service Quality*, *14*(2/3), 249–257. doi:10.1108/09604520410528662

Lupatini, M. (2005). *Relatório Setorial Preliminar: setor têxtil e vestuário. FINEP - Rede DPP*. Retrieved May 14, 2005, from www.finep.gov.br/portalDPP/relatório_setorial

Lurie, N., Rank, B., Parenti, C., Wooley, T., & Snoke, W. (1989). How Do House Officers Spend Their Nights? A

Time Study of Internal Medicine House Staff on Call. *The New England Journal of Medicine*, *320*, 1673–1677.

Lusch, R. F., & Vargo, S. L. (2006). Service-dominant logic: reactions, reflections and refinements. *Marketing Theory*, *6*(3), 281–288. doi:10.1177/1470593106066781

Lüthje, C. (2004). Characteristics of innovating users in a consumer goods field - An empirical study of sport-related product consumers. *Technovation*, *24*(9), 683–695. doi:10.1016/S0166-4972(02)00150-5

Lüthje, C., & Herstatt, C. (2004). The Lead User method: an outline of empirical findings and issues for future research. *R & D Management*, *34*(5), 553–568. doi:10.1111/j.1467-9310.2004.00362.x

Luttropp, C., & Lagerstedt, J. (2006). EcoDesign and The Ten Golden Rules: generic advice for merging environmental aspects into product development. *Journal of Cleaner Production*, *14*, 1396–1408. doi:10.1016/j.jclepro.2005.11.022

Lutz, C., & White, G. M. (1986). The Anthropology of Emotions. *Annual Review of Anthropology*, *15*, 405–436. doi:10.1146/annurev.an.15.100186.002201

Lynch, A., & Strauss, M. (2007). *Changing Fashion: a critical introduction to trend analysis and meaning*. New York: Berg.

Mace, R. (1997). *The Center for Universal Design, College of Design, NC State University*. Retrieved 12 April 2009 from http://www.design.ncsu.edu/cud/about_ud/about_ud.htm

Machlup, F. (1984). *Knowledge: Its Creation, Distribution and Economic Significance (3 Volumes)*. Princeton, NJ: Princeton University Press.

MacKenzie, D., & Wajcman, J. (Eds.). (1999). *The Social Shaping of Technology* (2nd ed.). Buckingham, UK: Open University Press.

Magee, C., & Frey, D. (2006). Experimentation and its role in engineering design: linking a student design exercise to new results from cognitive psychology. *International Journal of Engineering Education*, *22*(3).

Magnusson, P. R. (2009). Exploring the Contributions of Involving Ordinary Users in Ideation of Technology-Based Services. *Journal of Product Innovation Management*, *26*(5), 578–593. doi:10.1111/j.1540-5885.2009.00684.x

Magnusson, P. R., Kristensson, P., & Hipp, C. (2007, October, 8-10). *Exploring Ideation Patterns among Ordinary Users: The case of mobile telecommunication services.* Paper presented at the Mass Customization and Personalization Conference 2007, MIT, Cambridge, MA.

Magnusson, P. R., Matthing, J., & Kristensson, P. (2003). Managing User Involvement in Service Innovation. Experiments With Innovating End Users. *Journal of Service Research*, *6*(2), 111–124. doi:10.1177/1094670503257028

Mahlke, S. (2005). Understanding Users' Experience of Interaction. Source ACM International Conference Proceeding Series. In *Proceedings of the 2005 Annual Conference on European Association of Cognitive Ergonomics*.

Maldonado, T. (1976). *Disegno industriale: un riesame*. Milan, Italy: Feltrinelli.

Mankins, J. C. (1995). Technology Readiness Levels: A White Paper. In Technology ACO-OoSAa, (ed.), *Advanced Concepts Office of Space Access and Technology*. Washington DC: NASA

Mano, H., & Oliver, R. L. (1993). Assessing the Dimensionality and Structure of the Consumption Experience: Evaluation, Feeling, and Satisfaction. *The Journal of Consumer Research*, *20*(3), 451–466. doi:10.1086/209361

Manzini, E., & Jégou, F. (2003). *Sustainable everyday: scenarios of urban life*. Milano: Ambiente.

Manzini, E., & Vezzoli, C. (2003). *Product-service systems and sustainability: opportunities for sustainable solutions*. Paris: UNEP.

Maribondo, J. F. (2000). *Development of a Design Methodology for Modular Systems, Applied on Units Processing of Home Solid Residues*. (Thesis (Mechanical Engineering)). Federal University of Santa Catarina, Brazil. (in Portuguese)

Markides, C. C., & Geroski, P. A. (2005). *Fast Second: How Smart Companies Bypass Radical Innovation to Enter and Dominate New Markets*. New York: John Wiley & Sons.

Marsden, P. (2000, March). Brand Selection, Naturally Mapping Brand DNA, Engineering Mind Viruses and

Adapting to the Evolved Consumer. In *Proceedings of the Annual Market Research Society conference*. Brighton, 24-27. Retrieved December 5, 2004, from http://www.brandgenetics.com/archive/bsnat.htm

Martin, C. (2009). *Philippe Starck interview with Catherine Martin*. Vogue Living Australia.

Martin, C. R., & Horne, D. A. (1995). Level of success inputs for service innovations in the same firm. *International Journal of Service Industry Management*, *6*(4), 40–56. doi:10.1108/09564239510096894

Martindale, C. (1995). Creativity and connectionism. In Smith, S. M., Ward, T. B., & Finke, R. A. (Eds.), *The Creative Cognition approach* (pp. 249–268). Cambridge, MA: MIT Press.

Mascarenhas, J., & Esperto, L. (2006). Direct Metal Laser Sintering (in Portuguese). Ciência & Tecnologia dos Materiais. *Journal of Portuguese Materials Society*, *18*(3/4).

Maslach, C., & Jackson, S. (1989). Burnout in health professions: a social psychological analysis. In Sanders, G. S. J. (Ed.), *Social psychology of health and illness*. Hillsdale, NJ: Lawrence Erlbaum Associates.

Maslach, C., Schaufeli, W., & Leiter, M. (2001). Job Burnout. *Annual Review of Psychology*, *52*, 397–422. doi:10.1146/annurev.psych.52.1.397

Mather, M., & Carstensen, L. L. (2005). Aging and Motivated Cognition: The Positivity Effect in Attention and Memory. *Trends in Cognitive Sciences*, *9*(10), 496–502. doi:10.1016/j.tics.2005.08.005

Mattelmäki, T. (2007). *Probes*. Helsinki, Finland: University of Art and Design Helsinki.

Matthews, B., & Horst, W. (n.d.). *What can we learn from the probes? The role of interpretation in contribution to knowledge*. Demark: Mads Clausen Institute. *University of Southern Denmark*.

Mauss, M. (1954). *The Gift Forms and Functions of Exchange in Archaic Societies*. London: Routledge & Kegan Paul.

Mayer, R. E. (1999). Problem Solving. In Runco, M. A., & Pritzker, S. R. (Eds.), *Encyclopedia of Creativity* (*Vol. 2*, pp. 437–447). San Diego: Academic Press.

Maylor, H. (2001). Beyond the Gantt Chart: project management moving on. *European Management Journal*, 92–100. doi:10.1016/S0263-2373(00)00074-8

Mazur, G. H. (1994, June 12-14). *QFD for Small Business. A shortcut through the 'Maze of Matrices'*. Paper presented at the 6th Symposium on Quality Function Deployment, Novi, Michigan.

Mazuryk, T., & Gervautz, M. (1996). *Virtual Reality: History, Applications, Technology and Future*. Vienna, Austria: Vienna University of Technology.

McCracken, G. (1988). *The Long Interview*. Newbury Park, CA: Sage.

McDonagh, D. (2008). Do it until it hurts!: empathic design research. Design Principles and Practices. *International Journal (Toronto, Ont.)*, *2*(3), 103–110.

McDonough, W., & Braungart, M. (2002). *Cradle to cradle: remaking the way we make things* (1st ed.). New York: North Point Press.

McIntyre, A. (2008). *Participatory action research*. Los Angeles, CA: Sage.

Mckim, R. W. (1980). *Experiences in Visual Thinking* (2nd ed.). Monterey, CA: Brookes/Coles Publishing Co.

McKinley, M. (2009). *Re-Energizing the Fence: Designing a Desirable Electric Fencing System* (Unpublished Master Thesis). Massey University, Wellington.

Meadows, D. H. Club of Rome, & Potomac Associates. (1974). *The Limits to growth: a report for the Club of Rome's Project on the Predicament of Mankind* (2d ed.). New York: Universe Books.

Meadows, D. H., Richardson, J. M., & Bruckmann, G. (1982). *Groping in the dark: the first decade of global modelling*. New York: Wiley.

Medeiros, A. C. B., Crilly, N., & Clarkson, P. J. (2008). The Influence of Ageing on Product Experience. In *Proceedings of the 6th Design & Emotion Conference*. Hong Kong.

medGadget. (2009). MAC 800 Portable ECG System. *Internet Journal of Emerging medical Technologies*. Retrieved March 13, 2009, from http://medgadget.com/archives/2009/03/portable_ecg_system_with_sms_keyboard.html

Mehrabi, M. M Galip, G., Ulsoy, A., & Koren, Y. (2002). Manufacturing Systems and Their Design Principles, In Osita D.I. Nwokah and Yildirim Hurmuzlu (eds.), *The Mechanical systems design handbook: modeling, measurement, and control* (Vol. 1 – Manufacturing) Boca Raton, FL: CRC Press LLC.

Mehrhoff, W. A. (1999). *Community design: A team approach to dynamic community systems*. Thousand Oaks, CA: Sage.

Meijers, A. W. M. (2000). The relational ontology of technical artifacts. In Kroes, P. K., & Meijers, A. W. M. (Eds.), *The Empirical Turn in the philosophy of technology*. Oxford: Elsevier Science.

Merwe, J. van der (2007). The magic of Three. *Kybernetes*, (Cybernetics and Design), 36(9/10): 1436-1457.

Midgley, D. F., & Dowling, G. R. (1978). Innovativeness: The Concept and Its Measurement. *The Journal of Consumer Research*, *4*(4), 229–242. doi:10.1086/208701

Mikels, J. A., Larkin, G. R., Reuter-Lorenz, P. A., & Carstensen, L. L. (2005). Divergent Trajectories in the Aging Mind: Changes in Working Memory for Affective Versus Visual Information with Age. *Psychology and Aging*, *20*(4), 542–533. doi:10.1037/0882-7974.20.4.542

Miles, L. D. (1989). *Techniques of Value Analysis and Engineering*. London: Eleanor Miles Walker.

Milgram, P., & Kishino, F. (1994, December). A Taxonomy of Mixed Reality Visual Displays. *IEICE Transactions on Information and Systems*, *77*(12).

Millikan, R. G. (2004). *Varieties of Meaning: The 2002 Jean Nicod Lectures*. Cambridge, MA: MIT Press/ Bradford Books.

Minderhoud, S., & Fraser, P. (2005). Shifting paradigms of product development in fast and dynamic markets. *Reliability Engineering & System Safety*, *88*(2), 127–135. doi:10.1016/j.ress.2004.07.002

Models of Disability: keys to perspective (2001). *Models of Disability*. Retrieved September 13, 2007, from http://akmhcweb.org/ncarticles/models_of_disability.htm

Mohrman, S. A., Klein, J. A., & Finegold, D. (2001). *Managing the Global Knowledge-Creation Network: A Sense-Making perspective*. (Working Paper ESD-WP-2001-05). Cambridge, MA: MIT Engineering Systems Division.

Mok, H. S., Kim, H. J., & Moon, K. S. (1997). Disassemblability of mechanical parts in automobile for recycling. *Computers & Industrial Engineering*, *33*(3-4), 621–624. doi:10.1016/S0360-8352(97)00207-6

Montefusco, P. (1992). Interazione, non interfacce. In Anceschi, G. (Ed.), *Il progetto delle interfacce. Oggetti colloquiali e protesi virtuali* (pp. 97–131). Milan, Italy: Domus Academy.

Moore, G. A. (2002). *Crossing the Chasm. Marketing and Selling Disruptive Products to Mainstream Customers* (Revised ed.). New York: Harper Collins.

Morecroft, J. D. W., & Sterman, J. (1994). *Modeling for learning organizations*. Portland, OR: Productivity Press.

Morrow, S. L., & Smith, M. L. (2000). Qualitative Research for Counseling Psychology. In Brown, S. D., & Lent, R. W. (Eds.), *Handbook of Counseling Psychology* (3rd ed., pp. 199–230). Thousand Oaks, CA: Sage.

Mueller, J. (1999). *J. L. Mueller, Inc: Universal Design in products: case study - Oxo Good Grips*. Retrieved September 14, 2008, from http://home.earthlink.net/~jlminc/cases_oxo.html

Muir Wood, A., Moultrie, J., & Eckert, C. (2008). *Trends and Design: Relating literature to industry practice*. Paper presented at the Design Research Society Conference.

Muller, W. (2001). *Order and Meaning in Design*. Utrecht: Lemma Publishers.

Mullern, M. J., Wildeman, D. M., & White, E. A. (1993). Taxonomy of PD practices: a brief practitioner's guide. *Communications of the ACM*, *36*(4), 26–27.

Mullins, J. W. (2007). Discovering "Unk-Unks". *MIT Sloan Management Review*, *48*(4), 17–21.

Mumford, L. (1946). *Technics and Civilization*. London: George Routledge & Sons.

Murphy, C., Schubert, C. R., Cruickshanks, K. J., Klein, B. E. K., Klein, R., & Nondahl, D. M. (2002). Prevalence of Olfactory Impairment in Older Adults. *Journal of the American Medical Association*, *288*(18), 2307–2312. doi:10.1001/jama.288.18.2307

Nambisan, S. (2002). Designing virtual customer environments for new product development: Toward a theory. *Academy of Management. Academy of Management Review, 27*(3), 392–413. doi:10.2307/4134386

Nambisan, S., & Sawhney, M. (2007). Buyer's Guide to the Innovation Bazaar. *Harvard Business Review, 85*(6), 109–118.

Nanard, J., & Nanard, M. (1995). Hypertext design environments and the hypertext design process. *Communications of the ACM*, 49–56. doi:10.1145/208344.208347

NationMaster.com. (2009). *NationMaster.* New Zealand: Economy Retrieved March 2009, from http://www.nationmaster.com/country/nz-new-zealand/eco-economy

Neale, M. R., & Corkindale, D. R. (1998). Co-developing products: Involving customers earlier and more deeply. *Long Range Planning, 31*(3), 418–425. doi:10.1016/S0024-6301(98)80008-3

Neander, K. (2004). Teleological Theories of Mental Content. In Edward N. Zalta (ed.), *The Stanford Encyclopedia of Philosophy* (Summer 2004 Edition). Retrieved March 1, 2005, from http://plato.stanford.edu/archives/sum2004/entries/content-teleological/

Nelkin, D., Nelson, R., & Kiernan, C. (1987). Commentary: University-Industry alliances. *Science, Technology & Human Values, 12*(1), 65–74.

New Zealand Trade and Enterprise. (2004). *Creative Sector Engagement Strategy*. Retrieved February 2005, from http://www.nzte.govt.nz/common/files/ses-creative.pdf

Nielsen, J. (1993). *Usability Engineering*. San Francisco: Morgan Kaufmann.

Nieminen, E., & Järvinen. J. (2001). *Teollisen muotoilun teknologiaohjelman esiselvitys. Teknologiakatsaus 119/2001*. Helsinki, Finland: TEKES.

Noguchi, L. D. (2003). *The Goldsmith and the Skill: The Jewel Quality on the Interface Between Design and Production Process in the Handmade Jewellery*. (Thesis). Federal University of Minas Gerais, Belo Horizonte, Brazil. [in Portuguese].

Nonaka, I., & Takeuchi, H. (1995). *The Knowledge Creating Company*. New York: The Oxford University Press.

Norman, D. (1999). *Affordances, Conventions and Design (Part 2)*. Retrieved March 27, 2009, from http://www.jnd.org/dn.mss/affordance_conv.html

Norman, D. (1999). *Invisible Computer: Why Good Products Can Fail, the Personal Computer Is So Complex and Information Appliances Are the Solution*. Cambridge, MA: MIT Press.

Norman, D. (2004). *Emotional Design: Why we love (or hate) everyday things*. New York: Basic Books.

Norman, D. A. (2004). *Three levels of Design: Visceral, behavioral, and reflective. Emotional Design – Why we love or hate everyday things*. New York: Basic books.

Norman, D. A. (2008). *Design Emocional: por que adoramos (ou detestamos) os objetos do dia-a-dia. Rio de Janeiro/RJ*. Brazil: Rocco.

Normann, R. (2001). *Reframing Business: When the Map Changes the Landscape*. Chichester, UK: Wiley & Sons.

Normann, R., & Ramirez, R. (1993). From value chain to value constellation: Designing interactive strategy. *Harvard Business Review, 71*(4), 65–77.

Norretranders, T. (1999). *The user illusion: Cutting consciousness down to size*. New York: Penguin.

Nwokah, O. D. I., & Hurmuzlu, Y. (2002). *The Mechanical systems design handbook: modeling, measurement, and control*. Boca Raton: CRC Press LLC.

O'Connor, J., & McDermott, I. (1997). *The Art of Systems Thinking*. San Francisco: Thorsons.

O'Leary, Z. (2004). *The Essential Guide to Doing Research*. Thousand Oaks, CA: Sage Publications Ltd.

OCED (Organisation for Economic Co-Operation and Development). (2002). *High-Growth SMEs and Employment*. Paris: Organisation for Economic Co-Operation and Development.

Ochse, R. (1990). *Before the gates of excellence: The determinants of creative genius*. New York: Cambridge University Press.

Ogot, M., & Okudan, G. (2006). The five-factor model personality assessment for improved student design team performance. *European Journal of Engineering Education, 31*(5), 517–529. doi:10.1080/03043790600797335

Oliver, M. (2001). Disability issues in the postmodern world. In Barton, L. (Ed.), *Disability, politics and the struggle for change* (pp. 149–159). London: David Fulton Publishers.

Olson, E. L., & Bakke, G. (2001). Implementing the lead user method in a high technology firm: a longitudinal study of intentions versus actions. *Journal of Product Innovation Management*, *18*(6), 388–395. doi:10.1016/S0737-6782(01)00111-4

ORBIS. (2009). *ORBIS Hong Kong: Saving sight, rebuilding lives*. Retrieved June 22, 2010, from http://www.orbis.org/Default.aspx?cid=4870&lang=1

Osborne, A. F. (1979). *Creative collaboration by groups. Applied Imagination: Principles and Procedures of Creative Problem Solving*. New York: Charles Scribner's Sons.

Osgrove, B. (2005). *Comparison of Low Cost Rapid Prototyping Machines*. Retrieved from rpworld.net

Ostergaard, K., & Summers, J. (2007). Resistance Based Modeling of Collaborative Design. *Concurrent Engineering Research and Applications*, *15*(1), 21–32.

Østfold University College. (2009). Study Program for the Engineering Education: Product Design Engineering - course description. Retrieved from www.hiof.no/index.php?ID=1760&lang=eng

Otto, K. N., & Wood, K. L. (1998). *Product evolution: a reverse engineering and redesign methodology. Research in Engineering Design – Theory, Applications and Concurrent Engineering*. London: Springer.

Overbeek, K., Djajadiningrat, T., Hummels, C., & Wensveen, S. (2002). Beauty in Usability: Forget about Ease of Use! In Green, W. S., & Jordan, P. W. (Eds.), *Pleasure with products: Beyond Usability*. London: Taylor & Francis.

Owsley, C., Ball, K., McGwin, G., Sloane, M. E., Roenker, D. L., White, M. F., & Overley, E. T. (1998). Visual Processing Impairment and Risk of Motor Vehicle Crash among Older Adults. *Journal of the American Medical Association*, *279*(14), 1083–1088. doi:10.1001/jama.279.14.1083

Pahl, G., & Beitz, W. (1984). *Engineering Design*. London: Springer-Verlag.

Pahl, G., & Beitz, W. (1996). *Engineering Design: A Systematic Approach*. New York: Springer-Verlag.

Paivio, A. (1977). Images, Propositions, and Knowledge. In Nicholas, J. M. (Ed.), *Images, Perception, and Knowledge*. Boston: Reidel.

Panksepp, J. (2003). At the Interface of Affective, Behavioral, and Cognitive Neurosciences: Decoding the Emotional Feelings of the Brain. *Brain and Cognition*, *52*, 4–14. doi:10.1016/S0278-2626(03)00003-4

Papanek, V. (1985). *Design for the Real World: Human Ecology and Social Change* (2nd ed.). Chicago: Academy Publishers.

Pappas, E. (2002). *Creative Problem Solving in Engineering Design*. Paper presented at the ASEE Southeast Section Conference

Parkinson, S. (1982). The Role of the User in Successful New Product Development. *R & D Management*, *12*(3), 123–131. doi:10.1111/j.1467-9310.1982.tb00497.x

Pavitt, K. (1984). Sectoral patterns of technical change: Towards a taxonomy and a theory. *Research Policy*, *13*(6), 343–373. doi:10.1016/0048-7333(84)90018-0

Peças, P., Ribeiro, I., Folgado, R., & Henriques, E. (2009). A Life Cycle Engineering Approach for the Selection of Technology: a Case Study on Plastic Injection Moulds for Low Production Volumes. *International Journal of Cleaner Production*, *17*(9), 846–856. doi:10.1016/j.jclepro.2009.01.001

Pellegrini, A. D. (1984). The effects of exploration and play on young children's associative fluency. *Imagination, Cognition and Personality*, *4*, 29–40. doi:10.2190/0W5N-X8E8-WKWF-9A1R

Perelman, Ch., & Olbrechts-Tyteca, L. (1969). *The New Rhetoric – A Treatise on Argumentation*. South Bend, IN: University of Notre Dame Press.

Perkins, N. (1999). Women designers: making differences. In Rothschild, J., & Cheng, A. (Eds.), *Design and feminism: re-visioning spaces, places, and everyday things* (p. 121). Piscataway, NJ: Rutgers University Press.

Perrow, C. B. (1999). *Normal Accidents: Living With High Risk Technologie*. Princeton, NJ: Princeton University Press.

Phillips, L. W., & Sternthal, B. (1977). Age Differences in Information Processing: A Perspective on the Aged

Consumer. *JMR, Journal of Marketing Research*, *14*(4), 444–457. doi:10.2307/3151185

Pierson, P. (2004). *Politics in time: history, institutions, and social analysis*. Princeton, NJ: Princeton University Press.

Piller, F. T., & Walcher, D. (2006). Toolkits for idea competitions: a novel method to integrate users in new product development. *R & D Management*, *36*(3), 307–318. doi:10.1111/j.1467-9310.2006.00432.x

Piller, F. T., Schubert, P., Koch, M., & Möslein, K. (2005). Overcoming mass confusion: Collaborative customer co-design in online communities. *Journal of Computer-Mediated Communication*, *10*(4), 1–8.

Pinch, T. J., & Bijker, W. E. (1984). The Social Construction of Facts and Artefacts: or How the Sociology of Science and the Sociology of Technology Might Benefit Each Other. *Social Studies of Science*, *14*(3), 399–441. doi:10.1177/030631284014003004

Pisano, G. P., & Verganti, R. (2008). Which Kind of Collaboration Is Right for You? *Harvard Business Review*, *86*(12), 78–86.

Pitt, L. F., Watson, R. T., Berthon, P., Wynn, D., & Zinkhan, G. (2006). The Penguin's Window: Corporate Brands From an Open-Source Perspective. *Journal of the Academy of Marketing Science*, *34*(2), 115–127. doi:10.1177/0092070305284972

Pitta, D. A., Franzak, F., & Katsanis, L. P. (1996). Redefining new product development teams: learning to actualize consumer contributions. *Journal of Product and Brand Management*, *5*(6), 48–60. doi:10.1108/10610429610152831

Polya, G. (1965). *Mathematical Discovery. (V II)*. New York, London: John Wiley&Sons.

Post, T., & Cramer, K. (1989). Knowledge, Representation and Quantitative Thinking. In Reynolds, M. (Ed.), *Knowledge Base for the Beginning Teacher - Special publication of the AACTE* (pp. 221–231). Oxford, UK: Pergamon Press.

Pousa, C., Ribeiro, I., Folgado, R., Peças, P., & Henriques, E. (2009). LCC and LCA Simplified Models to Foster the Design of Sustainable Plastic Injection Moulds. In Waguih ElMaraghy (Ed.), *16th CIRP International Conference on Life Cycle Engineering – LCE 2009* (pp. 99-104). Cairo, Eqypt: University of Windsor.

Powers, M. (2008, May 19). Get Ready to Rethink What It Means to Be Green. *Wired, 16.06*.

Prahalad, C. K., & Krishnan, M. S. (2008). *The New Age of Innovation, Driving Cocreated Value Through Global Networks*. New York: McGraw Hill Books.

Prahalad, C. K., & Ramaswamy, V. (2003). *The Future of Competition: Co-Creating Unique Value with Customers*. Boston, MA: Harvard Business School Press.

Prahalad, C. K., & Ramaswamy, V. (2004). *The Future of Competition, Co-creating Unique Value with Customers*. Boston, MA: HBS Press.

Preiser, W. R. E., & Ostroff, E. (Eds.). (2001). *Universal Design handbook*. New York: McGraw-Hill.

Prensky, M. (2001). Digital Natives, Digital Immigrants. *On the Horizon. MCB University Press*, *9*(5), 1–6.

Prevent Blindness America. (2002). *Vision problems in the U.S.* Schaumburg, IL: Prevent Blindness America.

Pribram, K. H. (1977). Holonomy and Structure in the Organization of Perception. In Nicholas, J. M. (Ed.), *Images, Perception, and Knowledge*. Boston: Reidel.

Prigogine, I. (1980). *From Being to Becoming: Time and complexity in the physical sciences. San Fransisco: W.* H. Freeman.

Prochnik, V. (2002). *Estudo da Competitividade de Cadeias Produtivas no Brasil: impactos das zonas de livre comércio*. Cadeia Têxtil e Confecções (Nota técnica final), dez.

Purcell, A. T., & Gero, J. (1991). The Effects of Examples on the Results of Design Activity. In Gero, J. (Ed.), *Artificial Intelligence in Design AID91*. Oxford: Butterworth Heinemann.

Purcell, A. T., & Gero, J. (1996). Design and Other Types of Fixation. *Design Studies*, *17*(4), 363–383. doi:10.1016/S0142-694X(96)00023-3

Purcell, A. T., & Williams, P. (1993). Fixation Effects: Do they exist in design problem solving? *Environment and Planning. B, Planning & Design*, *20*(3), 333–345. doi:10.1068/b200333

Ramakers, R., & Bakker, G. (Eds.). (1998). *Droog design: Spirit of the nineties*. Rotterdam, The Netherlands: 010 Publishers.

Ramírez, R. (1999). Value co-production: intellectual origins and implications for practice and research. *Strategic Management Journal*, *20*(1), 49–65. doi:10.1002/(SICI)1097-0266(199901)20:1<49::AID-SMJ20>3.0.CO;2-2

Rao, M. N., Lo, A., & Fung, A. (2007). Nurturing students to think creatively in design education. In Katiyar, V. S., & Mehta, S. (Eds.), *Design Education, Tradition and Modernity*. Ahmedabad: National Institute of Design.

Rao, R. V. (2007). *Decision Making in the Manufacturing Environment: using graph theory and fuzzy multiple attribute decision making methods*. New York: Springer.

Rapaille, C. (2006). *The Culture Code: An Ingenious Way to Understand Why People around the World Live and Buy as They Do*. New York: Broadway.

Ray, N. (2009). *Jewellery Manufacturing*. Retrieved February 22, 2009, from Jewel Info 4 U, www.jewelinfo4u.com

Rebitzer, G., & Hunkeler, D. (2003). Life Cycle Costing in LCM: Ambitions, Opportunities, and Limitations. *International Journal of Life Cycle Assessment*, *8*(5), 253–256. doi:10.1007/BF02978913

Rech, S. R. (2002). *Moda: por um fio de qualidade. Florianópolis/SC*. Brazil: Editora da UDESC.

Rech, S. R. (2006). *Cadeia Produtiva da Moda: um modelo conceitual de análise da competitividade no elo confecção* (Unpublished PhD's thesis). Programa de Pós-Graduação em Engenharia de Produção, UFSC: Florianópolis/SC, Brazil.

Redelinghuys, C., & Bahill, A. (2006). A framework for the assessment of the creativity of product design teams. *Journal of Engineering Design*, *17*(2), 121–141. doi:10.1080/09544820500273136

Redström, J. (2006). Towards User Design? On the Shift from Object to User as the Subject of Design. *Design Studies*, *27*(2), 123–139. doi:10.1016/j.destud.2005.06.001

Reeve, J. (2005). *Understanding motivation and emotion*. New York: John Wiley & Sons.

Rehabilitation Alliance Hong Kong. (2004). *Research on the accessibility of public disabled toilets*. Hong Kong: Rehabilitation Alliance Hong Kong.

Reitmayr, G., & Schmalstieg, D. (2005). OpenTracker: A flexible software design for three-dimensional interaction. *Virtual Reality (Waltham Cross)*, 9.

Rey, J. D. (2008). Tapping the Community Pool Online Communities Let You See What's On Your Customers' Mind. *Inc*, *30*(7), 37–39.

Reymen, I. M., Hammer, D. K., Kroes, P. A., Aken, J. E., Dorst, C. H., & Bax, M. F. (2006). A domain-independent descriptive design model and its application to structured reflection on design processes. *Research in Engineering Design*, *16*(4), 147–173. doi:10.1007/s00163-006-0011-9

Ribeiro, I., Peças, P., Silva, A., & Henriques, E. (2008). Life cycle engineering methodology applied to material selection, a fender case study. *Journal of Cleaner Production*, *16*(17), 1887–1899. doi:10.1016/j.jclepro.2008.01.002

Riccò, D. (2008). *Sentire il design. Sinestesie del progetto di comunicazione*. Rome: Carocci.

Richards, A. R. (2008). The Art of Rhetoric as Self-Discipline: Interdisciplinarity, Inner Necessity, and the Construction of a Research Agenda. *Journal of Research Practice*, *4*(1). Retrieved from http://jrp.icaap.orh/index.php/jrp/article/view/115/104.

Richards, T. R., Lorrain, H., & Hilton, P. D. (2000). The Role of Rapid Tooling in Investment-Casting Applications. In Hilton PD. Jacobs PF (eds.), *Rapid Tooling: Technologies and Industrial Applications*. New York: Marcel Dekker.

Richardson, A. (1993). The death of the designer. *Design Issues*, *9*(2), 34–43. doi:10.2307/1511672

Rieple, A. (2004). Understanding Why Your New Design Ideas Get Blocked. *Design Management Review*, *15*(1), 36–42. doi:10.1111/j.1948-7169.2004.tb00148.x

Rittel, H., & Webber, M. (1973). Dilemmas in a General Theory of Planning. *Policy Sciences*, *4*, 155–169. doi:10.1007/BF01405730

Rittel, H., & Webber, M. (1979). Dilemmas in a General Theory of Planning. *Policy Sciences*, *4*, 155–169. doi:10.1007/BF01405730

Robertson, D., & Ulrich, K. (1998). Planning for product platforms. *Sloan Management Review*, *39*(4), 19–31.

Robinson, D. E. (1975). Style changes: cyclical, inexorable, and foreseeable. *Harvard Business Review*, *53*(6), 121–131.

Robson, C. (2002). *Real World Research*. Malden, MA: Blackwell Publishers Inc.

Rocha, V. C. (2007). *The Use of Information Systems on Jewellery Design and Production Process*. (Thesis). Federal University of Minas Gerais, Belo Horizonte, Brazil. [in Portuguese].

Rogers, E. M. (2003). *Diffusion of Innovation* (5th ed., *Vol. 5*). New York: Free Press.

Ron, A., & Penev, K. (1995). Disassembly and recycling of electronic consumer products: an overview. *Technovation*, *15*(6), 163–374.

Roozenburg, N. F. M., & Eekels, J. (1995). *Product Design: Fundamentals and Methods*. New York: John Wiley and Sons.

Rothwell, R. (1986). Innovation and Re-Innovation: A Role for the User. *Journal of Marketing Management*, *2*(2), 109–123. doi:10.1080/0267257X.1986.9964004

Rothwell, R., & Gardiner, P. (1983). The role of design in product and process change. *Design Studies*, *4*(3), 161–169. doi:10.1016/0142-694X(83)90025-X

Roy, R. (1994). The evolution of ecodesign. *Technovation*, *14*(6), 363–380. doi:10.1016/0166-4972(94)90016-7

Rugg, G., & McGeorge, P. (2003). Sorting and Categorization: Case Study. In Drake, M. A. (Ed.), *Encyclopedia of Library and Information Science: Pub-Zoo* (2nd ed., *Vol. 4*). Boca Raton, FL: CRC Press.

Rutter, B., & Agne, J. (1998). A Darwinian Theory of Good Design. *DMI Review*, *9*(4), 36–41.

Saffer, D. (2005). *The Role of Metaphor in Interaction Design*. (Thesis), Master of Design in Interaction Design, The School Of Design. Pittsburgh, PA: Carnegie Mellon University.

Salvador, F., Forza, C., & Rungtusanatham, M. (2002). How to mass customize: Product architectures, sourcing configurations. *Business Horizons*, *45*(4), 61–69. doi:10.1016/S0007-6813(02)00228-8

Sanders, E. B. N. (1999). *Postdesign and Participatory Culture. Useful and Critical: The Position of Research in Design*. Tuusula, Finland: University of Art and Design Helsinki.

Sanders, E. B. N. (2006). *Design Research in 2006. Design Research Quarterly*. Design Research Society.

Sanders, E. B. N. (2006). Scaffolds for Building Everyday Creativity. In Frascara, J. (Ed.), *Design for Effective Communications: Creating Contexts for Clarity and Meaning*. New York: Allworth Press.

Sanders, E. B. N., & Stappers, P. (2003) Generative tools for context mapping: tuning the tools from the Third International Conference on Design & Emotion. Boca Raton, FL: Taylor & Francis

Sanders, E. B. N., & Stappers, P. J. (2008). Co-creation and the new landscapes of design. *CoDesign*, *4*(1), 5–18. doi:10.1080/15710880701875068

Sanoff, H. (2000). *Community participation methods in design and planning* (p. 10). New York: Wiley.

Sanoff, H. (2005). Community participation in riverfront development. *CoDesign*, *1*(1), 61–78. doi:10.1080/15710880512331326022

Sant'anna, M. R. (2003). A Moda – o desafio de pensá-la além do produto. In: QUIRINO, S. F. S. (Ed.), *Modapalavra* (vol. 2, nº. 2, pp. 72-86). Florianópolis/SC, Brazil: UDESC/CEART.

Sant'anna, M. R. (2007). *Teoria da Moda: sociedade, imagem e consumo. Barueri/SP*. Brazil: Estação das Letras Editora.

Santamarina, J. C. (2002, February 21-22, 2003). *Creativity and Engineering - Education Strategies*. Paper presented at the International Conference/Workshop on Engineering Education Honouring Professor James T. P. Yao, College Station, Texas.

Sathikh, P. M. (1998). Singapore: A Healthy Environment for Innovation and Design. [). Tokyo, Japan]. *Axis Magazine*, *73*, 143.

Sawhney, M., & Prandelli, E. (2000). Communities of creation: Managing distributed innovation in turbulent markets. *California Management Review*, *42*(4), 24–54.

Scarry, E. (1985). *The body in pain: The making and unmaking of the world.* New York: Oxford University Press.

Schaber, F., & Thomas, V. (2007). Knowledge Transfer: Industry, Academia and the Global Gift Market. *Design Management Conference*. Seoul: Korea Institute of Design Promotion.

Scharmer, C. O. (2007). *Theory U – Leading from the Future as it Emerges*. Cambridge, Massachusetts: SoL.

Schifferstein, H. N. J., & Hekkert, P. (2007). *Product Experience*. London: Elsevier Science.

Schifferstein, H. N. J., & Hekkert, P. (2008). Introducing Product Experience. In Schifferstein, H. N. J., & Hekkert, P. (Eds.), *Product Experience*. San Diego: Elsevier.

Schoemaker, P. J. H. (1995). Scenario planning: A tool for strategic thinking. *Sloan Management Review*, *36*(2), 25–40.

Schon, D. A. (1979). *Generative Metaphor: A perspective on problem-setting as a social policy. Metaphor and Thought*. Cambridge, UK: Cambridge University Press.

Schön, D. A. (1983). *The Reflective Practitioner: How Professionals Think in Action*. New York: Basic Books.

Schön, D. A. (1992). Designing as reflective conversation with the materials of a design situation. *Knowledge-Based Systems*, *5*(1), 3–14. doi:10.1016/0950-7051(92)90020-G

Schreier, M., & Prügl, R. (2008). Extending Lead-User Theory: Antecedents and Consequences of Consumers' Lead Userness. *Journal of Product Innovation Management*, *25*(4), 331–346. doi:10.1111/j.1540-5885.2008.00305.x

Schuler, D., & Namioka, A. (Eds.). (1993). *Participatory design: Principles and practices*. Hillsdale, NJ: L. Erlbaum Associates.

Scrivener, S. A. R., Ball, L. J., & Woodcock, A. (Eds.). (2000). *Collaborative design*. New York: Springer.

Seamon, D., & Mugerauer, R. (2000). *Dwelling, place, and environment: towards a phenomenology of person and world.* Malabar, FL: Krieger Pub. Co.

Sedenkov, V. (2008). An attempt to answer perennial design questions. In Rohatynski, R. (Ed.), *Design Methods for Industrial Practice* (pp. 21–30). Poland, Gdynia-Zielona Gora.

Sedenkov, V. (2008). Design problem q-realization and its outcomes. In Vanek, V., Hosnedl, S., & Bartak, J. (Eds.), *AEDS 2008 Workshop* (pp. 91–98). Pilzen, Czech Republic: University of West Bohemia.

Sedenkov, V., & Guziuk, Z. (2004). Design machine: theory and implementation. In I. Horvath & P. Xirouchakis (Eds), *Tools and Methods of Competitive Engineering* (Vol. 2), (pp. 1119-1120, executive summary). (On CD-ROM). Rotterdam: Millpress.

Seidel, D., Crilly, N., Langdon, P. M., Matthews, F. E., Jagger, C., Brayne, C., Clarkson, P. J. & MRC CFAS (2009). Understanding Patterns of Capability Loss among Elderly Users. In *Proceedings of the 17th International Conference on Engineering Design*. ICED, Stanford.

Seidman, K. R. (2003). Understanding visual impairments. In Axel, E. S., & Levent, N. S. (Eds.), *Art beyond sight: A resource guide to art, creativity, and visual impairment* (pp. 50–57). New York: American Foundation for the Blind Press.

Senge, P. M., Carstedt, G., & Porter, P. L. (2001). Innovating Our Way to the Next Industrial Revolution. *MIT Sloan Management Review*, *42*(2), 24–38.

Seva, R. R., Duh, H. B.-L., & Helander, M. G. (2007). The marketing implications of affective product design. *Applied Ergonomics*, *38*, 723–731. doi:10.1016/j.apergo.2006.12.001

Seybold, P. B. (2006). *Outside Innovation: How Your Customers Will Co-Design Your Company's Future*. New York: Collins.

Shadbolt, N., & Burton, M. (1995). Knowledge elicitation: a systematic approach. In Wilson, J. R., & Corlett, E. N. (Eds.), *Evaluation of Human Work* (pp. 406–440). London: Taylor & Francis.

Shenoy, P. P. (2000). Valuation network representation and solution of asymmetric decision problems. *Eurobean Journal of Operational Research*, 579-608.

Shezi, A. (2005). Under the mattress or into the stokvel, SA's poor put money away for rainy day. *BusinessDay Online*. Retrieved May 24, 2009, from http://www.cssr.uct.ac.za/media/stokvels1.pdf

Shouse, E. (2005). Feeling, Emotion, Affect. *M/C Journal, 8*(6). Retrieved September 2, 2009, from http://journal.media-culture.org.au/0512/03-shouse.php

Sifonis, C., Chernoff, A., & Kolpasky, K. (2006). The Innovation Pipeline: Analogy as a tool for communicating about innovation. *International Journal of Innovation and Technology Management, 3*, 1–19. doi:10.1142/S0219877006000697

Sim, S. K., & Duffy, A. H. (2003). Towards an ontology of generic engineering design activities. *Research in Engineering Design*, 200–223. doi:10.1007/s00163-003-0037-1

SimaPro7 software. (n.d.). *PRe′ Consultants*. Retrieved from: http://www.pre.nl

Simoes, R., & Sampaio, A. M. (2008). Effect of technology-driven products in the future of product design and development. In *Proceedings of RPD 2008* (paper 8048). Oliveira de Azemeis, Portugal.

Simon, A. (1969). *Science of the Artificial*. Cambridge, MA: The MIT Press.

Simon, H. A., Kotovsky, K., & Cagan, J. (2001). *Formal Engineering Design Synthesis*. New York: Cambridge University Press.

Simonton, D. K. (1984). *Genius, Creativity and Leadership: Historimetric Enquiries*. Cambridge, MA: Harvard University Press.

Sinclair, M. A. (1995). Subjective assessment. In Wilson, J. R., & Corlett, E. N. (Eds.), *Evaluation of Human Work* (pp. 406–440). London: Taylor & Francis.

Singapore Economic Development Board. (2009). *Our History*. Retrieved from Singapore Economic Development Board. (2009). *Singapore Facts & Figures*. Retrieved from www.singstat.gov.sg/stats/keyind.html

Singer, T., Lindenberger, U., & Baltes, P. B. (2003). Plasticity of Memory for New Learning in Very Old Age: A Story of Major Loss? *Psychology and Aging, 18*(2). doi:10.1037/0882-7974.18.2.306

Siu, K. W. M. (2005a). Userfitness in design: A case study of public toilet facilities. In H. Pham & S. Yamada (Eds.), *2005 proceeding: Eleventh ISSAT international conference: Reliability and quality in design* (pp. 41-47). New Brunswick, NJ: International Society of Science and Applied Technologies.

Siu, K. W. M. (2005b). Pleasurable products: Public space furniture with userfitness. *Journal of Engineering Design, 16*(6), 545–555. doi:10.1080/09544820500273383

Siu, K. W. M. (2006a). Inclusive design: Public toilet designs for blind persons. *Annual Journal of SOPHEN, 2*(1), 75–77.

Siu, K. W. M. (2006b). FISH: Design quality of public toilets for visually impaired persons. In H. Pham & S. Yamada (Eds.), *2006 proceeding: Twelfth ISSAT international conference: Reliability and quality in design* (pp. 122-128). New Brunswick, NJ: International Society of Science and Applied Technologies.

Siu, K. W. M. (2007). *Accessible design: Devices for blind persons to access buses*. Hong Kong: The Hong Kong Polytechnic University.

Siu, K. W. M. (2008). Better design quality of public toilets for visually impaired persons: An all-round concept in design for the promotion of health. *The Journal of the Royal Society for the Promotion of Health, 128*(6), 313–319. doi:10.1177/1466424008092801

Siu, K. W. M., & Kwok, J. Y. C. (2004). Collective and democratic creativity: Participatory research and design. *Korean Journal of Thinking & Problem Solving, 14*(1), 11–27.

Siu, N. W. C., & Dilnot, C. (2001). The challenge of the codification of tacit knowledge in designing and making: a case study of CAD systems in the Hong Kong jewellery industry. *Automation in Construction, 10*, 701–714. doi:10.1016/S0926-5805(00)00088-1

Slappendel, C. (1996). Industrial design utilization in New Zealand firms. *Design Studies, 17*, 3–18. doi:10.1016/0142-694X(95)00004-B

Slater, S. F., & Narver, J. C. (1995). Market orientation and the learning organization. *Journal of Marketing, 59*(3), 63–74. doi:10.2307/1252120

Smith, E., Johnston, P., & Summers, J. D. (2007, September). Applying Lean Manufacturing Principles to Revolutionize Curbside Equipment and Collection Process. In E. Smith, P. Johnston, & J. Summers (Eds.), ASME-DETC2007, DFLM-35615. Las Vegas, NV: ASME.

Smith, P. G. (1998). Leverage the Cycle-Time Capability of Your Rapid Prototypes. *Time-Compression Technologies. North American Edition, 3*(7), 36-43.

Smith. N. (2009). From Mortarboard to Drawingboards. *The Institution of Engineering and Technology Knowledge Network*. Retrieved from http://kn.theiet.org/magazine/issues/0808/boards.cfm

Sobek, D. K., II. (2007). From Journals to Validated Tool: Results of Empirical Research on Student Designers. *International Conference on Engineering Design*. Paris: ICED.

Somalvico, M. (2003). Bodies and Robots. In Fortunati, L., Katz, J. E., & Riccini, R. (Eds.), *Mediating the Human Body: Technology, Communication, and Fashion* (pp. 31–38). Mahwah, NJ: Lawrence Erlbaum.

Soo, S. C., Yu, K. M., & Chiu, W. K. (2006). Modeling and fabrication of artistic products based on IFS fractal representation. *Computer Aided Design, 38*, 755–769. doi:10.1016/j.cad.2006.04.003

Sosa, M. E., Eppinger, S. D., & Rowles, C. M. (2003). Identifying Modular and Integrative Systems and Their Impact on Design Team Interactions. *Journal of Mechanical Design, 125*, 240–252. doi:10.1115/1.1564074

Spillers, F. (2004). Emotion as a Cognitive Artifact and the Design Implications for Products that are Perceived as Pleasurable. In *Proceedings of the 4th Conference on Design and Emotion*. Ankara Turkey.

Sproles, G. B., & Burns, L. D. (1994). *Changing Appearances*. New York: Fairchild Publications.

Stahl, J., Egan, M., & Goldman, J. (2005). *Introducing new technology into the OR: Measuring the impact on job performance and satisfaction*. Surgery.

Stahl, J., Goldman, J., Rattner, D., & Gazelle, G. (2007). Adapting to a new system of surgical technologies and perioperative processes among clinicians. *The Journal of Surgical Research, 139*(1), 61–67. doi:10.1016/j.jss.2006.08.030

Stahl, J., Holt, J., & Gagliano, N. (n.d.). Understanding Performance and Behavior of Tightly Coupled Outpatient Systems Using RFID: Initial Experience. *Journal of Medical Systems*.

Stake, R. (1994). Case Studies. In Denzin, N., & Lincoln, Y. (Eds.), *Handbook of Qualitative Research* (pp. 236–247). Thousand Oaks, CA: Sage.

Stamati, V., & Fudos, I. (2005). A parametric feature-based CAD system for reproducing traditional pierced jewellery. *Computer Aided Design, 37*, 431–449. doi:10.1016/j.cad.2004.07.004

Stamati, V., & Fudos, I. (2007). A feature based approach to re-engineering objects of freeform design by exploiting point cloud morphology In *Proceedings of SPM 2007: ACM Symposium on Solid and Physical Modeling*. Beijing, China.

Stamm, B. (2004). Innovation – What's Design Got to Do with It. *Design Management Review, 15*(1), 10–19. doi:10.1111/j.1948-7169.2004.tb00145.x

Stanford University. (2009). *Product Design - course description*. Retrieved from http://design.stanford.edu/PD/undergraduates.html

Staundiger, U. M., Cornelius, S. W., & Baltes, P. B. (1989). The Aging of Intelligence: Potential and Limits. *The Annals of the American Academy of Political and Social Science, 503*, 43–59. doi:10.1177/0002716289503001004

Stefik, M., & Stefik, B. (2005). The Prepared Mind Versus the Beginner's Mind. *Design Management Review, 16*(1), 10–16. doi:10.1111/j.1948-7169.2005.tb00002.x

Steiner, G. (2002). *Grammars of Creation*. London: Faber and Faber.

Stemple, J., & Schaub, P. (2002). Thinking in design teams – an analysis of team communication. *Design Studies, 23*(5), 473–496. doi:10.1016/S0142-694X(02)00004-2

Stephanidis, C., Akoumianakis, D., & Ziegler, J. (1997). Universal accessibility and standardisation: New opportunities and prospects. In Anogianakis, G., Bühler, C., & Soede, M. (Eds.), *Advancement of assistive technology* (pp. 39–43). Amsterdam: IOS Press.

Sterman, J. (2000). *Business dynamics: systems thinking and modeling for a complex world.* Boston: Irwin/McGraw-Hill.

Sternberg, R. J. (1999). The concept of creativity: prospects and paradigms. In Sternberg, R. J. (Ed.), *Handbook of Creativity*. New York: Reinhold.

Stouffer, W. B., Russell, J. S., & Oliva, M. G. (2004). *Making the Strange Familiar: Creativity and the Future of Engineering Education.* Paper presented at the 2004 American Society for Engineering Education Annual Conference and Exposition.

Strategies in emergency preparedness for transportation-dependent populations (2006). *Strategies in emergency preparedness.* Retrieved September 20, 2007, from http://www.dotcr.ost.dot.gov/Documents/Emergency/Emergency%20Preparedness%20Strategy%

Strauss, A., & Corbin, J. (1990). *Basics of qualitative research: Grounded theory procedures and techniques.* Newbury Park, CA: Sage.

Strickfaden, M., Devlieger, P., & Heylighen, A. (2009). Building empathy through dialogue. In Malins, J. (Ed.), *Design Connexity: Proceedings of the Eighth International Conference of the European Academy of Design* (pp. 448-452). Aberdeen, Scotland: Gray's School of Art

Sudjic, D. (2009). *The language of things: understanding the world of desirable objects* (pp 15, 20). London: W W Norton.

Suh, N. P. (1990). *The Principles of Design.* New York: Oxford University Press.

Suh, N. P. (2001). *Axiomatic Design.* New York: Oxford University Press.

Summers, J. D., & Ameri, F. (2008). *An algorithm for assessing design complexity through a connectivity view. Tools and Methods for Competitive Engineering.* Izmir, Turkey: TMCE.

Sutcliffe, A. (2004). Heuristic evaluation of virtual reality applications. *Interacting with Computers*, 16.

Swann, C. (2002). Action Research and the Practice of Design. *Design Issues*, *18*(2), 43–61.

Swinburne University of Technology. (2009). *Product Design Engineering - course description.* Melbourne, Australia: Swinburne University of Technology.

Takamura, J. H. (2005). *BrandX: Product Design's Influence on Brand Loyalty.* (Unpublished Master's Thesis), Arizona State University.

Takamura, J. H. (2007). The Product Design Genome: Mapping the Mysterious Interrelationships Between Products and Brands. In *Proceedings of IDSA Innovation,* (Summer 2007).

Takamura, J. H. (2007, December). *Molecular Branding: A Study on the Nature of the Relationship Between Product Design and Brand Loyalty.* In *Conference Proceedings of DeSForm.* Newcastle, UK: Northumbria University.

Tan, B. S. (Ed.). (2000). *Singapore Design Award 2000.* Singapore: Trade Development Board & Designers Association Singapore.

Tan, B. S. (Ed.). (2002). *Singapore Design Award 2002.* Singapore: Designers Association Singapore.

Tang, J. C. (1989). Toward an understanding of the use of shared workspaces by design teams. (Unpublished doctoral dissertation). Stanford, CA: Stanford University.

Tapscott, D., & Williams, A. D. (2006). *Wikinomics: How Mass Collaboration Changes Everything.* Portfolio Hardcover.

Teaching Company Scheme. (2009). *History.* Retrieved from http://www.ktponline.org.uk/strategy/history.aspx

Teece, D. J. (2006). *Managing intellectual capital: Organizational, strategic, and policy dimension.* Oxford, UK: Oxford University Press.

Teegavarapu, S., & Summers, J. D. (2008). *Case Study method for design research. Design Engineering Technical Conferences.* New York: ASME.

Teichert, T., von Wartburg, I., & Braterman, R. (2006). Tacit meaning in disguise: Hidden metaphors in new product development and market making. *Business Horizons*, *49*(6), 451–461. doi:10.1016/j.bushor.2006.02.006

Teixeira, F., & Piçarra, S. (2007). *Estado da arte em RFID.* Sybase. Retrieved February 20, 2008, from http://www.portalrfid.net

Thackara, J. (2006). *In the Bubble: Designing in a Complex World.* Cambridge, MA: MIT Press.

The Ministry of Trade and Industry. (2009). *Design Singapore Initiative: Summary Table from Preliminary Strategies and Recommendations*. Retrieved from http://app.mti.gov.sg/data/pages/507/doc/ERC_SVS_CRE_Annex3.2.pdf

The Ministry of Trade and Industry. (2009). *Media Release by Economic Review Committee, Subcommittee on Service Industries*. Retrieved from http://app.mti.gov.sg/data/pages/507/doc/ERC_SVS_CRE_PressStatement.pdf

The University of Northampton Sue Ryder Care Final Report. (2008). *Sue Ryder Care Final Report* (Unpublished). Northampton, UK: The University of Northampton.

The University of Northampton. (2008). *News Release 6 May 2008*. Retrieved from http://www.northampton.ac.uk/news/release/?id=UON8199

Thomas, V. (1984). *Gifts: The Designer's Role in the Commercialization of Gift Exchange* (Unpublished). London: Royal College of Art/Victoria & Albert Museum.

Thomke, S., & von Hippel, E. (2002). Customers as innovators: a new way to create value. *Harvard Business Review*, *80*(4), 74–81.

Thompson, C. J. (1998). Living the texts of everyday life. In Stern (Ed.), *Representing consumers. Voices, views and visions* (pp.130-136). London, New York: Routledge.

Tidd, J., & Bodley, K. (2002). The influence of project novelty on the new product development process. *R & D Management*, *32*(2), 127–138. doi:10.1111/1467-9310.00245

Tietz, R., Morrison, P. D., Luthje, C., & Herstatt, C. (2005). The process of user-innovation: a case study in a consumer goods setting. *International Journal of Product Development*, *2*(4), 321–338. doi:10.1504/IJPD.2005.008005

Tiger, L. (1992). *The pursuit of pleasure* (pp. 52–60). Boston: Little, Brown and Company.

Toffler, A. (1980). *The third wave.* New York: Bantam Books.

Tomes, A., Armstrong, P., & Clark, M. (1996). User groups in action: The management of user inputs in the NPD process. *Technovation*, *16*(10), 541–551. doi:10.1016/S0166-4972(97)82894-5

Tomiyama, T. (1990). Engineering design research in Japan. In *Proceedings of the ASME Design Theory and Methodology Conference – DTM'90* (pp. 219-224), Chicago: ASME.

Tomiyama, T., & Yoshikawa, H. (1987). *Extended General Design Theory*. In Design theory for CAD. North-Holland: Elsevier Science Publishers B. V.

Torrance, E. P. (1974). *The Torrance tests of creative thinking*. Bensenville, IL: Scholastic Testing Service.

Traktinsky, N., Katz, A. S., & Ikar, D. (2000). What is Beautiful is Usable. *Interacting with Computers*, *13*, 127–145. doi:10.1016/S0953-5438(00)00031-X

Treqaskis, C. (2004). *Constructions of disability: Researching the interface between disabled and non-disabled people.* London: Routledge. doi:10.4324/9780203299517

Tripathy, A., & Eppinger, S. D. (2007). A System Architecture Approach to Global Product Development. (Working Paper #4645-07). Cambridge, MA: MIT Sloan School of Management.

Tseng, H. E., Chang, C. C., & Li, J. D. (2008). Modular design to support green life-cycle engineering. *Expert Systems with Applications*, *34*, 2524–2537. doi:10.1016/j.eswa.2007.04.018

Tseng, I., Moss, J., Cagan, J., & Kotovsky, K. (2008). The Role of Timing and Analogical Similarity in the Stimulation of Idea generation in Design. *Design Studies*, *29*(3), 203–221. doi:10.1016/j.destud.2008.01.003

Tufte, E. R. (1986). *The visual display of quantitative information*. Cheshire, CT: Graphics Press.

Udo de Haes, H. A., & Heijungs, R. (2007). Life-cycle assessment for energy analysis and management. *Applied Energy*, *84*(7–8), 817–827.

Ullman, D. G., Dietterich, T. G., & Stauffer, L. A. (1988). A Model of the Mechanical Design Process Based on Empirical Data. *Artificial Intelligence for Engineering Design, Analysis and Manufacturing*, *2*(1), 33–52. doi:10.1017/S0890060400000536

Ullman, D. G., Wood, S., & Craig, D. (1990). The Importance of Drawing in the Mechanical Design process. *Computing and Graphics*, *14*(2), 263–274. doi:10.1016/0097-8493(90)90037-X

Ulrich, K. T., & Eppinger, S. D. (2008). *Product Design and Development* (4th ed.). New York: McGraw-Hill.

Un, C. A., Cuervo-Cazurra, A., & Asakawa, K. (2008). *R&D Collaborations and Product Innovation*. (Unpublished Paper accepted for publication in Journal of Product Innovation Management). University of South Carolina, Moore School of Business.

United Nations Enable. (n.d.). *World Programme of Action*. Retrieved April 12, 2009, from United Nations Enable - Rights and Dignity of Persons with Disabilities: http://www.un.org/esa/socdev/enable/diswpa04.htm

United Nations, W. C. o. E. a. D. (1987). Our Common Future. *Report of the World Commission on Environment and Development, General Assembly Document A/42/427*

United Nations. (1997). *A summary of United Nations agreements on human rights*. Retrieved June 22, 2010, from http://www.hrweb.org/legal/undocs.html/

United Nations. (2001). *World Population Ageing: 1950-2050*. Washington, DC: Department of Economics and Social Affairs, Population Division.

United Nations. (2008). *Rights and dignity of persons with disabilities*. Retrieved June 22, 2010, from http://www.un.org/disabilities

University of Strathclyde. (2009). *Product Design Engineering-course description 2009*. Retrieved from www.strath.ac.uk/dmem/studying/undergraduatecourses/choosingtherightcourseforyou/

Uotila, T., Harmaakorpi, V., & Melkas, H. (2006). A method for assessing absorptive capacity of regional innovation system. Fennia. *International Journal of Geography*, *184*(1), 49–58.

Upshaw, L. B. (1997). Transferable Truths of Brand Identity. *Design Management Journal, 8*(1).

Urban, G. L., & von Hippel, E. (1988). Lead User Analyses for the Development of New Industrial Products. *Management Science*, *34*(5), 569–582. doi:10.1287/mnsc.34.5.569

Urban, G. L., Hauser, J. R., Qualls, W. J., & Weinberg, B. D. (1997). Information acceleration: Validation and lessons from the field. *JMR, Journal of Marketing Research*, *34*(1), 143–153. doi:10.2307/3152071

Utterback, J. M. (1996). *Mastering dynamics of innovation*. Boston, MA: Harvard Business School Press.

Vaillant, G. E., & Mukamal, K. (2001). Successful Aging. *The American Journal of Psychiatry*, *158*, 839–847.

Vajna, S., Clement, S., Jordan, A., & Bercsey, T. (2005). The Autogenetic Design Theory: An Evolutionary view of the Design Process. *Journal of Engineering Design*, *16*(4), 423–440. doi:10.1080/09544820500267781

Valadez, J., Varga,s W., & DiPrete, L. (1990). Supervision of primary care in Costa Rica: time well spent? *Health Policy and Planning*, *5*, 118–125. doi:10.1093/heapol/5.2.118

Valkenburg, R., & Dorst, K. (1998). The reflective practice of design teams. *Design Studies*, *19*(3), 249–271. doi:10.1016/S0142-694X(98)00011-8

van Daelen, H. (2005). *Tools for Innovation through Customer Co-creation* (Unpublished Master Thesis). Tiburg University, Tilburg.

van der Merwe, J. (2008, July 10-12). Changing a phenomenal change: Reassembling the self through a new ethics of negotiation. In Proceedings of *Changing the Change*, an international conference on the role and potential of design research in the transition towards sustainability, Turin, Italy.

van Duijn, J. J. (1981). Fluctuations in innovations over time. *Futures*, *13*(4), 264–275. doi:10.1016/0016-3287(81)90143-9

van Kleef, E., van Trijp, H. C. M., & Luning, P. (2005). Consumer research in the early stages of new product development: a critical review of methods and techniques. *Food Quality and Preference*, *16*(3), 181–201. doi:10.1016/j.foodqual.2004.05.012

van Zanten, W. P. C. (1998). *Technische hulpmiddelen voor groepsbesluitvorming*. Baarn/Heerlen: Nelissen/Open Universiteit NL (in Dutch).

VanderWerf, P. A. (1992). Explaining downstream innovation by commodity suppliers with expected innovation benefit. *Research Policy*, *21*(4), 315–333. doi:10.1016/0048-7333(92)90031-X

Vargo, S. L., & Lusch, R. F. (2004). Evolving to a New Dominant Logic for Marketing. *Journal of Marketing*, *68*(1), 1–17. doi:10.1509/jmkg.68.1.1.24036

Vash, C. L., & Crewe, N. M. (2004). *Psychology of disability* (2nd ed.). New York: Springer Publishing.

VDI-Guideline 2221/1987: Systematic Approach to the Design of Technical Systems and Products. VDI, Düsseldorf.

Venkatesh, V., Morris, M. G., Davis, G. B., & Davis, F. D. (2003). User Acceptance of Information Technology: Toward a Unified View. *Management Information Systems Quarterly*, *27*(3), 425–478.

Verstijnen, I. M., & Hennessey, J. M. (1998). Sketching and Creative Discovery. *Design Studies*, *19*(4), 519–546. doi:10.1016/S0142-694X(98)00017-9

Vidal, R. V. V. (2006). *Creative and Participative Problem Solving – the art and the science*. Infromatics and Mathematical Modelling, Technical University of Denmark, on line publication.

Vihma, S. (1995). *Products as Representations: A Semiotic and Aesthetic Study of Design Products*. Helsinki: University of Art and Design.

Vincent, A. S., Decker, B. P., & Mumford, M. D. (2002). Divergent Thinking, Intelligence and Expertise: A Test of Alternative Models. *Creativity Research Journal*, *14*(2), 163–178. doi:10.1207/S15326934CRJ1402_4

Vincenti, W. G. (1998). *What engineers know and how they know it*. Baltimore, MD: The Johns Hopkins University Press.

Vincent-Ricard, F. (1989). *As Espirais da Moda. Rio de Janeiro/RJ*. Brazil: Paz e Terra.

von Glasersfeld, E. (2005). Thirty Years Radical Constructivism. *Constructivist Foundations*, *1*(1), 9–12.

von Hippel, E. (1986). Lead Users: A Source of Novel Product Concepts. *Management Science*, *32*(7), 791–806. doi:10.1287/mnsc.32.7.791

von Hippel, E. (1988). *The Sources of Innovation*. New York: Oxford University Press.

von Hippel, E. (1994). 'Sticky information' and the locus of problem solving: Implications of Innovation. *Management Science*, *40*(4), 429–439. doi:10.1287/mnsc.40.4.429

von Hippel, E. (2005). *Democratizing Innovation*. Cambridge, MA: MIT Press.

von Hippel, E., & Katz, R. (2002). Shifting innovation to users via toolkits. *Management Science*, *48*(7), 821–834. doi:10.1287/mnsc.48.7.821.2817

von Hippel, E., Thomke, S., & Sonnack, M. (1999). Creating breakthroughs at 3M. *Harvard Business Review*, *77*(5), 47–56.

von Krogh, G. (2006). Breakthrough Ideas for 2006: Customers Demand Their Slice of IP. *Harvard Business Review*, *84*(2), 45–46.

von Stamm, B. (2003). *Managing Innovation, Design and Creativity*. Chichester, UK: John Wiley & Sons.

Voss, C. A. (1985). Determinants of success in the development of applications software. *Journal of Product Innovation Management*, *2*(2), 122–129. doi:10.1016/0737-6782(85)90008-6

Walker, J. (1999). Writing to Build Brand Voice. *Public Relations Strategist*, *5*(1), 34–35.

Walker, S. (2006). *Sustainable by design: explorations in theory and practice*. London: Earthscan.

Wallach, M. A. (1970). Creativity. In Mussen, P. H. (Ed.), *Manual of child Psychology* (pp. 1211–1272). New York: Wiley.

Wallas, G. (1926). *The art of thought*. New York: Harcourt Brace.

Walton, T. (2005). Innovation as a necessity. *Design Management Review*, *16*(1), 6. doi:10.1111/j.1948-7169.2005.tb00001.x

Wang, Q. H., & Li, J. R. (2004). A desktop VR prototype for industrial training applications. *Virtual Reality (Waltham Cross)*, V7.

Wanyama, W., Ertas, A., Zhang, H. C., & Ekwaro-Osire, S. (2003). Life-cycle engineering: issues, tools and research. *International Journal of Computer Integrated Manufacturing*, *16*(4), 307–316. doi:10.1080/0951192031000089255

Warell, A. (2001). *Design Syntactics – A Functional Approach to Visual Product Form. Theory, Models, and Methods*. Gothenburg: Chalmers University of Technology.

Warell, A. (2008). *Modeling perceptual product experience – Towards a cohesive framework of presentation and representation in design"*. Paper presented at the 6th Design and Emotion Conference: Dare to Desire, Hong Kong Polytechnic University; School of Design, 6-9 October 2008.

Warell, A., Adank, R., Garrett, L., Goellner, M., & Parker, T. (2008). *Design for Desirability': Perceptions and Implications in New Zealand Export Industry*. Paper presented at the 15th International Product Development Management Conference, Hamburg: Helmut Schmidt University. June 29 – July 1.

Weaving designs in time and space to create a "sustainable service society": Ezio Manzini. (2004). *Future News*. Retrieved from www.changedesign.org/Students/Changes/Enabling/Enabling%20Docs/ffmanzini.pdf

Webb, E. J., Campbell, D. T., Schwartz, R. D., & Sechrest, L. (1966). *Unobtrusive measures* (pp. 112–141). Thousand Oaks, CA: Sage.

Weber, M. (2008, June 19-20). *Customer Co-creation in the Dutch Financial Services*. Paper presented at the IMCM-PETO 2008, Copenhagen, Denmark.

Weber, M. (2009, October, 5-6). *Customer Involved Open Innovation. Involving customers and end users in product and service innovation: the case of a coffee manufacturer.* Paper presented at the Mass Customization and Personalization Conference 2009 (MCPC), Helsinki, Finland.

Weightman, D., & McDonagh, D. (2003). People are doing it for themselves. In Forlizzi, J., Hanington, B., & Ayoob, E, (Eds.), *Proceedings of the International Conference on Designing Pleasurable Products and Interfaces*, (pp34-39). Pittsburgh, PA: ACM Press.

Weiser, M. (1991). The computer for the 21st Century. *Scientific American*, *265*(3), 94–104. doi:10.1038/scientificamerican0991-94

Wendel de Joode, R. (2005). *Understanding Open Source Communities. An organizational perspective.* Unpublished Dissertation, Technische Universiteit Delft, Delft.

West, J., & Gallagher, S. (2006). Challenges of open innovation: the paradox of firm investment in open-source software. *R & D Management*, *36*(3), 319–331. doi:10.1111/j.1467-9310.2006.00436.x

West, S. K., Munoz, B., Rubin, G. S., Schein, O. D., Bandeen-Roche, K., & Zeger, S. (1997). Function and Visual Impairment in a Population-Based Study of Older Adults. *Investigative Ophthalmology & Visual Science*, *38*(1), 72–81.

What is ID? (2004). *Industrial Designers Society of America*. Retrieved August 24, 2008, from http://www.idsa.org/webmodules/articles/articlefiles/what_is_id_brochure.pdf

Whitfield, R., Duffy, A., Coates, G., & Hills, W. (2002). Distributed Design Coordination. *Research in Engineering Design*, *13*(4), 243–252.

Whitney, D. E. (2004a). Product Architecture. In *Mechanical Assemblies, Their Design, Manufacture, and Role in Product Development*. New York, NY: Oxford University Press.

Whitney, D. E. (2004b). *The influence of architecture in engineering systems. ESD Symposium*. Cambridge, MA: MIT Press.

Wickremaratchi, M. M., & LLlewelyn, J. G. (2006). Effects of Ageing on Touch. *Postgraduate Medical Journal*, *82*, 301–304. doi:10.1136/pgmj.2005.039651

Wiklund, M. E. (2005). Making medical device interfaces more user-friendly. In Wiklund, M. E., & Wilcox, S. B. (Eds.), *Designing usability into medical products* (pp. 151–160). Boca Raton, FL: Taylor & Francis/CRC Press. doi:10.1201/9781420038088.ch16

Wilczynski, V., & Jennings, J. (2005). Creating Virtual Teams for Engineering Design. *International Journal of Engineering Education*, *19*(2), 316–327.

Wilkinson, M., & Sale, R. (2000). *Pressures for Change in the Education of Product Designers. Design Pedagogy: Basic Design and Academic Experiences*. Birmingham, UK: University of Birmingham.

Willems, C. G., Vlaskamp, F. J. M., & Knops, H. T. P. (1997). Use of technology in the support of independent living of people with special needs. In Anogianakis, G.,

Bühler, C., & Soede, M. (Eds.), *Advancement of assistive technology* (pp. 125–129). Amsterdam: IOS Press.

Willis, A.-M. (1999, May). Ontological Designing. In *Proceedings of the European Academy of Design conference, Design Cultures*, Sheffield.

Wind, Y. J., & Mahajan, V. J. (2002). *Convergence Marketing. Strategies for Reaching the New Hybrid Consumer*. New York: Prentice Hall.

Winograd, T., & Flores, F. (1988). *Understanding Computers and Cognition*. Reading, MA: Addison-Wesley.

Wise, D. A. (1997). Retirement against the Demographic Trend: More Older People Living Longer, Working Less, and Saving Less. *Demography. The Demography of Aging, 34*(1), 83–95. doi:10.2307/2061661

Wisner, A. *(1976).* Analyse de la Situation de Travail, Méthodes et Critères. *Paris: CNAN Ed. (in French)*

Wong, T. N., Cheung, C. H., & Lau, H. (1999). Decision support system for a jewellery manufacturer. *International Journal of Production Economics, 60-61*(20), 211–219. doi:10.1016/S0925-5273(98)00142-X

World Commission on Environment and Development. (1987). *Our common future*. New York: Oxford University Press.

World Health Organization. (2003). *Up to 45 million blind people globally – and growing*. Retrieved June 22, 2010, from http://www.who.int/mediacentre/news/releases/2003/pr73/en/

World Health Organization. (2009). *Visual impairment and blindness*. Retrieved June 22, 2010, from http://www.who.int/mediacentre/factsheets/fs282/en/

Woudhuysen, J. (1993). A Call for Transgenerational Design. *Applied Ergonomics, 24*(1), 44–46. doi:10.1016/0003-6870(93)90160-B

Wozniak, G. D. (1987). Human Capital, Information, and the Early Adoption of New Technology. *The Journal of Human Resources, 22*(1), 101–112. doi:10.2307/145869

Wunsch-Vincent, S., & Vickery, G. (2007). *Participative Web: User-Created Content* (No. JT03225396). Paris: Organisation for Economic Co-operation and Development.

Yamamoto, M., & Lambert, D. R. (1994). The Impact of Product Aesthetics on the Evaluation of Industrial Products. *Journal of Product Innovation Management, 11*(4), 309–324. doi:10.1016/0737-6782(94)90086-8

Yang, M. (2007). *Design Methods, Tools, and Outcome Measures: A Survey of Practioners. ASME International Design Engineering Technical Conferences (pp. DETC2007-35123)*. Las Vegas, NV: ASME.

Yin, R. K. (1994). *Case Study Research: Design and Methods* (2nd ed.). Newbury Park, CA: Sage Publications.

Yoon, E., & Lilien, G. L. (1988). Characteristics of the industrial distributor's innovation activities: An exploratory study. *Journal of Product Innovation Management, 5*(3), 227–240. doi:10.1016/0737-6782(88)90025-2

Yoshikawa, H. (1981). General Design Theory. In T. Sata & E. Warman (Eds), The IFIP WG 5.2-5.3 Working Conference (pp. 35-57). North Holland, Amsterdam.

Young, A. (1937). *Recurring Cycles of Fashion*. New York: Harper and Row.

Zachry, M. (2005). An interview with Donald A. Norman. *Technical Communication Quarterly, 14*(4), 469–487. doi:10.1207/s15427625tcq1404_5

Zaltman, G. (2003). *How Customers Think. Essential Insights into the Mind of the Market*. Boston, MA: Harvard Business School Press.

About the Contributors

Arlindo Silva, Instituto Superior Técnico. BSc in Mechanical Engineering at Instituto Superior Tecnico (IST), Lisbon, Portugal, in 1991, MSc in 1995 and PhD in 2001, both in Composite Materials. Assistant Professor at IST, Mechanical Engineering Department. He develops his research activities within the Institute of Mechanical Engineering, in Lisbon, Portugal. Co-author of three books (in Portuguese) on topics related to Mechanical Engineering and Manual Wheelchairs for the Impaired. Published numerous technical papers in refereed journals and conference proceedings, in the areas of rehabilitation of the impaired, composite materials, design ideation, product development and materials selection. Holds numerous patents, filed with his students on products developed in the courses taught at IST. His current research interests are in product development, engineering design methods and materials selection. Currently teaching Mechanical Design, Product Design and Development and Engineering Materials at IST. He is also involved in the MIT-Portugal Program through the Engineering Design and Advanced Manufacturing focus area, in the courses on Product Development and Technology Evaluation and Selection. He is the Vice-President of the CPIN - "Centre for the Promotion of Innovation and Business", a University-based non-profit organization providing integrated solutions for technology-based entrepreneurship, and fostering University spin-offs, currently holding a European Union label of Business Incubation Centre. He is also a member of the Product Development and Management Association, the International Council of Systems Engineering, the Design Research Society, the Design Society and the American Society of Engineering Education.

Ricardo Simões is an Associate Professor at the Polytechnic Institute of Cávado and Ave (IPCA), Barcelos, Portugal, teaching 3D Modelling and Physics of Materials. He conducts his research activities at the Institute for Polymers and Composites (IPC) of the University of Minho, Guimaraes, Portugal, and the Associate Laboratory I3N - Institute of Nanostructures, Nanomodelling and Nanofabrication. He cooperates with the Engineering Design and Advanced Manufacturing (EDAM) focus area of the MIT-Portugal program, teaching the Engineering and Manufacturing Systems course and supervising research projects. He has a 5-year bachelor degree in Polymer Engineering from the University of Minho (Portugal) and a PhD in Materials Science and Engineering from the University of North Texas (USA). He is currently coordinating projects in Mobile Health (in the framework of the MIT-Portugal program), RFID (SimeI&DT project), Automotive (with Ford Motor Company), Nanomaterials (with the US Air Force Research Laboratories, and also the Foundation for Science and Technology), and Medical Devices (with the Health Cluster Portugal). His main research interests are related to product design and development, through the application of engineering systems and materials science and engineering. He has over 40 scientific/technical publications, over 20 oral presentations at international conferences, and 8 invited lectures in several countries.

***

Rodney Adank's research interests are centred in product design/development. He has supervised numerous product innovation projects with New Zealand manufacturers. His work supports a strong connection between New Zealand industry and design research with a focus on affective design and sensory experience.

Eric E. Anderson is an independent consultant and a licensed professional Geophysicist with over 20 years of experience servicing both government and Fortune 500 companies. He specializes in preparedness planning and the development of risk management strategies to facilitate regional resilience through disaster risk reduction. He has a MS in Geosciences from the University of South Carolina, USA, a MS in Information Technology from Virginia Tech, USA and has been a Visiting Scholar to Ålborg University, DK. His research interests include: (1) the assessment of the vulnerability of communities, cities and regions and their critical life support infrastructure to disasters within the context of the globalization of people, information, markets and security, (2) the simulation of complex maladaptive human-environmental systems capable of creating disastrous cascading effects over space and time, (3) the design and use of web-based collaborative spatial decision support system tools to prioritize risk management solutions for investment.

Rina Bernabei is a Senior Lecturer in the Faculty of the Built Environment, University of New South Wales, Sydney. Rina Bernabei and Kelly Freeman are directors of the award winning product design company bernabeifreeman. Their design work and research is concerned with the emotional and storytelling potential in product design. Their design work has been published and exhibited widely both locally and internationally.

Fausto Brevi is a Senior Researcher and Assistant Professor in the Politecnico di Milano School of Design since 2002. He holds a degree cum laude in Architecture. Specialist in 3D computer modelling and visualization for product design, he spent several years working with some of the most important Italian Industrial Design departments mainly in car design industries. Recently he has focused on virtual prototyping and real-time rendering issues applied to industrial and interior design. He is the scientific supervisor of Indaco Department Virtual Prototyping Lab. He is member of the Academic board of PhD courses "Virtual Prototypes and Real Products" and "Industrial Design and Multimedia Communication". Vice-director of the Politecnico di Milano Specializing Master Course in Transportation & Automobile Design.

Gabrielle Stockey Chinchilha is affiliated with University of the State of Santa Catarina. Academic course of Bachelor of Fashion - University of Santa Catarina | Arts Center | Department of Fashion - Brazil. Member of the research group "Fashion Design & Technology" (UDESC / CNPq). Joined the research project "Future of the Present: space for observation, analysis and interpretation of signs" | Department of Fashion | Art Center - UDESC

P. John Clarkson holds a chair in Engineering Design and is director of the Engineering Design Centre at the University of Cambridge. His research interests are in the general area of engineering design, particularly the development of design methodologies to address specific design issues, for example, process management, change management, healthcare design and inclusive design. John is a Chartered Engineer, a Fellow of the Institution of Engineering and Technology and on the editorial board of the Journal of Engineering Design. As well as publishing over 400 papers, he has written a number of books on medical equipment design and inclusive design.

Nathan Crilly holds a lectureship in the Engineering Design at the University of Cambridge. His research interests are in the areas of design, creativity and communication. In particular, he focuses on how designers intend products to be understood and used, and on how they are subsequently experienced by users. More generally, Nathan has an interest in how different academic disciplines and professional practices can be used to inform the study of design. Nathan holds a bachelor's degree in Mechanical Engineering, a PhD in Product Aesthetics and has professional design experience in the aerospace and software industries.

Claudia Eckert is a senior lecturer of Design in the Design Group at the Open University. Formerly at the Engineering Design Centre at the University of Cambridge, she is interested both in understanding design in different design domains and different phases of the design process, and in developing techniques and computer tools for facilitating design activities. Most of her research is based on empirical studies in industry, but she is also interested in bringing theories and methods from different disciplines into design research. For example she is trying to apply insights from philosophy and sociology of science to design process modelling and from complexity theory to change prediction.

Eduardo Romeiro Filho, Graduate in Industrial Design at ESDI - Superior School of Industrial Design from UERJ - State University of Rio de Janeiro (1987), Master at Production Engineering (1993) and Ph.D. at Production Engineering from Federal University of Rio de Janeiro (1997). Has experience in Production Engineering and Industrial Design, focusing on Design Methodology, acting on the following subjects: Product Design, Ergonomics, Ecodesign, Computer Aided Design, Design Methodology and Concurrent Engineering. Author of books: "CAD in the Industries: Implementation and Management" (1997, in Portuguese), by Editora da UFRJ (Rio de Janeiro, Brazil) and "Product Design", by Elsevier Brasil (São Paulo, 2009, in Portuguese).

Luisa Fontana was born in Zurich (Switzerland). She graduated in architecture at Venice University and set-up her studio FONTANAtelier in Schio (Vicenza), Italy. Her designs are the result of advanced multidisciplinary research. She has entrusted the engineering aspects of her projects to ARUP in collaboration with whom she has developed for - runner projects in the sustainability field. She has collaborated as a lecturer for the department of architecture at the university of Venice, the university of Manitoba and the university of Montreal, Canada. She has received prestigious acknowledgements and articles about her work have been published in magazines and international architecture publications.

Davide Fornari is a researcher at SUPSI, University of Applied Sciences and Arts of Southern Switzerland in Lugano. He is a PhD candidate of the Design Sciences programme at University IUAV of Venice, with a thesis on humanoid interfaces. Honour degree in Architecture, master of Art and Culture Management, he studied at University IUAV of Venice, ETSA of Barcelona and MART Museo di Arte Moderna e Contemporanea di Rovereto (the Modern Art Museum in Rovereto). He is a junior member of AIAP, the Italian Association for Visual Communication Design, and a member of LISAV, the International Semiotic Laboratory in Venice.

Kelly Freeman graduated with Honors in Industrial Design from the University of New South Wales, Sydney and is a casual staff member of the Industrial Design program at the University of New South Wales, Sydney. Rina Bernabei and Kelly Freeman are directors of the award winning product design company bernabeifreeman. Their design work and research is concerned with the emotional and storytelling potential in product design. Their design work has been published and exhibited widely both locally and internationally.

Lyn Garrett is currently Undergraduate Programme Leader in Industrial Design at Massey University in Wellington, New Zealand. He has extensive professional design and lecturing experience in New Zealand and the UK. His professional awards include the 1988 Prince Philip Award for Industrial Design as part of the team that designed the Zaf chair range for leading New Zealand manufacturer Formway. Lyn is particularly interested in design research methodologies that connect product design and product experience, interface design, the integration of ergonomics factors into products and furniture, and ergonomics issues surrounding office and public furniture. He has a side interest in production design for science fiction movies and the relationship between industrial design, science fiction, and culture. His teaching specialities include ergonomics, design research, product design, furniture, and interface design. He is also a Research Associate within the Affect research centre, and is a Patron member of the New Zealand Ergonomics Society.

José Manuel Ferreira Gaspar received hisBachelor in Mechanical Engineering in 1998, Degree in Mechanical Engineering in 2003 and MSc in Engineering Design in 2007. Research member at the Institute of Science and Engineering Materials and Surfaces. Author and co-author in technical papers about creativity, innovation and product development in engineering design. Has a patent in the air conditioning field and is involved in the research, development and demonstration of the invention. Also involved as a PhD student in the MIT-Portugal Program through the Engineering Design and Advanced Manufacturing focus area. Currently is doing research on human and machine interfaces for the automotive audio panels.

Simone A.M. Geerts born in 1985, has studied industrial engineering and management science at the Eindhoven University of Technology, followed by a master in Innovation Management. Intrigued by innovating the innovation process, she has concluded her study with a master thesis on crowdsourcing, providing the opportunity to co-write this chapter on customer involved open innovation. Currently, she has started a traineeship at ABN AMRO, where she is also exploring possible applications of innovation management and crowdsourcing.

Mark Goellner is a Industrial Design Lecturer at Massey University in Wellington, New Zealand. He graduated as Industrial Designer (Braunschweig School of Art, Germany) and completed a PhD in Design Studies (University of Otago, 2005). He worked for different companies, like Hewlett Packard, and was awarded with the IF Product Design Award, in 2001 for the design of the AGILENT 2100 BiochipAnalyser. His teaching focuses on systematic design, design for desirability, R&D and CAD applications for industrial designers. His research at "AFFECT – The Centre for Affective Design Research" explores systematic design and affective design issues. Furthermore, he supports the Roofwater Research-Centre investigating the sustainable use of water.

Elsa Henriques has a graduation, a master and a doctorate in Mechanical Engineering at Instituto Superior Técnico. She is an associated professor at Instituto Superior Técnico in the Manufacturing Technologies and Industrial Management scientific area, being responsible for programmes in the "Mechanical Engineering" and "Engineering Design and Manufacturing" graduations and pos-graduations. During the last fifteen years she has participated and/or coordinated several national and European R&D projects in collaboration with the industry, mainly related to the design and manufacturing management based on value and lean principles. She has a large number of scientific and technical publications in national and international conferences and journals. She is a national delegate in the 7th Framework Programme of the EU.

Julie Holt is Project Manager for the Massachusetts General Hospital RFID in Outpatient Care Project. Prior to her experience at MGH she worked at PatientKeeper, Inc., a healthcare software company that specializes in mobile applications designed to improve physician workflow and patient safety. She has also worked in performance improvement, clinical research and outpatient care capacities at Tufts-New England Medical Center, Spaulding Rehabilitation Hospital, and the Boston Veterans Administration Research Institute. She holds a Master of Business Administration with a concentration in Healthcare Management and a Master of Public Health with a concentration in Health Management and Policy, both from Boston University. Her areas of interest include healthcare resource utilization, technology implementation, process improvement and health reform policy.

Tao Huang PhD, IDSA, is an assistant professor of Product Design in the Art + Design Department of Columbia College Chicago. She has earned her PhD in Architecture + Design Research from Virginia Tech. She holds a Bachelor of Science and a Master of Fine Arts degree both in Industrial Design from Beijing Institute of Technology and Guangzhou Academy of Fine Arts, respectively. She has practiced in multiple design fields for over 10 years, including product design, interior design, advertising, graphic design, and packaging design. Her doctoral dissertation, "Reforming Industrial Design Education in Mainland China for Sustainability", was published in 2009. Her current research projects include: Chinese contemporary design studies, product semantics and sustainable design, systems theory and its adaptation in design, design and disaster prevention, etc. She is also an influential blogger and a regular contributor for several Chinese newspapers and magazines.

Mirja Kälviäinen has graduated and worked as a designer in the 1980's turning into a design researched and manager during the 1990's. Her PhD. research approached the design professional assessment of products through sociological critical theory. The results showed how professionals evaluate the products through the socialization to the rules of their own profession. Her main research and practical interest during the last 10 years has been on user driven innovation methods and collaboration emphasizing the user value in product development instead of professional values. She works currently bridging research with practical consulting and company cases in the D'ART Design Resource Centre of the North Karelia University of Applied Sciences in Finland. In this work the user value and co creation emphasis has also led the applications of interdisciplinary innovation in the front end INNOstudio® practice.

Lydia Khuri, Psy.D., a Licensed Clinical Psychologist, received her Doctorate of Psychology from the Illinois School of Professional Psychology- Chicago in 1998. She has been in part-time private practice since 1998. At the University of Illinois at Urbana-Champaign, she has worked as a Clinical Counselor and with the Program on Intergroup Relations. She is currently a Program Coordinator for two multicultural Living-Learning Communities at the University of Illinois. She has published in the area of working with emotion in intergroup dialogue.

Rose Mary Rosa de Lima graduated at Electrical Engineering from Pontifícia Universidade Católica de Minas Gerais (1996), Master at Production Engineering from Federal University of Minas Gerais (2003). Professor at graduate and postgraduate courses on Production Engineering at Faculdade Pitágoras - FAP in Belo Horizonte, Brazil. Has experience in Production and Electrical Engineering, focusing on Production Engineering, Management, Design for Environment and Design for Recycling Methodology. Co-author of book "Tools on Environmental Management" (2008), by CEFET RN Publishing (Natal - Brazil, in Portuguese).

For ten years, **Michael Lye** has taught advanced Industrial Design studios and is currently assistant professor in the Industrial Design department at RISD. He is Project Lead for the Business Innovation Factory's Nursing Home of the Future program and co-directed the Business Innovation Factory/RISD Healthcare Innovation Project, which leveraged the skills of designers and design students to create a detailed visualization of the patients' experience of the primary care system in Rhode Island, and used this to define the boundary constraints for improvements in the primary care field. He has extensive experience teaching interdisciplinary studios, working with corporate and other partners in an academic setting. He has taught the NASA-supported Design for Extreme Environments studio and overseen the RISD/NASA collaboration to define future spacecraft and habitats for the moon. His consulting practice specializes in human centered design with an emphasis on design research and analysis, Universal Design issues, and the integration of design into the realm of healthcare.

Deana McDonagh, PhD., is an Associate Professor of Industrial Design within the School of Art + Design and a faculty member of the Beckman Institute of Advanced Science and Technology, the Institute of Aviation and Human Factors and the Academy of Entrepreneurial Leadership. She has focused on developing Empathic Design Research strategies to support new product development by responding to users' emotional needs beyond the functional (supra-functional). She joined the University of Illinois in 2004 from the Department of Design and Technology at Loughborough University (UK). She is the Co-Editor of the Design Journal, Fellow of the Design Research Society and the World Demographic Association. Recent University of Illinois Campus Research Grants have included "Trans-generational Learning: Communication through visual stimuli", "Disability + Relevant Design" and "Visual Data Capture".

Ana Cristina Medeiros is a third-year PhD research student in the Engineering Design Centre at the University of Cambridge. Ana's doctoral research is on the influence of ageing on user experience and the consequences this has for technology adoption. Her research interests are in product design, as well as in interaction and user experience at different life stages and among different cultures. Ana holds a bachelor's degree in Industrial Design and a Master's degree in Engineering, from universities in Brazil and Japan respectively.

For her achievements in design she has received four awards in Japan. Ana has a multicultural professional experience in product development and graphic design, having worked in Brazil, Japan and Germany.

Gavin Melles is Head of Interior and Industrial Design at Swinburne University. He currently supervises a range of students on topics around product and industrial design and education. He holds a Masters in Linguistics and a Doctorate in Education and continues to publish and research in design and education fields. He is supervising Ian de Vere's PhD on Product Design Education in the Faculty of Engineering, Swinburne University.

Johann van der Merwe teaches design thinking using various ways of looking at the hybrid world of objects and humans. His main research interests and influences are autopoiesis, cybernetics & systems, and phenomenology, all of which have been conceptually blended to form *A Grammatopology of Design Knowledge,* the title of the thesis he is currently finishing.

Stuart Miller is a project engineer with TTi, an international consumer product and hand tool manufacturing company, where he leads product development teams in creating the next generation of professional grade tools. Mr. Miller graduated with his MS degree from Clemson University in 2008 where his research was focused on developing case study research tools to aid design researchers in studying industry design projects. Previously, Mr. Miller graduated with his BS degree, also from Clemson University.

James Moultrie is a University Lecturer in Design Management. His research interests seek to improve the utilisation of design skills and increase design/innovation capability at project, firm and national levels. Specific interests include design management, design strategy and the role of the physical environment on enhancing innovation. His PhD focused on assessing and improving design capabilities in small firms. James is a Chartered Mechanical Engineer (IMechE) and has over ten years of industrial experience as a project manager, senior engineer and marketing product manager in the precision instruments sector. He has a 'Scientific and Technical Academy Award' and an Emmy for work on a range of lenses for professional 35mm cinematography.

Tony Parker is Associate Pro-Vice Chancellor of Massey University's College of Creative Arts. He is one of New Zealand's leading industrial designers. He studied at the internationally acclaimed Royal College of Art (MDes RCA), London and has a distinguished career in professional practice and design education spanning some thirty years. Tony has designed award winning, commercially successful products for international markets and continues to design for leading New Zealand manufacturers. His research interests focus around product identity and aesthetic innovation, styling and the human factors that influence the way designers give form to mass-produced products. Tony teaches in the industrial design programme with extensive experience in design studio subjects, drawing and rendering. His students have won numerous national and international design awards and many are now working in international companies as successful professional industrial designers.

Paulo Peças has a graduation and a doctorate in Mechanical Engineering at Instituto Superior Técnico (Technical University of Lisbon). He is an assistant professor at Instituto Superior Técnico in the Manufacturing Technologies and Industrial Management scientific area, being responsible for courses in the "Mechanical Engineering" and "Engineering Design and Manufacturing" graduations and pos-graduations. During the last ten years has participated and integrated coordination teams of R&DT projects in collaboration with the industry, at both national and European levels. The research activity has focused the development of models for diagnosis, comparative analysis and procedures implementation in the scope of lean and agile manufacturing. These models foster the build-up and implementation of solutions for productivity and competitiveness increasing in several industrial sectors, namely tools & dies, automotive suppliers, automotive assemblers and metal working. He has a large number of scientific and technical publications in national and international conferences and journals.

Feniosky Peña-Mora, Sc.D, is Dean and Morris A. and Alma Schapiro Professor of The Fu Foundation School of Engineering and Applied Science at Columbia University. Professor Peña-Mora earned a Master of Science degree in Civil Engineering and a Doctor of Science in Civil Engineering Systems from the Massachusetts Institute of Technology (MIT). Before joining Columbia in 2009, Professor Peña-Mora was an Associate Provost and an Edward William and Jane Marr Gutgsell Endowed Professor in the Department of Civil and Environmental Engineering, Center Affiliate at the National Center for Supercomputing Applications, and a Faculty Affiliate at the Beckman Institute at the University of Illinois. Previously, he was on the faculty of MIT's Civil and Environmental Engineering Department and was visiting professor at international universities, including Ecole Polytechnique Fédérale de Lausanne in Switzerland. He is the author or co-author of more than 100 scholarly publications and holds five patents.

Alice Pignatel holds a PhD in *Virtual Prototypes and Real Products* at Politecnico di Milano. She is involved in a research project that aims to improve the interaction with virtual prototypes in the Industrial Design field. Her main interest focus is on advanced interaction issues, in order to support and optimize the design process for industrial production. Recently, she extended her attention to the interior design field and she started to investigate the opportunities that virtual environments could offer to the interior design process. Since 2008 she is Adjunct Professor at the School of Design of Politecnico di Milano, teaching 3D CAID modelling and rendering techniques to both Bachelor of Science students and Specializing Master Course students.

António J. Pontes is Assistant Professor at the Department of Polymer Engineering, Universidade do Minho, since June of 2002. He develops scientific activities in the IPC - Institute of Polymer and Composites of Associated Laboratory I3N, Institute of Nanostructures, Nanomodelling and Nanofabrication of FCT. The main scientific interests are: (1) in the phenomenological aspects associated to functional system of injection moulds (2) in the incorporation of moulding blocks in alternative materials to the steel using Rapid Prototyping and Manufacturing techniques, and (3) in the manufacture of micro systems based on polymeric materials. He was the Director of Polymer Engineering Degree (Jan 2003 - Jan 2007) and Departmental Coordinator of Sócrates/Erasmus (Jan 2003 - Dec 2004). He is author or co-author of more than 30 scientific publications with referee and more than 60 communications in international conferences. Is the responsible or co-responsible in the Universidade do Minho of 1 European projects of 6th FP and 2 of 7th FP, and another of FCT in the scope of MIT-Portugal program. Is member of Professional Organization of Engineers since January 2000, in the College of Metallurgic Engineering and Materials, and associated of the Portuguese Society of Materials.

Jacqueline Power completed a degree in Interior Architecture, graduating with Honours Class 1. She is currently a PhD candidate in Interior Architecture at the University of New South Wales. Her PhD thesis is investigating Australian Indigenous interiors. Jacqueline was awarded an Australian Postgraduate Award to undertake her research. She is a causal staff member of the Interior Architecture program at the University of New South Wales and a research assistant. Jacqueline has co-authored a number of conference papers with industrial designers Rina Bernabei and Kelly Freeman.

Mamata N. Rao is presently an Associate Senior Faculty at the National Institute of Design (NID), R&D Campus, Bangalore, India. She has an architectural background with work experience in educational and private organisations in Hong Kong and Sydney in diverse fields such as architecture, creativity development, multimedia and development of web-based interfaces. She has a Masters in Design Science from the University of Sydney, Australia and Masters in Urban Design from School of Planning and Architecture, New Delhi, India. Before joining NID, she worked at the School of Design, the Hong Kong Polytechnic University, on projects in the area of creativity development for higher education and designing web applications for use in design education. Her areas of interest are creative thinking, spatial perception in retail environments, user experience design, and interactive media. She has various publications including has co-authored a book "Creative Tools" published by the School of Design, The Hong Kong Polytechnic University in November 2005. She believes in gaining insights for design from user behaviour/experiences and pursues the use of creative tools, together with the user-centred design process, for the development of design ideas to address user needs.

Sandra Regina Rech received Ph.D. (2006) in Production Engineering from Universidade Federal de Santa Catarina - Brazil. Associate professor at the University of Santa Catarina | Arts Center | Department of Fashion. Leader of the research group "Fashion Design & Technology" (UDESC / CNPq), she coordinates the research project of the "Future of the Present - a space for observation, analysis and interpretation of signs" (http://www.ceart.udesc.br/futurodopresente). Has experience in Design, with emphasis in Research and Fashion Management and develops projects and consultancy in the exploration of trends.

Inês Esteves Ribeiro has a degree in Mechanical Engineering from Instituto Superior Técnico, Universidade Técnica de Lisboa. She is currently a PhD student from MIT-Portugal EDAM program and in parallel she gives problems classes in the Manufacturing Technologies and Industrial Management scientific area at Instituto Superior Técnico. The research activity has been focused on Life Cycle approaches in the design stage of products. She has also some scientific and technical publications in international conferences and in journals.

Juan Carlos Campos Rúbio, Graduate in Mechanical Engineering at Federal University of Santa Maria (1988), master in Mechanical Engineering at Federal University of Santa Catarina (1992) and Ph.D. in Mechanical Engineering at Universidade de São Paulo, Brazil (2000), Postdoctorate at Universidade de Aveiro, Portugal (2007). Scholarship in Research Productivity 2 by CNPq, Brazilian Scientific National Council. Actually is professor and Doctorate Advisor at Federal University of Minas Gerais, Brazil. Editorial board member of the International Journal of Mechatronics and Manufacturing Systems (IJMMS) and of the Modern Manufacturing Technology (JMMT). Focus area: Processes of Manufacture, emphasizing Numerical Control, Precision Engineering, Industrial Automation, CNC Machine.

Álvaro Sampaio is an industrial designer and has a Masters degree (MSc) in Design, Materials and Product Management. He is currently pursuing his PhD in Polymer Engineering at the University of Minho, Portugal. He is a researcher at the Institute for Polymers and Composites of the University of Minho, and a collaborator at the I3N Associated Laboratory. He has devoted himself to the activities of product design and development, more specifically, in the field of socially responsible design, user-center design and medical devices sectors. He is an Assistant Professor at the Polytechnic Institute of Cávado and Ave, Portugal, teaching courses in several areas of Industrial Design, including Product Design and Development, Design Management and Design Methodologies, at the undergraduate level.

Peer M.Sathikh is an Assistant Professor at the School of Art, Design and Media at the Nanyang Technological University, Singapore and teaches product design in its BFA program. A Raffles Scholar, Peer obtained his Master of Design (MDesRCA) in Industrial Design from the Royal College of Art, London. He also has a Master of Design degree from the Indian Institute of Technology, Bombay, together with a Bachelor of Engineering (Hons) degree. A practicing Industrial Designer since 1986, Peer co-founded Inovasia Design, in 1996. After successfully merging Inovasia into a larger corporation, Peer left to pursue his interest in teaching design in 2008. Peer has been involved in design promotion and was the President of the Designers Association Singapore from 2003 till 2007. He is also the Chairman of the Advisory Committee for the diploma course on Product Design at the Ngee Ann Polytechnic, Singapore. As the Managing Director of Inovasia design, Peer was a participant and witness to the 'design renaissance' period in the 1990s, mentioned in this chapter.

Friedemann Schaber, Dipl. Des. (FH), studied at the University of Applied Science, Pforzheim, Germany. Currently, he is a Senior Lecturer in the Division of Design at the School of the Arts, The University of Northampton, UK. He has worked as an industrial designer in Europe and the Far East, which has led him to explore national differences in the use of materials, design education, government promotion and brand marketing. With Vicki Thomas, he researches the understanding of design in different societies, how it is taught and how ideas, visual and technical information are transferred between locations and markets. Knowledge transfer activities include conceptual models, prototyping systems for industrial and product design, and the introduction of live client projects into the design curriculum. This joint work with Randle Turner has been awarded the University's Teaching Fellowship in 2008.

Susann Heft Sears is a Disability Specialist with the University of Illinois' Division of Disability Resources and Educational Services (DRES). She determines reasonable accommodations for qualified students with physical and systemic disabilities who are enrolled at the university, as well as provides advising and support on career transition and study abroad issues for all students registered with DRES. She is currently a PhD student in Community Health with a Disability Studies Emphasis and her research interests include how campus / community social inclusion impacts the experiences of those with disabilities as well as investigating the career thoughts and perceptions of college students with disabilities.

Vladimir Sedenkov, a Senior lecturer of Belarusian State University. Research interests concern with fundamental and applied design theories, design methodology and education, developing computational tools for support design processes. He began his career as mechanical engineer and proceeded as a programmer and systems analyst of Computer Integrated Manufacturing at Leading Research Institute on Organizational and Management Engineering (Minsk). After postgraduate study at BSU, he combined research with faculty member duty. Current research projects include advanced technology of design computerization and development of an outline of Computer Urged Design (CUD) paradigm, which should succeed the CAD one. Teaching interests include software engineering, design automation, knowledge inference systems, students design projects.

Kin Wai Michael Siu is Professor and Leader of the Public Design Lab, The Hong Kong Polytechnic University. In 2001, he was invited by the University of Cambridge to carry out inclusive design research. From 2002 to 2003, he was Fulbright Scholar at MIT to research the reception of public spaces. From 2006-2007, he was awarded as ASIA Fellow by the Asian Scholarship Foundation and stayed at the National University of Singapore to study public spaces and facilities in a multicultural society. From 2007 to 2008, he was invited by UC Berkeley as Visiting Scholar to carry out research related to public design and contemporary urban cultures. His research and design focus is on both technological and social perspectives. He has provided consultancy services in public design. He has received international invention and design awards, and owns more than 40 design patents. His papers have been published in numerous leading journals and research outlets.

Kirsten Bonde Sørensen is a communication designer, MA in rhetoric and Ph.D. candidate. She is educated as a graphic designer at Kolding School of Design, Denmark and at The Utrecht School of the Arts, Holland (1995). Subsequently she was engaged as an art director in an international fashion company, based in Copenhagen, Denmark, responsible for the creative development of all international marketing material. In 2000 she set up her own company primarily working with graphic design and communication for cultural institutions, especially the music industry. In 2006 she completed her MA in Rhetoric at The University of Aarhus, Centre for Rhetoric. In 2007 she was hired at Danish Centre for Design Research and Kolding School of Design for a pre-doc study and in 2008 she was hired as a PhD candidate researching how to use design, design thinking and rhetoric in strategic design processes.

James Stahl, M.D., C.M., M.P.H. is a practicing board-certified internist and a Senior scientist at the MGH-Institute for Technology Assessment, Research Director for the MGH Outpatient RFID project and Read Trust Research Director for North Shore Medical Center. He is an outcomes researcher with expertise in decision analysis, health technology assessment, and simulation modeling. His current research examines clinical process redesign and the application of new technologies in the outpatient setting, transplant allocation and organ replacement technologies, and technology and ethics.

Joshua D. Summers, Associate Professor in Mechanical Engineering at Clemson University, co-directs the CEDAR Group (Clemson Engineering Design Applications and Research) which focuses on practicing engineering design in order to study how designers operate in order to develop needed design enabling tools. Dr. Summers earned his PhD in Mechanical Engineering from Arizona State University in 2004 researching design automation. Dr. Summers received his BSME and MSME from the University of Missouri-Columbia working on VR-based submarine design. Current research is supported by NASA, NIST, US Army Michelin, and BMW with past work supported by NSF, GM, Hartness International, Wright Metal Products, and others.

John Takamura has been instrumental in implementing brand and product development programs in both Asia and North America. John joined Sharp Corporation's elite Advanced Design Team in Tokyo where he worked on a variety of consumer electronic products. In 1997 John joined Fitch Inc. to establish and manage the Fitch office in Japan. Serving as Vice President of industrial design, John was involved in brand and product development research strategies and design. He served as manager of the Fitch Japan-based product innovation lab called the *MadLab* as well as Vice President and Director of Client Services for product design at the Fitch San Francisco office. With a major in industrial design, John received his BA in design from the UCLA and furthered his studies at Art Center Pasadena. He received his Master's degree in Human Factors in industrial design from Arizona State University and is currently an assistant professor in industrial design.

Joyce Thomas is an innovator and educator with a passion for design. As a professional industrial designer, she has worked across interdisciplinary boundaries and integrated marketing, engineering, and consumer needs into creative products for 30+ years. Joyce has invented products that have been awarded more than 50 US and international patents. She has employed the consumer as an active participant in the designing process, utilizing user-centered design, design ethnography, and empathic design research methods. Joyce is also a Visiting Assistant Professor of Industrial Design at the University of Illinois Urbana-Champaign where she teaches studio and technology courses. Her research interests include Design and Disability and Design for Disaster Preparedness and Recovery.

Vicki Thomas is a graduate of the London School of Economics and the Royal College of Art/Victoria and Albert Museum, in London. Her master dissertation was on the Designers' Role in the Gift Trade. She teaches on Product, Interior and Illustration courses at The University of Northampton; teaching primarily design in context, covering history, the social and economic context of design and design management. She is active researching cross-cultural aspects of design and is currently working on an illustration project funded by the Santander Bank. Outside the University she has run her own design practice, Vicki Thomas Associates, for twenty-five years, working with fifty illustrators and designers creating ranges for manufacturers and retailers both in the UK and abroad.

Randle Turner, (B.Tech(Hons), CENG, MIMechE, CertEd.) is the Divisional Leader for Design, within the School of the Arts. He studied Design initially at Loughborough University and then went on to work in the Aerospace, Automotive and IT industries within the UK. His past applied research portfolio has included: the development of a technical report commissioned by the EU concerning distance learning methodologies; the delivery of a programme to initiate distance learning technologies in the Philippines, commissioned by the UN Geneva. Randle has also been actively involved in Knowledge Transfer Partnerships (KTP), he has been the second academic for two KTPs, and lead academic for a KTP with "Sue Ryder Care". Currently he is lead academic for the scheme jointly set up between "BCE Distributers Ltd" and The University of Northampton. Randle's subject specialism includes the application of computer "Solid Modeling", linked to rapid prototyping technologies, which enable the production of high quality accurate models.

Ian de Vere is an industrial designer with extensive industry experience in product design and development within the design consultancy and manufacturing sectors. His design portfolio encompasses electronic, consumer and specialist medical products, design for the public domain, educational museum design for children, and commercial furniture for mass production. His academic roles include curriculum development, design teaching and the role of program coordinator for Product Design Engineering at Swinburne University of Technology, a course that integrates industrial design with mechanical engineering. His research examines the changing roles and societal responsibilities of engineering together with investigation of new curricula that develop responsible, human-centred and creative engineers through integration of 'designerly ways.' His work has been widely acknowledged with an Australian Design Award for consumer product design, a Swinburne Vice-Chancellors Award for his commitment to socially responsible design, and in 2009, an Erasmus Mundus Scholarship awarded by the European Commission of the EU.

Anders Warell holds an MSc in Mechanical Engineering (Ergonomic Design and Product Development) from Linköping University and a PhD in Industrial Design Engineering from Chalmers University of Technology. He has studied Industrial Design at HDK, the School of Design and Crafts, at Gothenburg University, and has worked as consultant designer since 1999. After finishing his PhD in 2002, he was Assistant Professor at Chalmers before joining Massey University School of Design in Wellington, New Zealand, in 2005. While at Massey, he led the research and educational programme as Associate Professor in Industrial Design, and initiated Affect, the Centre for Affective Design Research, which he headed as inaugural Research Director until 2008. His research interests centre on topics such as product experience, visual brand identity, design semantics, and strategic design management.

Marcel Weber, bornin 1953, Marcel Weber is nowadays a senior consultant at Altuition BV, a consultancy specialized in advising and supporting its clients in programs for obtaining customer insights, shaping customer interactions, developing new business and retaining customers. At the age of 16 he started a career as an air force officer by entering the Dutch Royal Military Academy in Breda, the Netherlands. After finishing the military academy, he served in several, maintenance and logistics positions in the Netherlands and abroad, with a specialization in total quality management and customer focus. After almost 30 years of service he joined Altuition in his present position. He has an MBA and a MSc in Strategy and Organisation Science. Presently he is an executive PhD student at Eindhoven University of Technology on the subject of Customer Involvement in Innovations.

Andrew Muir Wood, A doctoral researcher in his third year at the Design Management group in the Institute for Manufacturing at Cambridge University, Andrew is investigating the evolution of consumer products over time and the patterns of change that may occur. He is researching this area through a mixture of qualitative and quantitative methods, taking inspiration from a broad spectrum of academic disciplines from engineering design to consumer psychology. He has a first class MEng in Product Design and Manufacture from Loughborough University, and worked for a year as a concept engineer for Gillette. Once he has finished his PhD, he hopes to return to industry to translate his research into trend and product development processes.

Index

C

D

E

F

G

H

I

K

L

M

Q

R

S

T

U

V

W

Y

Z

Breinigsville, PA USA
21 September 2010
245510BV00003B/1/P